THE TOTAL SYNTHESIS OF NATURAL PRODUCTS

The Total Synthesis of Natural Products

VOLUME 1

Edited by

JOHN ApSimon

Department of Chemistry
Carleton University, Ottawa

WILEY-INTERSCIENCE, a Division of John Wiley & Sons, Inc.

New York · London · Sydney · Toronto

Library of Congress Cataloging in Publication Data:

ApSimon, John.
The total synthesis of natural products.

Includes bibliographical references.
1. Chemistry, Organic—Synthesis. I. Title.

QD262.A68 547′.2 72-4075
ISBN 0-471-03251-4

Printed in the United States of America

10 9 8 7 6 5 4 3 2 1

Contributors to Volume 1

U. Axen, Upjohn Company, Kalamazoo, Michigan
F. M. Dean, University of Liverpool, England
A. H. Jackson, University College, Cardiff, United Kingdom
F. Johnson, Dow Chemical Company, Wayland, Massachusetts
J. K. N. Jones, Queen's University, Kingston, Ontario
S. A. Narang, National Research Council of Canada, Ottawa
J. E. Pike, Upjohn Company, Kalamazoo, Michigan
W. P. Schneider, Upjohn Company, Kalamazoo, Michigan
K. M. Smith, University of Liverpool, England
W. A. Szarek, Queen's University, Kingston, Ontario
R. H. Wightman, Carleton University, Ottawa, Ontario

Preface

Throughout the history of organic chemistry we find that the study of natural products frequently has provided the impetus for great advances. This is certainly true in total synthesis, where the desire to construct intricate and complex molecules has led to the demonstration of the organic chemist's utmost ingenuity in the design of routes using established reactions or in the production of new methods in order to achieve a specific transformation.

These volumes draw together the reported total syntheses of various groups of natural products with commentary on the strategy involved with particular emphasis on any stereochemical control. No such compilation exists at present and we hope that these books will act as a definitive source book of the successful synthetic approaches reported to date. As such it will find use not only with the synthetic organic chemist but also perhaps with the organic chemist in general and the biochemist in his specific area of interest.

One of the most promising areas for the future development of organic chemistry is synthesis. The lessons learned from the synthetic challenges presented by various natural products can serve as a basis for this ever-developing area. It is hoped that this series will act as an inspiration for future challenges and outline the development of thought and concept in the area of organic synthesis.

The project started modestly with an experiment in literature searching by a group of graduate students about six years ago. Each student prepared a summary in equation form of the reported total syntheses of various groups of natural products. It was my intention to collate this material and possibly publish it. During a sabbatical leave in Strasbourg in the year 1968–1969, I attempted to prepare a manuscript, but it soon became apparent that if I was to also enjoy other benefits of a sabbatical leave, the task would take many years. Several colleagues suggested that the value of such a collection would be enhanced by commentary. The only way to encompass the amount of data collected and the inclusion of some words was to persuade experts in the various areas to contribute. I am grateful to all the authors for their efforts in producing stimulating and definitive accounts of the total syntheses described to date in their particular areas. I would like to thank those students who enthusiastically accepted my suggestion several years ago and produced valuable collections of reported syntheses. They are Dr. Bill Court, Dr. Ferial Haque, Dr. Norman Hunter, Dr. Russ King, Dr. Jack Rosenfeld, Dr. Bill Wilson, Mr. Douglas Heggart, Mr. George Holland, and Mr. Don Todd. I also thank Professor Guy Ourisson for his hospitality during the seminal phases of this venture.

JOHN APSIMON

Ottawa, Canada
February 1972

Contents

THE TOTAL SYNTHESIS OF NATURAL PRODUCTS

The Total Synthesis of Carbohydrates

J. K. N. JONES AND W. A. SZAREK

Department of Chemistry, Queen's University, Kingston, Ontario, Canada

1. INTRODUCTION

The carbohydrates comprise one of the major classes of naturally occurring organic compounds. Although the structures of carbohydrates appear to be quite complex, the chemistry of these compounds usually involves only two kinds of functional group, ketone or aldehyde carbonyls and hydroxyl groups. The carbonyl groups normally are not free but are combined with the hydroxyl groups in hemiacetal or acetal linkages; the carbon of the "masked" carbonyl is known as the anomeric center. When a free sugar is dissolved in an appropriate solvent, a dynamic equilibrium is achieved involving both anomerization and ring isomerization.

A wide variety of sugars has been found in nature and/or synthesized in the laboratory. These include not only the "classical" sugars but also derivatives such as amino, thio, halo, deoxy, branched-chain, and unsaturated sugars. Most synthetic sugars have been obtained by chemical transformations of naturally occurring sugars or their derivatives. In fact, the degree of achievement is such that the synthesis of a new mono-, di-, or trisaccharide can now be undertaken with a fair degree of confidence.

The total synthesis of sugars from noncarbohydrate precursors has also been achieved by many routes. Some methods are long, involved, are stereospecific and result in the formation of one or two sugars only; others are relatively simple but produce complex mixtures of carbohydrates which may resist fractionation. Practically all naturally occurring sugars are optically active. Most synthetic routes which employ noncarbohydrate precursors produce racemic mixtures of sugars which may be difficult to separate into the D and L isomers. However, if enzymes are used to effect condensation of fragments or to remove one or more of the components, optically pure isomers may be isolated. In this chapter the total synthesis of sugars and the related alditols and cyclitols, from noncarbohydrate substances by both specific and nonspecific methods, are discussed. Only compounds containing more than three carbon atoms are considered.

2. BASE-CATALYZED CONDENSATIONS WITH CARBON-CARBON BOND FORMATION. THE FORMOSE REACTION

The formose reaction has attracted the attention of biologists and chemists in recent years because it involves the self-condensation of formaldehyde to produce reducing sugars. This property is of interest in considering the problem of the origin of life on this planet, especially as formaldehyde has been detected in interstellar gases,[1] and also because of the feasibility of using carbon (as formaldehyde) as a possible source of sugars for the growth of microorganisms with the concomitant production of proteins and other complex organic compounds of importance to life and industry.[2]

The self-condensation of formaldehyde under the influence of base to yield a sugarlike syrup (methylenitan) was first observed by Butlerow[3] in 1861, when he treated trioxymethylene with calcium hydroxide solution. Calcium carbonate, magnesia, baryta, mineral clay, or even γ-radiation may also be used.[4] Fischer[5] showed that the sugarlike syrup called formose obtained from gaseous formaldehyde,[6] or methose prepared by the action of magnesium hydroxide suspension on formaldehyde,[7] contained monosaccharides, and that by reacting methose with phenylhydrazine acetate it was possible to obtain two hexose phenylosazones in low yield. Much higher yields were obtained from acrose (see below). Fischer named these derivatives α- and β-acrosazone (the corresponding sugars are α-acrose and β-acrose) and showed that α-acrosazone was DL-*arabino*-hexose phenylosazone (DL-glucose phenylosazone). In a remarkable series of experiments involving chemical and enzymic processes, Fischer was able to achieve the total synthesis from α-acrose of D- and L-glucose, D-gluconic acid, D- and L-mannose, D- and L-mannonic acids, D- and L-mannitol, and of D- and L-*arabino*-hexulose (D- and L-fructose), thus laying the basis for the synthesis of many other sugars. The transformations achieved are shown in Scheme 1.

β-Acrosazone was later identified as DL-*xylo*-hexose phenylosazone (DL-sorbose phenylosazone), which can be derived from DL-glucose, DL-idose, and DL-*xylo*-hexulose (DL-sorbose).[8] However, there has been a suggestion that β-acrosazone is really the phenylosazone of DL-dendroketose (see below). Fischer and Tafel[9] observed that 2,3-dibromopropionaldehyde (acrolein dibromide) when treated with dilute alkali yielded products which reacted like sugars (hence the name "acrose"). DL-Glyceraldehyde[10] also gave products that possessed the properties of sugars when treated similarly. Fischer and Tafel[9] explained the formation of acrose as an aldol-type condensation between DL-glyceraldehyde and 1,3-dihydroxypropanone (dihydroxyacetone), the latter compound being formed by a base-catalyzed isomerization from DL-glyceraldehyde (see Scheme 2).

H. O. L. Fischer and E. Baer[11] showed that D-glyceraldehyde and 1,3-dihydroxypropanone react in basic solution to yield D-*arabino*-hexulose and D-*xylo*-hexulose as major products. This reaction is a general reaction and novel sugars can be produced if D-glyceraldehyde is replaced by L-glyceraldehyde or by other aldehydes (see below). It is interesting that the biological origin of D-*arabino*-hexulose follows a similar route, but sugar phosphates and enzymes (aldolases) are involved.[12] In all cases the *threo* configuration is favored at the newly formed asymmetric centres but condensations involving enzyme-catalyzed reactions[13] usually yield the D-*threo* configuration only.

The conversion of formaldehyde to formose involves a complex series of reactions which have been rationalized by Breslow,[14] who suggested that two processes are involved in the formation of glycolaldehyde from formaldehyde. The first is a slow condensation of two molecules of formaldehyde to form glycolaldehyde, which then reacts rapidly with a further molecule of formaldehyde to produce glyceraldehyde. Part of this is then converted to

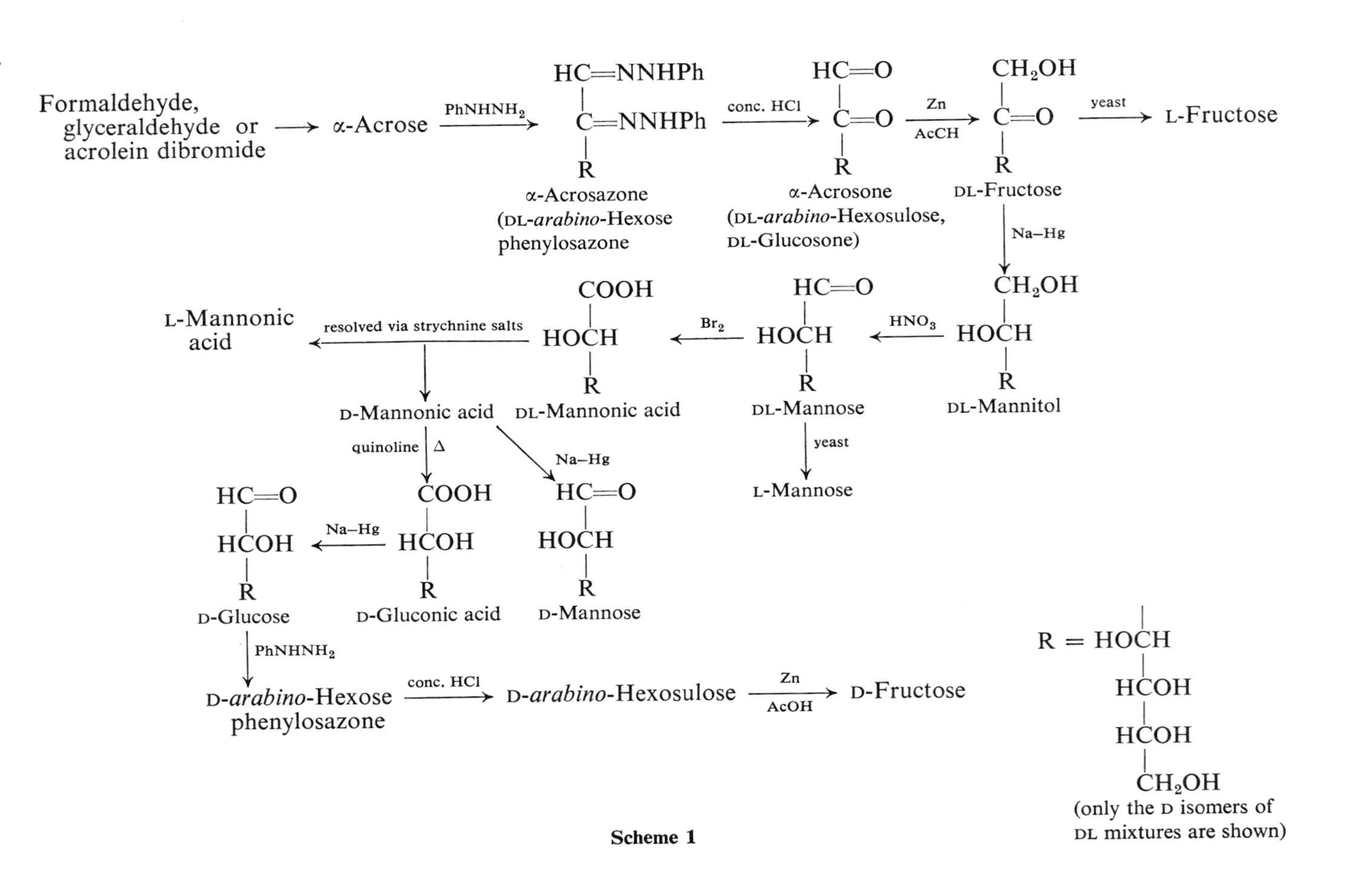
Formaldehyde, glyceraldehyde or acrolein dibromide
α-Acrose
PhNHNH₂
HC=NNHPh
C=NNHPh
R
α-Acrosazone (DL-arabino-Hexose phenylosazone
conc. HCl
HC=O
C=O
α-Acrosone (DL-arabino-Hexosulose, DL-Glucosone)
Zn
AcCH
CH₂OH
DL-Fructose
yeast
L-Fructose
Na–Hg
HOCH
DL-Mannitol
HNO₃
DL-Mannose
yeast
L-Mannose
Br₂
COOH
DL-Mannonic acid
resolved via strychnine salts
L-Mannonic acid
D-Mannonic acid
quinoline
Δ
Na–Hg
HCOH
D-Gluconic acid
D-Mannose
D-Glucose
PhNHNH₂
D-arabino-Hexose phenylosazone
conc. HCl
D-arabino-Hexosulose
Zn
AcOH
D-Fructose
R = HOCH
HCOH
HCOH
CH₂OH
(only the D isomers of DL mixtures are shown)

Scheme 1

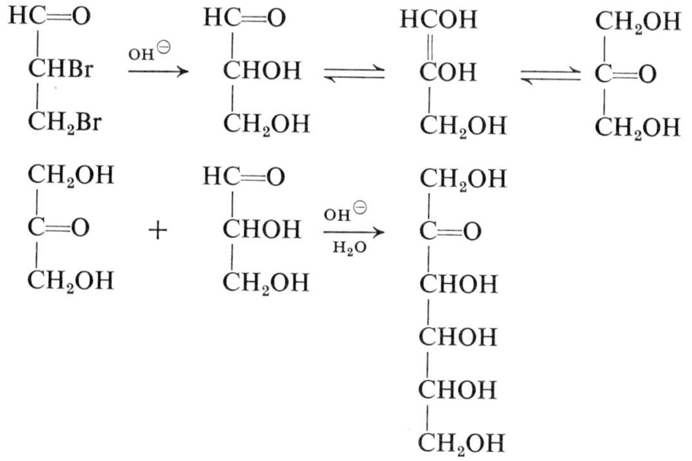

$$HC{=}O\text{—}CHBr\text{—}CH_2Br \xrightarrow{OH^{\ominus}} HC{=}O\text{—}CHOH\text{—}CH_2OH \rightleftharpoons HCOH{=}COH\text{—}CH_2OH \rightleftharpoons CH_2OH\text{—}C{=}O\text{—}CH_2OH$$

$$CH_2OH\text{—}C{=}O\text{—}CH_2OH + HC{=}O\text{—}CHOH\text{—}CH_2OH \xrightarrow[H_2O]{OH^{\ominus}} CH_2OH\text{—}C{=}O\text{—}CHOH\text{—}CHOH\text{—}CHOH\text{—}CH_2OH$$

Scheme 2

1,3-dihydroxypropanone,* which then rapidly reacts with formaldehyde to yield tetrulose and then tetrose, which then breaks down to two molecules of glycolaldehyde. The reaction is thus autocatalytic and is formulated as shown in Scheme 3. The rate of formose formation is dependent upon the metal cation of the base used. It is more rapid with those bases that form chelate compounds with enediols, which are intermediates in the foregoing reaction: thallium hydroxide > calcium hydroxide > sodium hydroxide. It follows that the composition of formose will depend upon the base used, the concentration of the reactants, and the temperature and time of reaction. Short periods of reaction favor the formation of lower molecular weight ketose sugars, longer periods of reaction yield more aldose sugars, while high concentration of alkali and long periods of heating yield saccharinic acids[16] and other products resulting from the decomposition of sugars by

* 1,3-Dihydroxypropanone has been prepared by Marei and Raphael[15] from nitromethane and formaldehyde:

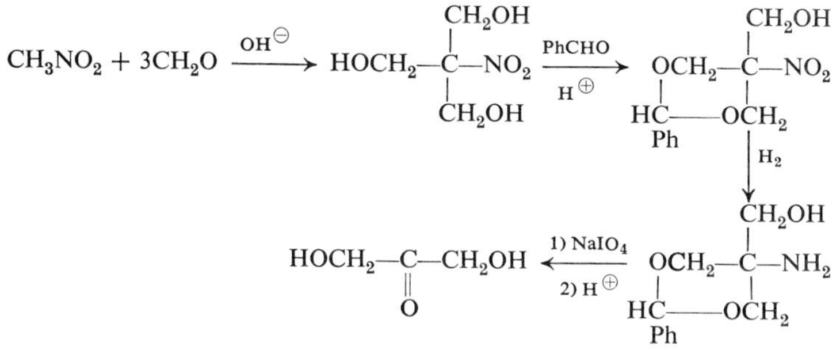

$$CH_3NO_2 + 3CH_2O \xrightarrow{OH^{\ominus}} HOCH_2\text{—}C(CH_2OH)_2\text{—}NO_2 \xrightarrow[H^{\oplus}]{PhCHO} \text{(PhCH acetal: } OCH_2\text{—}C(CH_2OH)(NO_2)\text{—}CH_2O\text{)} \xrightarrow{H_2} \text{(PhCH acetal: } OCH_2\text{—}C(CH_2OH)(NH_2)\text{—}CH_2O\text{)} \xrightarrow[2)\,H^{\oplus}]{1)\,NaIO_4} HOCH_2\text{—}C({=}O)\text{—}CH_2OH$$

$$CH_2O + HOCH_2\text{—}CHO \rightleftarrows HOCH_2\text{—}CHOH\text{—}CHO \rightleftarrows HOCH_2\text{—}CO\text{—}CH_2OH$$

$$CH_2O + HOCH_2\text{—}CO\text{—}CH_2OH \rightleftarrows HOCH_2\text{—}CHOH\text{—}CO\text{—}CH_2OH \rightleftarrows HOCH_2\text{—}CHOH\text{—}CHOH\text{—}CHO$$

$$HOCH_2\text{—}CHOH\text{—}CHOH\text{—}CHO \rightleftarrows 2\ HOCH_2\text{—}CHO$$

Scheme 3

alkali. With the advent of paper chromatography and gas-liquid chromatography, it has been possible to detect all eight aldohexose sugars, all four hexuloses, the four pentoses, two pentuloses, all possible tetroses, dendroketose, and three heptuloses.[17–19] Recently, sugars prepared by base-catalyzed condensation of formaldehyde were analyzed by combined gas-liquid chromatography and mass spectrometry; both branched and straight-chain products were detected.[19a] Cannizzaro reaction of formaldehyde proceeds in alkaline medium in conjunction with the formose reaction to produce aldoses and ketoses, and it has been shown[19b] that the extent of the two reactions is a function of the catalyst used. In a study with calcium hydroxide as catalyst, it was found that the ratio of branched-chain sugar derivatives, such as (hydroxymethyl)glyceraldehyde and apiose (see below), and straight-chain products could be controlled by manipulation of the reaction conditions. The branched products are very readily reduced by a crossed-Cannizzaro reaction with formaldehyde and large quantities of species such as (hydroxymethyl)glycerol are produced. Formose solutions are decomposed by microorganisms if allowed to stand in an open vessel in the laboratory.[20] Glycolaldehyde itself polymerizes under the influence of base to yield tetroses, hexoses, and other sugars.[21] Methoxyacetaldehyde polymerizes in aqueous potassium cyanide solution forming 2,4-dimethoxyaldotetroses:[22]

$$CH_3OCH_2\text{—}CHO + CH_3OCH_2\text{—}CHO \xrightarrow[\text{or}\ K_2CO_3]{KCN} CH_3OCH_2\text{—}CHOH\text{—}CHOCH_3\text{—}CHO$$

The polymerization of formaldehyde to yield sugars is, therefore, a very complicated process. For example, the formation of pentoses from formaldehyde may proceed via several routes. Glycolaldehyde and 1,3-dihydroxypropanone may react to form pentuloses, which subsequently are isomerized to pentoses, or formaldehyde and a tetrulose may combine to yield pent-3-uloses, which then isomerize to pentuloses and pentoses, or glyceraldehyde and glycolaldehyde may combine to form pentoses. To test these hypotheses, Hough and Jones[23] treated mixtures of glycolaldehyde and 1,3-dihydroxypropane and of glyceraldehyde and glycolaldehyde with lime water and found that pentoses were produced along with several other sugars. They were able to isolate arabinose, ribose, and xylose, as phenylhydrazones, from the complex mixture of sugars that results from the two reactions previously described.

Very recently, it was shown[23a] that it is possible, by making use of the hexokinase reaction, to extract some specific sugars from the complex synthetic formose sugars. The enzyme hexokinase is known to transfer the terminal phosphate of ATP to D-glucose:

$$\text{D-glucose} + \text{ATP} \xrightarrow{\text{hexokinase}} \text{D-glucose 6-phosphate} + \text{ADP}$$

However, the enzyme is not totally specific for glucose, other hexoses such as fructose and mannose being also susceptible to phosphorylation. The basis of the method of extraction involves phosphorylation of some hexoses by this means, which are then retained on a column of anion-exchange resin (together with unreacted ATP and formed ADP), while other unreacted, neutral components of the formose mixture pass through the column. The sugar phosphates are then eluted by a salt solution of appropriate concentration, and the unsubstituted hexoses are obtained by a phosphatase reaction.

The branched-chain sugar DL-dendroketose mentioned earlier was first isolated by Utkin,[24] who prepared it by adding sodium hydroxide to a solution of 1,3-dihydroxypropanone in water. It is formed so easily and in such high yield that it seems remarkable that it has not appeared in any natural product. Moreover, it is metabolized completely by baker's yeast.[25] Like the branched-chain sugar apiose,[26] hemiacetal formation results in the formation of a new optically active center with the possible formation of eight isomers from the D and L forms of dendroketose. Utkin was able to isolate D-dendroketose (4-*C*-hydroxymethyl-D-*glycero*-pentulose) when he observed that a microorganism which accidentally contaminated a solution of DL-dendroketose, metabolized the L-isomer only. He was able to prove the absolute configuration of the nonmetabolized material by relating it to D-apiose,[27] a sugar of known absolute configuration, by the series of reactions indicated in Scheme 4. It may be significant that D-dendroketose, which remained after fermentation of the DL mixture, possesses a potential L-*threo* disposition of hydroxyl groups at C-3 and C-4, while L-dendroketose which possesses a potential D-*threo* configuration at C-3 and C-4 is metabolized:

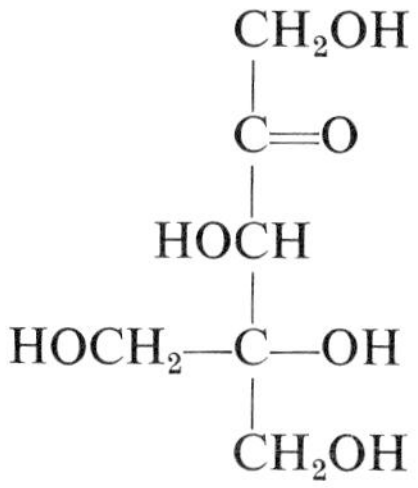

```
          CH2OH                     CH2OH
            |                         |
           C=O                       C=O
            |                         |
          HOCH                       HCOH
            |                         |
   HOCH2—C—OH                 HO—C—CH2OH
            |                         |
          CH2OH                     CH2OH
```

L-Dendroketose (metabolized)

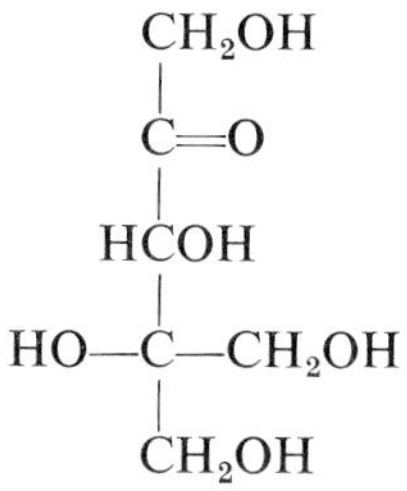

D-Dendroketose (not metabolized)

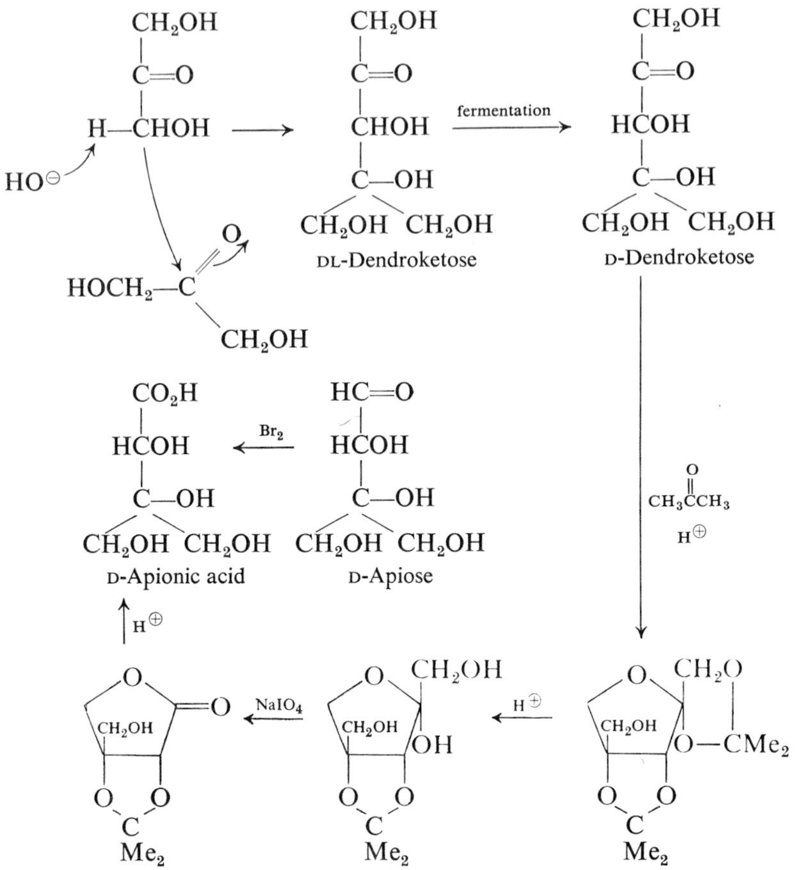

fermentation

DL-Dendroketose

D-Dendroketose

Br_2

D-Apionic acid

D-Apiose

CH_3COCH_3, $H^{\oplus}$

$NaIO_4$

$H^{\oplus}$

Scheme 4

3. SYNTHESES FROM ACETYLENIC AND OLEFINIC PRECURSORS

The directed synthesis of carbohydrates from noncarbohydrate precursors in most cases involves the preparation of compounds of acetylene. These acetylenic intermediates may be converted into *cis*- or *trans*-ethylenic derivatives dependent upon the mode of reduction of the acetylene. A further advantage of this approach is that the ethylene may then be hydroxylated in a *cis* or *trans* fashion, as decided by the mode of oxidation. In some cases steric effects may be used to force the predominant formation of one of the DL forms. This procedure is particularly effective when the hydroxylation of a ring compound is involved.

Several workers, chief among whom are Lespieau, Iwai, and Raphael, have synthesized carbohydrate derivatives from acetylenic and olefinic precursors. Stereochemical problems of hydroxylation were minimized either by *cis*-hydroxylation of double bonds of known stereochemistry using potassium permanganate or osmium tetroxide, or by epoxidation of double bonds of known stereochemistry followed by opening of the epoxide ring, with resulting *trans*-hydroxylation of the double bond.

Griner[28] appears to be one of the first to attempt the synthesis of sugar alcohols. He observed that when acrolein was hydrogenated by means of a zinc-copper couple and acetic acid, dimerization occurred and divinylglycol ($CH_2{=}CH{-}CHOH{-}CHOH{-}CH{=}CH_2$) resulted. This may exist in *meso* or DL modifications. Griner obtained the aid of LeBel to isolate a mold which would preferentially metabolize one of the isomers. In this, LeBel was successful. Griner had expected to obtain an optically active material but obtained a product devoid of activity and concluded that the *meso* form only was present. Lespieau[29] later showed this conclusion to be erroneous. Griner attempted to oxidize the divinylglycol, with permanganate solution, to a hexitol, but was unsuccessful. Later, in a brief note,[30] Griner stated that addition of two molecules of hypochlorous acid to divinylglycol gave a divinylglycol dichlorohydrin from which, after treatment with base, he was able to isolate DL-mannitol. Lespieau[31] repeated the attempted hydroxylation of divinylglycol but used osmium tetroxide-silver chlorate as the hydroxylating agent, and obtained allitol and DL-mannitol (see Scheme 5).

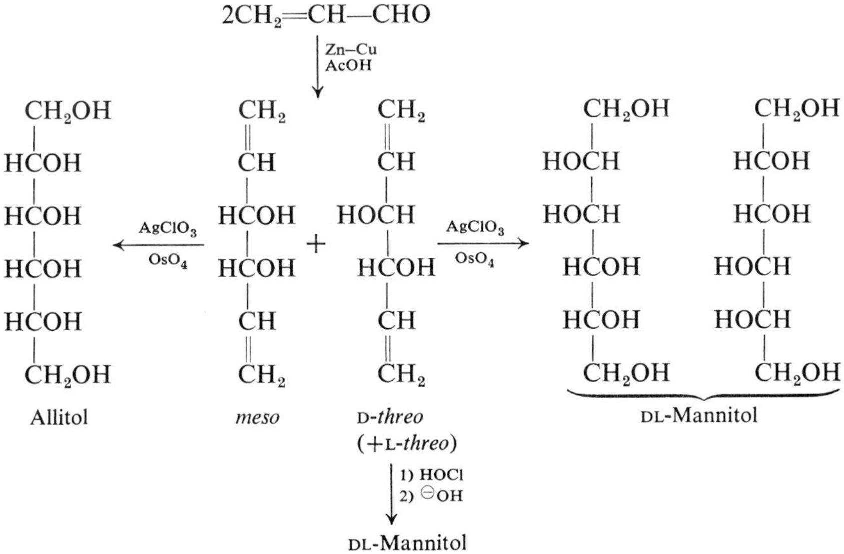

$$2CH_2{=}CH{-}CHO \xrightarrow{\text{Zn–Cu, AcOH}} meso + D\text{-}threo\ (+L\text{-}threo)$$

Allitol		*meso*		D-*threo* (+L-*threo*)		DL-Mannitol	
CH_2OH		CH_2		CH_2		CH_2OH	CH_2OH
HCOH		‖ CH		‖ CH		HOCH	HCOH
HCOH	← $AgClO_3$ / OsO_4	HCOH	+	HOCH	$AgClO_3$ / OsO_4 →	HOCH	HCOH
HCOH		HCOH		HCOH		HCOH	HOCH
HCOH		CH ‖		CH ‖		HCOH	HOCH
CH_2OH		CH_2		CH_2		CH_2OH	CH_2OH

D-*threo* (+L-*threo*) $\xrightarrow{\text{1) HOCl; 2) }^{\ominus}\text{OH}}$ DL-Mannitol

Scheme 5

Hence, assuming *cis* addition of the new hydroxyl groups, allitol arises from the *meso* compound and DL-mannitol from DL-divinylglycol. In a second method of synthesis,[32] involving the Grignard reagent derived from acetylene and chloroacetaldehyde, divinylacetylene dichlorohydrin

$$(CH_2Cl—CHOH—C{\equiv}C—CHOH—CH_2Cl),$$

was prepared, converted to the hexynetetrol, and reduced to the corresponding ethylene derivative. Hydroxylation of the product by means of osmium tetroxide-silver chlorate gave galactitol and allitol. The ethylene derivative, therefore, had the *meso* configuration (see Scheme 6).

Lespieau[33] also synthesized ribitol and DL-arabinitol using acrolein dichloride and acetylene as starting materials as shown in Scheme 7. Raphael[34] improved on these syntheses by using epichlorohydrin and acetylene as starting materials, and performic acid as the oxidizing agent (see Scheme 8).

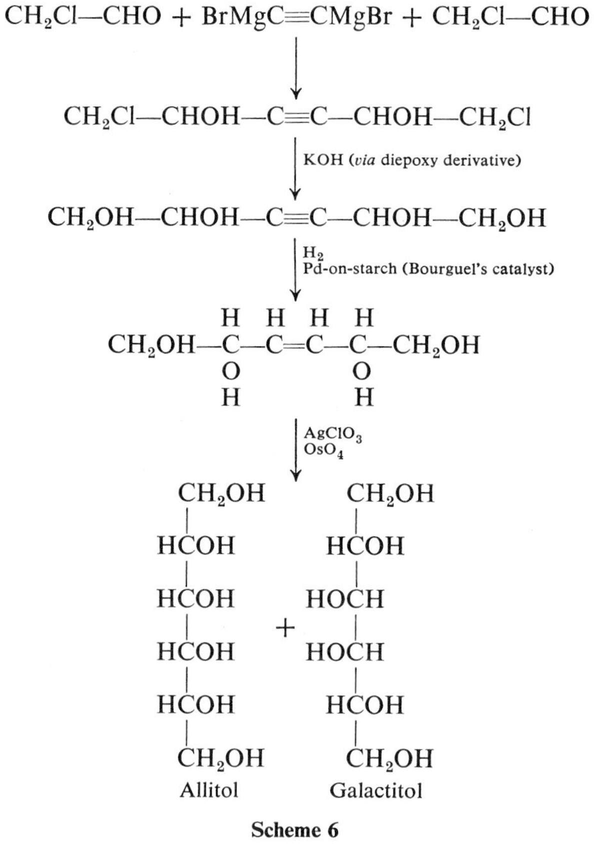

$$CH_2Cl—CHO + BrMgC{\equiv}CMgBr + CH_2Cl—CHO$$

$$\downarrow$$

$$CH_2Cl—CHOH—C{\equiv}C—CHOH—CH_2Cl$$

$$\downarrow \text{KOH (}via\text{ diepoxy derivative)}$$

$$CH_2OH—CHOH—C{\equiv}C—CHOH—CH_2OH$$

$$\downarrow H_2,\ \text{Pd-on-starch (Bourguel's catalyst)}$$

$$CH_2OH—C(H)(OH)—C(H){=}C(H)—C(H)(OH)—CH_2OH$$

$$\downarrow AgClO_3,\ OsO_4$$

Allitol		Galactitol
CH_2OH		CH_2OH
HCOH		HCOH
HCOH	+	HOCH
HCOH		HOCH
HCOH		HCOH
CH_2OH		CH_2OH

Scheme 6

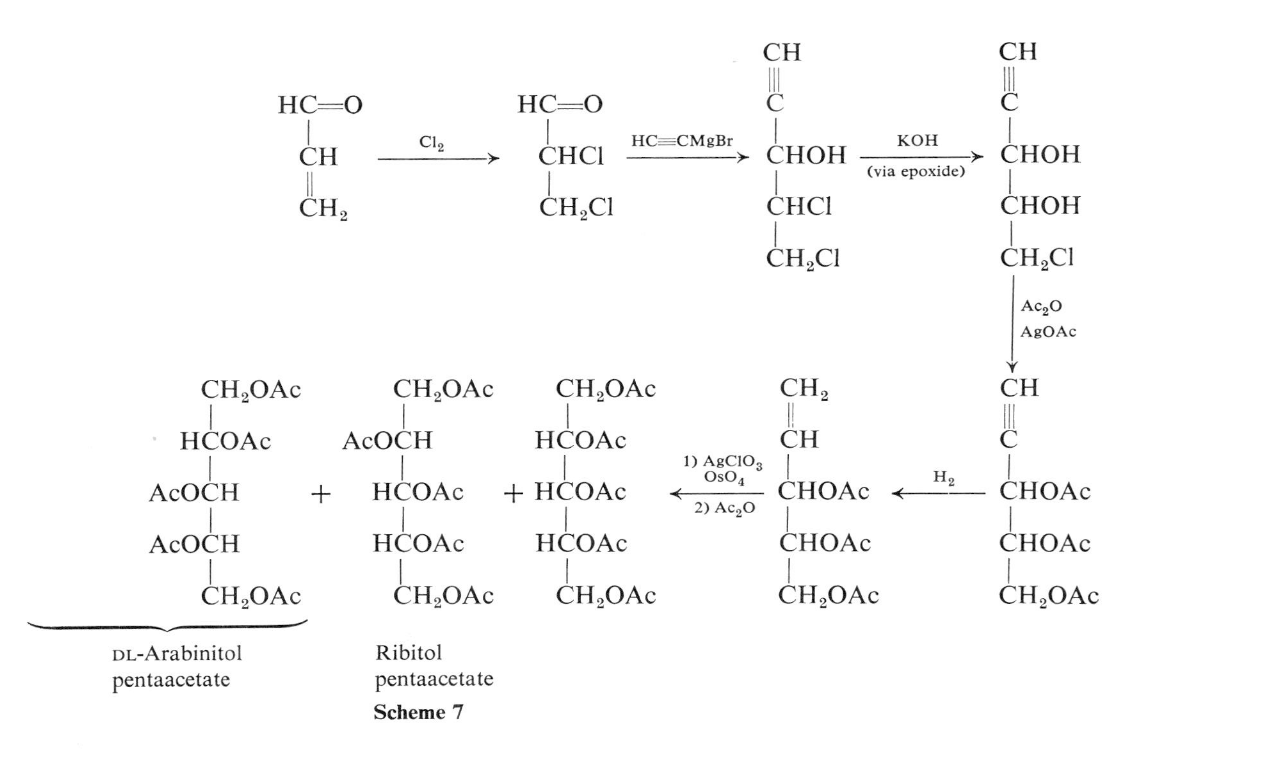
Cl$_2$
HC≡CMgBr
KOH
(via epoxide)
Ac$_2$O
AgOAc
H$_2$
1) AgClO$_3$
OsO$_4$
2) Ac$_2$O
DL-Arabinitol pentaacetate
Ribitol pentaacetate

Scheme 7

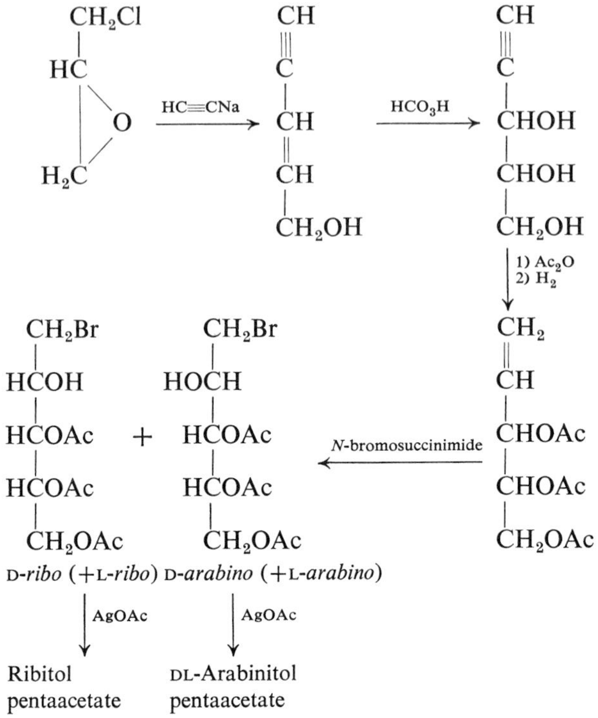

CH_2Cl–HC(–O–)H_2C $\xrightarrow{HC{\equiv}CNa}$ CH≡C–CH=CH–CH_2OH $\xrightarrow{HCO_3H}$ CH≡C–CHOH–CHOH–CH_2OH

$\xrightarrow{1)\ Ac_2O;\ 2)\ H_2}$ CH_2=CH–CHOAc–CHOAc–CH_2OAc

$\xrightarrow{N\text{-bromosuccinimide}}$ CH_2Br–HCOH–HCOAc–HCOAc–CH_2OAc + CH_2Br–HOCH–HCOAc–HCOAc–CH_2OAc

D-*ribo* (+L-*ribo*) D-*arabino* (+L-*arabino*)

AgOAc ↓ Ribitol pentaacetate; AgOAc ↓ DL-Arabinitol pentaacetate

Scheme 8

Iwai and his associates in Japan have achieved several total syntheses of pentose sugars using acetylenic compounds. Iwai and Iwashige[35] condensed the Grignard reagent derived from 3-(tetrahydropyranyl-2′-oxy)-propyne with 2,2-diethoxyacetaldehyde to yield 1,1-diethoxy-5-(tetrahydropyranyl-2′-oxy)pent-3-yn-2-ol, which, on reduction with lithium aluminum hydride, yields the *trans*-olefin. Catalytic hydrogenation, on the other hand, yields the *cis*-olefin. Acetylation of these products, followed by *cis* hydroxylation of the double bonds, affords products which, after hydrolysis of the acetal residues, yield the four DL-pentose sugars (see Scheme 9).

Iwai and Tomita have achieved a stereospecific synthesis of DL-arabinose[36] and a synthesis of a mixture[37] of DL-arabinose and DL-ribose as shown in Scheme 10.

DL-Ribose has been synthesized[38] stereospecifically by oxidative hydroxylation of 2-ethoxy-5-(tetrahydropyranyl-2′-oxy)methyl–2,5-dihydrofuran (see Scheme 11), which was obtained by hydrogenation of DL-1,1-diethoxy-5-(tetrahydropyranyl-2′-oxy)pent-2-yn-4-ol. This acetylenic compound was prepared by the Grignard reaction of (tetrahydropyranyl-2′-oxy)acetaldehyde with propargyl diethyl acetal magnesium bromide. One method for the

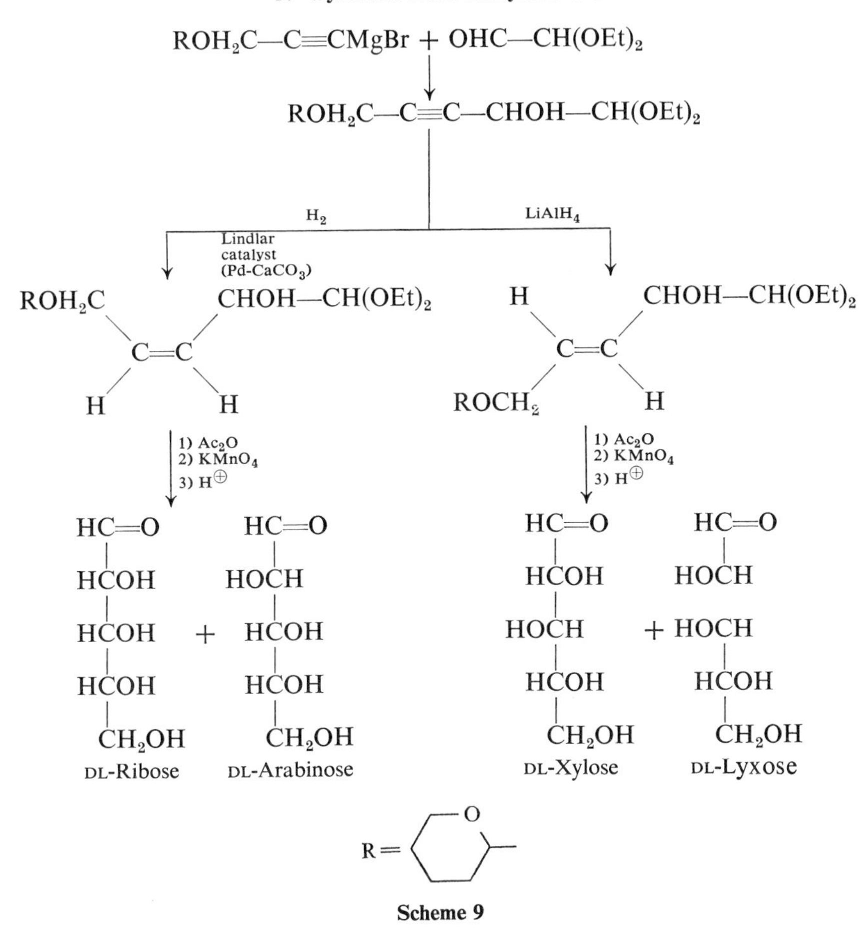

Scheme 9

preparation of (tetrahydropyranyl-2′-oxy)acetaldehyde involved ozonolysis of the tetrahydropyranyl ether of allyl alcohol. This synthesis of DL-ribose is the first example, in this chapter, of a total synthesis of a sugar, which involved a furan derivative; other examples are discussed in Section 6.

Total syntheses of deoxypentose sugars have also been reported. Hough[39] described a preparation of the biologically important 2-deoxy-D-*erythro*-pentose (2-deoxy-D-ribose) which involves the reaction of 2,3-*O*-isopropylidene–D-glyceraldehyde with allylmagnesium bromide. Hydroxylation of the resultant 5,6-*O*-isopropylidene–1-hexene-D-*erythro*-4,5,6-triol gave a mixture of products. Periodate oxidation of the hexitol derivatives, followed by hydrolysis, afforded almost exclusively 2-deoxy-D-*erythro*-pentose (Scheme 12). Another preparation of this sugar has also been achieved[40] using 2,3-*O*-isopropylidene–D-glyceraldehyde as a starting material, by condensation

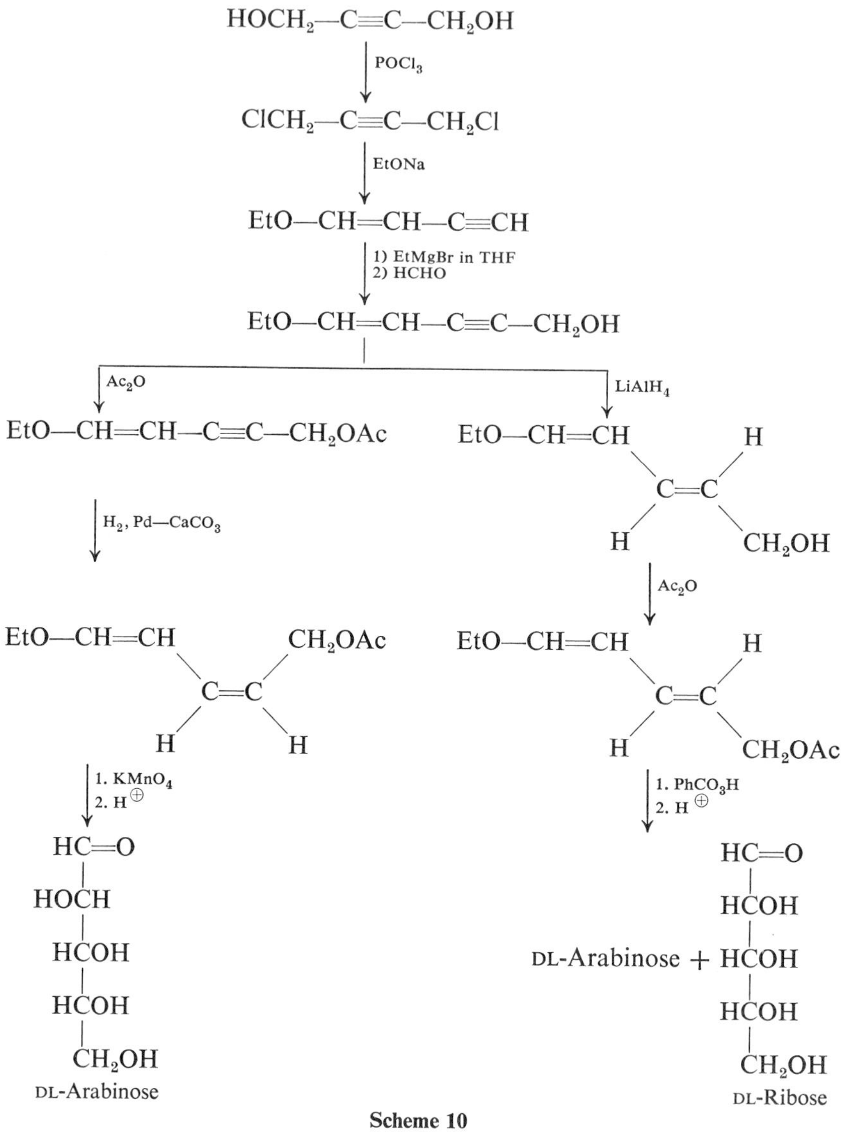

$HOCH_2—C{\equiv}C—CH_2OH$

↓ $POCl_3$

$ClCH_2—C{\equiv}C—CH_2Cl$

↓ EtONa

$EtO—CH{=}CH—C{\equiv}CH$

↓ 1) EtMgBr in THF 2) HCHO

$EtO—CH{=}CH—C{\equiv}C—CH_2OH$

↓ Ac_2O (left branch); ↓ $LiAlH_4$ (right branch)

$EtO—CH{=}CH—C{\equiv}C—CH_2OAc$; $EtO—CH{=}CH(H)C{=}C(H)CH_2OH$

↓ H_2, Pd—$CaCO_3$; ↓ Ac_2O

$EtO—CH{=}CH(H)C{=}C(H)CH_2OAc$; $EtO—CH{=}CH(H)C{=}C(H)CH_2OAc$

↓ 1. $KMnO_4$ 2. $H^{\oplus}$; ↓ 1. $PhCO_3H$ 2. $H^{\oplus}$

HC=O / HOCH / HCOH / HCOH / CH_2OH

DL-Arabinose

DL-Arabinose + HC=O / HCOH / HCOH / HCOH / CH_2OH

DL-Ribose

Scheme 10

with acetaldehyde in the presence of anhydrous potassium carbonate; 2-deoxy-D-xylose was also obtained.

Fraser and Raphael[41] have synthesized 2-deoxy-DL-*erythro*-pentose from but-2-yne-1,4-diol (Scheme 13). This compound was converted into 1-benzoyloxy-4-bromobut-2-yne (**1**) by monobenzoylation and treatment of the resultant half-ester with phosphorus tribromide. Condensation of

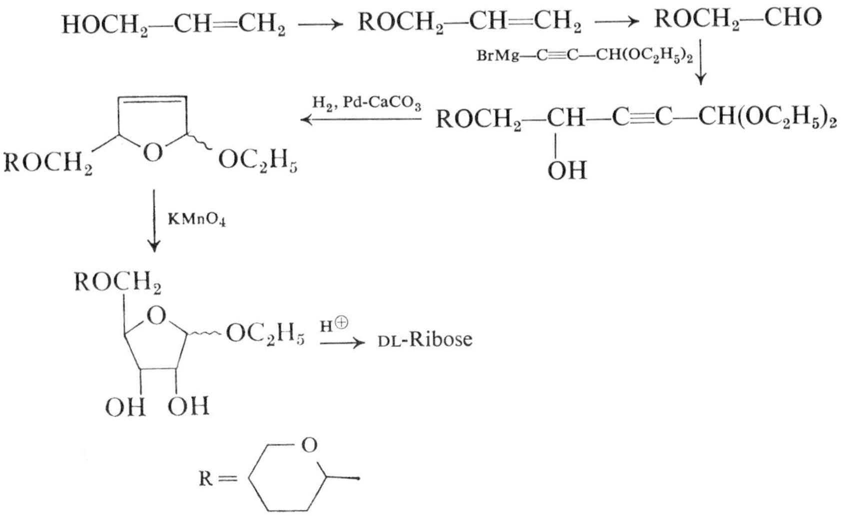

$HOCH_2—CH=CH_2 \longrightarrow ROCH_2—CH=CH_2 \longrightarrow ROCH_2—CHO$

$BrMg—C≡C—CH(OC_2H_5)_2$

$ROCH_2—CH(OH)—C≡C—CH(OC_2H_5)_2$

H_2, Pd-$CaCO_3$

$ROCH_2$ O OC_2H_5

$KMnO_4$

$ROCH_2$ O OC_2H_5 OH OH $\xrightarrow{H^{\oplus}}$ DL-Ribose

R = (tetrahydropyran-2-yl, O)

Scheme 11

1-benzoyloxy-4-bromobut-2-yne with ethyl sodiomalonate gave ethyl 5-benzoyloxypent-3-yne-1,1-dicarboyxlate (**2**), which was converted into the dihydrazide **3**. Compound **3** was then subjected to a double Curtius rearrangement; reaction with nitrous acid, followed by treatment of the resultant diazide with ethanol, afforded the acetylenic diurethane **4**. Catalytic hemihydrogenation of **4** gave the *cis*-ethylenic diurethane **5**. *cis*-Hydroxylation of **5**, followed by acid-catalyzed hydrolysis of the resultant *erythro*-triol, gave finally a small yield of 2-deoxy-DL-*erythro*-pentose.

Weygand and Leube[42] have also prepared 2-deoxy-DL-*erythro*-pentose (and 2-deoxy-DL-*threo*-pentose or 2-deoxy-DL-xylose) from an acetylenic precursor, 1-methoxy-1-buten-3-yne (**6**). Treatment of **6** with formaldehyde

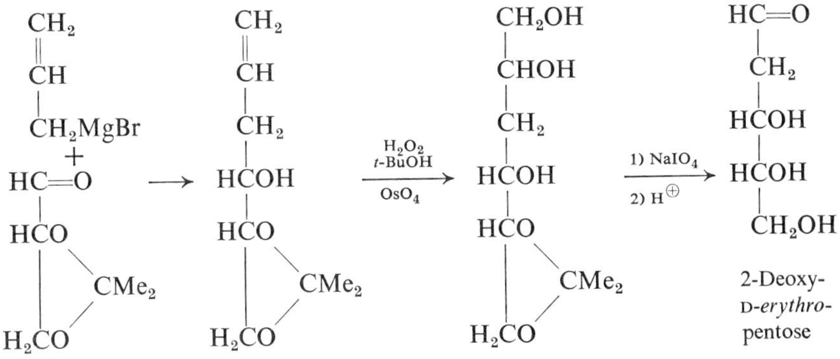

$CH_2=CH—CH_2MgBr$ + $HC=O$ / HCO / H_2CO (CMe_2) $\longrightarrow$ $CH_2=CH—CH_2—HCOH—HCO—H_2CO$ (CMe_2) $\xrightarrow[OsO_4]{H_2O_2,\ t\text{-BuOH}}$ $CH_2OH—CHOH—CH_2—HCOH—HCO—H_2CO$ (CMe_2) $\xrightarrow[2)\ H^{\oplus}]{1)\ NaIO_4}$ $HC=O—CH_2—HCOH—HCOH—CH_2OH$

2-Deoxy-D-*erythro*-pentose

Scheme 12

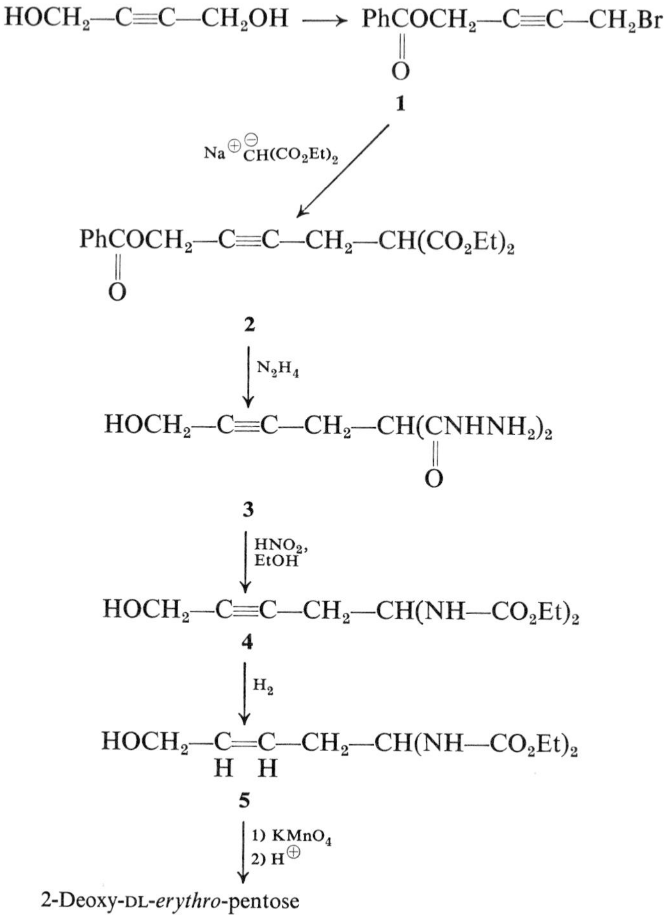

$$HOCH_2—C\equiv C—CH_2OH \longrightarrow \underset{\underset{O}{\|}}{PhC}OCH_2—C\equiv C—CH_2Br$$

1

$\downarrow$ $Na^{\oplus}\,\overset{\ominus}{C}H(CO_2Et)_2$

$$\underset{\underset{O}{\|}}{PhC}OCH_2—C\equiv C—CH_2—CH(CO_2Et)_2$$

2

$\downarrow$ N_2H_4

$$HOCH_2—C\equiv C—CH_2—CH(\underset{\underset{O}{\|}}{C}NHNH_2)_2$$

3

$\downarrow$ HNO_2, EtOH

$$HOCH_2—C\equiv C—CH_2—CH(NH—CO_2Et)_2$$

4

$\downarrow$ H_2

$$HOCH_2—\underset{H}{C}=\underset{H}{C}—CH_2—CH(NH—CO_2Et)_2$$

5

$\downarrow$ 1) $KMnO_4$ 2) $H^{\oplus}$

2-Deoxy-DL-*erythro*-pentose

Scheme 13

in methanol at 45–55° gives 1-methoxy-1-penten-3-yn-5-ol, but at 65–85° the dimethyl acetal **7** was produced. Hemihydrogenation of **7** over a Lindlar catalyst gave the corresponding ethylene **8**. Hydroxylation of **8** with osmium tetroxide and hydrogen peroxide in *t*-butanol, followed by acid-catalyzed hydrolysis, gave 2-deoxy-DL-*erythro*-pentose, whereas the use of peroxybenzoic acid afforded 2-deoxy-DL-*threo*-pentose (see Scheme 14).

A more recent synthesis of 2-deoxy-DL- and L-*erythro*-pentose has been reported by Nakaminami et al.[43] The first step (see Scheme 15) was a Reformatsky reaction of ethyl bromoacetate with acrolein to give the β-hydroxy ester **9**. Compound **9** was hydrolyzed by aqueous potassium hydroxide to give the DL-acid **10**, which was treated in an aqueous solution

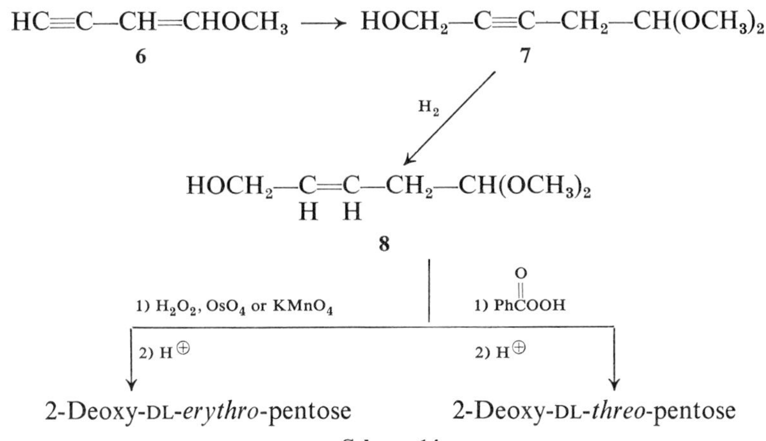

Scheme 14

with *N*-bromosuccinimide to afford the DL-bromolactone **11**. Successive basic and acidic hydrolysis gave "2-deoxy-DL-ribonolactone" (**12**). Treatment of **12** with disiamylborane [bis(3-methyl-2-butyl)borane], and hydrolysis of the resultant tris(disiamylborinate) ester, yielded 2-deoxy-DL-*erythro*-

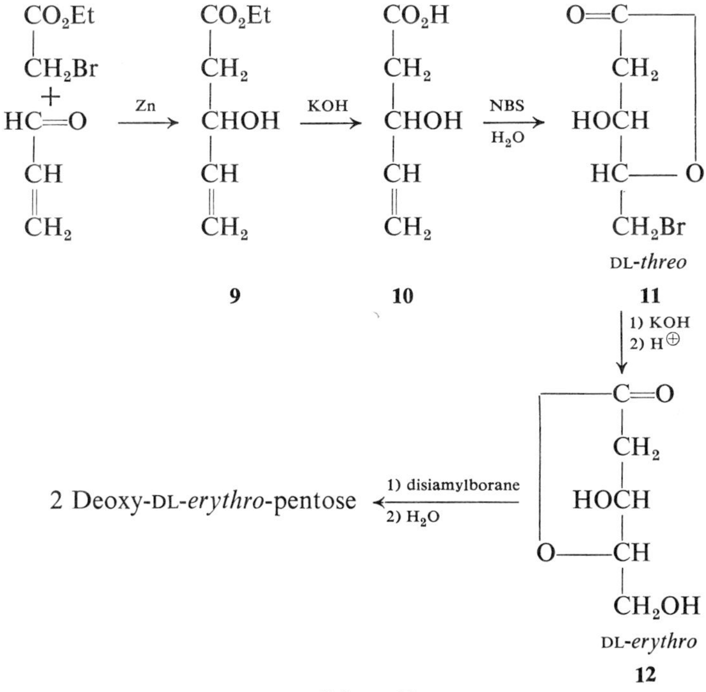

CO_2Et—CH_2Br + $HC{=}O$—$CH{=}CH_2$ $\xrightarrow{Zn}$ CO_2Et—CH_2—$CHOH$—$CH{=}CH_2$ (**9**) $\xrightarrow{KOH}$ CO_2H—CH_2—$CHOH$—$CH{=}CH_2$ (**10**) $\xrightarrow[H_2O]{NBS}$ $O{=}C$—CH_2—$HOCH$—HC(—O lactone)—CH_2Br, DL-*threo* (**11**)

11 $\xrightarrow{1)\ KOH;\ 2)\ H^{\oplus}}$ $C{=}O$—CH_2—$HOCH$—CH(—O lactone)—CH_2OH, DL-*erythro* (**12**)

12 $\xrightarrow{1)\ disiamylborane;\ 2)\ H_2O}$ 2 Deoxy-DL-*erythro*-pentose

Scheme 15

pentose. The preparation of 2-deoxy-L-*erythro*-pentose involved treatment of the racemic hydroxy acid **10** with a half equivalent of quinine and decomposition of the salt to yield (-)-**10**, which was then subjected to the same reactions as in the case of the racemic compounds.

2,3-Dideoxy-DL-pentose has been synthesized by Price and Balsley[44] by the Claisen rearrangement of allyl vinyl ether to 4-pentenal, conversion to the methyl acetal, and permanganate oxidation (see Scheme 16).

Total syntheses of tetroses and tetritols from olefinic precursors have also been achieved, using reactions which have already been described in this section. Thus Raphael[45] obtained erythritol tetraacetate by treatment of *trans*-2-butene-1,4-diol diacetate (**13**) with peroxyacetic acid, followed by complete acetylation, whereas similar treatment of the *cis* compound **14** produced DL-threitol tetraacetate:

$$\begin{array}{ccc} AcOCH_2 & & H \\ & C{=}C & \\ H & & CH_2OAc \end{array} \qquad \begin{array}{ccc} AcOCH_2 & & CH_2OAc \\ & C{=}C & \\ H & & H \end{array}$$

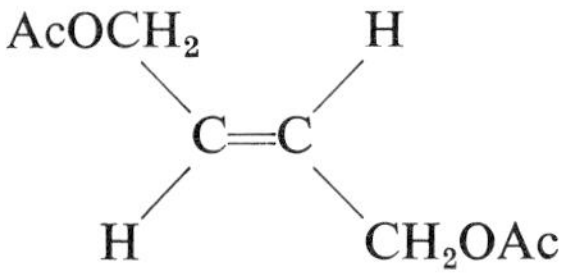

13 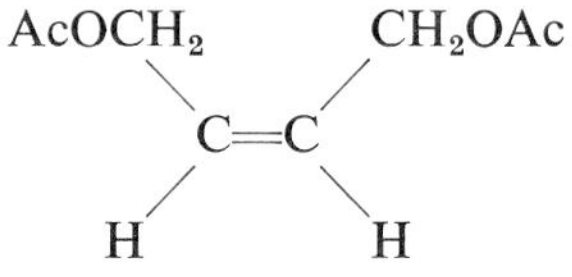 **14**

As usual, *cis*-hydroxylation of the *cis*-diacetate yielded erythritol tetraacetate, after complete acetylation, and the *trans*-diacetate gave DL-threitol tetraacetate, on treatment with osmium tetroxide and hydrogen peroxide in

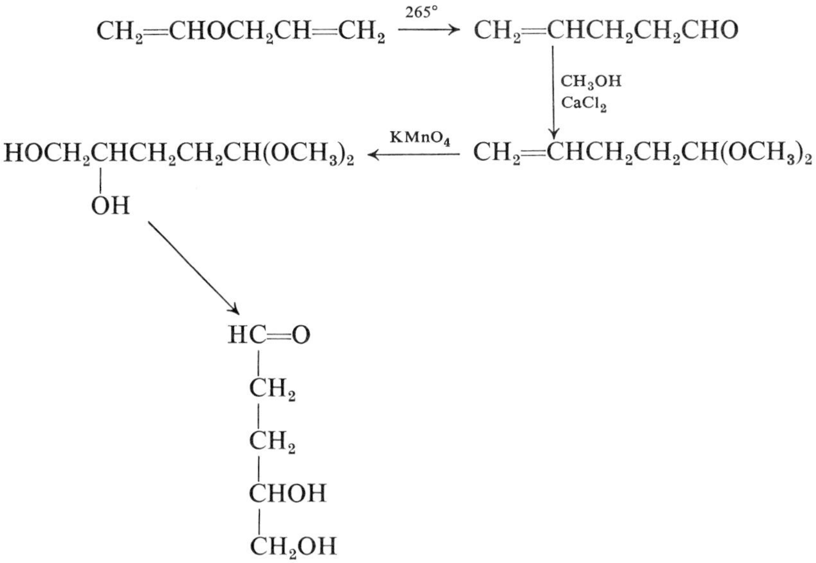

$$CH_2{=}CHOCH_2CH{=}CH_2 \xrightarrow{265^\circ} CH_2{=}CHCH_2CH_2CHO$$

$$\downarrow CH_3OH,\ CaCl_2$$

$$\underset{\displaystyle OH}{HOCH_2\underset{|}{C}HCH_2CH_2CH(OCH_3)_2} \xleftarrow{KMnO_4} CH_2{=}CHCH_2CH_2CH(OCH_3)_2$$

$$\downarrow$$

$$\begin{array}{l} HC{=}O \\ | \\ CH_2 \\ | \\ CH_2 \\ | \\ CHOH \\ | \\ CH_2OH \end{array}$$

Scheme 16

t-butanol and complete acetylation of the product. *trans*-Addition of hypobromous acid to the *cis* and *trans* compounds (**14** and **13**) was smoothly effected to give the *threo*- (**15**) and *erythro*-2-bromobutane-1,3,4-triol 1,4-diacetates (see Scheme 17). Chromium trioxide oxidation of either the *erythro*- or the *threo*-bromohydrin gave the same ketone, 2-bromo-1,4-diacetoxybutan-3-one (**16**), which, on treatment with silver acetate in acetic acid, yielded DL-*glycero*-tetrulose triacetate. Hydrolysis with baryta then gave DL-*glycero*-tetrulose. Other examples of the preparation of tetritol derivatives from olefinic precursors are known.[45a] Kiss and Sirokmán[45b] synthesized *erythro*-2-amino-1,3,4-trihydroxybutane stereospecifically from *trans*-1,4-dibromo-2-butene.

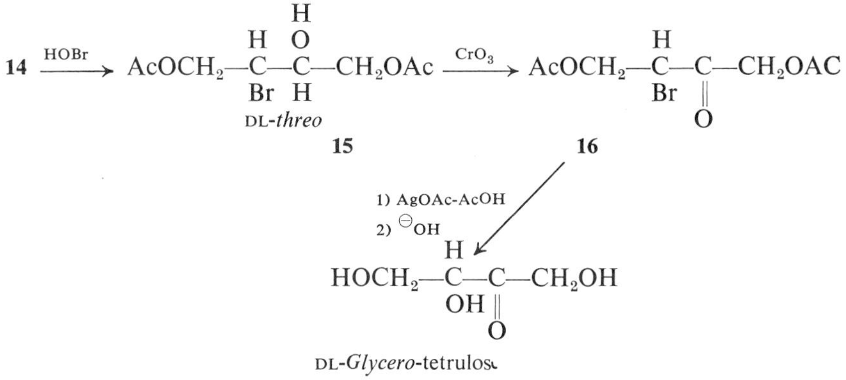

14 $\xrightarrow{HOBr}$ $AcOCH_2$—CH(Br)—CH(OH)—CH_2OAc (DL-*threo*, **15**) $\xrightarrow{CrO_3}$ $AcOCH_2$—CH(Br)—C(=O)—CH_2OAC (**16**)

16 $\xrightarrow{\text{1) AgOAc-AcOH; 2) } ^{\ominus}OH}$ $HOCH_2$—CH(OH)—C(=O)—CH_2OH

DL-*Glycero*-tetrulose

Scheme 17

Walton[46] has prepared D-threose and D-erythrose by a method similar to that employed by Hough[39] for the synthesis of 2-deoxy-D-*erythro*-pentose. Thus addition of vinylmagnesium chloride to 2,3-*O*-isopropylidene–*aldehydo*-D-glyceraldehyde gave a mixture of epimeric pentene derivatives, which were separated by gas-liquid chromatography. Ozonolysis followed by acid-catalyzed hydrolysis of each epimer afforded D-threose and D-erythrose, each in approximately 40% yield. A similar study has been made by Horton et al.[47] In that work, however, the first step was ethynylation of 2,3-*O*-isopropylidene–*aldehydo*-D-glyceraldehyde to give a 44:56 mixture of 4,5-*O*-isopropylidene–1-pentyne-D-*erythro* (and D-*threo*)-3,4,5-triol.

4. SYNTHESES FROM TARTARIC ACID AND OTHER NATURALLY OCCURRING ACIDS

The potential of the tartaric acids as possible precursors in the synthesis of tetroses and related compounds was recognized by Emil Fischer as early

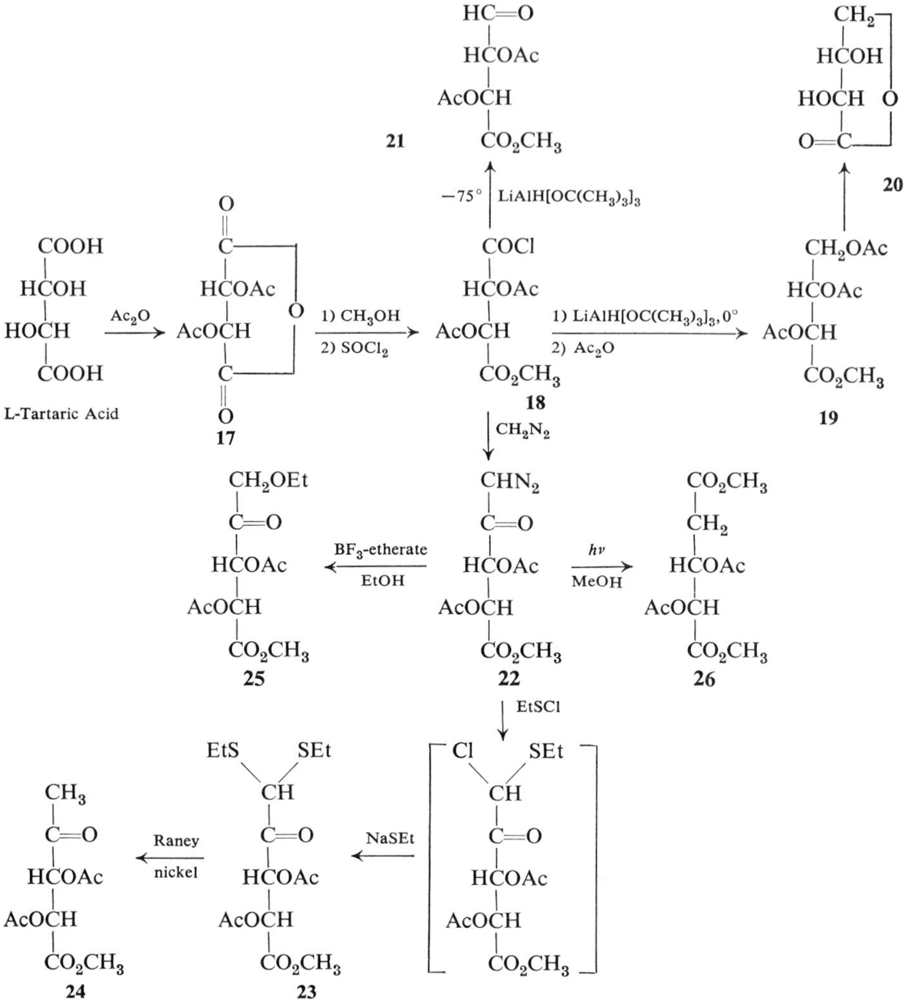

Scheme 18

as 1889;[48] however, he was unsuccessful in attempts to reduce tartaric acid. In 1941 Lucas and Baumgarten[49] reported a solution to this problem, and achieved a synthesis of L-threitol. More recently, Bestmann and Schmiechen[50] employed L-tartaric acid for the synthesis of a variety of tetrose and pentose derivatives (see Scheme 18). A key intermediate in that work[50] was the acid chloride of monomethyl di-*O*-acetyl-L-tartrate (**18**). Compound **18** was also an intermediate in the work of Lucas and Baumgarten.[49] Its preparation involved heating L-tartaric acid with acetic anhydride to give di-*O*-acetyl-tartaric anhydride (**17**), which reacts vigorously with methanol to give monomethyl di-*O*-acetyltartrate; the latter compound was then converted into **18** by treatment with thionyl chloride. The acid chloride group in **18** was reduced to a hydroxymethyl group by Bestmann and Schmiechen, on treatment with lithium tri-*t*-butoxyaluminum hydride at 0°; the reduced product was isolated as methyl tri-*O*-acetyl-L-threonate (**19**), which was converted into L-threono-1,4-lactone (**20**). When the reduction of **18** with lithium tri-*t*-butoxyaluminum hydride was performed at −75°, methyl 2,3-di-*O*-acetyl-L-threuronate (**21**) was produced; Lucus and Baumgarten[49] had obtained this compound by a Rosenmund reduction of **18**.*

The acid chloride **18** could be transformed with diazomethane into the diazoketone **22**.[51,52] Compound **22** was converted by Bestmann and Schmiechen[50] into the diethyl dithioacetal **23** by a reaction with ethylsulfenyl chloride, followed by treatment of the intermediate 1-chloro-1-ethylthio derivative with sodium thioethoxide. Desulfurization of **23** with Raney nickel gave methyl 2,3-di-*O*-acetyl–5-deoxy-L-*threo*-4-pentulosonate (**24**). Treatment of the diazoketone **22** with boron trifluoride-etherate in ethanol afforded methyl 2,3-di-*O*-acetyl–5-*O*-ethyl–L-*threo*-4-pentulosonate (**25**). Compound **25** had been obtained earlier by Ultée and Soons,[52] by treatment of **22** with cupric oxide in ethanol, instead of the expected Wolff rearrangement product. A Wolff rearrangement of the diazoketone **22** was achieved by Bestmann and Schmiechen by irradiation with ultraviolet light of a methanol solution of **22**; the product was di-*O*-acetyl–2-deoxy–L-*threo*-pentaric acid dimethyl ester (**26**).

The diazoketone **22** has also been utilized in a synthesis of the branched-chain sugar (see also Section 7C) L-apiose by Weygand and Schmiechen[51] (Scheme 19). Treatment of **22** with acetic acid in the presence of copper powder gave methyl 2,3,5-tri-*O*-acetyl–4-pentulosonate (**27**), which was converted into methyl 2,3,5-tri-*O*-acetyl–4,4′-anhydro-4-*C*-hydroxymethyl–L-*threo*-pentonate (**28**) with diazomethane. The opening of the epoxide ring, after saponification of **28** to give **29**, was achieved with a strongly acidic ion-exchange resin; the resultant product (**30**) was finally converted into L-apiose by a Ruff degradation procedure (oxidative decarboxylation of the calcium salt of **30** with hydrogen peroxide in the presence of ferric acetate).

Some deoxy sugar derivatives have been obtained by Lukes et al.[53] from

* Reduction of methyl 2,3-di-*O*-acetyl-L-threuronate (**21**) with sodium amalgam gave L-threonic acid, which was characterized as the brucine salt:[49]

$$\mathbf{21} \xrightarrow{\text{Na—Hg}} \begin{array}{c} CH_2OH \\ | \\ HCOH \\ | \\ HOCH \\ | \\ COOH \end{array}$$

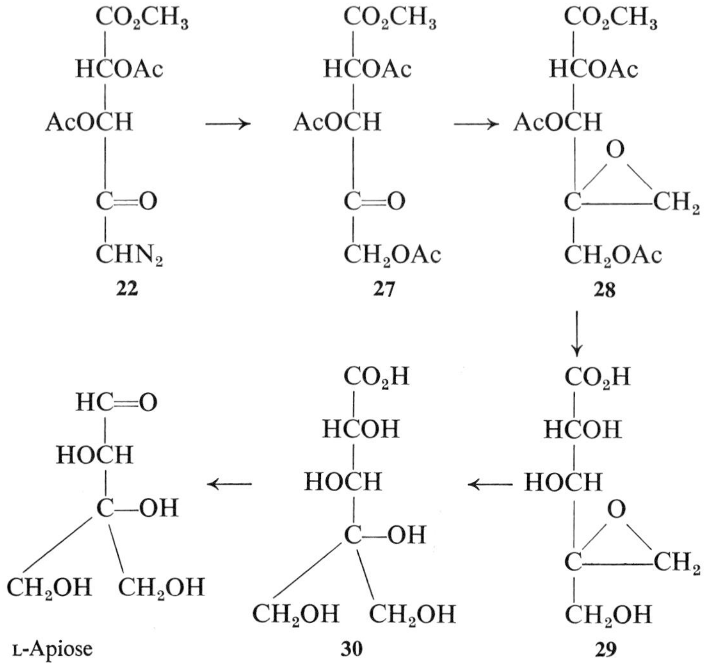

Scheme 19

L-parasorbic acid (**31**), which was isolated from *Sorbus aucuparia* berries. Hydroxylation of **31** with osmium tetroxide and sodium chlorate gave 4,6-dideoxy-L-*ribo*-hexonic acid 1,5-lactone (**32**). The calcium salt of the acid was then subjected to a Ruff degradation to afford 3,5-dideoxy-L-*erythro*-pentose (**33**) (Scheme 20).

A study, similar to the foregoing, is the stereospecific *trans*-hydroxylation of angelactinic acid (**34**).[54,55] Jary and Kefurt[55] found that hydroxylation of

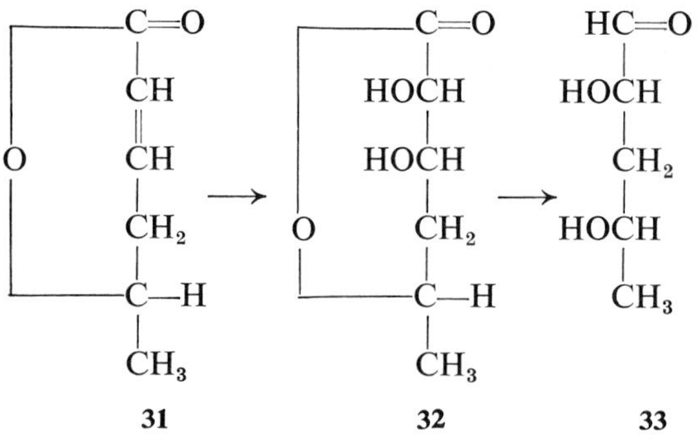

Scheme 20

34 with peroxyacetic acid gave the lactones of 5-deoxy-DL-arabinonic acid and 5-deoxy-DL-ribonic acid (**35** and **36**) in the ratio of 2.8: 1. These compounds were converted into 1-deoxy-DL-lyxitol (**37**) and 1-deoxy-DL-ribitol (**38**) by treatment of the lactones with lithium aluminum hydride in tetrahydrofuran (see Scheme 21; only one isomer of DL mixtures is shown).

Very recently, Koga et al.[56] described a new synthesis of D-ribose from L-glutamic acid (**39**) without the necessity of resolution at intermediate

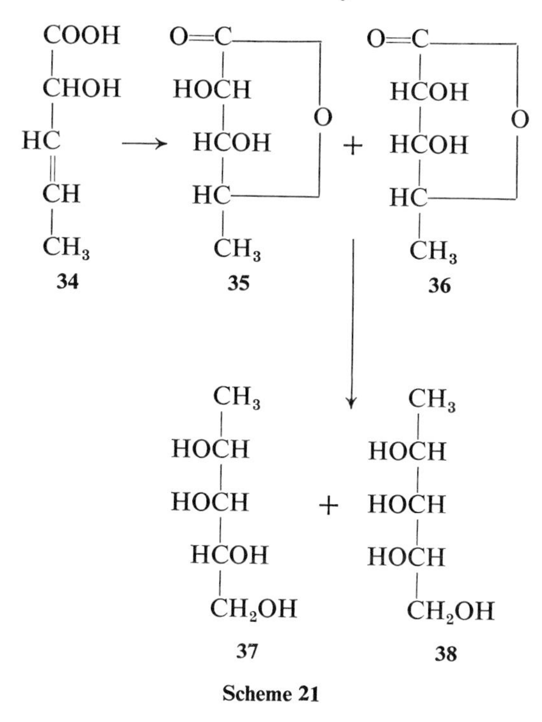

Scheme 21

stages; the asymmetric center of **39** became C-4 in D-ribose (see Scheme 22). The amino acid was deaminated to give, after esterification, the lactone ester **40**; this deamination was considered to proceed with full retention of configuration, because of the participation of the neighboring carboxyl group. Reduction of **40** with sodium borohydride in ethanol afforded the lactone alcohol **41**, which was converted into the benzyl ether **42**. Treatment of **42** with sodium and ethyl formate in ether gave the sodium salt **43**, which, on being heated in acidic aqueous dioxane, afforded 5-*O*-benzyl–2,3-dideoxy-D-pentofuranose (**44**), as a result of hydrolysis of the lactone ring, decarboxylation, and subsequent ring closure. Compound **44** was converted into a mixture of glycosides (see Section 6) which, on treatment with bromine and calcium carbonate, gave the monobromo derivative **45** as a mixture of diastereomers. Base-catalyzed dehydrobromination of **45** afforded the unsaturated derivative **46**. Surprisingly, hydroxylation of **46** with potassium permanganate or with osmium tetroxide gave a mixture of methyl 5-*O*-benzyl–β-D-ribofuranoside (**47**) and methyl 5-*O*-benzyl–α-D-lyxofuranoside (**48**). Compounds **47** and **48** could be separated as their acetonides or diacetates; alternatively, D-ribose could be isolated as its "anilide" by hydrogenation of the hydroxylation product to remove the benzyl group, followed by acid-catalyzed hydrolysis, and then treatment with aniline.

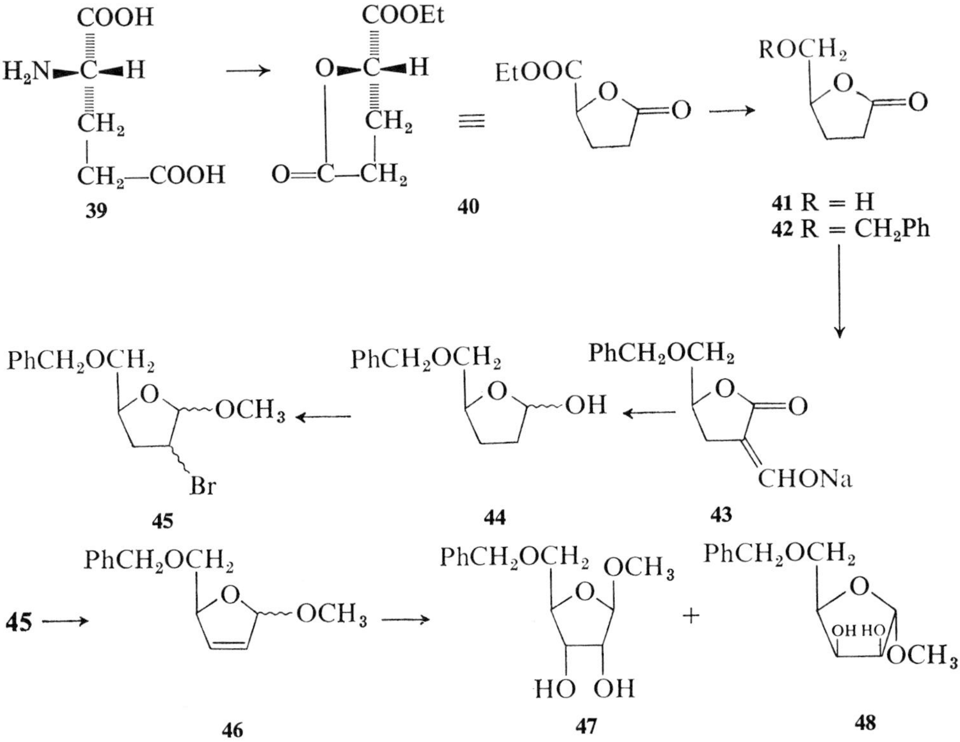

39 **40**

41 R = H
42 R = CH_2Ph

45 **44** **43**

46 **47** **48**

Scheme 22

Another synthesis of a sugar derivative from an amino acid (L-aspartic acid), and a synthesis involving both pyruvic acid and glycine, are discussed in Section 7A.

5. THE DIELS-ALDER REACTION

The Diels-Alder reaction has been employed by a number of workers for the preparation of dihydropyrans as substrates for the synthesis of a wide range of monosaccharides. These examples are discussed in Section 6, which is specifically concerned with the synthesis of sugars from pyran derivatives. In this section, the use of Diels-Alder condensations in two very elegant, total syntheses of novel carbohydrates is described.

The first example is that of Belleau and Au-Young,[57] whose objective was the total synthesis of amino sugars (Section 7A). They utilized the dienophilic properties of 1-chloro–1-nitrosocyclohexane and condensed it with methyl sorbate to yield *cis*-3-methyl–6-methoxycarbonyl–3,6-dihydro–1,2-oxazine hydrochloride (**49**). This Diels-Alder adduct is formed in a stereospecific manner, the *cis*-adducts only being formed when the diene has the *trans*, *trans* geometry, as is present in methyl sorbate. The adduct possesses a double bond at positions 4 and 5 and may therefore be hydroxylated to yield, after ring cleavage, 5-amino–5,6-dideoxy–DL-hexonic acids. The possible formation of a 2-amino–2-deoxy derivative was eliminated when the adduct, after hydrogenation and ring-opening, was shown to be an α-hydroxy acid and not an α-amino acid. When the *N*-benzoate of the adduct **50** was hydroxylated with osmium tetroxide-pyridine complex, attack of the reagent occurred from the least hindered side, and the diol-*N*-benzoyl-acid **51** resulted. Mild hydrolysis of **51**, followed by catalytic hydrogenation over Adam's catalyst, furnished 5-amino–5,6-dideoxy–DL-allonic acid (**52**) (see Scheme 23).

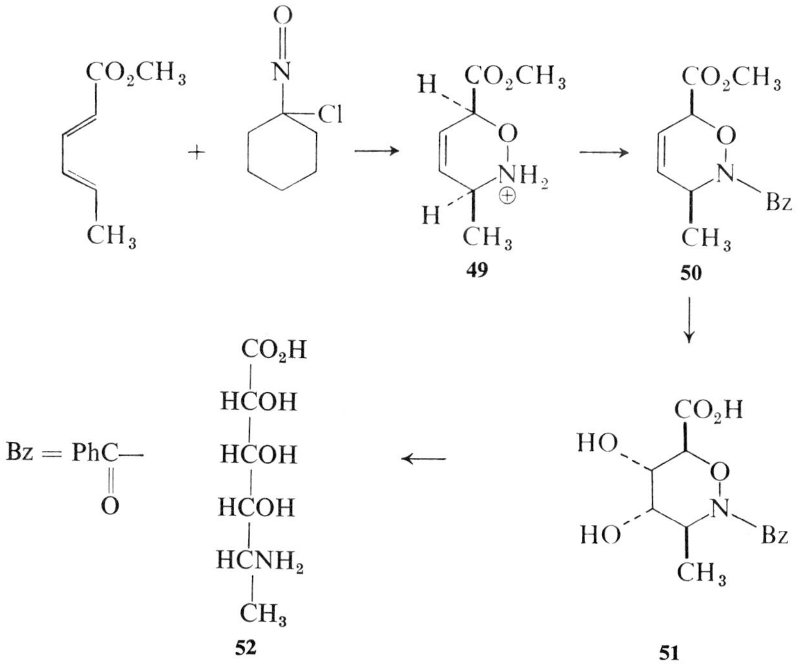

Bz = PhC(=O)—

Scheme 23

CO_2CH_3

O

N

Bz

CH_3

50

CO_2CH_3 O N Bz CH_3 **53**

CO_2CH_3 O N Bz CH_3 **54**

CO_2CH_3 HO O N HCO O Bz CH_3 **55**

O HCO CO_2CH_3 O N HO Bz CH_3 **56**

CO_2CH_3 HO O NH_2 ⊕ HO CH_3 **57**

O C=O OH N O ⊕H_2 CH_3 **58**

Scheme 24

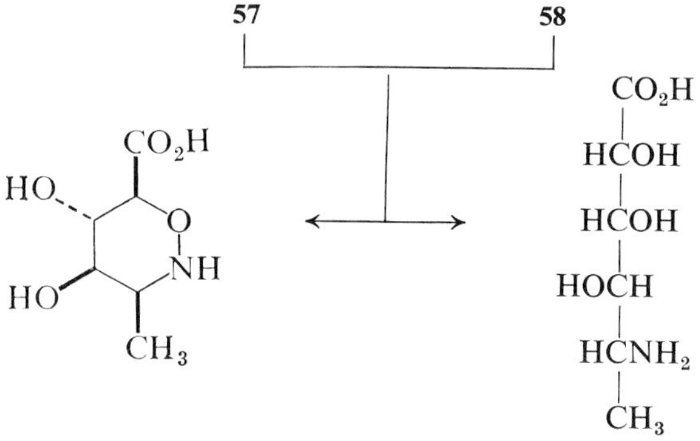

59 60

Scheme 24 (contd.)

The double bond in the *N*-benzoyl adduct **50** could be converted to the epoxide by reaction with peroxytrifluoroacetic acid. The reaction was not stereospecific; both possible isomers, **53** and **54**, were produced in equal amount and were separable by crystallization. Both epoxides, on reaction with formic acid, yielded a mixture of two *trans*-diol monoformates **55** and **56**. On treatment of this mixture with methanolic hydrogen chloride, a mixture of products **57** and **58** was obtained. Mild hydrolysis of this mixture gave a single tetrahydro-1,2-oxazine carboxylic acid (**59**), whereas catalytic hydrogenation of the mixture afforded 5-amino–5,6-dideoxy–DL-gulonic acid (**60**) (see Scheme 24).

In the early 1960s there was considerable interest in the synthesis of sugars in which the ring oxygen was replaced by other heteroatoms such as nitrogen or sulfur.[58] All of the syntheses were achieved by chemical modification of readily available monosaccharides. Very recently, Vyas and Hay[59] found that methyl cyanodithioformate possesses a very marked dienophilic activity and affords a facile one-step synthesis to a variety of unsaturated, deoxy, 1-thio sugars with sulfur in the ring by way of a Diels-Alder reaction. Thus with 1,3-butadiene, methyl cyanodithioformate affords a 75% yield of a crystalline Δ^3-dihydrothiopyran derivative **61*** after 24 hours in methylene chloride at room temperature. The adduct is a stable low-melting racemate:

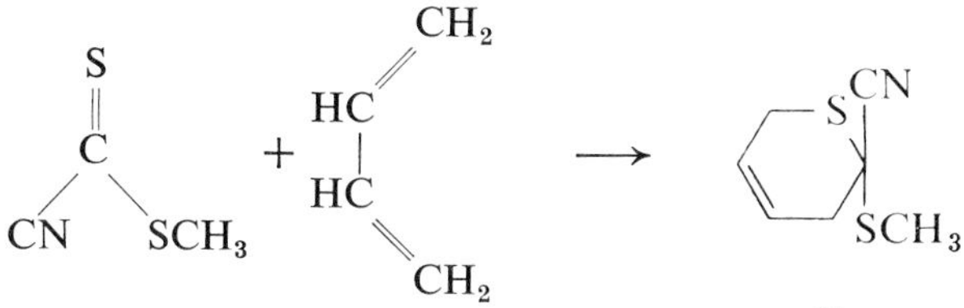

61

* Compound **61** is considered to be a carbohydrate derivative by virtue of the fact that it is an acetal, specifically, a dithioacetal. Compounds such as **61**, which have an *S*-alkyl (or *S*-aryl) group at C—1 of the sugar ring, are called 1-thioglycosides (see also Section 6).

2,3-Dimethyl–1,3-butadiene is more reactive and reacts with methyl cyanodithioformate in 5 min at 4° to give a racemic mixture of **62** and **63**:

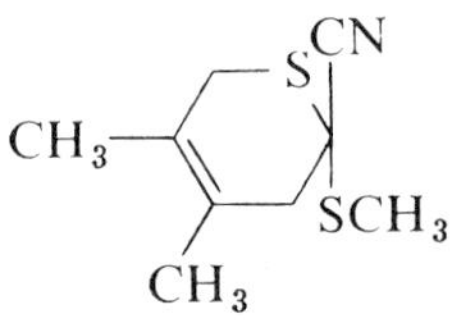

62 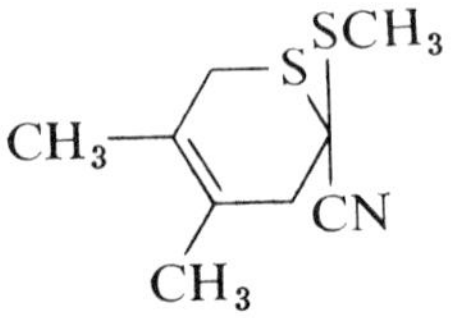 **63**

With cyclopentadiene, the reaction occurs instantaneously at 0° in methylene chloride to give an isomeric mixture of **64** and **65** in a 60:40 ratio, respectively. Isomer **64** crystallized, leaving a syrupy residue of **64** and **65**. The structure

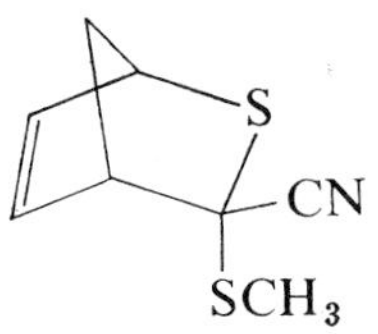

64 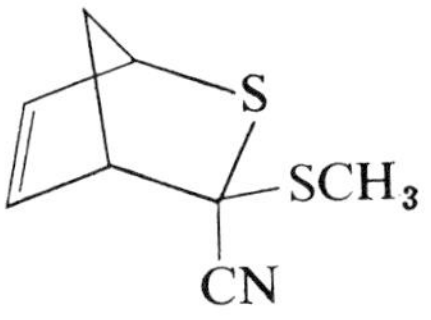 **65**

of 3-*endo*-thiomethyl–3-*exo*-cyano–2-thiobicyclo[2.2.1]hept-5-ene was assigned to **64** by nuclear magnetic resonance (NMR) spectroscopy. Isomerization of **64** into **65** was observed when **64** was kept in chloroform solution at room temperature. A stable crystalline adduct **66** has also been obtained from the reaction of *trans,trans*-1,4-diacetoxy–1,3-butadiene with methyl cyanodithioformate in refluxing chloroform:

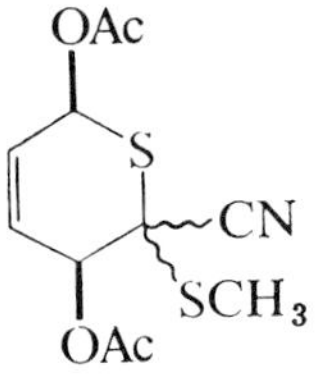

66

6. SYNTHESES FROM FURAN AND PYRAN DERIVATIVES

It is well known that 1,4- and 1,5-hydroxyaldehydes exist primarily as cyclic hemiacetals:[60]

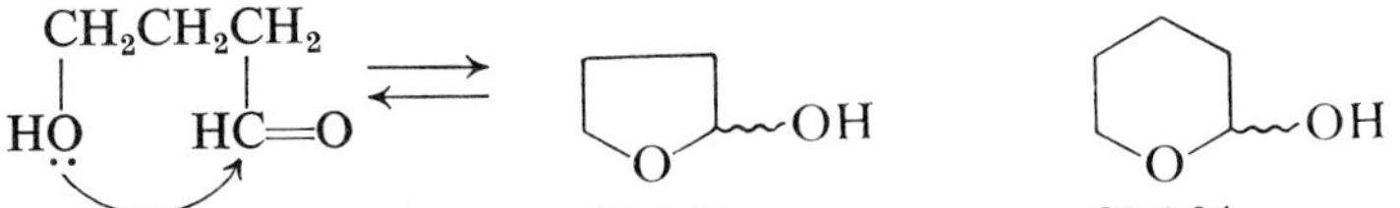

88.6% **67** 93.3% **68**

The two hemiacetals **67** and **68** may be regarded as carbohydrate models, namely, 2,3-dideoxytetrofuranose and 2,3,4-trideoxypentopyranose, respectively. A discussion of the compositions of the equilibrium mixtures of sugars in solution has been presented recently by Angyal.[61] Treatment of a cyclic hemiacetal with an alcohol in the presence of anhydrous acids yields an acetal. In the language of carbohydrate chemistry, the acetal is called a glycoside, and the conversion is said to involve the introduction of an aglycon at the anomeric center of the sugar:

ROH, $H^{\oplus}$ — Anomeric center — OR — Aglycon

Sugar (glycose) Glycoside

It is not surprising that furan and pyran derivatives themselves have been found to be useful substrates for the total synthesis of sugars.

Several syntheses of simple models of sugars have been achieved by addition reactions to the double bonds in 2,3-dihydrofuran (**69**) and 3,4-dihydro-2*H*-pyran (**70**) and their derivatives.* Compound **69** was first made,

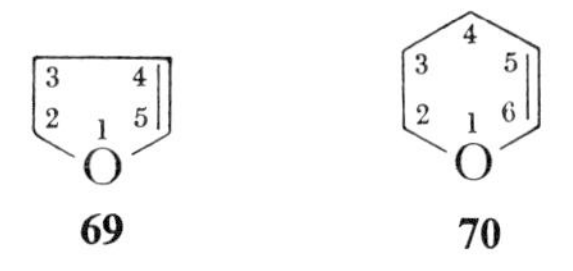

69 **70**

in 24% yield, by passing tetrahydrofurfuryl alcohol over a copper-nickel alloy;[62] higher yields have been obtained[63] by treatment of 3-chloro-2-alkoxytetrahydrofurans (made from tetrahydrofuran) with sodium. In another method,[64] butane-1,4-diol is dehydrogenated over a cobalt catalyst at 220° to give 2-hydroxytetrahydrofuran, which then eliminates water to afford 2,3-dihydrofuran in 80% yield. A convenient preparation is the rearrangement of the commercially available 2,5-dihydro isomer.[65]

3,4-Dihydro-2*H*-pyran (**70**) is made commercially by passing tetrahydrofurfuryl alcohol over alumina at 350°.[66] The mechanism of this ring-expansion has been followed with ^{14}C and found to proceed through the carbonium ion **71**:[67]

CH_2OH — Al_2O_3, Δ — $CH_2^{\oplus}$ —

71 **70**

* The numbering of furans and pyrans follows the universally adopted system: the heteroatom is called No. 1. In carbohydrate nomenclature, the anomeric center has been given this number. Compounds such as **69** and **70** in carbohydrate chemistry are called glycals.

Some α,β-unsaturated carbonyl compounds dimerize to 3,4-dihydro-2*H*-pyrans;[68] for example, acrolein, in the presence of a little hydroquinone, dimerizes to 2-formyl–3,4-dihydro–2*H*-pyran (**72**), and crotonaldehyde gives **73**:

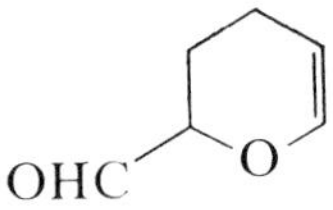

OHC 72 O

OHC 73 O H_3C CH_3

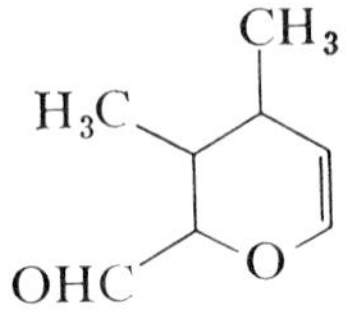

Several substituted 3,4-dihydro–2*H*-pyrans have been obtained by a Diels-Alder reaction, and these have served as substrates for the synthesis of sugars or were already simple models of sugars. Smith et al.[69] have added a wide variety of ethylenic compounds, such as vinyl ethers, unsaturated esters, methacrylonitrile, and olefins, to acrolein and other conjugated carbonyl compounds, such as methacrolein and crotonaldehyde:

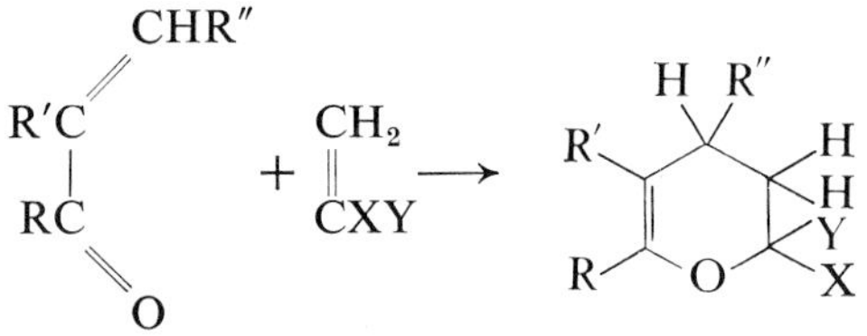

$$R'C(=CHR'')-C(R)=O + CH_2=CXY \longrightarrow$$ (dihydropyran bearing R, R′, H, R″, H, H, Y, X)

2,3-Dihydrofurans and 3,4-dihydro–2*H*-pyrans, which have the double bond in the α-position to the ring oxygen, undergo the usual reactions of vinyl ethers. Thus they react vigorously with water and other hydroxylic compounds in the presence of a trace of acid to give mainly 2-hydroxy- or 2-alkoxy–tetrahydro compounds or esters:

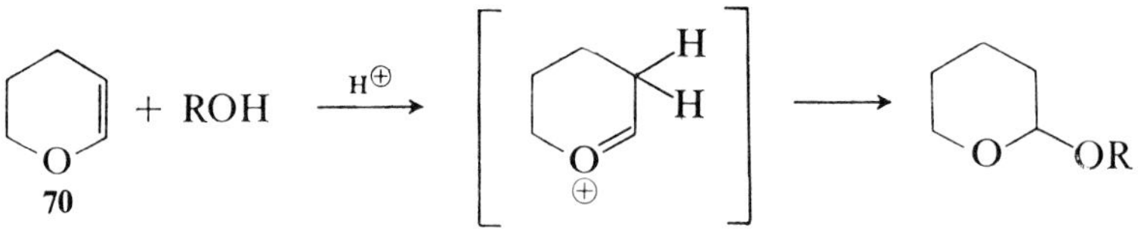

70 + ROH $\xrightarrow{H^{\oplus}}$ [oxocarbenium ion, O$^{\oplus}$, H, H] $\longrightarrow$ 2-OR tetrahydropyran

The acetal (or glycoside) shown in the preceding equation is unstable in the presence of aqueous acids, giving the alcohol and 2-hydroxytetrahydropyran (or free sugar). Many simple carbohydrates have been prepared by such addition reactions, and the compounds have been used as models for studies of glycoside hydrolysis[70] or for heterocyclic conformational analysis.[71] Of particular importance in the preparation of 2-oxy-substituted tetrahydropyrans is the anomeric effect[72] (or Edward-Lemieux effect[73]), by which term is meant the greater preference of an electron-withdrawing group for the axial position when it is located adjacent to a heteroatom in a ring than when it is located elsewhere. Thus, for example, it has been found[74] that acid-catalyzed addition of methanol to 2-methoxymethyl–2,3-dihydro–4*H*-pyran gave an equilibrium mixture of two isomers in the ratio of 70% *trans* to 30% *cis* (Scheme 25).

Various 2-tetrahydrofuranyl ethers (or furanosides) have been prepared[75] by the addition of alcohols to 2,3-dihydrofuran in the presence of acid. An alternative synthesis, from tetrahydrofuran and *t*-butyl perbenzoate in the presence of alcohols, is also available.[76]

2,3-Dihydroxy-tetrahydropyran and -tetrahydrofuran can be made from the dihydropyran **70** and the dihydrofuran **69** by reaction with osmium

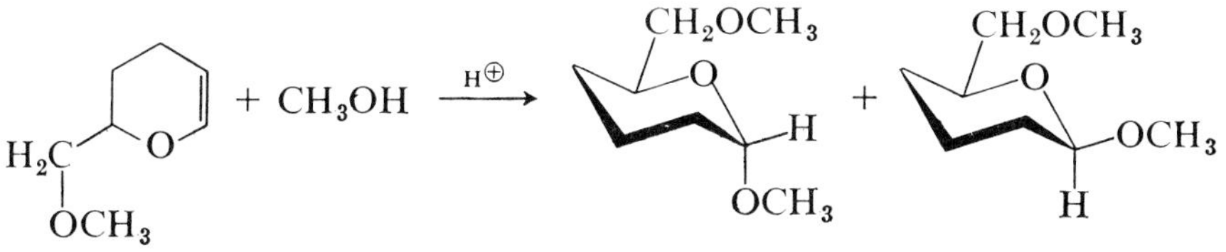

H_2C / OCH_3 (dihydropyran) + CH_3OH $\xrightarrow{H^\oplus}$ (CH_2OCH_3, O, H, OCH_3) + (CH_2OCH_3, O, OCH_3, H)

trans (70%) *cis* (30%)

Scheme 25

tetroxide and hydrogen peroxide in *t*-butanol[77] or with lead tetraacetate,[68] respectively. Both diols give 2,4-dinitrophenylosazones. In a more recent study,[78] **70** was treated with *m*-chloroperoxybenzoic acid in wet ether to give a diol (Scheme 26) whose NMR spectrum indicated it to be a mixture of the *cis* and *trans* isomers in the ratio of 30:70, respectively. The same ratio was obtained when the diol mixture was allowed to equilibrate in the presence of a small amount of *p*-toluenesulfonic acid. It therefore appears that if the reaction of the dihydropyran with the peroxy acid had taken place by way of an epoxy intermediate and produced initially the *trans*-diol stereospecifically, a rapid isomerization must have occurred to form the equilibrium mixture. Treatment of the crude diol with β-chloroethanol in the presence of *p*-toluenesulfonic acid produced 2-(β-chloroethoxy)–3-hydroxytetrahydropyran (**74**) as a mixture of the *cis* and *trans* isomers in a ratio of 69 : 31. Upon further treatment of the mixture with a catalytic amount of *p*-toluenesulfonic acid in β-chloroethanol, this ratio changed to the equilibrium mixture of 40% *cis:* 60% *trans*. Treatment of **74** with sodium hydride in 1,2-dimethoxyethane afforded both the *cis* and *trans* isomers of tetrahydropyrano[2,3-*b*]-1,4-dioxane. The results obtained with **74** indicate that the reaction of the diol with β-chloroethanol is highly stereospecific. A possible explanation[78] for this stereospecificity involves hydrogen bonding by the C-3

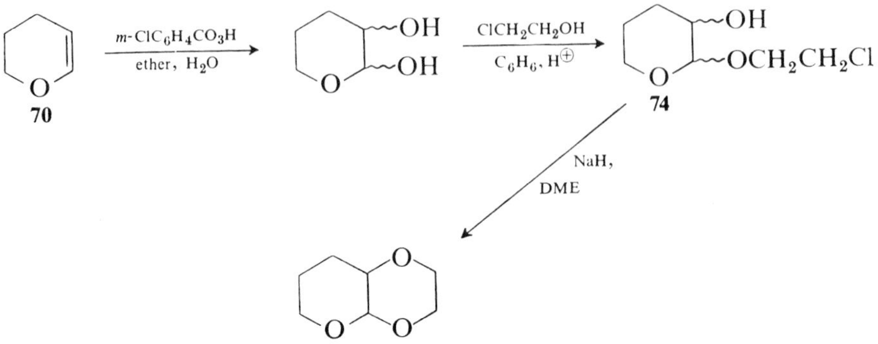

Scheme 26

hydroxyl group with the incoming alcohol, and hence a preference for attack by the β-chloroethanol (by way of an S_N2 or S_N1 mechanism) on the same side of the ring as is occupied by the C-3 hydroxyl. Partial isomerization would then yield the 69:31 ratio of *cis* to *trans* isomers (Scheme 27).

In a related study, Sweet and Brown[79] performed the oxidation of 2,3-dihydrofuran and 3,4-dihydro–2*H*-pyran with peroxybenzoic acid or *m*-chloroperoxybenzoic acid in the presence of an alcohol and obtained, respectively, *trans*-2-alkoxy–3-hydroxytetrahydrofurans and *trans*-2-alkoxy–3-hydroxytetrahydropyrans. Presumably, oxidation with the peroxy acids involves the formation of an epoxide intermediate or, alternatively, an "epoxidelike" transition state; the observed results are then in agreement with the known preferred *trans* opening of an epoxide ring. In the presence of acids, the acetals isomerized readily to given an equilibrium mixture of *cis* and *trans* isomers.

The reactions of 3,4-dihydro–2*H*-pyran (**70**) with halogens and halogen compounds have been well studied, and the products have proved to be useful intermediates for further syntheses. Compound **70** readily adds chlorine, bromine, hydrogen chloride, or hydrogen bromide[80] to give 2,3-dihalogeno- or 2-halogeno-tetrahydropyrans. The halogen atom at C-2 is removed as hydrogen halide, on distillation of the product at atmospheric pressure, to give 3-chloro- or 3-bromo-5,6-dihydro–4*H*-pyran (**75**, Scheme 28). The halogen atom at C-3 in all of these compounds is relatively inert, but that at C-2 resembles the halogen in α-chloro ethers. With alcohols or sodium salts of aliphatic acids, the dihalogeno compounds give 2-alkoxy or 2-acyloxy compounds, and with water substituted bis(tetrahydropyranyl) ethers are obtained.[81,82] Reaction of compound **70** or its derivatives with halogens in a hydroxylic solvent also gives the corresponding halogenated 2-hydroxy- or 2-alkoxy-tetrahydropyran; for example, 3-chloro–2-hydroxytetrahydropyran (**76**) can be made by chlorinating an emulsion of compound

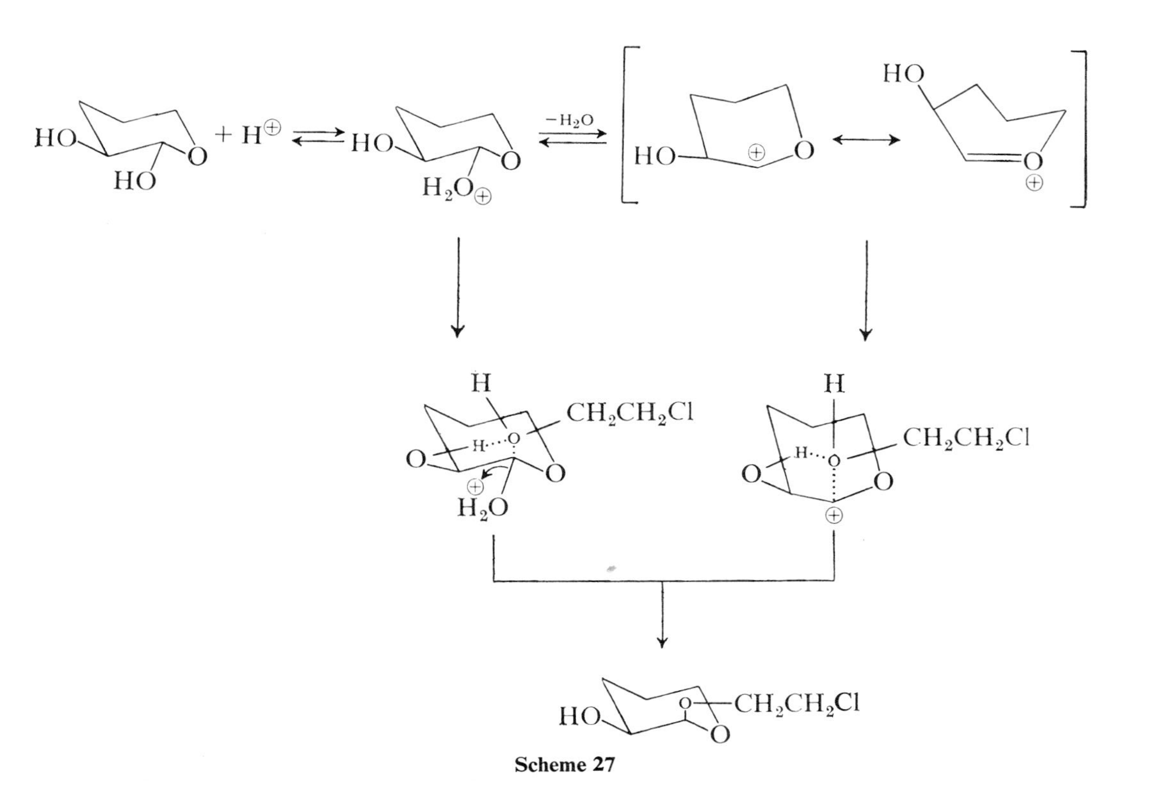

HO
HO
+ H⊕
HO
H_2O⊕
$-H_2O$
HO
⊕
O
HO
O
⊕
H
CH_2CH_2Cl
O
H
O
⊕
H_2O
H
CH_2CH_2Cl
O
H
O
⊕
HO
O
CH_2CH_2Cl
O

Scheme 27

70 in water. The mechanisms of halogenation and halogenomethoxylation of compound **70** have been discussed by Lemieux and Fraser-Reid.[83]

Addition of hydrogen chloride to 2,3-dihydrofuran (**69**) affords 2-chlorotetrahydrofuran.[63] 2,3-Dichlorotetrahydrofuran can be made by chlorinating compound **69** or, more conveniently, tetrahydrofuran itself. The chlorine atom at C-2 is as reactive as that in 2-chlorotetrahydropyran, and a corresponding series of hydroxy- and alkoxy-chlorotetrahydrofurans can be prepared.[63,84]

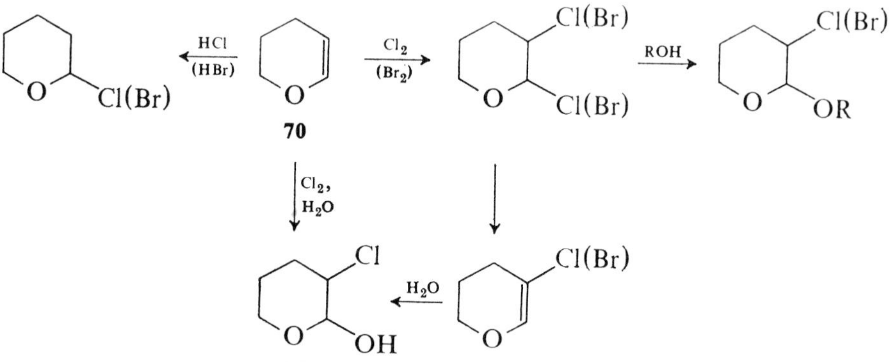

Scheme 28

The preparation of 3-alkythio–2-methoxy-tetrahydrofurans and -tetrahydropyrans has also been achieved. Senning and Lawesson[85] have described the synthesis of several 2-alkoxy–3-(trichloromethylthio)tetrahydrofurans by what was considered to be an "acid-catalyzed" addition of trichloromethylsulfenyl chloride to 2,3-dihydrofuran, followed by displacement of the halogen by an alkoxy group. More recently, Baldwin and Brown reported the results obtained when a similar procedure was applied to 3,4-dihydro–2*H*-pyran[86] and examined in detail the mechanism of the addition of ethanesulfenyl chloride.[87] It was found[86] that the reaction at $-20°$ of alkylsulfenyl chlorides with 3,4-dihydro–2*H*-pyran, followed by treatment of the product with sodium methoxide, gave, highly stereoselectively, *trans*-3-alkylthio–2-methoxytetrahydropyrans. Distillation under vacuum at temperatures above 55° resulted in the elimination of methyl alcohol to produce 5-alkylthio–3,4-dihydro–2*H*-pyrans in good yield. At lower temperatures, isomerization to a mixture of *cis*- and *trans*-3-alkylthio–2-methoxytetrahydropyrans occurred. Isomerization also occurred with acid catalysts. A mechanism for the formation of the 2-alkoxy–3-alkylthiotetrahydropyrans was proposed[86,87] and is shown in Scheme 29. The first step involved a nucleophilic attack of the α,β-unsaturated ether on the alkylsulfenyl chloride to displace the

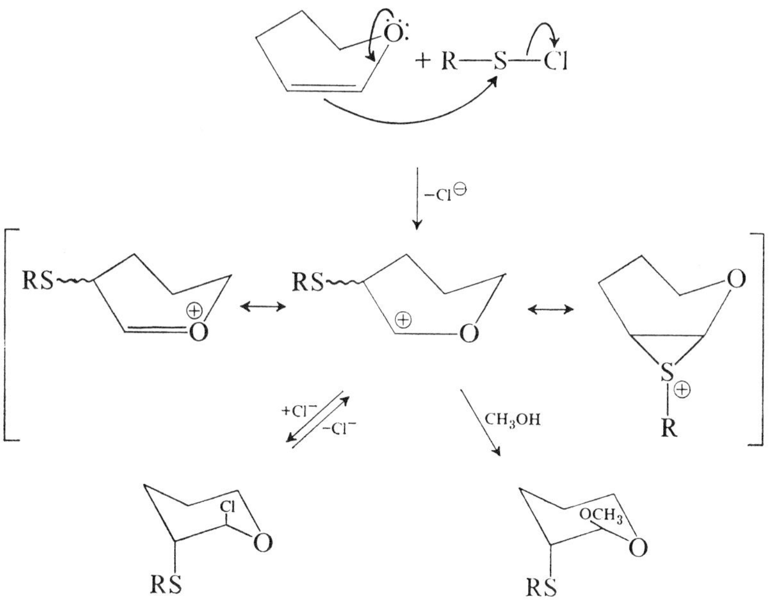

$R = CH_3, C_2H_5, PhCH_2$

Scheme 29

chloride ion and give an oxocarbonium-episulfonium ion in which the episulfonium contribution is considered to be paramount. Attack by chloride ion would then give, reversibly, *trans*-3-alkylthio–2-chlorotetrahydropyran. In the presence of an alcohol, the episulfonium ion, obtained by loss of the chloride ion, then gave, stereoselectively, the *trans*-3-alkylthio–2-alkoxy-tetrahydropyran.

In recent years, there has been considerable interest in the utility of 2-alkoxy–5,6-dihydro–2*H*-pyrans (**77**) and their derivatives as substrates for the synthesis of a wide range of monosaccharides. In addition to their use for the preparation of simple model carbohydrates, they have been intermediates in the total synthesis of some sugar moieties found in antibiotics (discussed later in this section). The 2-alkoxy–5,6-dihydro–2*H*-pyrans can be prepared by conversion of 3,4-dihydro–2*H*-pyran (**70**) into 2-alkoxy-3-halogenotetrahydropyrans followed by dehydrohalogenation.

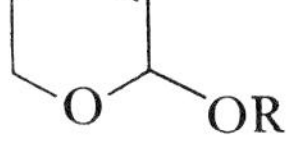

77

Sweet and Brown[88] have investigated the epoxidation of some 2-alkoxy-5,6-dihydro-2*H*-pyrans and the lithium aluminum hydride reduction of the resultant epoxides. Thus 2-methoxy- and 2-*t*-butoxy-5,6-dihydro-2*H*-pyran were found to react with *m*-chloroperoxybenzoic acid in ether solution, at 35°, to give mixtures of *trans*- and *cis*-2-alkoxy-3,4-epoxytetrahydropyrans in which the *trans* to *cis* ratio was 3:1 and 9:1 for the 2-methoxy and 2-*t*-butoxy compounds, respectively (Scheme 30). This selectivity can clearly be attributed to the steric effect of the 2-alkoxy group. Lithium aluminum hydride attacks the epoxide ring in these *cis* and *trans* compounds exclusively at the epoxide carbon remote from the alkoxy substituent to form only 2-alkoxy-3-hydroxytetrahydropyrans. This selectivity of hydride attack was

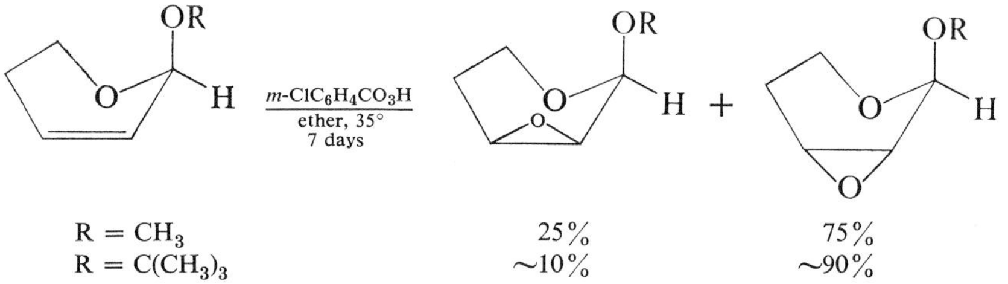

OR H $\xrightarrow[\text{ether, 35°, 7 days}]{m\text{-ClC}_6\text{H}_4\text{CO}_3\text{H}}$ OR H + OR H

R = CH_3 25% 75%
R = $C(CH_3)_3$ ~10% ~90%

Scheme 30

attributed to the polar effect of the two oxygen atoms at the anomeric carbon (C-2) in the tetrahydropyrans.

The foregoing approach was later employed by Sweet and Brown[89] to prepare a mixture of methyl 4-deoxy-3-*O*-methyl-α- and β-DL-*threo*-pentopyranosides starting with the *cis*-, *trans*-, or *cis-trans*-2-alkoxy-3,4-epoxytetrahydropyran (**78** and **79**). The mixture was obtained by a highly selective reaction of methanol with **78** and/or **79** in the presence of a catalytic amount of *p*-toluenesulfonic acid (Scheme 31). The result can be rationalized on the assumption that the first step was attack by methanol of the protonated epoxide exclusively at the carbon atom remote from the anomeric center. This was then followed by a slower isomerization to produce an equilibrium mixture of the α and β anomers of the DL mixture.

Sweet and Brown[90] have also found that acid-catalyzed methanolysis of 2-methoxy-5,6-dihydro-2*H*-pyran gave, in good yield, a 4:1 mixture of *trans*- and *cis*-2,4-dimethoxytetrahydropyran, respectively. The two isomers could be readily separated by gas-liquid chromatography. The formation essentially of only the two components shows that a highly selective reaction had occurred between methanol and 2-methoxy-5,6-dihydro-2*H*-pyran. A possible route is shown in Scheme 32. Because of the electron-withdrawing effect of the two oxygen atoms attached to the anomeric carbon, the more stable protonated state is that in which the carbon of the double bond more

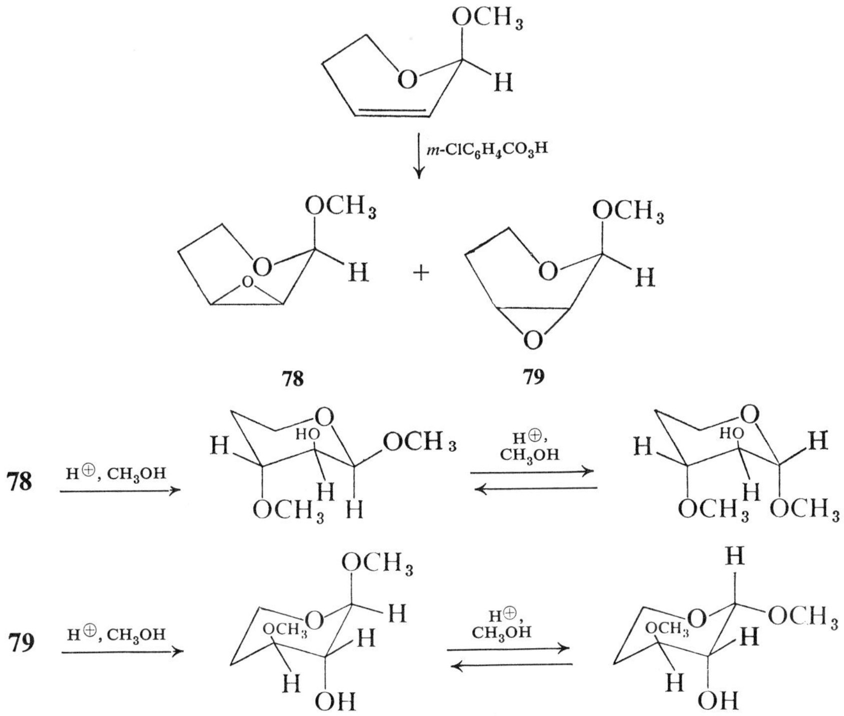

OCH_3 H

m-$ClC_6H_4CO_3H$

OCH_3 H + OCH_3 H

78 **79**

78 $H^{\oplus}$, CH_3OH → H HO OCH_3 OCH_3 H H ⇌ ($H^{\oplus}$, CH_3OH) H HO H OCH_3 H OCH_3

79 $H^{\oplus}$, CH_3OH → OCH_3 OCH_3 H H H OH ⇌ ($H^{\oplus}$, CH_3OH) H OCH_3 OCH_3 H H OH

Scheme 31

OCH_3 + $H^{\oplus}$ ⇌ OCH_3 ⊕ H H

CH_3OH
$-H^{\oplus}$

OCH_3 OCH_3 + OCH_3 OCH_3

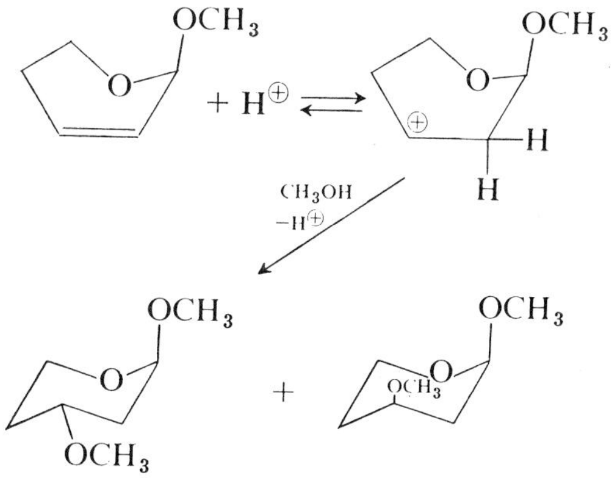

Scheme 32

remote from the anomeric center becomes positive. Another route to the formation of the 2,4-dimethoxytetrahydropyrans has also been considered by Sweet and Brown; this involves a preferred protonation of the oxygen atoms rather than the double bond; acetals are known to be unstable in an acidic medium. Treatment of a solution of 2,4-dimethoxytetrahydropyran, in a mixture of water and 1,2-dimethoxyethane with Amberlite IR-120 gave 2-hydroxy-4-methoxytetrahydropyran as an equilibrium mixture of *cis* and *trans* isomers in the ratio 1:1.[91]

The reactions of 2-methoxy-5,6-dihydro-2*H*-pyran with 1,3-dibromo-5,5-

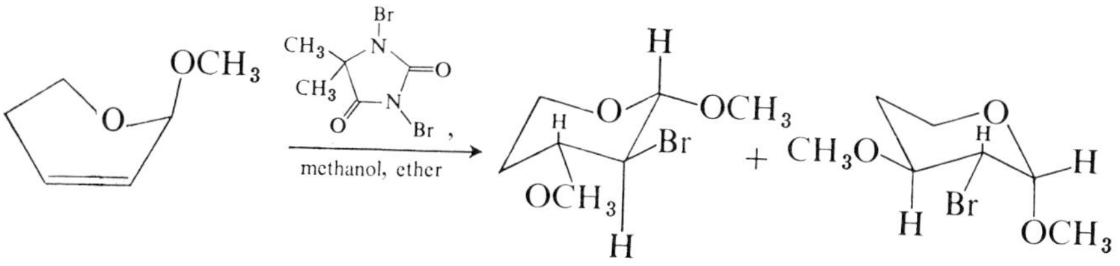

methanol, ether

80 **81**

Scheme 33

dimethylhydantoin in ether-methanol,[91] and with ethanesulfenyl chloride,[92] have been described by Baldwin and Brown. The former reaction gave a 2:1 mixture of the isomers 3β-bromo-2α,4α-dimethoxytetrahydropyran (**80**) and 3α-bromo-2α,4β-dimethoxytetrahydropyran (**81**), respectively. The structures and preferred conformations of the isomers are shown in Scheme 33. The reaction of ethanesulfenyl chloride with 2-methoxy-5,6-dihydro-2*H*-pyran gives only 4β-chloro-3α-ethylthio-2β-methoxytetrahydropyran (**82**). A proposed[92] route for this highly selective reaction is shown in Scheme 34. Because of the anomeric effect, the preferred conformer of 2-methoxy-5,6-dihydro-2*H*-pyran is considered to be that in which the C-2 methoxy group is *quasi* axial. The first step presumably is electrophilic attack of ethanesulfenyl chloride on the dihydropyran from the least hindered side of the molecule, namely, *trans* to the C-2 methoxy group, to give an episulfonium ion intermediate. The next step is attack by the chloride ion at either C-3 or C-4 with simultaneous opening of the episulfonium ring. It is believed that the steric effect of the *quasi* axial methoxy group, and its polar repulsion for the chloride ion, would strongly inhibit attack by the chloride ion at C-3. Moreover, the electron-withdrawing effect of the two oxygen atoms at C-2, which would destabilize an incipient positive charge at C-4 less than C-3, favors reaction of the chloride ion at C-4 rather than at C-3. The reaction of bromine with 2-ethoxy-5,6-dihydro-2*H*-pyran has also been studied[92a]; two isomeric dibromides were obtained, namely, 3α,4β-dibromo-2α-ethoxytetrahydropyran and 3α,4β-dibromo-2β-ethoxytetrahydropyran. Treatment of

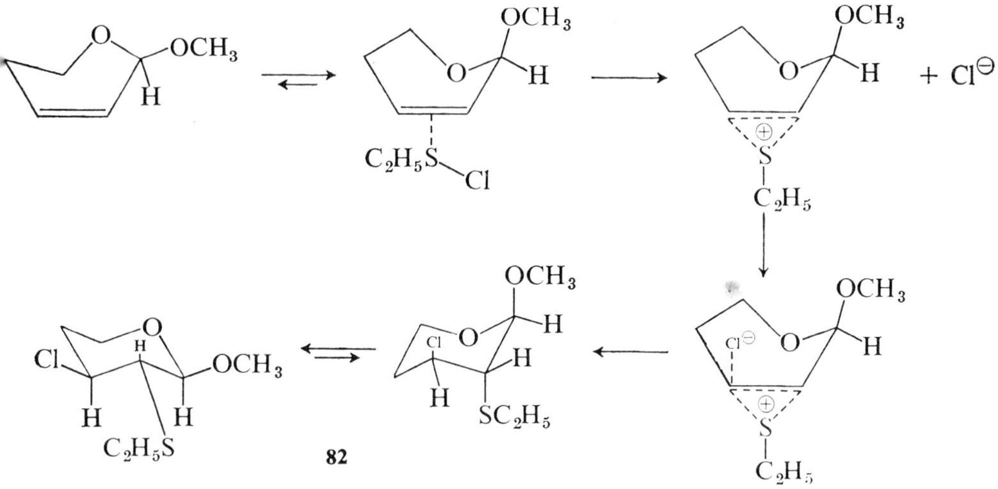

Scheme 34

these isomers with refluxing ethanolic sodium ethoxide afforded *trans*-5,6-diethoxy-5,6-dihydro-2*H*-pyran, *cis*-2,5-diethoxy-5,6-dihydro-2*H*-pyran, *trans*-2,5-diethoxy-5,6-dihydro-2*H*-pyran, and 3-bromo-2-ethoxy-5,6-dihydro-2*H*-pyran.[92b]

Cahu and Descotes[93] have also subjected 2-alkoxy–5,6-dihydro–2*H*-pyran to addition reactions such as hydrohalogenation, hydroxyhalogenation, epoxidation, and *cis*-hydroxylation, and the *cis*- and *trans*-2-alkoxy–3,4-epoxytetrahydropyrans, to acid hydrolysis, and addition of dimethylamine.

The potential scope of additions to 2-alkoxy–5,6-dihydro–2*H*-pyrans has recently been extended by the availability of some new derivatives. It has been shown[94,95] that the condensation of esters of glyoxylic[96] or mesoxalic acid with 1-alkoxy–1,3-butadienes affords esters of 2-alkoxy–5,6-dihydro–2*H*-pyran 6-carboxylic acid (**83**) and 2-alkoxy–5,6-dihydro–2*H*-pyran 6,6-dicarboxylic acid (**84**), respectively, in good yield. The butyl esters of **83** have been converted with lithium aluminum hydride into 2-alkoxy–5,6-dihydro–6-hydroxymethyl–2*H*-pyrans which, on hydrogenation over platinum, gave the corresponding 2-alkoxy–6-hydroxymethyltetrahydropyrans.[97] Treatment of some of the 6-substituted 2-alkoxy–5,6-dihydro–2*H*-pyrans with *m*-chloroperoxybenzoic acid yielded mixtures of stereoisomeric epoxides,

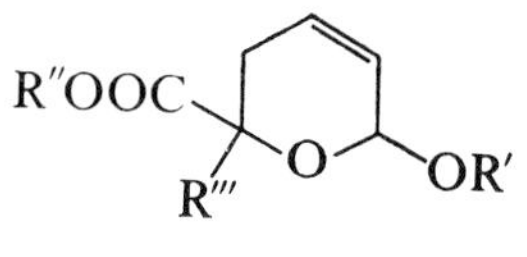

R′ = Me, Et, Pr, Bu
83 R″ = Et, Bu
R‴ = H
84 R″ = Et
R‴ = COOEt

which were separated by column chromatography.[98] Methanol has also been added, under acidic conditions, to 6-substituted 2-methoxy–5,6-dihydro–2*H*-pyrans to yield derivatives of 2,4-dimethoxytetrahydropyran.[99] The addition of bromine in ether solution to 6-substituted 2-methoxy-5,6-dihydro-2*H*-pyrans leads in the case of *trans* isomers to two stereoisomeric dibromo derivatives having the α-*xylo* and α-*arabino* configurations, whereas, *cis* isomers give β-*arabino* dibromo derivatives; in methanol solution, the treatment with bromine yields mixtures of dibromo and bromomethoxy derivatives.[99a] Stereospecific syntheses of four diastereomeric methyl 4-deoxy-DL-hexopyranosides have also been achieved from esters of 2-methoxy-5,6-dihydro-2*H*-pyran 6-carboxylic acid.[99b] Zwierzchowska and Zamojski[100] have demonstrated a "glycal-pseudoglycal" rearrangement[101] with some 6-substituted dihydropyran derivatives. Thus, for example, heating *trans*-**85** (a product of the condensation of 1-acetoxy-1,3-butadiene with butyl glyoxylate) in acetic anhydride gave two new products, the butyl esters of *trans*- and *cis*-4-acetoxy-5,6-dihydro-4*H*-pyran 6-carboxylic acid (**86**). The isomerization could be reversed, when *cis*- or *trans*-**86** were heated in acetic anhydride:

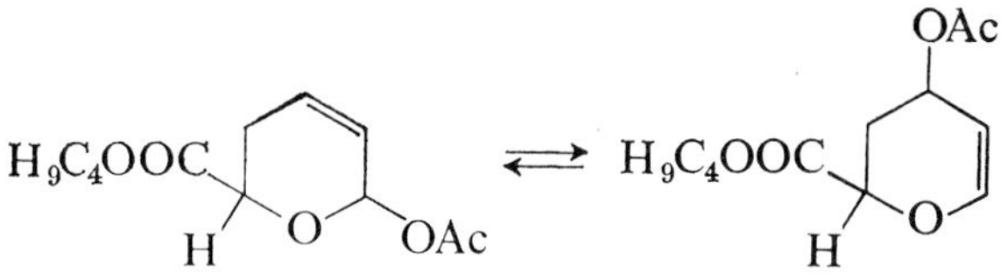

H_9C_4OOC H O OAc ⇄ OAc H_9C_4OOC H O

85 **86**

Much progress has been made in recent years by Brown and his associates in Canada toward the total syntheses, involving pyran intermediates, of monosaccharides. Some of this work has already been presented in this section. Other syntheses are now described which incorporate many of the reactions already discussed.

The synthesis of methyl 2,3-anhydro–4-deoxy–6-*O*-methyl–α-DL-*lyxo*-hexopyranoside (**91**) has been achieved[102] from 2-methoxymethyl–3,4-dihydro–2*H*-pyran (**87**). The sequence of reactions for this synthesis is shown in Scheme 35. Bromomethoxylation of **87** gave a 9:1 mixture of the two isomers **88** and **89**, respectively. This mixture was heated with sodium methoxide in refluxing methanol to give a product that contained at least 95% of the *trans*-2-methoxy–6-methoxymethyl–5,6-dihydro–2*H*-pyran, **90**. *m*-Chloroperoxybenzoic acid converted **90** into a mixture of methyl 2,3-anhydro–4-deoxy–6-*O*-methyl–α-DL-*lyxo*-hexopyranoside (**91**) (>95%) and methyl 2,3-anhydro–4-deoxy–6-*O*-methyl–α-DL-*ribo*-hexopyranoside (**92**) (<5%).

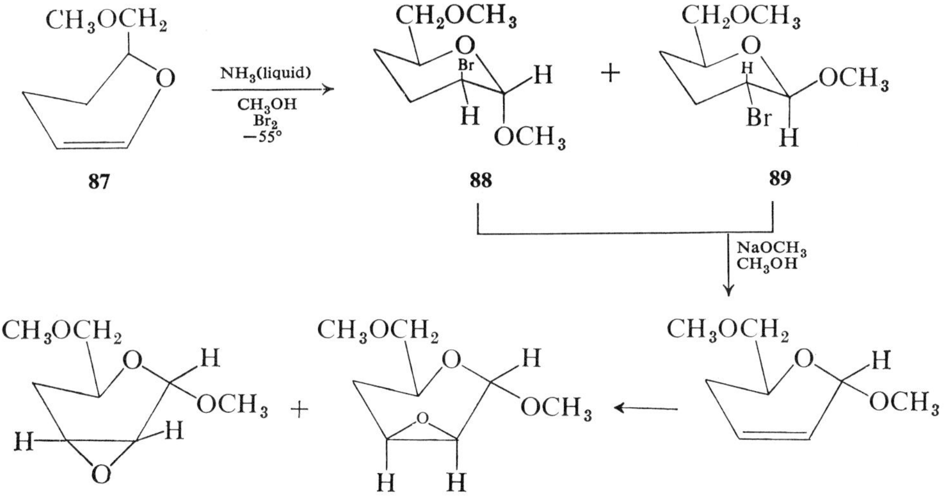

Scheme 35

Earlier in this section the preparation of *cis*- and *trans*-2,4-dimethoxytetrahydropyran was described. Recently, the synthesis of *cis*- and *trans*-2,5-dimethoxytetrahydropyran and the homolog *cis*- and *trans*-2,5-dimethoxy–6-methyltetrahydropyran were reported.[103] The starting material for the synthesis of *cis*- and *trans*-2,5-dimethoxytetrahydropyran (see Scheme 36) was 2 methoxy–3,4 dihydro–2*H* pyran (**93**), the condensation product of acrolein and methyl vinyl ether. Hydroboration and subsequent oxidation of **93** gave a 1:2 mixture of *cis*- and *trans*-5-hydroxy–2 methoxytetrahydropyran (**94**). From the methylated product **95**, *trans*-2,5-dimethoxytetrahydropyran could be obtained by gas-liquid chromatography. Distillation of the mixture of *cis*- and *trans*-2,5-dimethoxytetrahydropyran (**95**) over a catalytic amount of phosphoric pentoxide gave 3-methoxy–3,4-dihydro–2*H*-pyran (**96**). Bromomethoxylation of **96** with 1,3-dibromo-5,5-dimethylhydantoin in ether-methanol gave a mixture of three isomers, from which the major product, **97**, could be separated. Hydrogenation of **97** over palladium-on-charcoal afforded *cis*-2,5-dimethoxytetrahydropyran (**98**).

Both *cis*- and *trans*-2,5-dimethoxy–6-methyltetrahydropyran (**100** and **101**) were obtained by hydroboration and oxidation of 2-methoxy–6-methyl–3,4-dihydro–2*H*-pyran (**99**) (the condensation product of methyl vinyl ketone and methyl vinyl ether), followed by methylation of the resultant product (as shown on page 42).

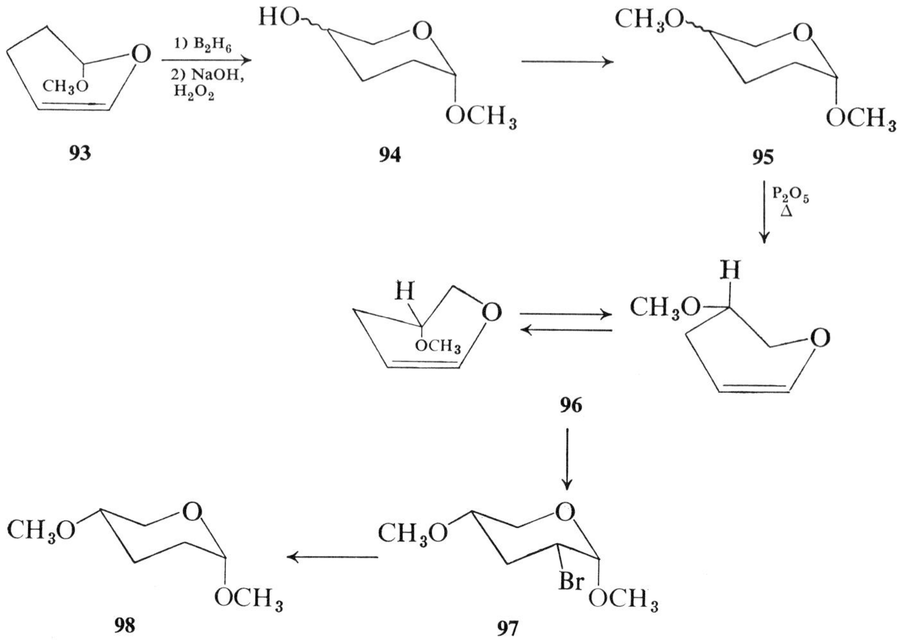

1) B_2H_6 2) NaOH, H_2O_2

93 94 95 96 97 98

Scheme 36

Srivastava and Brown[104] have utilized 3α-bromo–2α,5α-dimethoxytetrahydropyran (**97**), obtained by bromomethoxylation of 3-methoxy–3,4-dihydro–2*H*-pyran (**96**) as shown in Scheme 36, to prepare methyl 4-*O*-methyl–α-DL-arabinopyranoside (**104**) (see Scheme 37). Compound **97** was first dehydrobrominated with a boiling solution of potassium hydroxide in methanol to give *cis*-2,5-dimethoxy–5,6-dihydro–2*H*-pyran (**102**). Treatment of **102** with *m*-chloroperoxybenzoic acid in methylene chloride afforded, almost exclusively, the epoxide **103**, which was converted into methyl 4-*O*-methyl–α-DL-arabinopyranoside (**104**) with aqueous potassium hydroxide.

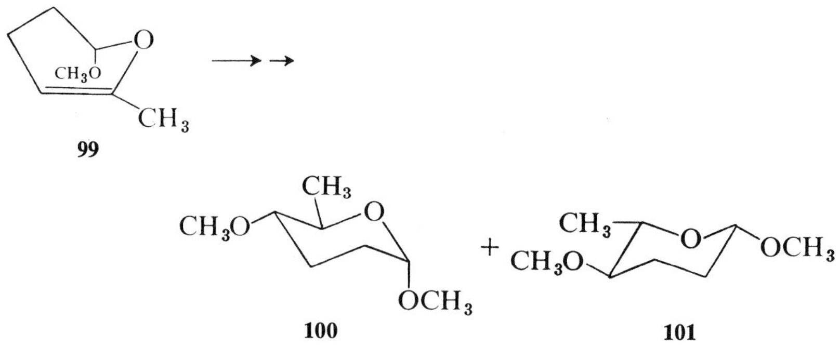

99 100 101

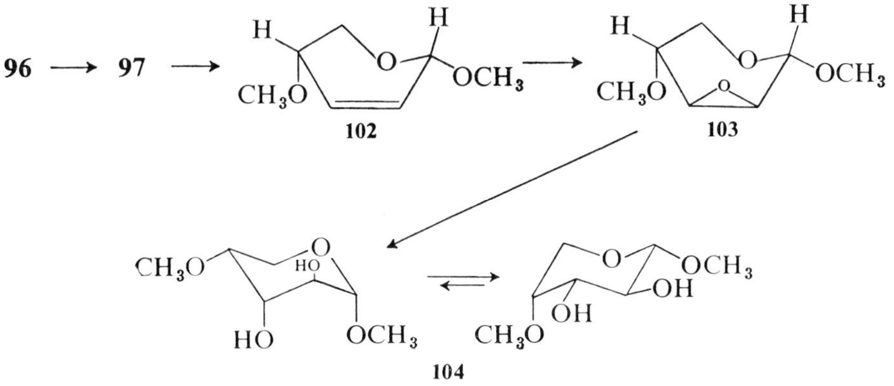

Scheme 37

A particularly significant development, for the total synthesis of monosaccharides, has been the conversion of acrolein dimer (**72**) into the olefins 6,8-dioxabicyclo[3.2.1]oct-3-ene (**109**) and 6,8-dioxabicyclo[3.2.1]oct-2-ene (**110**) by the sequence of reactions shown in Scheme 38.[105,106] Acrolein dimer was first reduced with sodium borohydride to give 2-hydroxymethyl–3,4-dihydro–2*H*-pyran (**105**). When the alcohol **105** was heated in refluxing benzene containing a catalytic amount of *p*-toluenesulfonic acid, 6,8-dioxabicyclo[3.2.1]octane (**106**) was formed. Treatment of **106** with bromine in carbon tetrachloride gave a mixture of two isomeric monobromides, considered to be *trans*- and *cis*-4-bromo–6,8-dioxabicyclo[3.2.1]octane (**107**

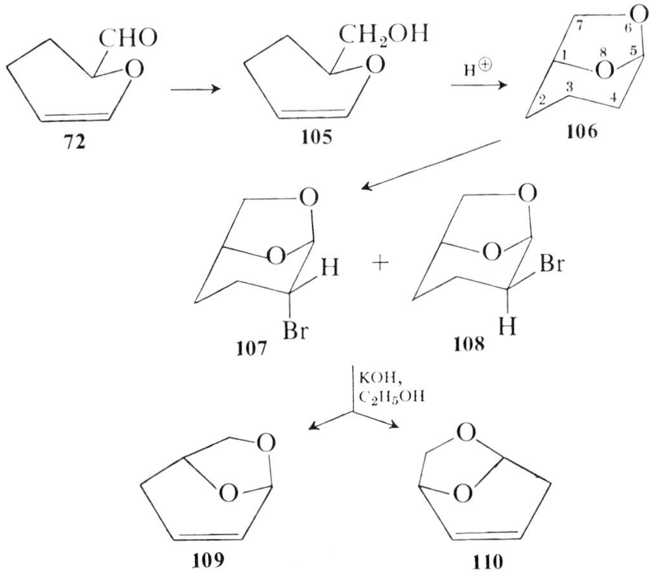

Scheme 38

and **108**). Heating the mixture of **107** and **108** in refluxing ethanolic potassium hydroxide gave the two olefins **109** and **110**, which were readily separated by gas-liquid chromatography; the proportion of the two isomeric olefins obtained depended upon the proportion of base to monobromide used during dehydrohalogenation.

6,8-Dioxabicyclo[3.2.1]oct-3-ene (**109**) and its isomer **110** have been used for the synthesis of several monosaccharides and their derivatives. Thus, for example, **109** has been converted into 1,6-anhydro-4-deoxy-β-DL-*xylo*-hexopyranose (**113**) by epoxidation with *m*-chloroperoxybenzoic acid to give 1,6:2,3-dianhydro-4-deoxy-β-DL-*ribo*-hexopyranose (**111**), contaminated with <5% of the isomeric 1,6:2,3-dianhydro-4-deoxy-β-DL-*lyxo*-hexopyranose (**112**), followed by treatment with aqueous potassium hydroxide (Scheme 39).[105]

The stereoselective synthesis of DL-glucose has been accomplished[107,107a] in 34% overall yield starting from 1,6:2,3-dianhydro-4-deoxy-β-DL-*ribo*-hexopyranose (**111**) (Scheme 40). Compound **111** was converted into 1,6-anhydro-3,4-dideoxy-β-DL-*erythro*-hex-3-enopyranose (**114**, R = H) with *n*-butyllithium. Treatment of **114** with *m*-chloroperoxybenzoic acid gave 1,6:3,4-dianhydro-β-DL-*allo*-hexopyranose (**115**), which with aqueous barium

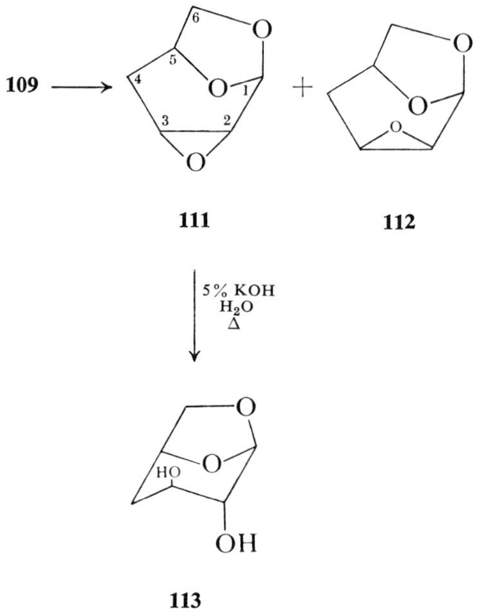

109 → **111** + **112**

5% KOH, H_2O, Δ

113

Scheme 39

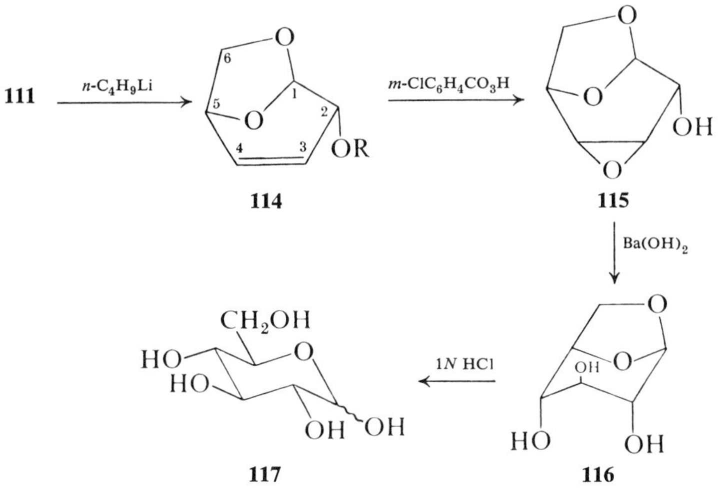

Scheme 40

hydroxide was converted into 1,6-anhydro-β-DL-*gluco*-hexopyranose (**116**). Acid-catalyzed hydrolysis of **116** afforded α,β-DL-glucose (**117**). 3-*O*-Methyl-DL-glucose and 3-deoxy-DL-*ribo*-hexopyranose have also been obtained[107a] from compound **115**. Thus, treatment of the oxirane **115** with sodium methoxide in methanol gave a very good yield of 1,6-anhydro-3-*O*-methyl-β-DL-*gluco*-hexopyranose as the only isolable product. Acid-catalyzed hydrolysis of the diacetate of this product afforded an excellent yield of 3-*O*-methyl-α,β-DL-glucopyranose. 1,6-Anhydro-3-*O*-methyl-β-DL-*gluco*-hexopyranose could also be obtained, along with methyl 3-*O*-methyl-α,β-DL-glucopyranoside, when **115** was heated in methanol containing *p*-toluenesulfonic acid monohydrate. Lithium aluminum hydride reacted with **115** to form 1,6-anhydro-3-deoxy-β-DL-*ribo*-hexopyranose, which was hydrolyzed readily by acid to 3-deoxy-DL-*ribo*-hexopyranose. In the above work of Singh and Brown,[107,107a] all of the products obtained from the oxirane **115** were those resulting only from *trans* diaxial opening of the oxirane ring.

Total syntheses of α,β-DL-allose (**118**) and α,β-DL-galactose (**119**) and their 2-*O*-methyl derivatives have also been accomplished[108] from **114** (R = H, CH_3— or CH_3CO—) by way of a stereoselective *cis*-hydroxylation by osmium tetroxide. The observation that the proportion of *allo* to *galacto* product obtained varied with the solvent (pyridine or dioxane) used in the hydroxylation reaction was interpreted as showing that the attack by osmic acid is subject to steric approach control.

Very recently, the total synthesis of several monodeoxy- and dideoxy-DL-hexopyranoses from the olefins **109** and **110** was reported.[109] The reactions

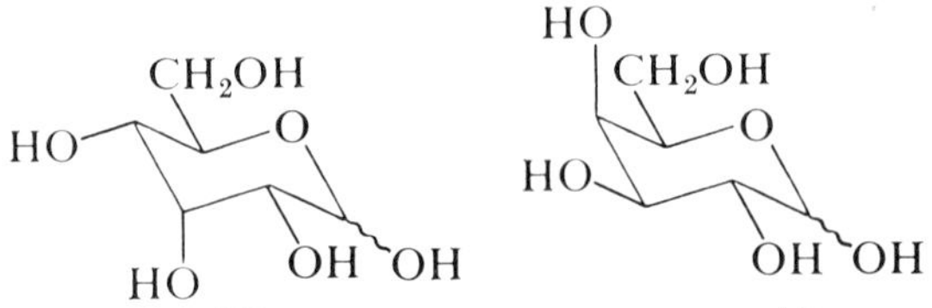

118 119

performed with olefin **109** are shown in Scheme 41. Reaction of osmic acid with **109** gave 1,6-anhydro–4-deoxy–β-DL-*ribo*-hexopyranose (**120**), which, on acid-catalyzed hydrolysis in aqueous dioxane, afforded 4-deoxy–α,β-DL-*ribo*-hexopyranose (**121**). Conversion of **109** into 1,6:2,3-dianhydro–4-deoxy–β-DL-*ribo*-hexopyranose (**111**), followed by treatment of **111** with lithium aluminum hydride, gave 1,6-anhydro–3,4-dideoxy–β-DL-*erythro*-hexopyranose (**122**). Compound **122** was hydrolyzed to 3,4-dideoxy–α,β-DL-*erythro*-hexopyranose (**123**). With olefin **110**, the same sequence of reactions led to 2-deoxy–αβ-DL-*ribo*-hexopyranose and, presumably, 2,3-dideoxy–α,β-DL-*erythro*-hexose, although the latter compound was not fully characterized.

Other examples of the stereoselective hydroxylation of double bonds by osmic acid were provided by the total synthesis of 4-*O*-methyl–DL-lyxose and 4-deoxy–DL-ribose[110] (see Scheme 42). The reaction of osmium tetroxide in pyridine with 2-methoxy–5,6-dihydro–2*H*-pyran and *cis*-2,5-dimethoxy–5,6-dihydro–2*H*-pyran afforded in excellent yield, as the sole isolable products, methyl 4-deoxy–β-DL-*erythro*-pentopyranoside (**124**) and methyl

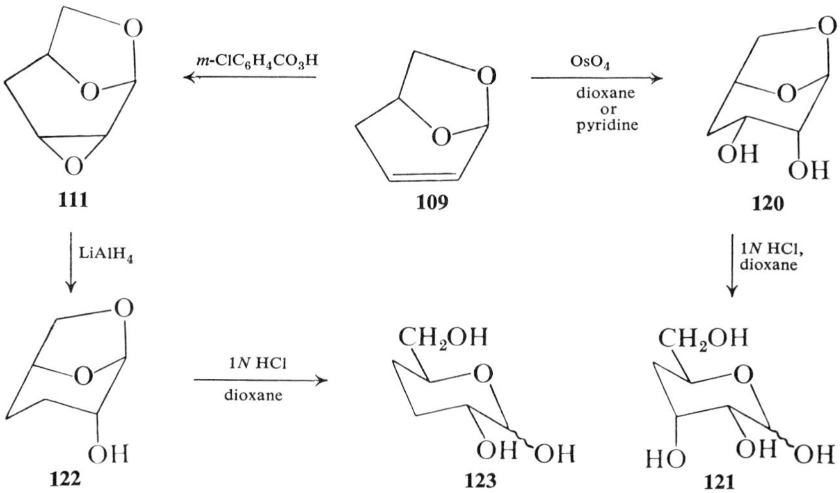

m-$ClC_6H_4CO_3H$; OsO_4 dioxane or pyridine ; $LiAlH_4$; 1N HCl, dioxane ; 1N HCl dioxane

111 109 120 122 123 121

Scheme 41

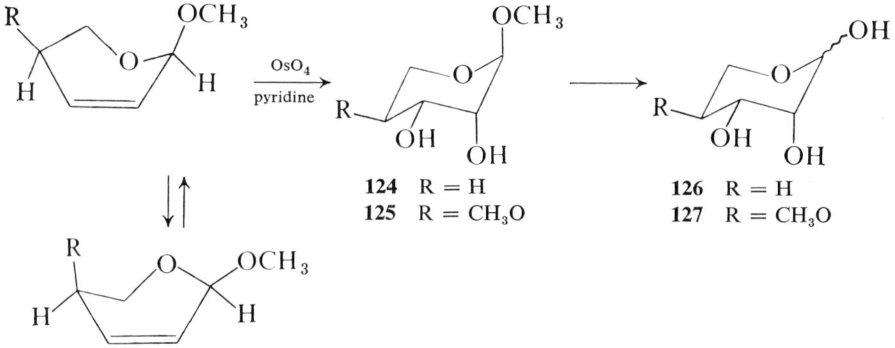

OsO_4 / pyridine

124 R = H
125 R = CH_3O

126 R = H
127 R = CH_3O

R = H
R = CH_3O

Scheme 42

4-*O*-methyl–α-DL-lyxopyranoside (**125**), respectively. Hydrolysis of **124** and **125** with aqueous dioxane containing sulfuric acid gave 4-deoxy–DL-*erythro*-pentose (4-deoxy–DL-ribose, **126**) and 4-*O*-methyl–DL-lyxose (**127**). The attack by osmic acid occurs nearly completely, if not exclusively, from the unhindered side, remote from the substituent. These results indicate that 2-methoxy–5,6-dihydro–2*H*-pyran must be predominantly in the conformation in which the C-2 alkoxy group is *quasi* axial.

Many antibiotics have been found to contain sugars of unusual structure, such as aminodeoxy, deoxy, and branched-chain sugars. During the 1960s, total syntheses, involving pyran intermediates of some of these carbohydrate moieties have been achieved. In 1962 Korte et al.[111] described a synthesis of DL-desosamine (DL-picrocin). Desosamine is a component of several macrolide antibiotics including erythromycin, oleandomycin, and narbomycin; its structure has been shown to be that of 3,4,6-trideoxy–3-dimethylamino–D-*xylo*-hexose (**128**). The starting material (see Scheme 43), δ-caprolactone (**129**), was reduced with lithium aluminum hydride in tetrahydrofuran to give 2-hydroxy-6-methyltetrahydropyran (**130**), which, on being heated in the presence of alumina, afforded 5,6-dihydro–6-methyl–4*H*-pyran (**131**). Compound **131** was then converted into 2,3-dibromo–6-methyltetrahydropyran (**132**) with bromine in carbon tetrachloride. Treatment

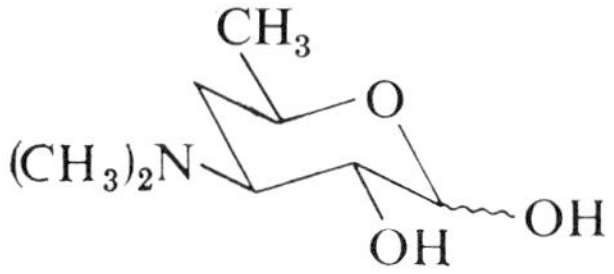

128

of **132** with ethanol saturated with ammonia gave the glycoside, 3-bromo–2-ethoxy–6-methyltetrahydropyran (**133**), which, when refluxed in ethanol containing sodium, yielded 5,6-dihydro–2-ethoxy–6-methyl–2*H*-pyran (**134**). Oxidation of **134** with peroxybenzoic acid afforded 3,4-epoxy–2-ethoxy–6-methyltetrahydropyran (**135**). Opening of the epoxide ring with aqueous dimethylamine gave 4-dimethylamino–2-ethoxy–3-hydroxy–6-methyltetra-

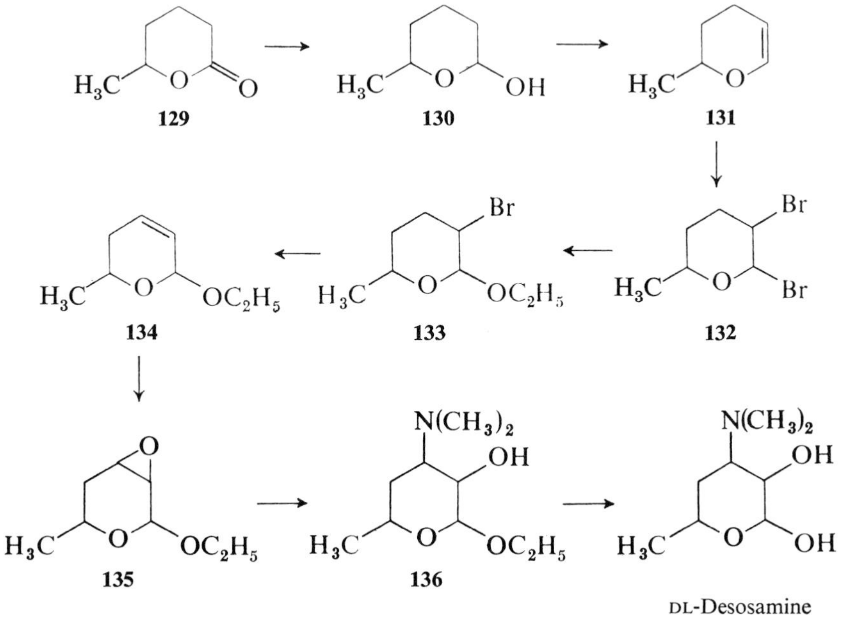

Scheme 43

hydropyran (**136**). Hydrolysis of **136** with aqueous hydrochloric acid finally afforded DL-desosamine hydrochloride.

Newman[112] has also achieved a synthesis of DL-desosamine by way of 5,6-dihydro–2-ethoxy–6-methyl–2*H*-pyran (**134**). This preparation of **134** is shown in Scheme 44. The lithium salt of propargyl aldehyde diethyl acetal reacted with propylene oxide to give 1,1-diethoxy–5-hydroxyhex-2-yne in 60% yield. This compound was converted into **134** by hydrogenation over 10% palladium-on-charcoal, followed by treatment with acid.

More recently, Mochalin et al.[113] have reported a stereospecific synthesis of desosamine by way of the reaction of **135**a with dimethylamine.

Yasuda and Matsumoto have recently described total syntheses of some sugars found in antibiotics. One of these syntheses was that of methyl DL-mycaminoside (**137**).[114] Mycaminose is a sugar component in the antibiotics magnamycin, spiramycin, and leucomycin. Hydroboration of

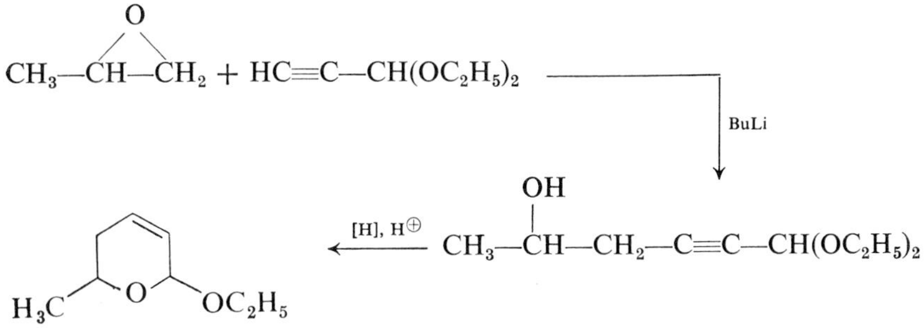

$CH_3—CH—CH_2$ (epoxide, O) + $HC≡C—CH(OC_2H_5)_2$ →(BuLi) $CH_3—CH(OH)—CH_2—C≡C—CH(OC_2H_5)_2$ →([H], $H^{\oplus}$) **134** (H_3C, O, OC_2H_5)

Scheme 44

3,4-dihydro–2-ethoxy–6-methyl–2*H*-pyran (**138**), followed by oxidation with hydrogen peroxide in alkaline solution, gave the alcohol **139** (see Scheme 45). Treatment of **139** with bromine in boiling methanol containing hydrogen chloride yielded three bromo compounds, **140**, **141**, and **142**. Each of these three compounds could be converted into the unsaturated alcohol **143**; **142**, for example, whose NMR spectrum suggested that the bromine and methoxyl groups are diaxial (one-proton doublet at τ 5.31, J 1 Hz), readily afforded **143** on treatment with sodium azide in *N*,*N*-dimethylformamide at 120–125°. Oxidation of **143** with peroxybenzoic acid gave the epoxide **144**, which afforded methyl DL-mycaminoside when treated with a saturated aqueous solution of dimethylamine.

Yasuda and Matsumoto have also prepared methyl DL-oleandroside (**145**) and its C-3 epimer, methyl DL-cymaroside (**146**).[115] Oleandrose is a sugar component of cardiac glycosides and of the antibiotic oleandomycin. A key intermediate in this work was the unsaturated alcohol **143**, whose preparation from 3,4-dihydro–2-ethoxy–6-methyl–2*H*-pyran has already been described. Compound **143** was first transformed into the benzyl ether **147** with benzyl

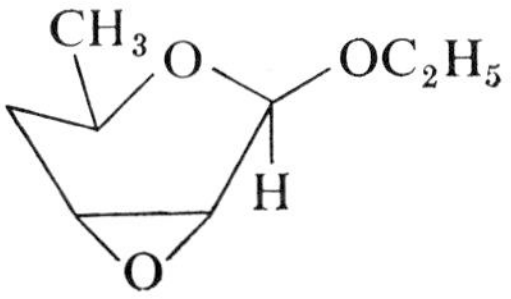

CH_3 O OC_2H_5 H O

135a

$N(CH_3)_2$ HO OH H_3CO O CH_3

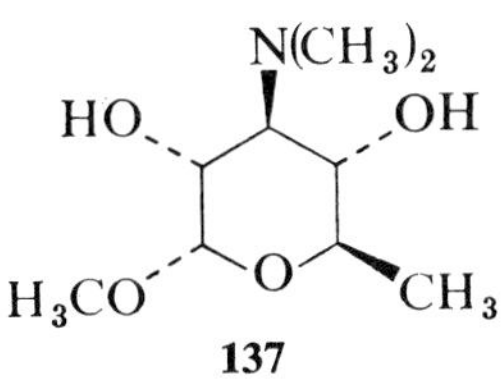

137

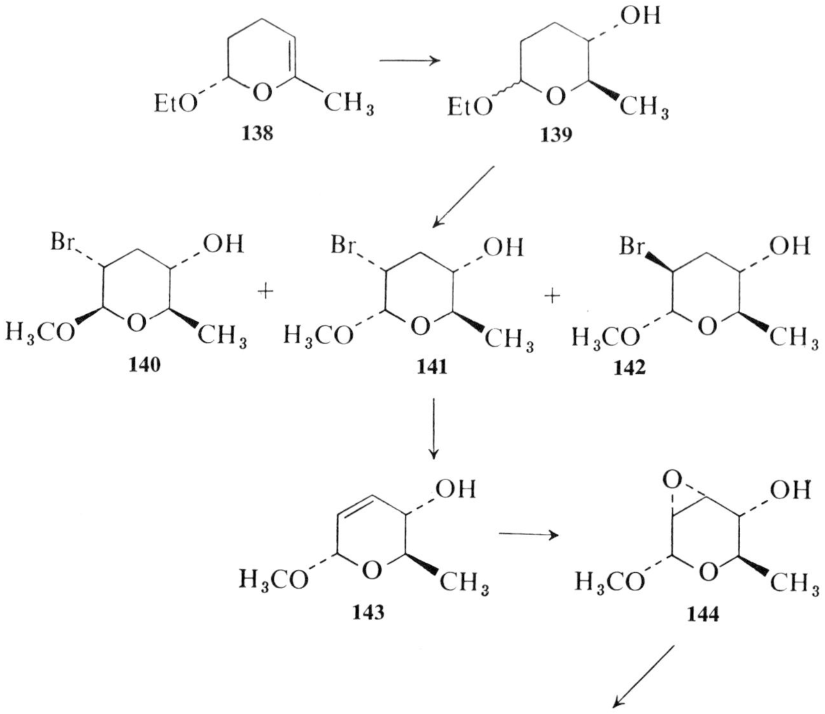

138 → 139 → 140 + 141 + 142

141 → 143 → 144 → 137

Methyl DL-mycaminoside

Scheme 45

chloride and sodium hydroxide (see Scheme 46). Treatment of **147** in refluxing methanol with a catalytic amount of *p*-toluenesulfonic acid gave two adducts, **148** and **149**. A similar addition of methanol to 5,6-dihydro–2-methoxy–2*H*-pyran has already been described in this section. Hydrogenolysis of **148** over palladium-on-carbon afforded methyl DL-oleandroside (**145**), whereas hydrogenolysis of **149** in methanol containing hydrogen chloride yielded an anomeric mixture of methyl DL-cymaroside (**146**).

The diamino sugar kasugamine **150** is a component of the antibiotic

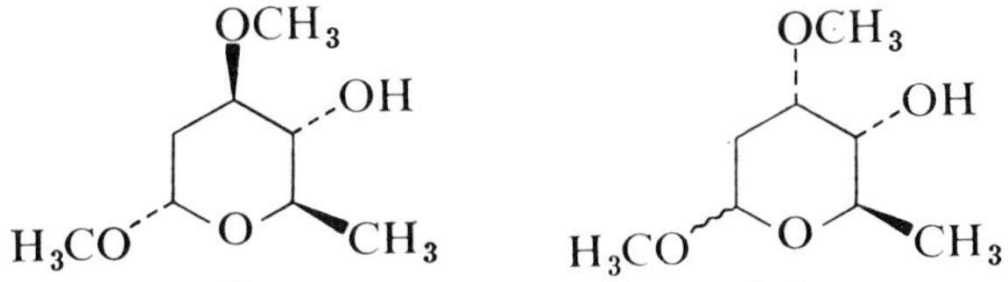

145 **146**

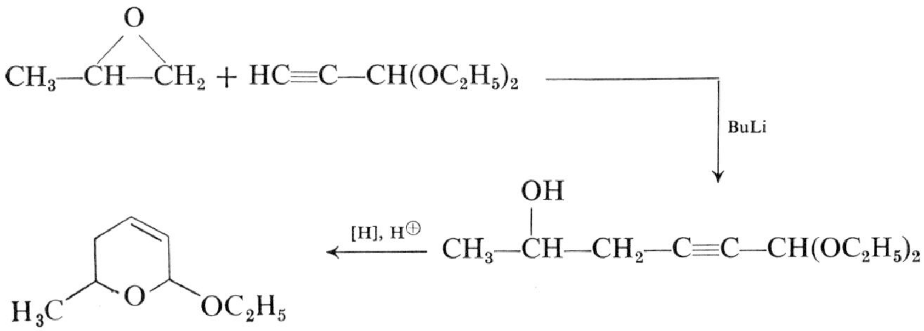

$$CH_3-\overset{O}{CH-CH_2} + HC\equiv C-CH(OC_2H_5)_2 \xrightarrow{BuLi} CH_3-\overset{OH}{\overset{|}{C}H}-CH_2-C\equiv C-CH(OC_2H_5)_2 \xrightarrow{[H],\ H^{\oplus}} \mathbf{134}$$

134

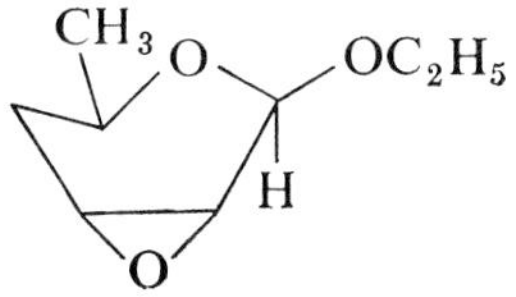

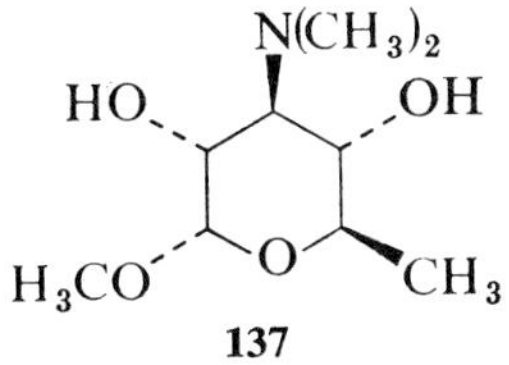

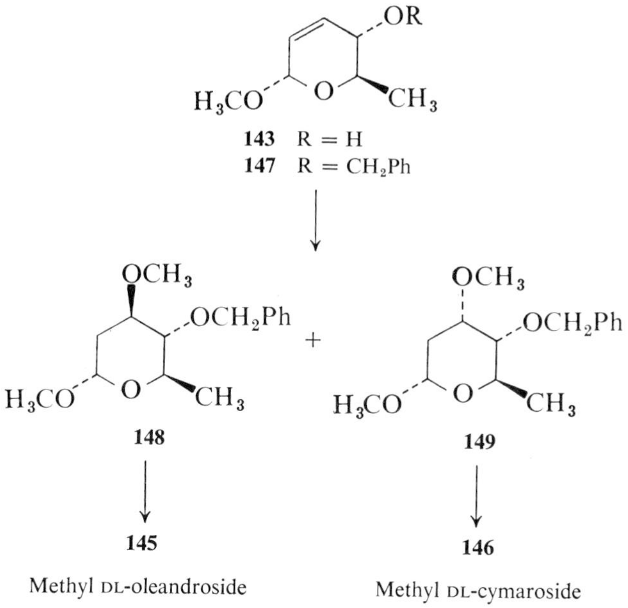

143 R = H
147 R = CH_2Ph

148 → **145** Methyl DL-oleandroside

149 → **146** Methyl DL-cymaroside

Scheme 46

kasugamycin. Two syntheses of derivatives of kasugamine involving pyran intermediates have recently been achieved.

The starting material for the first synthesis[116] (Scheme 47) was 3,4-dihydro-6-methyl-2*H*-pyran-2-one (**151**). Treatment of **151** with nitrosyl chloride in methylene chloride at −60° gave a dimer of 6-chloro-6-methyl-5-nitrosotetrahydropyran-2-one (**152**) in 97% yield. This dimer was easily hydrolyzed with water to give 4-oximino-5-oxohexanoic acid (**153**) which, on reduction with hydrogen over platinum, afforded stereoselectively DL-*erythro*-4-amino-5-hydroxyhexanoic acid (**154**). The acid **154** was lactonized by treatment with acetic anhydride at room temperature to give the *N*-acetylated lactone **155**. Compound **155** was reduced with lithium aluminum hydride to the hemiacetal **156** which, by refluxing with acetic anhydride and pyridine, gave the *N*-diacetyldihydropyran **157**. Treatment of **157** with nitrosyl chloride gave the chloronitroso dimer **158**; displacement with methanol in the presence of mercuric cyanide afforded the α-glycoside **159**. Methyl DL-kasugaminide

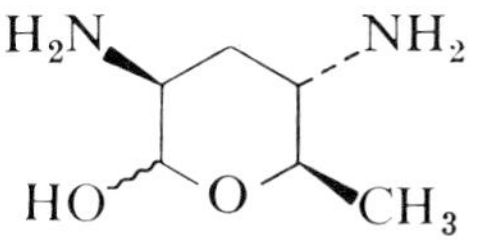

150

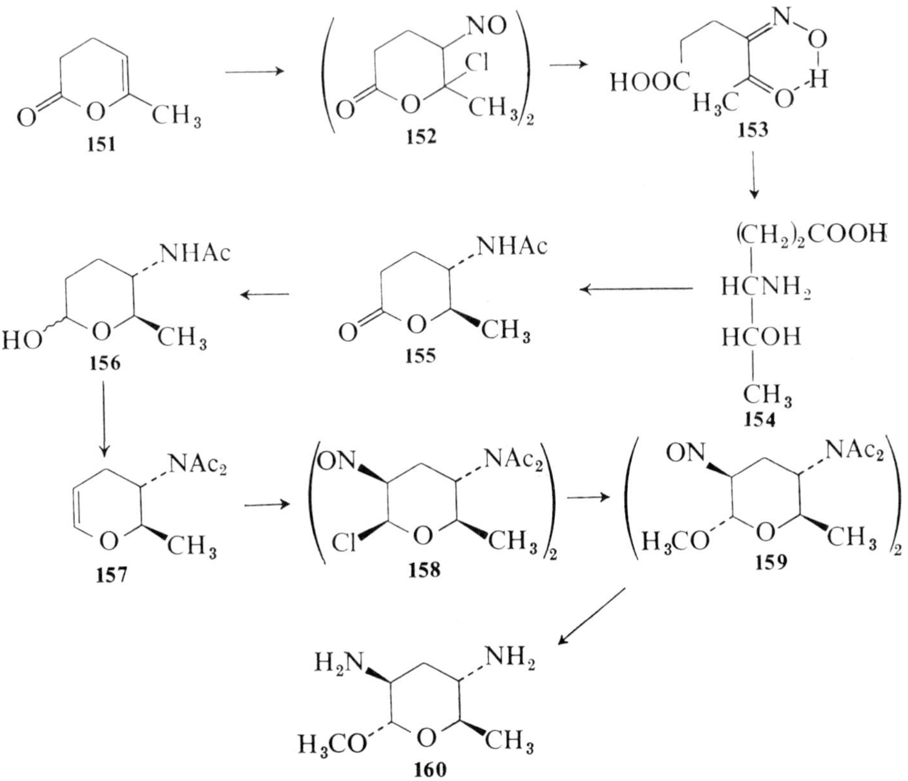

151 152 153 154 155 156 157 158 159 160

Scheme 47

(**160**) was finally obtained from **159** by reduction over platinum, followed by hydrolysis with barium hydroxide.

The *erythro* acid **154** has also been converted into the sugar DL-forosamine **161** in three steps. Forosamine has been obtained from the acid hydrolysate of spiramycin.

The second synthesis[117] of a derivative of kasugamine involved 3,4-dihydro-2-ethoxy-6-methyl-2*H*-pyran (**138**). Hydroboration of **138**, followed by treatment with chloramine, gave the amine **162**, which was isolated as the acetylated derivative **163** (see Scheme 48). Treatment of **163** with bromine

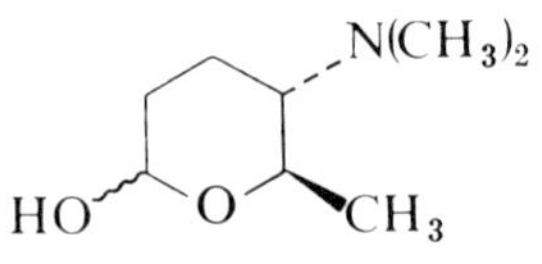

161

containing hydrogen chloride yielded three bromo compounds, **164**, **165**, and **166**; both of the isomers **165** and **166** could be converted into **164**. Compound **164** yielded the azide **167**, on treatment with sodium azide in methyl sulfoxide at 100–105°. Catalytic hydrogenation of **167** produced the amine **168**, which afforded the diacetyl derivative **169**. Optical resolution of

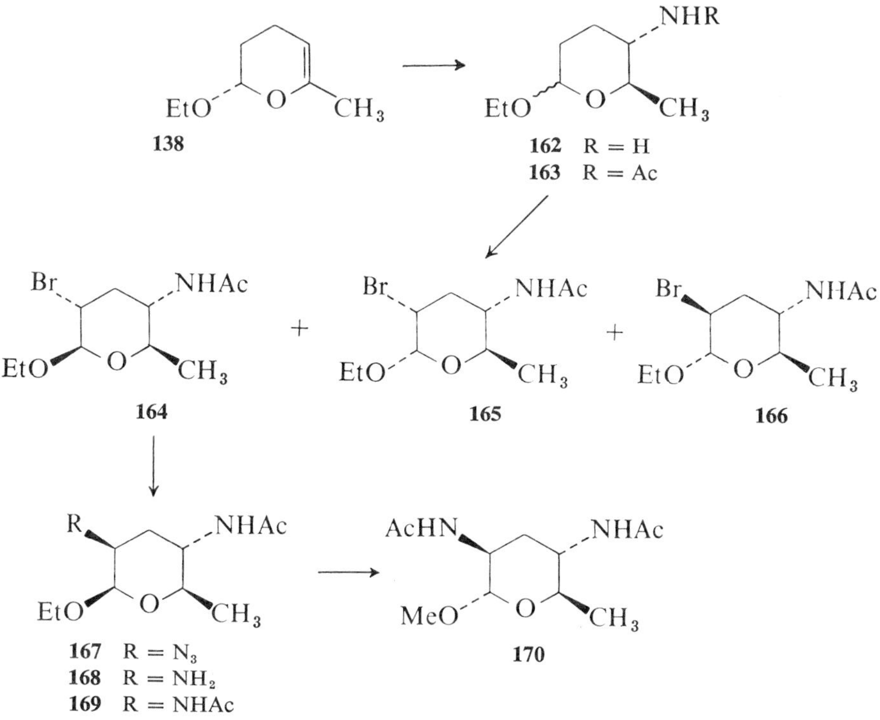

138

162 R = H
163 R = Ac

164 **165** **166**

167 R = N_3
168 R = NH_2
169 R = NHAc

170

Scheme 48

the amine **168** was effected with D-tartaric acid. The resolved amine was acetylated to give optically active **169**, which was converted into the methyl α-glycoside **170** in methanolic hydrogen chloride.

The synthesis of D-**170** is tantamount to the total synthesis of the antibiotic kasugamycin **171**, since D-**170** had already been condensed with the inositol derivative (see Section 9) to give kasuganobiosamine,[118] which, in turn, had been converted into kasugamycin.[119]

The final example of a synthesis involving a pyran intermediate of a sugar moiety of an antibiotic to be discussed is that of DL-chalcose from acrolein dimer.[120] Chalcose (lankavose) is a constituent of the antibiotics chalcomycin, lankamycin, and neutramycin; its structure has been shown to be that of

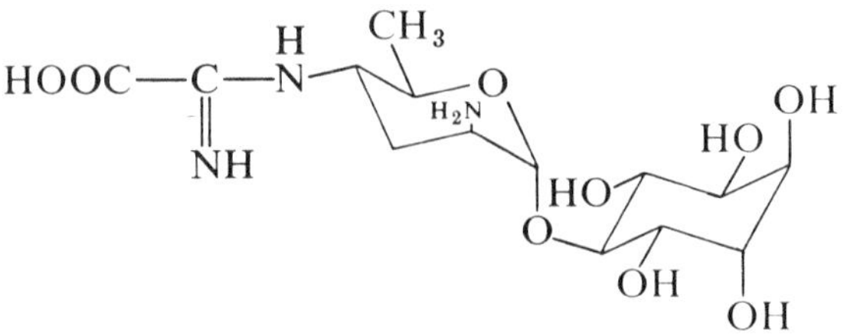

171

4,6-dideoxy–3-*O*-methyl–D-*xylo*-hexose (**172**). Acrolein dimer **72** was converted into 1,6:2,3-dianhydro–4-deoxy–β-DL-*ribo*-hexopyranose (**111**) as outlined in Schemes 38 and 39. Compound **111** was converted into 1,6-anhydro–4-deoxy–3-*O*-methyl–β-DL-*xylo*-hexopyranose (**173**) by reaction with acidic methanol or sodium methoxide in methanol (see Scheme 49). When heated in refluxing methanol containing Amberlite IR-120 (H^+), compound **173** gave a 2:1 mixture of the α and β isomers of methyl 4-deoxy–3-*O*-methyl–DL-*xylo*-hexopyranoside (**174**). Conversion of **174** into the corresponding di-*O*-*p*-toluenesulfonates **175**, followed by their selective reduction with lithium aluminum hydride, gave a mixture of the α and β isomers of methyl 4,6-dideoxy–3-*O*-methyl–2-*O*-*p*-tolylsulfonyl–DL-*xylo*-hexopyranoside (**176**). Saponification of **176** with sodium methoxide in methanol afforded a 2:1 mixture of the α and β isomers of methyl 4,6-dideoxy–3-*O*-methyl–DL-*xylo*-hexopyranoside (**177**) (methyl DL-chalcoside) from which the α isomer was obtained pure by column or gas-liquid chromatography. Acid-catalyzed hydrolysis of the α isomer in aqueous dioxane finally afforded a mixture of the α and β isomers of DL-chalcose.

Lukes et al.[121] have described the preparation of 5,6-dideoxy–DL-*ribo*-hexitol (**181**, Scheme 50) by way of a *cis*-hydroxylation of 2-ethyl–5-oxo–2,5-dihydrofuran (**179**). The starting material for this synthesis was homolevulinic acid which, on distillation with 85% phosphoric acid, gave 2-ethyl–5-oxo–4,5-dihydrofuran (**178**); **178** was isomerized to **179** with triethylamine. Hydroxylation of **179**, by treatment with osmium tetroxide and sodium hypochlorite, gave 5,6-dideoxy–DL-*ribo*-hexonic acid lactone (**180**). Reduction of **180** with lithium aluminum hydride in tetrahydrofuran afforded **181**.

A general approach to the total synthesis of monosaccharides from furan compounds has been developed by Achmatowicz et al.[122a] The method is

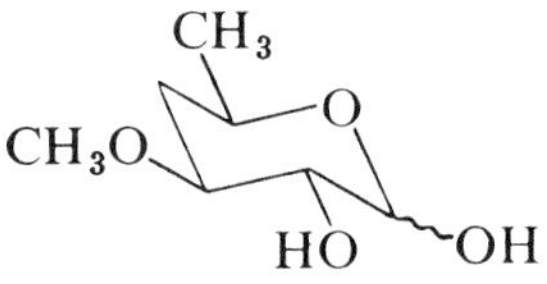

172

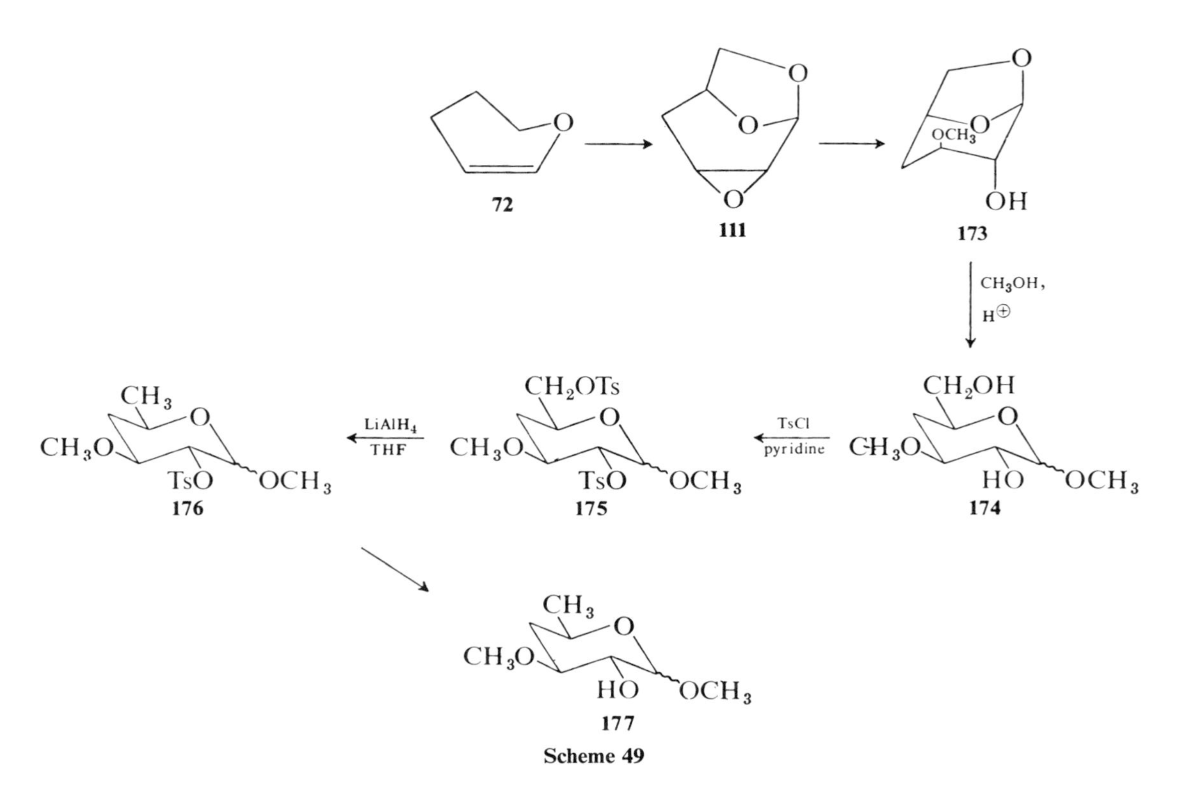
72
111
OCH3
OH
173
CH3OH,
H⊕
CH2OH
CH3O
HO
OCH3
174
TsCl
pyridine
CH2OTs
CH3O
TsO
OCH3
175
LiAlH4
THF
CH3
CH3O
TsO
OCH3
176
CH3
CH3O
HO
OCH3
177

Scheme 49

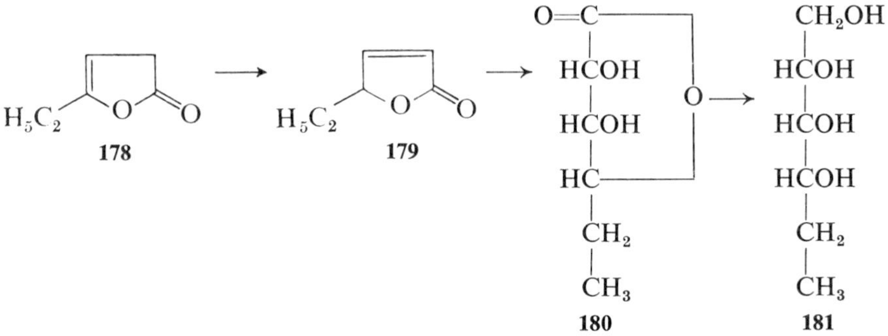

H_5C_2 178 → H_5C_2 179 → O=C, HCOH, HCOH, HC (ring O), CH_2, CH_3 180 → CH_2OH, HCOH, HCOH, HCOH, CH_2, CH_3 181

Scheme 50

outlined in Scheme 50a. A 2-furylcarbinol is converted, by treatment with bromine in methanol, into a mixture of *cis* and *trans* isomers of the corresponding 2,5-dimethoxy-2,5-dihydrofuran derivative. Mild acid hydrolysis results in a cleavage of the acetal bonds and formation of a dicarbonyl intermediate which cyclizes to the 2,3-dideoxy-DL-2-enopyranos-4-ulose; treatment with methyl orthoformate in the presence of a Lewis acid catalyst yields a mixture of methyl glycosides which can be separated by column or gas-liquid chromatography. Reduction with metal hydrides[122b] leads to stereoisomeric methyl 2,3-dideoxy-DL-2-enopyranosides. The method has been applied[122a] to furfuryl alcohol, 2-(1,2-*O*-isopropylidene-1,2-dihydroxyethyl)furan, 1-(2-furyl)ethanol, and 2-(2-furyl)glycerol 1,3-diacetate. The methyl 2,3-dideoxy-DL-2-enopyranosides can be converted, by way of hydroxylation or epoxidation and subsequent opening of the oxirane ring, into a variety of monosaccharides. If an optically active 2-furylcarbinol is used, it should be possible in some cases to obtain sugars of the D or L series in an optically pure state.[122c]

Making use of the general method outlined in Scheme 50a, Achmatowicz and Szechner[122d] have recently synthesized DL-cinerulose A (see Scheme 50b); cinerulose A has been isolated from the anthracycline antibiotic cinerubin A, and shown to have the structure of 2,3,6-trideoxy-L-*glycero*-hexopyranos-4-ulose. Treatment of 1-(2-furyl)ethanol with bromine in methanol gave 1-[2-(2,5-dimethoxy-2,5-dihydrofuryl)]ethanol which, on catalytic hydrogenation, afforded 1-[2-(2,5-dimethoxy-2,3,4,5-tetrahydrofuryl)]ethanol; the dihydro and tetrahydro derivatives were mixtures of *cis* and *trans* isomers. Acid-catalyzed hydrolysis of the tetrahydro derivative gave DL-cinerulose A as an anomeric mixture. Alternatively, DL-cinerulose A could be obtained by conversion of the dihydro derivative into 2,3,6-trideoxy-α-DL-hex-2-enopyranos-4-ulose, followed by catalytic hydrogenation.

A synthesis of 3-amino-3-deoxy-DL-pentofuranoside derivatives from the 2,5-dihydrofuran derivative **182** has been reported by Iwai et al.[122] The key

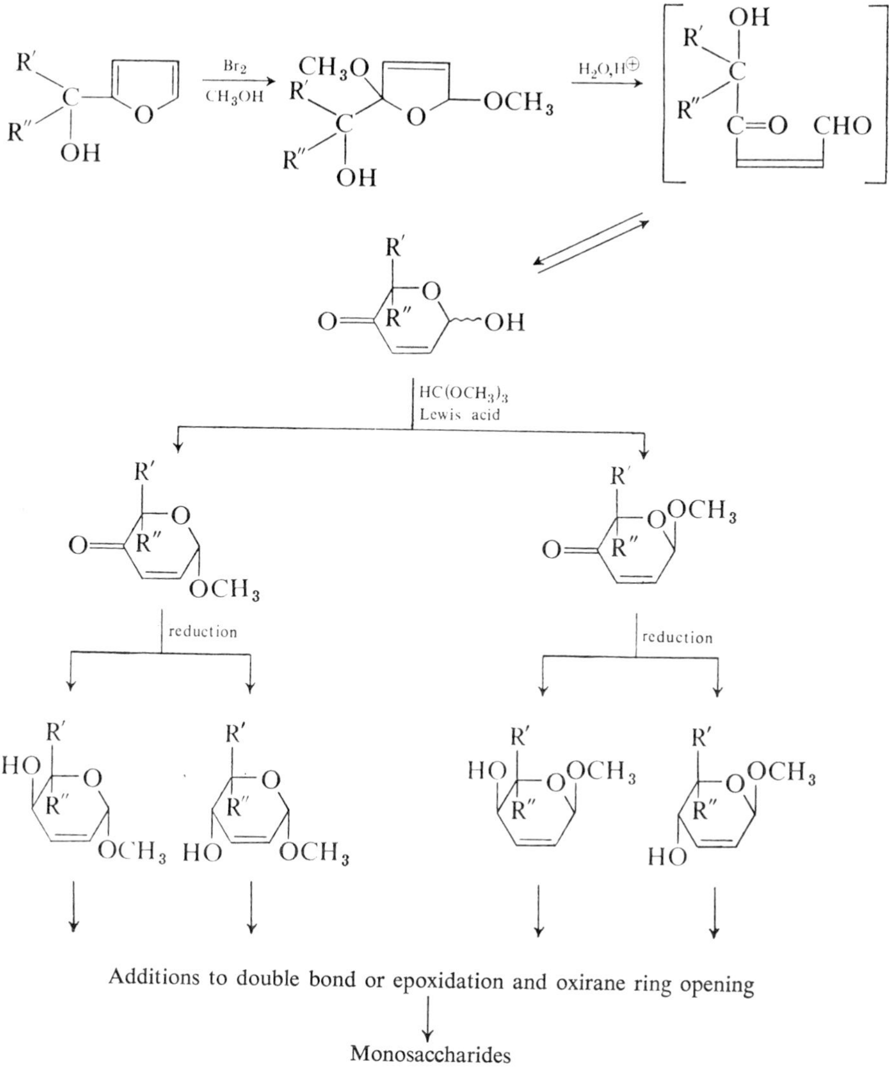

Br$_2$, CH_3OH; $H_2O, H^{\oplus}$; $HC(OCH_3)_3$, Lewis acid; reduction; reduction

Additions to double bond or epoxidation and oxirane ring opening

Monosaccharides

Scheme 50a

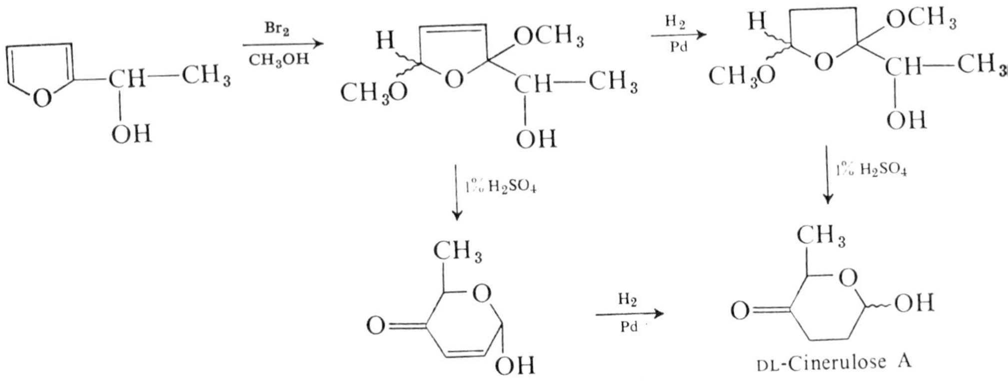

DL-Cinerulose A

Scheme 50b

steps were epoxidation of the double bond and opening of the epoxide ring with ethanolic ammonia. Other syntheses of amino sugars are discussed in detail in Section 7A.

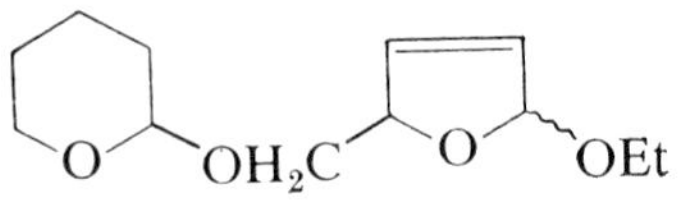

182

7. MISCELLANEOUS SYNTHESES

A. Amino Sugar Derivatives

Carbohydrates in which a hydroxyl group has been replaced by an amino substituent are called amino sugars (strictly, aminodeoxy sugars). These derivatives are of widespread occurrence in nature. They are found in many polysaccharides and mucopolysaccharides of microbiological and animal origin and in several antibiotics. The majority of amino sugars have been synthesized from readily available monosaccharide derivatives. However, syntheses from noncarbohydrate precursors have been reported.

David and Veyrières[123] have recently prepared 3-acetamido–2,3-dideoxy–D-tetrose (**188**) from L-aspartic acid by the route shown in Scheme 51. 3-Methoxycarbonyl–3-trifluoroacetamidopropionyl chloride[124] (**183**) was obtained from L-aspartic acid and converted into 3-methoxycarbonyl–3-trifluoroacetamidopropionaldehyde (**184**) by a Rosenmund reduction. Treatment of **184** with methanolic hydrogen chloride yielded methyl 2-amino–4,4-dimethoxybutyrate (**185**) which, on reduction with lithium

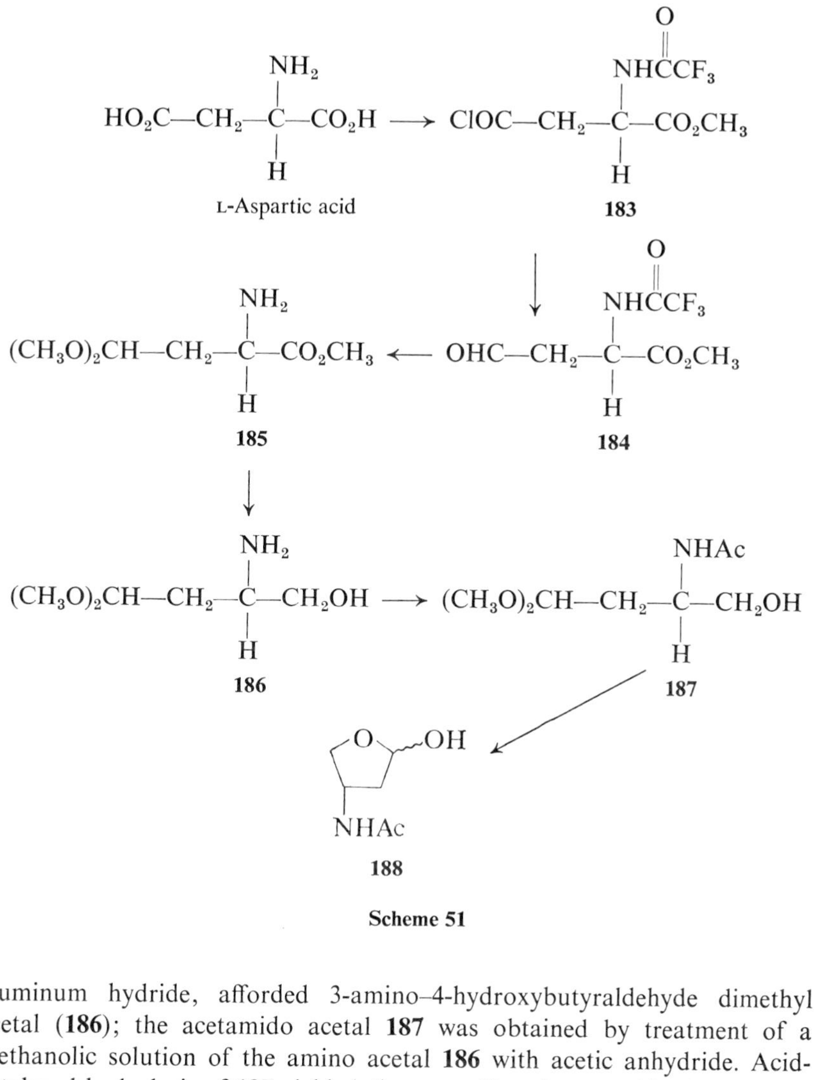

$$HO_2C{-}CH_2{-}C(NH_2)(H){-}CO_2H \longrightarrow ClOC{-}CH_2{-}C(NHCOCF_3)(H){-}CO_2CH_3$$

L-Aspartic acid **183**

$$(CH_3O)_2CH{-}CH_2{-}C(NH_2)(H){-}CO_2CH_3 \longleftarrow OHC{-}CH_2{-}C(NHCOCF_3)(H){-}CO_2CH_3$$

185 **184**

$$(CH_3O)_2CH{-}CH_2{-}C(NH_2)(H){-}CH_2OH \longrightarrow (CH_3O)_2CH{-}CH_2{-}C(NHAc)(H){-}CH_2OH$$

186 **187**

OH, NHAc

188

Scheme 51

aluminum hydride, afforded 3-amino–4-hydroxybutyraldehyde dimethyl acetal (**186**); the acetamido acetal **187** was obtained by treatment of a methanolic solution of the amino acetal **186** with acetic anhydride. Acid-catalyzed hydrolysis of **187** yielded the crystalline 3-acetamido–2,3-dideoxy D-tetrose (**188**).

The Akabori reaction[125] has been used by Ichikawa et al.[126] to prepare 2-amino–2-deoxyaldonic acids. The syntheses involved the base-catalyzed condensation of *N*-pyruvylideneglycinatoaquocopper(II) (**189**) with aldehydes to give the corresponding β-hydroxy amino acids; the reactions proceed at *p*H 8.0–9.5. Thus reaction of 2,3-*O*-isopropylidene-*aldehydo*-D-glyceraldehyde with **189**, followed by treatment of the mixture with sodium sulfide, gave a crystalline 2-amino–2-deoxy–4,5-*O*-isopropylidene–D-pentonic acid:

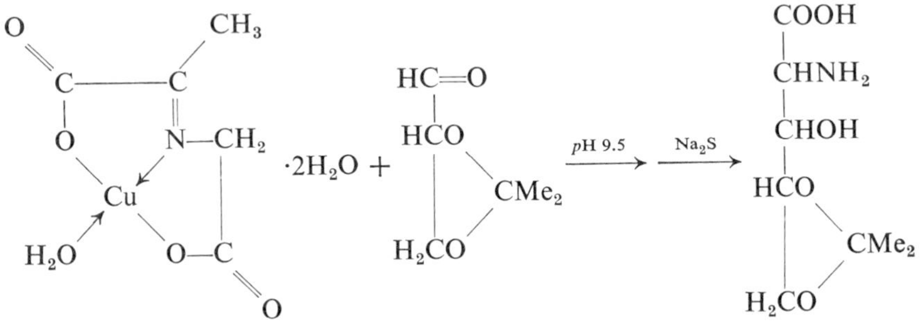

$$\mathbf{189}\cdot 2H_2O + HC{=}O,\ HCO,\ H_2CO\ (CMe_2) \xrightarrow{pH\ 9.5} \xrightarrow{Na_2S} COOH,\ CHNH_2,\ CHOH,\ HCO,\ H_2CO\ (CMe_2)$$

189

A branched-chain amino sugar has been prepared by Kuhn and Weiser.[127] Treatment of the aldehyde **190** (see Scheme 52) with benzylamine and hydrogen cyanide gave the amino nitrile **191**. Hydrogenation of **191** over prehydrogenated palladium oxide-on-barium sulfate afforded 2-amino–3,3-dimethyl–4-hydroxybutyraldehyde, the furanoid form (**192**) of which has been established.

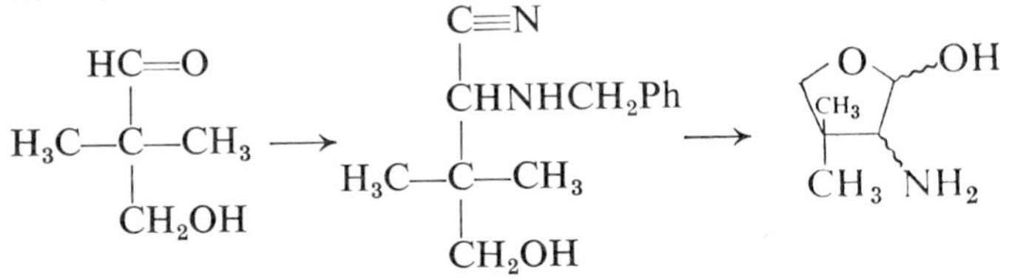

$$HC{=}O,\ H_3C{-}C{-}CH_3,\ CH_2OH\ (\mathbf{190}) \longrightarrow C{\equiv}N,\ CHNHCH_2Ph,\ H_3C{-}C{-}CH_3,\ CH_2OH\ (\mathbf{191}) \longrightarrow \mathbf{192}\ (O,\ OH,\ CH_3,\ CH_3,\ NH_2)$$

190 **191** **192**

Scheme 52

The total synthesis of 4-amino derivatives of ethyl 4-deoxy-α-DL-lyxo-pyranoside has been recently achieved by Mochalin et al.[127a] (see Scheme 52a) by way of a Cope reaction. Ethyl 3,4-dideoxy-3-(dimethylamino)-α-DL-*threo*-pentopyranoside was converted into the corresponding *N*-oxide by treatment with a 3% aqueous-methanolic solution of hydrogen peroxide; pyrolysis of the *N*-oxide gave 2-ethoxy-3,6-dihydro-2*H*-pyran-3-ol. Oxidation of the olefin with peroxybenzoic acid afforded ethyl 3,4-anhydro-β-DL-ribo-pyranoside. Treatment of the epoxide with aqueous solutions of ammonia, dimethylamine, and aniline at room temperature gave the 4-amino derivatives of ethyl 4-deoxy-α-DL-lyxopyranoside.

B. Deoxyfluoro Sugar Derivatives

For several years, there has been considerable interest in deoxyhalo sugars not only because of their potential intrinsic value in biochemistry or

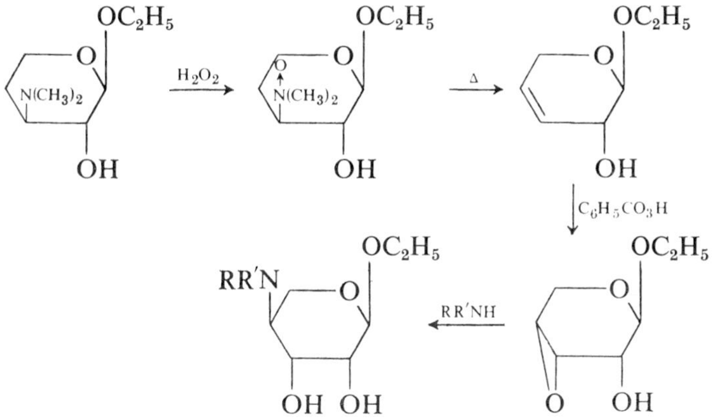

$R = R' = H; CH_3; \quad R = H, R' = Ph$

Scheme 52a

pharmacology but also because of their utility in the synthesis of other rare sugars such as deoxy and aminodeoxy sugars. The replacement of a hydroxyl group in a monosaccharide by a chlorine or a fluorine atom is particularly interesting to enzymologists, since the size of these halogen atoms is similar to that of a hydroxyl group, but they have very different capacities for forming covalent or van der Waals linkages. Until recently, the only way of producing sugar derivatives containing secondary fluorine was by total synthesis.[128] The first such derivatives to be prepared were fluoropolyols. Thus Claisen condensation of diethyl oxalate and ethyl fluoroacetate gave diethyl fluorooxalacetate, which, on reduction with lithium aluminum hydride[129] or sodium borohydride,[130] afforded two racemic fluorinated tetritols, 2-deoxy-2-fluoro-DL-erythritol and 2-deoxy-2-fluoro-DL-threitol (Scheme 53; only one isomer of each DL mixture is shown). In a similar way, condensation of methyl 2,3-*O*-isopropylidene-DL-glycerate with ethyl fluoroacetate has led to the preparation of 2-deoxy-2-fluoro-DL-ribitol.[131] By standard procedures, these 2-deoxy-2-fluoroalditols have provided fluoro derivatives of glyceraldehyde,[132] glycerol,[132] glyceric acid,[133] and erythronic acid.[134]

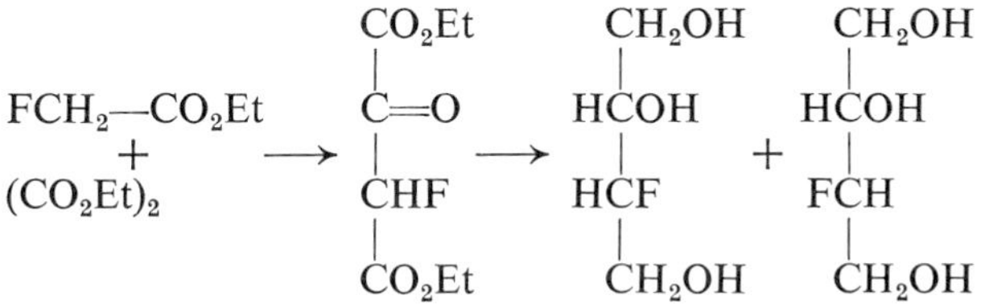

$FCH_2{-}CO_2Et + (CO_2Et)_2 \longrightarrow CO_2Et{-}C(=O){-}CHF{-}CO_2Et \longrightarrow CH_2OH{-}HCOH{-}HCF{-}CH_2OH + CH_2OH{-}HCOH{-}FCH{-}CH_2OH$

Scheme 53

C. Branched-Chain Sugars

The first branched-chain sugar was detected in 1901 by Vongerichten in the form of a glycosidic component in parsley. It was given the name apiose, and its structure was later found to be 3-*C*-hydroxymethyl–D-*glycero*-tetrose. A synthesis of L-apiose from a noncarbohydrate precursor has already been described (see Section 4, Scheme 19). In 1919 Emil Fischer and Freudenberg detected a branched-chain sugar in hamamelitannin from *Hamamelis virginiana* which they named hamamelose. It was later identified as 2-*C*-hydroxymethyl–D-ribose. For a long time, apiose and hamamelose remained the only examples of branched-chain sugars. With the advent of the "antibiotic era," however, several new branched-chain sugars were discovered. Many of these have been prepared from simple sugar derivatives. In this section syntheses of DL-mycarose and DL-epimycarose from noncarbohydrate precursors are described. Mycarose occurs as a component of a number of macrolide antibiotics. Its structure has been shown to be that of 2,6-dideoxy–3-*C*-methyl–L-*ribo*-hexose (**193**). Epimycarose is epimeric with mycarose at C-3.

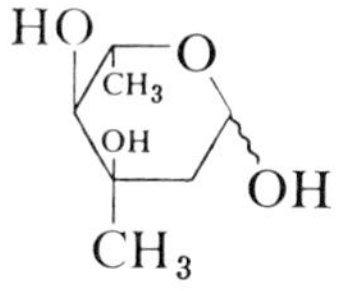

HO O CH_3 OH OH CH_3

193

Woodward and his associates[135] have synthesized both DL-mycarose and DL-epimycarose (see Scheme 54). Addition of the Grignard reagent derived from propyne to acetoacetaldehyde dimethyl acetal (**194**) gave the acetylenic alcohol **195**, which was converted into the *cis* olefin **196** by hemihydrogenation over a poisoned palladium catalyst. *cis*-Hydroxylation of **196** could be effected with osmium tetroxide or with dilute aqueous alkaline potassium permanganate. The mixture of triols **197** and **198** was cyclized in methanolic hydrogen chloride to give a mixture of the racemic glycosides **199** (methyl DL-mycaroside) and **200** (methyl DL-epimycaroside). The relative configurations of the separated glycosides were elucidated by chemical means, and the relationship of racemic **199** to mycarose was demonstrated by its conversion into the 4-*O*-*p*-tolysulfonyl derivative of the free sugar, which was spectroscopically indistinguishable from the corresponding natural sugar derivative. Synthetic racemic **199** was resolved by way of its bornanol-10-sulfonates.

Korte et al.[136] have prepared DL-mycarose and DL-epimycarose starting

CH_3

$(CH_3O)_2CH$ — C(=O) + $CH_3C{\equiv}CMgBr$

194

↓

H_3C, OH, $C{\equiv}CCH_3$, $(CH_3O)_2CH$

195

↓

H_3C, OH, H, $(CH_3O)_2CH$, H_3C, H

196

↙ ↘

H_3C, OH, OH, H, $(CH_3O)_2CH$, OH, H_3C, H

197

H_3C, OH, H, OH, $(CH_3O)_2CH$, H, H_3C, OH

198

↓

H_3C, OH, OH, H, H, H_3CO, O, CH_3

199

↓

H_3C, OH, H, OH, CH_3, H_3CO, O, H

200

Scheme 54

from 3-methylhex–3-eno-1 → 5-lactone (**201**):

201

The synthesis of DL-mycarose involved reduction of the lactone function with lithium aluminum hydride to the hemiacetal, which was converted into a glycopyranoside. *cis*-Hydroxylation of the double bond, followed by acid-catalyzed hydrolysis, afforded the free sugar. DL-Epimycarose was obtained by *trans*-hydroxylation of the appropriate intermediate.

Finally, a synthesis of DL-mycarose and DL-epimycarose was achieved by Grisebach et al.[137] from methyl 3-hydroxy–3-methyl–4-hexenoate (**202**). Oxidation of **202** with monoperoxyphthalic acid, followed by acid-catalyzed hydrolysis, gave a mixture of the 3-epimeric mycaric lactones (**203** and **204**),

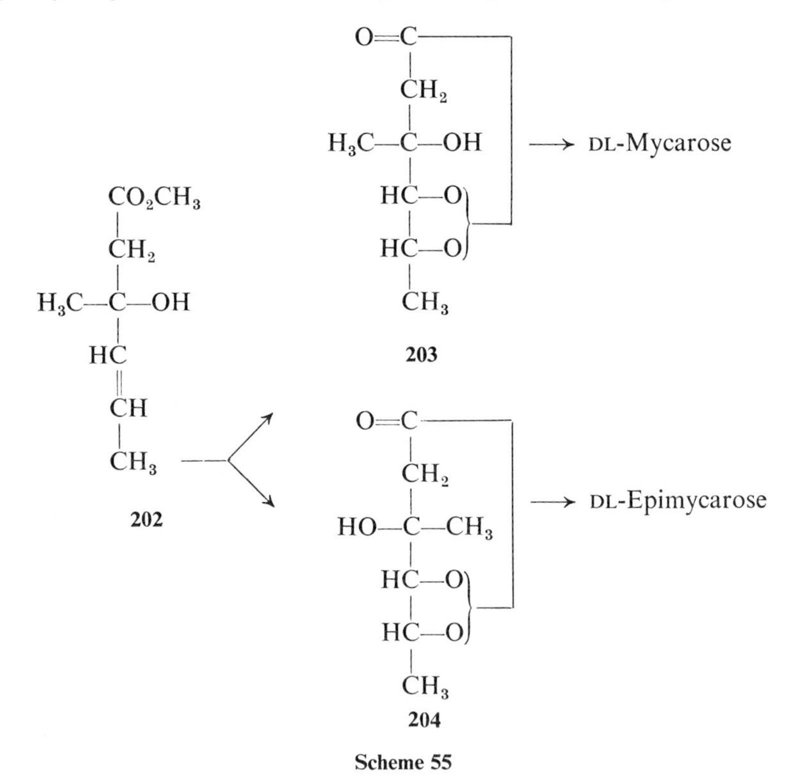

Scheme 55

which was reduced by bis(3-methyl-2-butyl)borane to a mixture of DL-mycarose and DL-epimycarose (Scheme 55). A separation of the epimeric sugars was achieved by preparative paper chromatography.

8. ENZYMIC SYNTHESES

Hough and Jones[138] speculated that, since dihydroxyacetone phosphate will react with D- or L-glyceraldehyde under the influence of the enzyme aldolase to yield D-fructose and L-sorbose 1-phosphates, respectively, then possibly other aldehydes would react in a similar manner. It was shown that glycolaldehyde in the presence of aldolase[139] produces D-*threo*-pentulose 1-phosphate, DL-lactaldehyde yields 6-deoxy-D-*arabino*- and L-*xylo*-hexulose 1-phosphates,[140] and that propionaldehyde gives 5,6-dideoxy-D-*threo*-hexulose 1-phosphate.[141] In these cases and in almost all other examples known, the two new hydroxyl groups produced have the D-*threo* configuration. In these examples the condensation leads in a stereospecific fashion to the formation in each case, of one sugar derivative only from noncarbohydrate material (see Scheme 56).

A recent example[142] of this type of condensation is the observation that pyruvate and formaldehyde, under the influence of an aldolase, yield 3-deoxy-2-tetrulosonic acid. Similarly, glyoxylate and pyruvate form a 3-deoxy-2-keto-D-pentaric acid (Scheme 57).

Uehara and Matsukawa[143] showed that D-ribose was produced exclusively

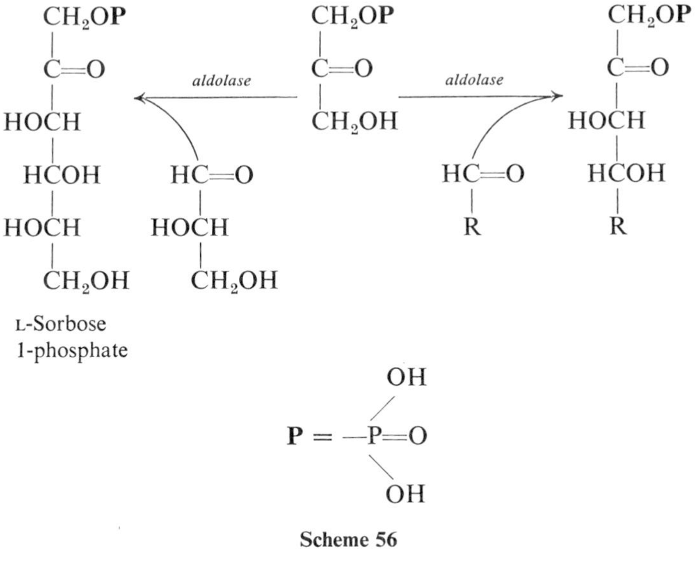

CH_2OP–C=O–HOCH–HCOH–HOCH–CH_2OH (L-Sorbose 1-phosphate) ←(aldolase; HC=O–HOCH–CH_2OH)— CH_2OP–C=O–CH_2OH —(aldolase; HC=O–R)→ CH_2OP–C=O–HOCH–HCOH–R

L-Sorbose
1-phosphate

P = —P(=O)(OH)(OH)

Scheme 56

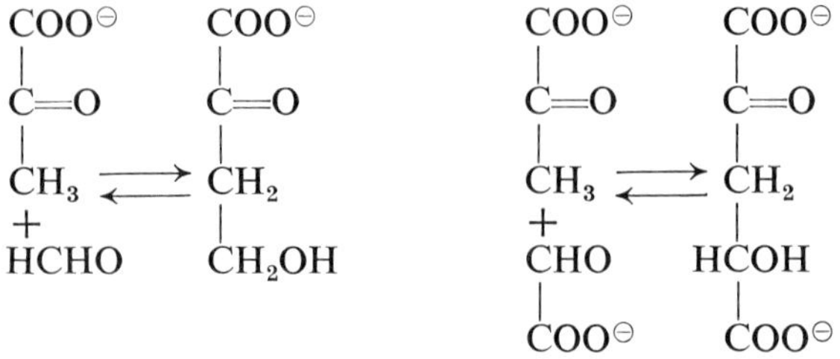

$$CH_3COCOO^{\ominus} + HCHO \rightleftharpoons HOCH_2CH_2COCOO^{\ominus}$$

$$CH_3COCOO^{\ominus} + OHC{-}COO^{\ominus} \rightleftharpoons {}^{\ominus}OOC{-}CO{-}CH_2{-}CHOH{-}COO^{\ominus}$$

Scheme 57

when baker's yeast was added to a solution of D-glyceraldehyde and hydroxypyruvic acid in a phosphate buffer at *p*H 6.8. Presumably, the α-keto acid, produced by an aldol-type condensation, is decarboxylated to yield the pentose sugar. It is interesting that in this case condensation leads to the production of two new hydroxyl groups with the D-*erythro* configuration (see Scheme 58).

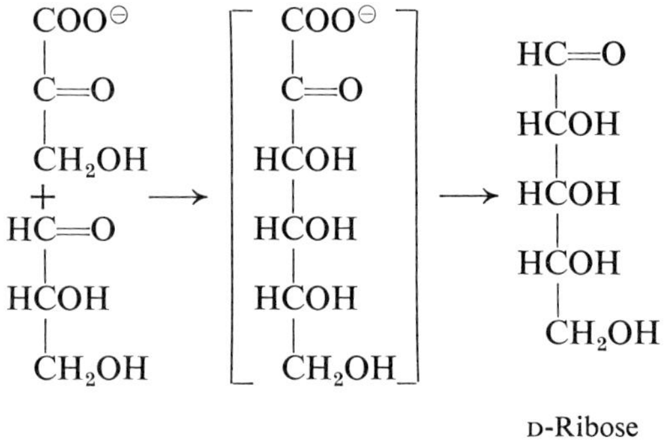

$$HOCH_2{-}CO{-}COO^{\ominus} + OHC{-}CHOH{-}CH_2OH \longrightarrow \left[{}^{\ominus}OOC{-}CO{-}CHOH{-}CHOH{-}CHOH{-}CH_2OH \right] \longrightarrow OHC{-}CHOH{-}CHOH{-}CHOH{-}CH_2OH$$

D-Ribose

Scheme 58

9. SYNTHESIS OF CYCLITOLS

The class of compounds called cyclitols, which includes the inositols (cyclohexanehexols), quercitols (cyclohexanepentols), inososes (pentahydroxycyclohexanones), inosamines (aminodeoxyinositols), and quinic and shikimic acids, has long been of interest to carbohydrate chemists. Many such compounds have been synthesized in nature and in the laboratory from sugar precursors. However, there is a significant literature on their synthesis from noncarbohydrate precursors; some representative examples are briefly discussed in this section.

Zelinski et al.[144] obtained the cyclohexanetetrol **205** by direct hydroxylation of 1,3-cyclohexadiene with permanganate:

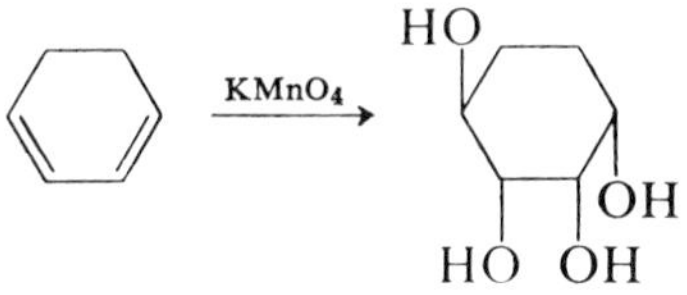

KMnO$_4$

HO OH HO OH

205

A mechanism for this *trans*-hydroxylation by permanganate has been proposed by Sable.[145] Compound **205** has also been prepared by a Diels-Alder synthesis (see Scheme 59).[146] Reaction of furan with vinylene carbonate gave the adduct **206**, which is called 1,4-anhydro–*cis*-conduritol carbonate. Successive acidic and basic hydrolysis afforded conduritol-*C;* or, after

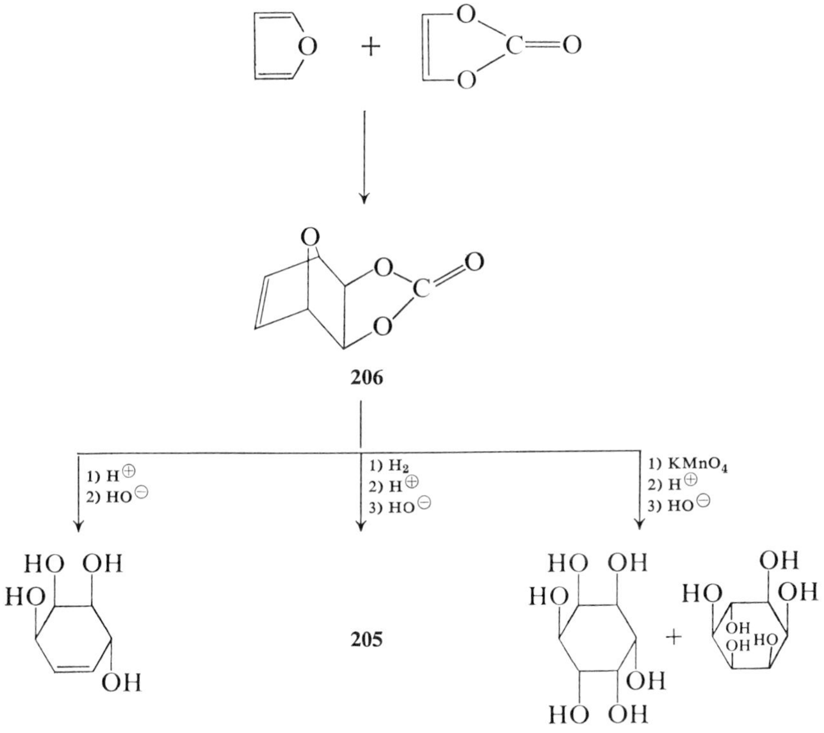

206

1) H$^{\oplus}$ 2) HO$^{\ominus}$

1) H$_2$ 2) H$^{\oplus}$ 3) HO$^{\ominus}$

1) KMnO$_4$ 2) H$^{\oplus}$ 3) HO$^{\ominus}$

205

Conduritol-*C*

neo-Inositol

epi-Inositol

Scheme 59

preliminary hydrogenation, the cyclohexanetetrol **205** was obtained. *cis*-Hydroxylation (*endo* or *exo*) of the adduct **206**, followed by acidic and basic hydrolysis, gave a mixture of *neo*- and *epi*-inositol.

The Diels-Alder reaction has also been used for the preparation of *allo*-inositol by Criegee and Becher[147] (see Scheme 60). Thus *trans*,*trans*-diacetoxybutadiene and vinylene carbonate condensed to give an adduct **207**, which, on hydroxylation by osmium tetroxide, followed by hydrolysis, gave *allo*-inositol. In this example the osmium tetroxide approaches from the un-

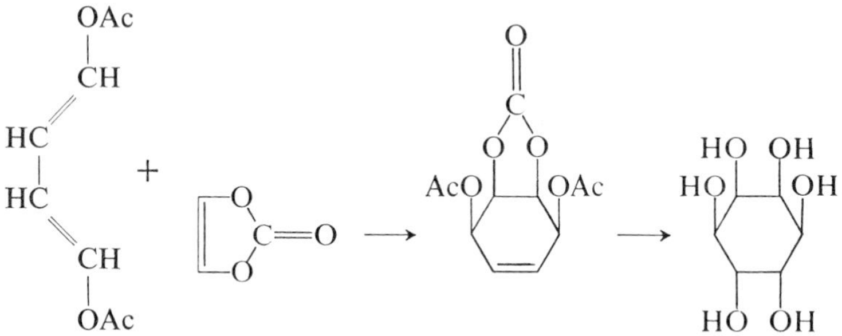

207 *allo*-Inositol

Scheme 60

hindered side of **207**, and the other possible product, *cis*-inositol, is not formed.

Posternak and Friedli[148] prepared five cyclohexanetetrols by the appropriate *cis*- or *trans*-hydroxylation of *cis*- or *trans*-3-cyclohexene–1,2-diol (see Scheme 61). *cis*-Hydroxylation may be achieved with permanganate or with silver chlorate-osmium tetroxide,[148] whereas *trans*-hydroxylation occurs with a peroxy acid,[148] or with silver benzoate-iodine (Prévost reagent).[149]

Nakajima and his associates[150] have also employed various hydroxylation methods in the synthesis of inositols from *cis*- and *trans*-3,5-cyclohexadiene–1,2-diols (**208** and **209**). These starting materials were obtained from α-3,4,5,6-tetrachlorocyclohexene by hydroxylation of the double bond, followed by removal of the chlorine atoms with zinc.

An early synthesis of inositols from a noncarbohydrate precursor involved hydrogenation of hexahydroxybenzene.[151] Kuhn et al.[152] repeated this study using a palladium catalyst and tetrahydroxybenzoquinone as starting material. The products of the hydrogenation were fully investigated by Angyal and McHugh[153] using cellulose chromatography. Five inositols, three quercitols, and one inosose (**210**) were isolated; *myo*-inositol (**211**) was the predominant product (**17%**). Some cyclohexane-tetrols and -triols were also formed.

(−)-Quinic acid, 1,3,4,5-tetrahydroxy–1-cyclohexanecarboxylic acid (**212**),

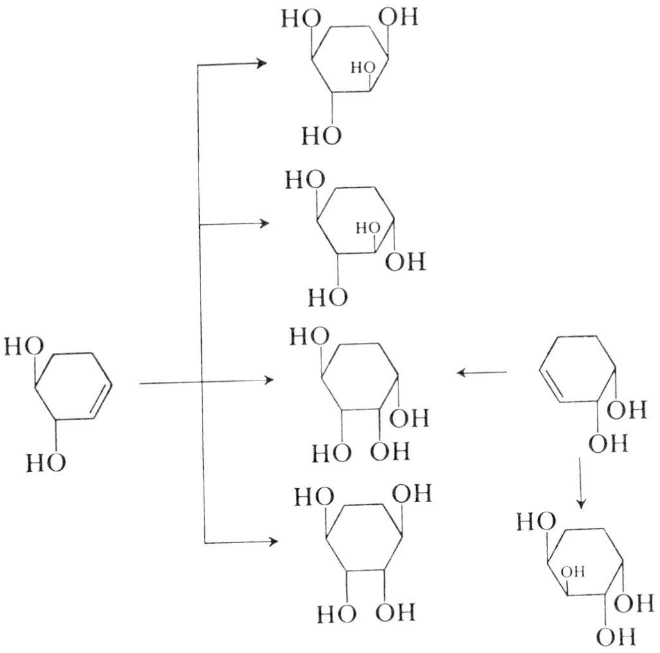

Scheme 61

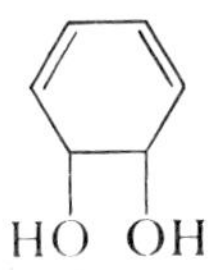

208

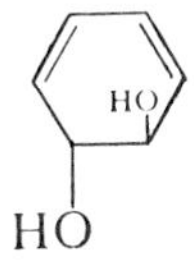

209

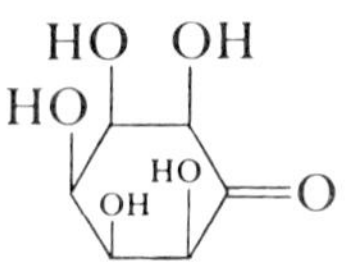

210
cis-Inosose

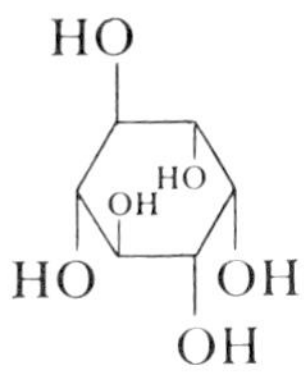

211
myo-Inositol

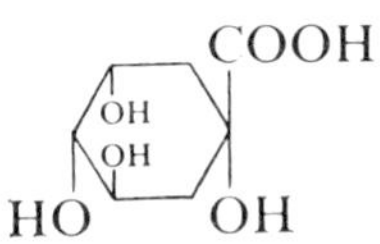

212

has long been known and occurs widely in the plant kingdom. Stereospecific syntheses of (±)-quinic acid from noncarbohydrate precursors have been achieved. One of these was by Smissman and Oxman,[154] and the sequence of reactions is shown in Scheme 62. A Diels-Alder reaction of *trans,trans*-1,4-

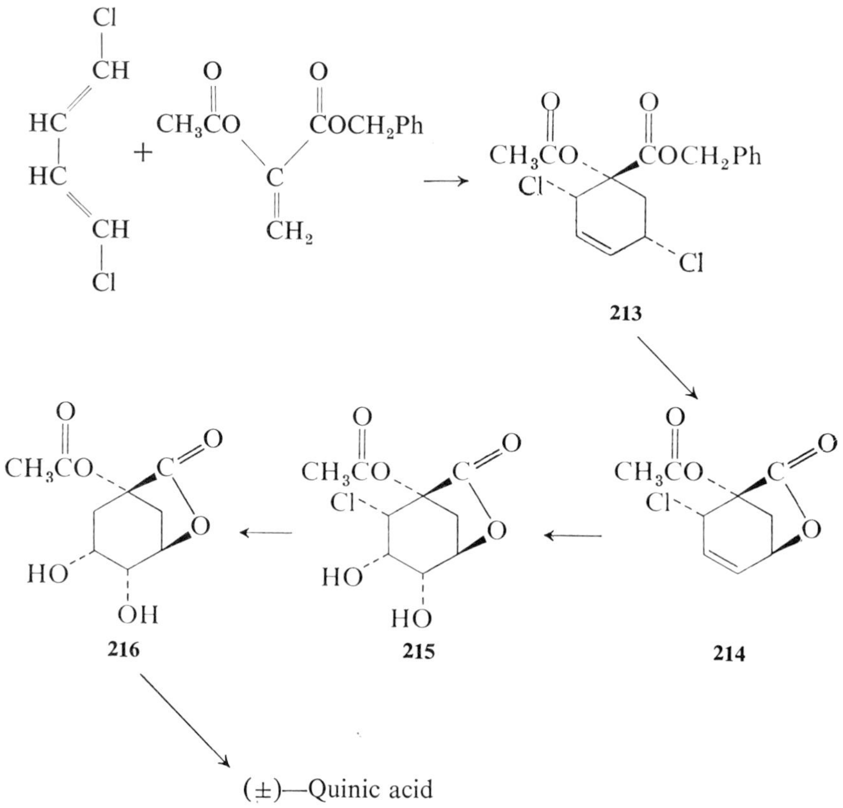

213

216 **215** **214**

(±)—Quinic acid

Scheme 62

dichlorobuta–1,3-diene with benzyl α-acetoxyacrylate (prepared from benzyl pyruvate by refluxing with acetic anhydride in the presence of *p*-toluenesulfonic acid) gave the adduct **213**, which was converted into the chloro lactone **214** on heating. Compound **214** was *cis* hydroxylated with osmium tetroxide to give 1-acetyl–6-chloroquinide (**215**). Hydrogenolysis of the chlorine atom was achieved with W-6 Raney nickel to yield (±)-1-acetylquinide (**216**), from which (±)-quinic acid could be obtained by heating in aqueous potassium hydroxide solution.

Wolinsky et al.[155] have also synthesized (±)-quinic acid (see Scheme 63). Methyl pyruvate was first converted into methyl α-acetoxyacrylate, which

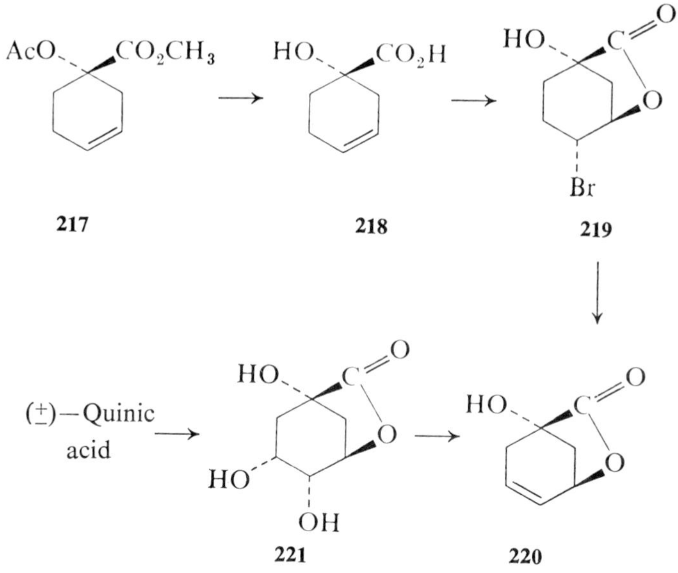

217 218 219

221 220

(±)—Quinic acid

Scheme 63

was then condensed with 1,3-butadiene to give methyl 1-acetoxy–3-cyclohexene–1-carboxylate (**217**); hydrolysis of the adduct **217** gave the hydroxy acid **218**. Bromo-lactonization of **218** afforded 1-hydroxy–5-bromo–3-oxabicyclo[3.2.1]octan-2-one (**219**). Dehydrobromination of **219** was effected by heating it at 130–180° with triethylamine in benzene. Hydroxylation of the unsaturated lactone **220** with osmium tetroxide gave (±)-quinide (**221**), which, on hydrolysis, afforded (±)-quinic acid.

Smissman et al.[156] have also achieved the first total chemical synthesis of shikimic acid (**222**). This compound has been shown to be an important intermediate in aromatic biosynthesis. *trans,trans*-1,4-Diacetoxybutadiene was condensed with methyl acrylate to produce *cis*-3,6-diacetoxycyclohexene–4-carboxylate (**223**) (Scheme 64). The double bond was *cis*-hydroxylated with osmium tetroxide to give methyl β-2,β-5-diacetoxy–α-3,α-4-dihydroxycyclohexylcarboxylate (**224**), which was converted into the 3,4-acetonide. The acetonide was pyrolyzed to give methyl (±)-3-*O*-acetyl–4,5-*O*-isopropylideneshikimate (**225**), which was converted into (±)-shikimic acid by successive acidic and basic hydrolysis.

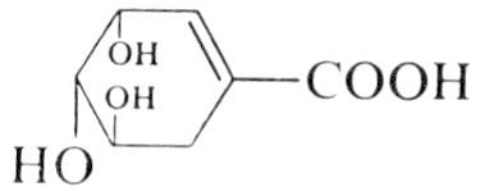

222

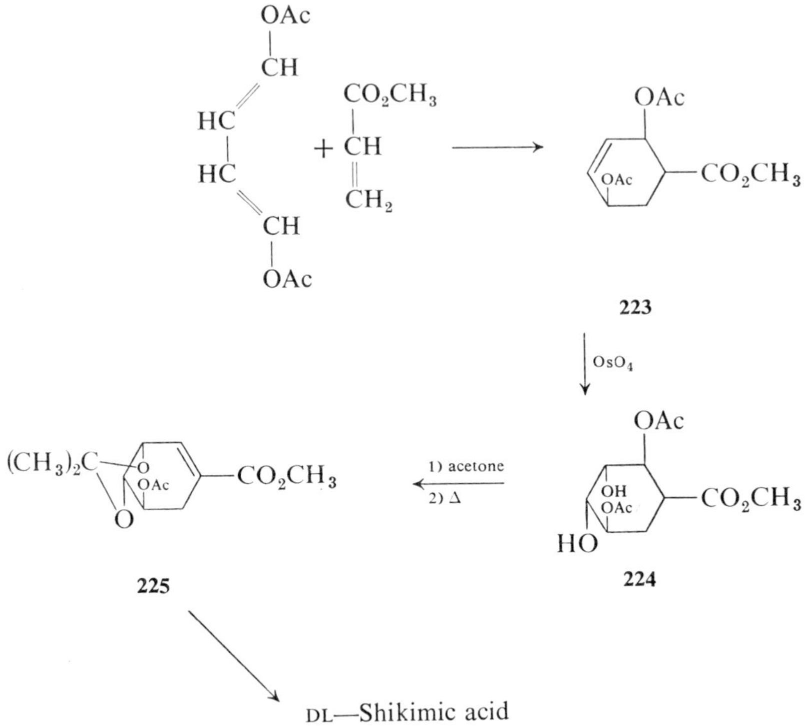

OAc CH HC HC CH OAc + CO_2CH_3 CH CH_2 → OAc OAc CO_2CH_3

223

OsO_4

OAc OH OAc CO_2CH_3 HO

224

1) acetone 2) Δ

$(CH_3)_2C$ O OAc O CO_2CH_3

225

DL—Shikimic acid

Scheme 64

A more recent synthesis of shikimic acid is by Grewe and Hinrichs[157] (Scheme 65). 1,4-Cyclohexadiene–1-carboxylic acid was prepared from 1,3-butadiene and propiolic acid, and converted into the methyl ester **226**. Compound 226 was *trans*-hydroxylated with peroxyformic acid, and the diol was acetylated to give **227**. The di-*O*-acetyl derivative was refluxed in carbon tetrachloride with *N*-bromosuccinimide, and the product was treated in acetic acid with silver acetate; removal of the acetyl groups finally gave methyl (±)-shikimate (**228**). Cleophax et al.[157a] have obtained methyl (−)-3,4,5-tri-*O*-benzoylshikimate by treatment of methyl (−)-3,4,5-tri-*O*-benzoylquinate with sulfuryl chloride-pyridine in anhydrous chloroform.

It has recently been shown that cyclitols can be converted into monosaccharides. The key intermediates in this work were seven-membered hemiacetal lactones. The first such synthesis was that of DL-allose from *myo*-inositol[158] (Scheme 66). DL-1,2:3,4-Di-*O*-isopropylidene–5,6-di-*O*-*p*-tolylsulfonyl–*epi*-inositol[159] was obtained from *myo*-inositol and treated with sodium methoxide in tetrahydrofuran to give DL-1,2:3,4-di-*O*-isopropylidene–6-*O*-methyl–*epi*-inositol (**229**). Oxidation of **229** with active manganese

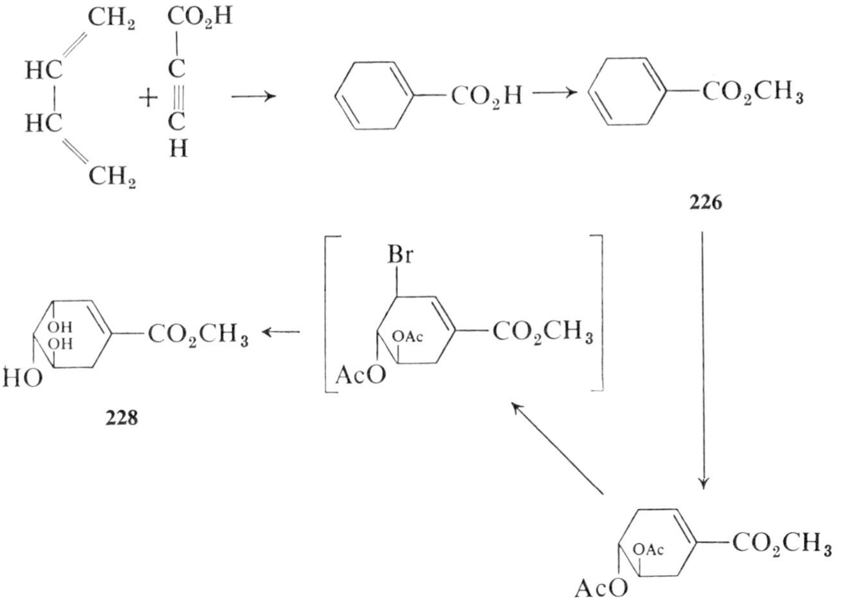

226

228

227

Scheme 65

dioxide gave the hemiacetal lactone **230**. A more convenient procedure for the preparation of **230** involved oxidation of **229** with the Pfitzner-Moffatt reagent (methyl sulfoxide-*N*,*N'*-dicyclohexylcarbodiimide-pyridinium phosphate) to give the *epi*-inosose derivative, which was then subjected to the Baeyer-Villiger reaction with peroxybenzoic acid. Compound **230** was heated in methanol containing a catalytic amount of sulfuric acid, and the resultant product was acetylated with acetic anhydride in pyridine; two components were isolated (**231** and **232**). Reduction of the monoacetate **231** with lithium aluminum hydride gave methyl 2,3-*O*-isopropylidene-β-DL-allofuranoside (**233**). Hydrolysis of **233** afforded DL-allose (**234**, only the D-isomer is shown). DL-Ribose (**235**) could also be obtained from **233** by successive oxidative cleavage by periodate, reduction with lithium aluminum hydride, and acid-catalyzed hydrolysis.

Fukami et al.[160] have used a similar approach to prepare 5-deoxy-DL-allose derivatives. In this case the seven-membered hemiacetal lactone **236** was employed. 4-Deoxy–DL-ribose (**238**) was also obtained from **236** by reduction with lithium aluminum hydride to give 2-deoxy–3,4-*O*-isopropylidene–DL-allitol (**237**), followed by oxidative cleavage with periodate and

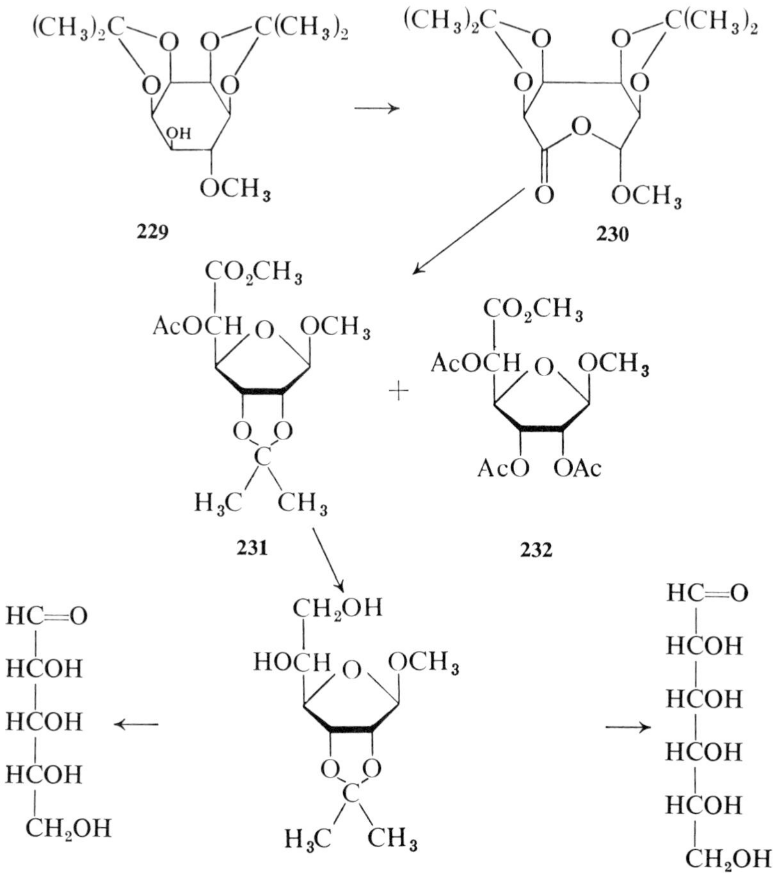

$(CH_3)_2C$ O O $C(CH_3)_2$ O O OH OCH_3

229

→

$(CH_3)_2C$ O O $C(CH_3)_2$ O O O O OCH_3

230

CO_2CH_3 AcOCH O OCH_3 O O C H_3C CH_3

231

+

CO_2CH_3 AcOCH O OCH_3 AcO OAc

232

HC=O HCOH HCOH HCOH CH_2OH

235

Ribose

←

CH_2OH HOCH O OCH_3 O O C H_3C CH_3

233

→

HC=O HCOH HCOH HCOH HCOH CH_2OH

234

Allose

Scheme 66

$(CH_3)_2C$ O O $C(CH_3)_2$ O O O O

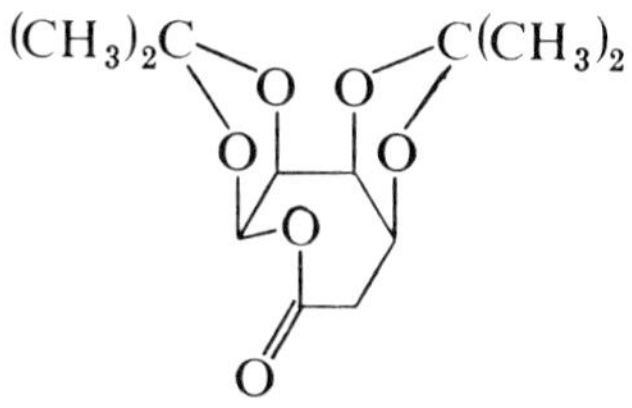

236

acid-catalyzed hydrolysis:

$$\mathbf{236} \xrightarrow{\text{LiAlH}_4} \begin{array}{l} CH_2OH \\ | \\ HCOH \\ | \\ HCO \diagdown \\ | \qquad C(CH_3)_2 \\ HCO \diagup \\ | \\ CH_2 \\ | \\ CH_2OH \end{array} \xrightarrow[2)\ H^{\oplus}]{1)\ IO_4^{\ominus}} \begin{array}{l} HC{=}O \\ | \\ HCOH \\ | \\ HCOH \\ | \\ CH_2 \\ | \\ CH_2OH \end{array}$$

237 **238**

REFERENCES

1. B. Zuckerman, D. Buhl, P. Palmer, and L. Snyder. *Astrophys. J.*, **160,** 485 (1969).
2. J. Shapira. *Chem. Eng. News*, **47** (41), 40 (1969); *Sci. Tech. Aerosp. Rep.*, **6,** 2597 (1968).
3. A. Butlerow, *Compt. rend.*, **53,** 145 (1861); A. Butlerow, *Ann.*, **120,** 295 (1861).
4. C. H. Riesz, *Sci. Tech. Rep*, **6,** 957 (1968); *Chem. Abs.*, 96815 (1969).
5. E. Fischer, *Ber.*, **21,** 989 (1888); **23,** 389, 2144 (1890).
6. O. Löw, *J. Prakt. Chem.*, [2] **33,** 321 (1886).
7. O. Löw, *Ber.*, **22,** 471 (1889).
8. E. Schmitz, *Ber.*, **46,** 2327 (1913); W. Küster and F. Schoder., *Hoppe-Seyler's Z. Physiol. Chem.*, **141,** 110 (1924).
9. E. Fischer and J. Tafel, *Ber.*, **20,** 1088, 2566 (1887).
10. E. Fischer and J. Tafel, *Ber.*, **20,** 3384 (1887).
11. H. O. L. Fischer and E. Baer, *Helv. Chim. Acta*, **19,** 519 (1935); **20,** 1213 (1937).
12. B. L. Horecker, *J. Cell. Comp. Physiol.*, **54** (Suppl. 1), 89 (1959); E. Grazi, T. Cheng, and B. L. Horecker, *Biochem. Biophys. Commun.*, **7,** 250 (1962).
13. W. G. Berl and C. E. Feazel, *J. Am. Chem. Soc.*, **73,** 2054 (1951); C. D. Gutsche, R. S. Buriks, K. Nowotony, and H. Grassner, *J. Am. Chem. Soc.*, **84,** 3775 (1962).
14. R. Breslow, *Tetrahedron Lett.*, **1959,** 22.
15. A. A. Marei and R. A. Raphael, *J. Chem. Soc.*, **1960,** 886.
16. E. Pfeil and H. Ruckert, *Ann.*, **641,** 121 (1961); J. K. N. Jones, personal observations.
17. H. Ruckert, E. Pfeil, and G. Scharf, *Ber.*, **98,** 2558 (1965).
18. L. Hough and J. K. N. Jones, *J. Chem. Soc.*, **1951,** 1126.
19. E. Mariani and G. Torraca, *Int. Sugar J.*, **55,** 309 (1953); *Chem. Abs.*, **48,** 4869 (1954).

19a. R. D. Partridge, A. H. Weiss, and D. Todd, Abstracts, 163rd National Meeting of the American Chemical Society, Boston, Mass., April 1972, CARB 2. See also A. H. Weiss and J. Shapira, *Hydrocarbon Process.*, **49**, 119 (1970); T. Mizuno, T. Mori, N. Shiomi, and H. Nakatsuji, *J. Agr. Chem. Soc. Jap.*, **44**, 324 (1970); T. Mizuno, M. Asai, A. Misaki, and Y. Fujihara, *ibid.*, **45**, 344 (1971).

19b. H. Tambawala and A. H. Weiss, Abstracts, 163rd National Meeting of the American Chemical Society, Boston, Mass., April 1972, PETR 8.

20. J. K. N. Jones, unpublished results.

21. C. Neuberg, *Ber.*, **35**, 2626 (1902).

22. C. D. Hurd and J. L. Abernethy, *J. Am. Chem. Soc.*, **63**, 1966 (1941).

23. L. Hough and J. K. N. Jones, *Nature*, **167**, 180 (1951); *J. Chem. Soc.*, **1951**, 1122, 3191.

23a. L. Tidwell, J. Lecocq, H. B. Chermside, and J. Shapira, *Proc. Western Pharmacol. Soc.*, **13**, 30 (1970).

24. L. M. Utkin, *Dokl. Akad. Nauk. S.S.S.R.*, **67**, 301 (1959).

25. A. Dmytraczenko, J. K. N. Jones, and W. A. Szarek, unpublished results.

26. F. Shafizadeh, *Adv. Carbohyd. Chem.*, **11**, 263 (1956).

27. L. M. Utkin, *Zh. Obshchei Khim.*, **25**, 530 (1955); *Chem. U.S.S.R.*, **25**, 499 (1955).

28. G. Griner, *Ann. Chim. Phys.* [6], **26**, 369 (1892).

29. R. Lespieau and J. Wiemann, *Bull. Soc. Chem.* [4], **53**, 1107 (1933); *J. Wiemann, Ann. Chim.* [11], **5**, 267 (1936).

30. Cf. R. Lespieau, *Adv. Carbohyd. Chem.*, **2**, 108 (1946).

31. R. Lespieau and J. Wiemann, *Compt. rend.*, **194**, 1946 (1932).

32. R. Lespieau, *Bull. Soc. Chim.* [4], **43**, 204 (1928).

33. R. Lespieau, *Bull. Soc. Chim.* [4], **43**, 657 (1928).

34. R. A. Raphael, *J. Chem. Soc.*, **1949**, S44.

35. I. Iwai and T. Iwashige, *Chem. Pharm. Bull. (Tokyo)*, **9**, 316 (1961); T. Iwashige, *Chem. Pharm. Bull. (Tokyo)*, **9**, 492 (1961).

36. I. Iwai and K. Tomita, *Chem. Pharm. Bull. (Tokyo)*, **9**, 976 (1961).

37. I. Iwai and K. Tomita, *Chem. Pharm. Bull. (Tokyo)*, **11**, 184 (1963).

38. I. Iwai, T. Iwashige, M. Asai, K. Tomita, T. Hiraoka, and J. Ide, *Chem. Pharm. Bull. (Tokyo)*, **11**, 188 (1963).

39. L. Hough, *Chem. Ind. (London)*, **1951**, 406.

40. W. G. Overend and M. Stacey, *J. Sci. Food Agr.*, **1**, 168 (1950).

41. M. M. Fraser and R. A. Raphael, *J. Chem. Soc.*, **1955**, 4280.

42. F. Weygand and H. Leube, *Ber.*, **89**, 1914 (1956).

43. G. Nakaminami, M. Nakagawa, S. Shioi, Y. Sugiyama, S. Isemura, and M. Shibuya, *Tetrahedron Lett.*, **1967**, 3983.

44. C. C. Price and R. B. Balsley, *J. Org. Chem.*, **31**, 3406 (1966).

45. R. A. Raphael, *J. Chem. Soc.*, **1952**, 401.

45a. W. F. Beech, *J. Chem. Soc.*, **1951**, 2483; B. W. Horrom and H. E. Zaugg, *J. Am. Chem. Soc.*, **79**, 1754 (1957).

45b. J. Kiss and F. Sirokmán, *Helv. Chim. Acta*, **43**, 334 (1960).

46. D. J. Walton, *Can. J. Chem.*, **45**, 2921 (1967).

47. D. Horton, J. B. Hughes, and J. K. Thomson, *J. Org. Chem.*, **33,** 728 (1968).
48. E. Fischer, *Ber.*, **22,** 2204 (1889).
49. H. J. Lucas and W. Baumgarten, *J. Am. Chem. Soc.*, **63,** 1653 (1941).
50. H. J. Bestmann and R. Schmiechen, *Ber.*, **94,** 751 (1961).
51. F. Weygand and R. Schmiechen, *Ber.*, **92,** 535 (1959).
52. A. J. Ultée and J. B. J. Soons, *Rec. Trav. Chim. Pays-Bas*, **71,** 565 (1952).
53. R. Lukes, J. Jary, and J. Nemec, *Coll. Czech. Chem. Commun.*, **27,** 735 (1962).
54. S. Sasaki, *J. Chem. Soc. Japan, Pure Chem. Sect.*, **78,** 1464 (1957); *Chem. Zentr.*, **1958,** 9467.
55. J. Jary and K. Kefurt, *Coll. Czech. Chem. Commun.*, **27,** 2561 (1962).
56. K. Koga, M. Taniguchi, and S. Yamada, *Tetrahedron Lett.*, **1971,** 263.
57. B. Belleau and Y.-K. Au-Young, *J. Am. Chem. Soc.*, **85,** 64 (1963).
58. For reviews see H. Paulsen, *Angew. Chem. Int. Ed.*, **5,** 495 (1966); H. Paulsen and K. Todt, *Adv. Carbohyd. Chem.*, **23,** 115 (1968).
59. D. M. Vyas and G. W. Hay, *Chem. Commun.*, **1971,** 1411; *Can. J. Chem.*, **49,** 3755 (1971).
60. For a recent review on acetals and hemiacetals see: E. Schmitz and I. Eichhorn, in S. Patai, Ed., *The Chemistry of the Ether Linkage*, Interscience, New York, 1967, Chapter 7.
61. S. J. Angyal, *Angew. Chem. Int. Ed.*, **8,** 157 (1969).
62. C. L. Wilson, *J. Chem. Soc.*, **1945,** 52.
63. H. Normant, *Compt. rend.*, **228,** 102 (1949).
64. P. Dimroth and H. Pasedach, *Angew. Chem.*, **72,** 865 (1960).
65. R. Paul, M. Fluchaire, and G. Collardeau, *Bull. Soc. Chim. Fr.*, **1950,** 668; M. L. A. Fluchaire and G. Collardeau, U.S. Patent 2,556,325 (1951); *Chem. Abs.*, **46,** 1046 (1952).
66. L. E. Schniepp and H. H. Geller, *J. Am. Chem. Soc.*, **68,** 1646 (1946).
67. W. J. Gensler and G. L. McLeod, *J. Org. Chem.*, **28,** 3194 (1963).
68. K. Alder, H. Oppermans, and E. Rüder, *Ber.*, **74,** 905, 920, 926 (1941).
69. C. W. Smith, D. G. Norton, and S. A. Ballard, *J. Am. Chem. Soc.*, **73,** 5267, 5270, 5273 (1951).
70. E. Dyer, C. P. J. Glaudemans, M. J. Koch, and R. H. Marchessault, *J. Chem. Soc.*, **1962,** 3361.
71. C. B. Anderson and D. T. Sepp, *Tetrahedron*, **24,** 1707 (1968); D. T. Sepp and C. B. Anderson, *Tetrahedron*, **24,** 6873 (1968).
72. R. U. Lemieux and N. J. Chu, Abstracts, 133rd National Meeting of the American Chemical Society, San Francisco, Calif., April 1958, p. 31N; E. L. Eliel, N. L. Allinger, S. J. Angyal, and G. A. Morrison, *Conformational Analysis*, Interscience, New York, 1965, p. 375.
73. S. Wolfe, A. Rauk, L. M. Tel, and I. G. Csizmadia, *J. Chem. Soc.*, (*B*), **1971,** 136.
74. U. E. Diner and R. K. Brown, *Can. J. Chem.*, **45,** 2547 (1967).
75. E. L. Eliel, B. E. Nowak, R. A. Daignault, and V. G. Badding, *J. Org. Chem.*, **30,** 2441 (1965).

76. S. O. Lawesson and C. Berglund, *Ark. Kemi*, **17**, 475 (1961); G. Sosnovsky, *J. Org. Chem.*, **25**, 874 (1960); *Tetrahedron*, **13**, 241 (1961).
77. C. D. Hurd and C. D. Kelso, *J. Am. Chem. Soc.*, **70**, 1484 (1948).
78. F. Sweet and R. K. Brown, *Can. J. Chem.*, **45**, 1007 (1967).
79. F. Sweet and R. K. Brown, *Can. J. Chem.*, **44**, 1571 (1966).
80. R. Paul, *Bull. Soc. Chim. Fr.*, **1** [V], 1403 (1934).
81. R. Paul, *Compt. rend.*, **218**, 122 (1944).
82. O. Riobé, *Bull. Soc. Chim. Fr.*, **1951**, 829.
83. R. U. Lemieux and B. Fraser-Reid, *Can. J. Chem.*, **43**, 1460 (1965).
84. W. Reppe and co-workers, *Ann.*, **596**, 86 (1955).
85. A. Senning and S.-O. Lawesson, *Tetrahedron*, **19**, 695 (1963).
86. M. J. Baldwin and R. K. Brown, *Can. J. Chem.*, **45**, 1195 (1967).
87. M. J. Baldwin and R. K. Brown, *Can. J. Chem.*, **46**, 1093 (1968).
88. F. Sweet and R. K. Brown, *Can. J. Chem.*, **46**, 707 (1968).
89. F. Sweet and R. K. Brown, *Can. J. Chem.*, **46**, 1592 (1968). See also A. Banaszek and A. Zamojski, *Roczniki Chem.*, **45**, 391 (1971).
90. F. Sweet and R. K. Brown, *Can. J. Chem.*, **46**, 1543 (1968).
91. M. J. Baldwin and R. K. Brown, *Can. J. Chem.*, **47**, 3099 (1969).
92. M. J. Baldwin and R. K. Brown, *Can. J. Chem.*, **47**, 3553 (1969).
92a. G. F. Woods and S. C. Temin, *J. Am. Chem. Soc.*, **72**, 139 (1950); R. M. Srivastava, F. Sweet, T. P. Murray, and R. K. Brown, *J. Org. Chem.*, **36**, 3633 (1971).
92b. R. M. Srivastava, F. Sweet, and R. K. Brown, *J. Org. Chem.*, **37**, 190 (1972).
93. M. Cahu and G. Descotes, *Bull. Soc. Chim. Fr.*, **1968**, 2975.
94. A. Konowal, J. Jurczak, and A. Zamojski, *Roczniki Chem.*, **42**, 2045 (1968).
95. A. Zamojski, A. Konowal, and J. Jurczak, *Roczniki Chem.*, **44**, 1981 (1970).
96. J. Jurczak and A. Zamojski, *Roczniki Chem.*, **44**, 2257 (1970).
97. J. Jurczak, A. Konowal, and A. Zamojski, *Roczniki Chem.*, **44**, 1587 (1970).
98. A. Konowal, A. Zamojski, M. Masojidkova, and J. Kohoutova, *Roczniki Chem.*, **44**, 1741 (1970). See also A. Banaszek and A. Zamojski, *ibid.*, **45**, 2089 (1971).
99. A. Zamojski, M. Chmielewski, and A. Konowal, *Tetrahedron*, **26**, 183 (1970).
99a. M. Chmielewski and A. Zamojski, *Roczniki Chem.*, **45**, 1689 (1971).
99b. A. Konowal and A. Zamojski, *Roczniki Chem.*, **45**, 859 (1971).
100. Z. Zwierzchowska and A. Zamojski, *Roczniki Chem.*, **44**, 1609 (1970).
101. R. J. Ferrier, N. Prasad, and G. H. Sankey, *J. Chem. Soc. (C)*, **1968**, 974.
102. F. Sweet and R. K. Brown, *Can. J. Chem.*, **46**, 2283 (1968).
103. R. M. Srivastava and R. K. Brown, *Can. J. Chem.*, **48**, 2334 (1970).
104. R. M. Srivastava and R. K. Brown, *Can. J. Chem.*, **48**, 2341 (1970).
105. F. Sweet and R. K. Brown, *Can. J. Chem.*, **46**, 2289 (1968).
106. T. P. Murray, C. S. Williams, and R. K. Brown, *J. Org. Chem.*, **36**, 1311 (1971).
107. U. P. Singh and R. K. Brown, *Can. J. Chem.*, **48**, 1791 (1970).
107a. U. P. Singh and R. K. Brown, *Can. J. Chem.*, **49**, 3342 (1971).
108. U. P. Singh and R. K. Brown, *Can. J. Chem.*, **49**, 1179 (1971).

109. T. P. Murray, U. P. Singh, and R. K. Brown, *Can. J. Chem.*, **49,** 2132 (1971).
110. R. M. Srivastava and R. K. Brown, *Can. J. Chem.*, **49,** 1339 (1971).
111. F. Korte, A. Bilow, and R. Heinz, *Tetrahedron*, **18,** 657 (1962).
112. H. Newman, *J. Org. Chem.*, **29,** 1461 (1964).
113. V. B. Mochalin, Yu. N. Porschnev, and G. I. Samokhvalov, *Zh. Obshch. Khim.*, **39,** 701 (1969); *Chem. Abs.*, **71,** 39346h (1969).
114. S. Yasuda and T. Matsumoto, *Tetrahedron Lett.*, **1969,** 4397.
115. S. Yasuda and T. Matsumoto, *Tetrahedron Lett.*, **1969,** 4393.
116. Y. Suhara, F. Sasaki, K. Maeda, H. Umezawa, and M. Ohno, *J. Am. Chem. Soc.*, **90,** 6559 (1968).
117. S. Yasuda, T. Ogasawara, S. Kawabata, I. Iwataki, and T. Matsumoto, *Tetrahedron Lett.*, **1969,** 3969.
118. M. Nakajima, H. Shibata, K. Kitahara, S. Takahashi, and A. Hasegawa, *Tetrahedron Lett.*, **1968,** 2271.
119. Y. Suhara, K. Maeda, and H. Umezawa, *J. Antibiotics*, **18A,** 187 (1965).
120. R. M. Srivastava and R. K. Brown, *Can. J. Chem.*, **48,** 830 (1970).
121. R. Lukes, M. Moll, A. Zobacova, and J. Jary, *Coll. Czech. Chem. Commun.*, **27,** 500 (1962).
122. I. Iwai, T. Iwashige, and M. Asai, *Chem. Abs.*, **65,** 3950 (1966).
122a. O. Achmatowicz, Jr., P. Bukowski, B. Szechner, Z. Zwierzchowska, and A. Zamojski, *Tetrahedron*, **27,** 1973 (1971).
122b. O. Achmatowicz, Jr., and B. Szechner, *Tetrahedron Lett.*, **1972,** 1205.
122c. O. Achmatowicz, Jr., and P. Bukowski, *Bull. Acad. Pol. Sci., Ser. Sci. Chim.*, **19,** 305 (1971).
122d. O. Achmatowicz, Jr., and B. Szechner, *Bull. Acad. Pol. Sci., Ser. Sci. Chim.*, **19,** 309 (1971).
123. S. David and A. Veyrières, *Carbohyd. Res.*, **13,** 203 (1970).
124. Y. Liwschitz, R. D. Irsay, and A. I. Vincze, *J. Chem. Soc.*, **1959,** 1308.
125. M. Sato, K. Okawa, and S. Akabori, *Bull. Chem. Soc. Jap.*, **30,** 937 (1957).
126. T. Ichikawa, T. Okamoto, S. Maeda, S. Ohdan, Y. Araki, and Y. Ishido, *Tetrahedron Lett.*, **1971,** 79.
127. R. Kuhn and D. Weiser, *Ann.*, **602,** 208 (1957).
127a. V. B. Mochalin, Z. I. Smolina, and B. V. Unkovskii, *Zh. Obshch. Khim.*, **41,** 1863 (1971).
128. For a recent review on fluorocarbohydrates see P. W. Kent, *Chem. Ind. (London)*, **1969,** 1128.
129. N. F. Taylor and P. W. Kent, *J. Chem. Soc.*, **1956,** 2150.
130. J. E. G. Barnett and P. W. Kent, *J. Chem. Soc.*, **1963,** 2743.
131. P. W. Kent and J. E. G. Barnett, *J. Chem. Soc.*, **1964,** 2497.
132. N. F. Taylor and P. W. Kent, *J. Chem. Soc.*, **1958,** 872.
133. A. Bekoe and H. M. Powell, *Proc. Roy. Soc.*, **250A,** 301 (1959).
134. R. C. Cherry and P. W. Kent, *J. Chem. Soc.*, **1962,** 2507.
135. D. M. Lemal, P. D. Pacht, and R. B. Woodward, *Tetrahedron*, **18,** 1275 (1962).

136. F. Korte, U. Claussen, and K. Göhring, *Tetrahedron*, **18,** 1257 (1962).
137. H. Grisebach, W. Hofheinz, and N. Doerr, *Ber.*, **96,** 1823 (1963).
138. L. Hough and J. K. N. Jones, *Adv. Carbohyd. Chem.*, **11,** 185 (1956).
139. L. Hough and J. K. N. Jones, *J. Chem. Soc.*, **1952,** 4047.
140. L. Hough and J. K. N. Jones, *J. Chem. Soc.*, **1952,** 4052.
141. T. C. Tung, K. H. Ling, W. L. Byrne, and H. A. Lardy, *Biochim. Biophys. Acta*, **14,** 488 (1954).
142. R. S. Lane, A. Shapley, and E. E. Dekker, *Biochem.*, **10,** 1353 (1971).
143. K. Uehara and T. Matsukawa, Jap. Patent 2279 (1959); *Chem. Abs.*, **54,** 13016 (1960).
144. N. D. Zelinski, J. I. Denisenko, and M. S. Eventova, *Compt. rend. Acad. Sci. URSS*, **1,** 313 (1935); *Chem. Zentr.*, **106, II,** 3765 (1935).
145. H. Z. Sable, Abstracts, 149th National Meeting of the American Chemical Society, Detroit, Mich., April 1965, p. 19C.
146. Y. K. Yurev and N. S. Zefirov, *Zh. Obsch. Khim.*, **31,** 685 (1961); *Chem. Abs.*, **55,** 24573 (1961). See also N. S. Zefirov, Y. Yurev, L. Prikazchilova, and M. Bykhovskay *Zh. Obsch. Khim.*, **33,** 2153 (1963).
147. R. Criegee and P. Becher, *Ber.*, **90,** 2516 (1957).
148. T. Posternak and H. Friedli, *Helv. Chim. Acta*, **36,** 251 (1953).
149. G. E. McCasland and E. C. Horswill, *J. Am. Chem. Soc.*, **76,** 1654 (1954).
150. S. Takai, M. Nakajima, and I. Tomida, *Ber.*, **89,** 263 (1956); M. Nakajima, I. Tomida, and S. Takei, *Ber.*, **90,** 246 (1957); **92,** 163 (1959); M. Nakajima, I. Tomida, N. Kurihara, and S. Takei, *Ber.*, **92,** 173 (1959).
151. H. Wieland and R. S. Wishart, *Ber.*, **47,** 2082 (1914).
152. R. Kuhn, G. Quadbeck, and E. Röhm, *Ann.*, **565,** 1 (1949).
153. S. J. Angyal and D. J. McHugh, *J. Chem. Soc.*, **1957,** 3682.
154. E. E. Smissman and M. Oxman, *J. Am. Chem. Soc.*, **86,** 2184 (1963).
155. J. Wolinsky, R. Novak, and R. Vasileff, *J. Org. Chem.*, **29,** 3596 (1964).
156. E. E. Smissman, J. T. Suh, M. Oxman, and R. Daniels, *J. Am. Chem. Soc.*, **81,** 2909 (1959).
157. R. Grewe and I. Hinrichs, *Ber.*, **97,** 443 (1964).
157a. J. Cleophax, D. Mercier, and S. D. Géro, *Angew. Chem. Int. Ed.*, **10,** 652 (1971).
158. H. Fukami, H.-S. Koh, T. Sakata, and M. Nakajima, *Tetrahedron Lett.*, **1967,** 4771.
159. S. J. Angyal and P. T. Gilham, *J. Chem. Soc.*, **1957,** 3691.
160. H. Fukami, H.-S. Koh, T. Sakata, and M. Nakajima, *Tetrahedron Lett.*, **1968,** 1701.

The Total Synthesis of Prostaglandins

U. AXEN, J. E. PIKE, AND W. P. SCHNEIDER

The Upjohn Company, Kalamazoo, Michigan

1. INTRODUCTION

The prostaglandins were discovered in 1933 to 1934 by Goldblatt in England and von Euler in Sweden. Smooth muscle stimulating and vasodepressor activities were found in extracts of human seminal fluid and sheep vesicular glands. von Euler showed that the activity associated with the lipid soluble fractions was not identical with any of the humoral agents known at that time and named the material prostaglandin. The next major development in the field was the isolation many years later of pure prostaglandin E_1 and $F_1\alpha$ by Bergstrom and his associates and later their classic work on the structure elucidation of these acidic lipids.[25,28] They are carboxylic acids with a 20-carbon structure incorporating a 5-membered ring and the hypothetical parent structure is designated prostanoic acid (Fig. 1).

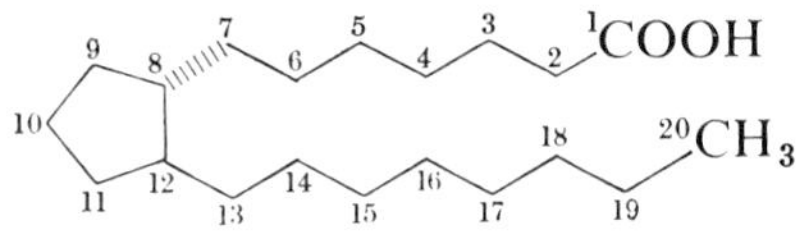

Figure 1. Prostanoic acid

The structures of naturally occurring materials are readily divided into four basic families: PGE, PGF, PGA, and PGB (Fig. 2). The various families include different structures, for example, $PGF_1\alpha$, $PGF_2\alpha$, and $PGF_3\alpha$ with, respectively, 13-*trans*, 5-*cis*–13-*trans*, and 5-*cis*–13-*trans*–17-*cis* double bonds. All six members of the prostaglandins of the E and F series are known as primary prostaglandins. The structures of the prostaglandins were established by Bergstrom and associates in Sweden by methods that included classical degradation and X-ray crystallographic studies on suitable derivatives.[1,2] A noteworthy feature of the structure elucidation was the extensive use of gas chromatography in combination with mass spectroscopy to identify the molecular fragments formed in various chemical degradations. The absolute configuration of PGE_1 has also been determined.[85] The presence of 19-hydroxy prostaglandins of the A and B classes was shown by Hamberg and Samuelsson in human seminal plasma.[63] Various earlier reviews cover the structural elucidation and associated chemistry.[13,20,95,99]

The chemical synthesis of the prostaglandins offers an unusually significant challenge to organic chemists for several reasons. It is now well established that the prostaglandins do not occur in male tissues only, but are found in low concentrations in nearly all organs.[29] The biological potency and diversity of the prostaglandins is remarkable and several areas are under intensive study to elucidate a possible physiological role for these materials.[68,97] Additionally, the possible clinical significance of prostaglandins is undergoing serious evaluation, particularly in the control of fertility.[67] The concentration of the natural prostaglandins in most tissues is less than 1 μg/g; one exception is in human seminal fluid where concentrations reach about 50–60 μg/ml. Clearly, then, extraction of various organs will not provide useful quantities of the agents. Until very recently prostaglandins had not been isolated from nonmammalian sources in amounts that would serve as potential precursors of the natural hormones. A report by Weinheimer and Spraggins that prostaglandins could be found in a horny coral or gorgonian

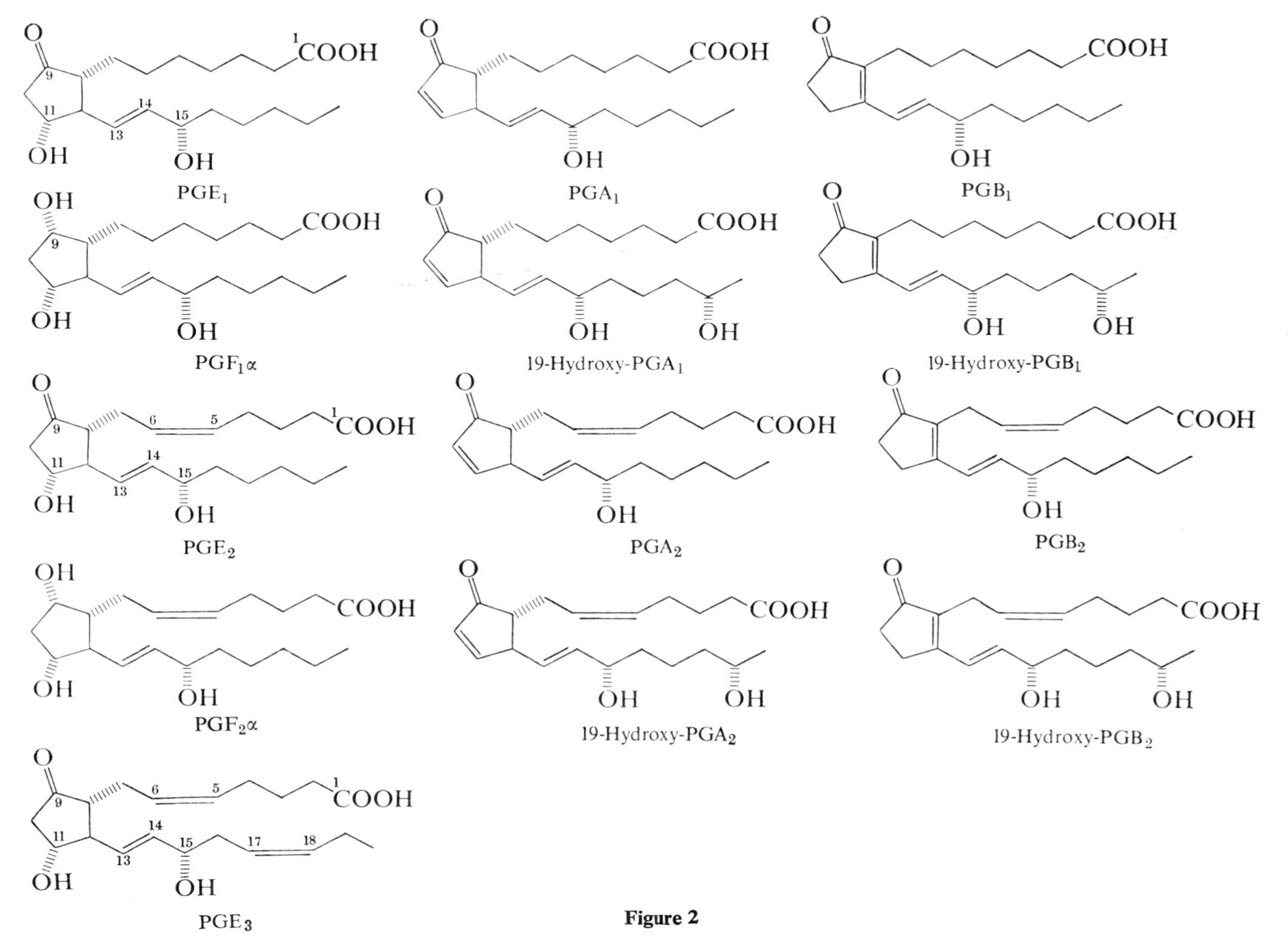

PGE$_1$
PGA$_1$
PGB$_1$
PGF$_1\alpha$
19-Hydroxy-PGA$_1$
19-Hydroxy-PGB$_1$
PGE$_2$
PGA$_2$
PGB$_2$
PGF$_2\alpha$
19-Hydroxy-PGA$_2$
19-Hydroxy-PGB$_2$
PGE$_3$

Figure 2

called *Plexaura homomalla*[118–120] has suggested a possible alternative source. Nevertheless, chemical total synthesis with its inherent flexibility seems to offer the best route to obtain these materials and possible structural analogs, which may be needed as therapeutic agents.

Before discussing the chemistry of the prostaglandins a brief review will be given of the biosynthesis and metabolism of prostaglandins. It was shown in 1964 that the essential fatty acids were enzymatic precursors of prostaglandins.[26,27,49,50] Bergstrom and co-workers in Stockholm and van Dorp and co-workers in Holland demonstrated that 8,11,14-eicosatrienoic acid was the biological precursor of PGE_1, arachidonic acid was the precursor of PGE_2, and 5,8,11,14,17-eicosapentaenoic acid led to PGE_3 (Fig. 3). In enzymes derived from guinea pig lung it was shown that arachidonic acid led to $PGF_2\alpha$.[8] The prostaglandin-synthesizing enzymes appear widely distributed in various animal tissues.[101,102] Studies on the mechanism of the biosynthesis, especially by Samuelsson, have established the mechanism indicated in Fig. 4 involving an intermediate endoperoxide.[102] The isolation of intermediates such as this have not yet been reported and would represent extremely intriguing synthetic objectives for an organic chemist. Experiments also have been described by workers at Unilever on the nonenzymatic conversion of 8,11,14-eicosatrienoic acid to PGE_1.[86] The present very low overall yield in this route suggests this is not a preferred way to obtain prostanoic acids. Chemical synthesis of many of the proposed intermediates in the biosynthesis, especially with specific C^{14} and tritium labeling, may be necessary in the future to delineate exactly the sequence of steps in the biosynthetic formation of prostaglandins.

Workers at the Unilever Laboratories have studied the substrate specificity of the enzyme system and these investigations have led to the synthesis of prostaglandins with varied side chain lengths and varied double bond isomers.[51,114] A particularly important finding was the correlation between the essential fatty acid activity of the precursor acids and the rate of formation of the biologically active prostaglandins. This work has strongly supported the idea that part of the essential function of the unsaturated fatty acids must be associated with their conversion to prostaglandins.

Another area of research which has attracted attention involving the biosynthesis has been the report of specific inhibitors of the synthetase. Particularly significant was the activity reported by Nugteren and co-workers for the 8-*cis*–12-*trans*–14-*cis*-eicosatrienoic acid[87] and by Downing for 5,8,11, 14-eicosatetraynoic acid.[3,4,52] Recent studies by Sih have concentrated on understanding the control mechanisms in the prostaglandin biosynthesis and factors regulating their formation.[111] Particularly intriguing is why the endoperoxide under differing physiological stimuli can produce either the PGE, PGF, or 11-dehydro-PGF structures.[112] Recently Pace-Asciak and

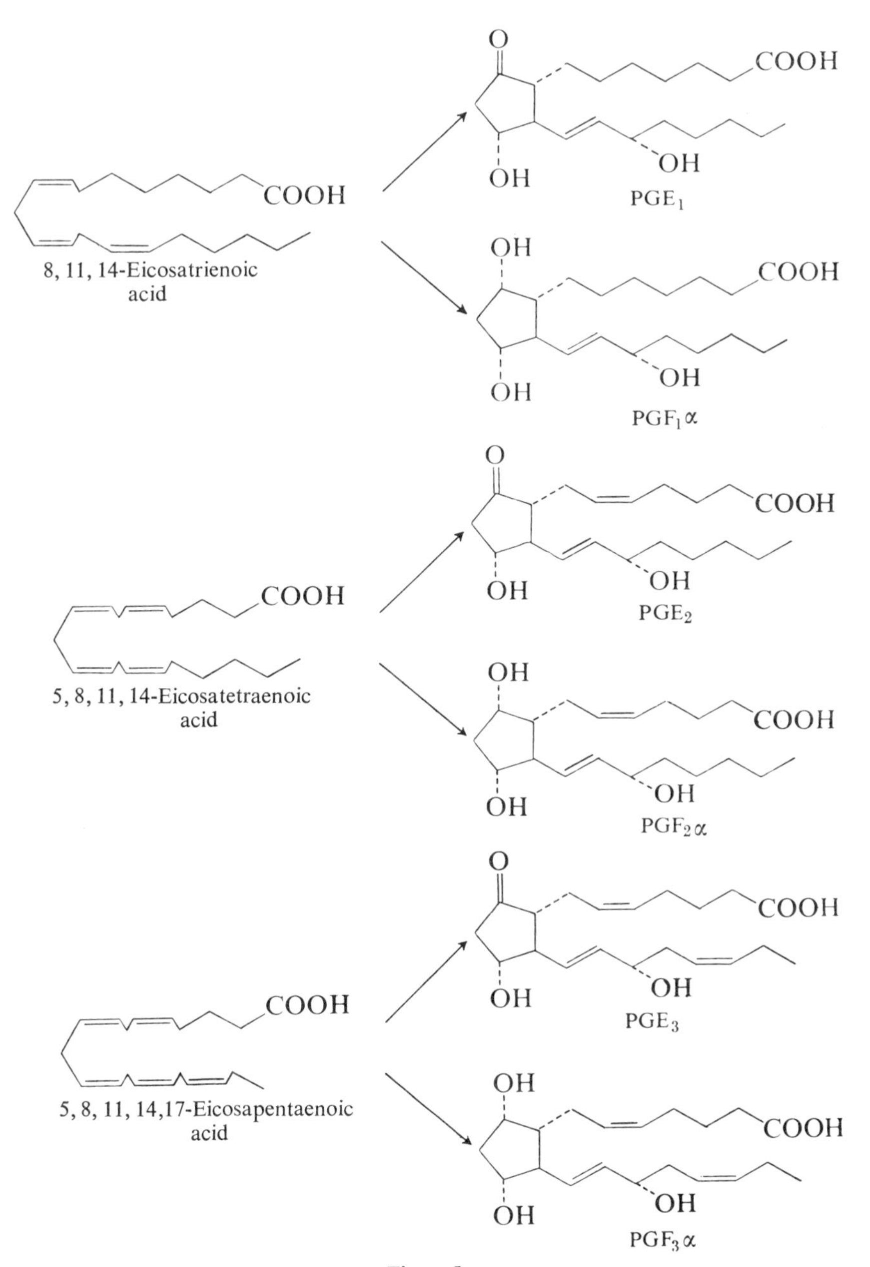

COOH
8, 11, 14-Eicosatrienoic acid
O
COOH
OH
OH
PGE$_1$
OH
COOH
OH
OH
PGF$_1$α
COOH
5, 8, 11, 14-Eicosatetraenoic acid
O
COOH
OH
OH
PGE$_2$
OH
COOH
OH
OH
PGF$_2$α
COOH
5, 8, 11, 14,17-Eicosapentaenoic acid
O
COOH
OH
OH
PGE$_3$
OH
COOH
OH
OH
PGF$_3$α

Figure 3

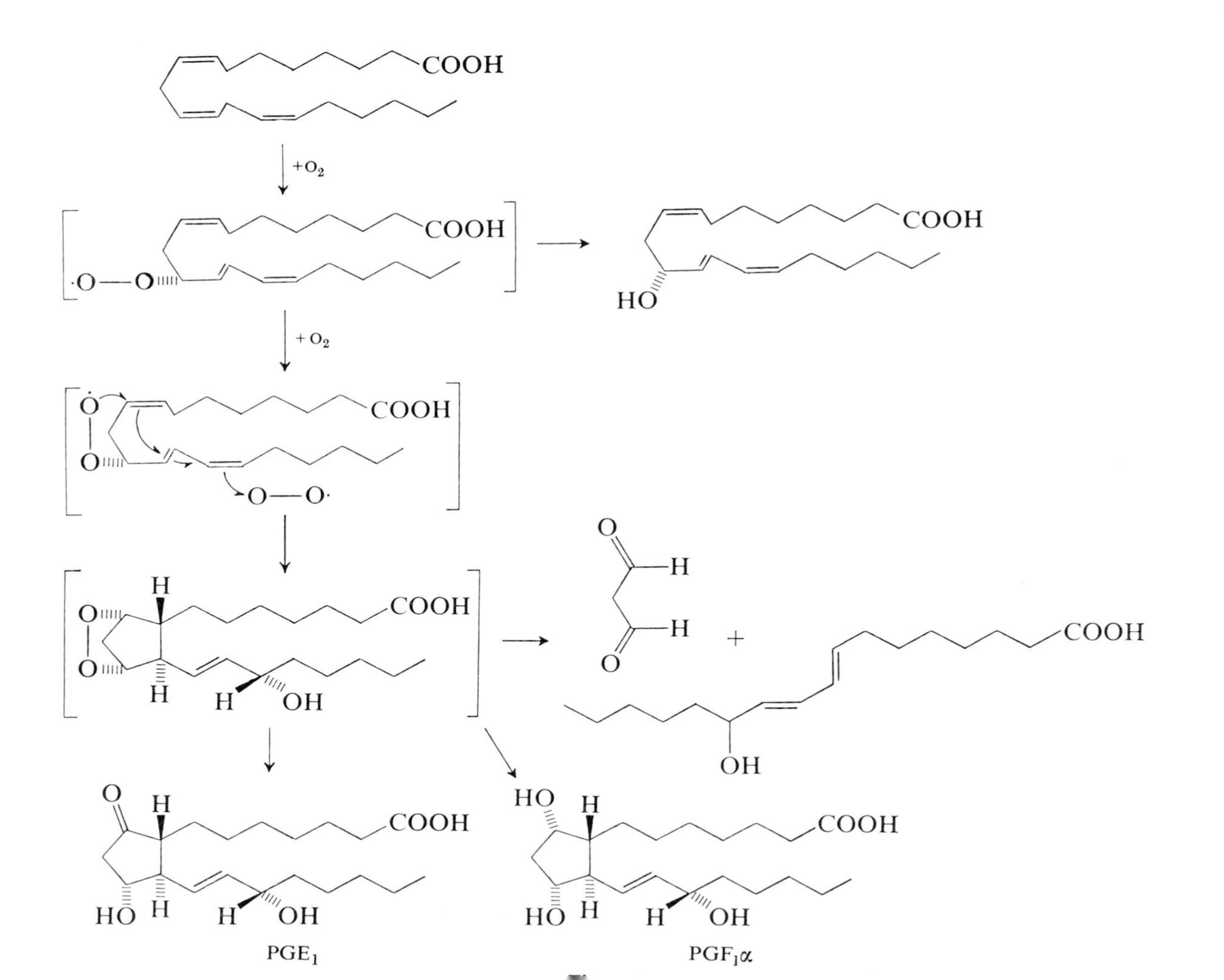
+O2
+O2
COOH
HO
OH
H
O
PGE1
PGF1α

Wolfe described some additional prostanoic acid by-products formed from arachidonic acid.[90,91]

Studies, particularly by Sameulsson, Änggård, and co-workers, have delineated the principal metabolic inactivation processes both *in vitro* and *in vivo* for the natural prostaglandins.[8] Especially important is the 15-hydroxydehydrogenase enzyme originally isolated from lung tissue.[9–12] This enzyme, which has been purified, has a high degree of structural specificity for prostaglandins.[82,83] The resulting 15-keto compounds appear to be much less biologically active than their 15(*S*)-hydroxy precursor.[10,93,109] Other metabolic transformations include the reduction of the 13,14-double bond of the Δ^{13}-15-keto metabolites, the β-hydroxylation and cleavage of the carboxy side chain, and the ω- and ω-1-hydroxylation of the alkyl (C_{13}–C_{20}) side chain.[101,102] The major urinary metabolites in man of both PGE_2 and $PGF_2\alpha$ have been identified (Fig. 5).[61,65]

More recently, Hamberg and Israelsson have described liver enzymes *in vitro* which convert the 9-keto prostaglandins both to the 9α- and 9β-hydroxy isomers.[64] This conversion of the PGE to the PGFα compounds *in vitro* is the first reported interconversion of these two series and the significance of this transformation *in vivo* remains to be evaluated. Enzymatic transformations have also been described which convert the saturated 15-keto compounds to the corresponding 15-hydroxy (*S* and *R*) metabolites.[64] It also now appears that the 8-iso compounds are formed in enzymatic transformations.[64] A particularly important aspect in the further understanding of prostaglandin metabolism is the development of very sensitive assays for prostaglandins. A recent publication by Levine has suggested that the radioimmunoassay technique may be applied to the prostaglandins.[77] Samuelsson and Sweeley have developed a new reverse isotope technique which promises to allow the GLC mass spectroscopic measurement of physiological levels of prostaglandins.[103]

Before discussing in detail the various synthetic approaches, some discussion is in order about the known chemistry of the prostaglandins, particularly the stereochemical features of the molecules. No really systematic studies have appeared detailing the basic chemistry of the molecule, although earlier work has covered general aspects of prostaglandin physicochemical properties. The stereochemistry of substituents has been designated α or β by analogy with steroid nomenclature; α-substituents are oriented on the same side of the five-membered ring as the carboxy (C_1 to C_8) side chain and β-substituents are above the plane of the cyclopentane ring and on the same side as the C_{13} to C_{20} side chain. The hydroxyl group at C_{15} has the 15(*S*) stereochemistry in the natural mammalian prostaglandins. In earlier publications this was also designated at 15α. The epimer of this hydroxyl group 15(*R*) has also been named 15β or 15-*epi*. The stereochemistry of the attached

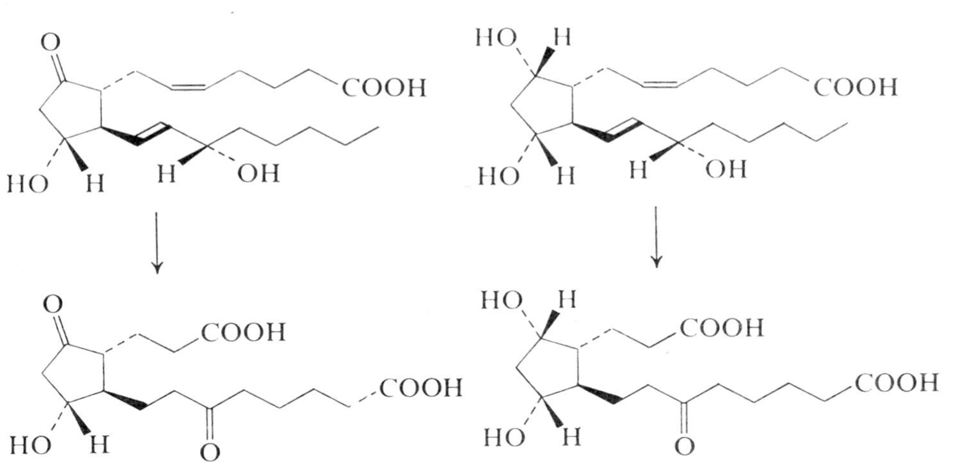

Figure 5. The major urinary metabolite of prostaglandin E_2 and $F_{2\alpha}$ in man.

side chain at C_8 and C_{12} was established principally by the earlier X-ray studies of Abrahamsson and co-workers.[1,2] The corresponding 8-*iso* compounds, that is, those with *cis*-oriented side chains, have also been described,[47] but the equilibrium occurs largely in favor of the natural configuration and under mild basic conditions the ratio is about 9:1. When the PGE or PGA compounds are treated with base, the PGB compounds or doubly unsaturated ketones are formed which have a characteristic UV absorption at 278 nm.[25] Under mild acidic treatment the PGE compounds can be converted to the PGA compounds.[46,93,94] Conditions which have been described include acetic acid and hydrochloric acid buffered with tetrahydrofuran. When prostaglandins are treated with stronger acids, for example, formic acid, epimerization of the C_{15}-hydroxyl occurs giving a mixture of both 15(*R*) and 15(*S*) derivatives.[94] Reductions of the ketone groups at C_9 and C_{15} have been described, and in both cases mixtures of hydroxyl epimers were obtained.[24,94]

Recently methods have been described for converting the PGA series back to the PGE and PGF prostaglandins via intermediate epoxy ketones (see below).[31]

The trichloroethyl ester of PGE_1 could be reconverted to the parent acid by treatment with zinc and acid under conditions which did not affect the β-hydroxy ketone system of the PGE compounds.[94] The 9-ketone of PGE_1 has been converted to an oxime from which the parent ketone could be regenerated using nitrous acid.[48,94] The reduction of the carboxyl group to a primary alcohol has been described for both the PGE[48,94] and PGF[89,93] series.

2. SYNTHETIC ROUTES TO PROSTAGLANDINS E AND F

A. General Strategy

Total synthesis of the prostaglandin molecule constitutes a formidable challenge to the organic chemist, not as much because of the nature of the different functional groups as because of the numerous asymmetric centers and the relative position of the functional groups to each other. PGE_1 (Fig. 2) has four asymmetric centers, an allylic alcohol with trans geometry of the double bond, and a β-ketol system which is sensitive to acid and base. Acid treatment causes elimination of water to form the α,β-unsaturated enone system, PGA_1, and base treatment causes elimination and double bond migration to the fully conjugated PGB system[25,93] (Fig. 6). Therefore the β-ketol system is generated at a late stage in all the syntheses of E-prostaglandins published

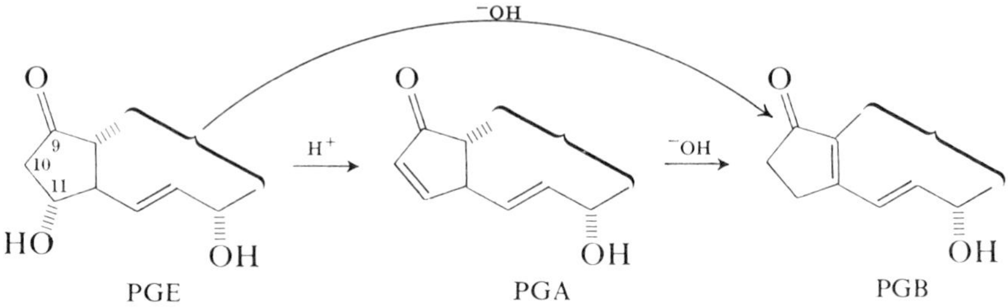

Figure 6

so far. Both possibilities have been realized: introduction of the 11-hydroxyl group at the end of the synthesis or generation of the 9-keto group as the last step.

$PGF_1\alpha$ (Fig. 2) is more stable than PGE_1 because the β-ketol unit is reduced to a 1,3-diol, but a new asymmetric center is introduced at C-9. PGE_2 and $PGF_2\alpha$ (Fig. 2) display an additional double bond at $C_{5,6}$; PGE_3 and $PGF_3\alpha$ (Fig. 2) display two additional double bonds at $C_{5,6}$ and $C_{17,18}$; all have cis geometry. All three E-prostaglandins have been converted to the corresponding PGF compounds by selective reduction of the keto function, but in these cases the corresponding 9β-isomers are formed in addition[24,25,62a] (Fig. 7).

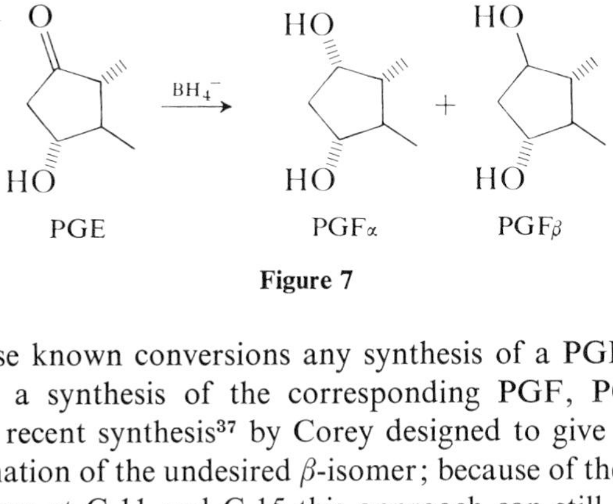

Figure 7

Because of these known conversions any synthesis of a PGE constitutes at the same time a synthesis of the corresponding PGF, PGA, and PGB compounds. A recent synthesis[37] by Corey designed to give $PGF_2\alpha$ directly avoids the formation of the undesired β-isomer; because of the use of suitable protecting groups at C-11 and C-15 this approach can still be used for the synthesis of PGE_2. In most other cases PGF's are obtained by reduction of the corresponding PGE's.

Most syntheses start with a suitably functionalized five-membered ring like cyclopentadiene, norbornadiene, or an indanol. Later formation of the ring by cyclization has been utilized, too, but no approach has yet been reported which generates the β-ketol unit and the ring simultaneously by aldol condensation of the corresponding keto-aldehyde.

B. Stereochemical Principles

Stereochemical control at C-8 and C-12 generally does not cause too many problems since the more stable *trans* configuration of the two side chains is the natural configuration. In most cases where the side chains are introduced subsequently the *trans* configuration is formed preferentially. Another possibility in the case of PGE's is the isomerization at C-8, which proceeds under basic conditions mild enough not to cause formation of PGA's or PGB's (Fig. 8).[47] More difficult is stereochemical control at C-9 of the PGF's.

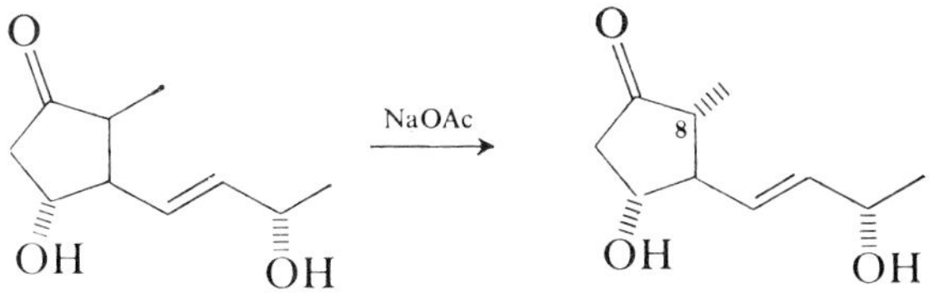

8-iso-PGE NaOAc → PGE

Figure 8

As mentioned, reduction of the keto-function produces mixtures of the desired α-isomer and the undesired β-isomer. Corey employed two different principles to obtain the desired cis configuration at C-8 and C-9: ketene addition to a double bond[41] (Fig. 9) and iodolactone formation[37] (Fig. 10). (Prostaglandin numbering is used for synthetic intermediates throughout this paper.) Both approaches not only assured the correct stereochemistry and functionality

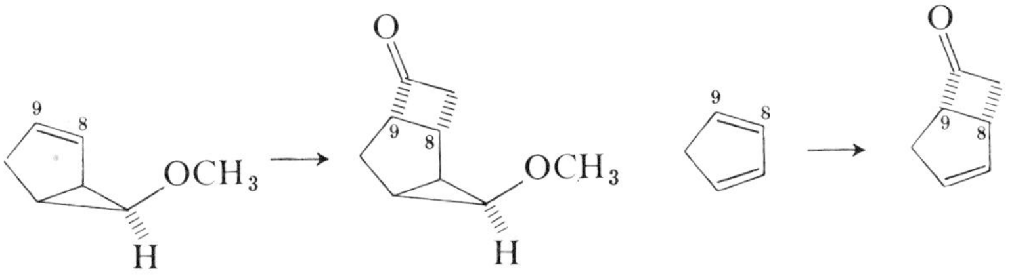

Figure 9

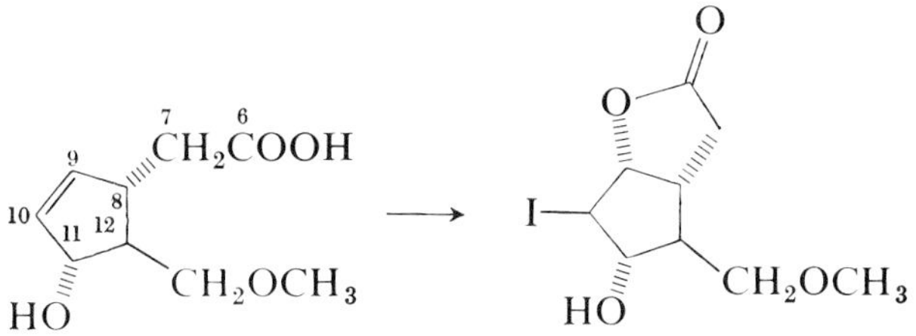

Figure 10

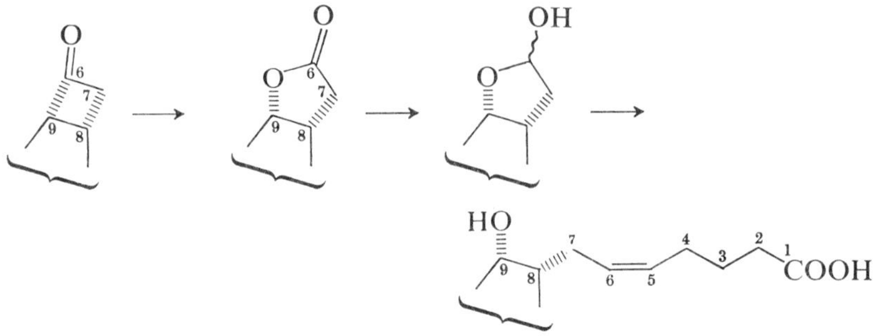

Figure 11

on the five-membered ring but also allowed a convenient introduction of the Δ^5 double bond in the synthesis of $PGF_2\alpha$: Baeyer-Villiger oxidation of the cyclobutanone produced a lactone similar to the one in Fig. 10, which was reduced to a hemiacetal and then subjected to a Wittig reaction (Fig. 11). The Wittig condensation was carried out under conditions which were known to produce selectively *cis*-olefins.[39] This strategy, construction of a γ-lactone fused to the cyclopentane ring at C-8 and C-9, proved to be very efficient since it allowed the introduction of the right functionalities with correct stereochemistry at three centers.

Introduction of the carboxy side chain by direct alkylation of the 9-ketone resulted in mixtures of 8α-and 8β-isomers[14,71,105] (Fig. 12). The thermodynamic equilibrium was found to be 65:35 in favor of the 8β-isomer when *exo* configuration at C-13 and 80:20 in favor of the 8α-isomer in the *endo* series. These results were surprising at first since the base-catalyzed equilibrium of PGE_1 and 8-*iso*-PGE_1 gave a 90:10 mixture in favor of the natural 8α-configuration, which was expected because *trans* configuration of the two side chains should be favored thermodynamically. The apparent preference

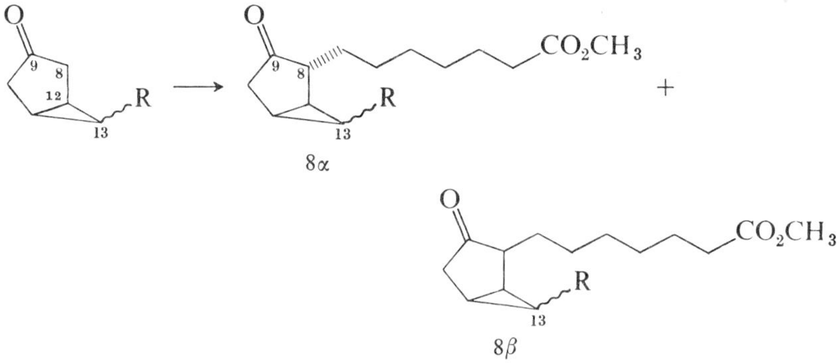

Figure 12

for cis substitution at C-8 and C-12 in the *exo*-bicyclohexane system is probably due to a boatlike conformation of the five-membered ring in bicyclo[3.1.0]hexanes (see, e.g., citations 8 and 13 in reference 71). In such a conformation the 8β-oriented side chain assumes an equatorial position. In the *endo*-bicyclohexane system, however, steric interaction with the substituent at C-13 causes preference of the 8α-position. In other syntheses the more stable *trans* configuration at C-8 and C-12 is obtained by thermodynamically controlled formation of the five-membered ring[35,36,38,45] or during generation of the asymmetric center at C-12 under basic conditions.[115]

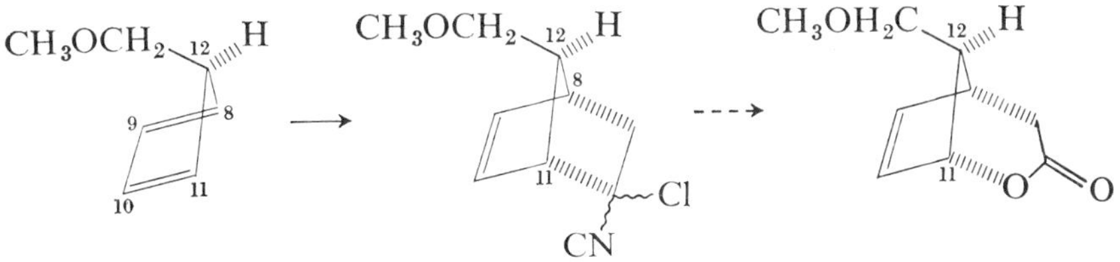

CH_3OCH_2 12 H, 9, 8, 10, 11 → CH_3OCH_2 12 H, 8, 11, Cl, CN ---→ CH_3OH_2C 12 H, 11, O, O

Figure 13

Stereochemical control at C-11 and C-12 proved to be not too difficult. In one of Corey's syntheses Diels-Alder addition assures the correct *trans-trans* relationship at C_8-C_{12}-C_{11} (Fig. 13).[37] When the hydroxyl-group at C-11 was introduced by opening of a cyclopropyl ring only the 11α isomer was formed[14,41,70,71,105,106] (Fig. 14). Whether this is due to mechanistic control, that is, the approach of the incoming hydroxyl group from the opposite side, or product control, the formation of the more stable *trans* configuration, depends on how much carbonium ion character one is willing to concede C-11 during opening of the cyclopropane ring.

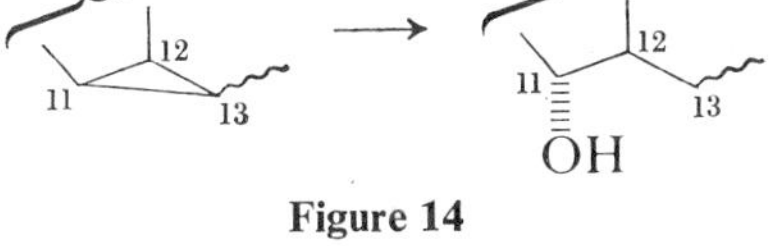

12, 11, 13 → 12, 11, 13, OH

Figure 14

In the Merck synthesis *trans* relationship between substituents at C-11 and C-12 was achieved by base-catalyzed epimerization of an 11-acetyl group.[115] Where the five-membered ring was formed via aldol condensation originally a mixture of 11-epimers was obtained[35,36] but later conditions were found which gave exclusively the desired α-isomer (Fig. 15).[38] Surprisingly, even direct epoxidation of 2-oxabicyclo[3.3.0]-oct-6-en-3-one yielded mainly the desired α-epoxide (Fig. 16).[42]

Two methods were used for formation of the allyl-alcohol system: (1) generation of an enone unit followed by reduction of the ketogroup[35–38,41,42,115] (Fig. 17); and (2) solvolysis of a bicyclo[3.1.0]-hexane system[14,71,105,106] (Fig. 18). In both cases only the desired *trans* olefin is formed. In the first

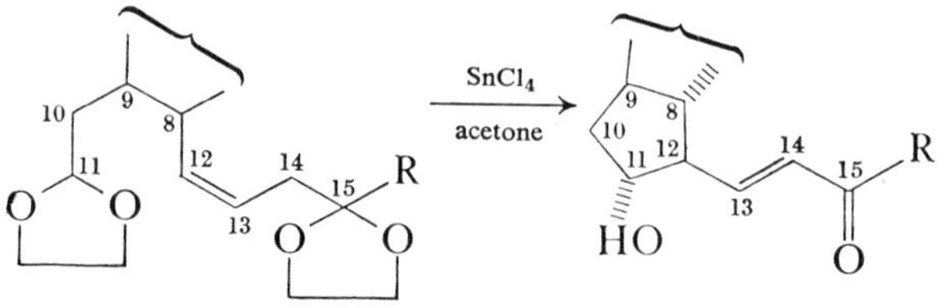

9 10 8 11 12 14 15 R O O 13 O O
$SnCl_4$
acetone
9 8 10 11 12 14 R 15 13 HO O

Figure 15

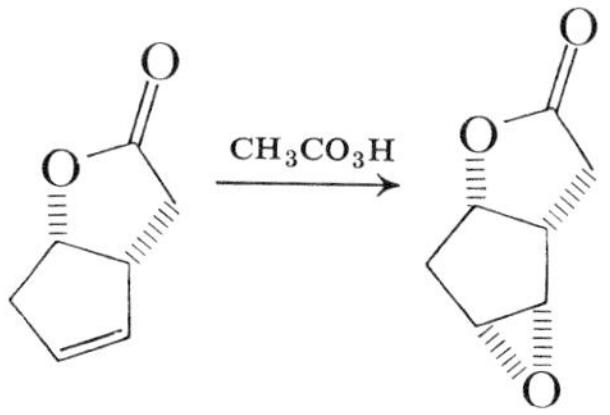

O O CH_3CO_3H O O O

Figure 16

12 14 15 13 O → 15 HO H + 15 H OH

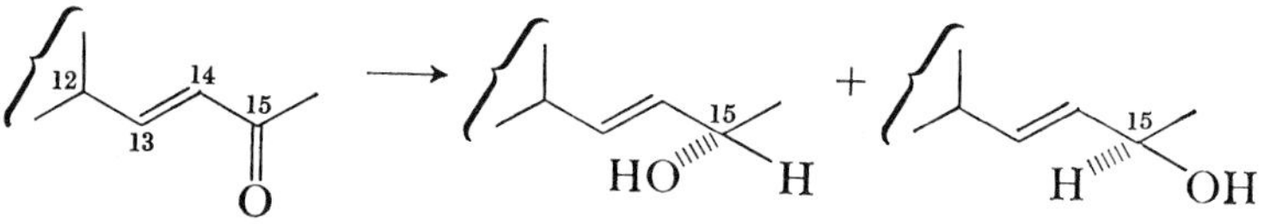

Figure 17

method *trans* geometry is assured because of the readily obtainable *trans*-enone; in the second method the activation energy for the rearrangement probably is high enough so that the more stable *trans* compound is formed without stopping at the energetically less stable *cis* olefin. The configuration at C-14 (Fig. 18) has no influence on stereochemistry or yield of the solvolysis reaction; both epimers give the same products. Unfortunately both methods result in mixtures of C-15 epimers, in the case of the latter again regardless of configuration at C-15 before solvolysis; so far in all prostaglandin syntheses the C-15 epimers have to be separated by column chromatography. The only exception has recently been disclosed by Corey;[45] in this variation a phosphonium salt with the correct 15(*S*) configuration is prepared in a five step synthesis from L(−) malic acid and then reacted with an aldehyde to afford exclusively the desired 15α-isomer (Fig. 19).

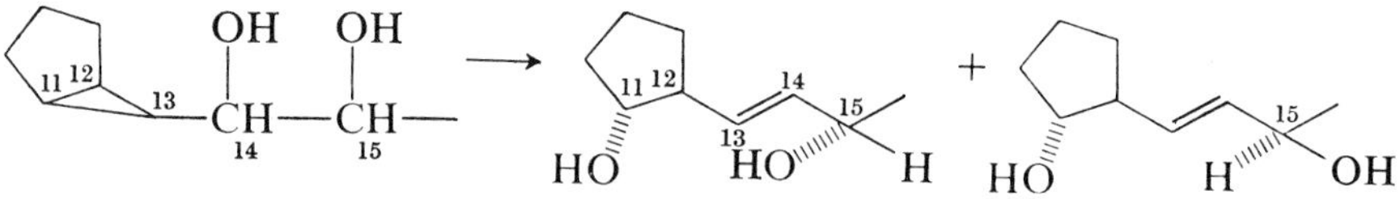

OH OH
11 12 13 —CH—CH—
14 15
→ 11 12 14 15 13 HO HO H + 15 HO H OH

Figure 18

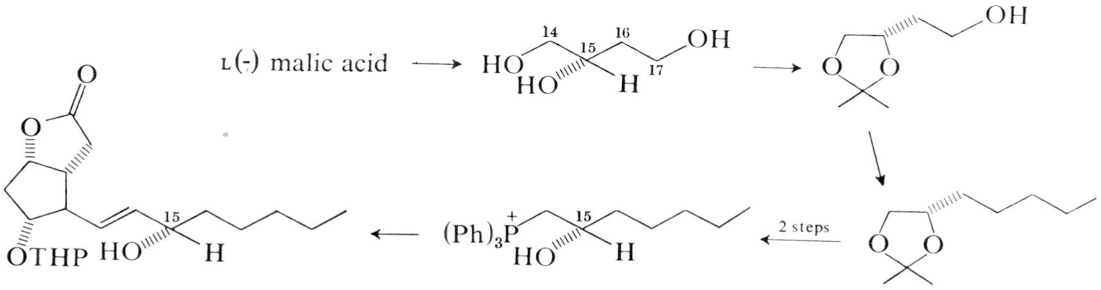

L(-) malic acid
HO
HO
H
OH
14
15
16
17
O
O
OH
2 steps
$(Ph)_3\overset{+}{P}$
HO
H
15
O
O
OTHP
HO
H
15

Figure 19

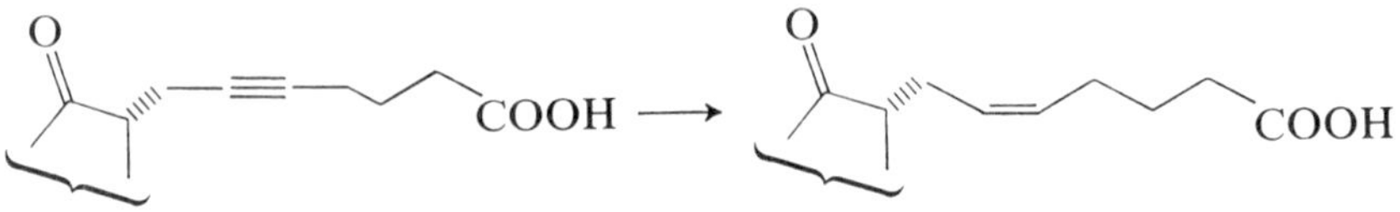

Figure 20

As mentioned, Corey et al. introduced the Δ^5-*cis* double bond in PGE_2 and $PGF_2\alpha$ by Wittig reaction under specific conditions. Another possibility was realized by Schneider,[104] who obtained the required *cis* geometry by hydrogenation of an acetylenic bond (Fig. 20). The same principle was utilized in the total synthesis of *d,l*-PGE_3 methylester, where both *cis* double bonds were generated simultaneously by hydrogenation of an acetylenic precursor (Fig. 21).[15]

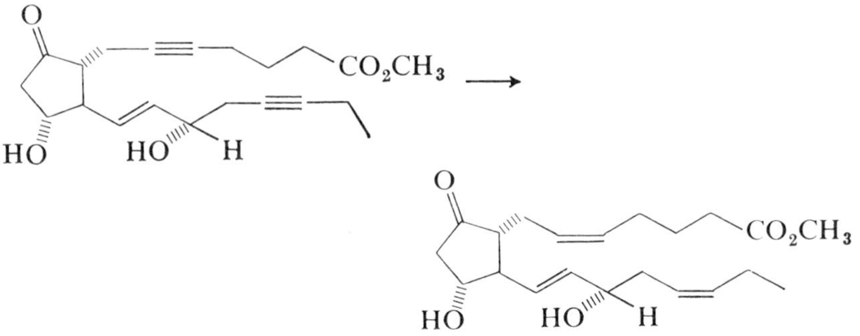

Figure 21

C. Individual Syntheses

How general strategy and stereochemical principles were implemented using the most efficient reaction sequences and best suitable protecting groups is outlined in the following charts. The figures describe the total syntheses of PGE's and PGF's completely, including the reagents used; we therefore limit the discussion, to pointing out the highlights, advantages, or disadvantages of the different syntheses.

Corey's early syntheses[35,36,38,40] have in common that the five-membered ring is formed by aldol condensation during the course of the synthesis and that the 9-keto group is generated at the end from an amine precursor (Figs. 22–25). The first synthesis,[35] leading to *d,l*-PGE_1, is outlined in Fig. 22. Noteworthy is the stereospecific Diels-Alder addition which gave adduct **4** as the major product and the position isomer only as a minor by-product. Base-catalyzed aldol cyclization of **9** yielded in addition to **10** its 11-epimer.

CH_2Br + Li, C_5H_{11}, S, S → C_5H_{11}, S, S; $O_2NCH{=}CH(CH_2)_6CN$ →

1 2 3

$NC(CH_2)_6$, C_5H_{11}, S, S, O_2N; Al–Hg →

4

$NC(CH_2)_6$, C_5H_{11}, S, S, H_2N; $CH_3C(=O)—O—C(=O)H$ →

5

$NC(CH_2)_6$, C_5H_{11}, S, S, OHCHN; $(CH_2OH)_2$, $HgCl_2$ →

6

$NC(CH_2)_6$, C_5H_{11}, O, O, OHCNH; OsO_4 →

7

$NC(CH_2)_6$, OH, C_5H_{11}, O, O, OHCHN, OH; $Pb(OAc)_4$ →

8

NHCHO, $(CH_2)_6CN$, C_5H_{11}, CHO, O, O, O; N, N; CH_2Cl_2 →

9

Figure 22

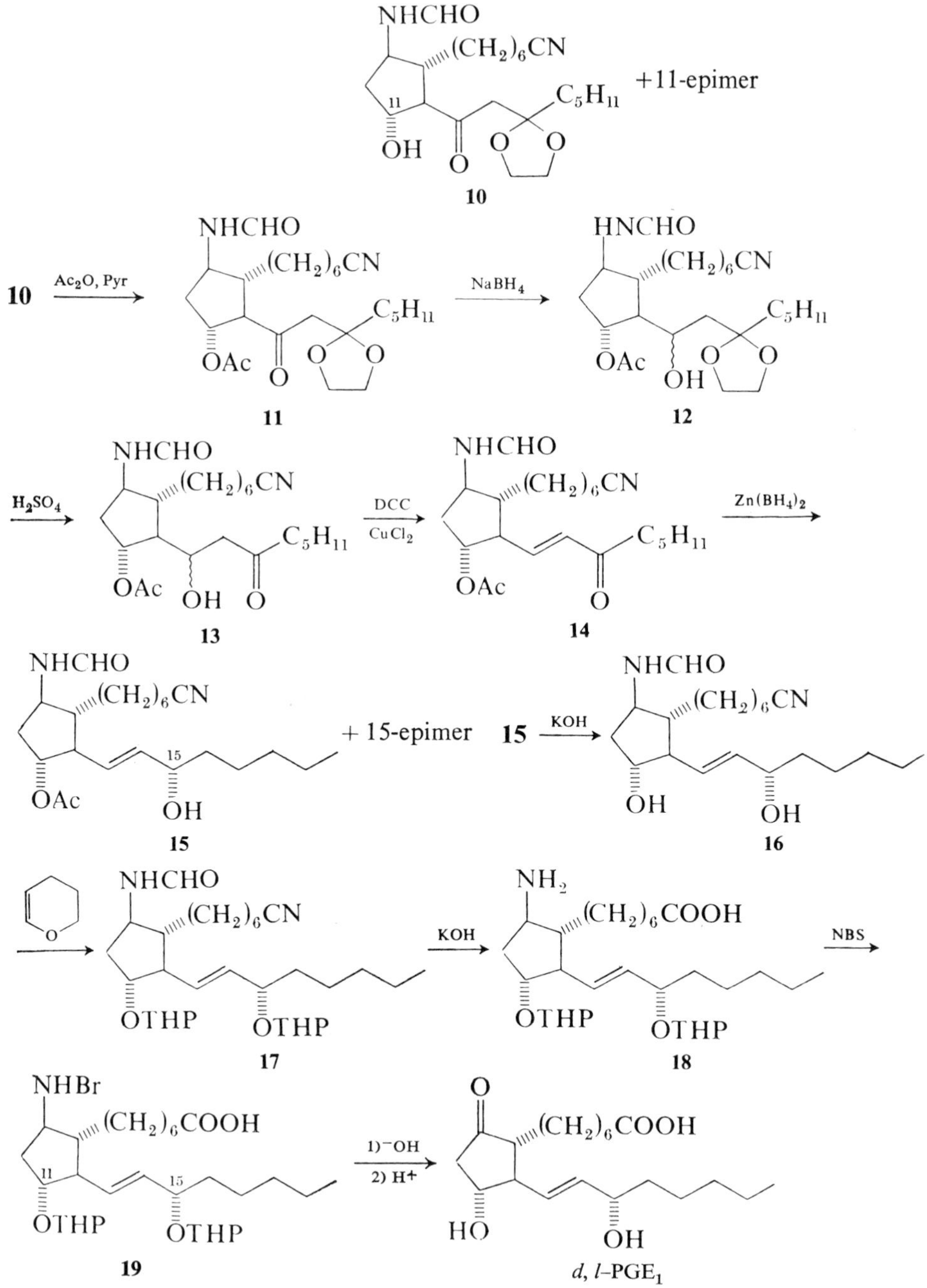

Figure 22 (contd.)

Important for the development of a later synthesis (Figs. 31, 32) was the finding that the tetrahydropyranyl-group at C-11 and C-15 could be removed under conditions which would not destroy the acid-sensitive β-ketol unit.

Another method[36] to prepare intermediate **14** (Fig. 23) involved acid-catalyzed aldol condensation. It was found that the Δ^{13}-double bond in **23** had to be transposed to the Δ^{12}-position for the cyclization; otherwise useless reaction products were obtained.[40] Using *p*-toluene sulfonic acid as catalyst, again a mixture of 11-epimers was obtained.

The formation of the undesired 11-epimer could be suppressed, however, by using stannic chloride as catalyst for the cyclization of intermediate **24**. In this variation[38] (Fig. 24) amine **28** was prepared and the amino group at C-9 could then be used as an handle for resolution via the α-bromocamphor-π-sulfonate. Conversion of resolved amine **28** by the procedures developed earlier (e.g., Fig. 22) led to the first total synthesis of resolved PGE_1. A very short route to intermediate **25** (Fig. 25) was disclosed recently by Corey;[45] unfortunately, this route is limited by low yield in the Wittig reaction (**31** → **32**).

Another approach, based on a concept by Just and Simonovitch[70] and developed by the Upjohn group,[14,15,71,104–106] involves solvolysis of a bicyclo[3.1.0]-hexane system as its key step. Intermediate **45** was prepared from Δ^3-cyclopentenol (Fig. 26). The original concept,[70] formolysis of epoxide **47**, was successful only in the PGF-series and only in very low yield.[71] Preparation of bismesylate **49** (Fig. 27) followed by solvolysis in acetone/water allowed adaption of this route to the synthesis of *d,l*-PGE_1 methyl-ester and *d,l*-PGE_1.[105,106] The yield of solvolysis products was in the order of 10% and was found to be independent of the configuration at C-14 and C-15 in bismesylate **49**. Minor isomers, however, isolated from this route and identified as C-13 isomers gave considerably higher yields of solvolysis products, which led to the development of a new synthesis, designed to produce only these *endo* isomers[14] (Fig. 28). Solvolysis of *endo*-bismesylates **65** gave solvolysis products in better than 40% yield. The higher yield in the *endo* series is probably due to steric hindrance at C-14, which makes the competing reaction, straight hydrolysis at C-14 without ring opening, less predominant than in the *exo* series. Modifications of the *endo*-bicyclohexane synthesis allowed the first total syntheses of *d,l*-PGE_2[104] (Fig. 29) and *d,l*-PGE_3 methylester[15] (Fig. 30). This approach, although adaptable to the synthesis of all prostaglandins known so far and a wide variety of analogous compounds, lacks stereocontrol except for the solvolysis reaction.

A stereocontrolled synthesis of $PGF_2\alpha$ by Corey et al.[37,43,44] is characterized by the build-up of a γ-lactone fused to the five-membered ring at C-8 and C-9, which is used for generation of the 9α-hydroxyl group and the Δ^5-*cis* double bond (Fig. 31). Noteworthy is the selective Baeyer-Villiger oxidation of ketone

NO_2 / $CH(OCH_3)_2$ (20) + OHC–CH=CH–$(CH_2)_6CN$ (21) ⟶

22 (NO_2, $(CH_2)_6CN$, CHO, CH(CH_3O)(OCH_3)) —$C_5H_{11}\overset{O}{\overset{\|}{C}}CH_2\overset{O}{\overset{\|}{P}}(OCH_3)_2$⟶

23 (NO_2, $(CH_2)_6CN$, 12, 13, 14, 15, C_5H_{11}, O, CH, CH_3O, OCH_3) —$(CH_2OH)_2$⟶

24 (NO_2, $(CH_2)_6CN$, t, 12, 13, 14, 15, C_5H_{11}) —Al–Hg⟶ —$CH_3\overset{O}{\overset{\|}{C}}OCHO$⟶

25 (NHCHO, $(CH_2)_6CN$, 12, 13, 14, 15, C_5H_{11}) —p-TsOH⟶

26 (NHCHO, $(CH_2)_6CN$, 11, 12, 13, 14, 15, OH, O) + 11-epimer

↓ Ac_2O, Pyr

14

Figure 23

NO_2 $(CH_2)_6CN$ C_5H_{11} *t*

24

$SnCl_4$ / acetone →

NO_2 $(CH_2)_6CN$ C_5H_{11} OH O

25

$Zn(BH_4)_2$ →

NO_2 $(CH_2)_6CN$ 15 OH OH + 15-epimer

26

26 $\xrightarrow{^{-}OH}$

NO_2 $(CH_2)_6CN$ OH OH

27

Al–Hg →

NH_2 9 $(CH_2)_6CN$ OH OH

28

Figure 24

NO_2 O S + $(CH_2)_5CN$ $CH_2P^+Ph_3$

30

$\xrightarrow{^{-}OH}$

NO_2 $(CH_2)_6CN$ O S $CH_2P^+(Ph)_3$

31

$OHCCH{=}C(C_5H_{11})SPh$ / PhLi →

NO_2 $(CH_2)_6CN$ O S C_5H_{11} SPh

32

$HgCl_2$ →

NO_2 CHO C_5H_{11} SPh

33

$SnCl_4$ →

NO_2 $(CH_2)_6CN$ C_5H_{11} HO O

25

Figure 25

OH

34

O

OTHP

35

EtO_2CCHN_2

Cu

OTHP

$CO_2C_2H_5$

36

NaOMe

OTHP

H

CO_2CH_3

37

$LiAlH_4$

OTHP

H

CH_2OH

38

Jones

OTHP

H

CHO

39

$Br^- Ph_3P^+(CH_2)_5CH_3$

OTHP

H

$CH{=}CH{-}C_5H_{11}$

40

$(COOH)_2$

OH

H

$CH{=}CH{-}C_5H_{11}$

41

Jones

O

H

$CH{=}CH{-}C_5H_{11}$

42

$I(CH_2)_6CO_2CH_3$

$KOBu^t$

O

$(CH_2)_6CO_2CH_3$

8

H

$CH{=}CH{-}C_5H_{11}$

43

+ 8-isomer

44

43 $\xrightarrow{NaBH_4}$

HO

$(CH_2)_6CO_2CH_3$

H

$CH{=}CH{-}C_5H_{11}$

45

+

OH

$(CH_2)_6CO_2CH_3$

H

$CH{=}CH{-}C_5H_{11}$

46

Figure 26

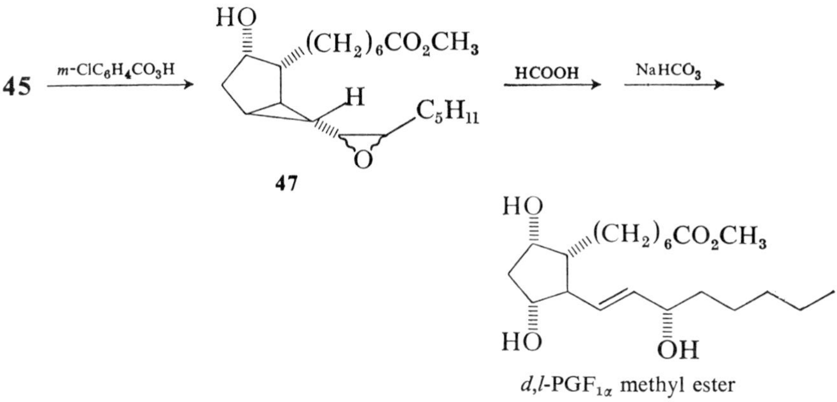

45 $\xrightarrow{m\text{-}ClC_6H_4CO_3H}$ 47 $\xrightarrow{HCOOH}$ $\xrightarrow{NaHCO_3}$ d,l-$PGF_{1\alpha}$ methyl ester

HO, $(CH_2)_6CO_2CH_3$, H, C_5H_{11}, O

47

HO, $(CH_2)_6CO_2CH_3$, HO, OH

d,l-$PGF_{1\alpha}$ methyl ester

Figure 26 (contd.)

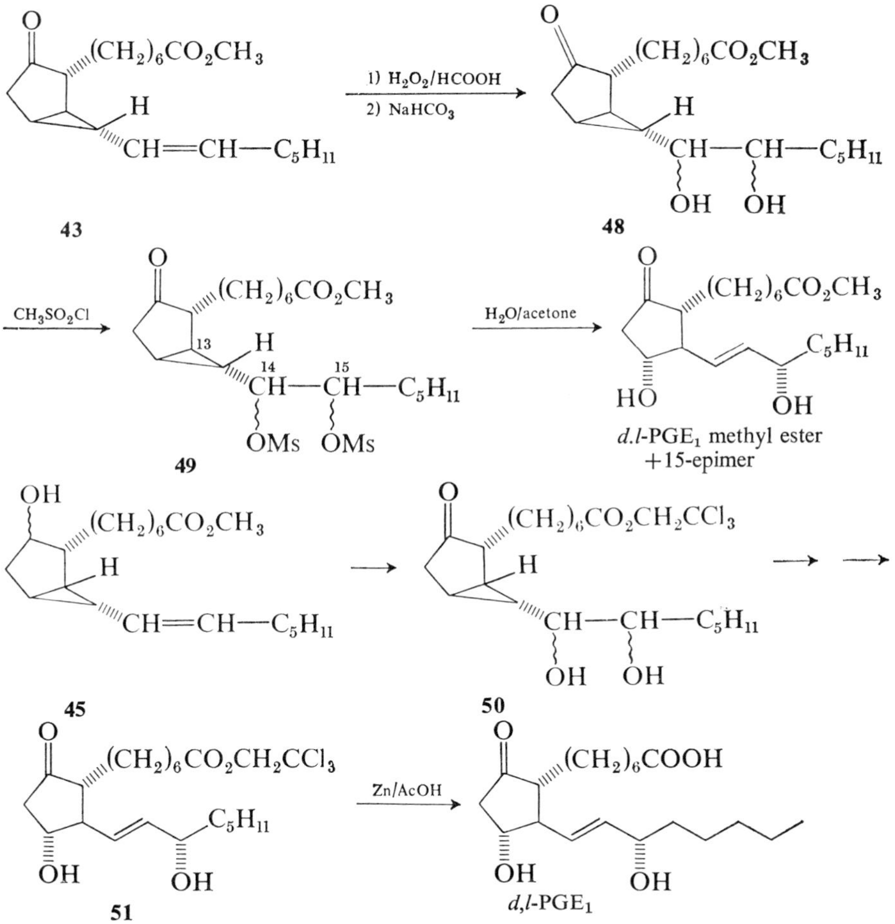

O, $(CH_2)_6CO_2CH_3$, H, $CH{=}CH{-}C_5H_{11}$

43

1) $H_2O_2/HCOOH$
2) $NaHCO_3$

O, $(CH_2)_6CO_2CH_3$, H, $CH{-}CH{-}C_5H_{11}$, OH, OH

48

CH_3SO_2Cl

O, $(CH_2)_6CO_2CH_3$, 13, H, 14, 15, $CH{-}CH{-}C_5H_{11}$, OMs, OMs

49

H_2O/acetone

O, $(CH_2)_6CO_2CH_3$, C_5H_{11}, HO, OH

d.l-PGE_1 methyl ester
+15-epimer

OH, $(CH_2)_6CO_2CH_3$, H, $CH{=}CH{-}C_5H_{11}$

45

O, $(CH_2)_6CO_2CH_2CCl_3$, H, $CH{-}CH{-}C_5H_{11}$, OH, OH

50

O, $(CH_2)_6CO_2CH_2CCl_3$, C_5H_{11}, OH, OH

51

Zn/AcOH

O, $(CH_2)_6COOH$, OH, OH

d,l-PGE_1

Figure 27

CO_2H H **52** $\xrightarrow{MeOH}$ CO_2CH_3 H **53** $\xrightarrow[2)\ H_2O_2]{1)\ B_2H_6}$

HO CO_2CH_3 H + HO CO_2CH_3 H

54 **55**

54 $\longrightarrow$ (dihydropyran, O) OTHP CO_2CH_3 H **56** $\xrightarrow{LiAlH_4}$ OTHP CH_2OH H **57** $\xrightarrow{Jones}$

OTHP CHO H **58** $\xrightarrow{Br^-Ph_3P^+(CH_2)_5CH_3}$ OTHP C_5H_{11} **59** $\xrightarrow{(COOH)_2}$

HO C_5H_{11} **60** $\xrightarrow{Jones}$ O C_5H_{11} **61** $\xrightarrow[KOBu^t]{I(CH_2)_6CO_2CH_3}$

O $(CH_2)_6CO_2CH_3$ C_5H_{11} **62** + 8-isomer **63** **62** $\xrightarrow{OsO_4}$ O $(CH_2)_6CO_2CH_3$ C_5H_{11} HO OH **64** $\xrightarrow{CH_3SO_2Cl}$

O $(CH_2)_6CO_2CH_3$ 12 13 14 15 C_5H_{11} MsO OMs **65** $\xrightarrow{H_2O/acetone}$ O $(CH_2)_6CO_2CH_3$ 15 HO OH *d,l*-PGE_1 methyl ester + 15-epimer

Figure 28

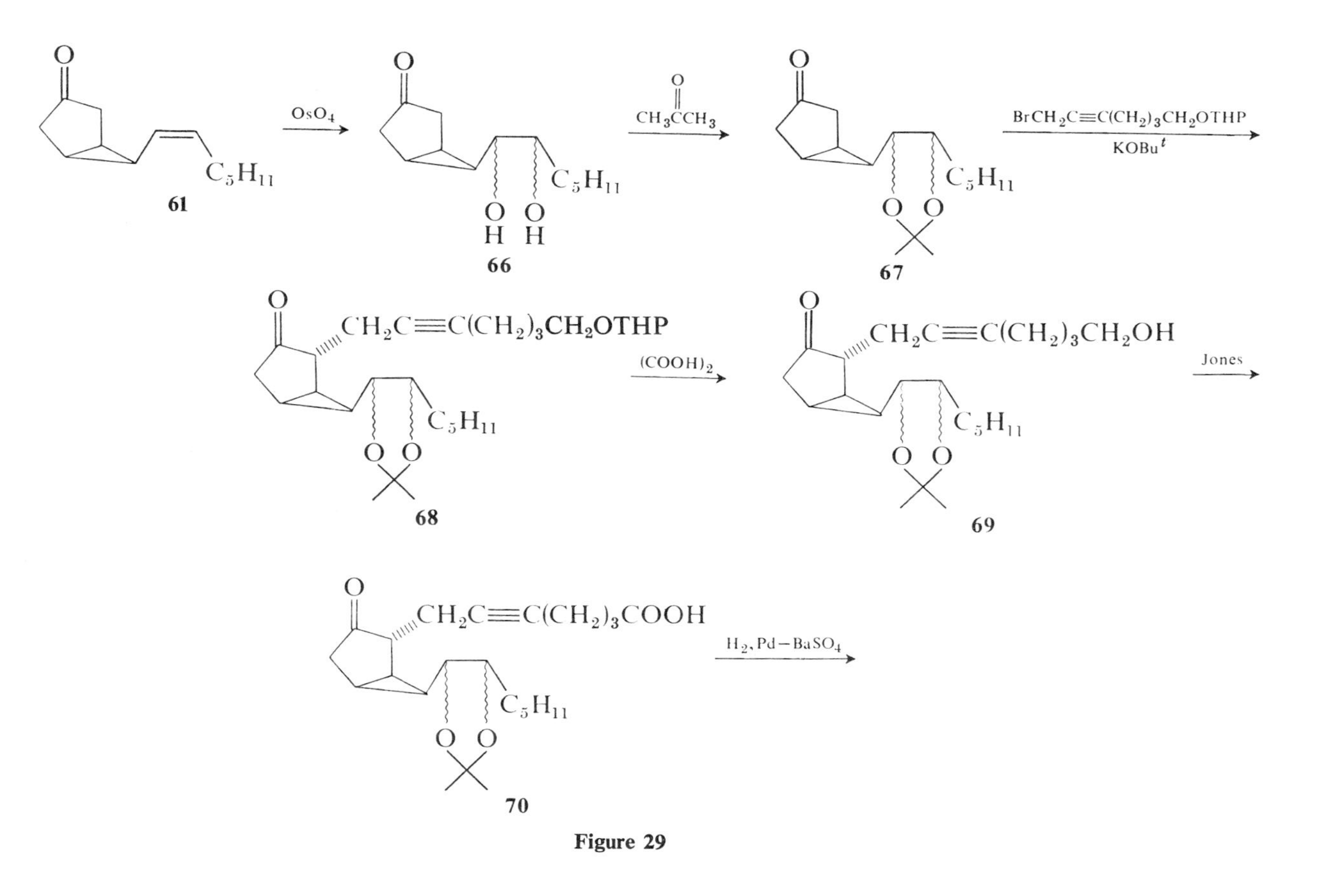

61
OsO4
66
CH3CCH3
67
BrCH2C≡C(CH2)3CH2OTHP
KOBu^t
CH2C≡C(CH2)3CH2OTHP
68
(COOH)2
CH2C≡C(CH2)3CH2OH
69
Jones
CH2C≡C(CH2)3COOH
70
H2, Pd−BaSO4

Figure 29

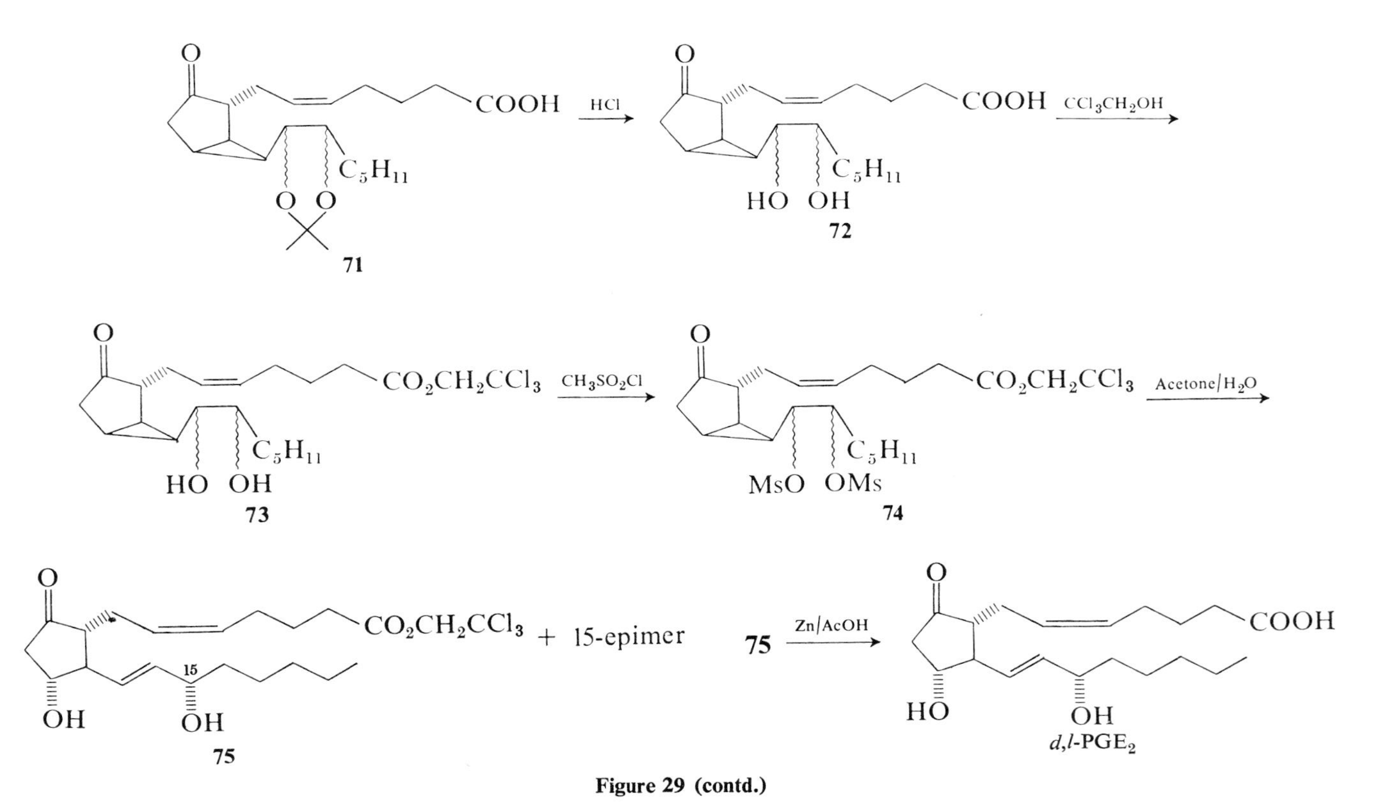

Figure 29 (contd.)

OTHP

CHO

H

58

$Br^- Ph_3P^+(CH_2)_2C{\equiv}CC_2H_5$

OTHP

H

76

$(COOH)_2$

OH

77

Jones

O

78

$BrCH_2C{\equiv}C(CH_2)_3CO_2CH_3$

$KOBu^t$

O

CO_2CH_3

79

OsO_4

O

CO_2CH_3

OH OH

80

CH_3SO_2Cl

O

CO_2CH_3

MsO OMs

81

H_2O/acetone

O

CO_2CH_3

+ 15-epimer

15

HO

OH

82

82 H_2, Lindlar

Quinoline

O

CO_2CH_3

HO

OH

d,l-PGE_3 methyl ester

Figure 30

CH_2OCH_3

Cl

$CH_2{=}C{-}CN$

1) Na

2) $ClCH_2OCH_3$

83

84

CH_3OCH_2

KOH

CH_3OCH_2

m-Cl-$C_6H_4CO_3H$

Cl

CN

O

85

86

CH_3OCH_2

NaOH

CH_2COOH

Resolution via ephedrine salt

O

O

CH_2OCH_3

OH

87

88

O

O

CH_2COOH

$KI{\cdot}I_2$

I

9 8 10 12 11

Ac₂O, py

CH_2OCH_3

CH_2OCH_3

OH

OH

89

90

O

O

O

O

Bu_3SnH

BBr_3

I

CH_2OCH_3

CH_2OCH_3

OAc

OAc

91

92

O

O

O

O

Collins reagent

O O

$C_5H_{11}CCH_2P(OCH_3)_2$

CH_2OH

CHO

OAc

OAc

93

94

Figure 31

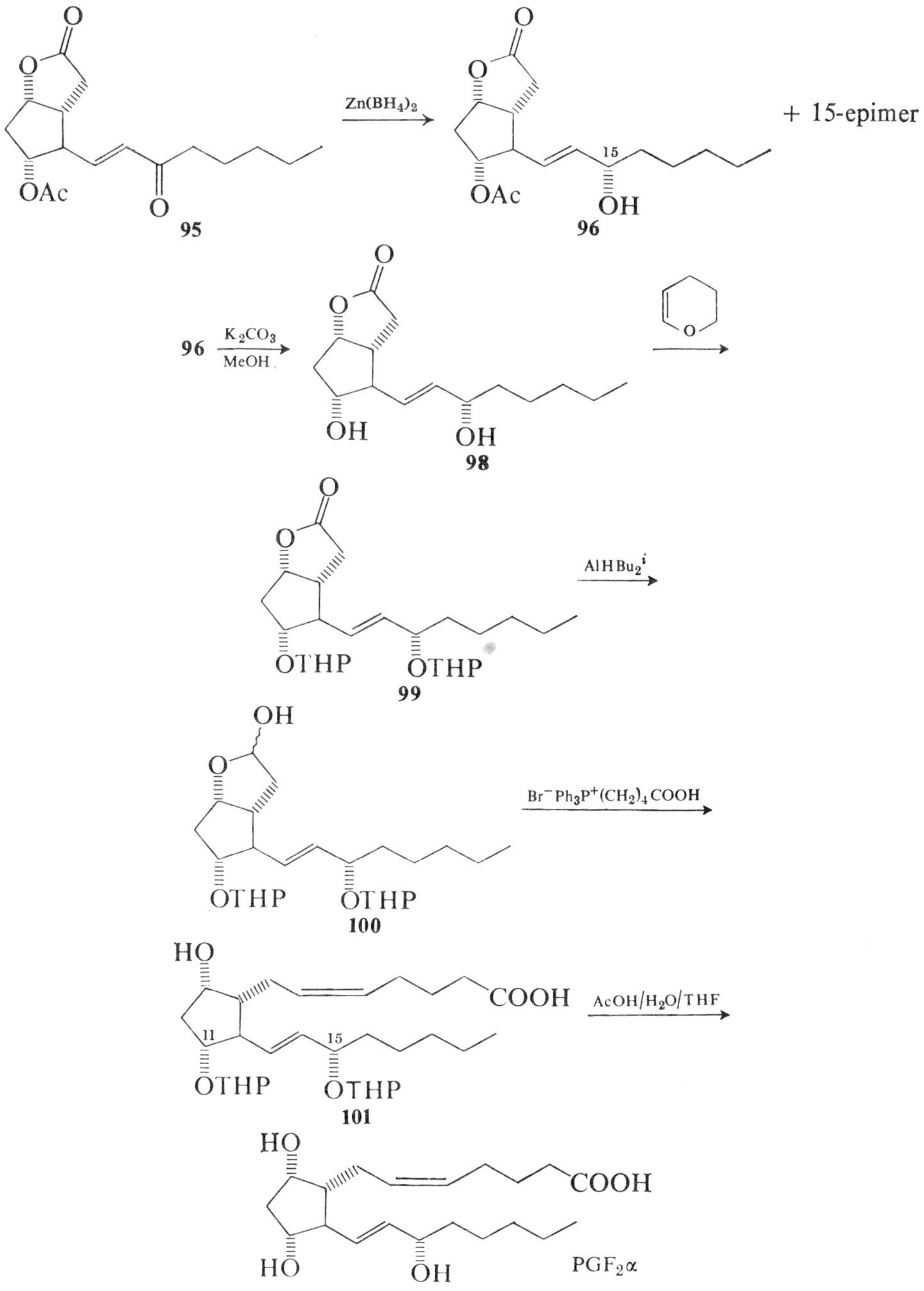

O
O
OAc
O
95

$Zn(BH_4)_2$

O
O
15
OAc
OH
96

+ 15-epimer

96 $\frac{K_2CO_3}{MeOH}$

O
O
OH
OH
98

O

O
O
OTHP
OTHP
99

$AlHBu_2^i$

OH
O
OTHP
OTHP
100

$Br^- Ph_3P^+(CH_2)_4COOH$

HO
COOH
11
15
OTHP
OTHP
101

AcOH/H_2O/THF

HO
COOH
HO
OH
$PGF_2\alpha$

Figure 31 (contd.)

86. Acid **88** was readily resolved via the ephedrine salt and iodolactonization of resolved acid **89** afforded key intermediate **90** with correct stereochemistry at all four asymmetric centers of the ring. The earlier finding[35] that 11,15-ditetrahydropyranyl ethers of PGE compounds can be hydrolyzed without β-elimination of the 11-hydroxy group permitted the synthesis of PGE_2 from intermediate **101** (Fig. 32). The same intermediate could also be selectively hydrogenated to give PGE_1 and $PGF_1\alpha$ (Fig. 32). A shorter approach[41] to intermediate **95** is handicapped by a very low yield in the ring-opening step **109** to **110** (Fig. 33). A third synthesis of $PGF_2\alpha$[42] using the γ-lactone principle is short and direct but lacks stereospecificity in the opening of epoxide **117** (Fig. 34).

A unique, if lengthy, synthesis of *d,l*-PGE_1 has recently been published by a group at Merck.[115] The intention was to exercise stereochemical control at all asymmetric centers (except C-15) and this has been achieved, although at the cost of additional steps; for example, the double bond in intermediate **136** is only hydrogenated after shift to the Δ^{10}-position under basic conditions to assure *trans* configuration at C-8 and C-12 (Fig. 35). The correct stereochemistry at C-11 is obtained in a similar manner by epimerization of intermediate **143** to the thermodynamically more stable isomer **144**.

Several of the syntheses outlined here provide practical routes to the pharmacologically important E and F prostaglandins and in several instances this has been achieved by developing new methodology which may be useful for other tasks.

3. PARTIAL SYNTHESES OF PGE_2 AND $PGF_2\alpha$ FROM 15(*R*)-PGA_2

In 1969, Weinheimer and Spraggins reported[119] the isolation of surprisingly large amounts of 15(*R*)-PGA_2 (**154**) and its 15-acetate, methyl ester (**155**) from a sea whip, *Plexaura homomalla*, a common gorgonian coral of Florida coastal waters. The combined weights of these two prostaglandins represented about 1.5% of the dry weight of the coral cortex. Before this finding, the highest concentration of prostaglandins found in nature was in mammalian semen, which would present some problems if considered as a source of large amounts of prostaglandins. These prostaglandins, **154–156**, the latter the result of partial hydrolysis of **155**, are not presently of major biological interest. They have, however, been converted[31] to the highly active prostaglandins, $PGF_2\alpha$ and PGE_2 methyl ester.

The two major problems in such a conversion involve inversion of configuration at C-15 and, formally, hydration of the 10,11-double bond. The second was accomplished by epoxidation with hydrogen peroxide and a small amount of potassium hydroxide in methanol, affording a mixture of epoxides. The ratio of α to β epoxide formed was 75:25 from **155**.

101 —Jones→ 102 —$AcOH/H_2O/THF$→ PGE_2

O, COOH, 11, 15, OTHP, OTHP

102

O, COOH, OH, OH

PGE_2

101 —H_2, Pd–C / MeOH→ 103 —$AcOH/H_2O/THF$→ $PGF_{1}\alpha$

HO, COOH, OTHP, OTHP

103

HO, COOH, OH, OH

$PGF_{1}\alpha$

103 —Jones→ 104 —$AcOH/H_2O/THF$→ PGE_1

O, COOH, OTHP, OTHP

104

O, COOH, OH, OH

PGE_1

Figure 32

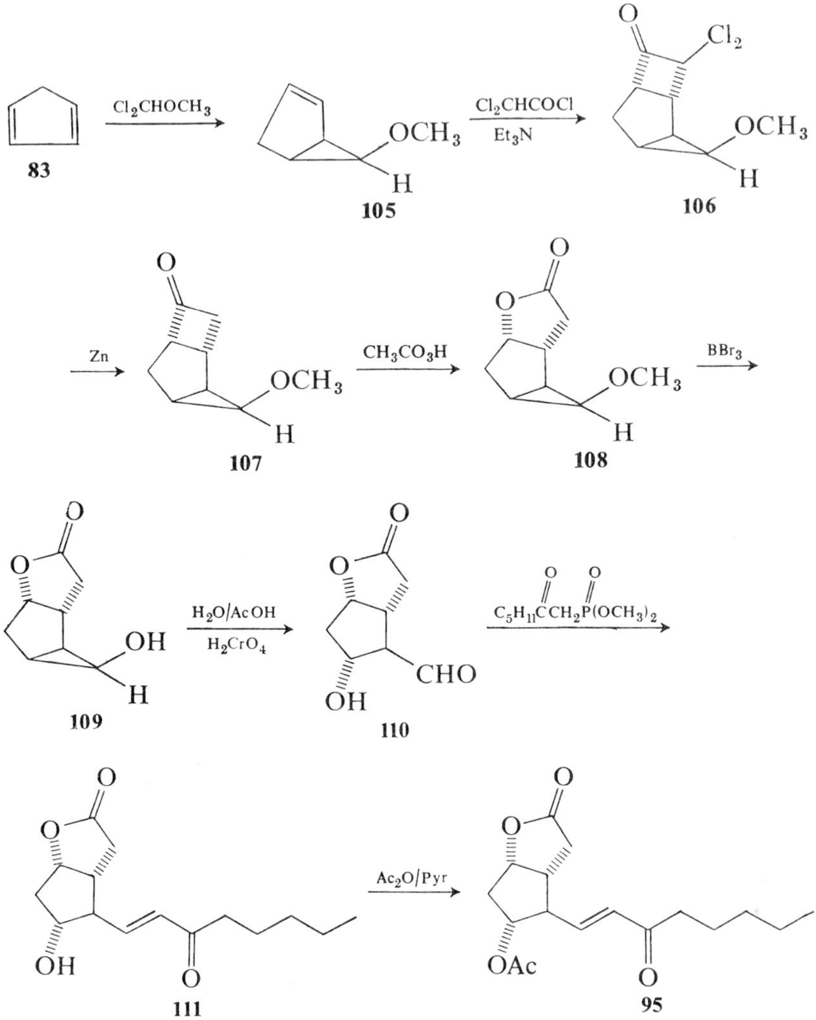

83 → Cl_2CHOCH_3 → 105 (OCH_3, H) → $Cl_2CHCOCl$ / Et_3N → 106 (O, Cl_2, OCH_3, H)

→ Zn → 107 (O, OCH_3, H) → CH_3CO_3H → 108 (O, O, OCH_3, H) → BBr_3 →

109 (O, O, OH, H) → $H_2O/AcOH$ / H_2CrO_4 → 110 (O, O, OH, CHO) → $C_5H_{11}\overset{O}{\overset{\|}{C}}CH_2\overset{O}{\overset{\|}{P}}(OCH_3)_2$ →

111 (O, O, OH, O) → Ac_2O/Pyr → 95 (O, O, OAc, O)

Figure 33

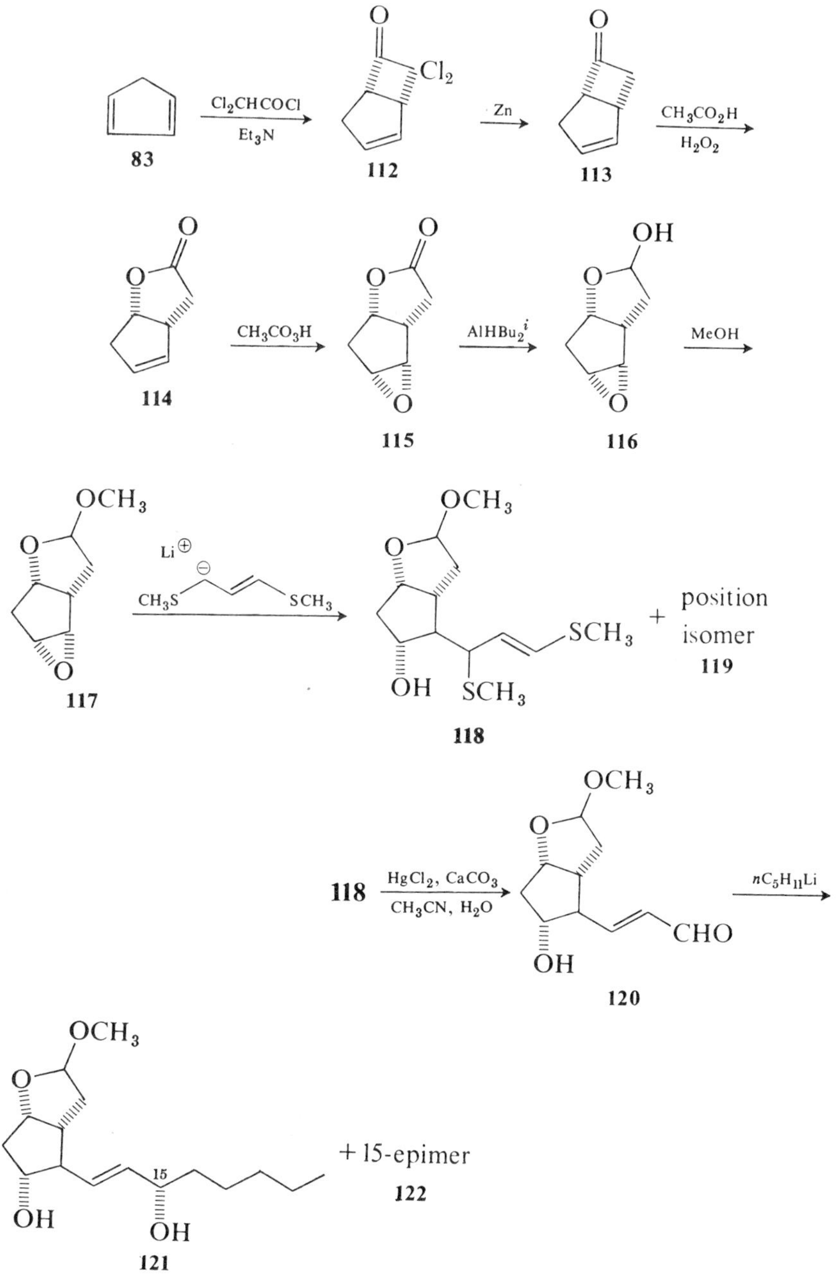

83 $\xrightarrow[Et_3N]{Cl_2CHCOCl}$ 112 $\xrightarrow{Zn}$ 113 $\xrightarrow[H_2O_2]{CH_3CO_2H}$

114 $\xrightarrow{CH_3CO_3H}$ 115 $\xrightarrow{AlHBu_2^i}$ 116 $\xrightarrow{MeOH}$

117 $\xrightarrow{Li^{\oplus}\ CH_3S\overset{\ominus}{C}H{-}CH{=}CH{-}SCH_3}$ 118 + position isomer 119

118 $\xrightarrow[CH_3CN,\ H_2O]{HgCl_2,\ CaCO_3}$ 120 $\xrightarrow{nC_5H_{11}Li}$

121 + 15-epimer 122

Figure 34

121 $\xrightarrow{HCl}$ 123 $\xrightarrow{Br^- Ph_3P^+(CH_2)_4COOH}$ d,l-$PGF_{2\alpha}$

OH, O, OH, OH, 123, HO, COOH, HO, OH

d,l-$PGF_{2\alpha}$

Figure 34 (contd.)

124 $\xrightarrow{Ph_3P \cdot HBr}$ 125 $\xrightarrow{OHC(CH_2)_5CO_2CH_3}$ 126 $\xrightarrow{CF_3COOH}$ 127 $\xrightarrow{OsO_4}$ 128 $\xrightarrow{p\text{-TsOH}}$

OH, OCH_3, 124

$^+Ph_3Br^-$, OCH_3, 125

$CH(CH_2)_5CO_2CH_3$, OCH_3, 126

$(CH_2)_6CO_2CH_3$, OCH_3, 127

OH, $(CH_2)_6CO_2CH_3$, OH, OCH_3, 128

Figure 35

$(CH_2)_6CO_2CH_3$ $(CH_2OH)_2$ $(CH_2)_6CO_2CH_3$ KOH

OCH_3 129 OCH_3 130

$(CH_2)_6COOH$ Li, NH_3 $(CH_2)_6COOH$ CH_2N_2

OCH_3 131 OCH_3 132

$(CH_2)_6CO_2CH_3$ HOAc

OCH_3 133

$(CH_2)_6CO_2CH_3$ MeI, Ph_3CLi $(CH_2)_6CO_2CH_3$ $LiAlH(OBu^t)_3$

H_3C O 134 O 135

$(CH_2)_6CO_2CH_3$ 9 8 10 12 11 $HClO_4$

H_3C OH 136

Figure 35 (contd.)

$(CH_2)_6CO_2CH_3$ $(CH_2)_6CO_2CH_3$

NaOMe

H_2 Pd/C

10 9 8 11 12

H_3C OH H_3C OH

137 **138**

$(CH_2)_6CO_2CH_3$

$MeSO_2Cl$

H_3C OH

139

$(CH_2)_6CO_2CH_3$ $(CH_2)_6CO_2CH_3$

Me_2SO

100°C

$(CH_2OH)_2$

H_3C OSO_2CH_3 H_3C

140 **141**

$(CH_2)_6CO_2CH_3$ $(CH_2)_6CO_2CH_3$

$KMnO_4-NaIO_4$

NaOMe

COOH

H_3C H_3C O

142 **143**

$(CH_2)_6CO_2CH_3$ $(CH_2)_6CO_2CH_3$

$PhCHN_2$

CF_3CO_3H

COOH CO_2CH_2Ph

H_3C O H_3C O

144 **145**

Figure 35 (contd.)

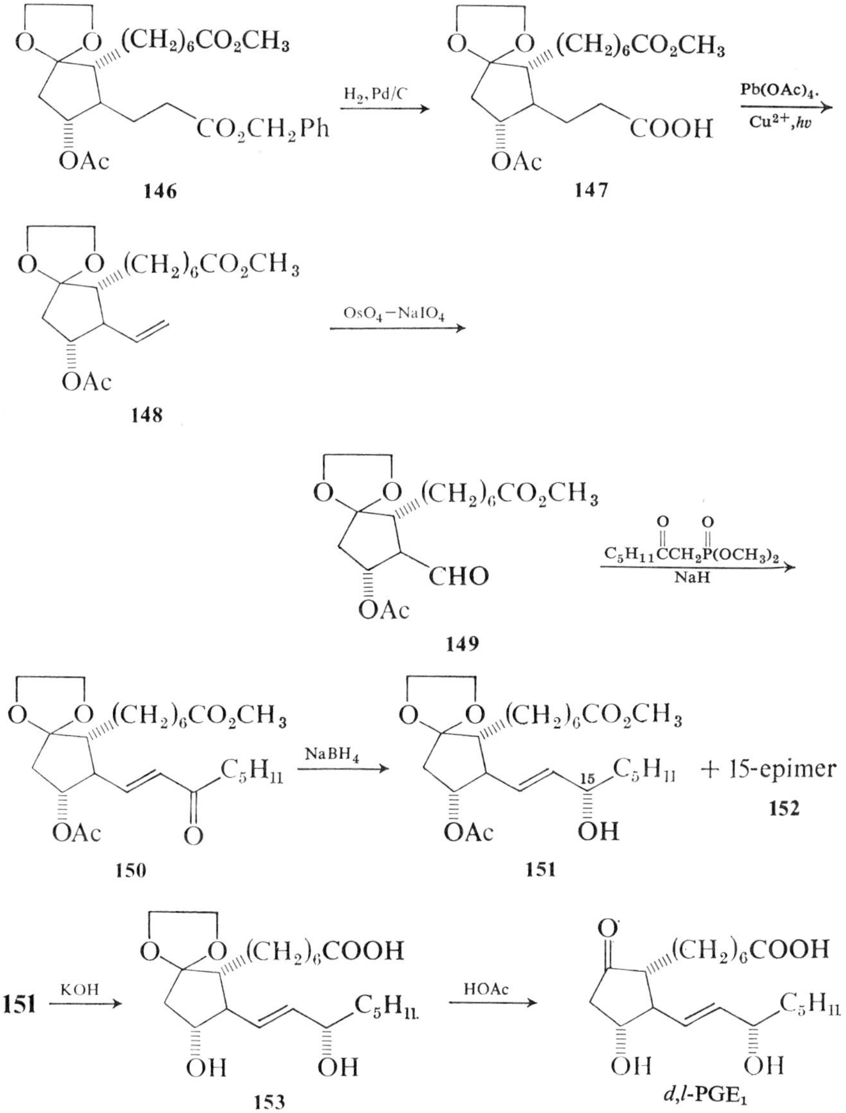

Figure 35 (contd.)

Reductive opening of the epoxide mixture by chromous acetate in aqueous acetic acid[32,108] gave 15(*R*)-PGE_2, 15-acetate, methyl ester (**159**), which was separated chromatographically from the 11β-isomer. This was converted to its trimethylsilyl ether and reduced with sodium borohydride to give 9α-hydroxy compound **160** (55% yield from **159**), separated chromatographically from the corresponding 9β-isomer. The 9α:9β ratio was considerably less if the silylation step was omitted. The purified **160** was hydrolyzed to **161** and selectively oxidized with 2,3-dichloro–5,6 dicyano–1,4-benzoquinone in dioxane to **162**. Reduction of the 15-ketone with zinc borohydride in dimethoxyethane[37] gave a mixture of $PGF_{2}\alpha$ (**163**) and its 15-epimer. Again, epimer ratio was more favorable [73:27 = 15(*S*):15(*R*)] when the hydroxyl groups were first protected as trimethylsilyl ethers.

A modification of the preceding route allowed the preparation of PGE_2 methyl ester from the mono ester **156**. Inversion at C_{15} was accomplished in fair yield by forming the 15-methanesulfonate and solvolyzing it in acetone-water. The desired product **164** was accompanied by starting material (**156**) and several by-products. The introduction of the 11α-hydroxyl group into **164** was accomplished as in the previous example; epoxidation to **165** and chromous acetate reduction gave a mixture of 11-epimers from which PGE_2 methyl ester (**166**) was separated by chromatography.

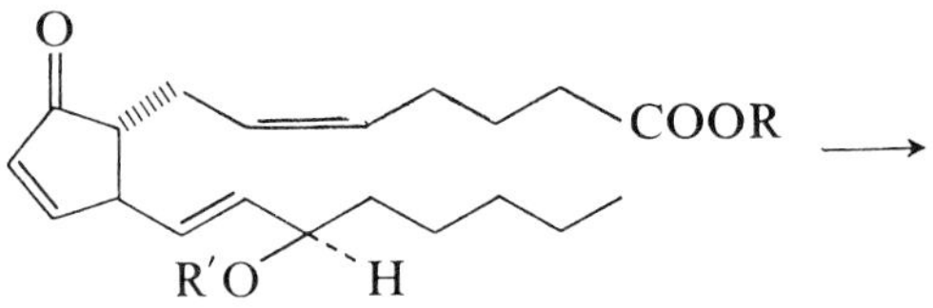

O
COOR
R'O H

154 R = R' = H
155 R = CH_3, R' = Ac
156 R = CH_3, R' = H

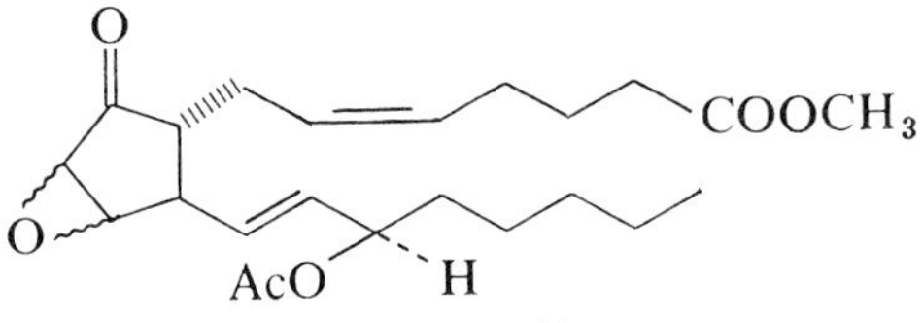

O
COOCH$_3$
O
AcO H

157 α-epoxide
158 β-epoxide

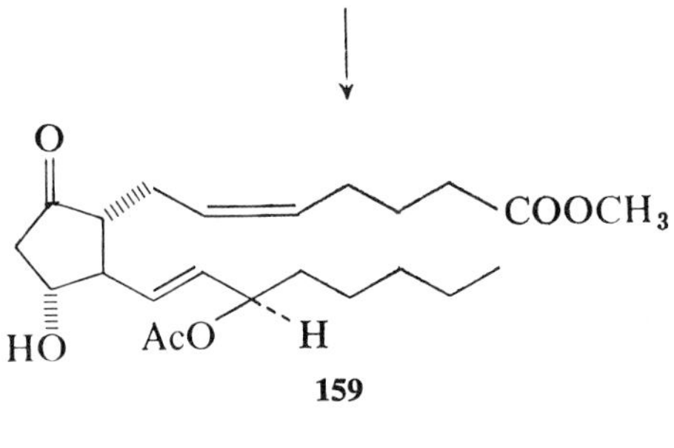

O
COOCH$_3$
HO AcO H

159

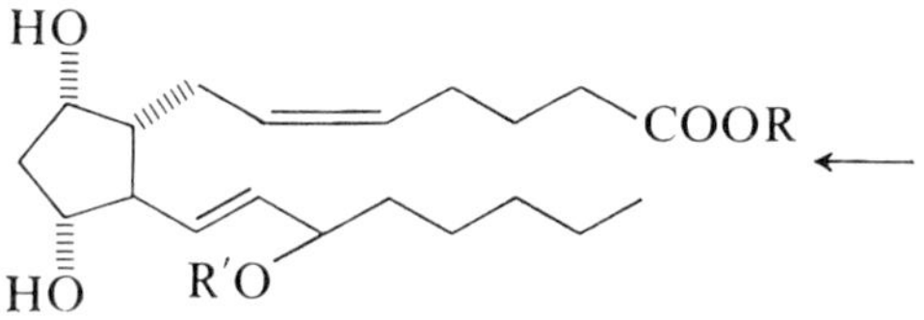

HO

COOR

HO R′O

160 R = CH_3, R′ = Ac
161 R = R′ = H

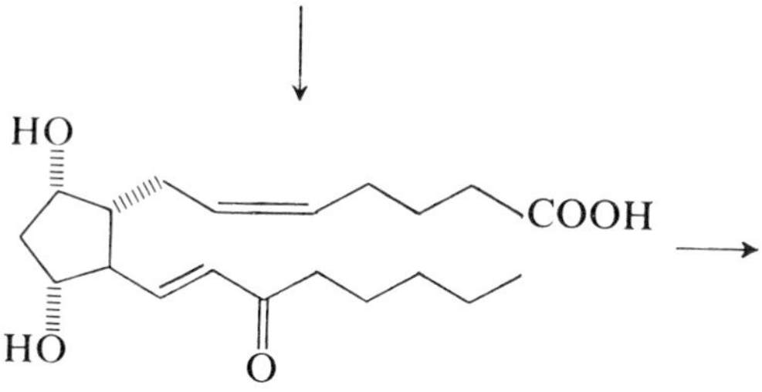

HO

COOH

HO O

162

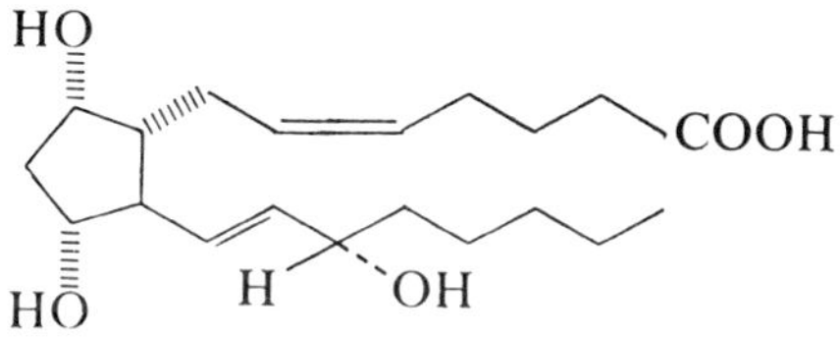

HO

COOH

HO H OH

163

156

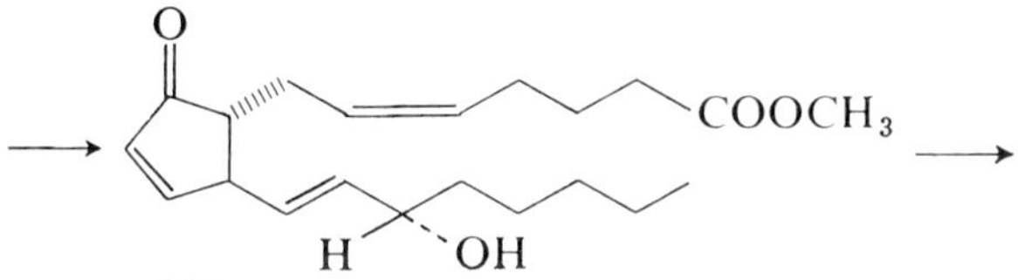

O

$COOCH_3$

H OH

164

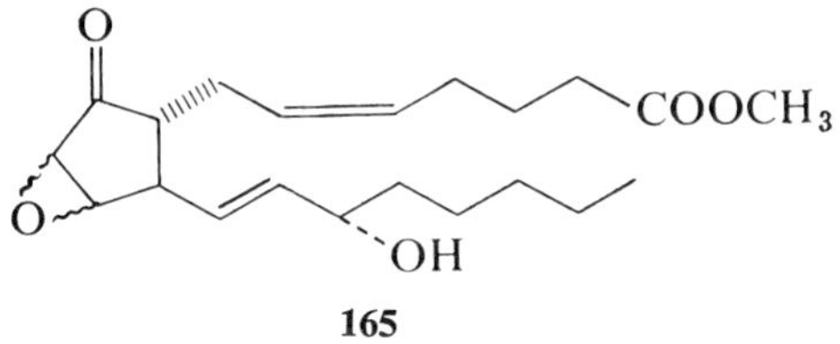

O

$COOCH_3$

O OH

165

O

$COOCH_3$

HO OH

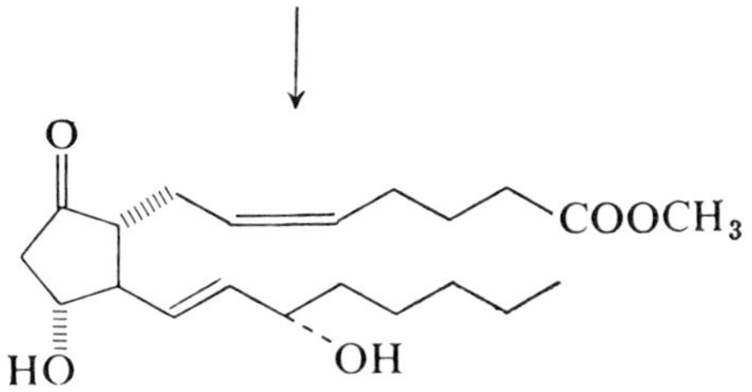

166

4. SYNTHETIC ROUTES TO STRUCTURALLY SIMPLIFIED PROSTANOIC ACIDS

The restraints placed upon available synthetic methods by the variety of functional groups of natural prostaglandins are, of course, lessened by making simplified analogs the synthetic goal. For example, the preparation of completely saturated prostaglandins allows one to use a wide range of reducing agents not possible if PGE_2 is the objective of synthesis. In some of the following syntheses, stereochemical control has also been sacrificed; isomers were not always separated, making evaluation of the work by biological or other activity difficult. However, a number of these syntheses make use of novel reactions or unusual protective groups and so have intrinsic value for synthetic organic chemistry.

A. *d,l*-13,14-Dihydro-PGE_1

13,14-Dihydro-PGE_1 is a naturally occurring, biologically active metabolite of PGE_1.[7] The synthesis, in 1966, of the racemic ethyl ester **172** of this material by Beal, Babcock, and Lincoln[22,23] represents the first synthesis of a biologically active natural prostaglandin. Their route illustrates a novel use of the Wittig olefination reaction and represents an extremely short and efficient synthesis of the prostanoic acid skeleton.

The sodium enolate of 5-formyl–3-ethoxy–2-cyclopentenone **167** was reacted with the phosphonium bromide **168** derived from ethyl 6-bromosorbate and triphenylphosphine to give **169**, via the *in situ* formed ylid. Catalylic hydrogenation and reformylation was followed by a second, similar Wittig reaction, this time using *n*-hexanoylmethyltriphenylphosphonium chloride **173** to give **170**. In this product, the prostaglandin skeleton and oxygen substitution pattern is already established. A series of reductive steps, after changing ethyl enol ether to benzyl enol ether, gave **172** as a mixture of stereoisomers. Chromatographic separation gave material having biological activity and the same polarity and spectral properties as authentic natural dihydro-PGE_1 ethyl ester. In addition, an isotope dilution method was used to demonstrate the presence of at least 22% of the natural isomer. This synthesis suffers from a low yield only in the final reduction of the benzyl enol ether system to the β-hydroxy-ketone **171** → **172**.

Klok, Pabon, and van Dorp[74] have also reported a synthesis of a mixture of steroisomers of dihydro-PGE_1 as an extension of their synthesis of *d,l*-PGB_1 and *d,l*-PGE_1-237 (**174**) (see below). After acetylation of the hydroxyl group of **174**, allylic bromination followed by displacement of bromide with silver acetate gave **175**. The ester functions were hydrolyzed and

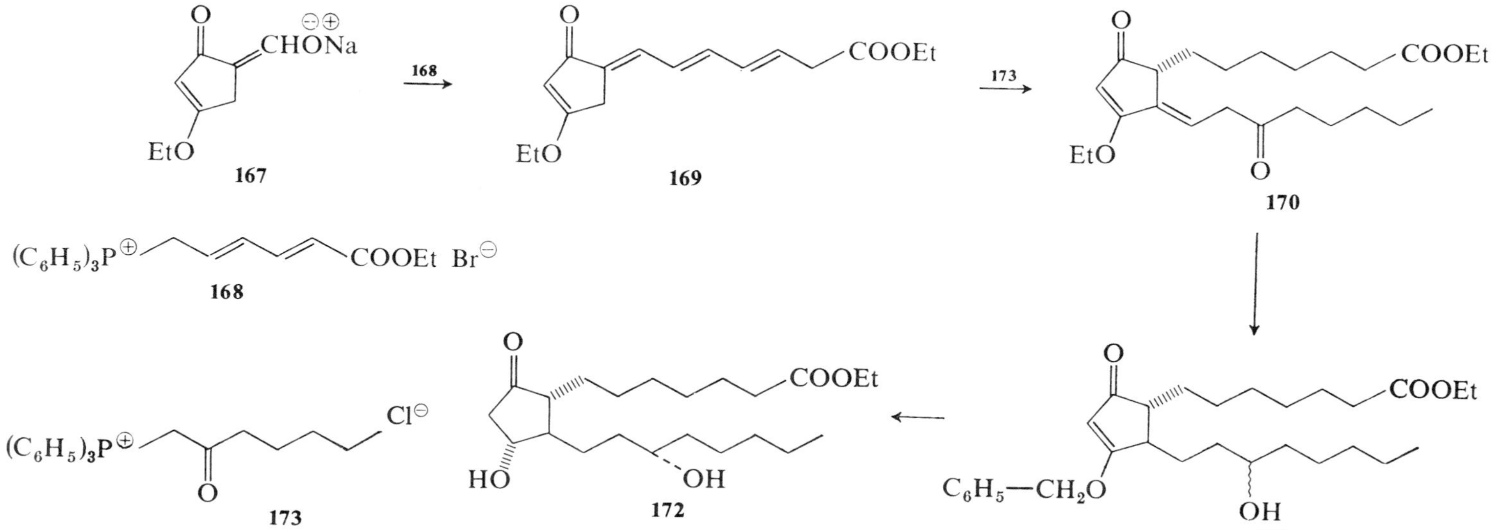

167 168 169 170 171 172 173

the resulting **176** reduced over rhodium-charcoal catalyst to give a mixture of **177** and **178**, the latter unfortunately predominating. The diol **177** was separated by chromatography and was shown to be spectrally identical to 13,14-dihydro-PGE_1. It had about 15% of the biological activity of the natural material.

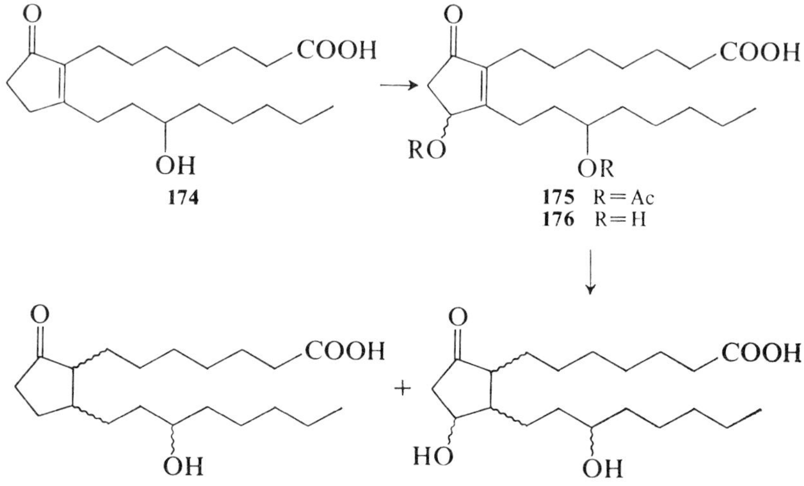

174

175 R=Ac
176 R=H

178 + **177**

A third synthesis of the same PGE_1 metabolite, again as a mixture of stereoisomers, has recently been reported by Strike and Smith.[113] A series of alkylation steps was used to build up the acyclic structure **179**. After removal of the *t*-butoxycarbonyl group, the acetylenic side chain was partially reduced to give **180**. Ozonolysis converted this chain to an aldehyde which underwent aldolization with base to **181**. None of the desired intermediate β-hydroxyketone was obtained. Epoxidation (alkaline hydrogen peroxide), catalytic hydrogenation over palladium-charcoal, and acidic removal of the tetrahydropyranyl group gave a stereoisomeric mixture having spectral and chromatographic properties consistent with structure **183**. It was not shown how much of the natural stereoisomer was present in the mixture, which did have considerable biological activity.

B. 15-Dehydro-PGE_1

Miyano and Dorn[79] carried out a short and efficient synthesis of a mixture of stereoisomers said to be spectrally indistinguishable from 15-dehydro-PGE_1 (187). Condensation of 3-keto-undecan–1,11-dioic acid with styrylglyoxal gave crystalline **184**, which was cyclized to the cyclopentenone **185**. Cleavage

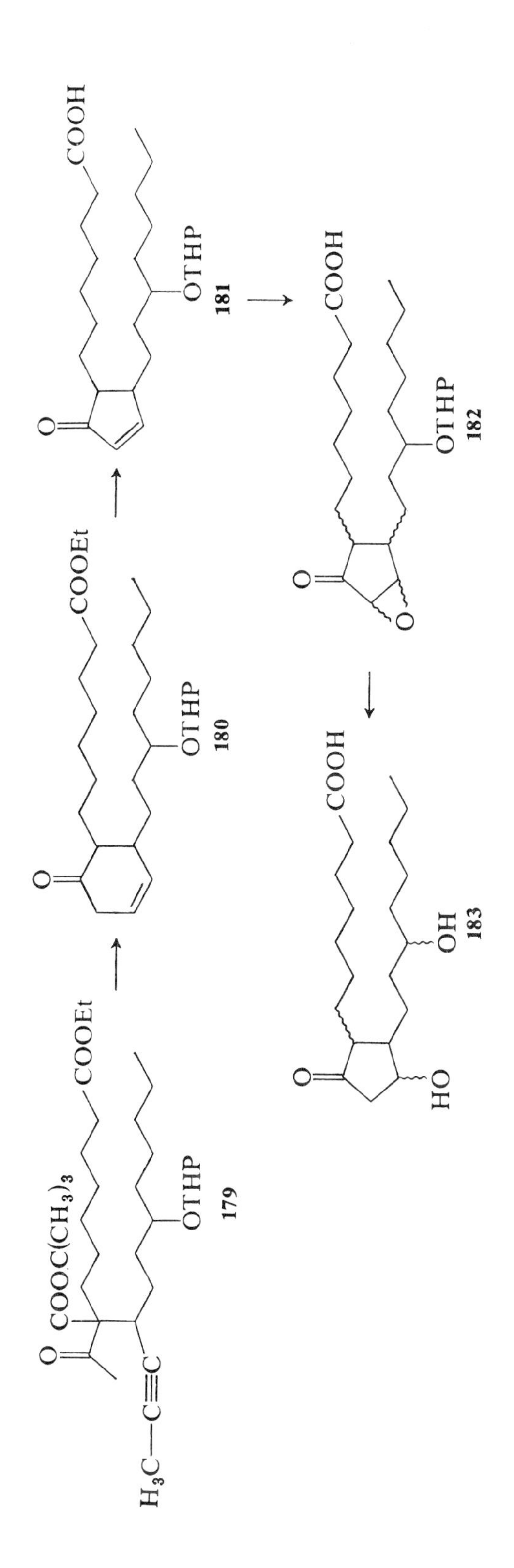
179
180
181
182
183
COOC(CH$_3$)$_3$
COOEt
OTHP
COOH
OH
HO
H_3C—C≡C

at the styryl double bond gave aldehyde **186** in which the conjugated double bond was then reduced with zinc. A Wittig olefination with *n*-hexanoyl-methylenetriphenylphosphorane gave a mixture of stereoisomers of structure **187**. Evidently selective reduction of the 15-ketone to PGE_1 was unsuccessful. Borohydride reduction of both keto groups to $PGF_1\alpha$ was not mentioned, but this would have presumably given proof of structure and stereochemistry of their product.

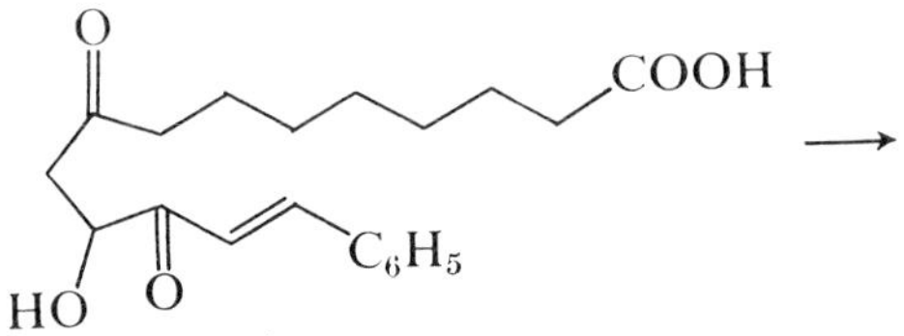

COOH
C_6H_5
HO O
184

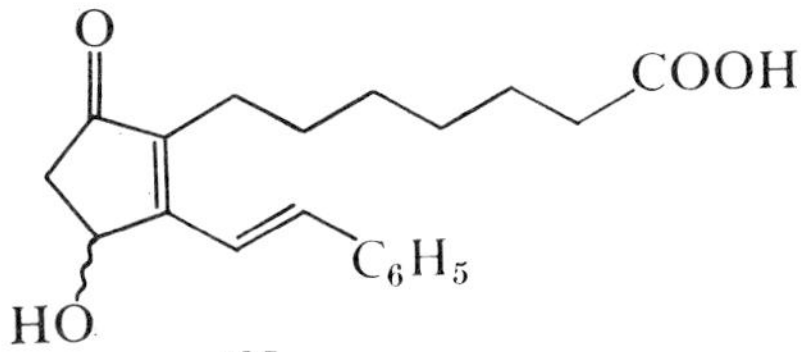

COOH
C_6H_5
HO
185

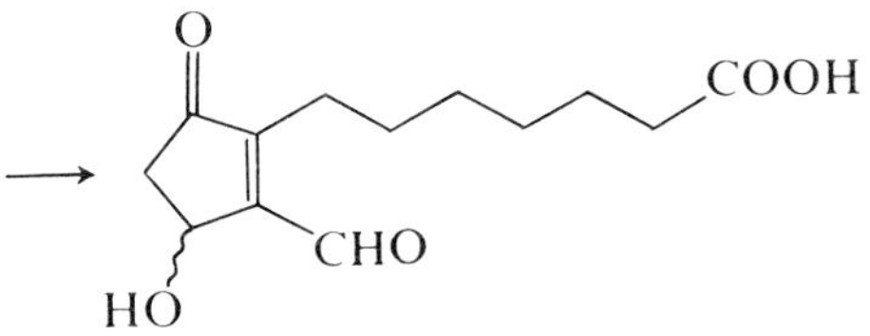

COOH
CHO
HO
186

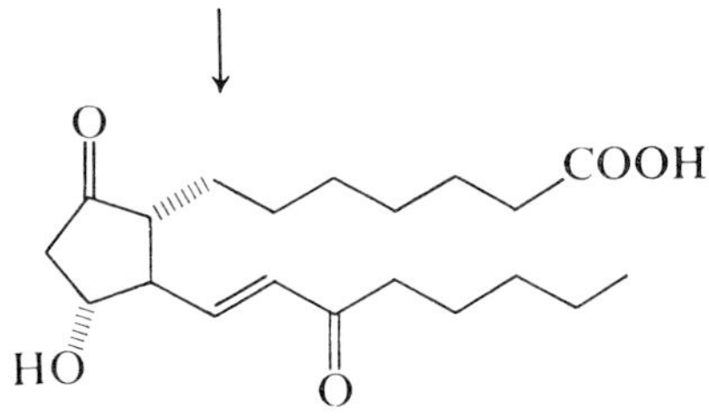

COOH
HO O
187

C. 11-Desoxy Prostaglandins

Several 11-*desoxy* prostaglandins were prepared by Bagli and Bogri[16] of Ayerst in 1966 and these unnatural materials were shown to retain prostaglandin-like activity. These authors subsequently[17] improved the synthesis and proved the stereochemical structure of their products. In the later work, the substituted cyclopentenone **189** was prepared by alkylation of ethyl 2-cyclopentanone carboxylate (**188**), bromination, and treatment with ethanolic sulfuric acid. Addition of HCN to **189** and subsequent hydrolysis gave keto acid **190**, which was monoesterified with *p*-toluenesulfonic acid in methanol (**191**). The acid was converted to its acid chloride **193**, which added

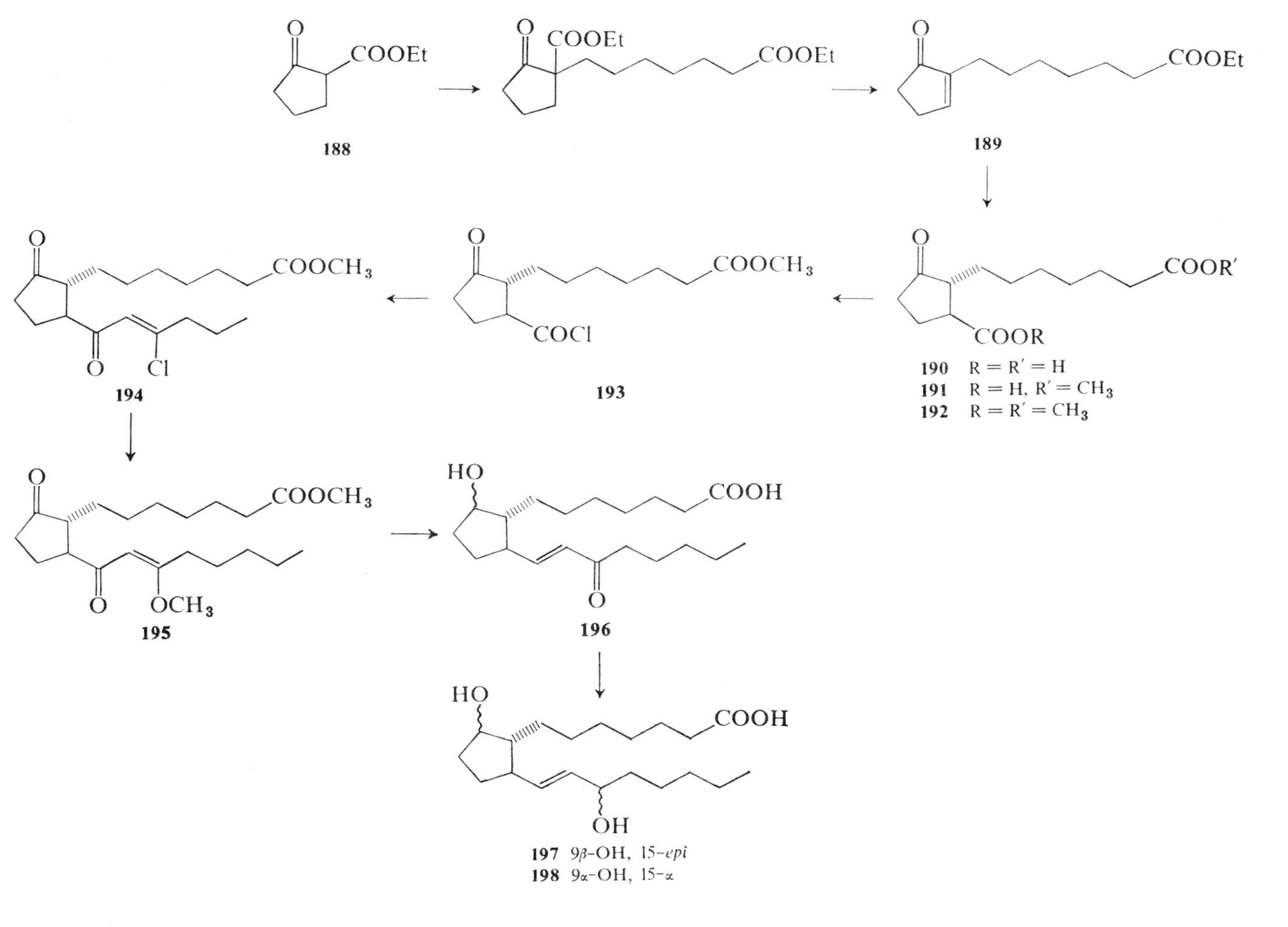
COOEt
188
COOEt
COOEt
COOEt
189
COOR′
COOR
190 R = R′ = H
191 R = H, R′ = CH3
192 R = R′ = CH3
COOCH3
COCl
193
COOCH3
Cl
194
COOCH3
OCH3
195
HO
COOH
196
HO
COOH
OH
197 9β-OH, 15-epi
198 9α-OH, 15-α

to 1-heptyne in the presence of aluminum chloride to give **194**. This chlorovinyl ketone was converted with methanolic sodium hydroxide to enol ether **195**. Sodium borohydride reduction of both ketones and acidification gave the 13,14-unsaturated-15-ketone **196**. Further borohydride reduction produced *d,l*-11-*desoxy*–$PGF_1\beta$ and its 15-epimer **197**, which were not separated. Presumably *d,l*-11-*desoxy*–$F_1\alpha$ **198** was also produced as a minor product, since these authors noted that borohydride reduction of the diester **19** gave two isomeric 9-alcohols, $9\alpha:9\beta = 15:85$. In view of the later observations of Ramwell et al.,[96] it might be surmised that much of the biological activity mentioned by Bagli et al. for the 9β-ol **197** may be due to the *ent*-11-*desoxy*–15-*epi*–$PGF_1\beta$ and/or 11-*desoxy*–$F_1\alpha$ present in the mixture, rather than to *nat*-11-*desoxy*–$F_1\beta$.

A novel approach to prostanoic acids via a photochemically prepared bicyclo[3,2,0] heptanone has also been used by Bagli and Bogri.[18] The same substituted cyclopentenone **189** used above was irradiated (high-pressure mercury lamp, Pyrex filter) with the chlorovinyl ketone **199** to give adduct **200**. When this photoadduct was refluxed with zinc in acetic acid, the major product isolated (44%) was the methyl prostanoate diketone **201**. The minor product was simply dechlorinated adduct. Sodium borohydride reduction of the diketone gave a mixture of diols which was separated into two fractions by chromatography, reported to be the 9-epimers **202** and **203**, both of which had biological activity.

D. Cyclopentenone Prostanoic Acids

The substituted cyclopentenone **207** [the 15(*S*)-epimer], or prostaglandin B_1, is the end-product of a series of base catalyzed steps involving dehydration of PGE_1 and double-bond migrations.[99] Racemic and also optically active PGB_1 have been synthesized by a number of groups, not all of which had this product as their initial goal. These syntheses give some valuable methods for cyclopentene substitution.

The first reported synthesis of *d,l*-PGB_1 was by Hardegger and associates.[66] The substituted cyclopentenone **189** was condensed with 3-*t*-butoxyoctynyl magnesium bromide. Acid-catalyzed allylic rearrangement of the initial product **204** and oxidation gave the disubstituted cyclopentenone **206**. Partial catalytic reduction of the triple bond, removal of the *t*-butyl ether, and base hydrolysis completed the synthesis of *d,l*-PGB_1 **207**. The basic hydrolysis step also served to isomerize the initially *cis*-13,14 double bond to *trans*. Complete reduction of the triple bond of **206** and removal of protective groups gave the racemic form of PGE-237 **(208)**, also previously obtained from PGE_1.[99] A very similar synthesis of **207** and **208** was also reported by Klok, Pabon, and van Dorp.[73]

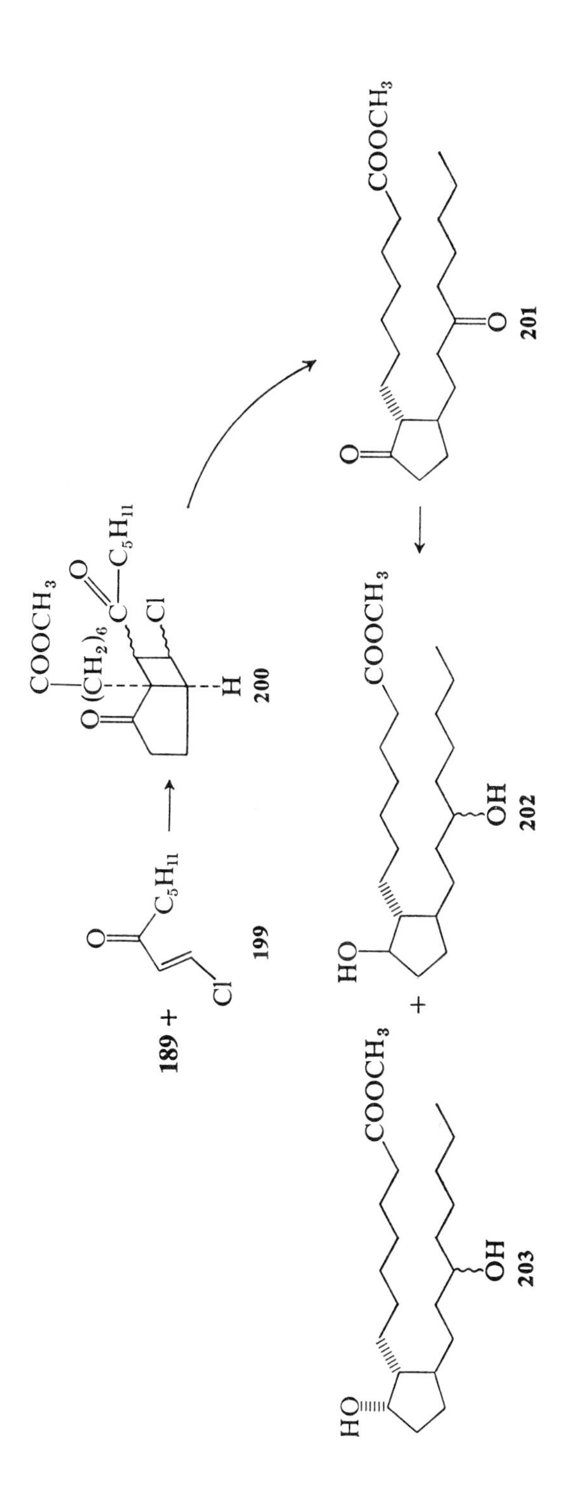
189 +
C5H11
Cl
199
COOCH3
(CH2)6
C5H11
Cl
H
200
COOCH3
201
COOCH3
HO
OH
202
COOCH3
HO
OH
203

189 → **204** → **205** → **206** → **207**, **208**

204: OH, $C{\equiv}C{-}CH{-}C_5H_{11}$, $OC(CH_3)_3$, COOEt

205: HO, $C{\equiv}C{-}CH{-}C_5H_{11}$, $O{-}C(CH_3)_3$, COOEt

206: O, $C{\equiv}C{-}CH{-}C_5H_{11}$, $O{-}C(CH_3)_3$, COOEt

207: O, OH, COOH

208: O, OH, COOH

Three other groups[34,72,121] prepared PGB_1 from the 2-substituted cyclopentan-1,3-dione enol ether **209**. This starting material was prepared by Collins, Jung, and Pappo from the hydroxydione **212**, acid-catalyzed hydrogenolysis and enol ether formation of which gave **209**. Yura and Ide cyclized the keto acid **213** using aluminum chloride and propionyl chloride at 80°. Katsube and Matsui prepared **209** from ethyl 9-oxodecanoate. All three groups then reacted **209** with the acetylenic magnesium bromide from

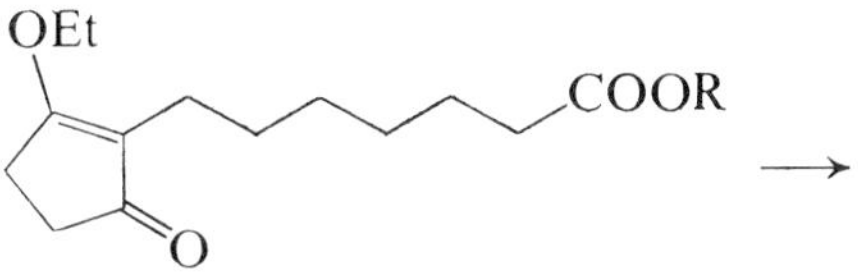

209 R = H or Et

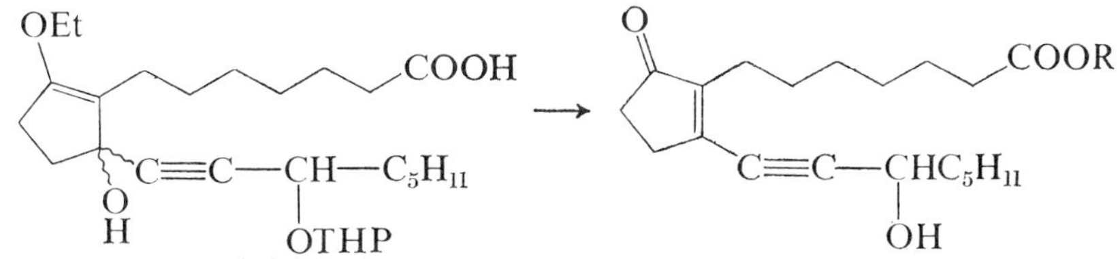

210

211

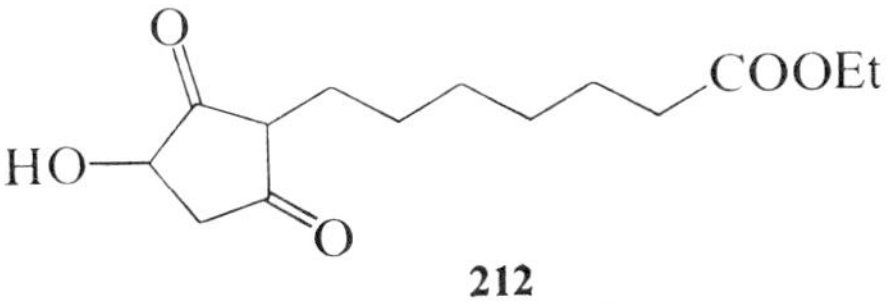

212

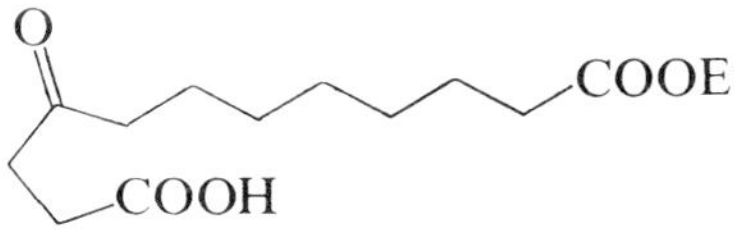

213

1-octyn–3-ol, tetrahydropyranyl ether to give adduct **210**. Acid hydrolysis of the enol ether and concomitant dehydration gave the cyclopentenone **211** which was converted to PGB_1 as in the preceding route. Pappo[92] recently reported the resolution of the intermediate 1-octyn–3-ol and the use of the (*S*)-enantiomer for the synthesis of natural 15(*S*)-PGB_1. The method of Dale, Dull, and Mosher[45a] involving an NMR analysis of the derived (−)α-methoxy–α-trifluoromethylphenyl acetate ester was used to determine the enantiomeric purity of the resolved 1-octyn–3-ols. This method has also been used[107] to assay mixtures of 15(*R*) and 15(*S*)-PGB_2.

In the same report,[92] Pappo described the conversion of the triketo acid **214** to the enol ether **215**. This, with difficulty, condensed with the same octynyl magnesium bromide to give **216**, which is the 11-hydroxy analog of **211**, opening the way to synthesis of 11-hydroxy prostaglandins. In this connection, Vandewalle, Sipido, and DeWilde[116] carried out some similar studies with the simpler triketone **217**. They converted it in a series of steps to **218**, which on lithium-ammonia reduction gave some **219**, stereochemistry

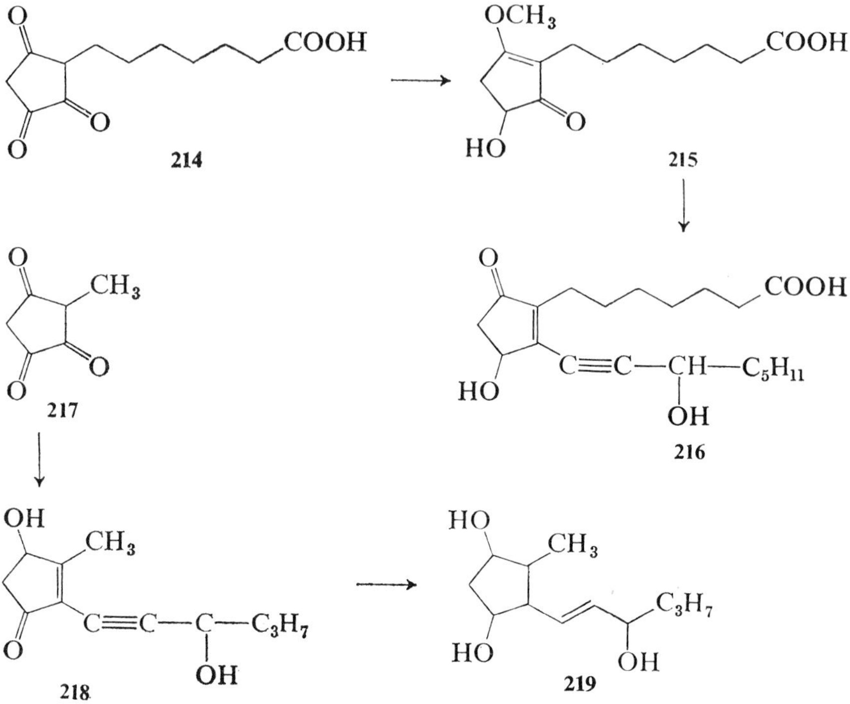

214 **215** **216** **217** **218** **219**

unspecified, among other products. This is being investigated further as a potential $PGF_1\alpha$ synthesis.

An attempt to repeat the Just PGE_1 synthesis by Holden et al.[67a] gave *d,l*-PGB_1 instead. Morin et al.[81] also describe a synthesis via the aromatic intermediate **220**, which gave a poorly characterized mixture said to contain **221** and the methyl ester of *d,l*-PGB_1.

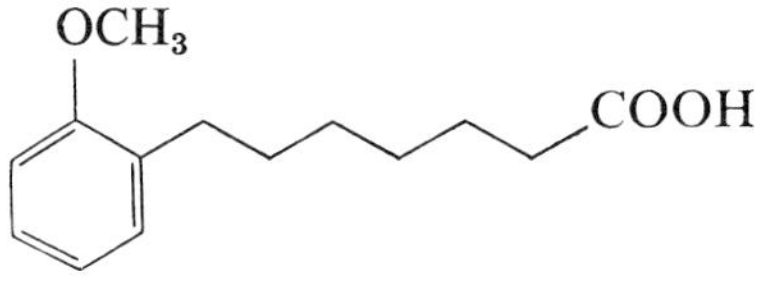

220

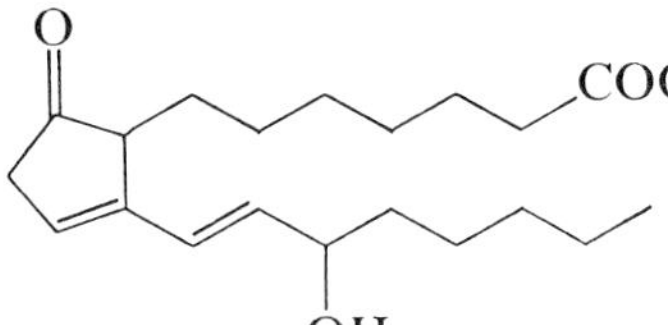

221

A synthesis of the 15-keto analog of racemic PGB_1 methyl ester by Miyano[78,80] illustrates the unusual use of a bicyclo[2,2,1]heptene as a protecting group and precursor of a *trans* carbon-carbon double bond. The starting material **222**, prepared from 5-norbornene–2,3-endodicarboxylic anhydride, was condensed with the sodio-enolate of dimethyl 3-oxoundecan–1,11-dioate to afford the triketo diester **223**. This was cyclized to the diketo diacid **224** with base. Decarboxylation (copper-quinoline) and esterification gave **225**, which was pyrolyzed, producing 15-dehydro–PGB_1 methyl ester **226** by a reverse Diels-Alder reaction. The new double bond was shown to be

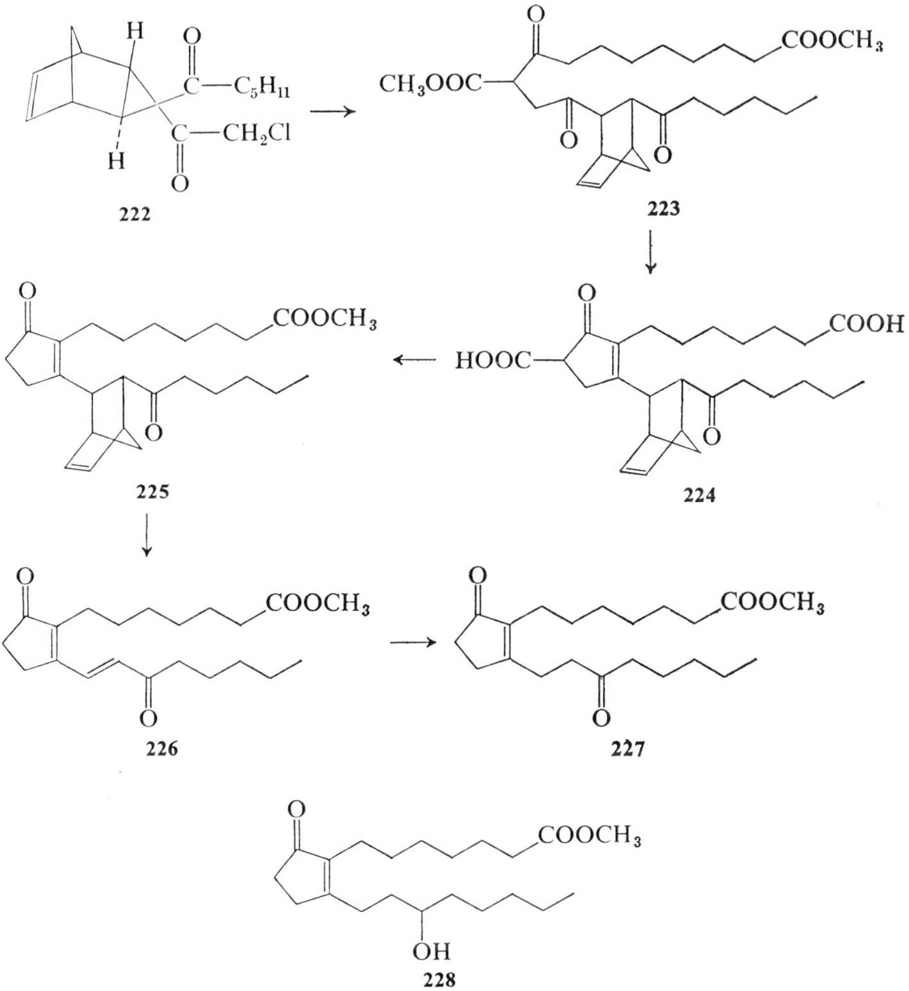

222 **223** **224** **225** **226** **227** **228**

trans, and could be selectively reduced with zinc in acetic acid to give **227**, which is 15-dehydro PGE-237. Some *d,l*-PGE-237 (**228**) was obtained by catalytic hydrogenation of **226**, along with **227**.

The earliest synthesis of a cyclopentenone prostanoic acid was by Samuelsson and Stallberg.[98] The two routes shown lead to a chromatographically separable mixture of **229** and **230**, the latter being identical to a degradation product of PGE_1.[99]

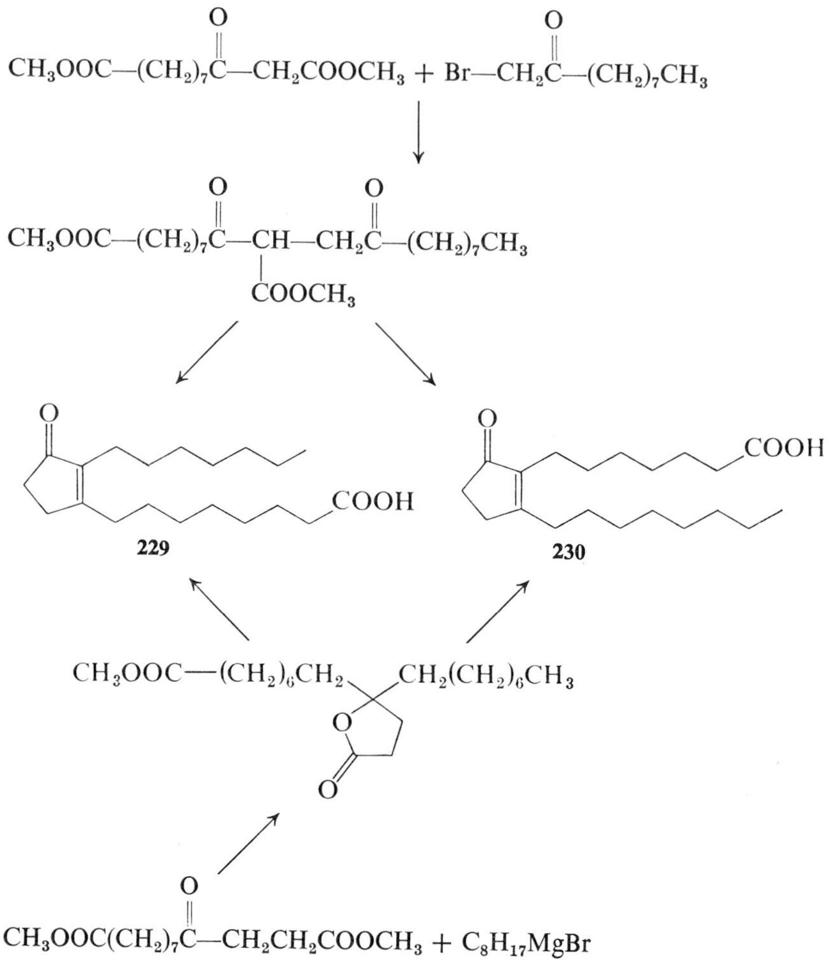

$$CH_3OOC{-}(CH_2)_7\overset{O}{\overset{\|}{C}}{-}CH_2COOCH_3 + Br{-}CH_2\overset{O}{\overset{\|}{C}}{-}(CH_2)_7CH_3$$

$$\downarrow$$

$$CH_3OOC{-}(CH_2)_7\overset{O}{\overset{\|}{C}}{-}\underset{COOCH_3}{\underset{|}{CH}}{-}CH_2\overset{O}{\overset{\|}{C}}{-}(CH_2)_7CH_3$$

O, COOH

229

O, COOH

230

$CH_3OOC{-}(CH_2)_6CH_2$ $CH_2(CH_2)_6CH_3$ (lactone: O, O)

$$CH_3OOC(CH_2)_7\overset{O}{\overset{\|}{C}}{-}CH_2CH_2COOCH_3 + C_8H_{17}MgBr$$

E. *d,l*-PGE$_1$ Methoxime

In a synthesis involving cyclopentenone intermediates similar to some of those in Section 4.D, Finch and Fitt[53] have succeeded in preparing *d,l*-PGE$_1$ methoxime. Although in a simple model system the methoxime of a β-hydroxyketone could be hydrolyzed to the β-hydroxyketone, these authors were unable to isolate pure *d,l*-PGE$_1$ from hydrolysis attempts from their synthetic methoxime. They mention a new oxime reagent they developed which can be removed under very mild conditions, but this has not yet been disclosed.

The cyclopentenone diester **231** was allylically brominated and the resulting bromide displaced by silver acetate to give acetate dimethyl ester **232**.

Methanolysis of the acetate gave alcohol **233**, which was silylated and then reduced with hydrogen on Raney nickel. Silylation was said to direct the *cis*-hydrogenation to the side of the ring opposite the bulky silyloxy group, producing the all *cis* stereochemistry as shown in **234**. Formation of the 9-methoxime and hydrolysis of the silyl ether gave **235**, which was much more resistant to dehydration than the corresponding β-hydroketone. Saponification (methanolic K_2CO_3) and reesterification gave the new diester **236** in which epimerization α to the carbomethoxy group was assumed to have taken place. Epimerization α to the methoxime was considered unlikely based on model studies. The tetrahydropyranyl ether of **236** was prepared and treated with sodium borohydride in ethanol, giving alcohol **237** (40% yield). It was thought that complexing of borohydride with the tetrahydropyranyl group (or methoxime?) helped to give the desired selective reduction. Oxidation of alcohol **237** to the aldehyde and Wittig olefination with tributylphosphoranylidene-2-heptanone gave unsaturated ketone **238**. Borohydride reduction of **238** followed by hydrolysis of tetrahydropyranyl ether and methyl ester produced a mixture of diol acids from which a crystalline isomer **239** (m.p. 97–99°) was separated. This racemic material was compared with one of the isomers (syn or anti?) of the methoxime (m.p. 55–57°) of natural PGE_1 by the usual criteria, but a satisfactory hydrolysis of this to *d*,*l*-PGE_1 was not achieved.

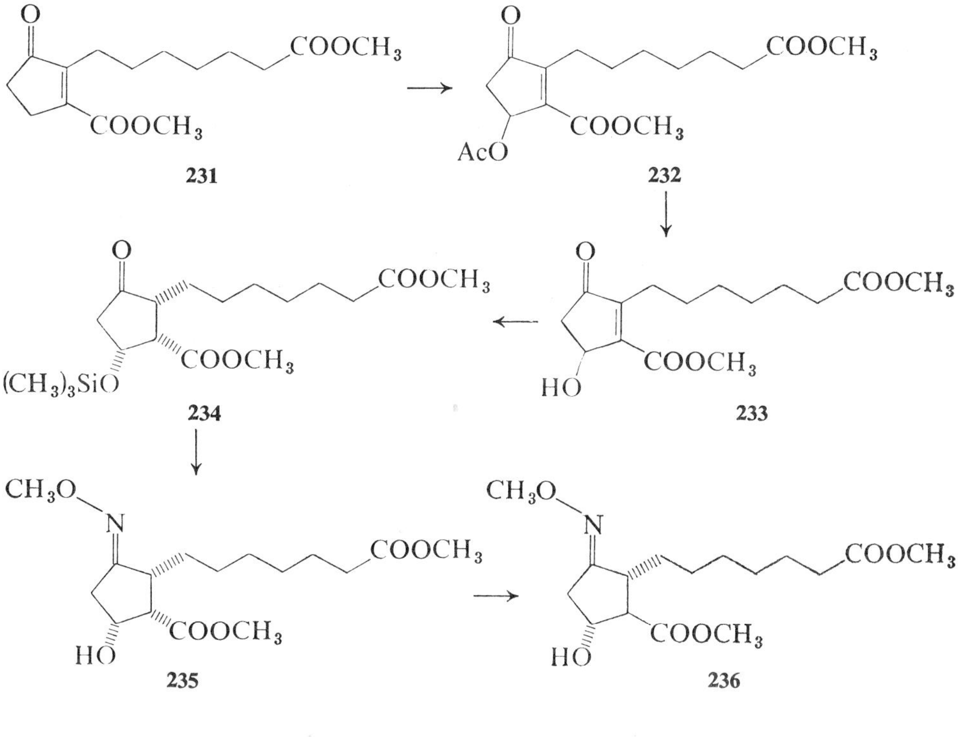

O COOCH$_3$ COOCH$_3$ **231** → O COOCH$_3$ AcO COOCH$_3$ **232**

↓

O COOCH$_3$ HO COOCH$_3$ **233** → O COOCH$_3$ $(CH_3)_3SiO$ COOCH$_3$ **234**

↓

CH_3O N COOCH$_3$ HO COOCH$_3$ **235** → CH_3O N COOCH$_3$ HO COOCH$_3$ **236**

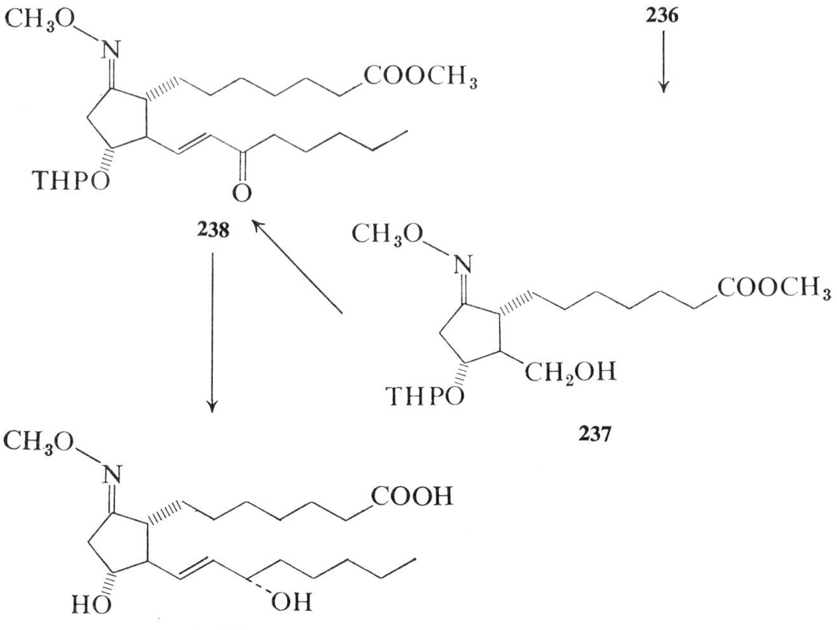

236 → 237 → 238 → 239

5. OXAPROSTAGLANDINS

Fried and co-workers[56] have carried out a well-planned synthesis of 7-oxaprostaglandins in which the *trans* opening of an epoxide is used to insure the correct steric relationships of the four substituents on the cyclopentane ring. The all *cis*-1,2-epoxycyclopentan–3,5-diol **240** was dibenzylated and the resulting diether **241** reacted with diethyl-1-octynylalane, giving **242**. The dialkyl alkynyl aluminum reagent gives nearly quantitative yields of **242** at room temperature, contrasting with the failure of alkali or magnesium acetylides to react even at elevated temperatures. The alane can be generated *in situ*[30] and employed directly in toluene solution. The resulting alcohol was then alkylated with

$$Et_3Al \longleftarrow NEt_3 + HC{\equiv}C{-}R \longrightarrow Et_2Al(C{\equiv}C{-}R)(\leftarrow NEt_3) + C_2H_6$$

t-butyl 6-iodohexanoate (with dimsyl sodium in DMSO) to give ester **243** in 65% yield. Removal of the *t*-butyl group with trifluoroacetic acid gave acid

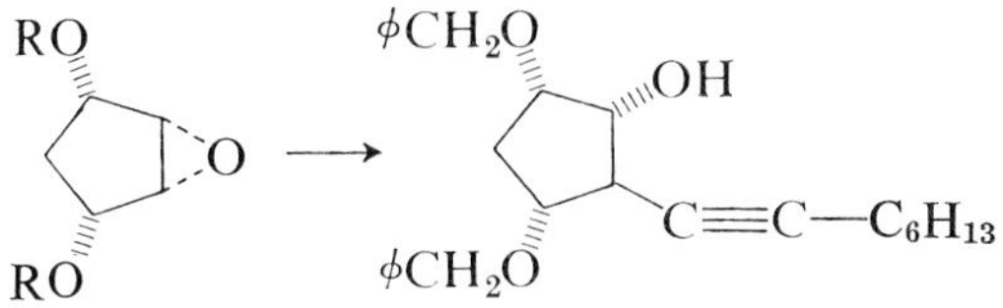

240 R = H—
241 R = ϕCH_2—

242

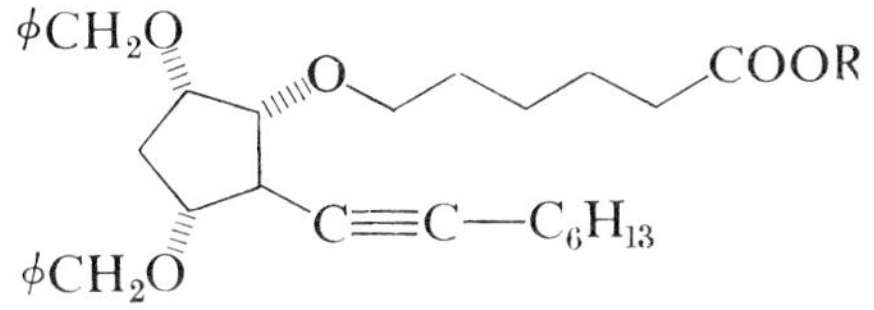

243 R = t-butyl
244 R = H

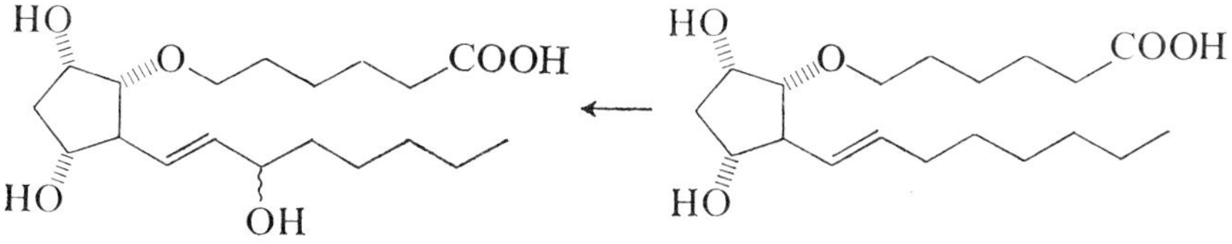

246

245

244 which was reduced with lithium in methylamine. This removed the benzyl groups as well as reducing the acetylenic bond to a *trans* double bond, producing **245**. The synthesis of 7-oxa–$PGF_1\alpha$ (**246**) was completed by selenium dioxide oxidation, which was nonstereospecific.

It was found subsequently[60] that intermediate **242** could be resolved via reaction with (+) or (−)-α-phenethyl isocyanate and recrystallization of the urethanes. The absolute configuration of the resolved (+) **242** was determined by catalytic hydrogenation to triol **247**, acetonide formation (**248**) and oxidation to ketone **249**. Optical rotatory dispersion studies of this ketone were definitive in assigning configuration.

The ability to form acetonides from such *vic*-glycols was used to accomplish a synthesis of 7-oxa-PGE_1. The required triol **251** was obtained from **250** by reaction as above with an acetylenic alane and hydrolysis of silyl groups. Now, acetonide formation (**252**) and benzylation gave **253**, which could be

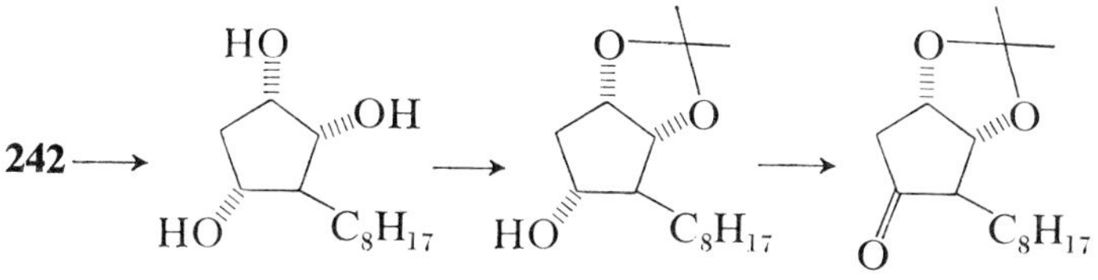

242 →

247

248

249

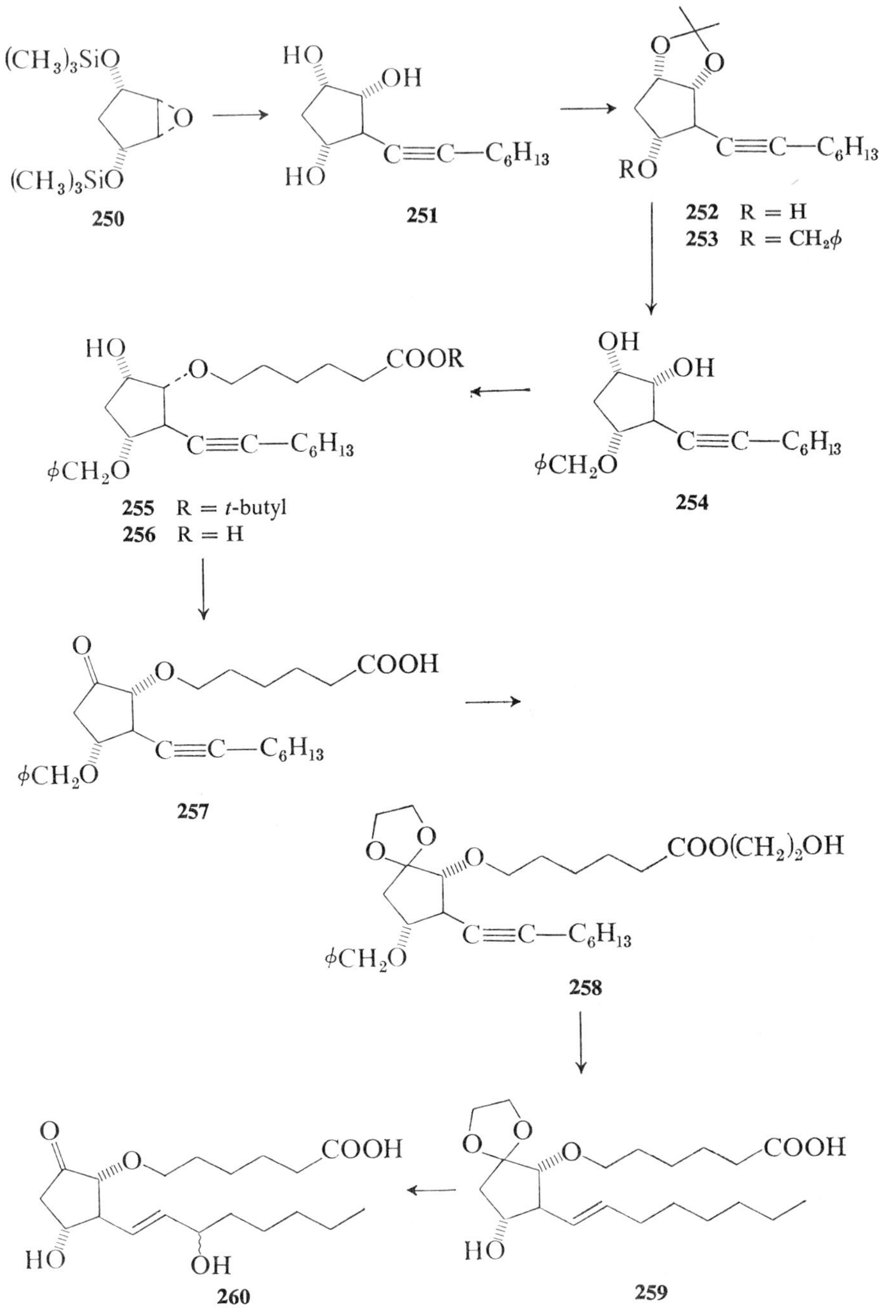

$(CH_3)_3SiO$

$(CH_3)_3SiO$

250

HO OH

HO $C{\equiv}C{-}C_6H_{13}$

251

RO $C{\equiv}C{-}C_6H_{13}$

252 R = H
253 R = $CH_2\phi$

OH OH

ϕCH_2O $C{\equiv}C{-}C_6H_{13}$

254

HO O COOR

ϕCH_2O $C{\equiv}C{-}C_6H_{13}$

255 R = *t*-butyl
256 R = H

O O COOH

ϕCH_2O $C{\equiv}C{-}C_6H_{13}$

257

O O O $COO(CH_2)_2OH$

ϕCH_2O $C{\equiv}C{-}C_6H_{13}$

258

O O O COOH

HO

259

O O COOH

HO OH

260

selectively hydrolyzed (aqueous trifluoroacetic acid) to glycol **254**. Alkylation as above unexpectedly gave 65% of the desired product **255** and only 20% of the other possible monoalkylated product.

Hydrolysis of the *t*-butyl ester and oxidation gave keto-acid **257**. The keto group was protected from reduction by ketal formation, which simultaneously esterified the acid. Saponification of **258** followed by lithium-methylamine reduction gave **259**, which was allylically hydroxylated with selenium dioxide. Removal of the ketal group with trifluoroacetic acid at 0° gave 7-oxa-PGE_1 (**260**), found to be sensitive to dehydration of the β-hydroxy-ketone system.

Application of similar synthetic principles by Fried has led to a number of other oxa-prostanoic acids.[60] A further extension is proposed as a synthesis of $PGF_1\alpha$. The isomeric "*trans*" expoxy dibenzyl ether **261** was used as starting material. Reaction as before with an acetylenic alane reagent gave **262**, where now the acetylenic side chain is to be used as the precursor of the carboxylic acid side chain of the prostanoic acid. The derived diol **263** could be resolved via a diurethane from (+)-α-phenethyl isocyanate. Tritylation of the primary hydroxyl and tosylation of the secondary gave **266**, which was simultaneously debenzylated, detritylated, and the triple bond saturated by hydrogenation over palladium catalyst. Base treatment of the resulting **265** resulted in epoxide formation **266**. Silylation of the diol and reaction with another acetylenic alane gave a mixture of two isomeric products, **267** and **269**. The desired prostanol **267**, which is unfortunately the minor product of the epoxide opening, was reduced to **270**. Further proposed transformations of **270** involving protection of secondary alcohols and oxidation of the primary alcohol should give $PGF_1\alpha$.

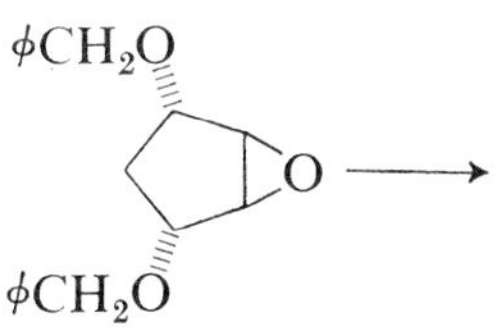

ϕCH_2O

O →

ϕCH_2O

261

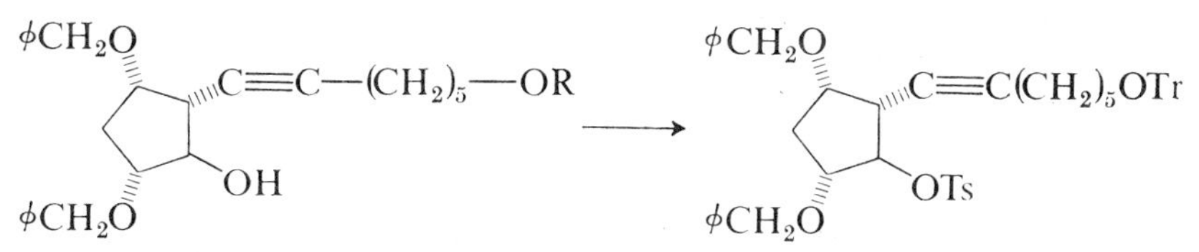

ϕCH_2O C≡C—$(CH_2)_5$—OR

OH

ϕCH_2O

262 R = THP

263 R = H

→

ϕCH_2O C≡C$(CH_2)_5$OTr

OTs

ϕCH_2O

264

↓

264

HO, OH, OTs, HO

265

HO, OH, O

266

HO, OH, HO, $C{\equiv}C{-}CHC_5H_{11}$, OR

267 R = THP
268 R = H

HO, OH, OH, C_8

269

HO, OH, HO, OH

270

REFERENCES

1. S. Abrahamsson, S. Bergström, and B. Samuelsson. *Proc. chem. Soc.*, **1962,** 332.
2. S. Abrahamsson. *Acta Crystallog.*, **16,** 409 (1963).
3. D. G. Ahern and D. T. Downing. *Biochim. Biophys. Acta*, **210,** 456 (1970).
4. D. G. Ahern and D. T. Downing. *Fed. Proc. Fed. Am. Socs. exp. Biol.*, **29,** 854 (1970). (Abstract).
5. N. H. Andersen. *J. Lipid Res.*, **10,** 316 (1969).
6. N. H. Andersen. *J. Lipid Res.*, **10,** 320 (1969).
7. E. Änggård and B. Samuelsson. *J. Biol. Chem.*, **239,** 4097 (1964).
8. E. Änggård and B. Samuelsson. *J. Biol. Chem.*, **240,** 3518 (1965).
9. E. Änggård and B. Samuelsson. *Ark. Kemi*, **25,** 293 (1966).
10. E. Änggård and B. Samuelsson. *Mem. Soc. Endocr.*, **14,** 107 (1966).
11. E. Änggård and B. Samuelsson. *Acta Physiol. Scand.*, **68** (Suppl. 277), 232 (1966).
12. E. Änggård and B. Samuelsson. *Meth. Enzymol.*, **14,** 215 (1969).

13. U. Axen. In C. K. Cain, Ed., *Annual Reports in Medicinal Chemistry 1967*, Academic Press, New York, 1968, pp. 290–296.

14. U. Axen, F. H. Lincoln, and J. L. Thompson. *Chem. Commun.*, **1969,** 303.

15. U. Axen, J. L. Thompson, and J. E. Pike. *Chem. Commun.*, **1970,** 602.

16. J. F. Bagli, T. Bogri, R. Deghenghi, and K. Wiesner. *Tetrahedron Lett.*, **1966,** 465.

17. J. F. Bagli and T. Bogri. *Tetrahedron Lett.*, **1967,** 5.

18. J. F. Bagli and T. Bogri. *Tetrahedron Lett.*, **1969,** 1639.

19. J. F. Bagli and T. Bogri. Abstracts, 158th Meeting Am. Chem. Soc., New York, 8–12 September 1969, MEDI 4.

20. J. F. Bagli. In C. K. Cain, Ed., *Annual Reports in Medicinal Chemistry*, Academic Press, New York, 1970, pp. 170–179.

21. J. F. Bagli and T. Bogri. Abstracts, 5th Middle Atlantic Regional Meeting, Am. Chem. Soc., Newark, Del., 1–3 April 1970, p. 59.

22. P. F. Beal, III, J. C. Babcock, and F. H. Lincoln. *J. Am. Chem. Soc.*, **88,** 3131 (1966).

23. P. F. Beal, J. C. Babcock, and F. H. Lincoln. In S. Bergstrom and B. Samuelsson, Eds., *Nobel Symposium 2*, *Prostaglandins*, Almqvist and Wiksell, Stockholm, 1967, pp. 219–230.

24. S. Bergström, L. Krabisch, B. Samuelsson, and J. Sjövall. *Acta. Chem. Scand.*, **16,** 969 (1962).

25. S. Bergström, R. Ryhage, B. Samuelsson, and J. Sjövall. *J. Biol. Chem.*, **238,** 3555 (1963).

26. S. Bergström, H. Danielsson, D. Klenberg, and B. Samuelsson. *J. Biol. Chem.*, **239,** PC4006 (1964).

27. S. Bergström, H. Danielsson, and B. Samuelsson. *Biochim. Biophys. Acta*, **90,** 207 (1964).

28. S. Bergström. *Science*, **157,** 382 (1967).

29. S. Bergström, L. A. Carlson, and J. R. Weeks. *Pharmac. Rev.*, **20,** 1 (1968).

30. P. Binger. *Angew. chem.*, **75,** 918 (1963).

31. G. L. Bundy, F. H. Lincoln, N. A. Nelson, J. E. Pike, and W. P. Schneider. *Ann. N.Y. Acad. Sci.*, **180,** 76 (1971).

32. W. Cole and P. J. Julian. *J. Org. Chem.*, **19,** 131 (1954).

33. A. Collet and J. Jacques. *Chim. ther.*, **5,** 163 (1970).

34. P. Collins, C. J. Jung, and R. Pappo. *Israel J. Chem.*, **6,** 839 (1968).

35. E. J. Corey, N. H. Andersen, R. M. Carlson, J. Paust, E. Vedejs, I. Vlattas, and R. E. K. Winter. *J. Am. Chem. Soc.*, **90,** 3245 (1968).

36. E. J. Corey, I. Vlattas, N. H. Andersen, and K. Harding. *J. Am. Chem. Soc.*, **90,** 3247 (1968) [see erratum **90,** 5947 (1968)].

37. E. J. Corey, N. M. Weinshenker, T. K. Schaaf, and W. Huber. *J. Am. Chem. Soc.*, **91,** 5675 (1969).

38. E. J. Corey, I. Vlattas, and K. Harding. *J. Am. Chem. Soc.*, **91,** 535 (1969).

39. E. J. Corey and E. Hamanaka. *J. Am. Chem. Soc.*, **89,** 2758 (1967).

40. E. J. Corey. In W. O. Milligan, Ed., *Proceedings of the Robert A. Welch Foundation Conferences on Chemical Research.* XII. *Organic Synthesis*, Houston, Texas, 1969, pp. 51–79.

41. E. J. Corey, Z. Arnold, and J. Hutton. *Tetrahedron Lett.*, **1970,** 307.
42. E. J. Corey and R. Noyori. *Tetrahedron Lett.*, **1970,** 311.
43. E. J. Corey, R. Noyori, and T. K. Schaaf. *J. Am. Chem. Soc.*, **92,** 2586 (1970).
44. E. J. Corey, T. K. Schaaf, W. Huber, U. Koelliker, and N. M. Weinshenker. *J. Am. Chem. Soc.*, **92,** 397 (1970).
45. E. J. Corey. *Ann. N.Y. Acad. Sci.*, **180,** 24 (1971).
45a. J. A. Dale, D. L. Dull, and H. S. Mosher. *J. Org. Chem.*, **34,** 2543 (1969).
46. E. G. Daniels, J. W. Hinman, B. A. Johnson, F. P. Kupiecki, J. W. Nelson, and J. E. Pike. *Biochem. Biophys. Res. Commun.*, **21,** 413 (1965).
47. E. G. Daniels, W. C. Krueger, F. P. Kupiecki, J. E. Pike, and W. P. Schneider. *J. Am. Chem. Soc.*, **90,** 5894 (1968).
48. E. G. Daniels and J. E. Pike. In P. W. Ramwell and J. E. Shaw, Eds., *Prostaglandin Symposium of the Worcester Foundation for Experimental Biology*, Interscience, New York, 1968, pp. 379–387.
49. D. A. van Dorp, R. K. Beerthuis, D. H. Nugteren, and H. Vonkeman. *Nature, London*, **203,** 839 (1964).
50. D. A. van Dorp, R. K. Beerthuis, D. H. Nugteren, and H. Vonkeman. *Biochim. Biophys. Acta*, **90,** 204 (1964).
51. D. A. van Dorp. *Naturwiss.*, **56,** 124 (1969).
52. D. T. Downing, D. G. Ahern, and M. Bachta. *Biochem. Biophys. Res. Commun.*, **40,** 218 (1970).
53. N. Finch and J. J. Fitt. *Tetrahedron Lett.*, **1969,** 4639.
54. S. H. Ford and J. Fried. *Life Sci.*, **8** (part 1), 983 (1969).
55. J. Fried, S. Heim, P. Sunder-Plassman, S. J. Etheredge, T. S. Santhanakrishnan, and J. Himizu. In P. W. Ramwell and J. E. Shaw, Eds., *Prostaglandin Symposium of the Worcester Foundation for Experimental Biology*, Interscience, New York, 1968, pp. 351–363.
56. J. Fried, S. Heim, S. J. Etheredge, P. Sunder-Plassman, T. S. Santhanakrishnan, J. Himizu, and C. H. Lin. *Chem. Commun.*, **1968,** 634.
57. J. Fried, T. S. Santhanakrishnan, J. Himizu, C. H. Lin, S. H. Ford, B. Rubin, and E. O. Grigas. *Nature, London*, **223,** 208 (1969).
58. J. Fried, M. M. Mehra, W. Kao, and C. H. Lin. Abstracts, 5th Middle Atlantic Regional Meeting, Am. Chem. Soc., Newark, Del., 1–3 April 1970, p. 60.
59. J. Fried, M. M. Mehra, W. L. Kao, and C. H. Lin. *Tetrahedron Lett.*, **1970,** 2695.
60. J. Fried, C. H. Lin, M. M. Mehra, W. L. Kao, and P. Dahren. *Ann. N.Y. Acad. Sci.*, **180,** 38 (1971).
61. E. Granström and B. Samuelsson. *J. Am. Chem. Soc.*, **91,** 3398 (1969).
62. K. Gréen. *Chem. Phys. Lipids*, **3,** 254 (1969).
62a. K. Gréen and B. Samuelsson. *J. Lipid Res.*, **5,** 117 (1964).
63. M. Hamberg. *Eur. J. Biochem.*, **6,** 147 (1968).
64. M. Hamberg and U. Israelsson. *J. Biol. Chem.*, **245,** 5107 (1970).
65. M. Hamberg and B. Samuelsson. *J. Am. Chem. Soc.*, **91,** 2177 (1969).
66. E. Hardegger, H. P. Schenk, und E. Borger. *Helv. Chim. Acta.*, **50,** 2501 (1967). Other work including model studies by this group is found in Dissertations. Nr.

3794, W. Graf; 3796, H. A. Kindler, 3870, P. Müller; 3942, F. Naf; 3943, E. A. Broger; 4003, H. Schenk; and 4006, J. Vonarburg, E. T. H. Zurich.

67. J. W. Hinman. *Postgrad. Med. J.*, **46,** 562 (1970).

67a. K. G. Holden, B. Hwang, K. R. Williams, J. Weinstock, M. Harman, and J. A. Weisbach. *Tetrahedron Lett.*, **1968,** 1569.

68. E. W. Horton. *Physiol. Rev.*, **49,** 122 (1969).

69. W. Jubiz and J. Frailey. *Clin. Res.*, **19,** 127 (1971).

70. G. Just and C. Simonovitch. *Tetrahedron Lett.*, **1967,** 2093.

71. G. Just, C. Simonovitch, F. H. Lincoln, W. P. Schneider, U. Axen, G. B. Spero, J. E. Pike. *J. Am. Chem. Soc.*, **91,** 5364 (1969).

72. J. Katsube and M. Matsui. *Ag. Biol. Chem.*, **33,** 1078 (1969).

73. R. Klok, H. J. J. Pabon, and D. A. van Dorp. *Rec. Trav. chim. Pays-Bas Belg.*, **87,** 813 (1968).

74. R. Klok, H. J. J. Pabon, and D. A. van Dorp. *Rec. Trav. chim. Pays-Bas Belg.*, **89,** 1043 (1970).

75. O. Korver. *Rec. Trav. chim. Pays-Bas Belg.*, **88,** 1070 (1969).

76. C. Larsson and E. Änggård. *Acta Pharmacol. Tox.*, **28** (Suppl. 1), 61 (1970).

77. L. Levine and H. Van Vunakis. *Biochem. Biophys. Res. Commun.*, **41,** 1171 (1970).

78. M. Miyano. *Tetrahedron Lett.*, **1969,** 2771.

79. M. Miyano and C. R. Dorn. *Tetrahedron Lett.*, **1969,** 1615.

80. M. Miyano. *J. Org. Chem.*, **35,** 2314 (1970).

81. R. B. Morin, D. O. Spry, K. L. Hauser, and R. A. Mueller. *Tetrahedron Lett.*, **1968,** 6023.

82. J. Nakano, E. Änggård, and B. Samuelsson. *Eur. J. Biochem.*, **11,** 386 (1969).

83. J. Nakano, E. Änggård, and B. Samuelsson. *Pharmacologist*, **11,** 238 (1969).

84. D. H. Nugteren, R. K. Beerthuis, and D. A. van Dorp. *Rec. Trav. chim. Pays-Bas Belg.*, **85,** 405 (1966).

85. D. H. Nugteren, D. A. van Dorp, S. Bergström, M. Hamberg, and B. Samuelsson. *Nature, London*, **212,** 38 (1966).

86. D. H. Nugteren, H. Vonkeman, and D. A. van Dorp. *Rec. Trav. chim. Pays-Bas Belg.*, **86,** 1237 (1967).

87. D. H. Nugteren. *Biochim. Biophys. Acta*, **210,** 171 (1970).

88. D. E. Orr and F. B. Johnson. *Can. J. Chem.*, **47,** 47 (1969).

89. H. J. J. Pabon, L. van der Wolf, and D. A. van Dorp. *Rec. Trav. chim. Pays-Bas Belg.*, **85,** 1251 (1966).

90. C. Pace-Asciak and L. S. Wolfe. *Chem. Commun.*, **1970,** 1234.

91. C. Pace-Asciak and L. S. Wolfe. *Chem. Commun.*, **1970,** 1235.

92. R. Pappo, P. W. Collins, and C. J. Jung. *Ann. N.Y. Acad. Sci.*, **180,** 64 (1971).

93. J. E. Pike, F. P. Kupiecki, and J. R. Weeks. In S. Bergstrom and B. Samuelsson, Eds., Nobel Symposium 2, Prostaglandins, Almqvist and Wiksell, Stockholm, 1967, pp. 161–171.

94. J. E. Pike, F. H. Lincoln, and W. P. Schneider. *J. Org. Chem.*, **34,** 3552 (1969).

95. P. W. Ramwell, J. E. Shaw, G. B. Clarke, M. F. Grostic, D. G. Kaiser, and J. E.

Pike. In R. T. Holman, Ed., *Progress in the Chemistry of Fats and Other Lipids*, Vol. 9, Pergamon Press, Oxford, 1968, pp. 231–273.

96. P. W. Ramwell, J. E. Shaw, E. J. Corey, and N. Andersen. *Nature, London*, **221,** 1251 (1969).
97. P. W. Ramwell and J. E. Shaw. *Recent Prog. Horm. Res.*, **26,** 139 (1970).
98. B. Samuelsson and G. Ställberg. *Acta Chem. Scand.*, **17,** 810 (1963).
99. B. Samuelsson. *Angew. Chem. Int. Ed.*, **4,** 410 (1965).
100. B. Samuelsson. *J. Am. Chem. Soc.*, **85,** 1878 (1963).
101. B. Samuelsson. Abstract, 4th Int. Congr. Pharmac., Basle, 14–18 July 1969, p. 11.
102. B. Samuelsson. Proc. 4th Int. Congr. Pharmac., Basle, 14–18 July 1969, Vol. 4, pp. 12–31, Schwabe, Basle, 1970.
103. B. Samuelsson, M. Hamberg, and C. C. Sweeley. *Anal. Biochem.*, **38,** 301 (1970).
104. W. P. Schneider. *Chem. Commun.*, **1969,** 304.
105. W. P. Schneider, U. Axen, F. H. Lincoln, J. E. Pike, and J. L. Thompson. *J. Am. Chem. Soc.*, **90,** 5895–5896 (1968) [see erratum **91,** 1043 (1969)].
106. W. P. Schneider, U. Axen, F. H. Lincoln, J. E. Pike, and J. L. Thompson. *J. Am. Chem. Soc.*, **91,** 5372 (1969).
107. W. P. Schneider and J. Muenzer. Unpublished data. Useful differences were seen in both the proton and fluorine magnetic resonance spectra. It has been suggested by G. Slomp of The Upjohn Company that the relevant peaks in the fluorine spectra would be sharper and more easily integrated if the corresponding tri-deuteromethoxy reagent were used.
108. V. Schwarz. *Coll. Czech. Chem. Comm.*, **26,** 1207 (1961).
109. H. Shio, N. H. Andersen, E. J. Corey, and P. W. Ramwell. Abstracts, 4th Int. Congr. Pharmac. Basle, 14–18 July 1969, p. 100.
110. H. Shio, P. W. Ramwell, N. H. Andersen, and E. J. Corey. *Experientia*, **26, 355** (1970).
111. C. J. Sih, G. Ambrus, P. Foss, and C. J. Lai. *J. Am. Chem. Soc.*, **91,** 3685 (1969).
112. C. J. Sih, C. Takeguchi, and P. Foss. *J. Am. Chem. Soc.*, **92,** 6670 (1970).
113. D. P. Strike and H. Smith. *Tetrahedron Lett.*, **1970,** 4393.
114. C. B. Struijk, R. K. Beerthuis, H. J. J. Pabon, and D. A. van Dorp. *Rec. Trav. chim Pays-Bas Belg.*, **85,** 1233 (1966).
115. D. Taub, R. D. Hoffsommer, C. H. Kuo, H. L. Slates, Z. S. Zelawski, and N. L. Wendler. *Chem. Commun.*, **1970,** 1258.
116. M. Vandewalle, V. Sipido, and H. DeWilde. *Bull. Soc. Chim. Belg.*, **79,** 403 (1970).
117. H. Vonkeman, D. H. Nugteren, and D. A. van Dorp. *Biochim. Biophys. Acta*, **187,** 581 (1969).
118. A. J. Weinheimer and R. L. Spraggins. Abstracts, 158th Meeting Am. Chem. Soc., New York, 8–12 September 1969, MEDI 41.
119. A. J. Weinheimer and R. L. Spraggins. *Tetrahedron Lett.*, **1969,** 5185.
120. A. J. Weinheimer and R. L. Spraggins. In H. W. Youngken, Jr., Ed., *Food-Drugs from the Sea*, Proc. Marine Technology Society 1969, 1970, pp. 311–318.
121. Y. Yura and J. Ide. *Chem. Pharm. Bull., Tokyo*, **17,** 408 (1969).

The Total Synthesis of Pyrrole Pigments

A. H. JACKSON

Department of Chemistry, University College, Cardiff

AND

K. M. SMITH

Robert Robinson Laboratories, University of Liverpool

1. INTRODUCTION

In this essay we review the main methods currently available for the synthesis of naturally occurring porphyrins and bile pigments, as well as the structurally and biosynthetically related vitamin B_{12} (cobalamin). A short section is devoted to the small group of tripyrrolic bacterial pigments of which prodigiosin is the characteristic example.

Heme and chlorophylls *a* and *b* are the most widespread natural pigments and perform a complementary role in nature, being associated with the oxidative and energy-liberating processes of plant and animal metabolism on the one hand and the reduction and energy-trapping processes of photosynthesis on the other. As well as being the oxygen carrier of hemoglobin, heme is also the prosthetic group in many of the cytochromes, catalases, and peroxidases, and modified hemes occur in the prosthetic group of cytochromes *a*, and a_3, and *c*. Other important naturally occurring porphyrins include the algal chlorophylls *c*, *d*, and *e*, the bacterial chlorophylls such as bacteriochlorophyll, and the *Chlorobium* chlorophylls (650) and (660). With the notable exceptions of heme and chlorophylls *a* and *b* most of these pigments have not yet been synthesized and their structures rest upon comparisons of their degradation products, with other porphyrins of importance in relation to heme and chlorophyll biosynthesis. Some of the more important earlier reviews in this area are given in references 1–8.

2. NOMENCLATURE

Two systems are currently in use for numbering the porphyrin nucleus and related open-chain polypyrroles, and their relative merits are now under active discussion with a view to choosing a unified convention. The earlier system[2] devised by Hans Fischer is shown in **1** and the new system devised by the IUPAC Nomenclature Committee,[9] to include vitamin B_{12} and related macrocycles in the same scheme as porphyrins, is shown in **2**.

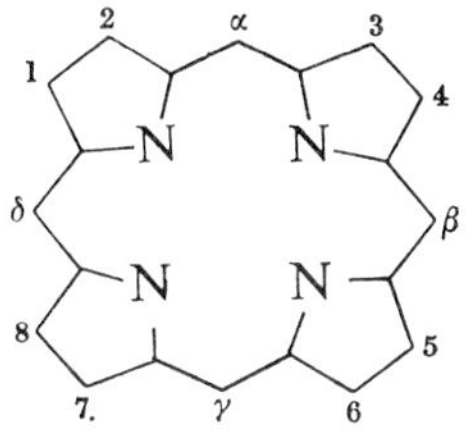

1

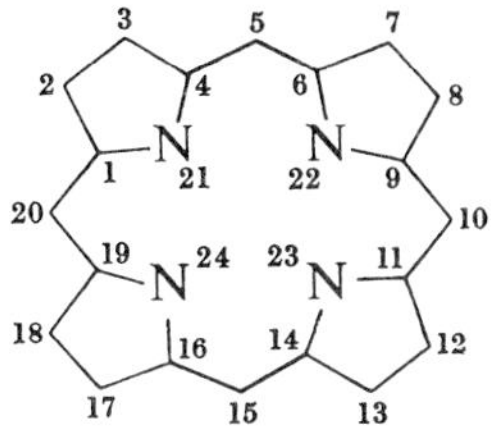

2

The Fischer system is used in this review for porphyrins, to provide continuity with the massive chemical and biological literature of the past and because it is still used by the majority of chemists and biochemists working in the field today. However, the IUPAC system (**2**) is used for vitamin B_{12} and other "corrinoid" compounds. The chlorophylls present in most green plants and the *Chlorobium* chlorophylls are dihydroporphyrins or *chlorins*, in which one of the peripheral double bonds has been reduced; conventionally,[3] their structures have always been written with the *D* ring partially reduced (**3**). Bacteriochlorophyll, a tetrahydroporphyrin, is usually written with the peripheral linkages in rings *B* and *D* reduced. The recently discovered dihydroporphyrin system (**4**), in which hydrogen is added to one of the *meso* positions and to nitrogen, has been named *phlorin*.[10]

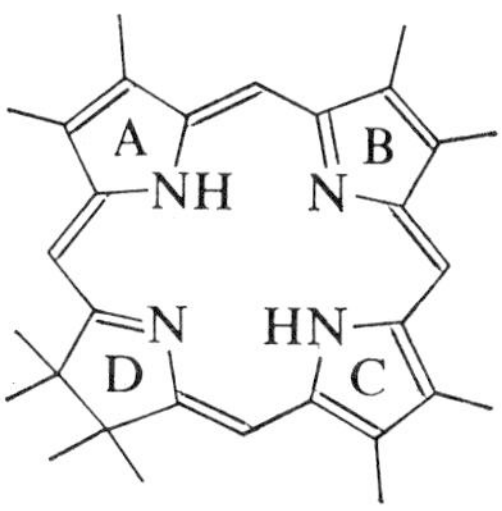

3

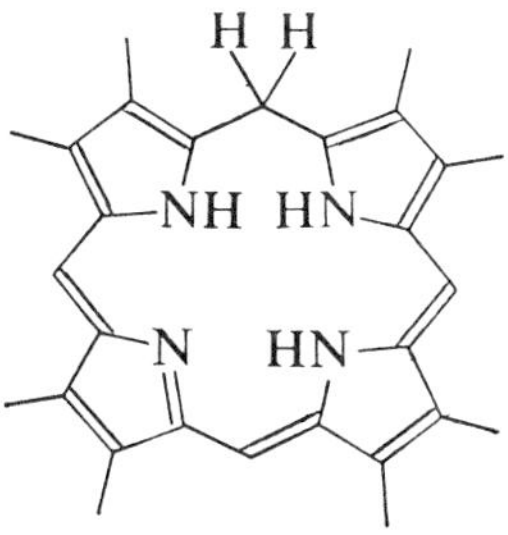

4

The nomenclature for di-, tri-, and tetrapyrrolic compounds in which the pyrrole rings are linked together by methane (—CH_2—) (**5a**), methene (—CH=) (**5b**), or carbonyl (—CO—) (**5c**) bridges has not yet been fully systematized. The accepted numbering system[1,2] for the dipyrrolic compounds is shown in **5**, and these compounds should strictly be referred to as 2,2′-

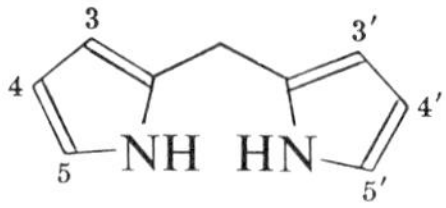

5a 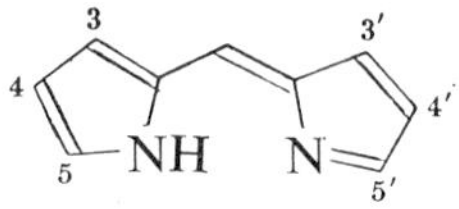 **5b** 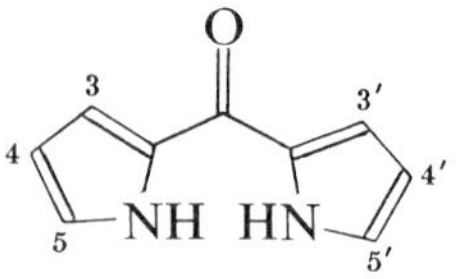 **5c**

dipyrrolyl methanes, methenes, or ketones. These names frequently are abbreviated to dipyrromethane, dipyrromethene, and dipyrroketone, or to the even simpler *pyrromethane*, *pyrromethene*, and *pyrroketone;* the latter convention is used in this review. (This does not normally cause any confusion because few of the 2,3′- and 3,3′-analogs are known, and moreover they are not known to have any biological significance.)

The numbering system[2] shown in **6** is used for the tetrapyrrolic compounds, although some authors[5] have used an alternative system based on the IUPAC nomenclature for vitamin B_{12} and porphyrins. The generic name *bilane* was originally applied[2] to the tetrapyrrolic bile pigments (**6**) with three methylene (CH_2) bridges. Subsequently, Lemberg[11] used this name for the corresponding 1′,8′-dioxygenated tetrapyrrolic bile pigments, but many synthetic non-

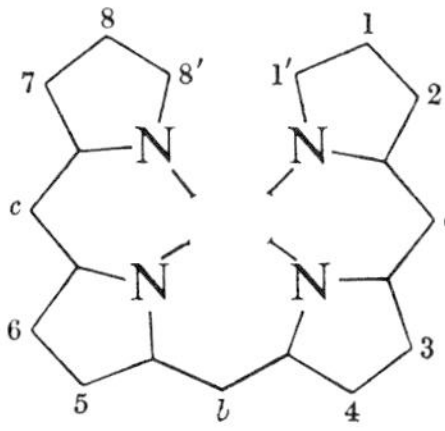

6

oxygenated analogs are now known and to refer to these as "dideoxybilanes" seems rather cumbersome. For this reason the name *bilane* is used generically in this review for all the compounds with the carbon nitrogen skeleton shown in **6**, and in accord with Lemberg[11] the related tetrapyrroles with one or more methene linkages are referred to as *bilenes*, *biladienes*, and *bilatrienes*. These compounds are often referred to in the literature as "linear" tetrapyrroles, but the term "open-chain" tetrapyrroles is to be preferred, because space filling models clearly show that the β-substituents in the individual pyrrole nuclei tend to force the adoption of a coiled conformation. For this reason also the structures are written as in **5** and **6** to emphasize this point.

As with other natural products, a wide variety of trivial names are in current use for both porphyrins and bile pigments.[2,3] A further complication arises because of the possibilities for isomerism about such a symmetrical nucleus as that in porphyrins, and Fischer[2] devised a system for numbering isomeric compounds (Table 1), especially the simpler degradation products of the

Table 1. Trivial Names of Porphyrins Related to Heme and Chlorophyll

Trivial Name	Substituent Combinations in the β-Positions 1–8	Number of Isomers
Etio-porphyrins	4(Me, Et)	4
Copro-porphyrins	4(Me, P)	4
Uro-porphyrins	4(A, P)	4
Meso-porphyrins	2(Me, Et), 2(Me, P)	15
Proto-porphyrins	2(Me, V), 2(Me, P)	15
Deutero-porphyrins	2(Me, H), 2(Me, P)	15
Pyrro-porphyrins	1(Me, H), 2(Me, Et), 1(Me, P)	25
Rhodo-porphyrins	1(Me, CO_2H), 2(Me, Et), 1(Me, P)	25

A = CH_2CO_2H, P = $CH_2CH_2CO_2H$, V = $CH{=}CH_2$

natural materials. There are four possible *etioporphyrins*, for example, in which each pyrrole ring bears one methyl and one ethyl substituent, and the isomer designated as etioporphyrin-III can not only be obtained by reduction of heme but is also closely related to the chlorophylls. Nearly all naturally

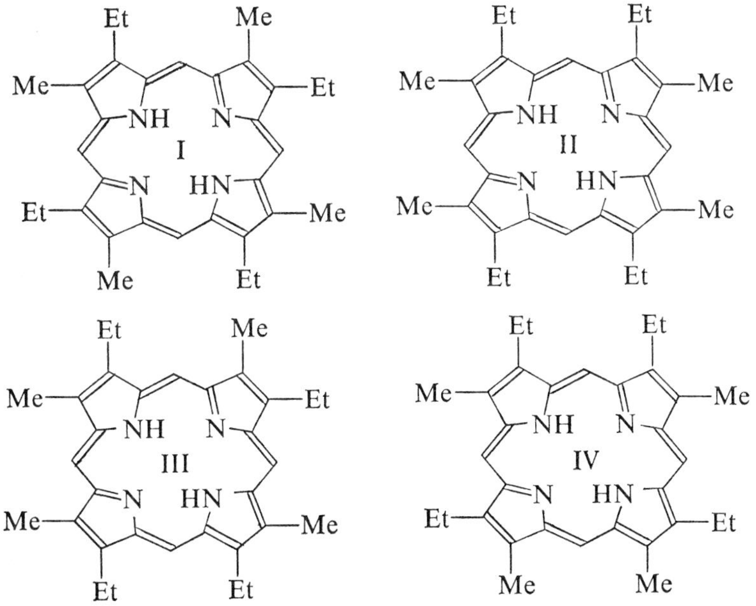

Etioporphyrin type isomers.

occurring porphyrins and the corrin chromophore of vitamin B_{12} can be related, at least formally, to etioporphyrin-III and are often referred to as type III compounds. Small amounts of type I porphyrins occur in certain rare pathological conditions, but type II and IV porphyrins have not been found in nature.

There are also four isomers of the *copro-* and *uro*-porphyrins, but when there are three different substituents as in the related *meso-*, *proto-*, and *deutero*-porphyrins, then 15 isomers are possible; with the *pyrro-* and *rhodo*-porphyrins which have four different substituents, the number of positional isomers rises to 25 (Table 1). The substitution pattern in the type III isomers of *meso-*, *proto-*, *deutero-*, *pyrro-*, and *rhodo*-porphyrins is indicated schematically in 7.

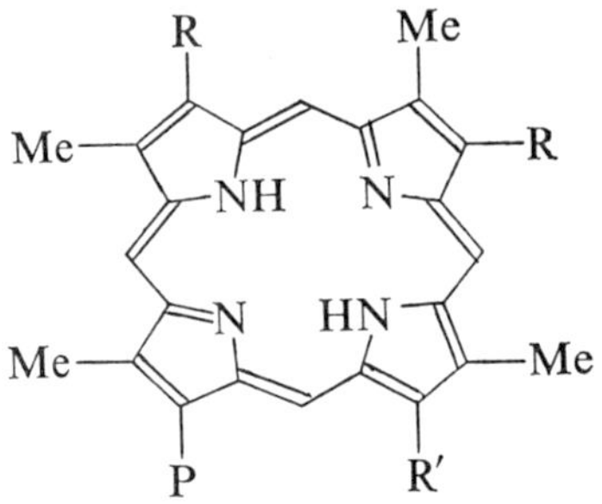

copro-III	R = R′ = P
meso-IX	R = Et; R′ = P
proto-IX	R = V; R′ = P
deutero-IX	R = H; R′ = P
pyrro-XV	R = Et; R′ = H
rhodo-XV	R = Et; R′ = CO_2H

7

3. STRATEGY OF PORPHYRIN SYNTHESIS

In reviewing recent progress[5,7,8] made in the synthesis of porphyrins it seems appropriate to consider the general philosophy and strategy underlying the various methods now available and to cite individual naturally occurring porphyrins as examples. The most widely used porphyrin synthesis to date has been the fusion of two pyrromethene units;[2] Fischer's classical synthesis of heme is perhaps the most well-known example. However, in recent years the pre-eminence of this method has been challenged by the development of new "stepwise" methods for the rational synthesis of open-chain (or "linear") tetrapyrroles, which can then be cyclized to porphyrins. This approach has proved increasingly attractive since Corwin and Coolidge's original preparation and cyclization of an open-chain bilane to etioporphyrin-II nearly 20 years ago,[12] although recent work has cast some doubts on the isomeric purity of this product (see below). Mild methods are now available for the preparation of bilenes,[13] biladienes,[14] and *a*- and *b*-oxobilanes,[15] all of which can be cyclized to porphyrins under mild conditions. None of these methods is truly stepwise, for they all involve the coupling of two dipyrrolic units, but there do not seem to be any advantages in synthesizing the tripyrrolic compounds as intermediates;[16] indeed, there may be positive disadvantages if the tri- or tetrapyrrolic compounds are relatively unstable.

One other attraction of these new stepwise methods is that they generally involve much milder reaction conditions than the Fischer pyrromethene synthesis,[2] and consequently some of the more labile side chains in the desired porphyrin can be carried through from the pyrrole stage, either directly or in a modified or protected form. This is generally more efficient, but as will be seen it is still occasionally necessary to introduce a substituent at a late stage in this synthesis after macrocycle formation, and the position of insertion may require blocking during the mono-, di, and tetrapyrrolic stages, with, for example, bromine or iodine, which can subsequently be removed by hydrogenolysis.

The synthesis of chlorins (dihydroporphyrins) and tetrahydroporphyrins such as the chlorophylls and the bacteriochlorophylls presents additional problems. Extra hydrogen atoms can be introduced by reduction of porphyrins in various ways, but the stereochemistry and position of addition are difficult to control.

Since pyrroles are the basic building blocks for all the present porphyrin syntheses we first discuss briefly the synthesis of monopyrroles and their coupling to give the various dipyrrolic intermediates, pyrromethanes, pyrromethenes, and pyrroketones. The direct synthesis of porphyrins from the mono- and dipyrrolic compounds is also discussed, followed by the coupling of dipyrrolic compounds to form tetrapyrrolic intermediates and their conversion into porphyrins.

4. PYRROLES

In choosing a synthetic approach to a given pyrrole, symmetry factors within the molecule are of primary importance, and a choice is made from one of the general routes[17] outlined in this section. The large majority of the methods discussed are from noncyclic precursors, and so the synthetic plan is variable with respect to the points at which the pyrrole ring (**8**) may be closed.

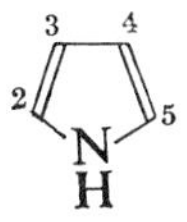

3 4
2 5
N
H
8

A. Formation of C—N Bonds Only

Typically, this method involves the much used reaction of a primary amine, or ammonia, with a 2,5-diketone:

$$\mathrm{R{-}CO{-}CH_2{-}CH_2{-}CO{-}R^1} + \mathrm{R^2NH_2} \longrightarrow \text{(pyrrole: 2-R, 5-}\mathrm{R^1}\text{, N-}\mathrm{R^2}\text{, 3,4-H)}$$

Known in general terms as the Paal-Knorr method, this synthesis can give very good yields of pyrroles and is limited mainly by the inaccessibility of suitable diketones. It has not found a great deal of application in the preparation of pyrroles for contemporary porphyrin synthesis.[5] An example of this general approach is the original, and still useful, synthesis of pyrrole (**8**) by the pyrolysis of ammonium mucate (**9**) and pyrrole itself is still much

$$\mathrm{NH_4^{+}\,{}^{-}O_2C{-}CH(OH){-}CH(OH){-}CH(OH){-}CH(OH){-}CO_2^{-}\,{}^{+}NH_4} \xrightarrow{\Delta} \text{pyrrole}$$

9 **8**

used in porphyrin synthesis because when it is polymerized in the presence of aldehydes (the Rothemund[18] reaction) it furnishes the corresponding *meso*-$\alpha,\beta,\gamma,\delta$-tetrasubstituted porphyrin.

B. Formation of 3—4 and C—N Bonds

This approach includes the classical (and without doubt most frequently used) avenue to pyrroles, the Knorr synthesis. In its most popular form this method gives an excellent route to 2,4-dialkylpyrrole–3,5-dicarboxylates [e.g., "Knorr's pyrrole" (**14**)]. Thus condensation of an α-aminoketone (**12**) with a ketone (**13**) possessing an active methylene group (usually a β-ketoester) in acetic acid buffered with sodium acetate furnishes pyrroles (e.g., **14**) in good yield. The α-aminoketones (e.g., **12**) are prepared *in situ* by zinc (or sodium dithionite) reduction of the corresponding oximinone **11** prepared earlier from the appropriate ketone [or β-ketoester (**10**)] with nitrous acid. The synthesis is of wide application, which accounts for its popularity through the years.

The comparatively recent introduction of *t*-butyl and benzyl esters as protecting groups has increased the versatility of this synthesis even further,

10 →(HONO) 11 →(Zn, HOAc)

[12] + 13 → 14

since differential protection of the two carboxylates expedites the specific manipulation of these functions at later stages. The 5-substituent can be hydrogen, but yields are better when it is an acyl or carboxylate group; yields are poor when the 3-substituent is alkyl. The most important limitation to Knorr's synthesis is the inherent symmetry of the products (e.g., **14**).

The Hantzsch synthesis involves the condensation of a β-ketoester (**15**) and an α-chloroketone (**17**) or α-chloroacetaldehyde in the presence of ammonia; it seems likely that 2-aminocrotonate (**16**) is an intermediate in

15 →(NH_3) 16 + 17 →

this reaction. It has long been thought to be far less generally applicable than the Knorr synthesis, but recently the Hantzsch synthesis has been extended by MacDonald,[19] who showed that α-halo derivatives of aldehydes other than acetaldehyde may be used, and also that in certain cases benzyl and *t*-butyl pyrrole-4-carboxylates are accessible. Hitherto, only eight or nine different pyrroles had been made by the Hantzsch method, but the extension to the use of aldehydes (which are much more easily halogenated specifically than are ketones) has increased the general applicability of this route.

C. Formation of 2—3 and C—N Bonds

In attempting to modify the Knorr synthesis, Fischer and Fink[20] discovered that condensation of β-ketoaldehydes [in the form of their acetals (**19**)] and the usual Knorr oximinones (e.g., **18**) in the presence of zinc and buffered acetic acid gave acceptable yields (~40%) of pyrroles of type **20**. An acetyl group is lost in the process, showing that, rather unexpectedly, the 2—3 and

MeCO, EtO_2C, C=N–OH (**18**) + (MeO)(OMe)HC–CH_2–C(=O)Me (**19**) —Zn/HOAc→ EtO_2C-pyrrole(H, H)-Me, NH (**20**)

C—N bonds had been formed. Kleinspehn[21] found that this method could be modified by use of oximinomalonic ester or 2-oximino-1,3-diketones rather than the oximinoketoesters (**18**). The route was then modified even further by Johnson,[22] who recommended the use of oximinoacetoacetic esters (**21**) (particularly the benzyl and *t*-butyl esters) or α-oximino ketones in condensation with 1,3-diketones (**22**). This modification is probably one of the most

MeCO, RO_2C, C=N–OH (**21**) + Me–C(=O)–CH(R^1)–C(=O)Me (**22**) —Zn/HOAc→ RO_2C-pyrrole(Me, R^1)-Me, NH

frequently used routes to pyrroles in modern synthesis, due to the easy accessibility of suitable 2-substituted 1,3-diketones (**22**).

A further method developed by Fischer and Fink[20] utilized the hydroxymethylenemethyl ethyl ketone **23** and oximinoacetic ester **18**, furnishing the pyrrole **24** after the usual zinc reduction.

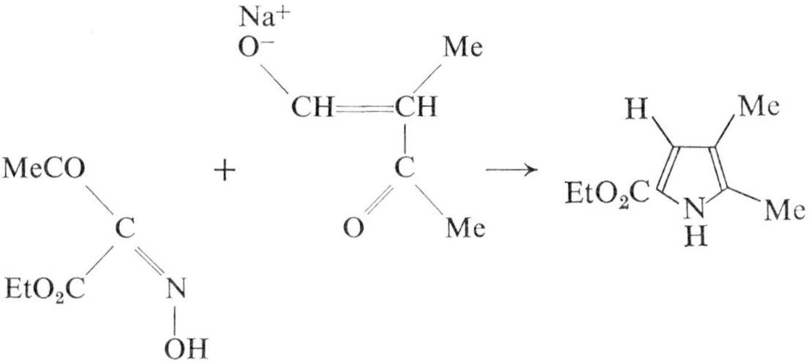

18 23 24

Kenner and co-workers have reported a pyrrole synthesis utilizing *N*-tosyl glycine derivatives.[23] Thus base-catalyzed addition of *N*-tosyl glycine (**25**) to

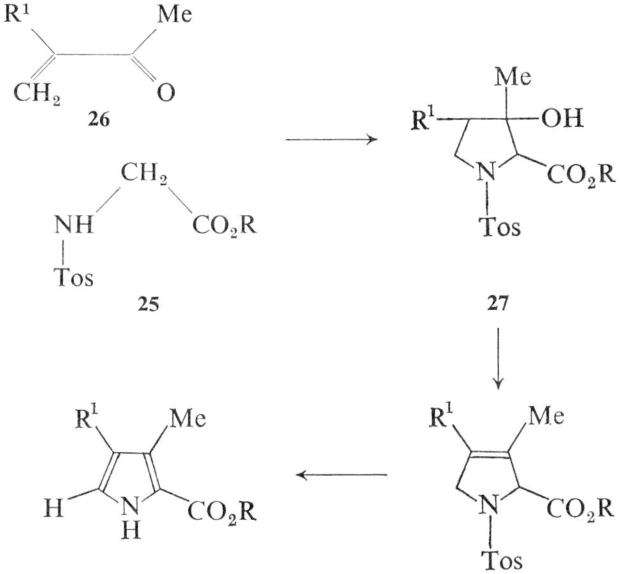

26 25 27 28 29

R^1 = H or Me
R = Et, Bu^t, or CH_2Ph

an α,β-unsaturated ketone (**26**) gave **27**, which, after dehydration to the Δ^3-pyrroline **28**, afforded the corresponding pyrrole-2-carboxylate **29**.

D. Formation of 2—3 and 4—5 Bonds

Very few useful examples of this general method exist. Perhaps the best is the preparation of *N*-substituted pyrroles (**32**) by condensation of a suitable tri-substituted amine (**30**) with an α-diketone (**31**). As might be expected, the

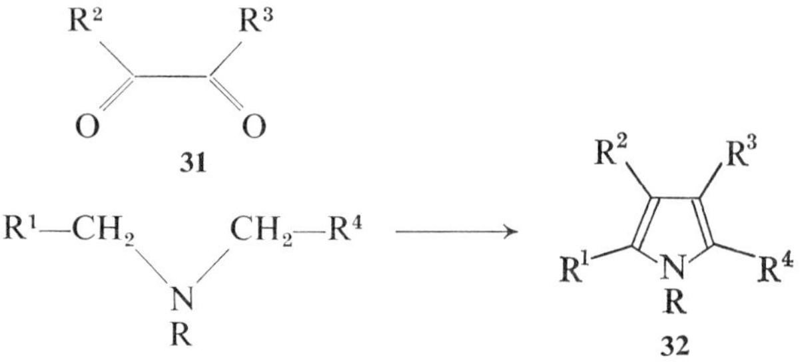

31

30

32

pyrroles obtained directly from this method, because of their *N*-substitution. are of little use as intermediates in porphyrin synthesis.

E. Pyrroles from Other Heterocycles

The best known method[24] of this type is the treatment of furans (e.g., **33**) with ammonia or substituted amines, at temperatures between 300 and 500°. The moderate yields, limitations due to availability of furans, and the drastic

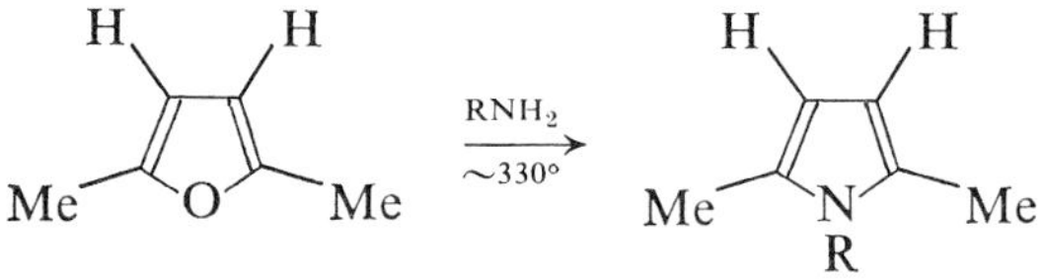

33

conditions all combine to make this method of limited value for preparation of the diverse pyrroles required for porphyrin synthesis.

F. Modification of Pyrroles for Use as Intermediates

Even with the large number of general methods for pyrrole synthesis, which have been outlined briefly, there are still many cases in which a required pyrrole cannot be obtained directly by ring synthesis. In such cases, pyrroles obtained from the Knorr synthesis or one of its modifications are usually used as substrates for further elaboration.

"Reductive C-alkylation" is a recently reported method[25] by which pyrroles from existing syntheses can be modified to give access to pyrroles which, due to deficiencies in these very syntheses, were not readily available. When a pyrrole is treated with hydriodic acid and paraformaldehyde at 100° it is *C*-methylated, alkoxycarbonyl and acetyl groups being lost. In hydriodic acid at lower temperatures (15–45°) typical pyrroles retain their labile groups and all free positions are *C*-alkylated (methylated if formaldehyde is used or else otherwise alkylated depending on the carbonyl agent). Apart from providing a route to diverse alkylated pyrroles, it seems that this method may also be applicable to the degradative structure determination of porphyrins.[26]

The "rational" synthesis of all four uroporphyrin isomers was not achieved until recently by MacDonald,[27–30] due to difficulties[31] involved in the preparation of the required pyrroles (e.g., **34**) rather than the construction of the

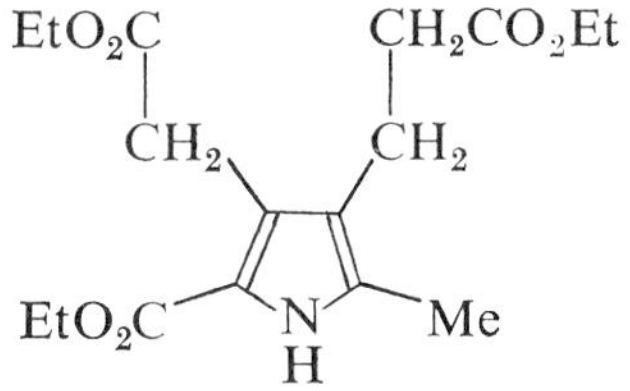

34

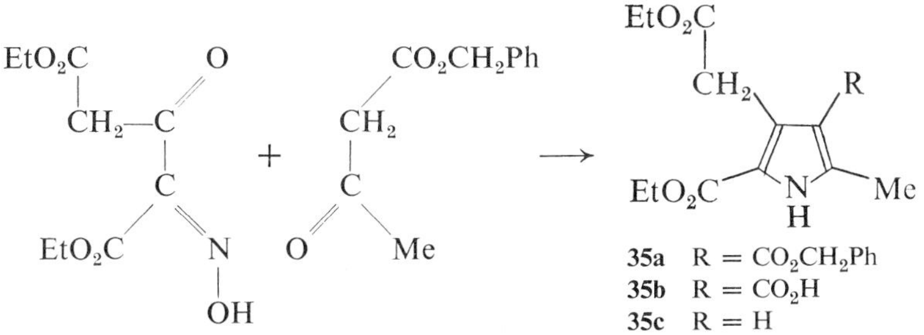

35a R = CO_2CH_2Ph
35b R = CO_2H
35c R = H
35d R = CHO
35e R = $CH{=}CHCO_2H$

porphyrin macrocycle itself. The problem was overcome by first synthesizing the pyrrole **35a** from benzyl acetoacetate and oximinoacetone dicarboxylic ester (a typical example of the Knorr method). Catalytic hydrogenation gave the β-carboxylic acid **35b**, which was thermally decarboxylated to the corresponding β-free pyrrole **35c**. Formylation gave **35d**, which was elaborated to the acrylic pyrrole **35e** by means of a Knoevenagel condensation with malonic acid. Catalytic hydrogenation and esterification gave the pyrrole **34**, an invaluable intermediate in the synthesis of the uroporphyrin isomers.

Without doubt, the most important and synthetically desirable pyrrole in recent times has been porphobilinogen (PBG) (**36**). Since its isolation[32] from the urine of patients suffering with acute porphyria, it has been found to be a direct precursor[33] in the biogenesis of the blood and plant pigments as well as the biliproteinoids and vitamin B_{12}. Any truly useful synthesis of PBG

HO_2C CH_2CO_2H
CH_2 CH_2
NH_2CH_2 N H
H

36

must be capable of modification to allow introduction of isotopic labels at specific sites within the molecule, for exploitation in biosynthetic investigations. Already, much work has been done with PBG obtained enzymatically from labeled δ-aminolevulinic acid, but further investigations, particularly the examination of the pathway by which PBG is polymerized to the unsymmetrical type III uroporphyrinogen, require more refined labeling at given

EtO_2C CH_2CO_2Et
CH_2 CH_2
R N CO_2Et
H

37a $R = CH_3$
37b $R = CHO$
37c $R = CH{=}NOH$
37d $R = CH_2NH_2$

HO_2C CH_2CO_2H
CH_2 CH_2
NH_2CH_2 N CO_2H
H

38

CH_2CO_2H
O
CH_2
HN
N R
H

39a $R = CO_2H$
39b $R = H$

HO_2C CH_2CO_2H
CH_2 CH_2
OHC N R
H

40a $R = CO_2H$
40b $R = I$

HO_2C CH_2CO_2H
CH_2 CH_2
HC N H
H
N
HO

41

positions, and such substrates can be obtained only by total synthesis. In a series of papers, MacDonald and co-workers[34] described the synthesis of PBG, with gradually increasing efficiency, based on a rather classical overall plan of attack. Thus the pyrrole **37a** was dichlorinated with sulfuryl chloride and the 2-formylpyrrole **37b** obtained by hydrolysis. This was converted to the oxime **37c** by standard methods from which the aminomethylpyrrole **37d** was accessible by catalytic hydrogenation. Hydrolysis gave the tricarboxylic acid **38**, which was decarboxylated most readily as the lactam **39a** to PBG lactam (**39b**). Alkaline hydrolysis gave PBG (**36**) in an overall yield of 5% from the pyrrole **37a**. This was later improved by hydrolysis of the formyl triester **37b** to the corresponding tricarboxylic acid **40a** followed by iodinative decarboxylation to **40b**; treatment with hydroxylamine afforded the pyrrole oxime **41** with concomitant loss of the iodine. Alternatively the formyl acid (**40a**), could be converted directly into the oxime (**41**) by treatment with hydroxylamine. Catalytic hydrogenation of **41** gave porphobilinogen (**36**) in 29% overall yield from the pyrrole **37a**.

Plieninger, Hess, and Ruppert[35] recently improved the early stages of the MacDonald synthesis of porphobilinogen by a more efficient synthesis of **37a**.

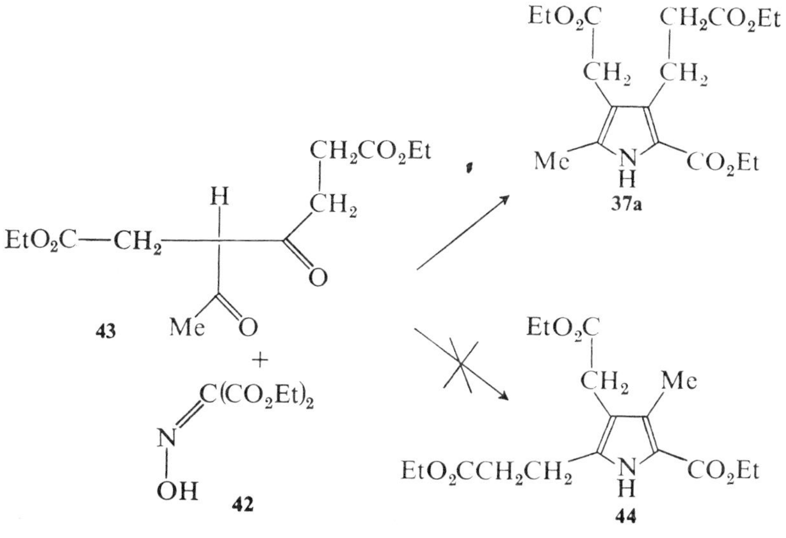

Their condensation of diethyl oximinomalonate (**42**) with the 2,4-diketone **43** in presence of zinc and sodium acetate gave **37a** in 41% yield, with none of the unwanted isomer (**44**) being detected.

A most ingeniously conceived synthesis of PBG based on its readily reversible conversion to the lactam **39b** has recently been reported by Rapoport and co-workers.[36] The starting material, a pyridine (**45**), was converted to PBG

(**36**) in an overall yield of 19%, a considerable improvement on existing routes. In addition, the totally different concept in the synthetic approach outlined below provides entirely different opportunities for incorporation of radioactive tracers; the high overall yield also makes it possible to contemplate the incorporation of tracers at unusually early stages. Thus the trisubstituted pyridine **45** was chosen as the starting material, the nitrogen atom of the nitro group being destined to become the nitrogen atom of the pyrrole ring in PBG. After suitable modification, the pyridine ring itself was to become the lactam ring in PBG lactam (**39b**), with the pyridine 4-methyl group and its acidic hydrogens being the site for initial connection of the future pyrrole ring.

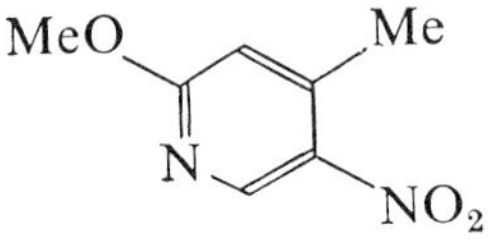

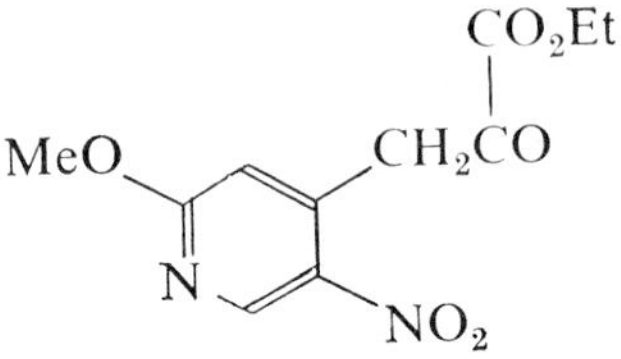

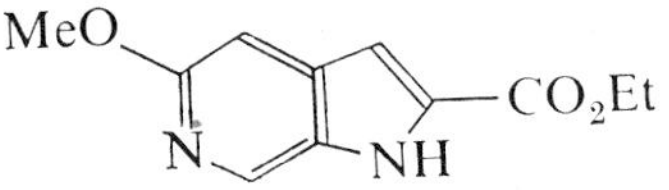

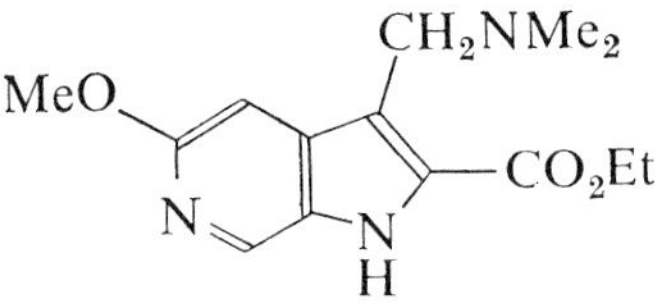

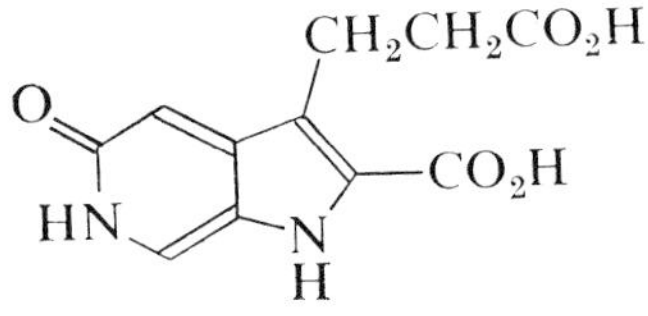

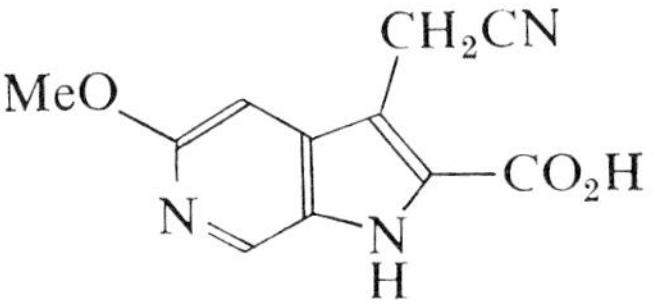

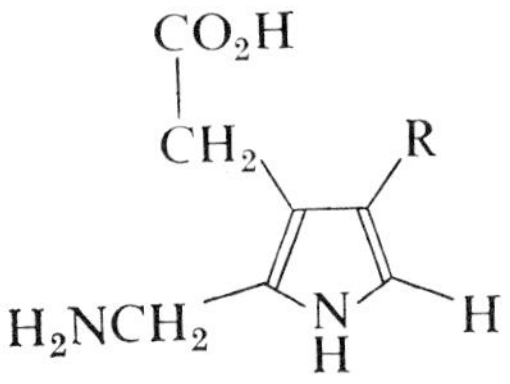

Treatment of **45** with diethyl oxalate and sodium ethoxide gave **46**, which cyclized to the azaindole **47** on reduction of the nitro group. The β-position of indoles is well known for its vulnerability to electrophiles, and so the required propionate side chain was constructed by means of a Mannich reaction (to give **48**) followed by treatment with diethyl sodiomalonate, furnishing **49** after exhaustive hydrolysis. Catalytic reduction of the newly created pyridone ring gave the pyrrole **39a**, an intermediate in MacDonald's approach, which could be decarboxylated to **39b** and then hydrolyzed to PBG. This general approach has been extended to the synthesis of PBG analogs such as **51**; R = H and **51**; R = CH_2COOH by direct transformation of the azaindoles, **47** and **50**, respectively; the latter was accessible by treatment of the quarternary salt of the Mannich base **48** with sodium or potassium cyanide.

Without exception, the methods for producing polypyrroles utilize ionic reactions, and the pyrrole ring is the nucleophile. Because of the "π-excessive" nature of pyrrole,[37] the most common nucleophilic species are 2-unsubstituted pyrroles (**52**). In certain cases, and particularly when 3-, 4-, and 5-substituents are not electron-withdrawing functionalities, the corresponding 2-carboxylic acid **53** may be used instead. Such pyrroles (**53**) are readily accessible from the corresponding esters by hydrolysis or hydrogenolysis. It should

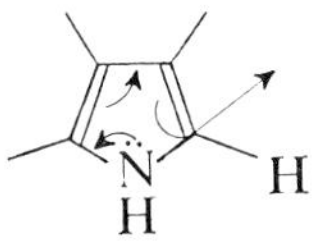

N H H

52

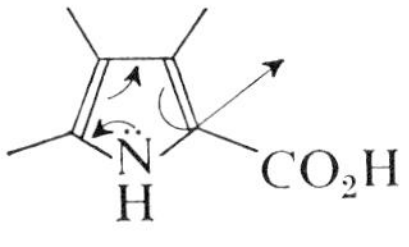

N H CO_2H

53

be noted, however, that *t*-butyl esters of pyrroles, when deesterified with, for example, trifluoroacetic acid, suffer concomitant decarboxylation to **52** and, indeed, 2-unsubstituted pyrroles are often accessible from the corresponding carboxylic acid by either acid-catalyzed or thermal decarboxylation. Pyrroles (e.g., **54**) bearing electron-withdrawing groups usually cannot be decarboxylated with such ease and may require "iodinative decarboxylation" involving iodination to **55** followed by hydrogenolysis of the halogen atom over Adams

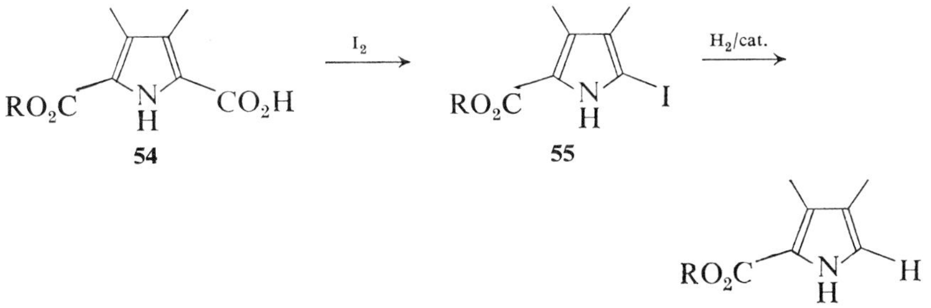

RO_2C N H CO_2H $\xrightarrow{I_2}$ RO_2C N H I $\xrightarrow{H_2/cat.}$ RO_2C N H H

54 **55**

catalyst.[38] Pyrroles such as **54** are available from 5-methylpyrrole–2-carboxylates (**56**), (the most common type of pyrrole produced from the classical syntheses,) by trichlorination to **57** with sulfuryl chloride, followed by hydrolysis (see Scheme 1). Mono-halogenation of pyrroles (**56**) furnishes the α-halomethylpyrroles **58**, a useful electrophilic species in pyrromethane synthesis. Further chlorination to the dichloromethylpyrroles **59** followed by hydrolysis furnishes the corresponding α-formylpyrroles **60**, which are used extensively in pyrromethene synthesis. Alternatively, acetoxylation of **56** with lead tetraacetate furnishes the mono- and diacetoxymethyl analogs of **58**

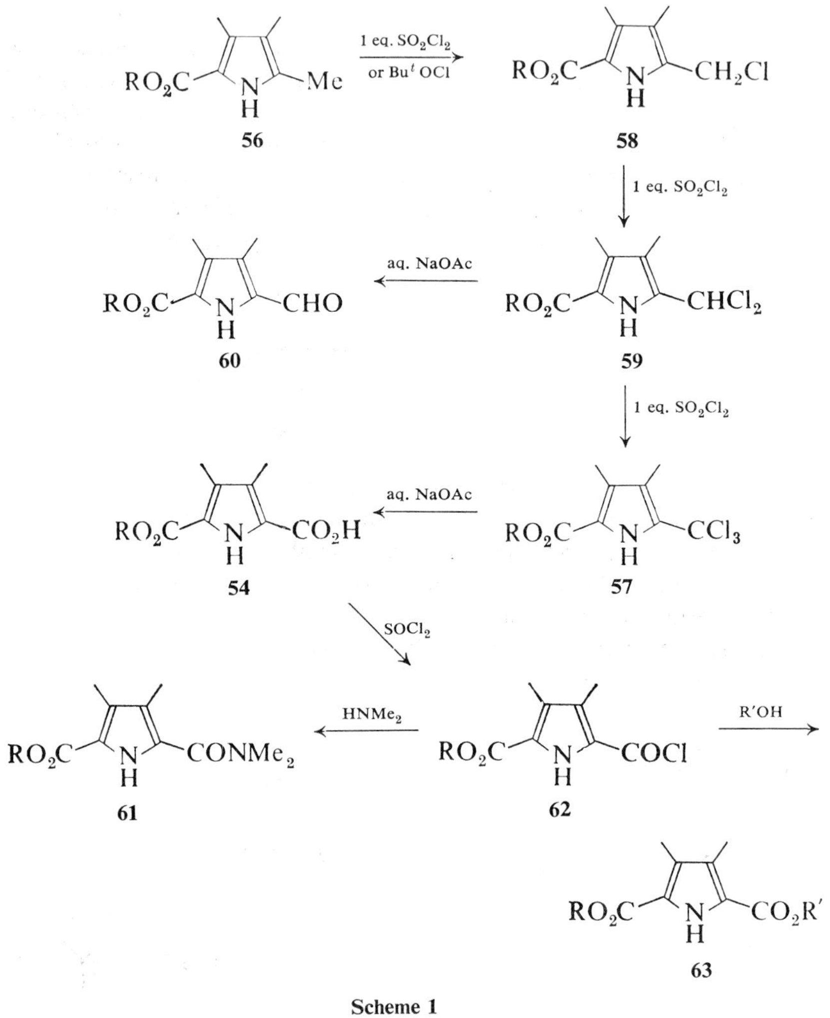

RO_2C–pyrrole(NH)–Me (**56**) —[1 eq. SO_2Cl_2 or Bu^tOCl]→ RO_2C–pyrrole(NH)–CH_2Cl (**58**)

58 —[1 eq. SO_2Cl_2]→ RO_2C–pyrrole(NH)–$CHCl_2$ (**59**)

59 —[aq. NaOAc]→ RO_2C–pyrrole(NH)–CHO (**60**)

59 —[1 eq. SO_2Cl_2]→ RO_2C–pyrrole(NH)–CCl_3 (**57**)

57 —[aq. NaOAc]→ RO_2C–pyrrole(NH)–CO_2H (**54**)

54 —[$SOCl_2$]→ RO_2C–pyrrole(NH)–COCl (**62**)

62 —[$HNMe_2$]→ RO_2C–pyrrole(NH)–$CONMe_2$ (**61**)

62 —[R'OH]→ RO_2C–pyrrole(NH)–CO_2R' (**63**)

Scheme 1

and **59**; the latter can also be hydrolyzed to the formylpyrrole **60**. *NN*-Dimethyl carboxamidopyrroles (**61**) have found much application in the synthesis of pyrroketones and are obtained[39] either by treatment of α-trichloromethylpyrroles (**57**) with dimethylamine followed by hydrolysis or else by treatment of pyrrole α-acid chlorides (**62**) [obtained from the corresponding α-carboxylic acid (**54**) with thionyl chloride] with dimethylamine. If the acid chloride **62** is treated with an alcohol R^1OH in presence of a base (e.g., *N*,*N*-dimethylaniline), the corresponding 2,5-dicarboxylate (**63**) results.

5. DIPYRROLIC COMPOUNDS

A. Pyrromethanes (Dipyrrylmethanes)

These compounds (**5a**), until fairly recent times, were thought to be unstable and not suitable as intermediates in porphyrin synthesis for a variety of reasons. However, a highly successful route to porphyrins devised by MacDonald[29] has shown that the supposed limitations of pyrromethanes are not as serious as had been suggested, for example, by Fischer's almost total reliance on pyrromethenes.[2]

Symmetrically substituted pyrromethanes (**64**) can be synthesized in several ways. Treatment of a 2-unsubstituted pyrrole (**65**) with formaldehyde furnishes pyrromethanes (**64**) in good yield. The same product can be obtained in even higher yield by refluxing the corresponding 2-bromomethylpyrrole **66a** in alcoholic solvents for short periods; a recent modification, due to

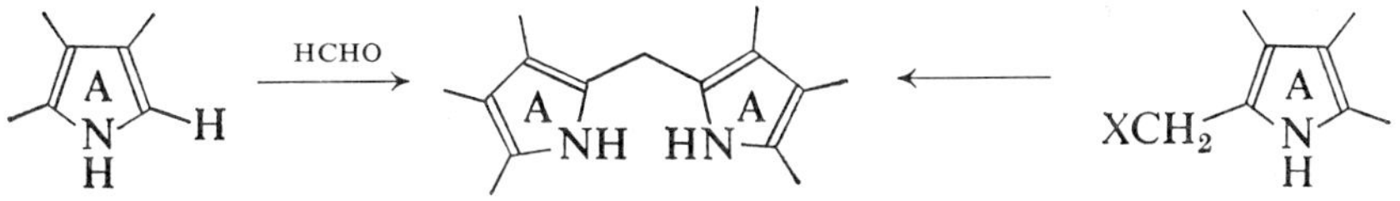

HCHO

65 **64**

66a X = Br
66b X = OAc

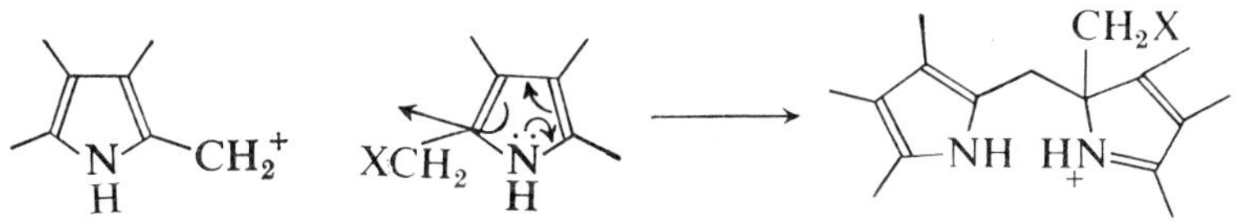

67

66a X = Br
66b X = OAc

Russian workers,[40] utilizes the heating of 2-acetoxymethylpyrroles (**66b**) in hydrochloric acid/methanol and can furnish the symmetrical pyrromethane **64** in yields as high as 90%. The mechanism of this type of self-condensation presumably involves initial formation of the carbonium ion **67**, which is attacked by a further molecule of the pyrrole **66**. Elimination of the one carbon fragment (probably as formaldehyde) then furnishes the required product.

Unsymmetrically substituted pyrromethanes (e.g., **70**) are frequently required in porphyrin synthesis, and compounds of this type pose a considerably more complex problem than their symmetrical counterparts. An early solution[41] to the problem employed the treatment of a pyrrole Grignard reagent (**68**) with a pyrrole (**69**) bearing a potential carbonium ion. Pyrrole

A N H MgBr

68

EtMgBr + XH_2C B N H → A NH HN B

69 **70**

X = Cl, Br, OMe, OAc

A N H H

65

Grignard reagents are easily obtained by treatment of the corresponding pyrrole **65** with ethyl magnesium bromide. Perhaps the most successful avenue to unsymmetrical pyrromethanes is that developed by MacDonald and co-workers.[30] Thus brief heating of 2-unsubstituted pyrroles (**65**) with 2-substituted pyrroles (**69**) in refluxing glacial acetic acid buffered with sodium acetate gives good yields of the corresponding unsymmetrically substituted pyrromethanes (**70**). When the bromomethylpyrrole (**66a**) is used as the electrophilic species, it seems likely that this is first converted, in the buffered acetic acid, to the corresponding 2-acetoxymethylpyrrole (**66b**), quantities of which are often recovered from the reaction. Even when the 2-unsubstituted pyrrole **65** bears an electron-withdrawing function (e.g., CO_2R), it is still sufficiently nucleophilic to react to give pyrromethane. The reaction conditions are not so drastic as to cleave *t*-butyl esters, which gives an added degree of selectivity.

In an alternative approach to unsymmetrical pyrromethanes (**70**), the pyridinium salt **71** of a bromomethyl pyrrole (**66a**) and the lithium salt **72** of a pyrrole α-carboxylic acid are heated in polar solvents (such as methanol/water or formamide) and good yields of the appropriate unsymmetrically

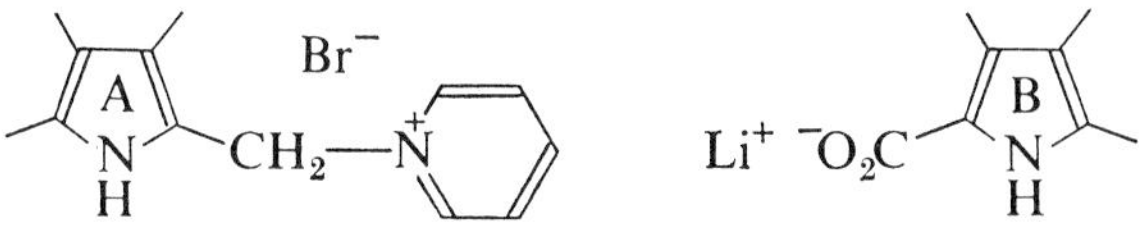

substituted pyrromethane (**70**) are obtained.[15,42] A major limitation to this method is the low nucleophilicity of pyrroles (**72**) having strongly electron-withdrawing substituents.

Borohydride reduction of pyrromethenes has found some application, but this is limited by reactions of some substituents with sodium borohydride, and also by the range of synthetically available pyrromethenes.

B. Pyrromethenes (Dipyrrylmethenes)

Pyrromethenes are highly colored compounds (free base **73**, λ_{max} 450 nm; salt **74**, λ_{max}, 500 nm) which have found much use in the synthesis of

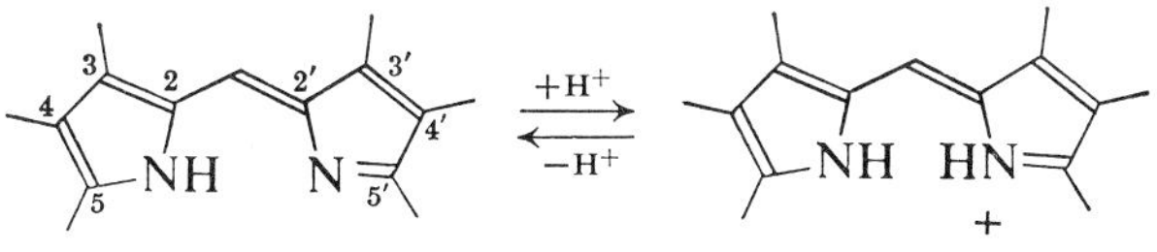

porphyrins.[2] They are particularly difficult to purify, probably because of their facile protonation and deprotonation; chromatography rarely produces any improvement in purity, and it is therefore a prerequisite that any synthesis must be highly efficient to enable convenient isolation of the pyrromethene. The classical, and still most favored method for pyrromethene synthesis is that due to Fischer,[2] i.e. acid-catalyzed condensation of a 2-unsubstituted pyrrole (**65**) with a 2-formylpyrrole (**75**), which gives excellent yields of the corresponding unsymmetrically substituted pyrromethene salt (**74**), and phosphoryl chloride has been shown to be a satisfactory reagent for this condensation in certain cases.[43]

In some cases, the initially formed pyrromethene (**74**) may undergo further reaction with the 2-unsubstituted pyrrole to form a tripyrrylmethane (**76**), disproportionation of which may occur in a different way to give a symmetrical pyrromethene (**64**) rather than the anticipated unsymmetrical product (**74**). However, this rarely occurs, and the high degree of success with this general method has deterred further developments to find alternative

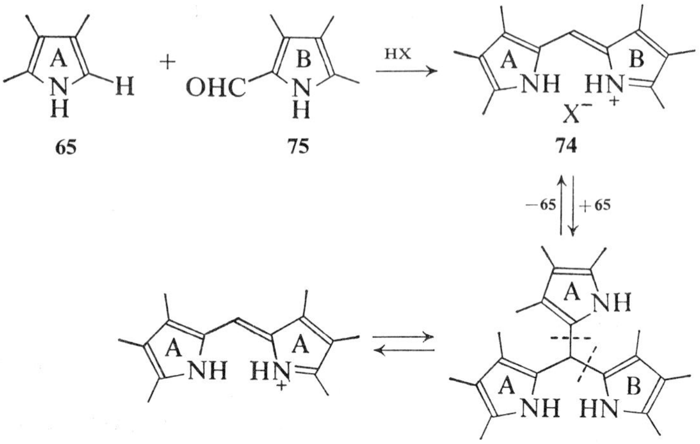

65 75 74

64 76

routes to pyrromethenes. The treatment of α-unsubstituted pyrroles with formic acid and hydrobromic acid has found application, but this only provides a route to symmetrical pyrromethenes. The controlled oxidation (e.g., with bromine) of pyrromethanes provides a further, but limited, route to pyrromethenes.

C. Pyrroketones (2,2′-Dipyrrylketones)

These compounds have found much use in the synthesis of porphyrins and oxophlorins.[15] They are bisvinylogs of amides and do not show any normal

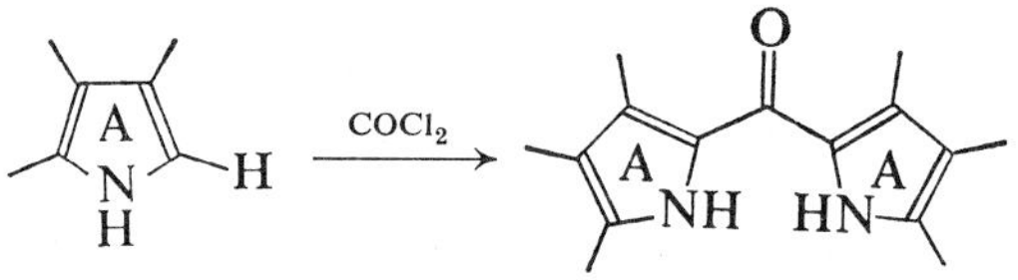

65 77

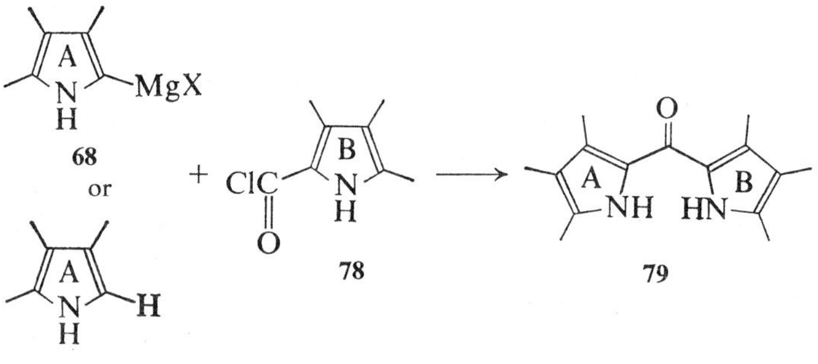

68 or 65 78 79

ketonic behavior (e.g., toward phenylhydrazine or toward reduction with borohydride). They can, however, be reduced[39] to the corresponding pyrromethanes with diborane; this is an important reaction in their adaptation as intermediates in porphyrin synthesis.

Symmetrical pyrroketones are available by treatment of an α-unsubstituted pyrrole (**65**) with phosgene, whereas unsymmetrical ones are available from the same nucleophile (**65**) or its Grignard derivative (**68**) with a 2-chlorocarbonyl pyrrole (**78**). A further method of limited applicability is the oxidation of pyrromethanes with lead dioxide and lead tetraacetate.[44] A reexamination of this method has been reported and it has been found that the same transformation can be brought about more efficiently using bromine followed by sulfuryl chloride, or in some cases by sulfuryl chloride alone.[45]

A very satisfactory general method for the preparation of pyrroketones based on the Vilsmeier-Haack procedure has been reported.[39] Thus treatment of 2-*N*,*N*-dimethylamidopyrroles (**80**) with phosphoryl chloride furnishes the

CONMe$_2$ N H **80** $POCl_3$ N H C=$\overset{+}{N}Me_2$ $OPOCl_2$ Cl^- **81** H N H **82**

Cl^-

O NH HN **79** H_2O/OAc^- Cl^- $\overset{+}{N}Me_2$ NH HN **83**

corresponding complexes (**81**), which are strong electrophiles and react with 2-unsubstituted pyrroles (**82**) to give imine salts (**83**). Hydrolysis gives pyrroketones (**79**) in overall yields often in excess of 80%.

6. PORPHYRINS FROM MONO- AND DIPYRROLIC PRECURSORS

A. From Pyrroles

Polymerization of 2,5-unsubstituted pyrroles (**84**) with formic acid,[46] aldehydes,[18] etc., or self-condensation of suitable substituted pyrroles (**85**), followed by aerial oxidation often affords porphyrins in quite good yields,[22,47–56] but unless the β-substituents in the pyrroles **84** or **85** are identical, a "random" mixture of the four possible porphyrin isomers is always obtained.[55,57,58] These methods are therefore clearly of very limited usefulness for the preparation of the unsymmetrical naturally occurring porphyrins, although partial separation of isomeric mixtures such as the coproporphyrins has been achieved by chromatographic methods.[57,59] However, the mechanism of self-condensation of pyrroles of type **85** is of great interest in relation to the enzymically controlled polymerization of

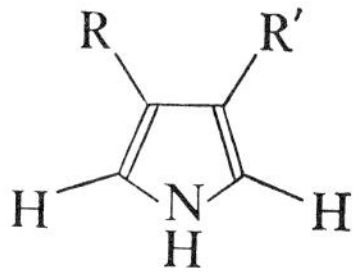

R R′ H N H H

84

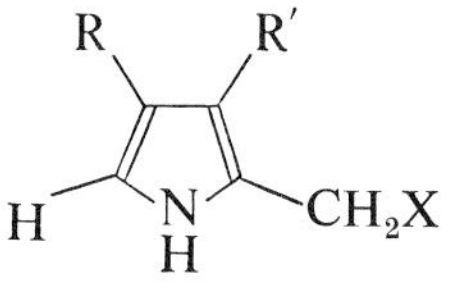

R R′ H N H CH_2X

85

X = Cl, Br, OAc, NH_2, etc.

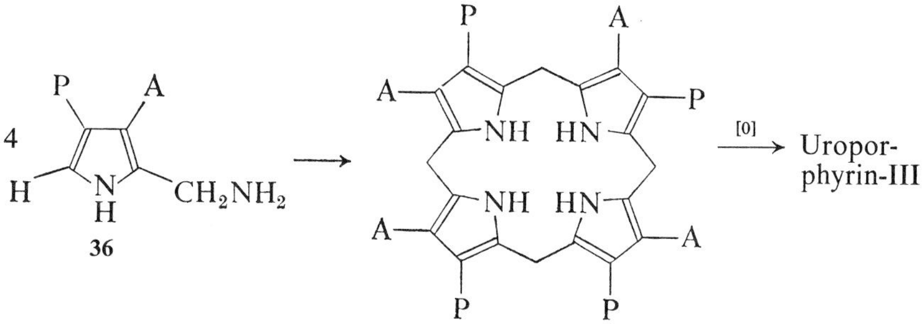

4 P A H N H CH_2NH_2 **36** ⟶ (P A A P NH HN NH HN A A P P) $\xrightarrow{[O]}$ Uropor-phyrin-III

86

Uroporphyrinogen-III

A = CH_2CO_2H

P = $CH_2CH_2CO_2H$

porphobilinogen (**36**), which normally leads to uroporphyrinogen-III (**86**) (although the centrosymmetrical uroporphyrinogen-I is occasionally observed in rare pathological conditions). Many hypotheses have been advanced to account for the almost ubiquitous occurrence of type III porphyrins in nature (of which uroporphyrinogen-III is the precursor), but no experimental evidence to show how one ring is reversed relative to the other three has yet been provided.[7,22,34,58,60–64]

B. From Pyrromethanes

One of the earliest reported porphyrin syntheses[65] from pyrromethanes involved the condensation of the pyrromethane **87a** with formic acid in presence of air to give etioporphyrin-II (**88a**); more recent applications

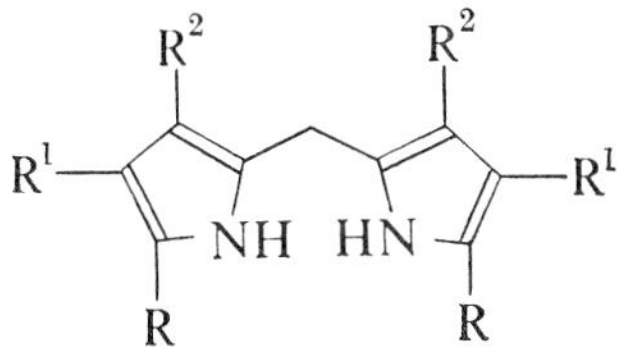

87

R = H or CO_2H

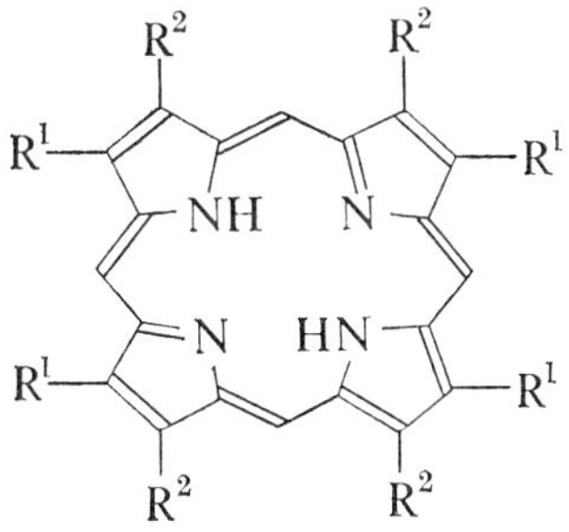

88

a R^1 = Me; R^2 = Et
b R^1 = Me; R^2 = P
c R^1 = A; R^2 = P

include the synthesis of copro- and uroporphyrins-II (**88b** and **88c**).[29,66] However, mixtures of porphyrins are sometimes obtained because under the acidic conditions of the reaction, cleavage and recondensation reactions[57,67] may occur at the methane bridges, presumably by the mechanism outlined in Scheme 2. A milder modification developed in Corwin's laboratory[68] was the

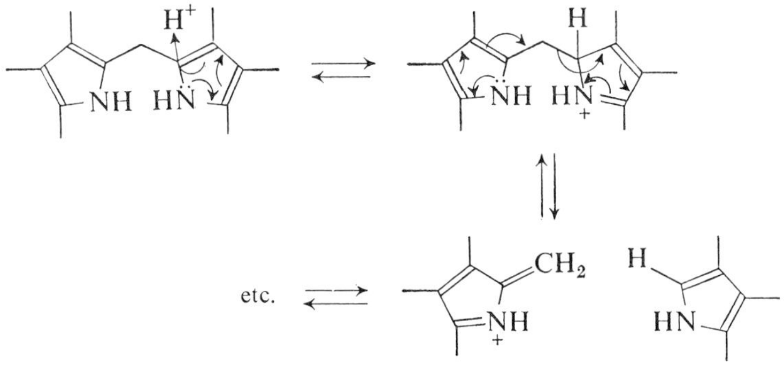

etc.

Scheme 2

use of the *N*-methyl pyrrole aldehyde **89** instead of formic acid to provide the bridge carbon atoms presumably via tripyrrylmethane-type intermediates (**90**). However, both procedures are clearly limited to the preparation of symmetrical porphyrins.

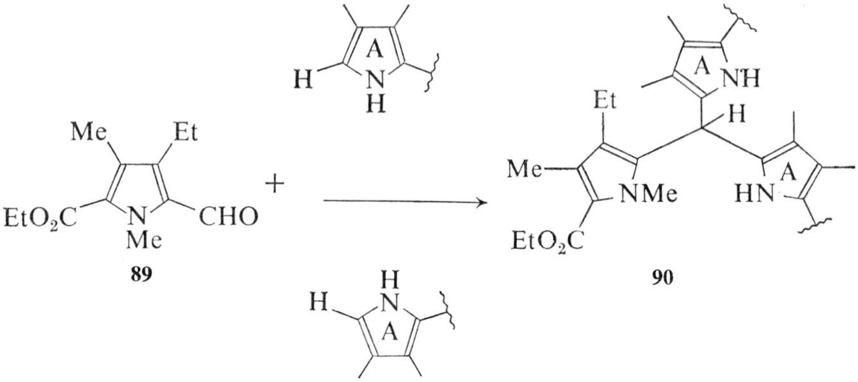

89 **90**

A much more versatile method introduced by MacDonald and his colleagues[29] in Canada is the mild acid-catalyzed condensation of 5,5′-diformylpyrromethanes (**91**) with 5,5′-unsubstituted pyrromethanes (**92**). The best

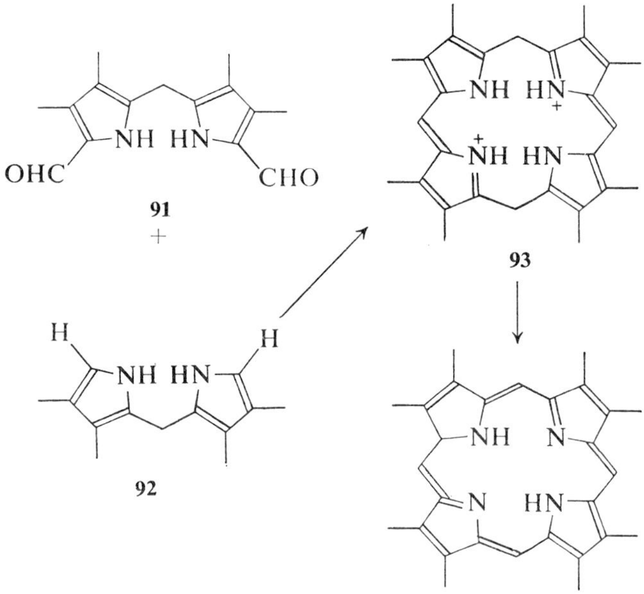

91 + **92** **93** **94**

conditions for the reaction involve the use of a very dilute solution of hydriodic acid in acetic acid and the intermediate porphodimethene **93** formed is oxidized to porphyrin **94** by aeration in a buffered solution. In the preparation of uroporphyrins-II, III, and IV, yields of up to 65% of porphyrin were obtained,[29] but other groups[69,70] have since applied the method in syntheses of a rhodoporphyrin,[16] coproporphyrin-II, and etioporphyrin-II, among others. Preliminary investigations[71] of a route to porphyrin-*a*, which were aimed at applying the MacDonald method, have also been reported. However, the MacDonald method still suffers from the inherent limitation that one of the two pyrromethanes must be symmetrical, or otherwise two porphyrins could be formed. This limitation has been overcome in one notable example, the Woodward synthesis of chlorin-e_6[72] (constituting a formal total synthesis of chlorophyll-*a*); in the crucial stage two unsymmetrical pyrromethanes are coupled together in a unique manner, by means of an intermediate Schiff's base.

MacDonald has also condensed a 5,5′-di-unsubstituted pyrromethane (**95**) with a 5,5′-di(bromomethyl)pyrromethene (**96**) to give a 33% yield of porphyrin, isolated as the hexamethyl ester **97**.[29] No other porphyrins were

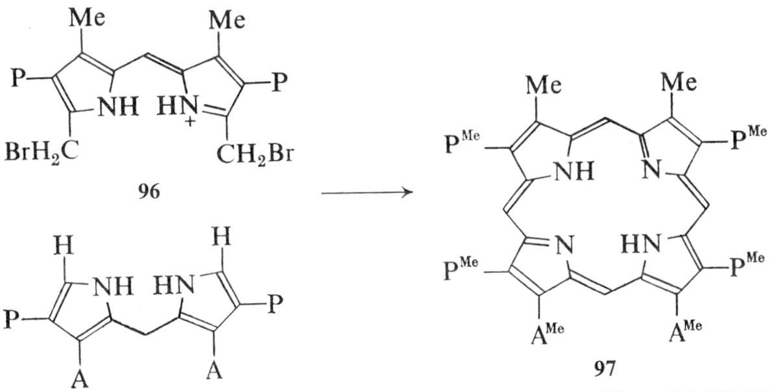

Me Me P P NH HN+ BrH_2C CH_2Br

96

H H NH HN P P A A

95

Me Me P^{Me} P^{Me} NH N N HN P^{Me} P^{Me} A^{Me} A^{Me}

97

$A^{Me} = CH_2CO_2Me$; $P^{Me} = CH_2CH_2CO_2Me$

formed in this reaction, thus further demonstrating the utility of pyrromethanes in porphyrin synthesis.

C. From Pyrromethenes

The earliest examples of the classical Fischer synthesis[73,74] of porphyrins involved the self-condensation of 5-bromo-5′-methylpyrromethenes (**98a**) in a melt of succinic or tartaric acids at temperatures in the range 160–200°. In this way Fischer obtained good yields of symmetrical porphyrins such as etioporphyrin-I (**99a**) and coproporphyrin-I (**99b**), even better yields could be obtained if the corresponding 5′-bromomethylpyrromethenes (**98b**) were

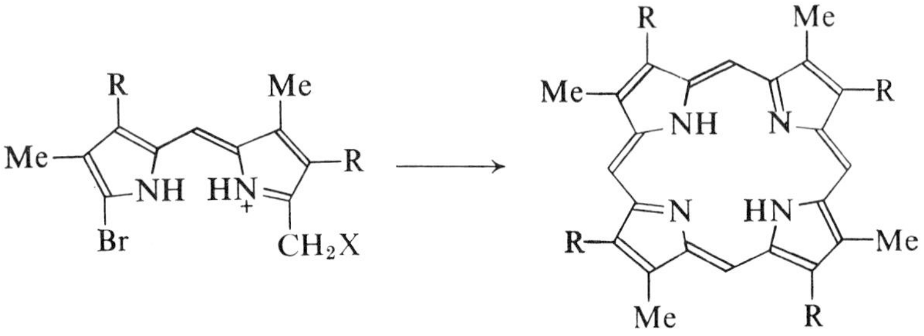

98a X = H
98b X = Br

99a R = Et
99b R = P

heated at 100° in formic acid.[2] This type of synthesis is largely limited to centrosymmetrical porphyrins because condensation of two different pyrromethenes can lead to three different porphyrins, two by self-condensation and a third by cross-condensation. However, in many cases, the porphyrin arising from cross-condensation has been successfully separated from the mixture.[75]

A more general approach[65b] subsequently developed by the Munich School involved the condensation of 5,5′-dibromopyrromethenes (**100**) with 5,5′-dimethyl- or 5,5′-di(bromomethyl)pyrromethenes (**101**) in an organic acid melt. Fischer's synthesis of deuteroporphyrin-IX (**102**), a crucial intermediate in his synthesis of protoporphyrin-IX (**103**) and hemin, is a classic example

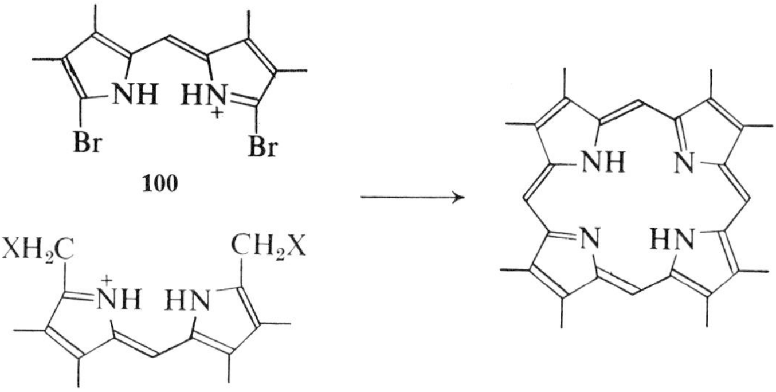

100

101a X = H
101b X = Br

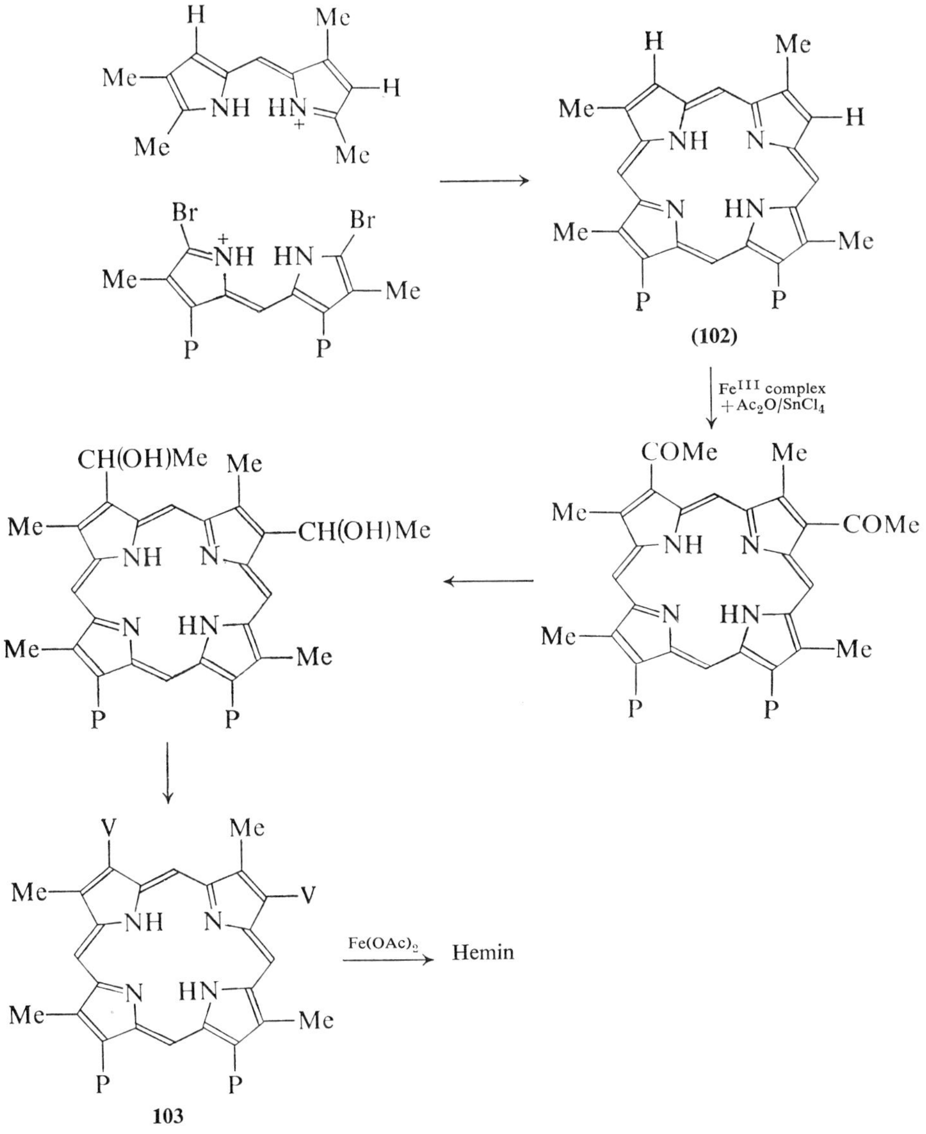

Me, H, Me, H, Me, Me, NH, HN, +

Br, Br, Me, Me, P, P, NH, HN, +

(102)

Fe^{III} complex $+Ac_2O/SnCl_4$

COMe, Me, Me, COMe, Me, Me, P, P

CH(OH)Me, Me, Me, CH(OH)Me, Me, Me, P, P

V, Me, Me, V, Me, Me, P, P

103

$\xrightarrow{Fe(OAc)_2}$ Hemin

Scheme 3

104; 7%

+ Etioporphyrin
R = H, Br

Meso-V (main product)

106a R = R′ = Et
106b R = V; R′ = Et
106c R = R′ = V.
(**106a** yield ≪ 1%)

+ trace of another porphyrin

(Scheme 3).[2,76] Acetylation of the deuteroporphyrin-IX was carried out on the iron complex, because Friedel-Crafts type reactions are unsuccessful with free porphyrins, presumably owing to formation of dication species.

Apart from symmetry considerations, other limitations to the usefulness of the pyrromethene fusion method for porphyrin synthesis are that labile substituents may not survive the drastic experimental conditions, and, moreover, yields may be extremely low. However, in spite of all these difficulties, the Fischer method has been of immense utility in the synthesis of a wide variety of different porphyrins; for example, in the work leading up to the determination of the structure and synthesis of hemin,[76] Fischer synthesized the four etioporphyrins, and 12 of the 15 isomeric mesoporphyrins.[2] The total synthesis of many of the degradation products of chlorophylls *a* and *b*, such as pyrroporphyrin-XV[77] (**104**), phylloporphyrin-XV[78] (**105**), rhodoporphyrin-XV[79] (**106**) (Scheme 4) enabled the Munich School to deduce their structures almost completely.

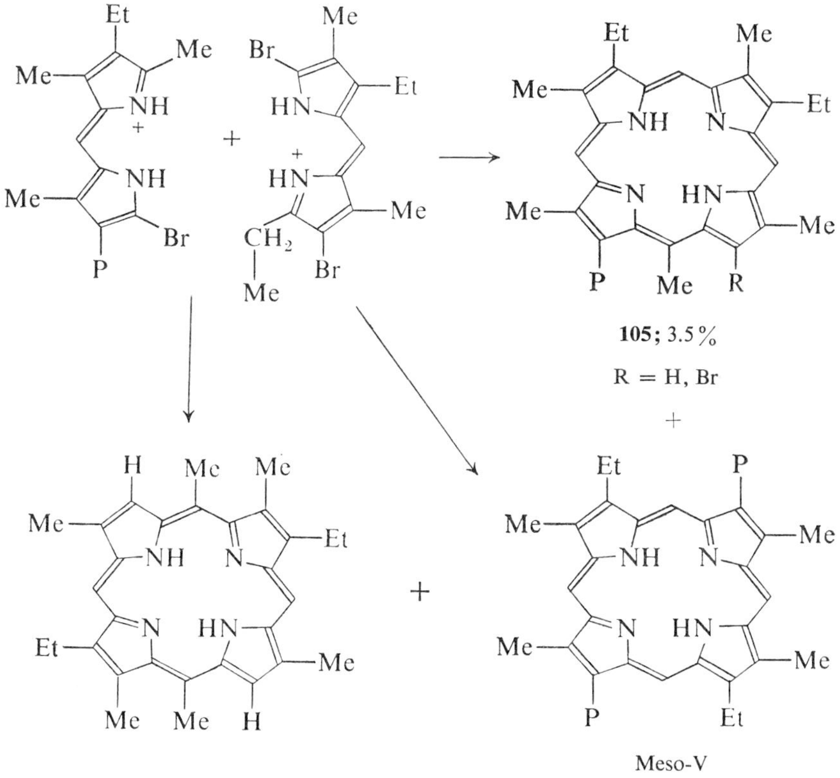

105; 3.5%

R = H, Br

Meso-V

Scheme 4

In recent years perhaps the most notable applications of the Fischer method have been the achievements of MacDonald's group in Ottawa. Their syntheses of all four uroporphyrins[27,28,72] and of the corresponding coproporphyrins[80] showed definitively that the naturally occurring isomers were exclusively of types I and III. The identity of the compounds produced by the Fischer method, with those they also obtained from pyrromethanes, and with the natural products was rigorously confirmed by m.p., X-ray, and chromotographic comparisons. (As MacDonald has pointed out, m.p. or X-ray comparisons are not always sufficiently reliable by themselves for confirmation of identity owing to the polymorphic behavior of many porphyrins;[80,81] however, it must also be said that the massive edifice of porphyrin structural chemistry erected by Fischer on m.p. comparisons has well withstood the test of time and the impact of the more sophisticated spectroscopic and chromatographic techniques now available.)

Other important applications of the Fischer method by the Canadian group include the synthesis of a number of degradation products[82] of heme-*a*, and of the *Chlorobium* chlorophylls (650) and (660) which have considerably helped in structural studies.[81] For example, resorcinol fusion of heme-*a* affords a "*des*-methyl deuteroporphyrin" and MacDonald and his colleagues therefore synthesized all four possible des-methyl analogs of deuteroporphyrin-IX. The 8-*des*-methyl compound **107**, synthesized as shown, was identified with

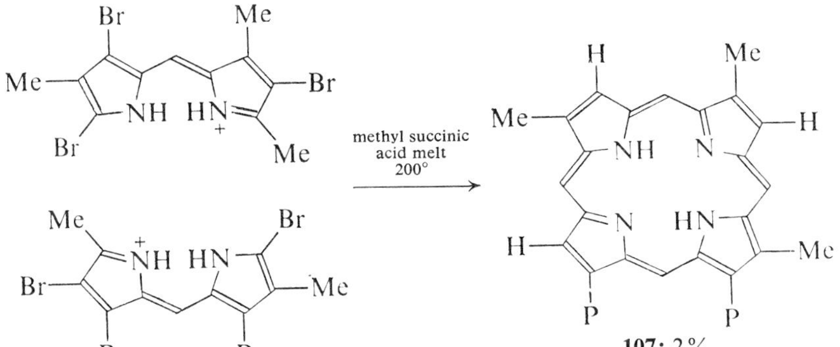

methyl succinic acid melt 200°

107; 2%

+ other porphyrins

the resorcinol fusion product, and this provided good evidence for the location of the three labile groups in porphyrin-*a*, whose currently disputed structures (**108**) are shown here.

The structures of the *Chlorobium* chlorophylls (650) (**109**) have been established beyond reasonable doubt by degradative methods[83] and by MacDonald's synthesis of the derived pyrroporphyrins (**111**) from pyrromethanes.[81]

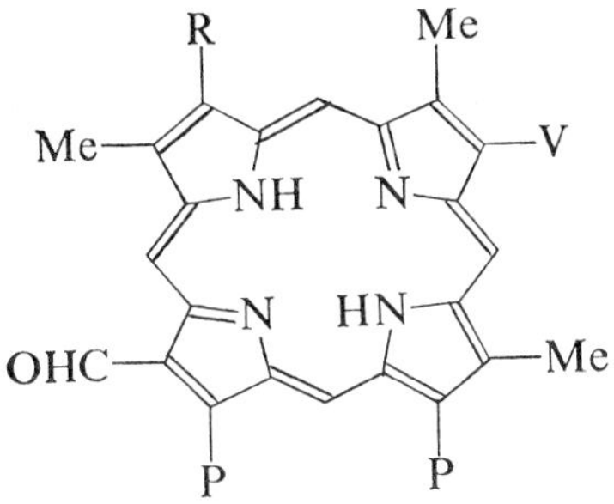

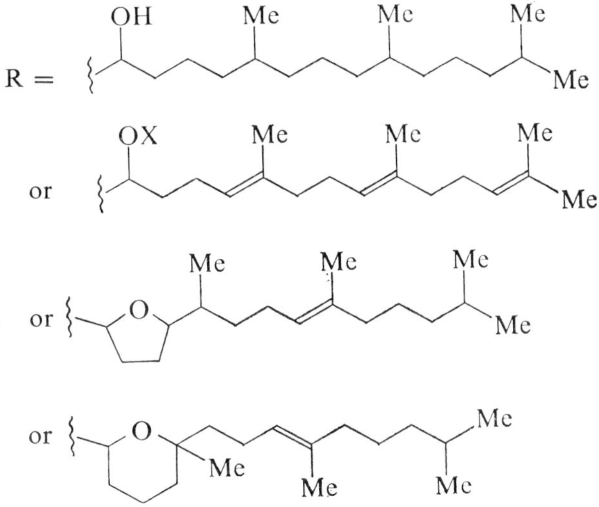

108

However, there is much less certainty about the *Chlorobium* chlorophylls (660). The currently accepted structures[84] (**110**) are based on degradation to maleimides and to the corresponding phylloporphyrins (**112**) as well as mass

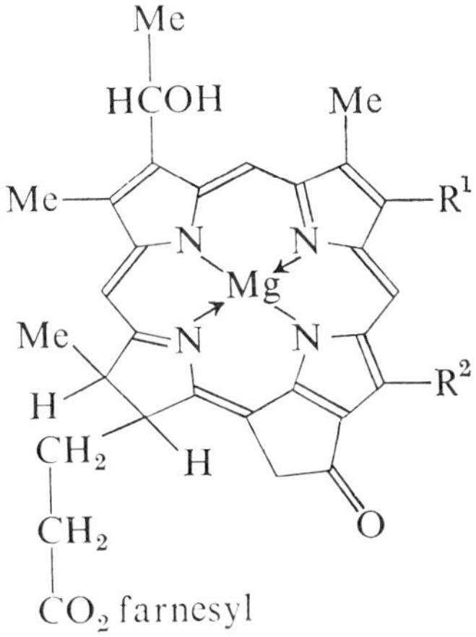

109

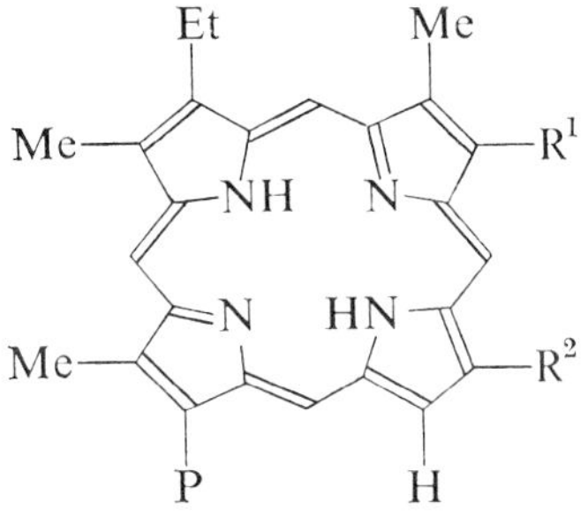

111

Fraction	R^1	R^2
1	Bu^i	Et
2	Pr^n	Et
3	Bu^i	Me
4	Et	Et
5	Pr^n	Me
6	Et	Me

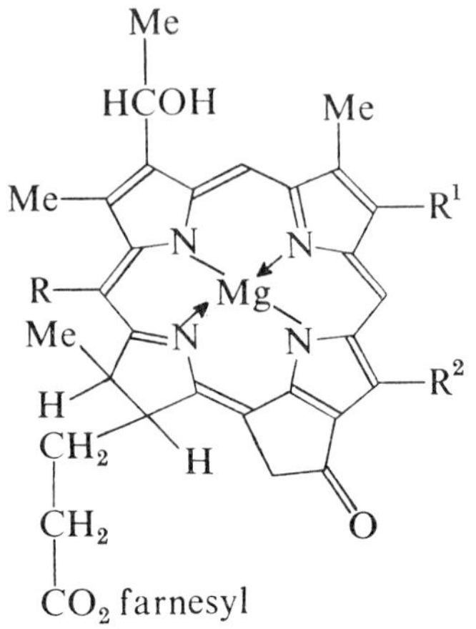

110

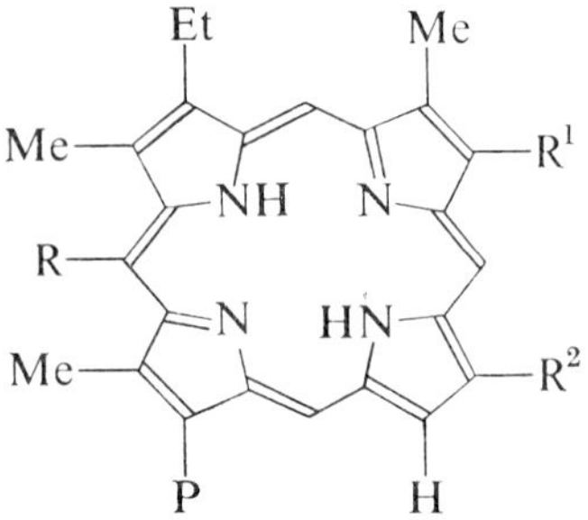

112

Fraction	R^1	R^2	R
1	Bu^i	Et	Et
2	Bu^i	Et	Me
3	Pr^n	Et	Et
4	Pr^n	Et	Me
5	Et	Et	Me
6	Et	Me	Me

spectral determinations.[85] MacDonald's group have since synthesized[81] the phylloporphyrins from fractions 5 and 6 by the pyrromethene method; unfortunately, a similar synthesis of the phylloporphyrin from fraction 3 gave very low yields and comparisons with naturally derived material were inconclusive. There is still some doubt about the location* and nature of the *meso*-substituents, for Mathewson, Richards, and Rapoport[86] have reported NMR evidence favoring the α-position (rather than the δ-); the Liverpool group's results have also cast some doubt on the existence of any *meso*-ethylated *Chlorobium* chlorophylls.

D. From Pyrroketones

Until recently pyrroketones were little more than chemical curiosities, but since 1966 their tetrapyrrolic analogs, the *a*- and *b*-oxobilanes,[15] have found

* Reductive *C*-methylation (cf. **25** and **26**) has recently confirmed the siting of the *meso*-alkyl groups at the γ-positions (S. F. MacDonald, private communication).

extensive use in the synthesis of porphyrins and oxophlorins. However, the direct synthesis of oxophlorins from pyrroketones has also attracted attention recently and Clezy's group in Australia[88–94] prepared a number of oxophlorins (**115**) and acetoxyporphyrins (**116**) from 5,5′-diformylpyrroketones (**113**) and 5,5′-unsubstituted pyrromethanes (**114a**) or their dicarboxylic acid analogs (**114b**).

O
NH HN
OHC CHO
113

R R
NH HN
114a R = H
114b R = CO_2H

H^+, air →

O
NH HN
N HN
115

↓

OAc
NH N
N HN
116

meso-Unsubstituted porphyrin ← 1) H_2/Pd-C 2) air, I_2, or DDQ

The diformyl pyrroketones are usually prepared by direct oxidation of the corresponding 5,5′-dimethylpyrroketones with lead tetracetate.[88,91,95] The cyclization is of course analogous to the MacDonald method with pyrromethanes, but the formyl groups which are to form the bridging carbon atoms must be situated in the pyrroketone moiety, because the *oxo*-function in a pyrroketone deactivates the 5- and 5′-positions toward electrophilic attack.[39]

The primary products of the reaction are oxophlorins (**115**) (oxyporphyrins), but these are usually converted directly to the corresponding *meso*-acetoxyporphyrins (**116**); it has been shown by other workers that the acetoxy substituent may be removed by hydrogenolysis and the porphyrinogen formed can be reoxidized to porphyrin.[96] Like the analogous MacDonald synthesis from pyrromethanes, the Clezy procedure is, however, somewhat limited in scope for preparing porphyrins because one of the two components **113** and **114** must be symmetrical, or two isomeric oxophlorins will be formed. On the other hand, it represents a relatively rapid approach to macrocycle synthesis, and has considerable potential, for a wide variety of oxophlorins have now been synthesized including a number with electronegative substituents (e.g., bromine,[92] acetyl,[93,94] and ethoxycarbonyl[93]). However, perhaps Clezy's most noteworthy achievement has been his recent synthesis[94] of α-benzyloxy- and α-acetoxy–protoporphyin-IX dimethyl esters (**119a** and **119b**, respectively); 4-acetyl–α-hydroxydeuteroporphyrin-IX* (**117**) was first prepared from a diformyl monobromomonoacetyl pyrroketone, and then acetylation of the acetoxy derivative with acetic anhydride (or ether/acetyl chloride) in presence of stannic bromide gave the 2,4-diacetyl–α-acetoxy deuteroporphyrin-IX **118**. Hydrolysis to the corresponding oxophlorin followed by borohydride reduction afforded the *bis*(hydroxyethyl) oxophlorin, which was dehydrated and benzoylated by treatment with benzoylchloride to give the desired α-benzoyloxyporphyrin **119a**. The corresponding α-acetoxy porphyrin **119b** was obtained in a similar manner; however, attempts to hydrolyze either compound were fraught with problems, but passage through

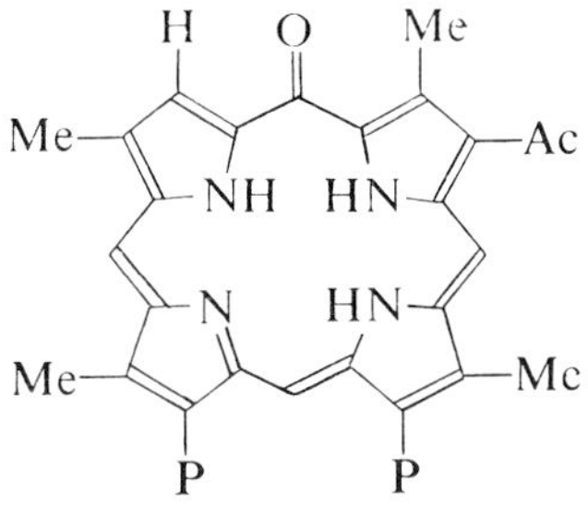

117

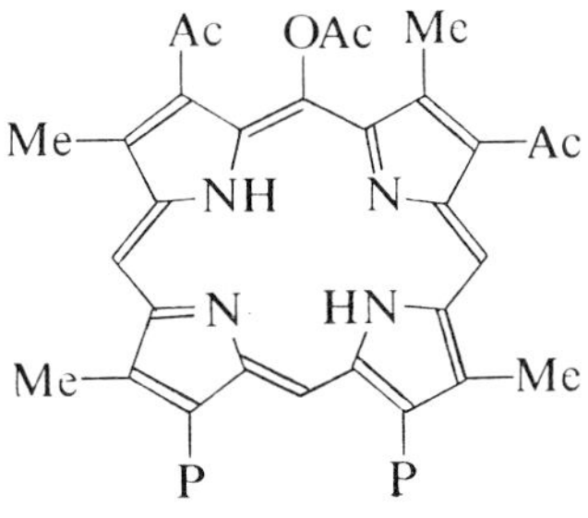

118

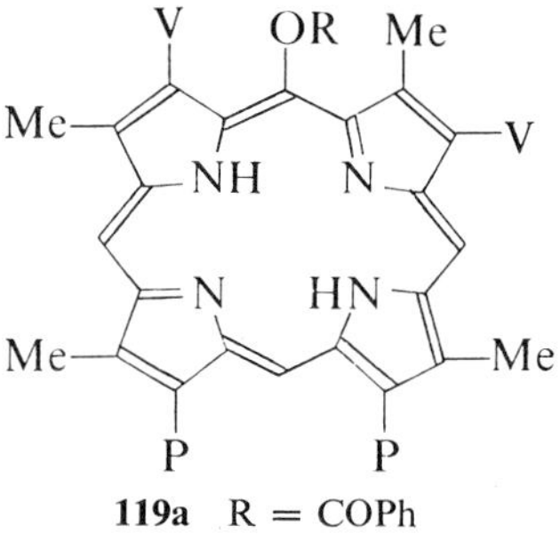

119a R = COPh
119b R = Ac

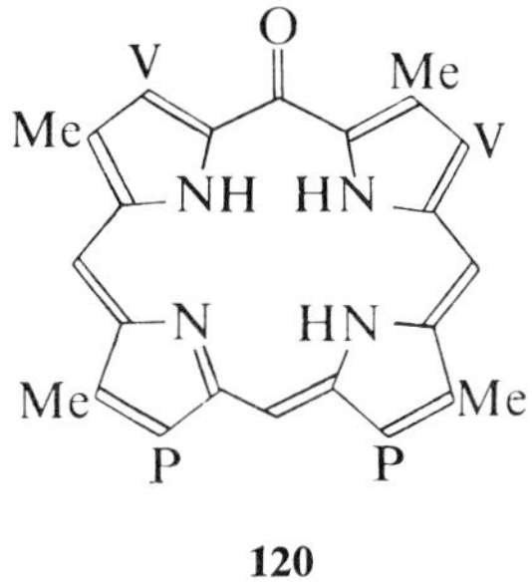

120

* For systematic nomenclature purposes it is convenient to name and number oxophlorins as their hydroxyporphyrin tautomers.

a basic alumina column gave a solution of the unstable blue-green "α-oxyprotoporphyrin-IX" **120**. Owing to shortage of material, **120** was not fully characterized, although Clezy and Liepa[94] were able to demonstrate the conversion of its ferric complex by aerial oxidation and subsequent hydrolysis to a blue product which was identical with biliverdin dimethylester in its visible spectroscopic and TLC behavior. This result is of great interest in relation to the probable metabolic route from heme to bile pigments.

7. PORPHYRIN SYNTHESIS FROM OPEN-CHAIN TETRAPYRROLIC INTERMEDIATES

A. From *a,c*-Biladienes

In one variant of the pyrromethane fusion method, Fischer and Schormüller[97] condensed the 5-bromomethylpyrromethene **121** with the 5-unsubstituted pyrromethene **122** to give a mixture of etioporphyrin-I **124** and pyrroporphyrin-XVIII **125**. These were easily separated because **125** has a side-chain

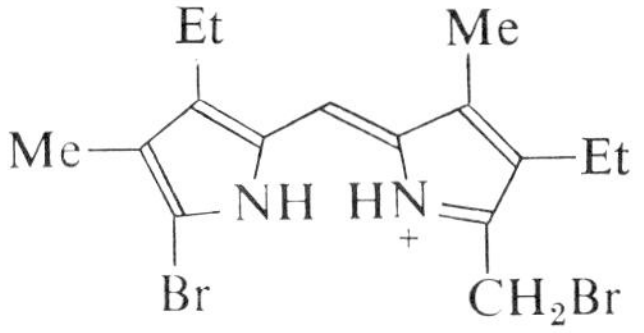

Et Me Me Et NH HN + Br CH_2Br

121

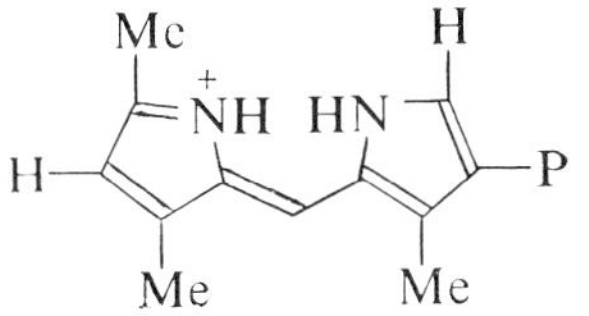

Me H + NH HN H P Me Me

122

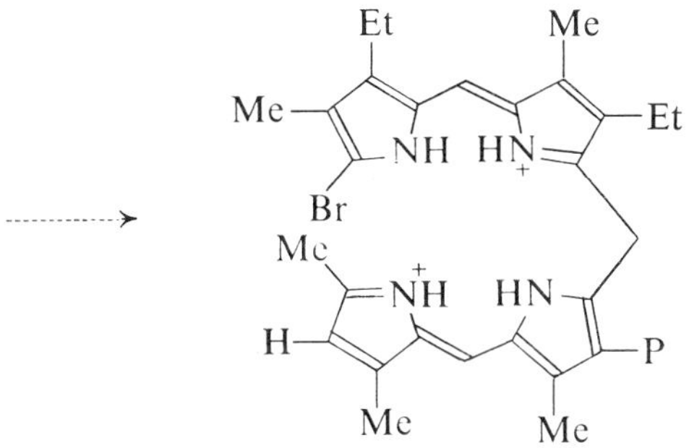

Et Me Me Et NH HN + Br Me + NH HN H P Me Me

123

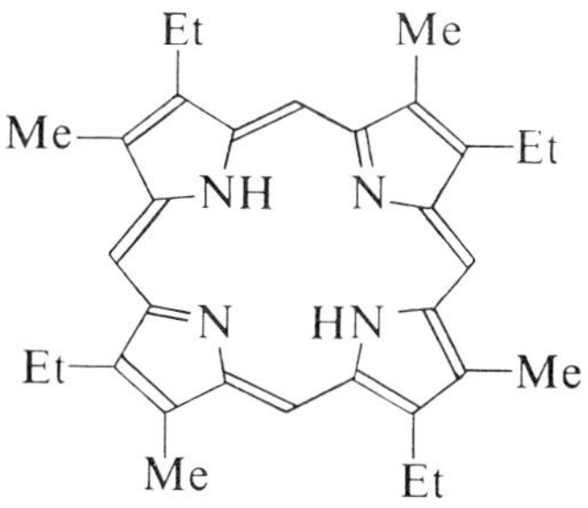

Et Me Me Et NH N N HN Et Me Me Et

124

Et Me Me Et NH N N HN H P Me Me

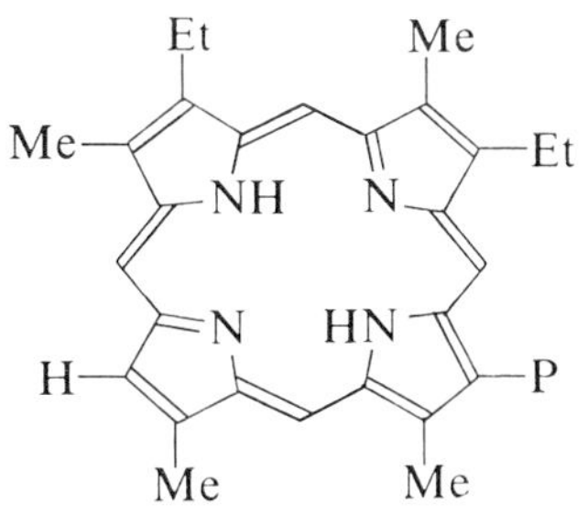

125

carboxylic acid group, and moreover a third porphyrin is not formed in this case because the pyrromethene **122** does not carry a 5-bromosubstituent. The intermediate tetrapyrrole formed *en route* to pyrroporphyrin-XVIII must presumably be the *a*,*c*-biladiene **123**, and a logical extension of the Fischer synthesis was clearly to prepare *a*,*c*-biladienes of this type under mild conditions and then cyclize to porphyrin in a second step. This has now been achieved by members of the Nottingham school,[14] who showed that 5-bromo-5′-bromomethylpyrromethenes (**126**) could be coupled with a 5-unsubstituted-5′-methylpyrromethene (**127**) at room temperature in presence of stannic chloride. Treatment of the reaction product with methanolic hydrogen bromide gave the relatively insoluble *a*,*c*-biladiene hydrobromides (**128**), which were then usually cyclized directly to porphyrins (**129**) without further purification by heating in boiling *o*-dichlorobenzene for a few minutes. Some

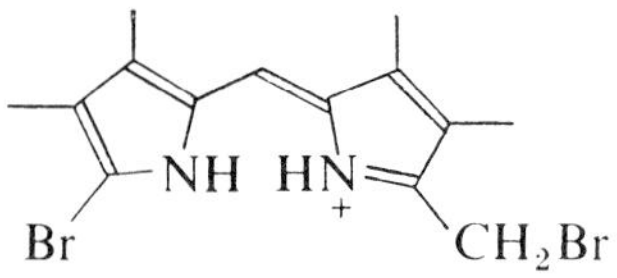

126

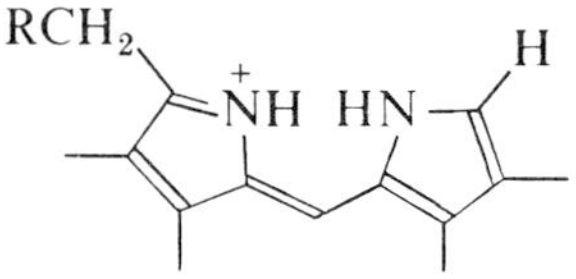

1) $SnCl_4$
2) HBr

127

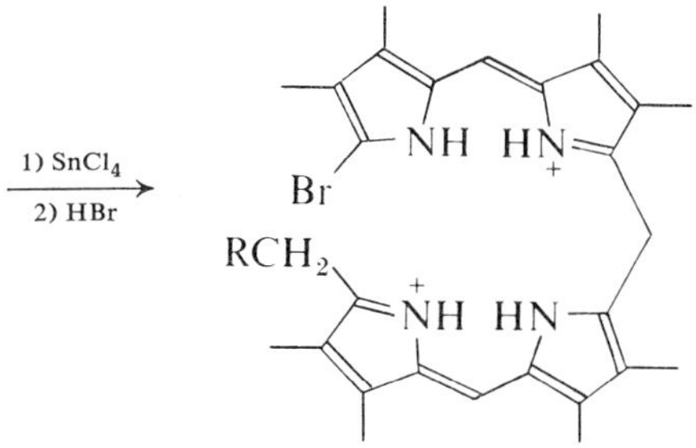

128

hot o-$C_6H_4Cl_2$

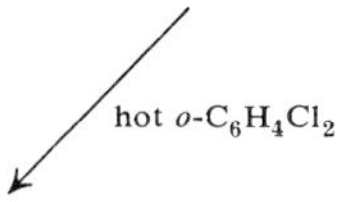

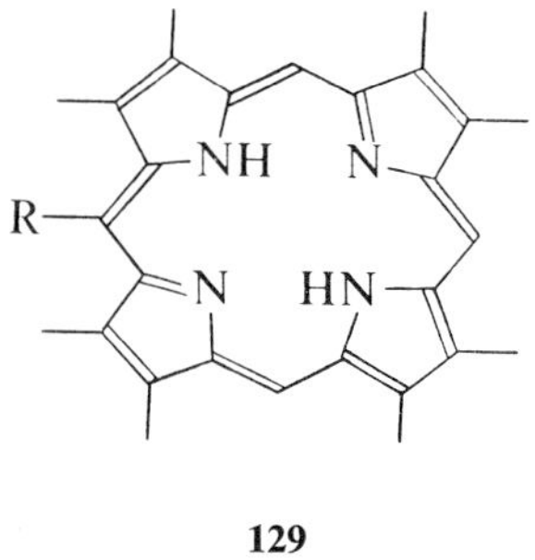

129

R = H or Alkyl

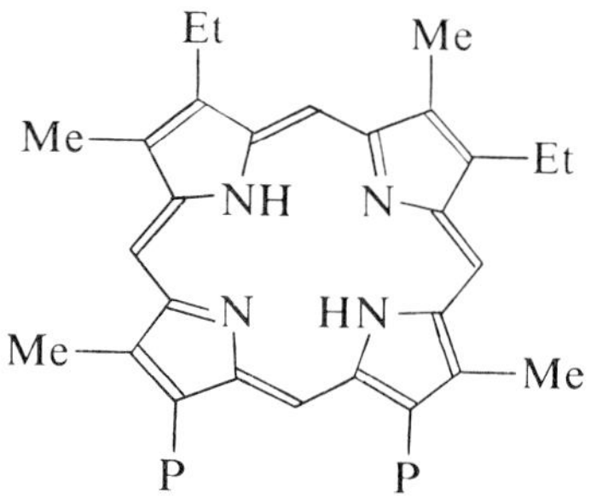

130

twenty or more porphyrins have now been prepared by this method, yields of up to 80% being obtained in the first stage and up to 90% in the second stage.[14,98–100] A useful modification of the reaction conditions for the cyclization, which gives improved yields in some cases, is to dissolve the *a,c*-biladiene in dimethyl sulfoxide and pyridine at room temperature with exclusion of light.[98] After about two days porphyrin formation is complete, and mesoporphyrin-IX (**130**), for example, has been prepared in 78% yield in this way.

Lower yields of the intermediate *a,c*-biladienes and of the porphyrins are often obtained in the synthesis of *meso*-substituted porphyrins or those containing electron-withdrawing groups[99] (e.g., CO_2R). However, the yields of rhodoporphyrin-XV (**106**) (31%) and of γ-phylloporphyrin-XV (**105**) (29%), two key degradation products from chlorophyll, still represent a very substantial improvement on the earlier Fischer syntheses. A number of

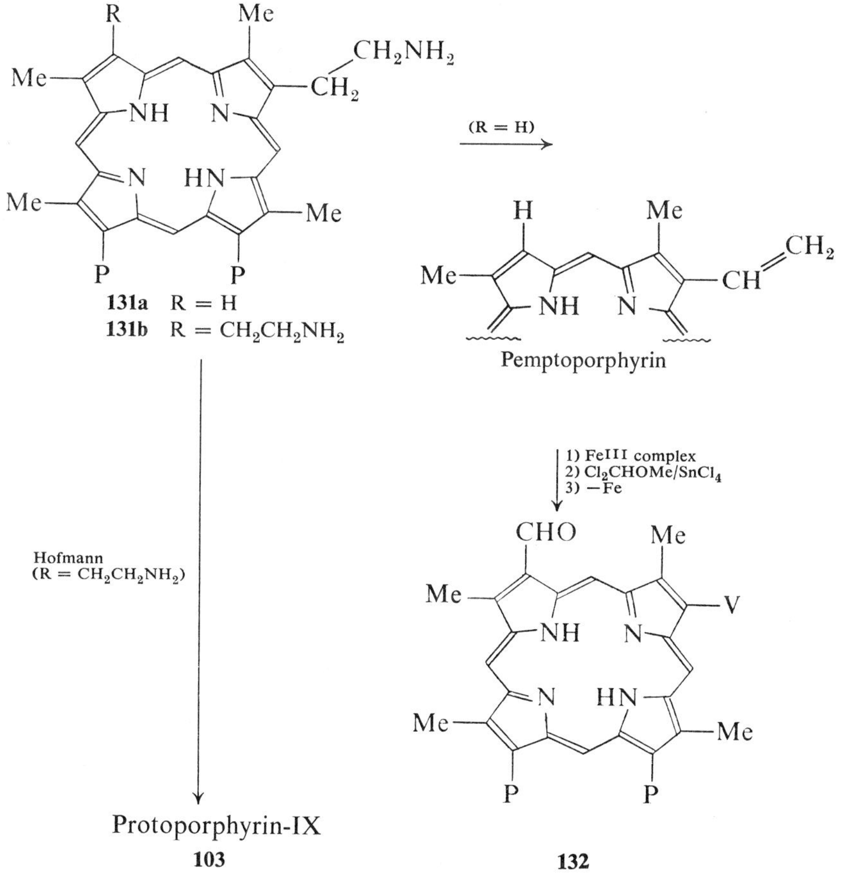

131a R = H
131b R = $CH_2CH_2NH_2$

(R = H) →

Pemptoporphyrin

1) FeIII complex
2) $Cl_2CHOMe/SnCl_4$
3) −Fe

Hofmann
(R = $CH_2CH_2NH_2$)

Protoporphyrin-IX
103

132

porphyrin esters, including rhodoporphyrin esters, were in fact synthesized in order to evaluate the effect of the proximity of the ester group to the position of cyclization.[99]

More recent examples of the biladiene method include the synthesis of pemptoporphyrin[101], chlorocruoroporphyrin[101] (**132**) and protoporphyrin-IX[102] (**103**). Pemptoporphyrin has recently been isolated from human feces, while the heme from chlorocruoro (*Spirographis*) porphyrin is the oxygen-carrying pigment of the polychete worm *Spirographis Spallanzanii*. In these cases the vinyl groups were introduced by Hofmann degradation of aminoethyl side chains carried through from the pyrrolic precursors, and the formyl group of chlorocruoroporphyrin was introduced by direct formylation of a metal complex of pemptoporphyrin.

An interesting feature of the behavior of *a,c*-biladienes is that they are readily converted into the corresponding *a,b,c*-bilatriene *mono*-salts (**133**), for example, on heating in dimethyl sulfoxide; the loss of proton from the *b-meso*-position is facilitated by an electron-withdrawing substituent in the neighboring rings. Addition of base to the dimethyl sulfoxide solution causes

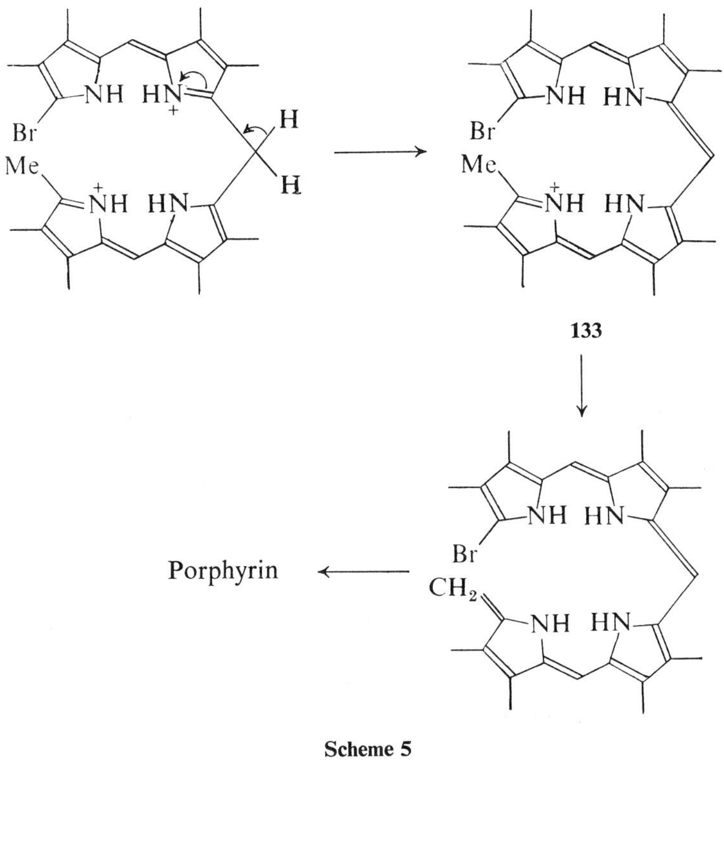

133

Porphyrin

Scheme 5

rapid conversion to porphyrin through a transient green color (the *a,b,c*-bilatriene free base?).[98] This observation has formed the basis for a further modification of the biladiene synthesis in which solutions of the salts in dimethyl sulfoxide were treated with various one-electron oxidizing agents—and yields of up to 46% of porphyrin were obtained within 1 or 2 hours.[103] The overall mechanism suggested for the oxidative cyclization is shown in Scheme 5, and it may well be that the cyclizations in hot *o*-dichlorobenzene or

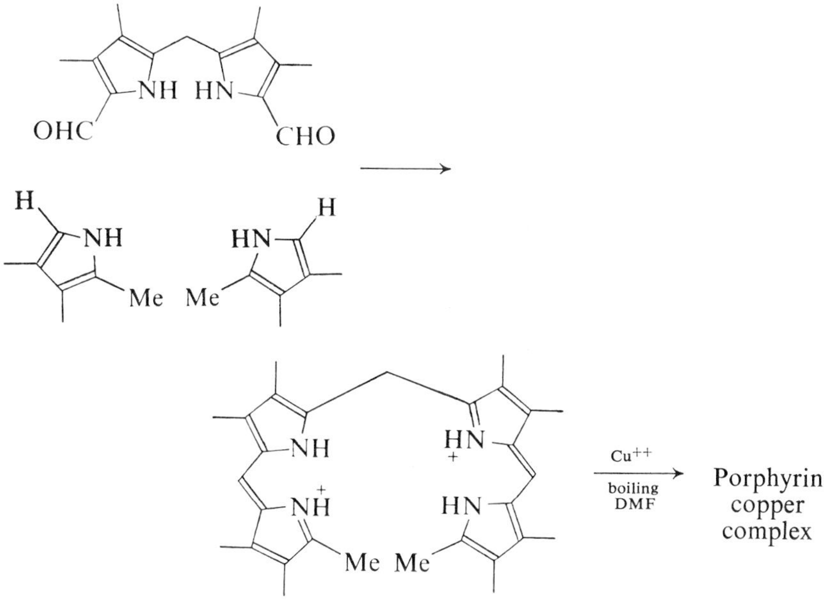

OHC NH HN CHO

H NH HN H

Me Me

NH HN+ NH+ HN Me Me

$\xrightarrow[\text{boiling DMF}]{Cu^{++}}$ Porphyrin copper complex

134

dimethylsulfoxide and pyridine also take place by a similar mechanism, the oxidizing agent being aerial oxygen.

This mild and rapid method is of potential use in the preparation of naturally occurring porphyrins bearing labile side-chains, and another synthesis[103] which proceeds under mild experimental conditions is the oxidative cyclization of 1′,8′-dimethyl-*a,c*-biladienes (**134**) with an excess of a cupric salt in boiling *N,N*-dimethylformamide. High yields of porphyrin were obtained after heating for only 2 minutes and the effects of the nature and concentration of the cupric salt were also studied. This method has considerable potential, particularly in the synthesis of more symmetrical porphyrins, because the 1′,8′-*a,c*-biladienes (**134**) can be readily prepared by condensation of α-free pyrroles with 5,5′-diformylpyrromethanes. Attempts to adapt this method to the synthesis of porphyrins via *a,c*-biladienes bearing acetyl substituents in the terminal rings were less satisfactory, very low yields being obtained in the cyclization step.[104]

In general it can be said that the Johnson two-stage version of the pyrromethene fusion synthesis is such a very great improvement on the classical Fischer method that it can be regarded as a new method in its own right. Its versatility is shown by the ability to synthesize not only unsymmetrical alkylated porphyrins but also those bearing labile substituents such as alkoxycarbonyl or *meso*-alkyl substituents. As with the other currently available porphyrin syntheses, however, the introduction of vinyl, formyl, and acetyl groups must usually be effected at the porphyrin stage. Apart from these minor limitations the only other problems encountered lie in the synthesis of the intermediate pyrromethenes and their coupling to biladienes. In some instances it has proved impossible[105] to brominate the 5-bromo–5′-methylpyrromethene **135a** to the 5′-bromomethyl derivative **135b**. In two

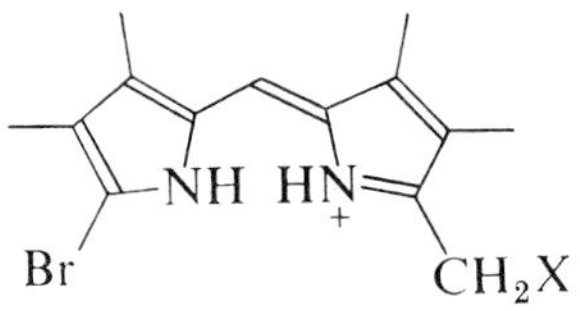

NH HN
+
Br
CH_2X

135a X = H
135b X = Br

other cases the stannic chloride complex of the brominated *a,c*-biladienes could not be decomposed in acid without destroying the biladienes.[101] Direct cyclization gave low yields of the stannic porphyrins, from which it was difficult to remove the metal, and thus protection of vacant β-positions in the intermediate pyrromethenes and their precursors seems to present difficulties.

It should also be mentioned (although it does not directly fall within the scope of this review) that *a,c*-biladienes have proved to be useful intermediates in the synthesis of azaporphyrins, and especially of corroles and tetradehydrocorrins, whose carbon skeleton is present in vitamin B_{12}.[100]

B. From *a*-Oxobilanes

In contrast to the Fischer porphyrin synthesis, and the later modifications involving isolation of intermediate *a,c*-biladienes, the biosynthesis of porphyrins presumably involves stepwise coupling of four porphobilinogen units followed by cyclization to uroporphyrinogen-III; the bridges between the rings in the latter and also in its open-chain tetrapyrrolic precursor are all methylene groups. Unfortunately, however, such tetrapyrroles (bilanes) cannot be prepared in the laboratory in a completely stepwise fashion; moreover, when prepared in other ways they are not only very susceptible to oxidation but readily undergo acid-catalyzed rearrangements, so that attempts at cyclization lead to mixtures of porphyrinogens. This was in accord with Mauzerall's finding that the porphyrinogens themselves (prepared by reduction of porphyrins) readily isomerize in acidic media.

In consequence of these findings the Liverpool group turned their attention to the synthesis of open-chain tetrapyrroles partially stabilized by carbonyl linkages between the rings.[15] The initial stages of this work relied upon the development of a new pyrroketone synthesis[39] (described earlier) and the required *a*-oxobilanes (**138a**) bearing protective benzyl ester groups in both terminal rings (at the 1′- and 8′-positions) were prepared by coupling 5′-pyridiniummethyl pyrroketones (**136c**) with the sodium salts of pyrromethane-5-carboxylic acids (**137**).[106]

The pyridiniummethyl pyrroketones **136c** were readily available from the corresponding 5-methylpyrroketones **136a** by chlorination (with sulfuryl chloride or *t*-butyl hypochlorite) followed by treatment with pyridine. The *a*-oxobilane-1′,8′-dicarboxylic acids **138b** prepared by hydrogenolysis of the benzyl ester groups could not, however, be cyclized to porphyrins by condensation with a variety of one-carbon units owing to the relative inertness of pyrroketones toward electrophilic attack (discussed earlier). The highly crystalline *a*-oxobilanes were therefore reduced with diborane to the corresponding bilanes **139a** and hydrogenolyzed to the di-acids **139b**. (The carbonyl function in pyrroketones is fairly resistant to hydrogenation or to borohydride reduction owing to its highly polar character, but this in turn makes it readily reducible by diborane, a highly electrophilic reagent.)

The bilane dicarboxylic acids **139b**, give mixtures of porphyrins on cyclization and aeration. However, oxidation (with one molar equivalent of *t*-butyl hypochlorite), followed by cyclization with trimethyl orthoformate in dichloromethane in presence of trichloroacetic acid, and aeration, gave isomerically pure porphyrin; in the first example studied, mesoporphyrin-IX dimethylester was isolated by chromatography and shown to be completely homogenous by thin-layer chromatography, NMR, and mixed m.p. comparisons with naturally derived material.[106] It was concluded that the major oxidation product from the bilane di-acid **139b** was the corresponding *b*-bilene **140** and that this had then cyclized to porphyrin, without rearrangement owing to the protective effects of the unsaturated linkage and of the terminal carboxyl in preventing acid-catalyzed cleavages of the central and the end rings respectively. Moreover, the terminal carboxyl groups could be expected to protect (at least partially) the *a*- and *c*-methylene groups of the bilane from oxidation, and even if formed the *a*- and *c*-bilenes would not cyclize to porphyrin under the mild conditions of the reaction.[106]

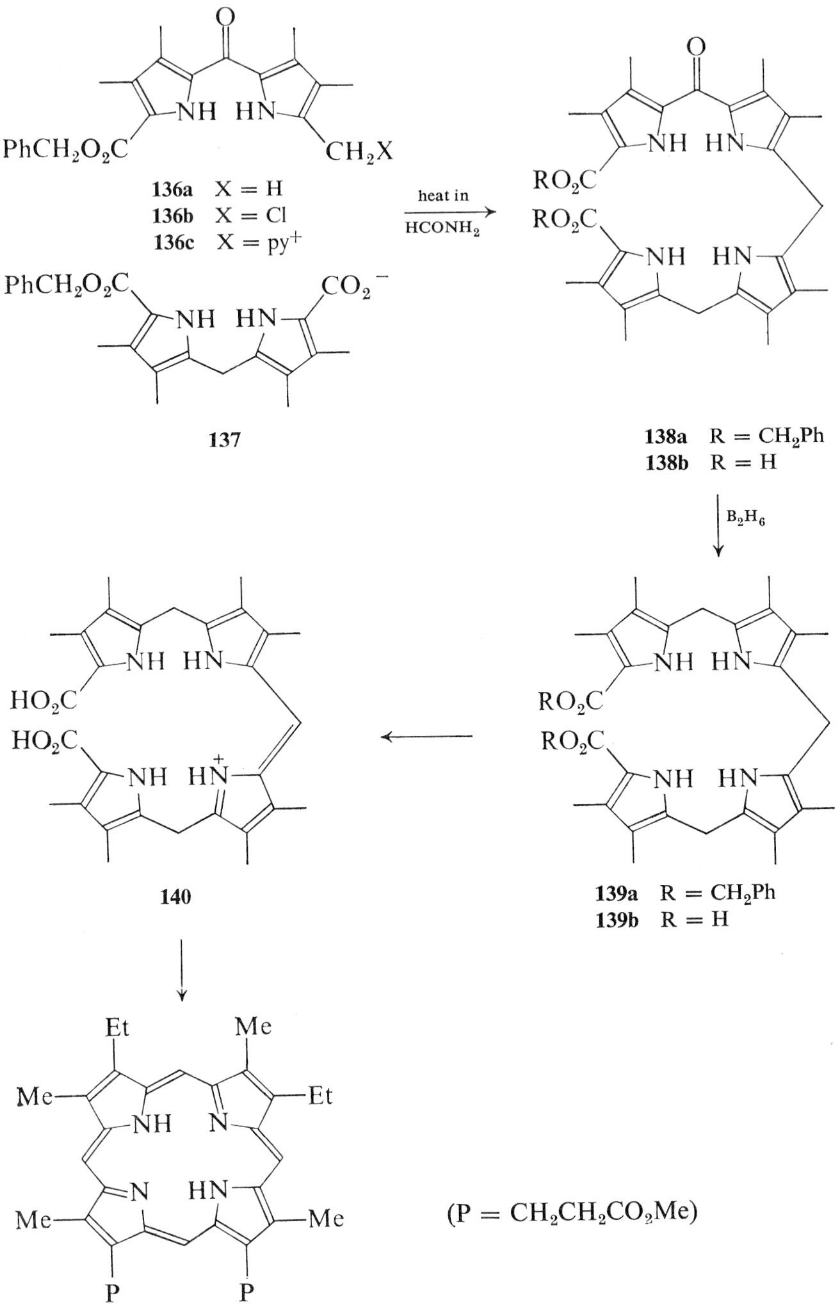

$PhCH_2O_2C$ NH HN CH_2X O

136a X = H
136b X = Cl
136c X = py^+

$PhCH_2O_2C$ NH HN CO_2^-

137

heat in $HCONH_2$ →

RO_2C RO_2C NH HN NH HN O

138a R = CH_2Ph
138b R = H

↓ B_2H_6

RO_2C RO_2C NH HN NH HN

139a R = CH_2Ph
139b R = H

←

HO_2C HO_2C NH HN NH $H\overset{+}{N}$

140

↓

Et Me Me Et NH N N HN Me Me P P

141

(P = $CH_2CH_2CO_2Me$)

Although the overall conversion of *a*-oxobilane **138a** to porphyrin **141** involves five stages, yields of porphyrin were in the 25–30% range, and there was no evidence of formation of mixtures. It was not necessary to isolate the various intermediates, but each of the individual stages could be followed and checked spectroscopically. The method has been applied to the synthesis of a number of other porphyrins, including coproporphyrins-III and IV,[107] protoporphyrin-IX,[108] and 2-vinyldeuteroporphyrin-IX[107,109] (isopempto-porphyrin) methyl esters. The free 4-position in the latter was protected by a bromine substituent until the porphyrin stage, while the vinyl group was introduced by transformation of an acetoxyethyl substituent (which had been carried through from the pyrrole stage).

Earlier examples of the synthesis suffered from the limitation that the pyrromethane carboxylate **137** was derived by partial hydrogenolysis of a symmetrical dibenzyl ester. This disadvantage has now been overcome[110] by use of a pentachlorophenyl ester (as an alternative protecting group) which can be removed by mild alkaline hydrolysis while retaining a nuclear benzyl ester group. The completely unsymmetrical mesoporphyrin-XI dimethyl ester (**141**) has now been synthesized in this way,[110] as has protoporphyrin-IX dimethyl ester.[108]

C. From *b*-Oxobilanes and Oxophlorins

A logical corollary[96] to the *a*-oxobilane method was to synthesize the analogous *b*-oxobilanes and study their cyclization. Coupling of the phosphoryl chloride complexes of the pyrromethane amides (**142**) with the 5-unsubstituted pyrromethane **143a** followed by hydrolysis of the intermediate imine salts (**144a**) afforded the *b*-oxobilanes (**144b**) in good yield. Hydrogenolysis of **144b** followed by cyclization with trimethyl orthoformate in dichloromethane in presence of trichloroacetic acid affords "oxomethenes" (**145**) which, upon aerial oxidation, give the oxophlorins[96] ("oxyporphyrins") (**146**) in up to 70% yield from the *b*-oxobilanes (**144b**). The oxophlorins can be isolated as deep blue crystalline solids, but in solution they are unstable and undergo photooxidation to form ill-defined red pigments.[111] For conversion to porphyrins it is usually more convenient to treat the crude reaction product with acetic anhydride and pyridine. The resulting *meso*-acetoxy-porphyrin (**147**) can then be catalytically reduced to the *meso*-unsubstituted porphyrinogen and reoxidized to form the corresponding *meso*-unsubstituted porphyrin **148**, preferably by aerial oxidation or with DDQ.[96] (Dilute solutions of iodine have also been used, but if there are any free β-positions in the porphyrinogen, they may be iodinated and iodoporphyrins will be formed.)

The earliest examples of the use of this method were the syntheses of mesoporphyrin-IX,[96] coproporphyrin-III,[96] and protoporphyrin-IX[108] methyl

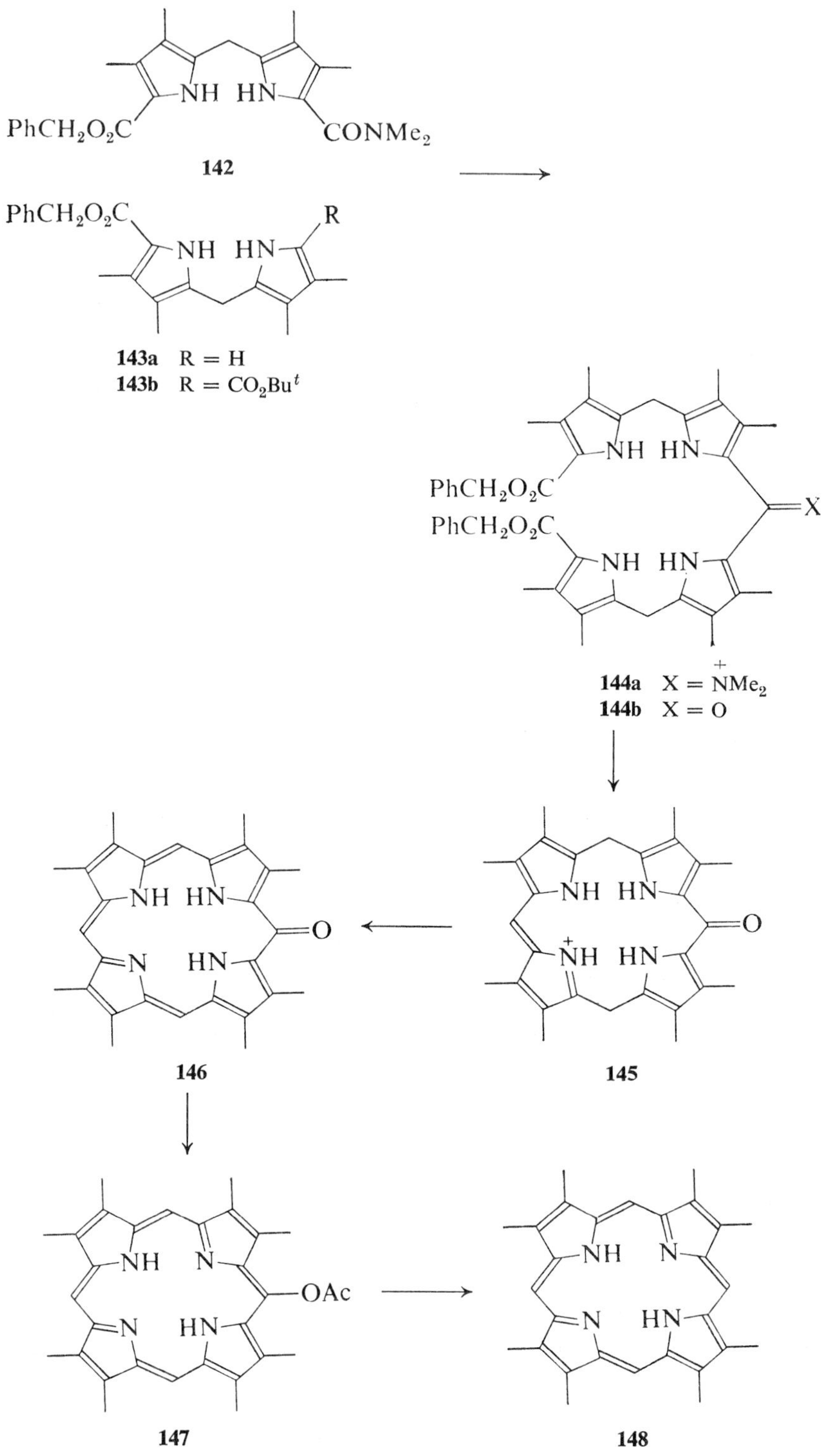

$PhCH_2O_2C$ NH HN $CONMe_2$

142

$PhCH_2O_2C$ NH HN R

143a R = H
143b R = CO_2Bu^t

$PhCH_2O_2C$ NH HN =X
$PhCH_2O_2C$ NH HN

144a X = $\overset{+}{N}Me_2$
144b X = O

NH HN =O $\overset{+}{N}H$ HN

145

NH HN =O N HN

146

NH N –OAc N HN

147

NH N N HN

148

esters; as with the *a*-oxobilane route the 5-unsubstituted pyrromethane **143a** was derived from a symmetrical dibenzyl ester by partial hydrogenolysis, followed by decarboxylation. More recently this limitation has been overcome by use of pyrromethane 5-benzyl, 5′-*t*-butyl esters (**143b**).[110] In this way completely unsymmetrical pyrromethanes were prepared and the *t*-butyl esters removed selectively by hydrolysis and decarboxylation with neat trifluoroacetic acid. The resulting 5-unsubstituted pyrromethane benzyl esters (**143a**) were then used directly without purification for coupling with pyrromethane amides (**142**). The vinyl side-chains of protoporphyrin-IX were introduced via acetoxyethyl side chains (as in the *a*-oxobilane route), which were modified at the porphyrin stage.[108]

Unlike the crystalline *a*-oxobilanes, most of the *b*-oxobilanes so far prepared have only been obtained as "foams" or gums, but the homogeneity of these intermediates after chromatographic purification has been shown by TLC and NMR (and to a lesser extent by mass spectrometry, since their high molecular weights render them unstable and molecular ions are not often observed). A useful modification of the work-up procedure is to chromatograph the intermediate imine salts (**144a**) before hydrolysis; starting materials and neutral impurities may be readily removed by elution with benzene and ethyl acetate, leaving the polar imine salts on the column, from which they can then be stripped with methanol and ethyl acetate. The method has considerable potential for synthesis of a wide variety of porphyrins, and in addition to the examples already mentioned it has been applied to the synthesis of the methyl esters of 4-vinyl-deuteroporphyrin-IX[109] (pemptoporphyrin) (**150**), rhodoporphyrin-XV (**106a**) and its 2-vinyl (**106b**) and 2,4-divinyl (**106c**) analogs,[112,113] as well as intermediates required in approaches to the synthesis of porphyrin-*a*.[114]

The 4-vinyl deuteroporphyrin dimethyl ester was shown to be identical with the dimethyl ester of pemptoporphyrin, a recently isolated fecal metabolite, by mixed m.p., mass, and NMR spectral comparisions. The synthetic 2-vinyl isomer prepared by the *a*-oxobilane route was shown to be different from the natural material by mixed m.p. comparisons as well as by differences in NMR spectra.[107,109]

The intermediate acetoxyethylporphyrin **149a** (*en route* to pemptoporphyrin **150**) was also formylated via its iron complex, and the product **149b** was converted by the sequence shown in Scheme 6 to 2-formyl–4-vinyl deuteroporphyrin-IX dimethylester **132**. The latter was identical in all respects with the porphyrin from chlorocruoroheme (spirographis heme), the oxygen-carrying pigment of the polychete worm *Spirographis spallanzanii* and with one of the two formyl porphyrins obtained indirectly by photooxidation of protoporphyrin.[115]

The synthesis of rhodoporphyrin-XV dimethyl ester (**106a**) is outlined in

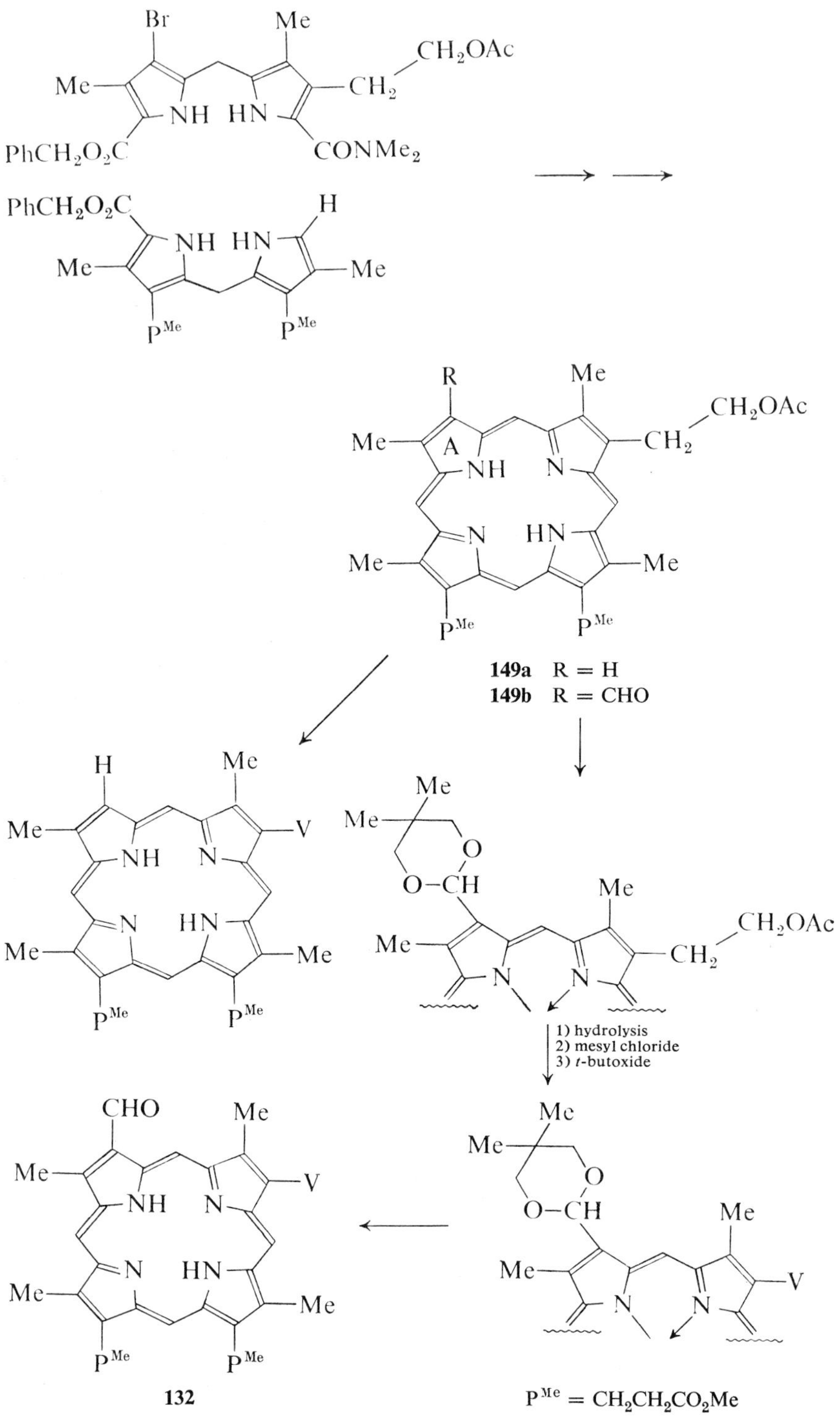

149a R = H
149b R = CHO

1) hydrolysis
2) mesyl chloride
3) *t*-butoxide

132

$P^{Me} = CH_2CH_2CO_2Me$

Scheme 6

Scheme 7, and the corresponding mono- and divinyl analogs were prepared

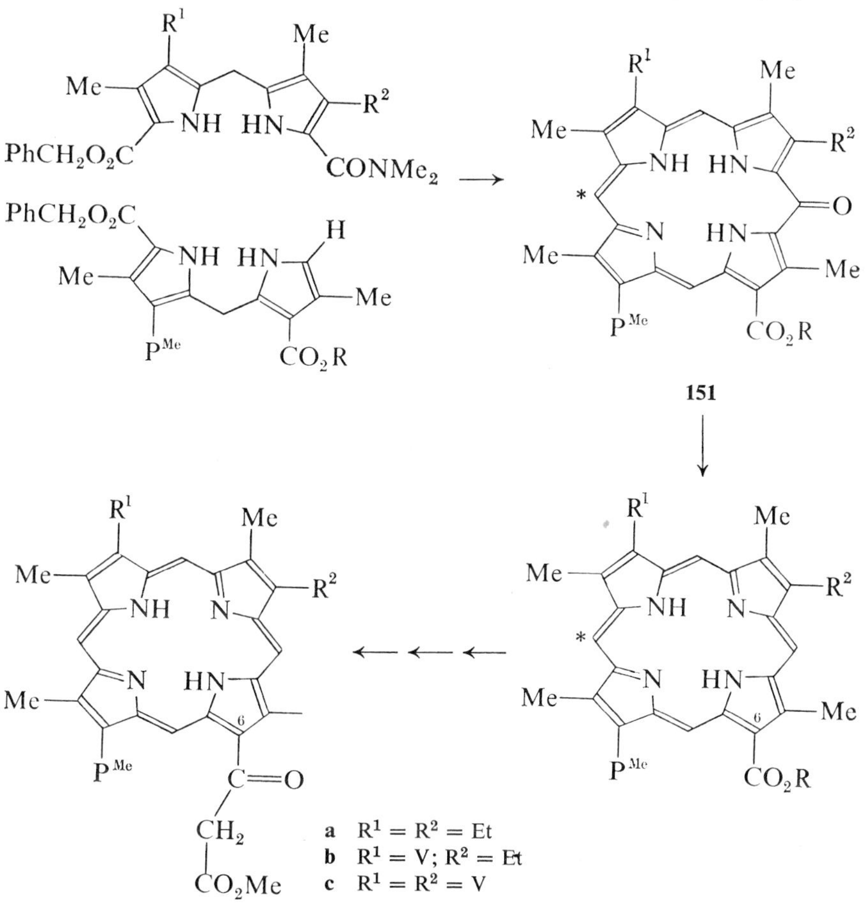

R^1 Me Me NH HN R^2 $PhCH_2O_2C$ $CONMe_2$ →
$PhCH_2O_2C$ H Me NH HN Me P^{Me} CO_2R

R^1 Me Me NH HN R^2 * O Me N HN Me P^{Me} CO_2R

151

↓

R^1 Me Me NH N R^2 * Me N HN 6 Me P^{Me} CO_2R

← ← ←

R^1 Me Me NH N R^2 Me N HN 6 P^{Me} C=O CH_2 CO_2Me

a $R^1 = R^2 = Et$
b $R^1 = V; R^2 = Et$
c $R^1 = R^2 = V$

152 **106**

Scheme 7

in the same way utilizing acetoxyethyl side chains as precursors of the vinyl groups.[112,113] Modification of the nuclear ester function at position-6 as shown below then affords the corresponding 6-β-ketoesters[113] (**152**), whose magnesium complexes are of great interest as possible intermediates in the biosynthesis of chlorophylls. Indeed, the capacity of oxophlorins (such as

$$\text{Por-6-CO}_2\text{R} \longrightarrow \text{Por-CO}_2\text{H} \longrightarrow \text{Por-COCl} \longrightarrow \text{Por-COCH}(\text{CO}_2\text{Me})(\text{CO}_2\text{Bu}^t) \longrightarrow \text{Por-COCH}_2\text{CO}_2\text{Me}$$

151) to undergo hydrogen exchange at the *meso*-position opposite the *oxo*-function[111] (see below) has made it possible to synthesize tritium-labeled rhodoporphyrins (**106**) and their β-ketoester analogs (**152**). Furthermore, ^{14}C labels can be introduced readily into the ketoester side chain by use of appropriate malonate derivatives, and consequently single- and double-labeled biosynthetic experiments are now feasible. Preliminary results[116] have shown that an isolated chloroplast system obtained from bean leaves can convert *meso*-tritiated magnesium protoporphyrin-IX (prepared by the *b*-oxobilane route) into chlorophyll-*a*, and the stage is thus set for definitive experiments on the role of porphyrin β-ketoesters in chlorophyll formation. *In vitro* experiments[113] have already shown that oxidation of the magnesium

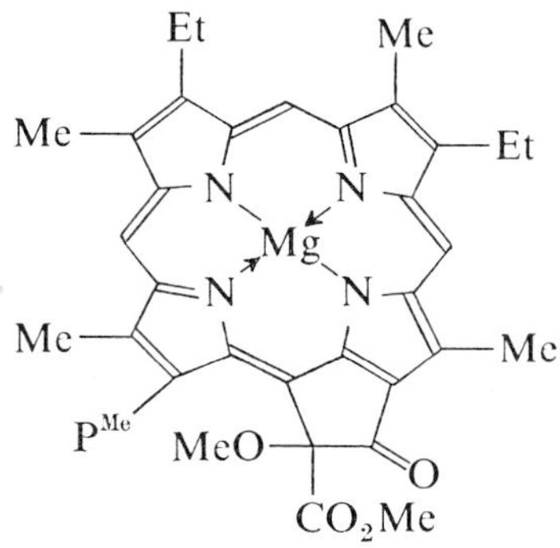

153

complex of ketoester **152a** with iodine in methanol in presence of carbonate effects cyclization to the pheoporphyrin derivative **153**.

A further recent example[117] of the *b*-oxobilane method was the synthesis and proof of structure of Harderoporphyrin, a monovinyl tricarboxylic acid porphyrin isolated from the Harderian glands of the rat. Both the dimethyl ester **154c** and its isomer **154d** were synthesized, and **154c** was shown to be identical with the natural product by m.p., NMR spectral, and counter-current comparisons.

In addition to its usefulness in the synthesis of porphyrins, the *b*-oxobilane route also brings with it the added bonus of making rationally synthesized unsymmetrical oxophlorins readily available. These compounds were originally prepared by Lemberg, Fischer, and their co-workers by oxidation of the pyridine complexes of hemins.[118,119] Although this method leads to mixtures, there is no doubt about the spectroscopic identity of the chromophores obtained with those prepared more recently by ring synthesis[96,111] and with other more symmetrical oxophlorins recently obtained directly from diformyl pyrroketones.[88–95]

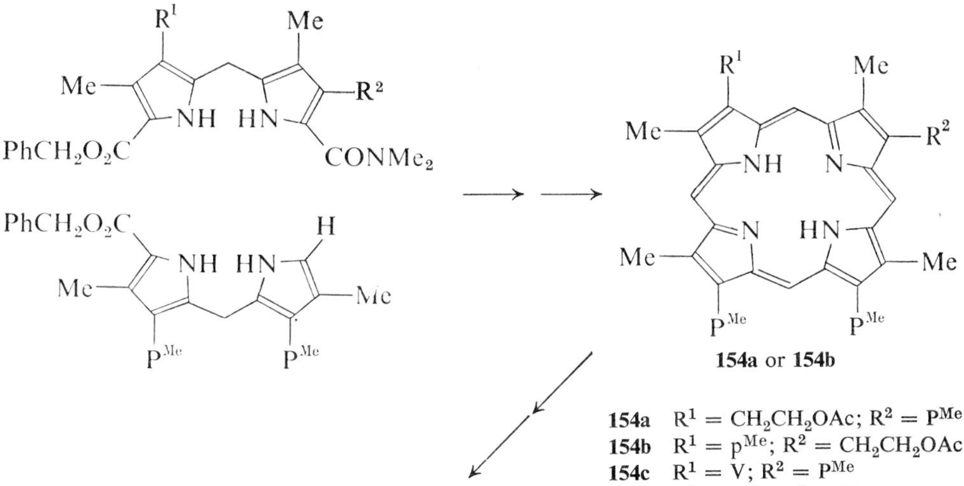

154a or **154b**

154c or **154d**

154a $R^1 = CH_2CH_2OAc$; $R^2 = P^{Me}$
154b $R^1 = P^{Me}$; $R^2 = CH_2CH_2OAc$
154c $R^1 = V$; $R^2 = P^{Me}$
154d $R^1 = P^{Me}$; $R^2 = V$

Interest in oxophlorin synthesis has arisen not only because of their use as intermediates in porphyrin synthesis, but more importantly because it has long been thought[120] that the α-hydroxy derivative of hemin is an intermediate in the breakdown of hemin to bile pigments. The red metal complexes of oxophlorins (or "oxyporphyrins" as they used to be known) must exist in the tautomeric *meso*-hydroxy form unlike the blue free bases (λ_{max}, 401, 588, and 635 nm) in which the oxygen function is clearly in the *oxo*-form[111] (as shown by their visible spectra and the carbonyl absorption at 1560 cm^{-1}, which is closely similar to that observed for simple pyrroketones). The NMR spectra of the free bases are somewhat ill-defined owing to partial free radical character, but the spectra of the salts are quite well resolved.[89,95,111] The free radical character has been attributed to a low energy triplet state, and Bonnett has recently shown that the free electrons are largely associated with the methine bridges rather than the peripheral positions.[121] The oxophlorins are very strong bases, being readily converted into their monocations (λ_{max}, 407, 500, 537, 584) even by acetic acid; stronger acid is, however, required to convert them into their violet red dications (λ_{max}, 416, 560, 615 nm).[111]

The monocations readily undergo exchange of the *meso*-proton opposite to the *oxo*-group in mildly acidic media and this provides a useful method of tritium labeling.[111] The ferric complex of α-oxymesoporphyrin-IX (**155**) labeled in this manner (asterisk in formula) has recently been shown in the rat to be degraded to mesobilirubin and this provides good circumstantial evidence for the involvement of the α-oxy derivative of hemin in the normal

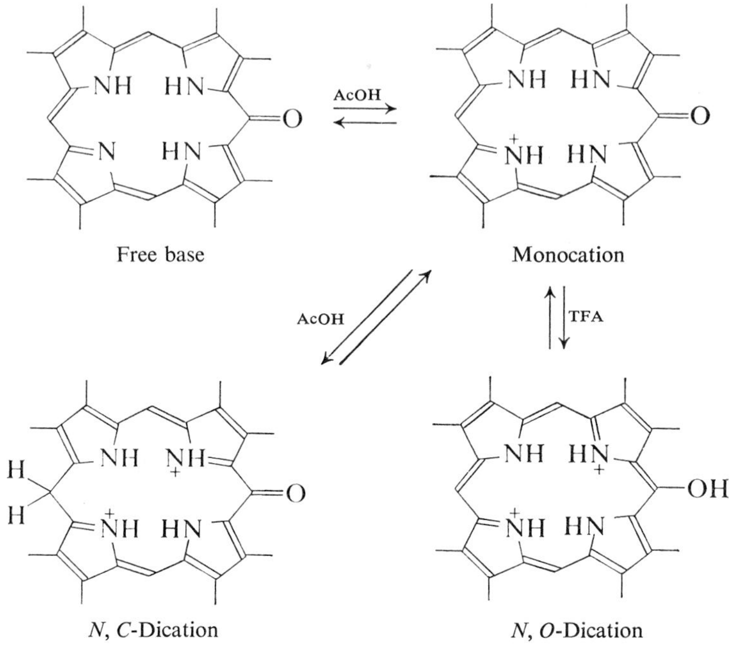

Free base

Monocation

AcOH

TFA

N, *C*-Dication

N, *O*-Dication

breakdown of haem to bile pigments[122] (see also below). The ferric complex of the β-oxy isomer is not, however, degraded to the corresponding mesobilirubin *in vivo*.

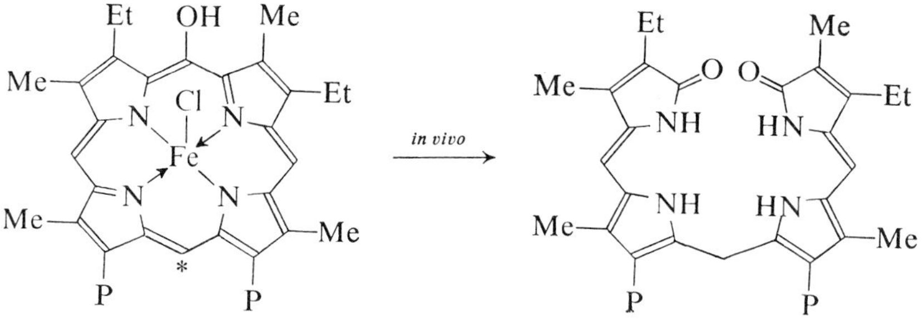

in vivo

155

Mesobilirubin-IXα

D. From *b*-Bilenes

In the synthesis of porphyrins from *a*-oxobilanes the penultimate stage is the cyclization of a *b*-bilene dicarboxylic acid. A more direct route[13,123] to such *b*-bilenes is the condensation of an α-formylpyrromethane **(157)** with an α-unsubstituted pyrromethane (**156**; R = H) or the corresponding α-carboxylic acid (**156**; R′ = CO_2H). Initially attempts were made to utilize *b*-bilene dibenzyl esters (**158a**), but removal of the benzyl groups by hydrogenolysis could not be accomplished without concomitant hydrogenation of the methene linkage. However, the di-*t*-butyl esters (**158b**) were readily cleaved by cold trifluoroacetic acid and the resulting 1′,8′-di-unsubstituted

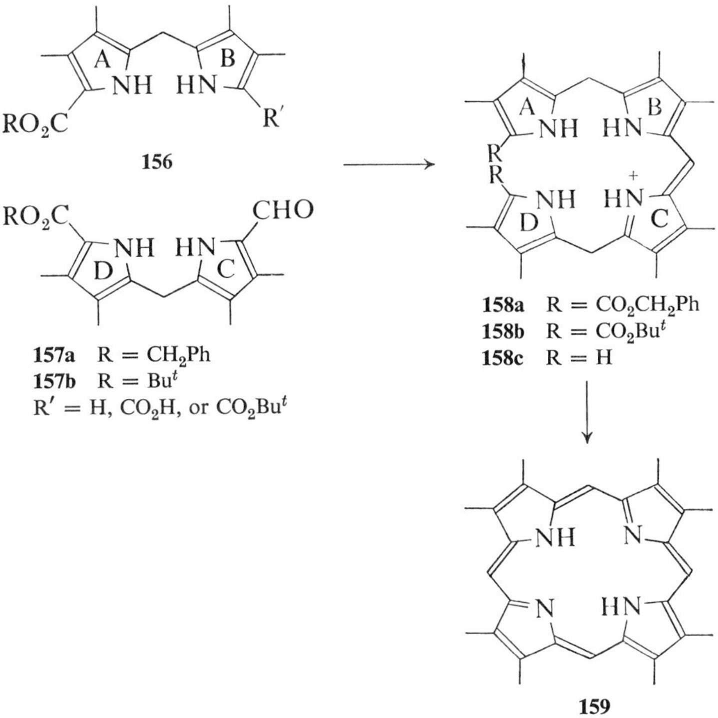

156

157a R = CH_2Ph
157b R = Bu^t
R′ = H, CO_2H, or CO_2Bu^t

158a R = CO_2CH_2Ph
158b R = CO_2Bu^t
158c R = H

159

b-bilenes (**158c**) could by cyclized to porphyrin **159** by treatment with trimethyl orthoformate in dichloromethane in presence of trichloroacetic acid. Porphyrin formation was complete in a few minutes and, for example, the dimethyl esters of mesoporphyrins-IV, X, and XIII were prepared in 39, 47, and 57% yields, respectively, from appropriate *b*-bilenes.

Later attempts to synthesize rhodoporphyrins (i.e., porphyrins bearing nuclear carboxylic ester substituents) by cyclization of *b*-bilenes bearing a methoxycarbonyl group at position-3 (i.e., in the *B*-ring) gave rise to mixtures of porphyrins.[85] On the other hand, a similar synthesis involving cyclization of a *b*-bilene bearing a 2-methoxycarbonyl group (i.e., in the *A*-ring) gave a moderate yield of a rhodo-type porphyrin; this was isomerically pure and

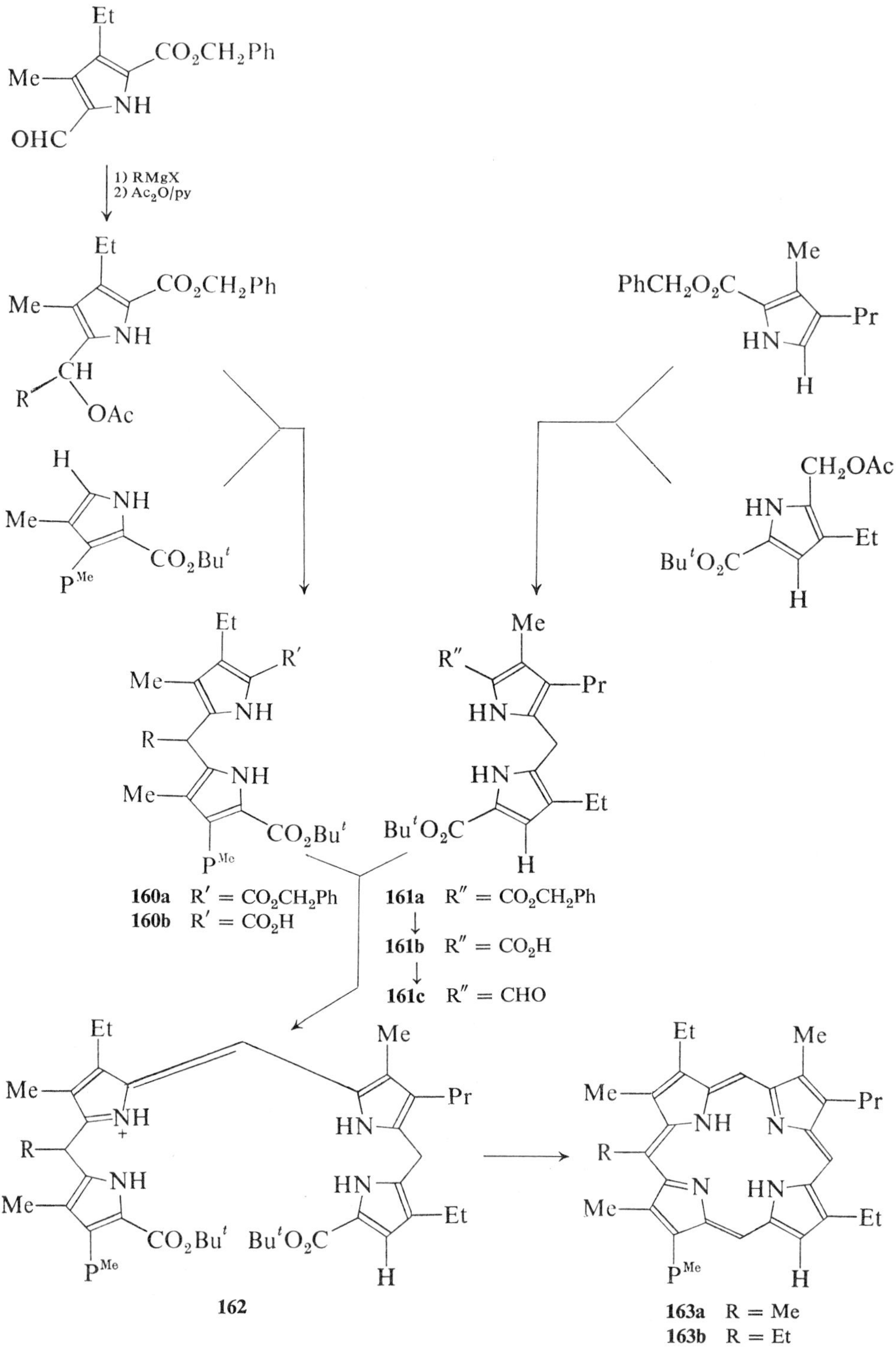

1) RMgX
2) Ac_2O/py

160a $R' = CO_2CH_2Ph$
160b $R' = CO_2H$

161a $R'' = CO_2CH_2Ph$
↓
161b $R'' = CO_2H$
↓
161c $R'' = CHO$

162

163a R = Me
163b R = Et

Scheme 8

uncontaminated with other porphyrins, although cyclization was much slower, as expected, owing to deactivation by the nuclear ester group.[85,114] The failure of the *b*-bilene method to give a single isomerically pure porphyrin when the methoxycarbonyl group was in the *B*-ring may be attributed to the increased susceptibility of the methene carbon atom in this case toward nucleophilic attack, for example, by the terminal pyrrole ring of the intermediate decarboxylated *b*-bilenes. If this occurred, then a tripyrrylmethane-type intermediate would be formed and this could break down in three different ways (discussed earlier), and hence mixtures of porphyrins would be formed. More recent work has also shown that in certain cases having complex unsymmetrically disposed side chains small amounts of porphyrin by-products are obtained in cyclization of pure *b*-bilenes.[114] However, the method has been successfully applied to the synthesis of *meso*-substituted porphyrins related to the *Chlorobium* chlorophylls[13,123] and to desoxophylloerythroetioporphyrin.[124]

The *meso*-substituted pyrromethanes (**160a**) were first prepared as shown in Scheme 8 and the corresponding carboxylic acids (**160b**) were condensed with the formyl pyrromethanes (**161c**) to give the *b*-bilenes (**162**). Subsequent cyclization of *b*-bilenes followed the usual procedure, deesterification and decarboxylation with trifluoroacetic acid followed by cyclization with trimethylorthoformate-trichloroacetic acid and aerial oxidation. However, the yields of the *meso*-methyl and *meso*-ethylporphyrins (**163a**) and (**163b**) were only about 5% compared with the 50% observed in the early work. The low yields may be attributable to the influence of the *meso*-substituent insofar as it may sterically hinder attainment of the conformation required for cyclization; alternatively, the presence of an unsubstituted β-position may be the cause, since similar results have been obtained when a β-position has been free but there has been no *meso*-substituent.[114]

The two synthetic porphyrins **163a** and **163b** and their copper complexes were compared with the phylloporphyrins derived from fractions 3 and 4 of *Chlorobium* chlorophylls(660). The synthetic *meso*-methyl porphyrin (**163a**) was shown[13,123] to be identical with the naturally derived porphyrin and with an earlier synthetic sample prepared by MacDonald[81] (utilizing the Fischer method) by means of X-ray powder photographs and m.p. comparisons. However, the synthetic δ-ethyl porphyrin **163b** was not identical with any of the phylloporphyrins derived from the *Chlorobium* chlorophylls(660) and this result in combination with other evidence has cast considerable doubts on the occurrence of δ-ethyl substituents in this series.[123] More degradative and synthetic work is clearly needed, however, before a definitive answer to the outstanding structural problems can be obtained.

Rapoport approached the synthesis of porphyrins by the *b*-bilene route independently of the Liverpool group, and this work culminated in an elegant

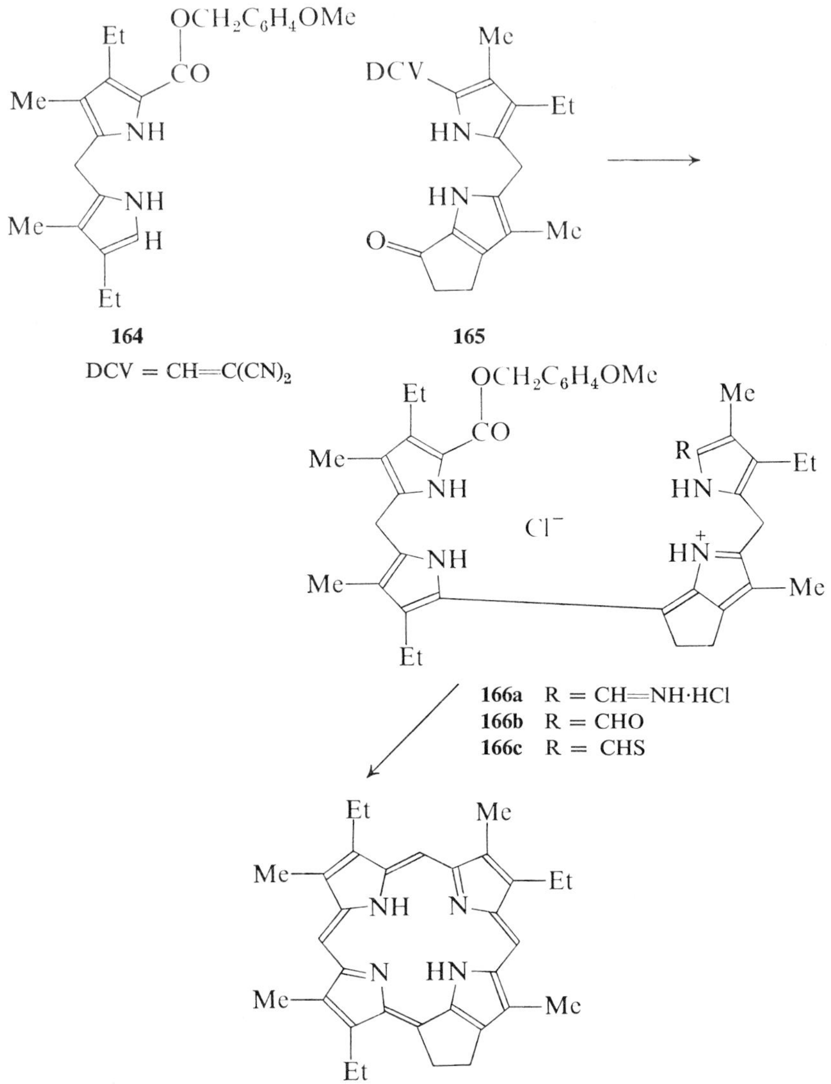

164

165

DCV = $CH{=}C(CN)_2$

166a R = $CH{=}NH \cdot HCl$
166b R = CHO
166c R = CHS

167

synthesis of desoxophylloerythroetioporphyrin (**167**), the predominant porphyrin present in petroleum. This synthesis utilizes a variety of protecting groups; for example, the acid labile anisyl protecting group was used for the pyrromethane **164** corresponding to rings *D* and *A* of the final porphyrin, while the formyl group of the other pyrromethene (**165**) was protected initially as a dicyanovinyl group (during pyrromethane formation) and later as an aldimine hydrochloride (during *b*-bilene formation). The formimino *meso*-substituted *b*-bilene dihydrochloride **166a** was hydrolyzed to the formyl-*b*-bilene **166b** and converted to the thioformyl analog **166c**. Either of these two *b*-bilenes could be cyclized with 2.5% hydriodic acid in acetic acid followed by air oxidation to give desoxophylloerythroetioporphyrin (**167**) (conditions which are similar to those employed in the MacDonald method). The yield was only 6%, but this was attributed to the steric effect of the isocyclic ring. This view was supported by results obtained by applying the *ac*-biladiene route; the required biladiene was obtained in excellent yield, but its cyclization in DMSO-pyridine gave less than 3% of desoxophylloerythroetioporphyrin.[124]

In conclusion therefore it can be said that the *b*-bilene route is useful for the preparation of isomerically pure but fairly simple porphyrins; severe constraints may occur if electron-withdrawing groups, vacant β-positions, or *meso*-substituents are present. It is essentially a two-stage version of the MacDonald synthesis enabling it to be applied to the synthesis of less symmetrically substituted porphyrins; a *b*-bilene was also a fleeting intermediate in the Harvard synthesis[72] of chlorophyll.

8. CHLORINS AND OTHER PARTIALLY REDUCED PORPHYRINS

Although the chemistry of chlorophylls *a* and *b* (**168**) has been studied extensively,[3,6,125] there is as yet no direct rational stepwise synthesis of the dihydroporphyrin (or chlorin) ring which they contain. Most of the synthetic chlorins to date have been prepared either by degradation of chlorophylls or by reduction of porphyrins.

The earliest preparation of chlorins from porphyrins was Fischer's reduction of porphyrin iron complexes with sodium in amyl alcohol.[3] More recently Inhoffen and his colleagues[126] have shown that diborane reduction of octaethylporphyrin in tetrahydrofuran affords a mixture of *cis*- and *trans*-octaethylchlorin (ratio 5:1, respectively) whereas diimide in pyridine reduction is stereoselective affording the *cis*-isomer only;[127] in contrast, the sodium in alcohol reduction affords the *trans*-isomer. In more complex cases, it would be difficult to control the position of reduction and so these methods are of limited utility for chlorin synthesis; however, Fischer has claimed[128] that reduction of γ-phylloporphyrin-XV (**105**) affords the 7,8-dihydro derivative (**169**), presumably because reduction of the 7,8-double bond relieves the steric strain[129] between the γ-methyl group and the 7-propionate side chain.

Simple symmetrically substituted chlorins have been prepared by self-condensation of 2-dimethylaminomethyl pyrroles in presence of ethyl magnesium bromide in boiling xylene;[130] like other chlorins they can be

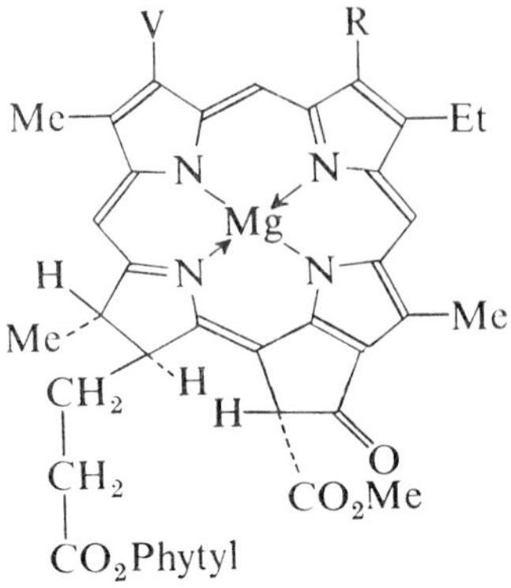

168a R = Me
168b R = CHO

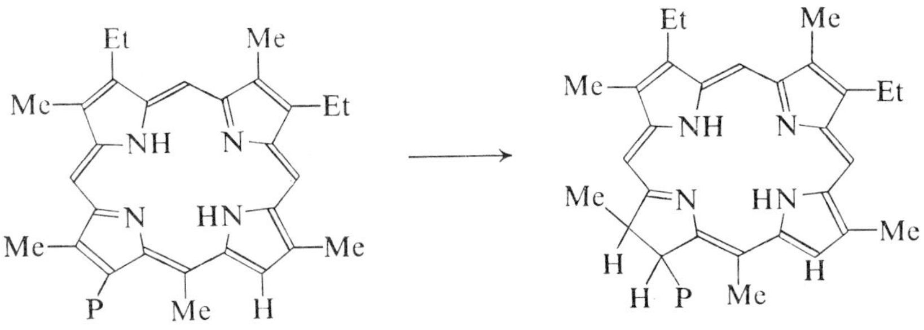

105

169

readily dehydrogenated by high potential quinones to the corresponding porphyrins.[131] Simple tetrahydroporphyrins[130] (analogs of the bacteriochlorophylls) have also been prepared.

During the course of work on the synthesis[72] of chlorophyll (see below) Woodward[10] and his colleagues discovered the phlorins (**170**), a new type of dihydroporphyrin formed as intermediates in the synthesis of chlorophyll-*a*. The phlorins which have additional hydrogens on nitrogen and on one of the *meso*-carbon bridges can be prepared by reduction of the porphyrin ring system in various ways—by addition of thiol acids, photoreduction, and other one-electron processes.[10,132,133] Their blue color (λ_{max}, 620 mm) is very characteristic, and they also form green monocations (λ_{max}, 725 mm) by protonation on the remaining nitrogen; in strong acid further protonation occurs on carbon to give a "*bis*-porphomethene" (**171**).[10]

Phlorins are very readily reoxidized to porphyrins by mild oxidizing agents such as air or iodine; further reduction affords porphomethenes and porphyrinogens.[10]

An interesting new synthesis of simple chlorins and their reduced porphyrin derivatives has been explored recently by the Braunschweig school.[134] Oxidation of octaethylporphyrin with hydrogen peroxide in sulfuric acid or with osmium tetroxide affords the diol **172**, which on treatment with acid undergoes a pinacol-type rearrangement to the chlorin, or "gemini ketone" **173**. Repetition of the same process leads to di- and even triketones. This

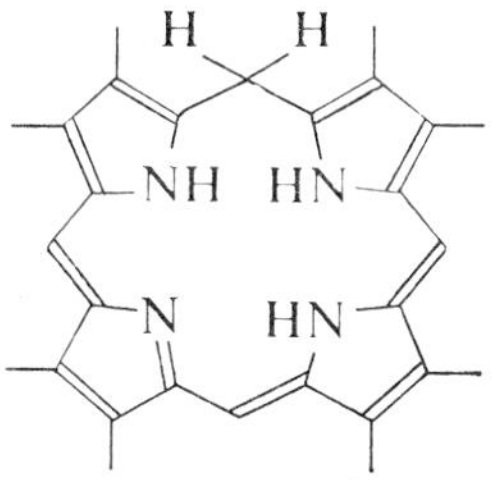

170

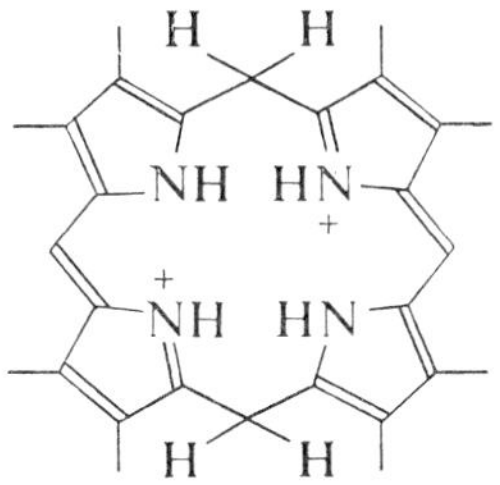

171

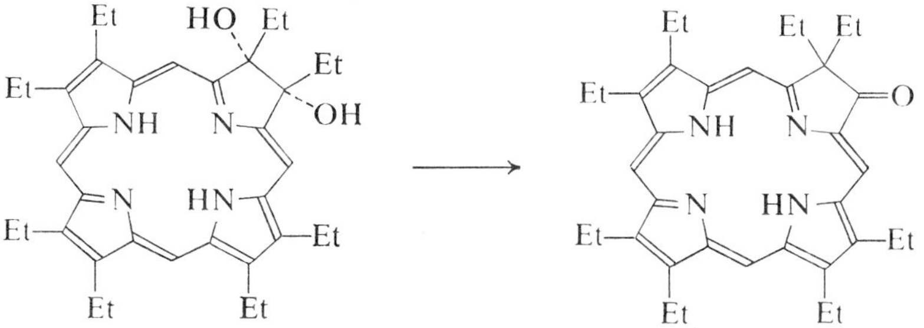

172

173

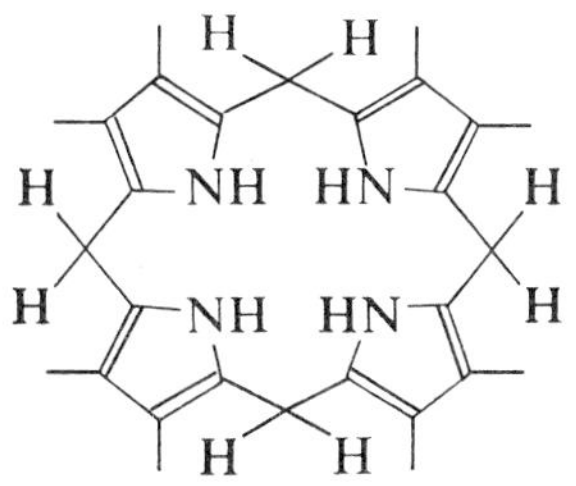

174

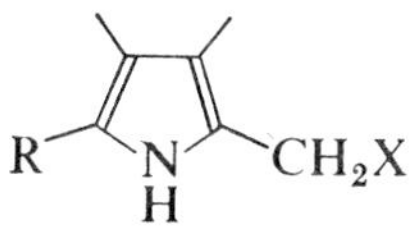

85

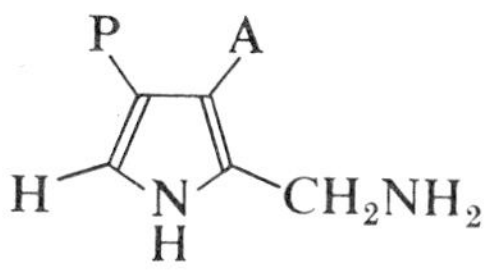

R = H or CO_2H

36

A = CH_2CO_2H
P = $CH_2CH_2CO_2H$

approach would, however, be difficult to apply to more complex naturally occurring chlorins, although it is potentially useful in corrin and Vitamin B_{12} chemistry.

Porphyrinogens (**174**), the colorless hexahydro derivatives of porphyrins, are readily prepared by reduction of porphyrins by using a variety of reducing agents such as sodium amalgam, sodium borohydride, zinc dust and alkali, and hydrogenation (over platinum). They are readily reoxidized by air, iodine, quinones, or other oxidizing agents to the parent porphyrin. Porphyrinogens can also be obtained[135] by polymerization of simple pyrroles of type **85** under anaerobic conditions, but unless the substituents in the β-positions are identical, mixtures of isomers are obtained. However, enzymic polymerization of porphobilinogen **36** normally leads specifically[33] to uroporphyrin-III except in rather rare pathological conditions where uroporphyrin-I is formed.

Woodward's synthesis of chlorin-e_6, accomplished in 1961, still represents the only synthesis of a completely unsymmetrical chlorin from pyrrolic intermediates.[72] It involves first the preparation of an unsymmetrical porphyrin, followed by a series of further reactions in which the two additional hydrogen atoms are introduced into the *D*-ring in a stereospecific manner. The two pyrromethanes **175** and **176** were first prepared as shown from pyrrolic intermediates and then condensed together in an ammonium acetate/acetic acid buffer. Under these conditions the highly reactive thioformyl group of the pyrromethane **176** formed a Schiff's base with the side chain amino group of the pyrromethane **175**, the ketonic moiety in the other pyrrole ring of **176** being insufficiently reactive under these conditions to interfere. The unstable product **177**, in which the Schiff's base served to hold the two methanes together in the desired manner, was then treated immediately with hot methanolic 12*M* hydrogen chloride and cyclized to the phlorin **178**, presumably by condensation first to form the α-bridge and then the γ-bridge. The phlorin was oxidized to porphyrin **179** with iodine, and after acetylation of the side chain amino group, further oxidation by air gave the *meso*-acrylic ester derivative **180**. The latter could then be equilibrated with the purpurin **181** because of the steric interactions between the *meso*-acrylic ester and the substituents at the 6- and 7-positions in the porphyrin ring. The acetamidoethyl side chain in the purpurin was then hydrolyzed with methanolic hydrogen chloride and subjected to Hofmann degradation to generate the 2-vinyl group. Subsequently, photooxidation cleaved the double bond in the isocyclic ring and the resulting 7-oxaloyl chlorin **182** underwent fission of the oxaloyl group on treatment with dilute methanolic alkali to afford a *meso*-formyl chlorin **183**; this immediately cyclized with the 6-methoxycarbonyl substituent to form the chlorin lactone **184**. Mild alkaline hydrolysis followed by reesterification afforded the racemic chlorin-e_5 (**185**), which was resolved by means of its quinine salt. The (+)-enantiomer was identical with that obtained by degradation of chlorophyll and the γ-formyl group was then transformed into methoxycarbonylmethyl via the cyano-lactone **186** and the cyanomethyl acid **187**; the optically active chlorin-e_6 (**188**) was identical with a sample derived from natural chlorophyll-*a*. This completed the formal total

CH_2NH_2 CH_2 H Me NH H NC CN C HC Me Et HN CH_2Cl

+

OHC NH Me H P^{Me} 1) H^+ 2) $NaBH_4$

O HN H Me C CH_2 CO_2Et CH_2 CO_2Me

NC CN C Me HC Et NH HN O Me C CH_2 CO_2Et CH_2 CO_2Me

CH_2NH_2 CH_2 H Me NH NH Me H P^{Me}

175

SHC Me Et NH HN O Me C CH_2 CO_2Et CH_2 CO_2Me

176

NEt Me HC Et HN

Me OHC Et HN

177

178

180

179

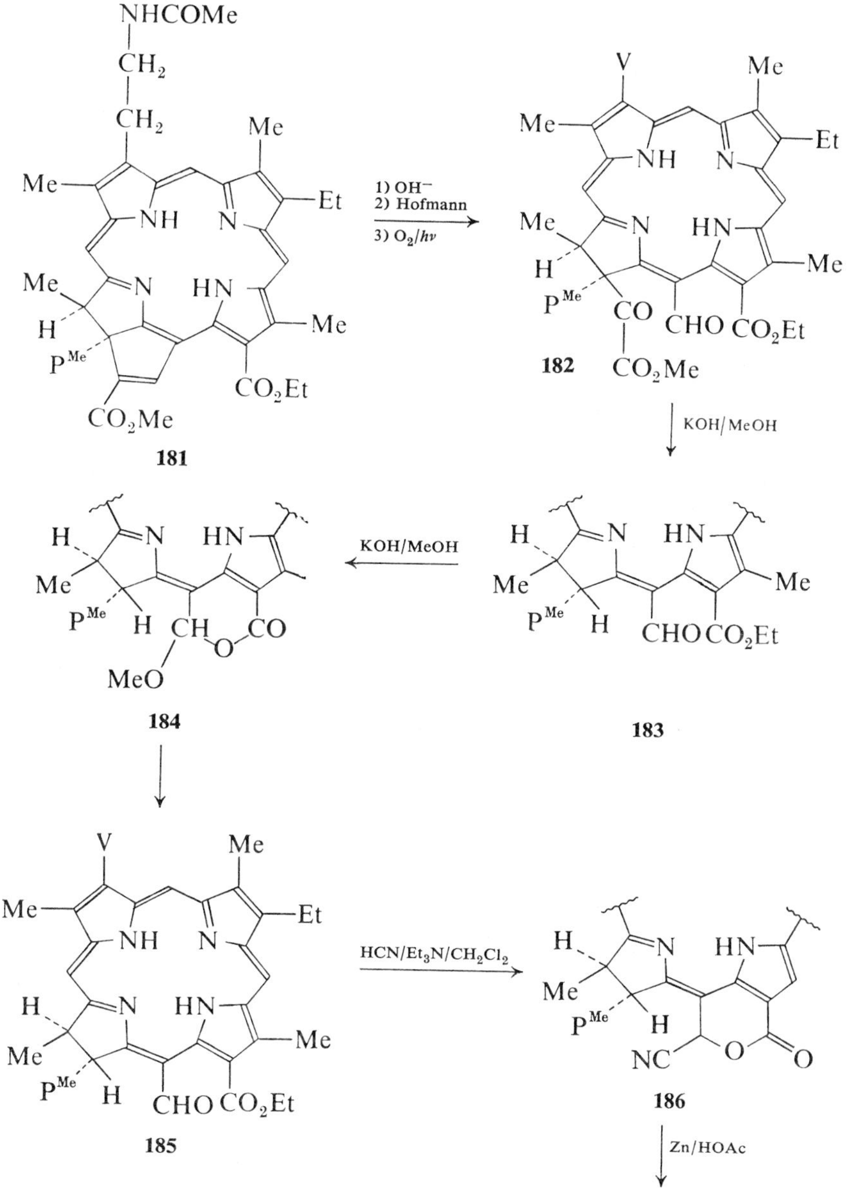
NHCOMe
CH2
CH2
Me
Me
Et
NH
N
Me
H
N
HN
Me
PMe
CO2Et
CO2Me
181
1) OH−
2) Hofmann
3) O2/hν
V
Me
Me
Et
NH
N
Me
H
N
HN
Me
PMe
CO
CHO
CO2Et
182
CO2Me
KOH/MeOH
H
N
HN
Me
Me
PMe
H
CHO
CO2Et
183
KOH/MeOH
H
N
HN
Me
PMe
H
CH
CO
O
MeO
184
V
Me
Me
Et
NH
N
H
N
HN
Me
Me
PMe
H
CHO
CO2Et
185
HCN/Et3N/CH2Cl2
H
N
HN
Me
PMe
H
NC
O
O
186
Zn/HOAc

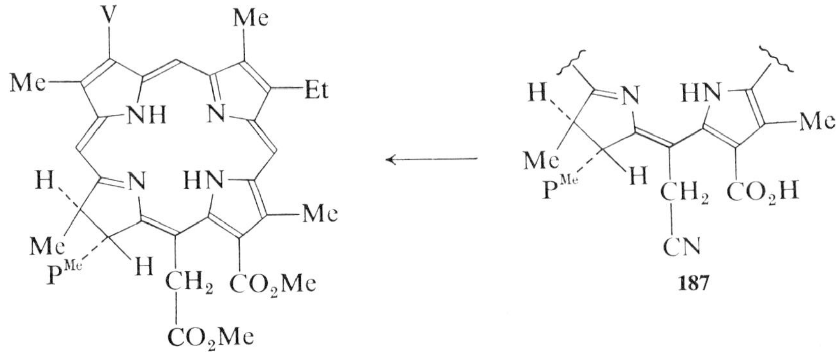

187

188 Chlorin-e_6

synthesis of chlorophyll-*a* because chlorin-e_6 had been previously converted[125a] back into chlorophyll, although this is far from easy and has not since been repeated.

Almost simultaneously with the appearance of the Harvard synthesis, Strell and Kalojanoff[136] reported the completion of work on the synthesis of phaeophorbide-*a* begun many years previously in the Munich school under the aegis of Hans Fischer. The phylloporphyrin **189** was synthesized[139] in very low yield from pyrromethenes and its iron complex reduced by sodium and amyl alcohol to give a product (**190a**) which was thought to be reduced specifically in the *D*-ring. Transformation of the γ-methyl group into γ-methoxycarbonylmethyl giving **190b** was then effected,[138] and the copper derivative acetylated to give the diacetyl derivative **191**. Partial deacylation then gave the 2-acetyl chlorin **192**. Inhoffen has since criticized[6,139] these results, and he was able to obtain only a 2-acetyl compound by modifying the experimental conditions. Borohydride reduction of the acetyl group followed by treatment of the iron complex with dichloromethylmethyl ether then gave a chlorin (**193**) containing the isocyclic ring. Dehydration of the 2-hydroxyethyl group and oxidation of the secondary hydroxyl function in the isocyclic ring then afforded pheophorbide-*a* (**194**). This formally completed the total synthesis of chlorophyll, since pheophorbide-*a* had been converted back into chlorophyll by insertion of magnesium and phytylation of the propionic acid side chain.

Further work on this synthesis is probably very desirable and because of the low yields obtained in many of the stages the racemic compounds were not resolved and, furthermore, the 2-acetylchlorin (**192**) obtained by degradation of natural chlorophyll was used as a relay.

Since these syntheses were completed the absolute stereochemistry of chlorophyll[140] (**168**) and of bacteriochlorophyll[140,141] (**195**) have been determined. Epimers have also been shown to exist (at C-10), and they are

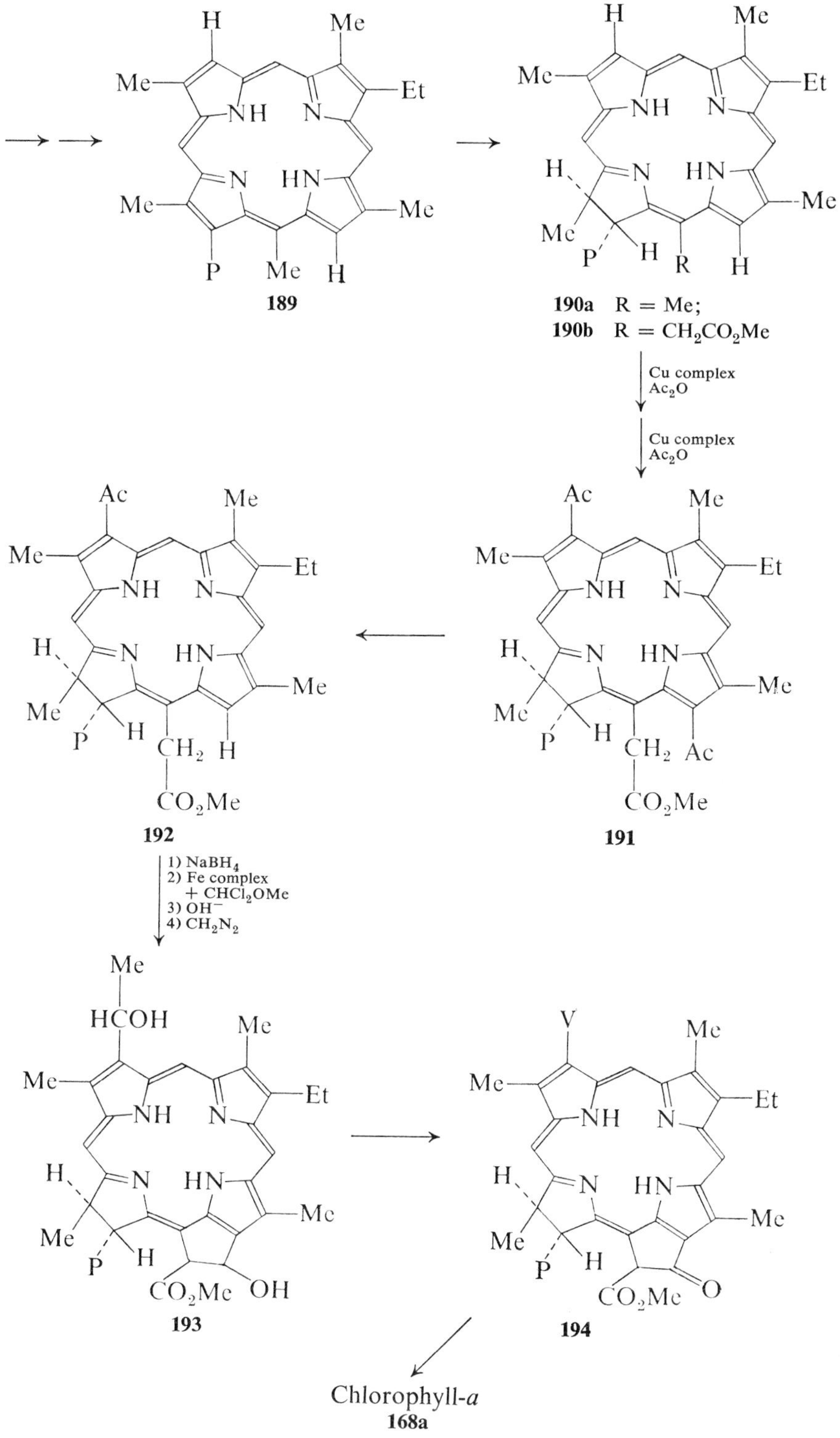

189

190a R = Me;
190b R = CH_2CO_2Me

Cu complex
Ac_2O

Cu complex
Ac_2O

191

192

1) $NaBH_4$
2) Fe complex
+ $CHCl_2OMe$
3) OH^-
4) CH_2N_2

193

194

Chlorophyll-*a*
168a

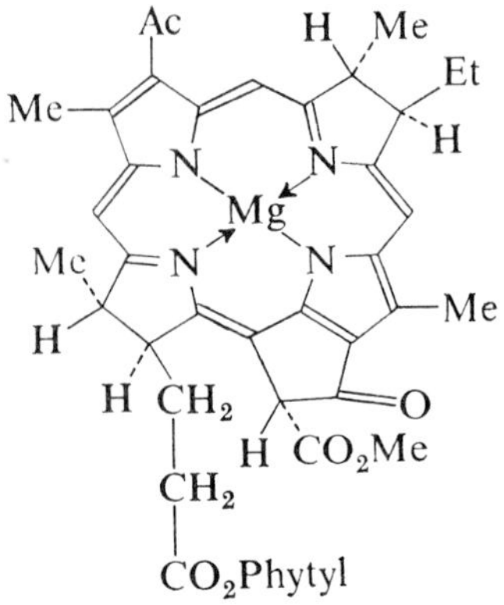

195

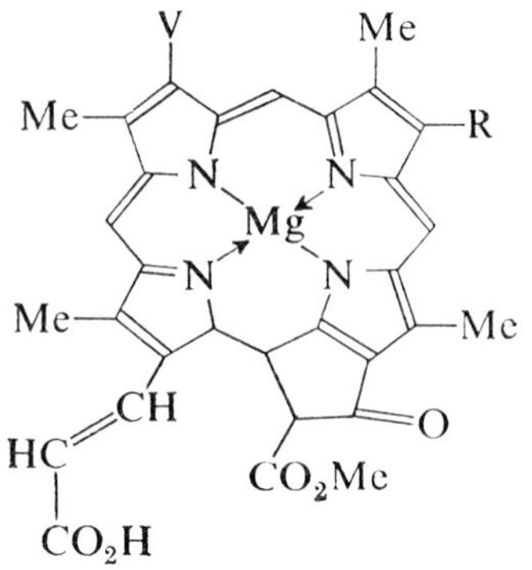

196 R = Et or V

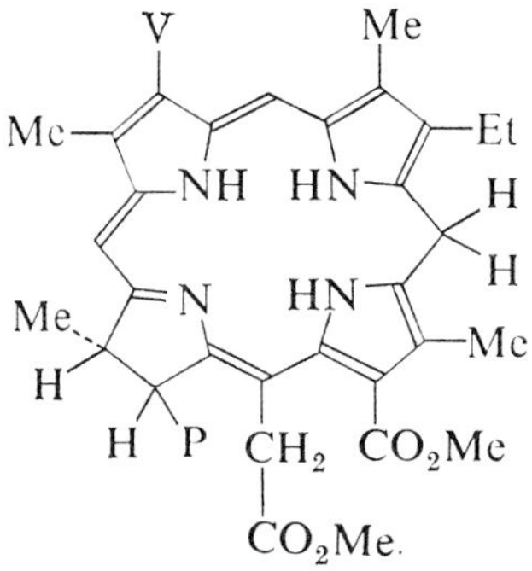

197

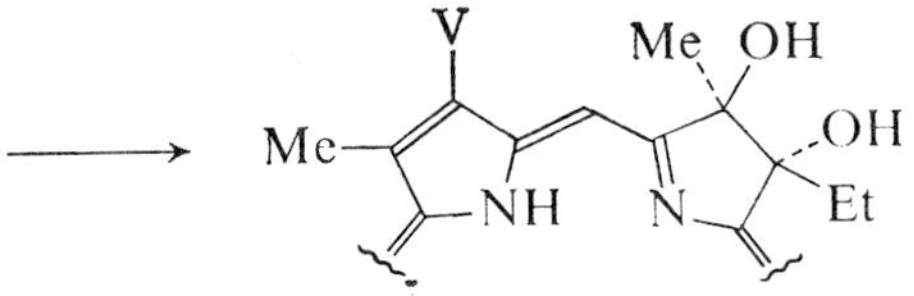

198

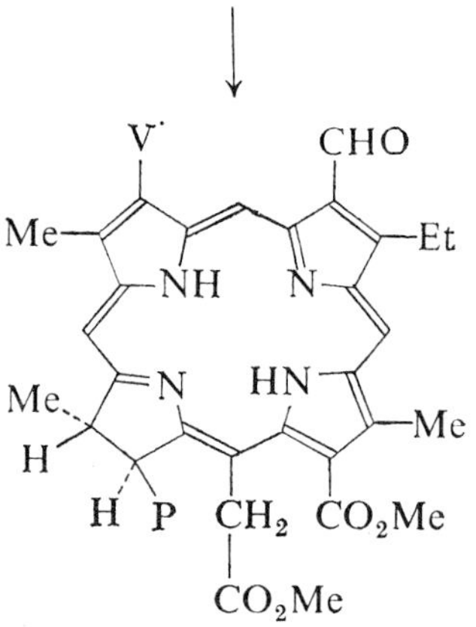

199

Chlorophyll-*b*
168b

formed in small amounts by slow equilibration of the major epimer.[142] The structures of the chlorophylls-c[143] (**196**) (which, it should be noted, are porphyrins rather than chlorins) and of the *Chlorobium* chlorophylls (**109** and **110**) have also been investigated[81,83,84] recently but they have not yet been synthesized.

The partial synthesis of rhodin-g_7 (**199**), a degradation product of chlorophyll-b, has recently been reported by Inhoffen.[134] Electrolysis of natural chlorin-e_6 (**188**) gives the "phlorin-chlorin" **197**, which undergoes photooxidation in dioxan-water to the *trans*-diol **198** (which is a bacteriochlorin). The latter was then transformed into rhodin-g_7-trimethylester **199**, and since chlorophyll-b is accessible from **199**, this constitutes a formal total synthesis of chlorophyll-b, since chlorin-e_6 was synthesized by Woodward.

9. BILE PIGMENTS

The bile pigments are open-chain tetrapyrrolic compounds derived in nature by oxidative metabolism and ring opening of the prosthetic groups of hemoproteins.[11,120] Biliverdin (**200**) is the first open-chain tetrapyrrole to be formed and is then rapidly reduced to bilirubin (**201**). With only one exception, all known bile pigments of natural origin are derived by cleavage of the

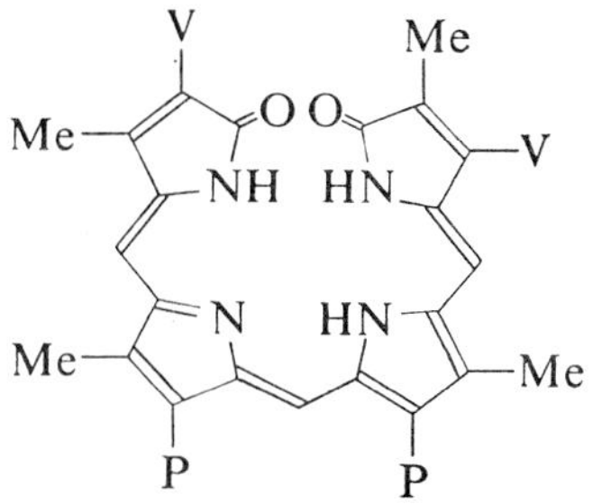

200

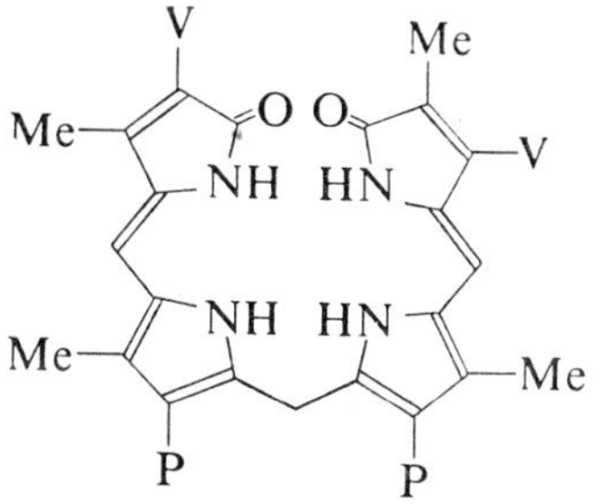

201

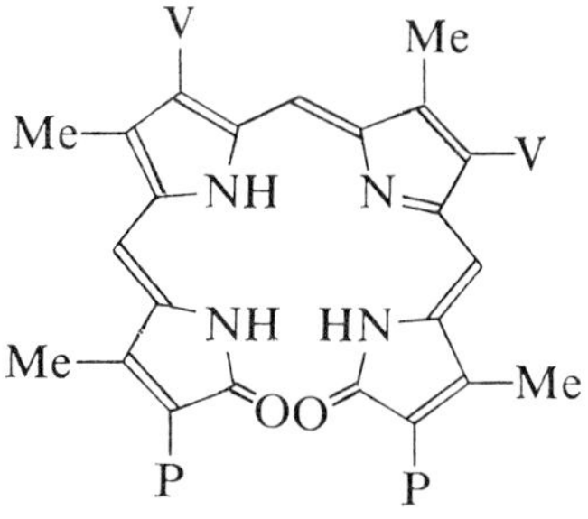

202

heme at the α-bridge and thus belong to the so-called IXα- series. The exception is the blue-green tegumental pigment of the caterpillar of the cabbage butterfly, which has recently been shown[144] to be biliverdin-IXγ (**202**) and is presumably formed by ring opening of heme at the γ-position.

The chemical transformation of porphyrins to bile pigments was first studied in the 1930s by Fischer, Lemberg, and their co-workers.[118–120] They showed that coupled oxidation of porphyrin iron complexes with hydrogen peroxide in presence of reducing agents such as ascorbic acid and hydrazine followed by hydrolysis afforded biliverdin-type pigments. Some of the earlier workers suggested that these chemical oxidations were specific and led like the natural process to IXα- pigments. However, more recent work has clearly demonstrated that an almost random oxidation of the pyridine hemochrome results in a mixture of all four possible biliverdins.[144,145] Interestingly, however, O'Carra[146] has now shown that *in vitro* coupled oxidation of myoglobin leads specifically to biliverdin-IXα, and *in vitro* oxidation of hemoglobin affords mainly biliverdin-IXα but with some of the IXβ- isomer; on the other hand, myoglobin denatured by 8*M* urea gives a random mixture of biliverdins. The X-ray evidence[147] shows that the α-bridge is the least accessible, being at the bottom of the heme crevice in the protein, whereas the γ-bridge is

1) $(PhCO_2)_2$
2) removal of metal

OCOPh

NH N N HN

203

O

NH HN N HN

204

1) $Fe(OAc)_2$
2) pyridine

Hemochromogen

1) O_2/pyridine
2) hydrolysis

Biliverdins

exposed, and O'Carra therefore suggests that the heme-binding site must have a positive effect on the position of the natural cleavage.

Intermediates in the chemical oxidation of heme and bile pigments are the oxyporphyrins (or oxophlorins as they are now known) and these can be degraded by further oxidation to biliverdins. Simple oxophlorins (**204**) can now be more conveniently prepared by oxidation of porphyrin metal complexes with benzoyl peroxide[95] followed by hydrolysis of the resulting benzoates (**203**), but unless the porphyrin is symmetrically substituted, the products will clearly be mixtures. Another, indirect, method[148] of preparing oxophlorins is by the oxidation of porphyrins with lead dioxide in acetic acid to xanthoporphyrinogens (**205**) followed by reduction with hydrogen bromide

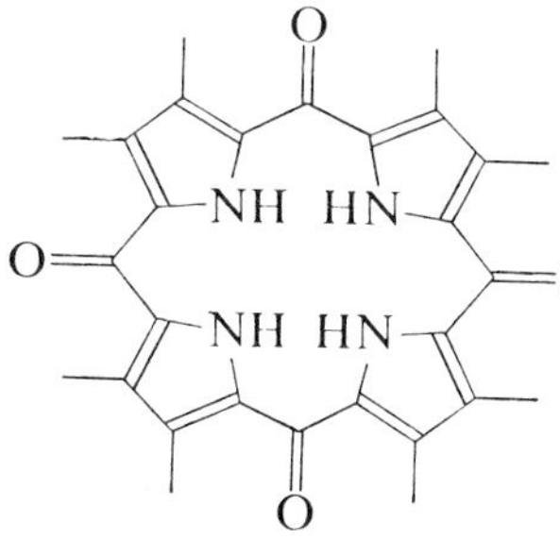

205

in acetic acid, but this method has little general synthetic utility.

The best methods for synthesizing isomerically pure oxophlorins, which can then be converted to bile pigments, are the ring syntheses described earlier, involving *b*-oxobilane intermediates, or the condensation of diformyl pyrroketones with pyrromethanes. Oxophlorins can be ruptured to give bile pigments following Lemberg's and Fischer's original procedures;[118,119] the oxophlorin pyridine hemochrome **206** is oxidized by atmospheric oxygen to give first an intermediate verdoheme **207**, which is then hydrolyzed to give

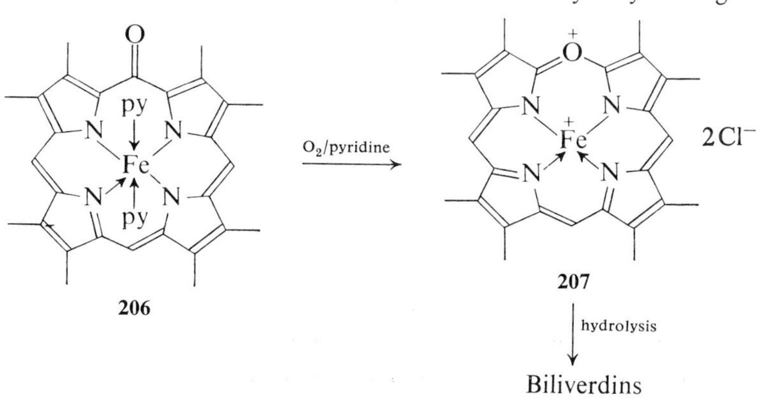

206 $\xrightarrow{O_2/\text{pyridine}}$ **207** $2Cl^-$

hydrolysis

Biliverdins

the biliverdin. In this way, for example, members of the Liverpool school have synthesized mesobiliverdin-IXα[110] and mesobiliverdin-IXβ[111] and Clezy has prepared biliverdin-IXα (**200**) itself, albeit in rather low yield.[94]

There is now little doubt that the long-held view[120] that an oxoporphyrin iron complex is an intermediate in the catabolism of heme is correct, for tritium-labeled α-oxymesoporphyrin ferriheme was shown to be converted into mesobilirubin in the rat;[122] moreover, the enzymic degradation was shown to be specific for the α-isomer, for the corresponding β-oxymesohemin was not converted into a bilirubin.[122] Oxophlorins are thus clearly of con siderable significance in their own right as well as being useful intermediates in the chemical synthesis of bile pigments.

Useful as the foregoing methods are for bile pigment synthesis, however, they would be difficult to apply directly to compounds other than biliverdins or bilirubins for many of the bile pigments recently characterized or discovered are much further reduced; for example, stercobilinogen (**208**), the final, colorless product of bacterial reduction in the gut, is a dodecahydro derivative of bilirubin. Phycoerythrobilin[149] (**209**) and phycocyanobilin[150]

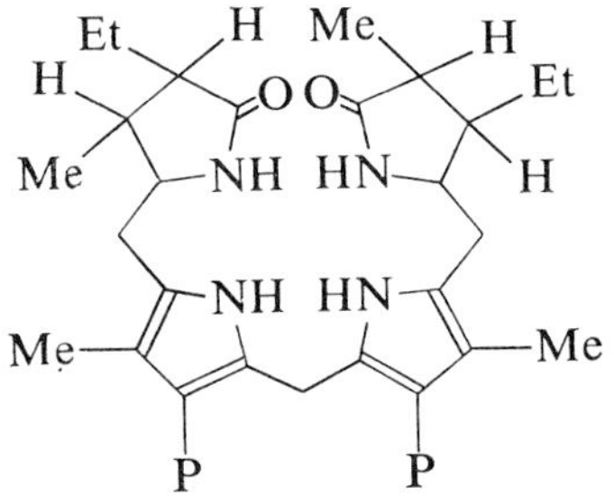

208

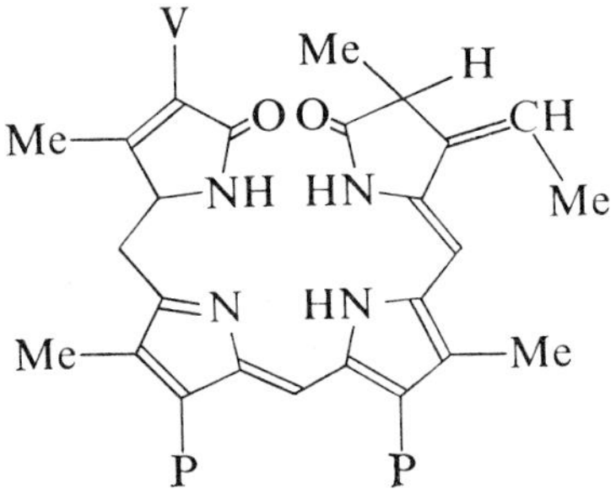

209

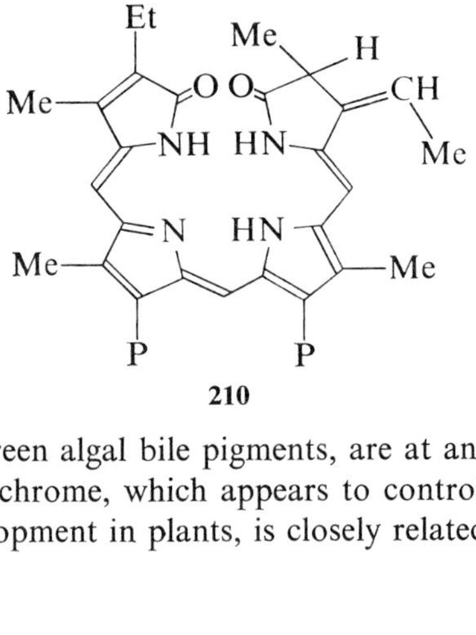

210

(**210**), the red and green algal bile pigments, are at an intermediate level of reduction and phytochrome, which appears to control the photoregulation of growth and development in plants, is closely related to phycocyanobilin, although its precise structure (**211**) is still uncertain.[149] For bile pigments of this kind the stepwise coupling of pyrrole or reduced-type units appears

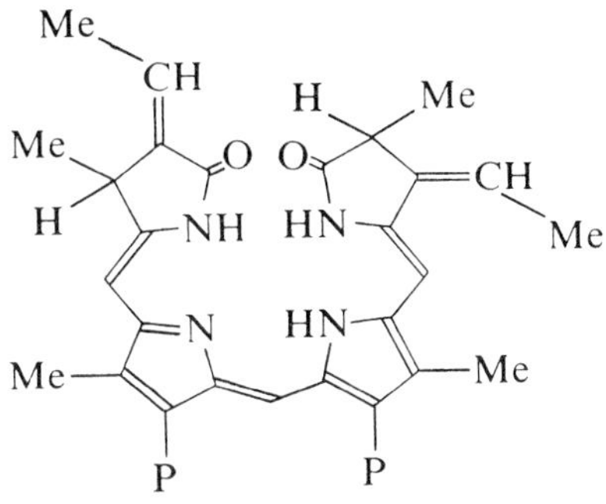

211

to be the most useful synthetic method. This is the approach that was originally developed by the Munich school and successfully continued in recent years by Plieninger's group in Heidelberg (see below).

Perhaps the main achievements in the earlier work were the confirmation of the structures of bilirubin **201** and biliverdin **200** by synthesis.[151,152] Mild hydriodic acid reduction of bilirubin affords a mixture of bilirubic acid **212a** and its isomer **213a** and oxidation of these by alkaline permanganate leads to the corresponding orange yellow methenes, xanthobilirubic acid (**214a**), and isoxanthobilirubic acid (**215a**).[153–155] The closely related neo- and isoneoxanthobilirubic acids **214b** and **215b** were obtained[156] by resorcinol

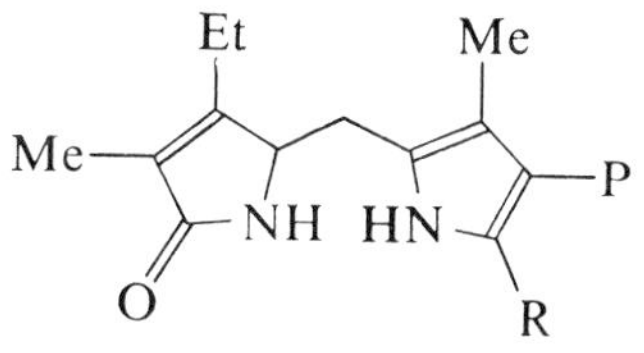

212 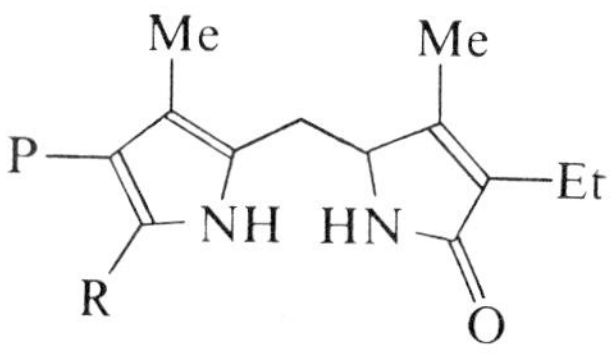 **213**

a R = Me
b R = H
c R = CHO
d R = CH_2OH

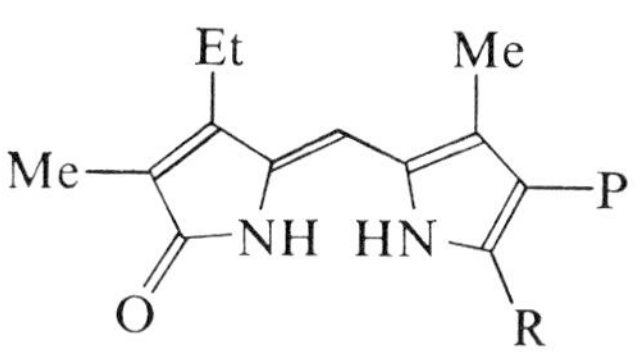

214 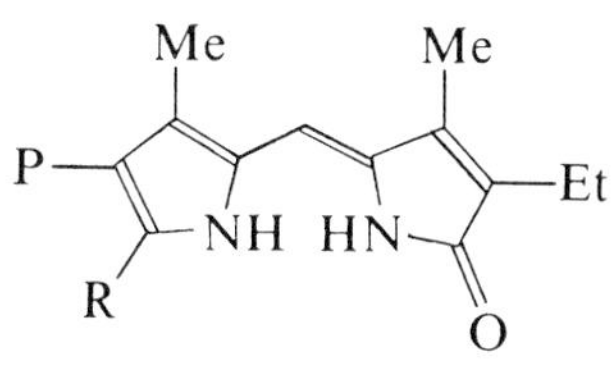 **215**

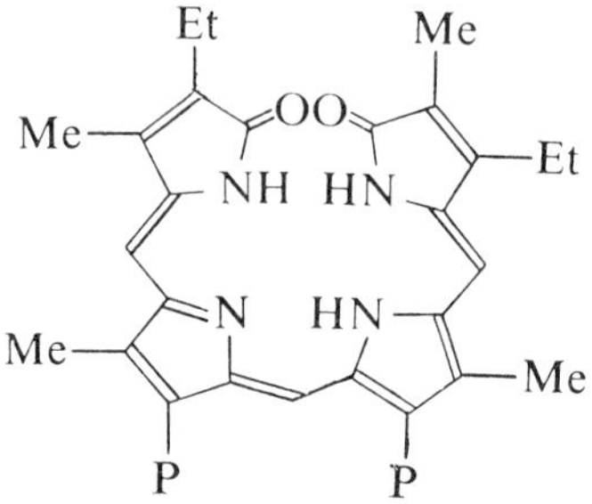

216

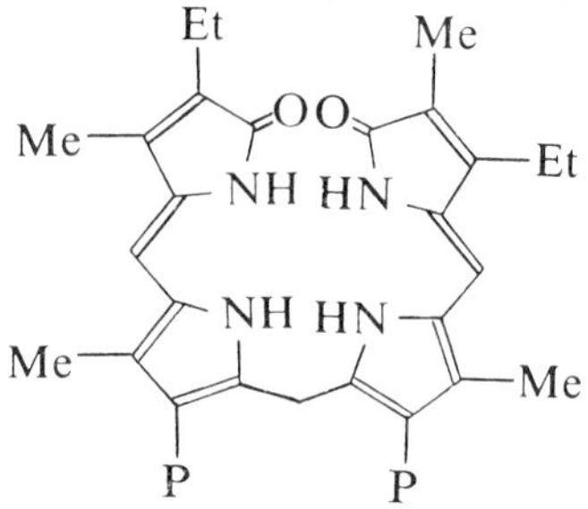

217

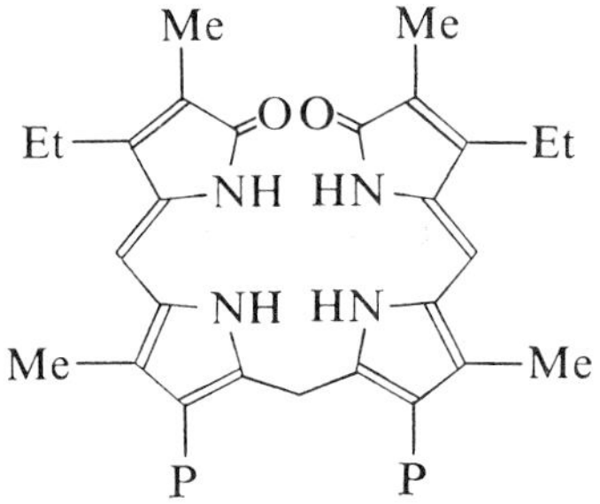

218

fusion of bilirubin, and it was shown that condensation of formyl neoxanthobilirubic acid **214c** with isoneoxanthobilirubic acid **215b** gave mesobiliverdin **216**.[157]

Reduction of the formyl acid to the corresponding hydroxymethyl derivative **214d** followed by coupling with the isoneoxanthobilirubic acid **215b** similarly afforded[151] mesobilirubin-IXα (**217**), whereas direct coupling[158] of two moles of neoxanthobilirubic acid with formaldehyde gave mesobilirubin-XIIIα (**218**).

The dipyrrolic acids were also synthesized directly from monopyrrolic precursors; for example, neoxanthobilirubic acid was obtained[159] by condensation of a bromopyrrole aldehyde with an α-free pyrrole as shown in Scheme 9, and the ring oxygen function was introduced into the intermediate pyrromethene by alkaline hydrolysis. A synthesis[159] of xanthobilirubic acid involved the oxidation of 2,4-dimethyl–3-ethyl pyrrole with peroxide to the corresponding pyrrolinone, which was then brominated in the α-methyl group and coupled with 2,4-dimethylpyrrole–3-propionic acid. The other two acids, iso-, and isoneoxanthobilirubic acids have also been synthesized by variants of the first method.[160–162] The synthesis of bilirubin and biliverdin was somewhat more complex owing to the necessity to protect the sensitive

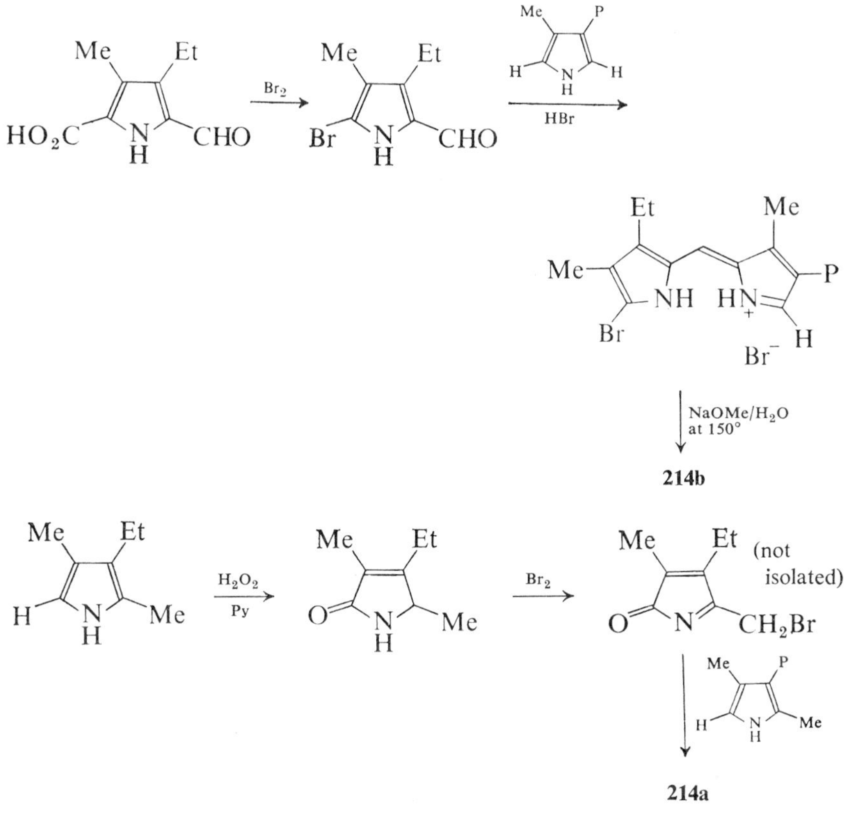

Scheme 9

vinyl groups during the intermediate stages.[152] This was effected by the use of urethane derivatives, conveniently generated from propionic acid side chains by Schmidt degradation. Thus condensation of the formyl methene **220** with the α-free methene **219** afforded the bilatriene **221**, which by hydrolysis and exhaustive methylation was converted into biliverdin-IXα (**200**). Reduction of the latter by hydrosulfite then gave bilirubin (**201**).[152]

In a similar manner formyl neobilirubic acid (**212c**) and isoneobilirubic acid (**213b**) gave[163] urobilin-IXα (**222**), and the mesobilirubin isomers "mesobilirhodin" (**223**) and "mesobiliviolin" (**224**) were also synthesized as shown.[164]

In more recent years the peroxide oxidation method for converting α-free pyrroles to pyrrolinones has been exploited by the Nottingham school,[165] particularly in relation to the synthesis of pyrrolylmethylpyrrolinones such as **226a** from pyrromethanes (**225**). Further oxidation with manganese

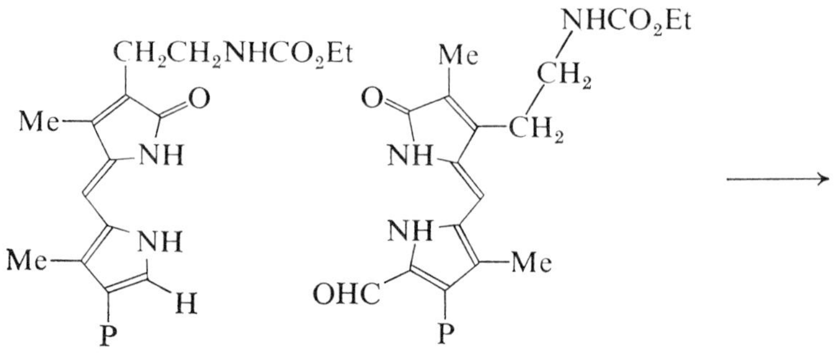

219 + **220** ⟶

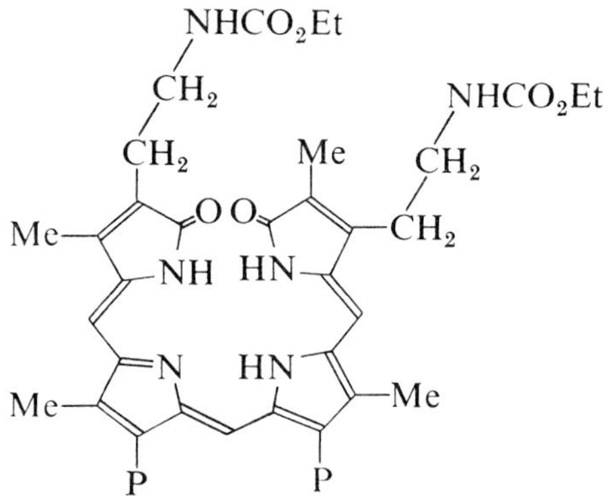

221

↓ Me_2SO_4/OH^-

Bilirubin-IXα $\xleftarrow{Na_2S_2O_4}$ Biliverdin-IXα

201 **200**

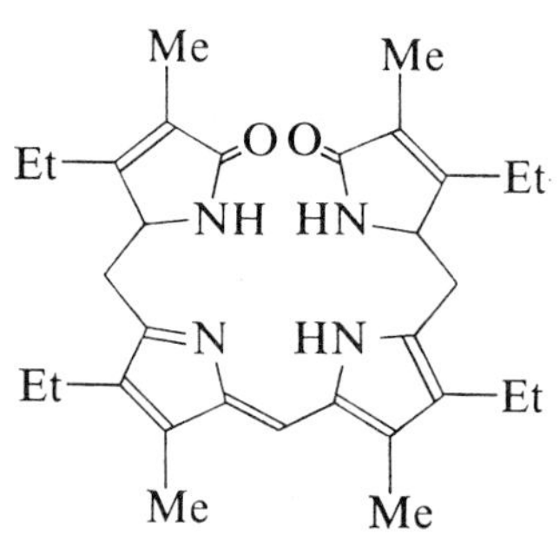

222

212c + **215b** ⟶ "Mesobilirhodin"
223

214c + **213b** ⟶ "Mesobiliviolin"
224

dioxide or dehydrogenation with palladium-charcoal affords the corresponding methylene pyrrolinones **227a**. These pyrrolinones were then hydrolyzed

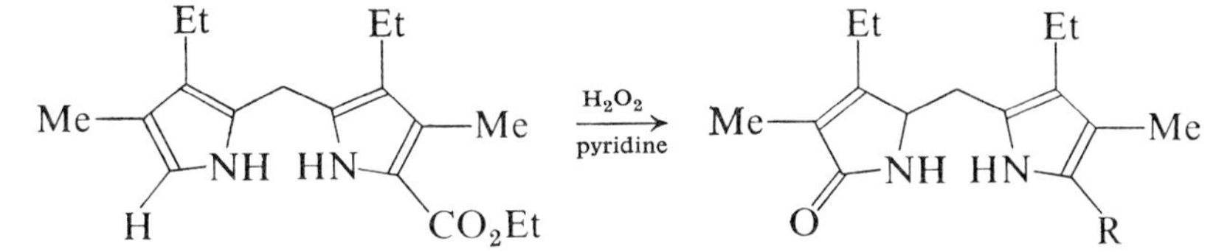

H_2O_2, pyridine

225 **226**

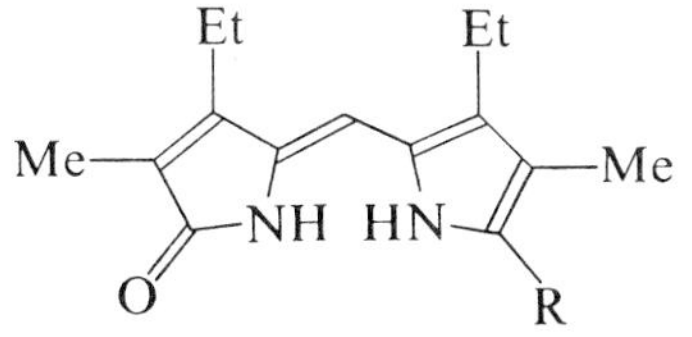

227

a R = CO_2Et
b R = H
c R = CHO

and decarboxylated to **226b** and **227b** and formylation by the Vilsmeier-Haack procedure gave the corresponding aldehydes **226c** and **227c**. Condensation of **226b** with **226c** then gave the *b*-bilene **228**, and the corresponding *a*,*b*-biladiene **229** was prepared in a similar manner from **226c** and **227b**.

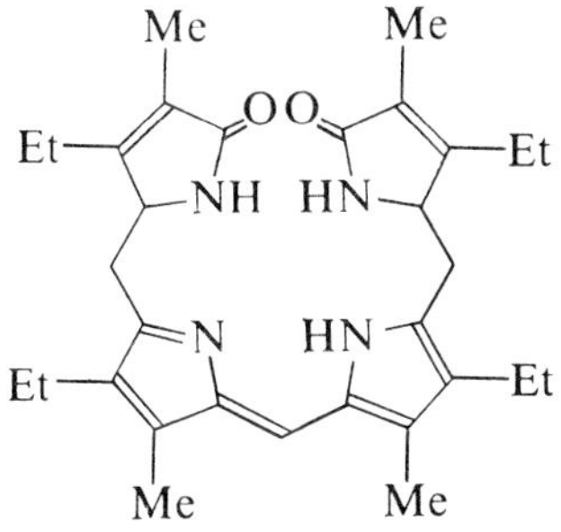

228 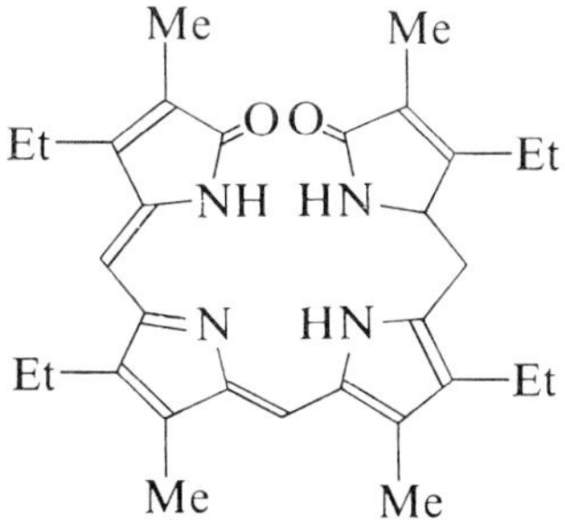 **229**

Hydrogenation of the pyrrolylmethylpyrrolinone **226a** afforded the pyrrolidone **230a** (obtained as a mixture of stereoisomers), and this was converted into the aldehyde **230c** as before; condensation of **230b** and **230c** then gave the tetrapyrrole **231**. The preparation of the tetrapyrroles **228**, **229**, and **231** provides models for the synthesis of various naturally occurring bile pigments

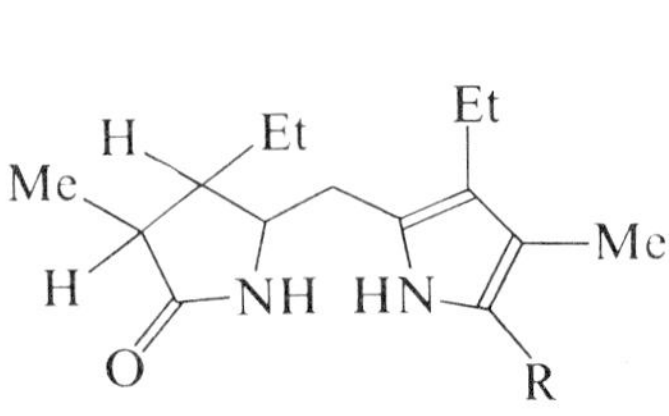

230

230a R = CO_2Et
230b R = H
230c R = CHO

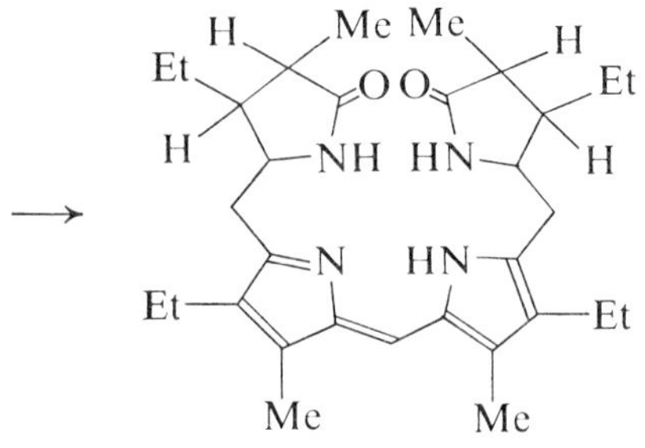

231

such as *i*-urobilin, mesobiliviolin (**224**), and stercobilin. A number of other similar model syntheses were carried by the Nottingham workers, and the use of pyrrolylmethylpyrrolinones in the synthesis of macrocyclic tetrapyrrolic structures of the dehydrocorrin type[165] was also investigated.

A variant of the peroxide oxidation procedure—direct oxidation of pyrrole and pyrromethane α-carboxylic acids—has been investigated in Liverpool,[166] for the carboxylic acids are generally more readily accessible (e.g., by catalytic hydrogenation of benzyl esters). However, mixtures of products were obtained and yields of the desired pyrrolinones were low.

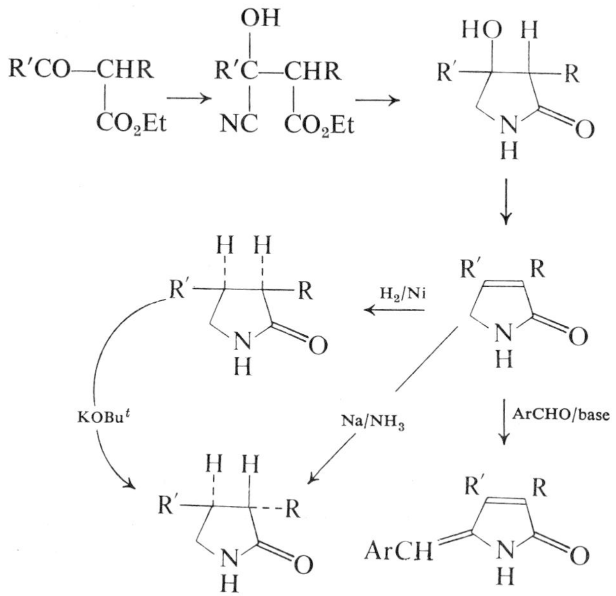

232

H_2/Ni

233

KOBu^t

Na/NH_3

ArCHO/base

234

Perhaps the most generally useful method of preparing pyrrolinones now available is Plieninger's new ring synthesis.[167] Catalytic hydrogenation of the cyanhydrins of β-ketoesters with Raney nickel in acetic acid yields a diastereoisomeric mixture of hydroxpyrrolidones (**232**), which can then be dehydrated to the desired pyrrolinones (**233**). Model experiments showed that the 5-methylene groups in **233** rapidly reacted with aromatic aldehydes in presence of base to give arylidene pyrrolinones (**234**).[167] Later it was shown that catalytic reduction of the pyrrolinones gave mainly the *cis* addition products, whereas sodium in liquid ammonia afforded the *trans* isomers; the *cis* compounds could also be isomerized to the *trans* form by treatment with potassium *t*-butoxide.[168]

Following these model experiments on simple pyrrolinones, neoxanthobilirubic acid (**214b**) and isoneoxanthobilirubic acid(**215b**) were synthesized by

Me Et (**233a**) ⟶ **214b**

Et Me (**233b**) + OHC–(Me, P pyrrole) ⟶ **215b**

condensation of 4-ethyl–3-methylpyrrolinone and 3-ethyl–4-methylpyrrolinone with formylopsopyrrole carboxylic acid (i.e., 2-formyl–3-methylpyrrole–4-propionic acid). Sodium amalgam reduction of **214b** gave the dihydroderivate **212b**, which was esterified with diazomethane (to facilitate purification) and then reduced catalytically at 115–120° over Raney nickel. The product was a mixture of stereoisomeric dihydroneobilirubic esters (**235**), which were separated by fractional crystallization. The higher melting isomer **235a** (m.p. 163–165°) condensed with ethyl orthoformate in presence of borontrifluoride etherate to give a stercobilin-XIIIα hydrochloride (**236**), m.p. 120°.[168]

For the synthesis[168] of stercobilin-IXα, the two dihydroneobilirubic acid methyl esters (**235**) were separately formylated with hydrogen cyanide under Gatterman conditions, but owing to the alkaline work-up the methyl substituent on the pyrrolidone ring underwent epimerization to give the *trans* arrangement of alkyl substituents on this ring (**237**). The other half of the stercobilin was similarly prepared from isoneoxanthobilirubic acid by sodium amalgam reduction followed by hydrogenation of the sodium salt under pressure. Owing to the alkaline conditions of the hydrogenation, the ethyl group neighboring the pyrrolidone ring carbonyl group underwent epimerization and the product obtained was an inseparable mixture of epimeric acids (**238**). This mixture was condensed with each of the formyl dihydroneobilirubic

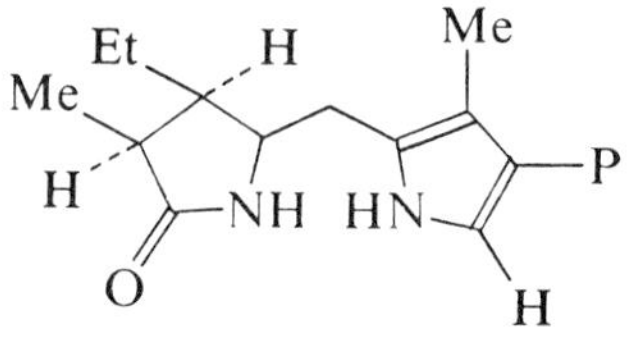

235a
235b

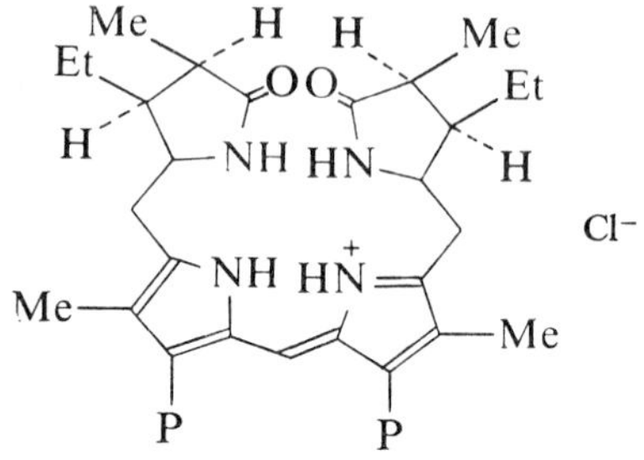

236

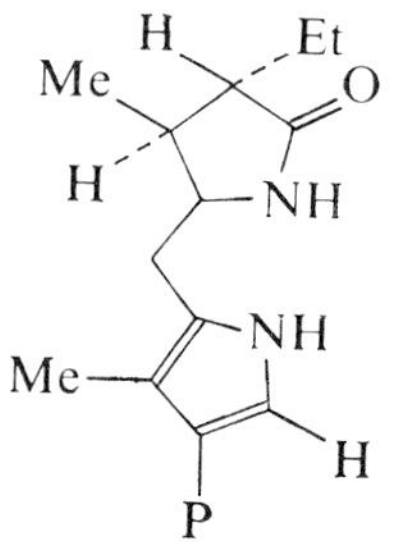

238

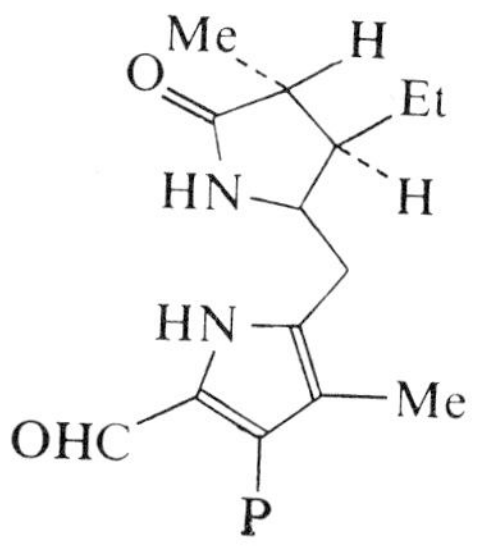

237
237a
237b

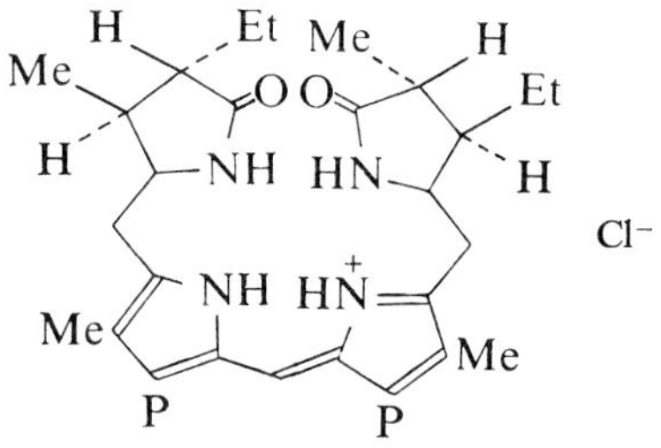

239a
239b

esters **237a** and **237b** in acetic acid containing hydrogen bromide, giving two stercobilins (**239**) which were isolated as their hydrochlorides, m.p. 208–209° and 166–169°, respectively. The IR, UV, and chromatographic properties of these two compounds were practically indistinguishable from those of natural stercobilin hydrochloride.[168] The product obtained earlier by perhydrogenation of bilirubin over palladium-charcoal to stercobilinogen followed by reoxidation with ferric chloride also had very similar properties.

In more recent work, Plieninger and Ruppert have synthesized[169] the dipyrrolic dicarboxylic acids **240a** and **241a** by condensation of 2-ethoxycarbonyl–3-ethoxycarbonylethyl–4-methyl–5-formylpyrrole with the two pyrrolinones **233a** and **233b** in 40% sodium hydroxide. (These dicarboxylic acids on heating with methanolic sulfuric acid gave neo- and isoneoxanthobilirubic acid methyl esters, **214b** and **215b**, respectively.) If the condensation was carried out with sodium ethoxide in ethanol then the corresponding esters, **240b** and **241b**, were obtained. Hydrogenation of the oxodipyrromethenes **240b** and **241b** over palladium on barium sulfate at 1 atm and 25°

gave the dihydrodiesters **242** and **243**, respectively, in ~60% yields. Hydrogenation of **241b** over Raney nickel at 120° and 130 atm gave an oily tetrahydro derivative which was fractionally crystallized to give a 60% yield of the all-*cis* diester **246**. The epimer **247**, m.p. 125°, was isolated in 10% yield from the mother liquors, and the structures of the two products were confirmed by NMR assignments. Similarly, high-pressure hydrogenation of the isomeric

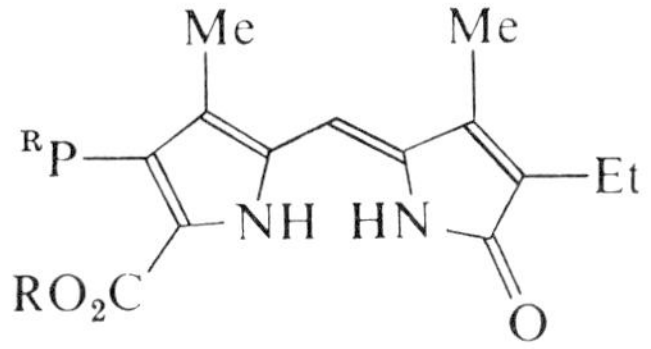

a R = H
b R = Et

241

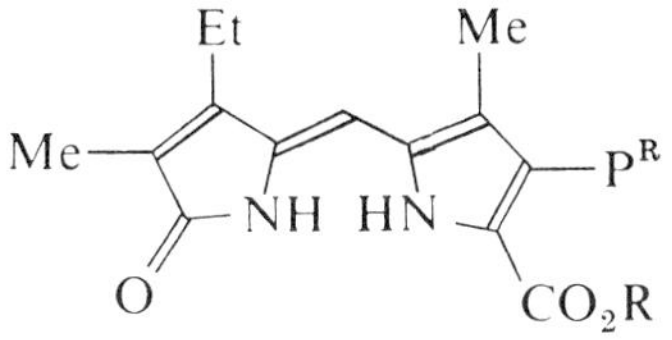

240

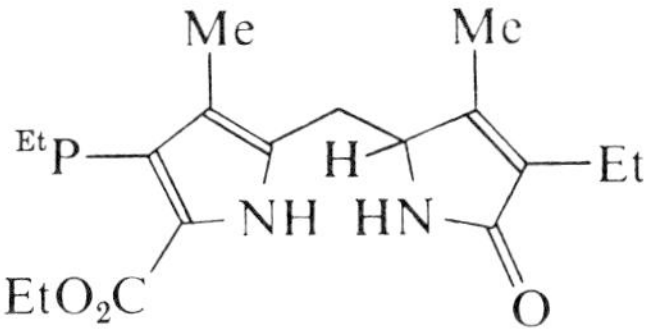

243

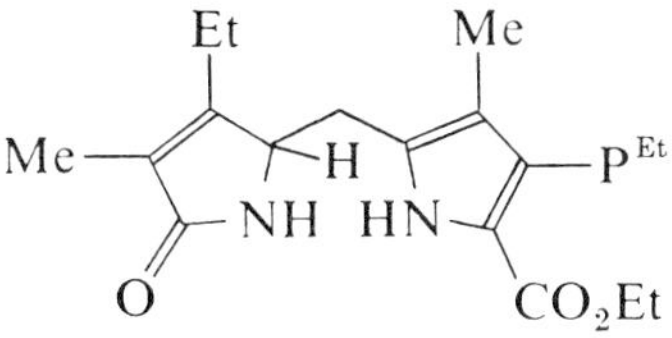

242

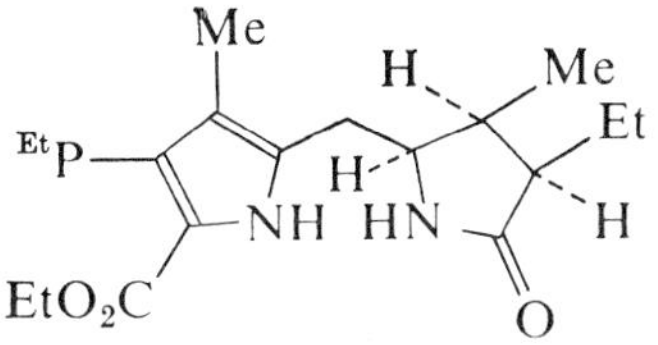

246

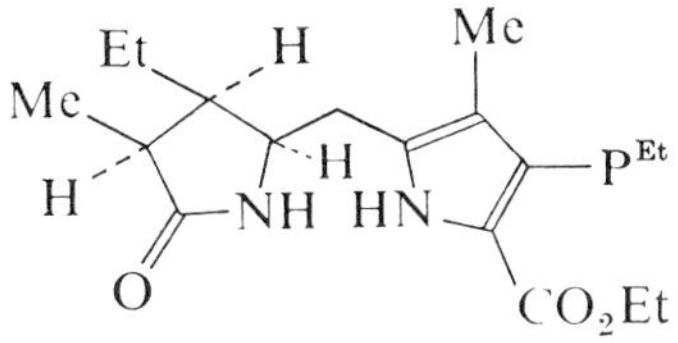

244

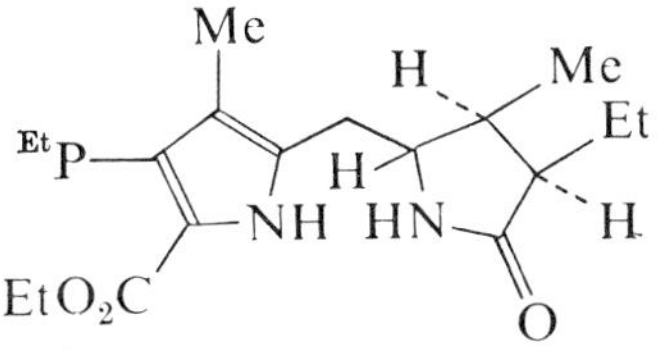

247

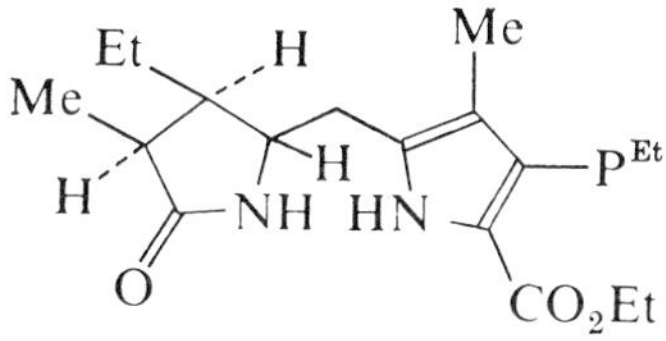

245

diester **240b** gave a mixture of stereoisomeric tetrahydro derivatives **244** and **245**, which were separated by fractional crystallization, and again assigned structures on the basis of their NMR spectra. The chemical shifts of the α-proton in the pyrrolidone ring were ~0.2 ppm to lower field in the all-*cis* compounds **244** and **246** than in their isomers **245** and **247**, respectively, and the m.p. of the all-*cis* compounds were also higher than their isomers.[169]

Alkaline hydrolysis of three of these diesters (**244**, **245**, and **246**) gave the dicarboxylic acids **248a**, **249a**, and **250a** with epimerization of the alkyl groups on the carbon atoms neighboring the pyrrolidone carbonyl groups. These dicarboxylic acids were then resolved as their brucine salts. Thermal decarboxylation of (+) and (−) **250a** and sublimation afforded the corresponding (+) and (−) monocarboxylic acids (**250b**) in high yield, the progress of the decarboxylation being followed by the diminution in the absorption maximum at 282 nm due to the pyrrole α-carboxylic acid chromophore.

Condensation of the (+)-dihydroisoneobilirubic acid (**250b**) with orthoformic ester in hydrogen bromide-acetic acid at 50° then gave (+)-stercobilin-IIIα (**251**), isolated as its hydrochloride $[\alpha]_D^{20} = 2830°$ in chloroform, raised

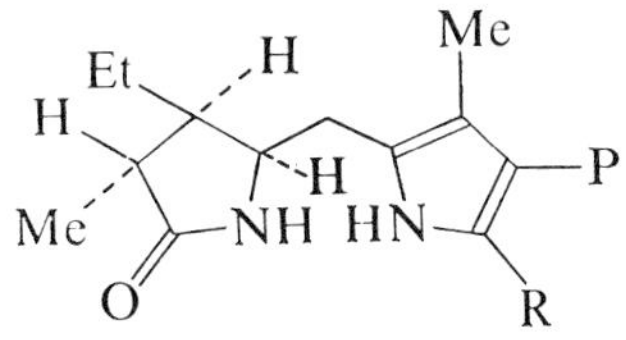

248

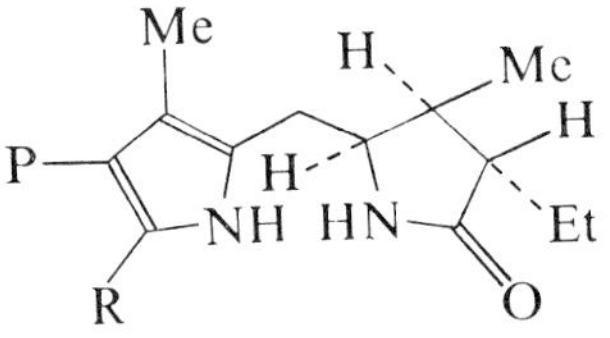

249

(+) and (−) forms

250a R = CO_2H
250b R = H
250c R = CHO

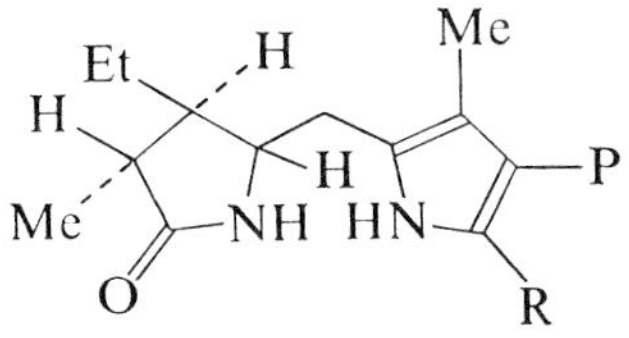

250b + $HC(OEt)_3$ $\xrightarrow{HBr/HOAc}$

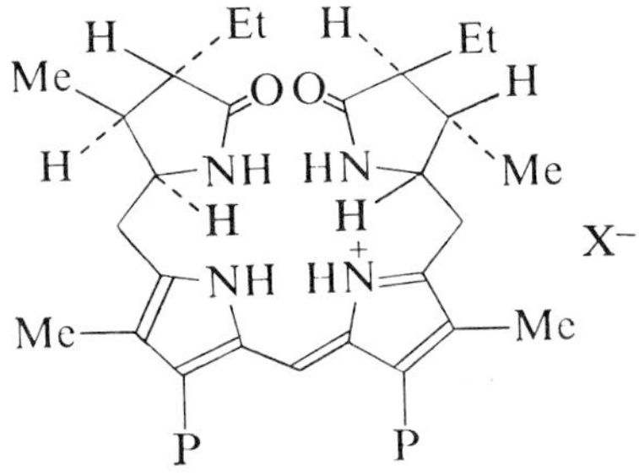

251

(+) and (−) forms

to +3460° after three crystallizations from chloroform-acetone. (−)-Stercobilin-IIIα hydrochloride (**251**) $[\alpha]_D^{20} = -3570°$ was also prepared in a similar manner from the (−)-acid **250**.[169]

Unfortunately only the dicarboxylic acid **250a** could be resolved, and so the other two racemic acids **248a** and **249a** were thermally decarboxylated and formylated (with HCN/HCl) to give the epimeric aldehydes **248c** and **249c**. Condensation of (−)-**250b** with the racemic aldehyde (**248c**) in acetic acid containing hydrogen bromide gave a diastereoisomeric mixture of the two products **252** and **253**. Similar condensation of the other aldehyde (**249c**) with (−)-dihydroisoneobilirubic acid (**250b**) gave a diastereoisomeric mixture of the two further products **254** and **255**.[169]

The crude product **253** from the first reaction had an optical activity $[\alpha]_D^{20} = -1040°$, raised by five recrystallizations from chloroform and one crystallization from methanol-ethyl acetate to −3570°. The mother liquors from the crude product, and the first recrystallization, showed no optical activity, because the other product (**252**) is a *quasi-meso* type structure. (The

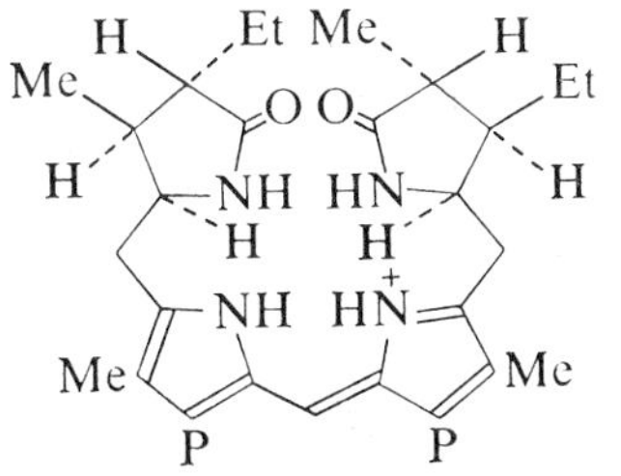

252

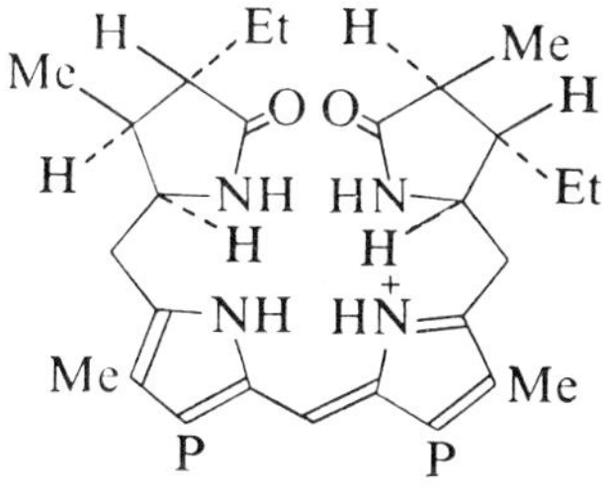

253

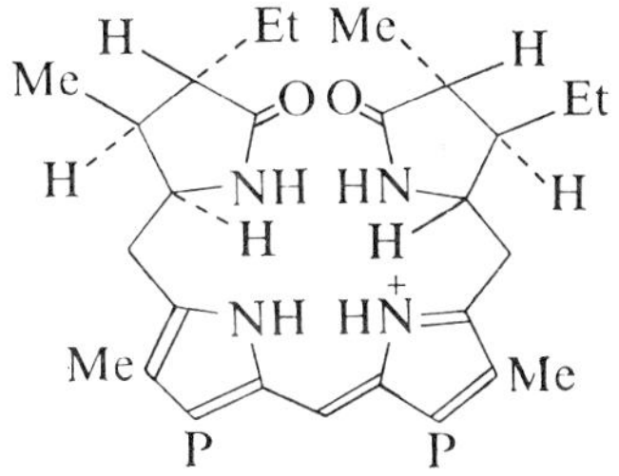

254

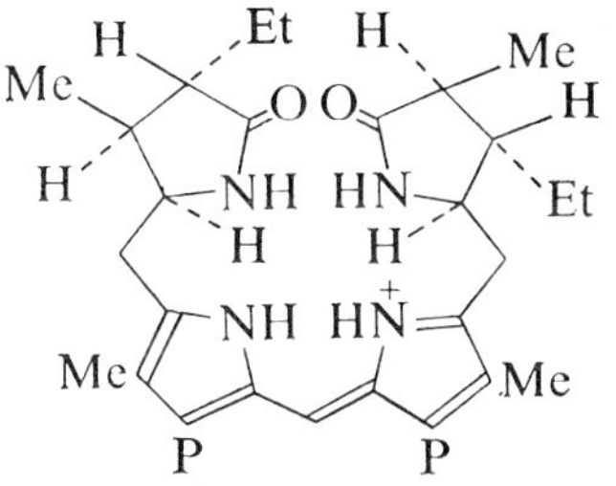

255

two end rings are not identically substituted, and so it is not strictly a *meso*-compound). The optically active (−)-stercobilin hydrochloride **253** prepared[169] in this way slowly decomposed above 160° and melted at 175–177°, whereas the natural product was reported to decompose at 157–162°.

One of the second pair of products is also a quasi-*meso* form (**255**) and the other after isolation and repeated recrystallization gave another (−)-stercobilin hydrochloride (**254**) $[\alpha]_D^{20} = -4200$, which decomposed slowly above 158–160°, and melted at 177–183°. The two synthetic (−)-stercobilins had electronic spectra and chromatographic behavior identical with natural stercobilin.[169]

However, comparison of X-ray powder photographs showed that the (−)-stercobilin hydrochloride **253** (prepared from the aldehyde **248c**) and (−)-dihydroisoneobilirubic acid **250b** was identical with the natural material.[169] Thus the relative configurations of the two end rings are now known, but their absolute configurations remain to be determined.*

Plieninger and Ruppert also synthesized (+)-stercobilin hydrochloride by condensation of (+)-dihydroisoneobilirubic acid (**250b**) with the racemic aldehyde **248c**; the crude product had $[\alpha]_D^{20} = +2800°$ raised to $+3430°$ by recrystallization from chloroform-acetone.

The precise nature of the intermediates between bilirubin **217** and the colorless stercobilinogen **209** and the order in which they are formed in nature is still not entirely clear. (+)-*d*-Urobilin, for example, is thought[121] to be the hexahydrobilirubin derivative **256a** or **256b** containing one vinyl group, while *i*-urobilin (**257**) is the corresponding octahydro analog in which both the original vinyl groups of bilirubin are reduced.[171] The latter, whether

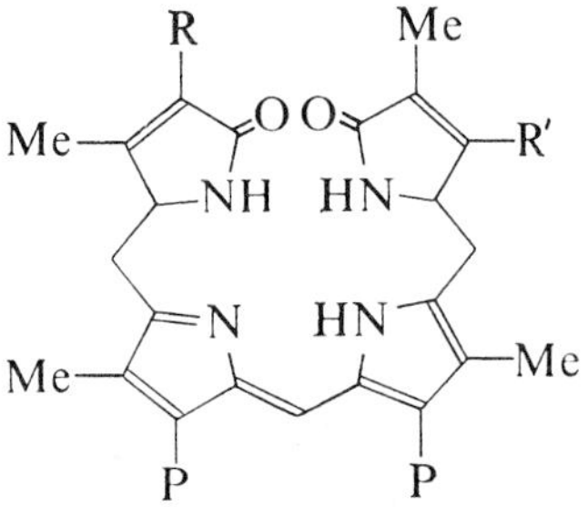

256a R = Et; R′ = V
256b R = V; R′ = Et

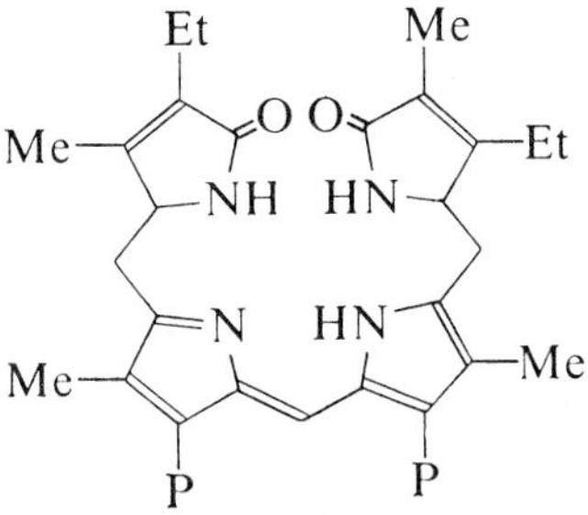

257

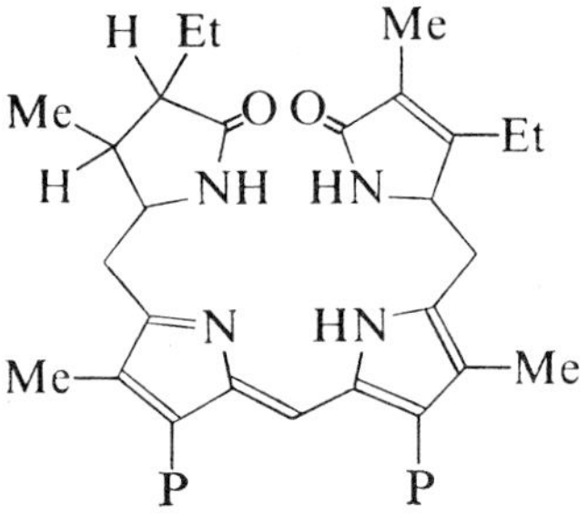

258

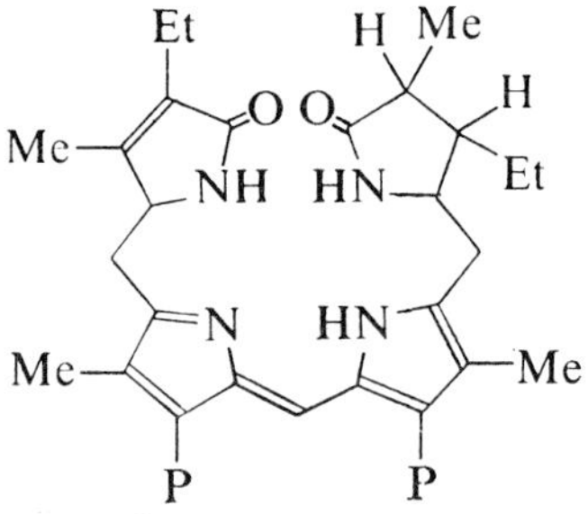

259

* Added in proof: Degradations of the natural and synthetic products to methylethylmaleimides by Dr. Brockmann have now shown that the terminal rings have the RR and SS configurations respectively: hence the natural product has the RRS . . . SRR configuration and the synthetic material the SSS . . . SSS configuration. (Prof. H. Plieninger, private communication).

isolated from urine or feces or of synthetic origin, is optically inactive. A further reduction step affords a "half-stercobilin,"[172] **258** or **259**.

Some of these compounds (e.g., *d*-urobilin analogs) have now been synthesized by Plieninger and co-workers.[173] The formyl pyrrole **260** was condensed in presence of methanolic sodium hydroxide with the protected aminoethyl pyrrolinones **261** and **262** to give the yellow oxodipyrromethenes **263** and **264**. Catalytic hydrogenation of the benzyloxycarbonyl protecting groups and reduction of the pyrromethenes afforded the aminoethyl oxodipyrromethanes **265a** and **266a**. Exhaustive methylation followed by brief heating with alkali under argon then gave the desired monovinyloxopyrromethanes **265b** and **266b**. These products showed a tendency to disproportionate to neo- and isoneoxanthobilirubic acids **214b** and **215b** by rearrangement of hydrogen from the methane bridges to the vinyl groups, and hence were used directly in the next stages of the syntheses without purification. The formyl neo- and isoneobilirubic acids **212c** and **213c** were prepared by the Vilsmeier procedure and condensed in acetic acid/hydrobromic acid with the vinyl pyrromethanes **265b** and **266b**, respectively, to give the

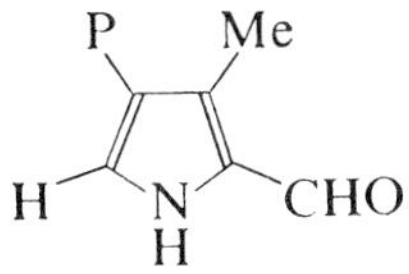

260 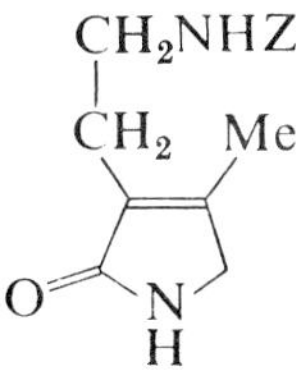 **261** 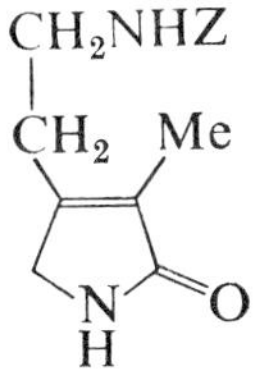 **262**

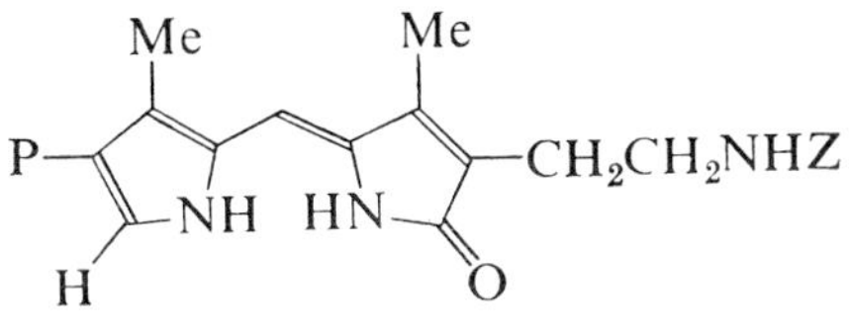

263 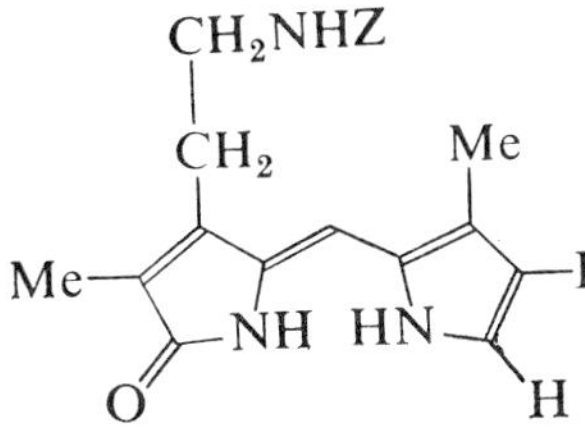 **264**

$Z = CO_2CH_2Ph$

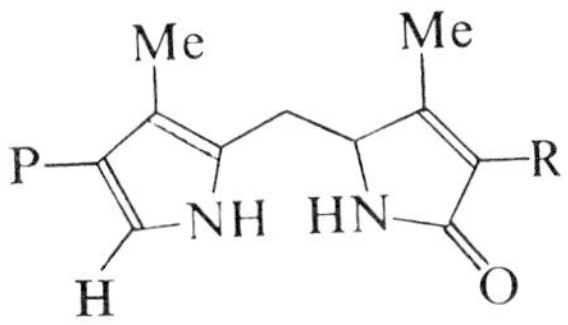

265a $R = CH_2CH_2NH_2$
265b $R = V$

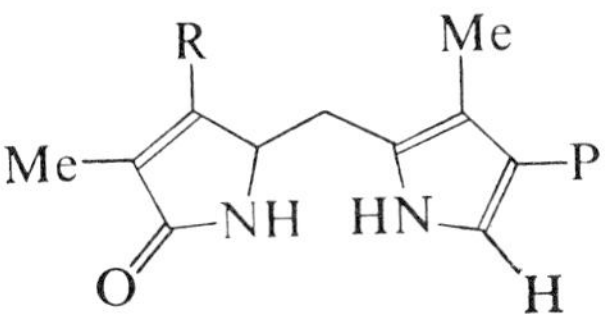

266a $R = CH_2CH_2NH_2$
266b $R = V$

racemic monovinyl urobilins **256a** and **256b** (isolated as their hydrochlorides in 6 and 7% yields, respectively).

These synthetic products were difficult to compare[173] directly with the natural *d*-urobilin, but the UV spectrum of **256a** showed a band at 230 nm (also occurring in that of the related monocyclic vinylpyrrolinone) which was absent in the UV spectrum of **256b** and in that of the natural product. Mass spectral studies were complicated by the tendency of this type of bile pigment to undergo disproportionation in the mass spectrometer.[174]

Plieninger has also described[175] syntheses of optically active urobilins of the III, XIII, and IX series. Optically active neo- and isoneobilirubic acids, **212b** and **213b**, were prepared by resolution of the corresponding dicarboxylates and decarboxylation either by heating their barium salts in water, or by

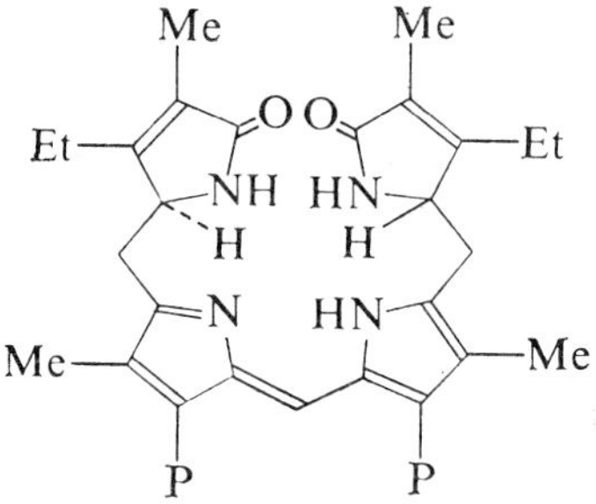

267

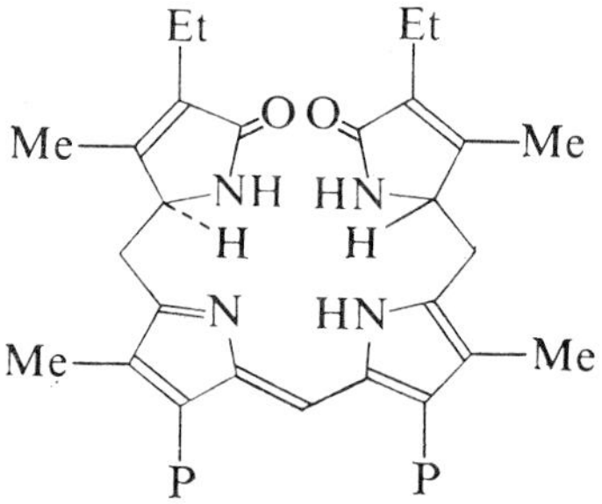

268

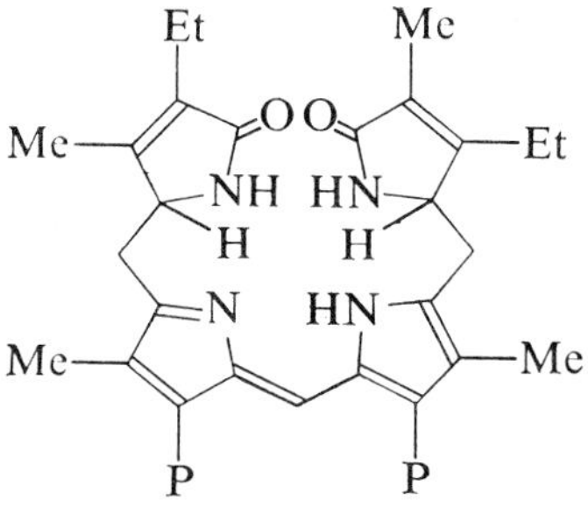

269a

Enantiomer = **269b**

warming the quinine salts in acetic acid, a better method which avoids any racemization. The (+)-acids **212b** and **213b** on self-condensation with orthoformate ester in hydrogen bromide/acetic acid, gave (+)-urobilin-XIII (**267**) and (+)-urobilin-III (**268**), respectively, ($[\alpha]_D^{20} = +4{,}500 \pm 100$ for both products). In the synthesis of optically active urobilin-IXα the racemic formylisoneobilirubic acid (**213c**) was condensed separately with both the (+)- and (−)-enantiomers of neobilirubic acid (**212b**). Each diasterenisomeric mixture produced[175] was crystallized three times to give two urobilin-IX hydrochlorides, the first with $[\alpha]_D^{20} = +4500°$ and the other with $[\alpha]_D^{20} = -4800°$. These were presumably the RR and SS enantiomers (**269**) and showed pronounced Cotton effects at λ_{max}, 499 nm, positive for the dextrorotatory form, and negative for the levorotatory form.

A new and potentially very useful route to pyrrolinones (e.g., **233**) has recently been reported,[176] which makes use of an internal Emmons reaction. Typically a 2,2-diethoxyalkylamine (**270**) and an α-diethyl phosphonalkanoic

R′—C(OEt)(OEt)—CH₂—NH₂ + H—C(R)(CO₂H)—P(=O)(OEt)₂ —[1) DCC; 2) H⁺]→

270 **271**

R′—CO—CH₂—NH—CO—CH(R)—P(=O)(OEt)₂ —[NaH]→ pyrrolinone (R′, R; N–H; C=O)

272 **233**

acid (**271**) in the presence of DCC gave, after acidic work-up, the keto-amide **272**. The latter was cyclized, using sodium hydride, to the required pyrrolinone **233** in an overall yield of about 50%. Applications of this procedure to bile pigment synthesis are awaited with interest.

10. PRODIGIOSIN AND RELATED COMPOUNDS

Prodigiosin is a tripyrrolic red pigment occurring in the bacterium *Serratia marcescens* formerly known as *Bacillus prodigiosus*. This organism occurs widely in both soil and water, and prodigiosin itself has considerable antibiotic and antifungal activity, but it is too toxic for therapeutic use. A number of possible structures for prodigiosin were put forward by Wrede[177] in 1933, but the one which he later favored[178] contained a tripyrrylmethene moiety. However, some ten years ago Rapoport[179] showed, by synthesis, that the true structure (**273**) was a pyrrolylpyrromethene, two of the pyrrole rings being joined together by a direct link. More recently, several analogs have been discovered in other bacteria, including nonylprodigiosin[180] (**274a**), undecylprodigiosin[181] (**274b**), metacycloprodigiosin[182] (**275**), and a cyclic nonylprodigiosin[183] (**276**).

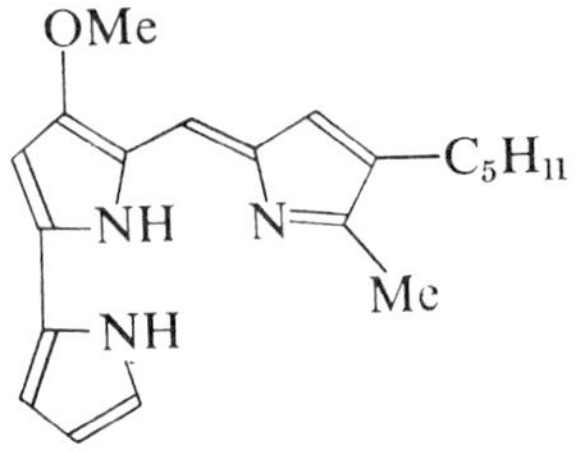

273

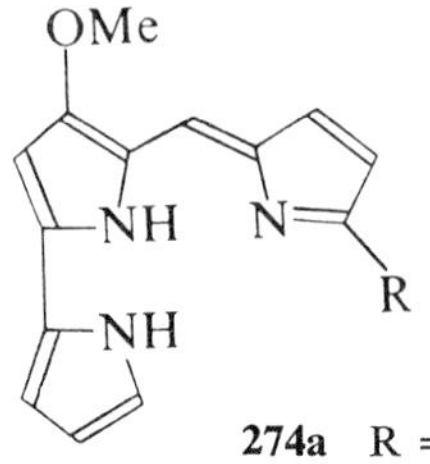

274a R = C_9H_{19}
274b R = $C_{11}H_{23}$

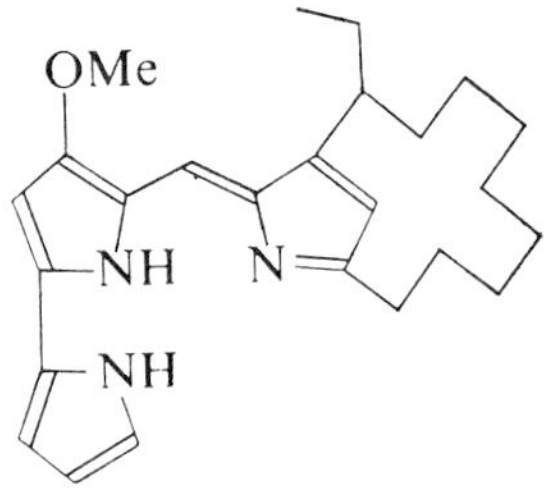

275

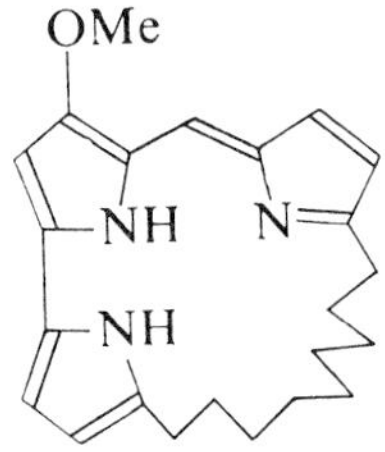

276

The crucial piece of evidence in the structure determination of prodigiosin was the isolation[184] of a dipyrrolic compound, $C_{10}H_{10}N_2O_2$, from a mutant strain of *S. marcescens*. This compound was shown to be a true precursor of prodigiosin by tracer experiments;[184] moreover, it could be converted[185] into prodigiosin by acid-catalyzed condensation with 2-methyl–3-*n*-amylpyrrole. Rapoport soon afterward showed by synthesis[179] that the C_{10}-precursor was the formyl bipyrrole **277** rather than a dipyrroketone as would have been expected on the tripyrrylmethene formulation.

The synthesis[179] of **277** required the development of new methods for preparing 3-methoxypyrroles and 2,2′-bipyrroles (other methoxy pyrroles were also prepared in the course of this work for comparative purposes).[179,186]

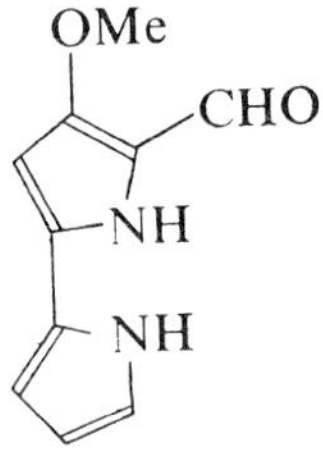

277

EtO_2C CO_2Et EtO CH_2CO_2Et HN CO_2Et $\xrightarrow[\text{xylene}]{\text{Na}}$

EtO_2C O N CO_2Et CO_2Et $\rightleftharpoons$ EtO_2C OH N CO_2Et CO_2Et

EtO_2C OMe N Me CO_2Et + EtO_2C OMe N H CO_2Et $\xleftarrow{CH_2N_2}$ EtO_2C OH N H CO_2Et

278

1) conc. H_2SO_4
2) heat

OMe N H CO_2Et

279

279 + N → OMe CO_2Et NH NH → OMe CO_2Et NH NH

280 **281**

C_5H_{11} N H Me

273 $\xleftarrow{MeOH/HCl}$ **277**

Base-catalyzed condensation of ethyl *N*-ethoxycarbonylglycinate and diethyl ethoxymethylene malonate afforded the diester **278**, which was then methylated, selectively hydrolyzed with concentrated sulfuric acid, and decarboxylated to give the 3-methoxypyrrole–2-carboxylate **279**.

Previous syntheses of bipyrroles were of little use in the prodigiosin work because they generally led to highly substituted symmetrical products, and drastic conditions were used for Ullman-type coupling reactions. However, an earlier report that Δ′-pyrroline would couple with pyrrole on heating to give pyrrolidinylpyrrole led Rapoport and Holden[179] (after conducting model experiments) to heat the ester **279** with Δ′-pyrroline. The resulting pyrrolidinylpyrrole **280** was then dehydrogenated by refluxing in xylene over 5% palladium on carbon to the bipyrryl ester **281**. Conversion of the ester function to an aldehyde was achieved by the McFadyen-Stevens reduction[188] (because model experiments had revealed that such esters were somewhat inert to aluminum hydrides). The product **277** was found to be identical with the dipyrrolic compound isolated from *S. marcescens*, and its condensation with 2-methyl–3-amylpyrrole then gave a red pigment identical in all respects to natural prodigiosin (**273**).

Rapoport also investigated other methods for synthesizing bipyrroles,[189] because these compounds are clearly of importance not only in relation to prodigiosin but also to the synthesis of vitamin B_{12} and the corrins.[100] The most useful outcome of this work was the Vilsmeier-type condensation of 2-pyrrolidone with pyrrole followed by dehydrogenation to give bipyrrole **282** itself. In later work Rapoport showed that the dehydrogenation step could be avoided by the use of 2-pyrrolinones (**233**);[189b] other methods for synthesizing bipyrroles have also since been described by Johnson and Grigg.[190]

The synthesis of undecylprodigiosin (**274b**) was readily accomplished[181] by condensation of the formyl bipyrrole **277** with 2-undecyclpyrrole in ethanolic

O NH POCl$_3$ → NH OPOCl$_2$ $\overset{+}{N}$H → NH N → NH NH R R′

282

O NH R R′

233

hydrogen chloride; this confirmed the structural assignments and also served as a formal total synthesis.

Metacycloprodigiosin presents a more formidable synthetic objective owing to the necessity to construct a "meta"-bridged cycloalkylpyrrole (**290**). However, this was achieved by a multistage procedure starting from cyclo-dodecanone.[182b] Sodamide-catalyzed alkylation in glyme gave the 2-ethyl derivative, which was converted into its ketal and brominated with pyridine hydrobromide perbromide. The resulting bromo-ketal (**283**) was dehydro-brominated with 1,5-diazabicyclo(4,3,0)-non-5-ene to give the unsaturated ketal **284**, which was hydrolyzed to the ethyl cyclodocenone **285**. Epoxidation of the double bond gave a mixture of diastereoisomers, which, on treatment with hydrazine in an aqueous ethanol containing a catalytic amount of acetic

1) $EtBr/NaNH_2$
2) $(CH_2OH)_2$ TosOH
3) $PyHBr/Br_2$

283

$-HBr$

284

H_3O^+

285

1) H_2O_2/OH^-
2) N_2H_4

286

$Na_2Cr_2O_7/H^+$

287

KCN/NH_4Cl

288

1) $(CH_2OH)_2$ TosOH
2) $(iBu)_2AlH$

289

1) H_3O^+
2) $(NH_4)_2CO_3$ in H_2O/DMF

290

EtOH/HCl + **277**

275

acid, afforded 4-ethyl-2-cyclododecenol (**286**). Oxidation of **286** with acid dichromate gave the dodecenone **287**, which underwent 1,4-addition of cyanide ion. The cyanoketone **288** was converted into its ketal and reduced to the formyl ketal **289** with diisobutyl aluminum hydride. Hydrolysis followed by treatment of the resulting keto-aldehyde with ammonium carbonate then gave the *dl*-metacyclopyrrole **290**. The latter was identical with a pyrolysis product of natural (−)-metacycloprodigiosin and was converted[182b] into *dl*-metacycloprodigiosin by condensation with the known formylbipyrrole (**277**). The product had UV, visible, IR, NMR, and mass spectra identical with those of the natural prodigiosin. The cyclic nonyl prodigiosin[183] **276** has not yet been synthesized but a number of "unnatural" prodigiosin analogs[191] as well as related bipyrroles and pyrromethenes[192] have been synthesized for studies of their chemotherapeutic activity.

11. CORRINS AND VITAMIN B_{12}

The corrin ring system **291** is the parent ligand of the structure of vitamin B_{12}

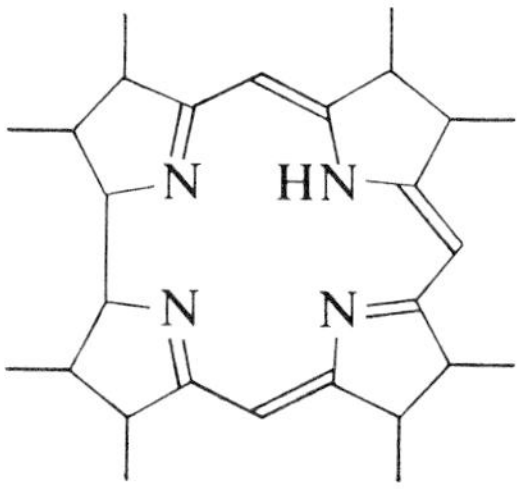

N HN
N N

291

(**292a**) and its coenzyme (**292b**). Little of the details of the biosynthesis of vitamin B_{12} is known, but there is no doubt that it is "pyrrole-derived" since it has been shown that porphobilinogen (**36**) is a precursor, although the exact point of divergence from the porphyrin pathway is unknown. However, many corrins have been synthesized from pyrrolic intermediates;[100] for example, Johnson and co-workers[191] found that tetradehydrocorrin salts (**293**) obtained by cyclization of *ac*-biladienes can be hydrogenated over Raney nickel at 160° and 100 atm to give a mixture of corrin epimers (**294**) (obtained as the perchlorates). It is clear that the lack of stereochemical control inherent in this approach makes it of little value for application to vitamin B_{12} synthesis. It would seem that all routes involving reduction of pyrrolic compounds are equally doomed to failure in view of the presence of no less than nine chiral centers in the vitamin B_{12} ligand (**292**).

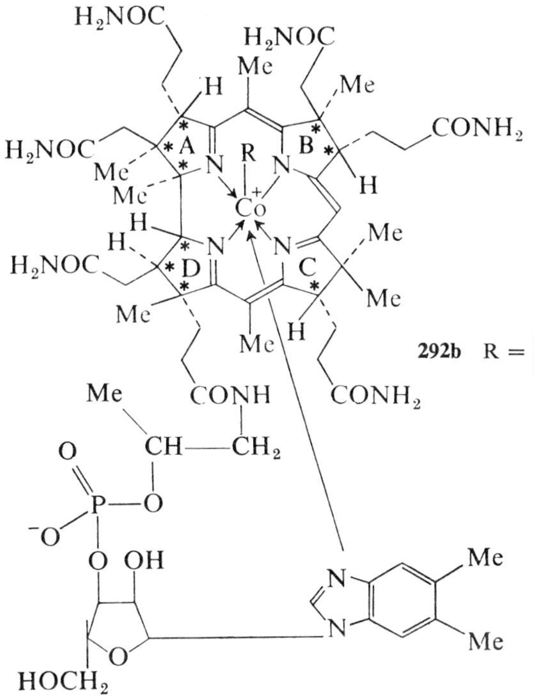

292a R = CN

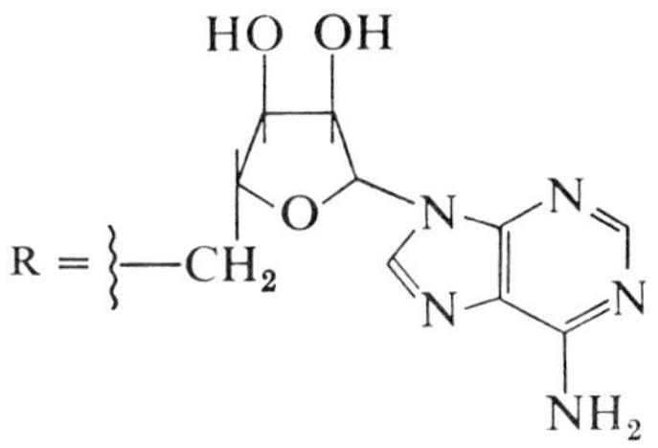

292b

Alternative and more highly controlled syntheses of corrins were therefore required, and in any synthesis of the vitamin itself, it was clearly necessary to pay great attention to stereochemical detail from the outset. Three different approaches to the synthesis of vitamin B_{12} are currently under investigation, (a) Eschenmoser's study of general syntheses of corrinoid compounds, (b) the joint attempt by the Eschenmoser and Woodward groups to develop a specific synthesis of cobyric acid, a vitamin B_{12} degradation product, and (c) Cornforth's approach via isoxazole intermediates.

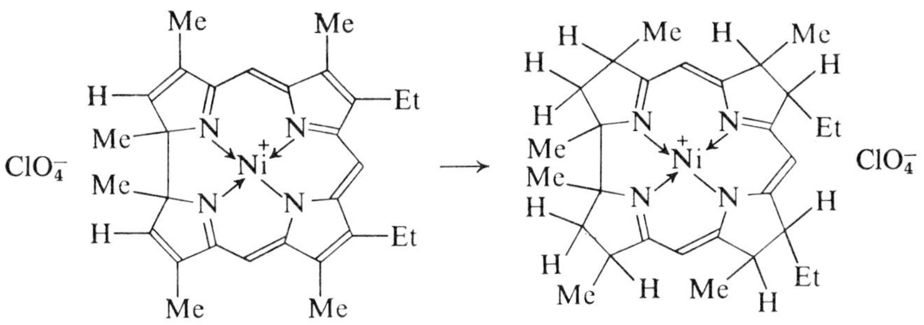

293 → **294**

A. Eschenmoser's Approach

The first objective of Eschenmoser and his co-workers[194] was the corrin **295,** and their achievement was described in one of the most notable synthetic papers of 1964. Although the target corrin **295** required little attention to stereochemical detail, the route chosen was clearly capable of modification so that it could be used as a basis for the approach to vitamin B_{12} itself.

The chromophore of the corrin macrocycle (**291**) is nominally made up of three interacting vinylogous amidine systems (**296**). In view of this fact,

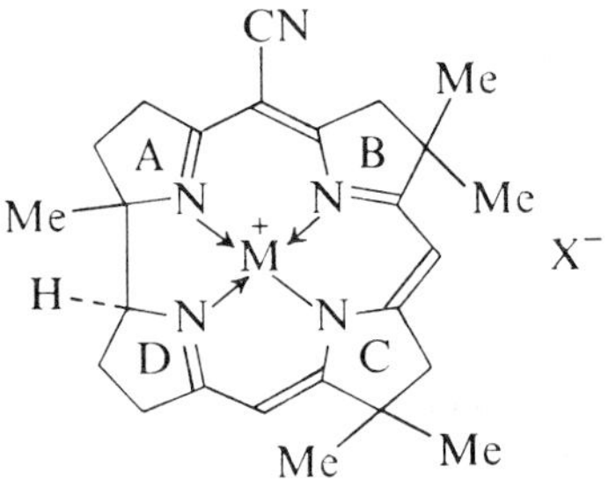

295 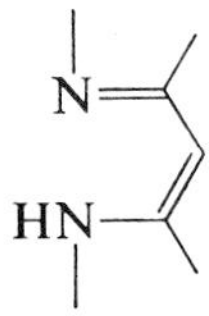 **296**

Eschenmoser based his strategy on the concept of carbon-carbon bond formation through condensations involving amide groups (**297**) activated as their corresponding imino ethers (**298**) by means of the hitherto little used Meerwein trialkyloxonium salts (see Scheme 10).[195]

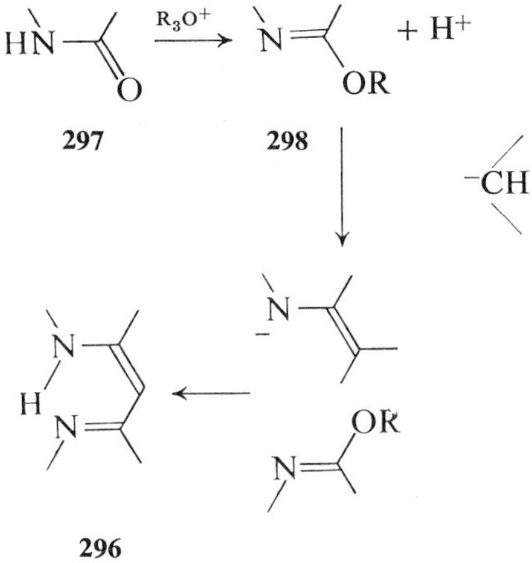

297 **298** **296**

Scheme 10

The best way to accommodate this plan in terms of the "dislocation"[196] of the molecule was by straightforward "east-west" retrosynthetic division, giving **299** and **300** as the immediate synthetic objectives.

Scheme 11 shows the synthetic sequence leading to the western (*A–D*) half (**299**). Thus Diels-Alder cycloaddition of isoprene and tetramethyl ethylene

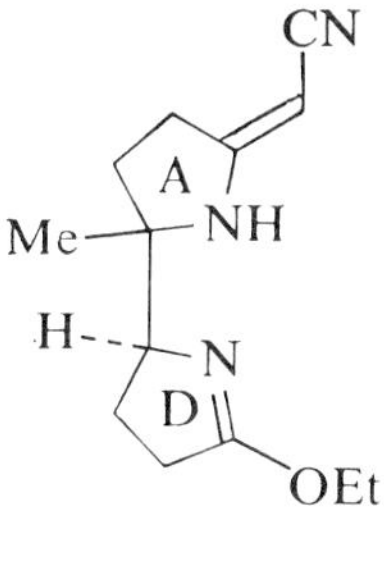

299 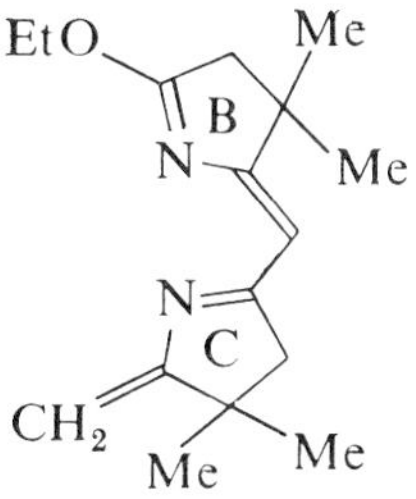 **300**

tetracarboxylate gave the cyclic tetraester **301**, which was converted to the acyclic material **302** by ring cleavage with sodium in liquid ammonia. This was converted to the diamide **303** before creation of the aziridine ring and transformation to the diester **304**; the *cis* configuration was assigned on the basis of mechanistic postulates, the validity of which were later confirmed by degradation of **304** to known substances. The second nitrogen atom in the prospective product **299** was introduced by azidolysis of the aziridine ring, the azide ion attacking at the least hindered carbon atom; the pyrrolidone **305** was obtained by cyclization of this aliphatic intermediate during 15–20 days. The stereochemistry of **305** is predictable from the *trans* opening of the aziridine ring by the incoming azide ion. Meerwein's reagent served to activate the lactam function of **305** while still retaining the azide grouping. Condensation with *t*-butyl cyanoacetate then gave **306**, the *cis* orientation of the amino and ester groups being established on the basis of IR comparisons with model compounds of defined geometry. The azide group in **306** was reduced catalytically, giving the bicyclic monolactam **307**, which was transformed to **308** with trifluoroacetic acid and from thence activated to the required *A–D* component **299** with triethyloxonium tetrafluoroborate.

The synthesis of the eastern (*B–C*) half (**300**) proceeded along the lines indicated in Scheme 12. The ethyl ester (**309**) of β,β-dimethyllevulinic acid was converted efficiently to β,β-dimethylmethylenebutyrolactam (**310**) by reaction with hot ethanolic ammonia followed by pyrolysis. Treatment of β,β-dimethyllevulinic acid with thionyl chloride gave the butyrolactone derivative **311**, whose sodium salt reacted with the lactam **310** to give the corresponding

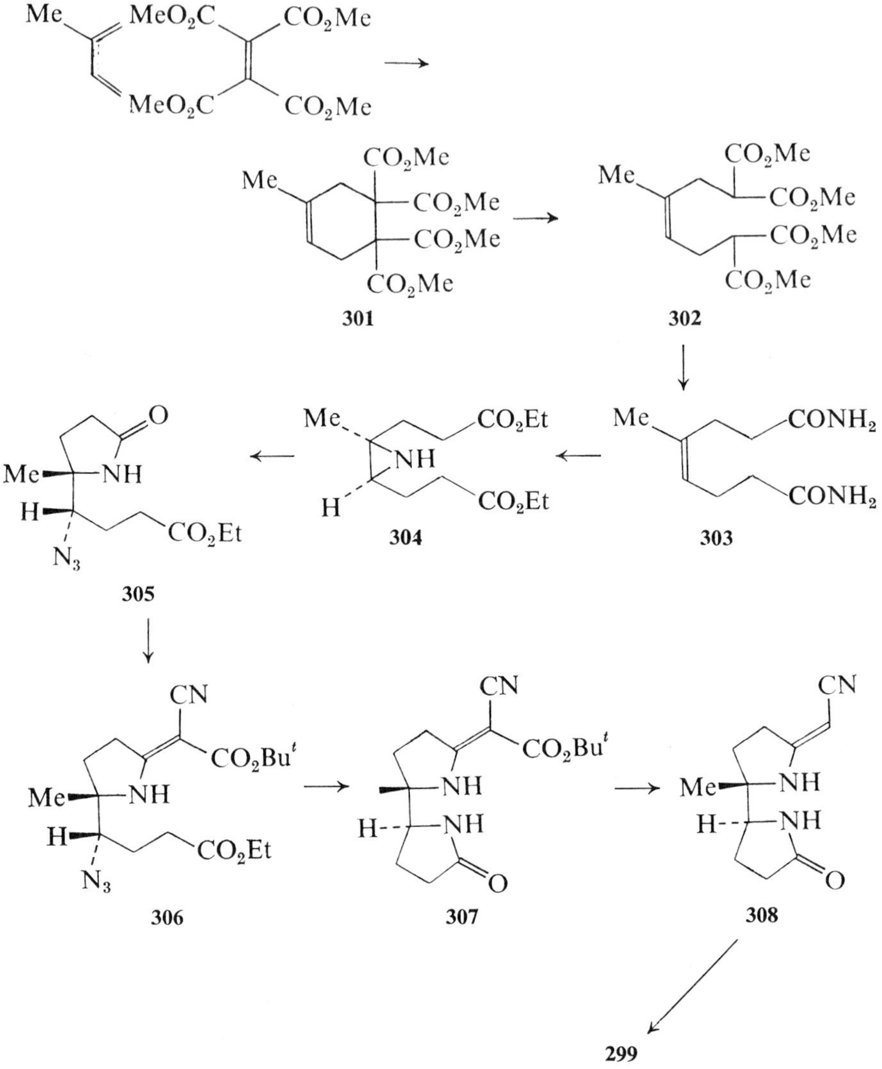

Me, MeO_2C, CO_2Me, MeO_2C, CO_2Me →

301 → 302 → 303 → 304 → 305 → 306 → 307 → 308 → 299

Scheme 11

N-acylated substance **312**; an ingenious photochemically induced migration of the acyl group to carbon in a constant flow system gave **313**, which was converted to the bicyclic material **314** on treatment with methanolic ammonia. Pyrolytic or base-promoted dehydration gave **315**, which was converted to the sensitive iminoether derivative **300** through the agency of triethyloxonium tetrafluoroborate.

The crucial problem of combination of **299** and **300** in the required sense

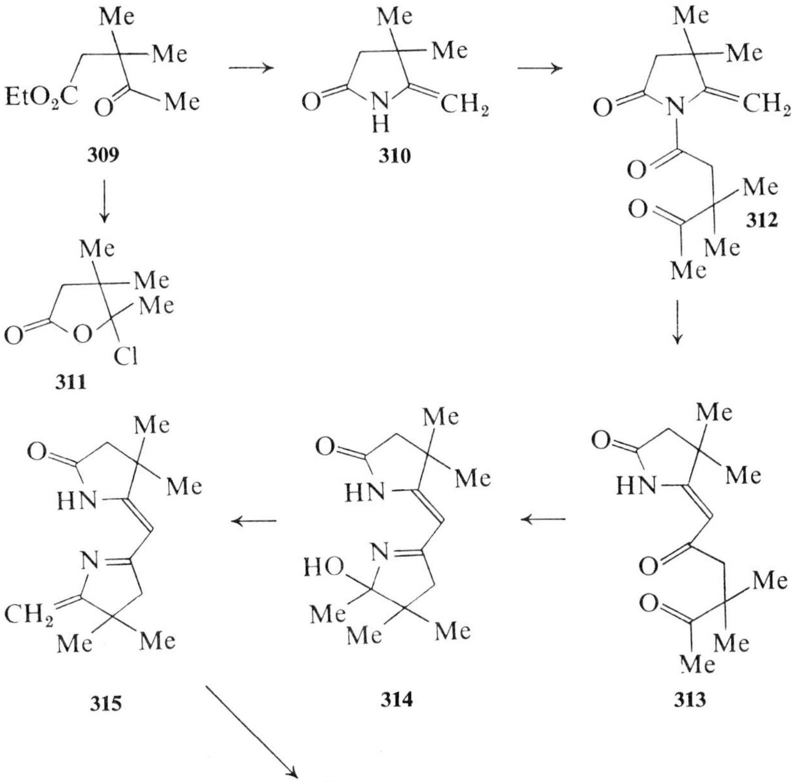

309 310 312 311 315 314 313 300

Scheme 12

was solved in the beautiful series of reactions outlined below. The sodium salt of **299** condensed specifically with the iminoether function of **300** to give the open-chain material **316**, isolated as its sodium salt; the greater electrophilic reactivity of the conjugated iminoether of **300** compared with the isolated function in **299** meant that none of the unwanted dimer **317** or its further

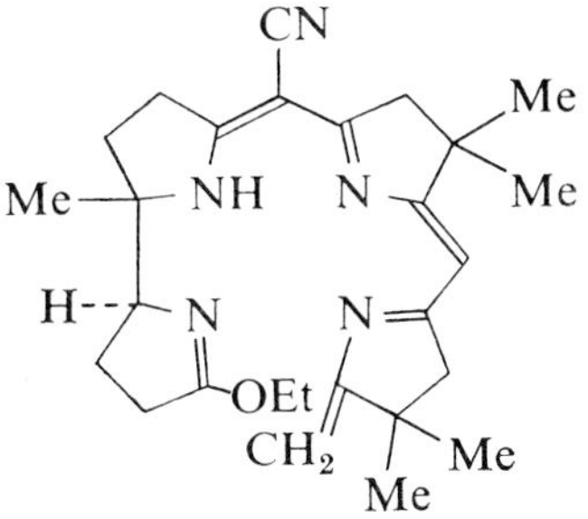

316

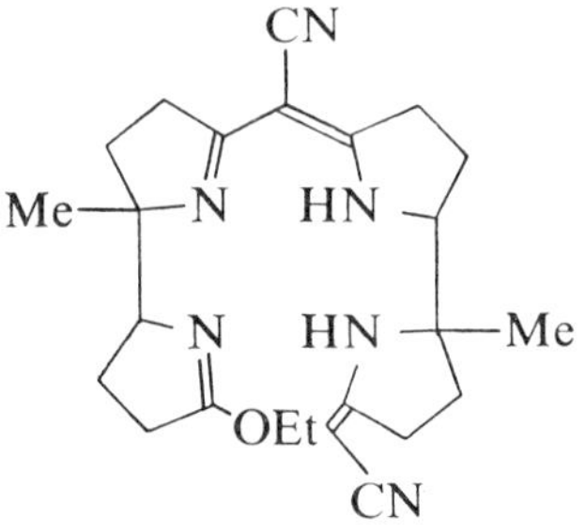

317

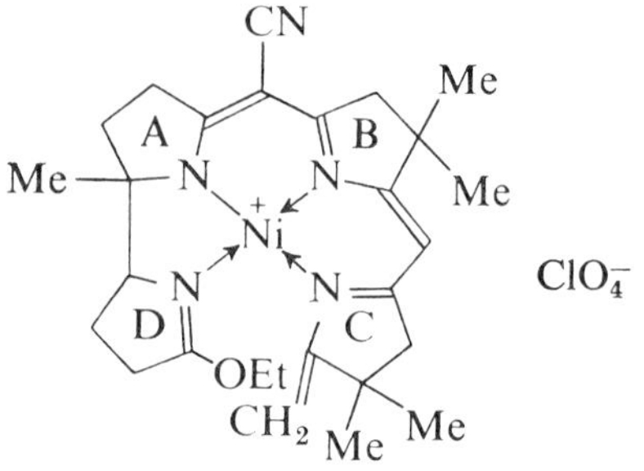

318

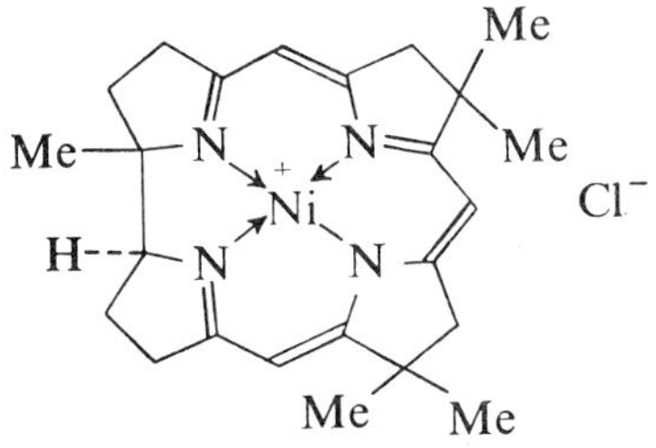

319

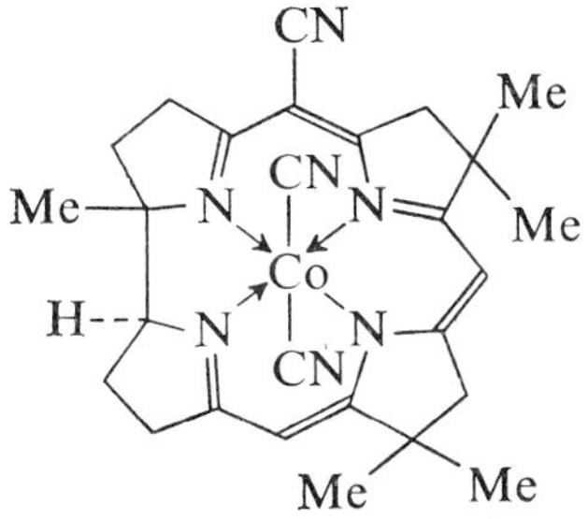

320

reaction products was obtained. Nickel(II) perchlorate in acetonitrile furnished the chelate **318**, largely solving the problem of closure of **316** to a corrin due to the enforced proximity of the appropriate functional groups in rings *C* and *D*. The corrin perchlorate (**295**; X = ClO_4) or hydrochloride (**295**; X = Cl) was obtained in excellent yield merely by treatment of the pre-corrinoid chelate **318** with potassium *t*-butoxide followed by the appropriate mineral acid. The cyano group could be removed by hydrolysis and decarboxylation in 0.1*M* hydrochloric acid at 220° in a greater than 90% yield to give the corrin chloride **319**. The structure of the corrin **295** was confirmed in all details by a full three-dimensional X-ray analysis.[194,197]

The obvious extension of this synthesis to other metal complexes of corrins has been carried out;[198] in particular the cobalt(III) chelate (**320**) has been synthesized and its properties examined.

More recent developments[199] from Eschenmoser's group have involved the preparation of an *A–D* moiety by modification of the enamide **310**, which had earlier been used only as a precursor for the *B–C* half (**300**). Thus this enamide (**310**) reacted readily with nitromethane in the presence of a catalytic amount of potassium *t*-butoxide to give the enantiomers **321** which were subjected to Michael addition with methyl acrylate, furnishing the diastereoisomers **322**.

After catalytic hydrogenation these gave the corresponding diastereoisomers of the bicyclic material **323**, which were separated, and the lower melting

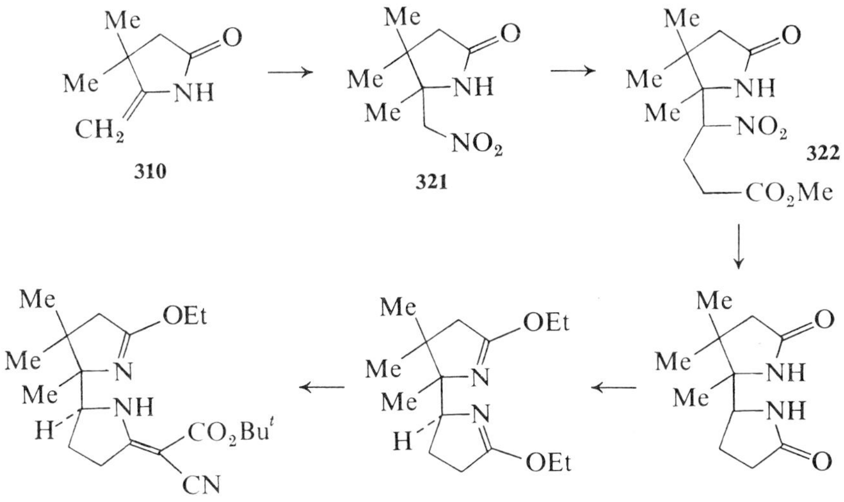

310 **321** **322** **323** **324** **325**

(undoubtedly *anti-*) isomer carried through the reaction sequence described below. Treatment with Meerwein's salt gave the *bis*-iminoether **324**, which was condensed with *t*-butyl cyanoacetate in presence of triethylamine to give **325**, the most sterically favored adduct, from which the *A–D* intermediate **326**

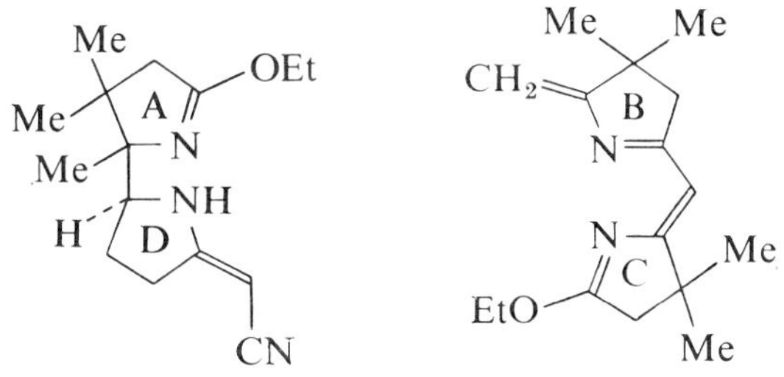

326 **300**

was accessible by treatment with dry trifluoroacetic acid. Condensation with **300** under previously defined conditions[198] led to the required racemic dicyanocorrin cobalt(III) complex (**328**; R = CN) or its hydrolysis and decarboxylation product (**328**; R = H) (Scheme 13).

A major disadvantage of all of the corrin syntheses discussed so far is the essential presence of a metal ion in the product. In most cases this has been

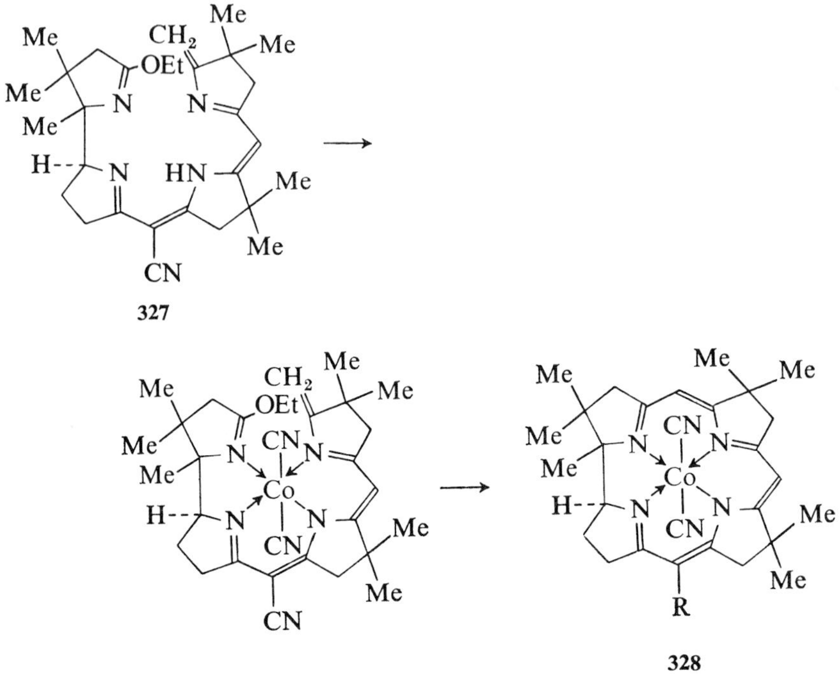

327

328

Scheme 13

cobalt(III) or nickel(II) and the vital templating effect of these ions in corrin synthesis has been alluded to. Currently there exists no satisfactory method for removal of such ions to give the metal-free ligand, though such compounds have been an attractive objective, both esthetically and theoretically, for some time. This flaw in existing synthetic strategy has been highlighted even further by reports[200] of the isolation and characterization of vitamin B_{12} compounds lacking cobalt and of the quite remarkable *p*H dependence of their visible absorption spectra. In addition, certain of their chemical properties are surprising, particularly the relatively drastic conditions required to insert cobalt into the corrin ligand, which had hitherto been regarded as a potent scavenger for various metal ions. Apart from the biosynthetic implications of the natural occurrence of metal-free vitamin B_{12} compounds, it was clear that an efficient approach to the free corrin ligand was desirable so that the macrocycle could be studied in detail; the accessibility of such a ligand would also be a good source of many presently unavailable metal complexes. The relatively small quantities of natural materials from photosynthetic bacteria[200] obviously would not suffice for this purpose.

The answer to this new problem was supplied by Eschenmoser and Fischli,[201] largely as an extension of the already monumental contributions to corrin chemistry made by the Zurich group. The usual method of templating the ligand into the required geometry was discarded in favor of "sulfide contraction" (Scheme 14). In this method the sulfur atom of a thiolactam (accessible by P_2S_5 treatment of the corresponding lactam), which is sterically unhindered due to the long length of the carbon-sulfur bond, is linked to the methylidene carbon atom of the enamide (Scheme 14) to give a sulfur-bridged intermediate (**329**). Collapse to the episulfide derivative **330**, which

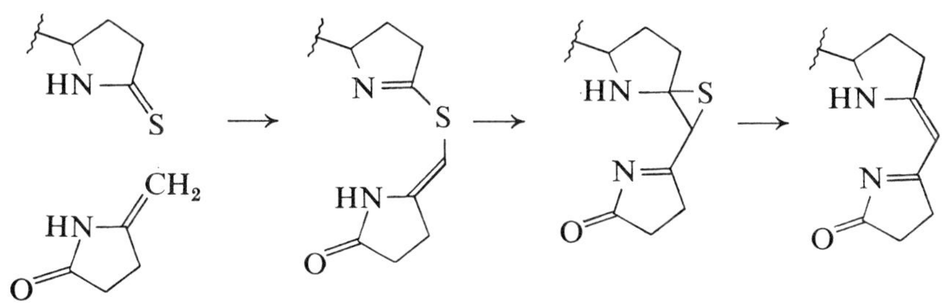

329 **330** **331**

Scheme 14

in the presence of a suitable sulfur acceptor suffers sulfur extrusion[201] to give the required product **331**. This sequence highlights an important principle in synthetic chemistry[202]:

> "*Whenever in the synthesis of complex organic molecules one is confronted with a situation where the success of an intermolecular synthetic process is thwarted by any type of kinetically controlled lack of reactivity, one should look for opportunities of altering the structural stage in such a way that the critical synthetic step can proceed intramolecularly rather than intermolecularly.*"

Thus the sodium salt of the precorrinoid ligand **327** was treated with hydrogen sulfide in the presence of trifluoroacetic acid in order to obtain the corresponding ring-*A* thiolactam derivative. In the event, the compound obtained after "loose" complexation with zinc(II) was the cyclic isomer **332**, which, though unexpected, was adapted for sulfide contraction by treatment with benzoyl peroxide in presence of trifluoroacetic acid giving the ring expanded compound **333** in **72%** yield after contact with methanol; the probable mechanistic pathway of this reaction has been outlined[202] and need not be repeated here. Contraction with trifluoroacetic acid in hot dimethylformamide furnished the corrin complexes **334** (R = H) and **334** (R = SH) in about 80% yield, the former already having suffered loss of sulfur. The major product (**334**; R = SH) was smoothly desulfurized with triphenyl phosphine in the presence of a catalytic amount of trifluoroacetic acid to give **334** (R = H) from which the free ligand **335** was liberated with excess trifluoroacetic acid in acetonitrile. The spectroscopic properties of the free corrin have been found to mimic closely those of Toohey's naturally occurring compound,[200] and a variety of metal complexes of **335** have been obtained in high yield and their properties examined.[201]

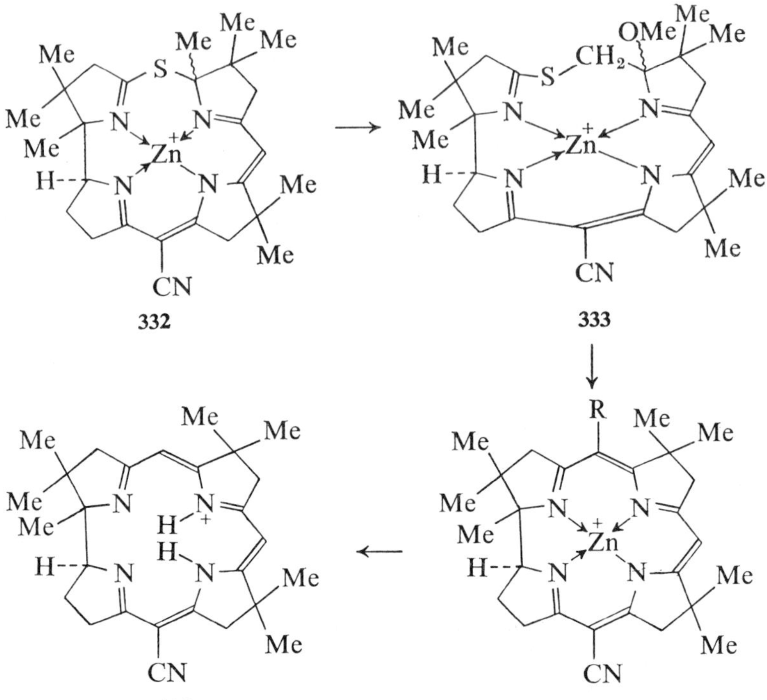

As if to emphasize the deep interest of Eschenmoser's group in corrins of all types, a totally new route to this macrocycle was devised concurrently with the developments just described, named "The New Road"[202,203] and based on a Woodward-Hoffmann type cyclization as the *last* stage in the synthetic sequence. As a matter of general synthetic principle, it is by far the safest course of action to accommodate any "doubtful" transformations at the earliest possible opportunity, thus providing the maximum amount of latitude for circumventing difficulties. It seems clear in Eschenmoser's new road that failure of the final cyclization step with its concomitant loss of effort was not considered seriously since the reaction had been deemed "allowed" by the Woodward-Hoffmann rules.[204]

The ubiquitous enamide **310** was treated with aqueous potassium cyanide giving the derivative **336** with the strongly nucleophilic double bond protected. This was treated with P_2S_5, furnishing the thiolactam **337** which was oxidized with benzoyl peroxide to the reactive disulfide **338**; in the presence of the enamide **310** this was smoothly transformed into the bicyclo thio-bridged compound **339**. Contraction to the vinylogous amidine **340** was accomplished

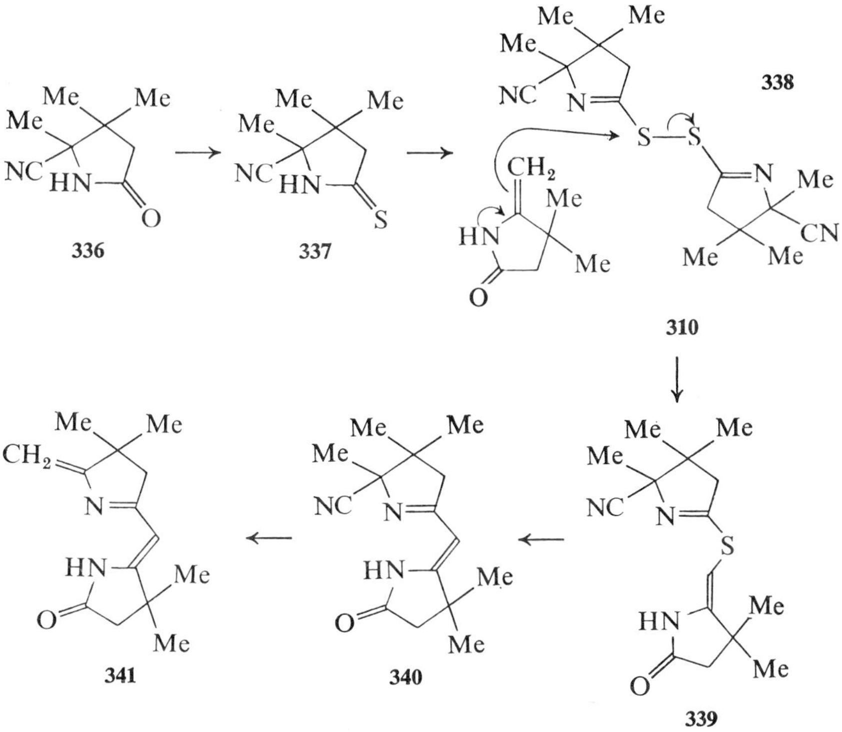

336 **337** **338** **310** **339** **340** **341**

with triphenylphosphine, and the exocyclic double bond regenerated by treatment with potassium *t*-butoxide, giving the deprotected intermediate **341**. Repetition of this series of reactions with **341** and more of the disulfide **338** furnished the tricyclic material **342**. In this series of operations, the sulfide contraction was catalyzed with boron trifluoride. The *silver* complex **343** was alkylated with Meerwein's salt, so avoiding the complications of indiscriminate *O*- and *N*-alkylation which were apparent with the metal free compound **342**, and direct treatment with the enamine **344** gave the required open-chain tetracyclic material, isolated as its nickel(II) chelate (**345**; M = Ni^+). Excess of cyanide ion served to remove the metal from the ligand, laying open a pathway to other chelates [e.g., **345**; M = Pd^+, Pt^+, $Co(CN)_2$]. The cyanide-protecting group in such compounds (**345**) is easily removed with potassium *t*-butoxide, to give the *A*/*D* *seco*-corrinoid systems [**346**; M = Ni^+, Pd^+, $Co(CN)_2$]; the corresponding complexes (**346**; M = Zn^+, Mg^+) are also accessible at this stage from (**346**; M = Ni^+).

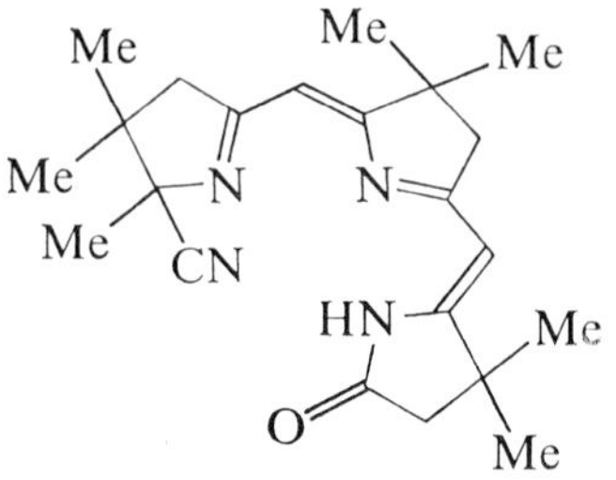

342

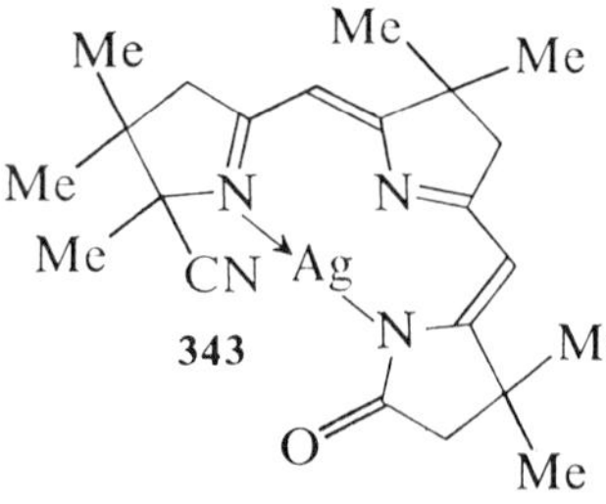

343

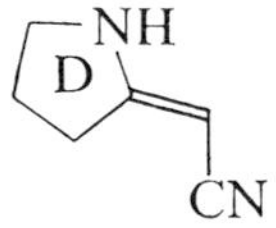

344

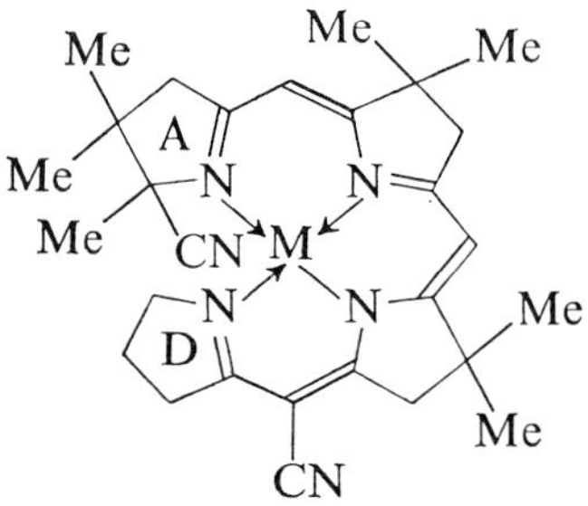

345

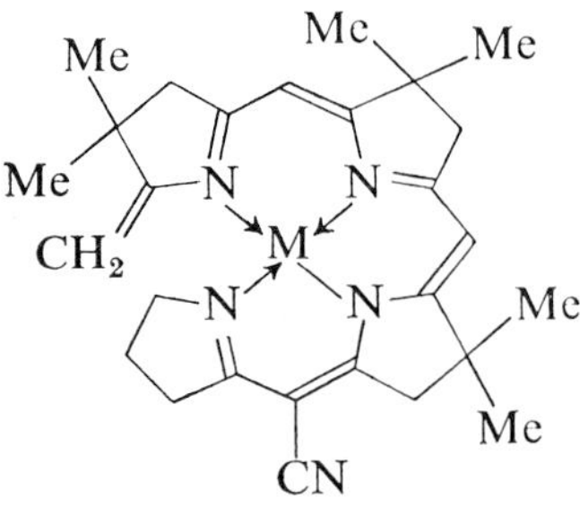

346

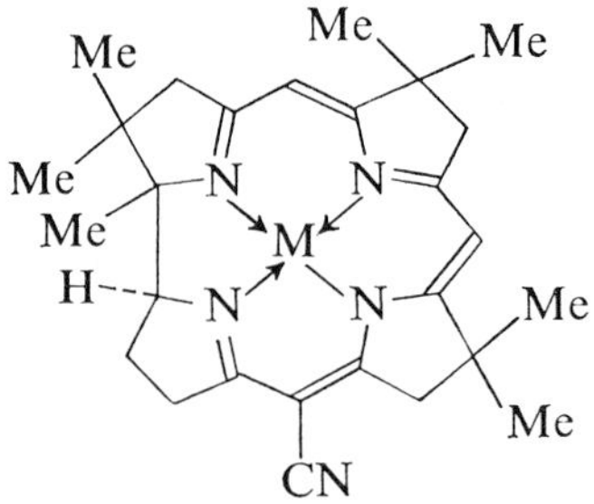

347

The critical cyclization of the *seco*-corrinoids (**346**) to corrins appears to depend in an absolutely all-or-nothing sense on the precise nature of the central metal atom. Thus the chelates (**346**; M = Pd^+, Pt^+, Zn^+, Mg^+) all cyclize to the corresponding corrins (**347**; M = Pd^+, Pt^+, Zn^+, Mg^+) in yields greater than 90% by a remarkable, symmetry-allowed, photochemically induced (and formally classified) antarafacial sigmatropic 1,16-hydrogen transfer and an antarafacial electrocyclic 1,15-π,σ-isomerization; no cyclization was observed with the cases [**346**; M = Ni^+, $Co(CN)_2$]. These observations have prompted a great deal of new theoretical and experimental investigation, and Eschenmoser has already outlined briefly his proposed route to the vitamin B_{12} degradation product cobyric acid (**348**).[202]

B. The Woodward-Eschenmoser Approach

The simplest known natural vitamin B_{12} derivative[205] is cobyric acid (**348**), the synthesis of which would constitute a formal total synthesis of the vitamin since it has been transformed successfully to vitamin B_{12} by earlier workers. A seemingly trivial, but nonetheless vitally important point of detail is the fact that the 17-propionic side chain must be distinguishable from the remaining side chains so that it may be manipulated specifically in order to exploit Bernhauer's avenue to the vitamin itself.[206]

A considerable amount of information has already been published concerning the joint approach by the Woodward and Eschenmoser groups at Harvard and Zurich and has been given in numerous lectures.[207] The strategy of Woodward and Eschenmoser's approach lies in the retrosynthesis of the target cobyric acid (**348**) to two molecules, the western (*A–D*) (**349**) and eastern (*B–C*) (**350**) halves. The *A–D* molecule **349**, allocated to the Harvard group, is the more synthetically demanding of the two halves, with its "crowded concatenation of six contiguous asymmetric centers,"[207] but any Swiss effort spared by this division of labor was to be made up in full by the tremendous wealth of expertise and experience available to the Zurich workers as a result of their several years of successful endeavor in the corrin field. In particular, the method of sulfide contraction, discussed earlier (Scheme 14), which was developed primarily for the coupling of bulkily substituted rings *B* and *C*, was to prove invaluable also in the coupling of the *A–D* and *C–B* components.

Thus the *B–C* half was divided into the two monocyclic units **351** and **352**. The latter, bearing only one chiral center, was obtained from (+)-camphorquinone (**353**). Ring *B* (**351**) was synthesized from β-methyl–β-acetylacrylic acid (**354**), the two required chiral centers being specifically created in the required sense by Diels-Alder cycloaddition with butadiene in the presence of stannic chloride; the enantiomeric mixture so produced was resolved by fractional crystallization of the diastereomeric α-phenethylamine salts and eventually furnished the required monocyclic material (**355**). Chromic acid oxidation served to rupture the double bond and further reaction of one of the newly created carboxyl groups with the ketonic carbonyl group furnished the dilactone acid **356**. Homologation of the acetate side chain to propionate

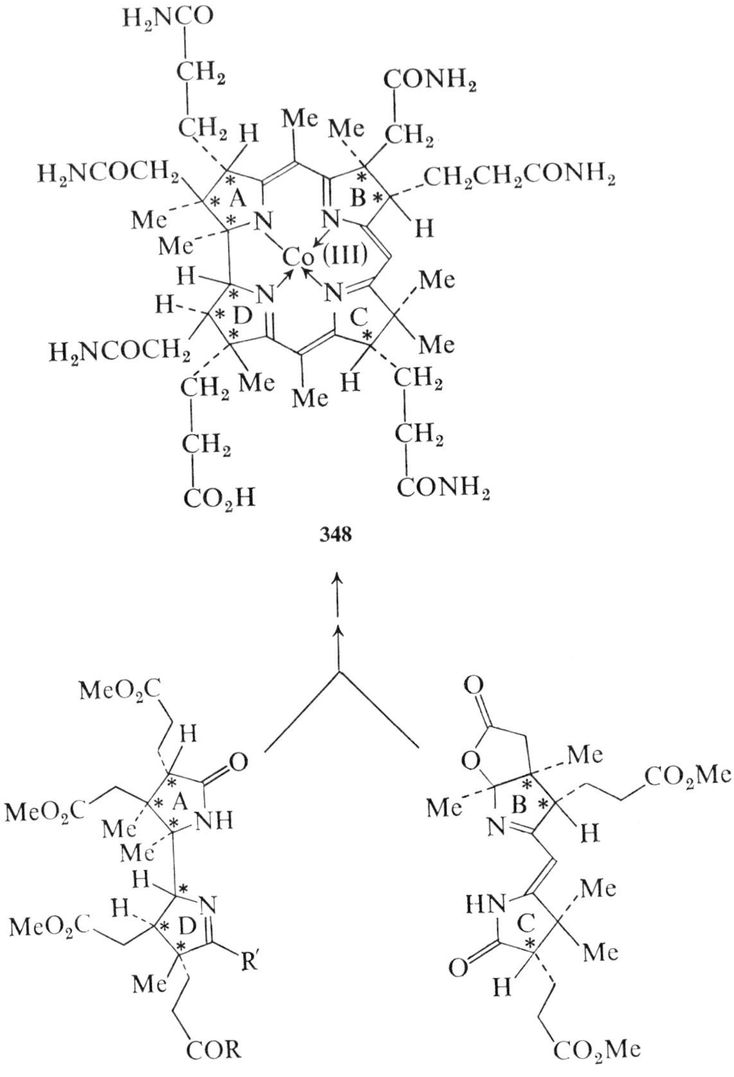

348

349 **350**

by the Arndt-Eistert procedure gave **357**, from which the lactone-lactam **358** was obtained on treatment with ammonia in methanol. Treatment of this material with P_2S_5 furnished the required thiolactam **351**, having its acetate side chain concealed as a lactone function. This was to facilitate future activation of the methyl group adjacent to nitrogen as an enamine and the thiolactam sulfur atom was to be utilized in the sulfur contraction method

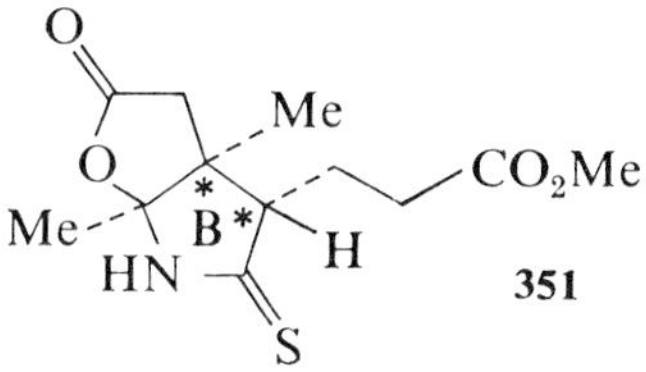

351

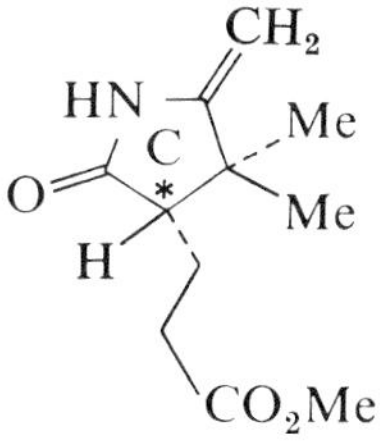

352

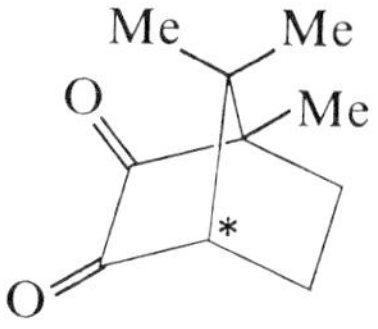

353

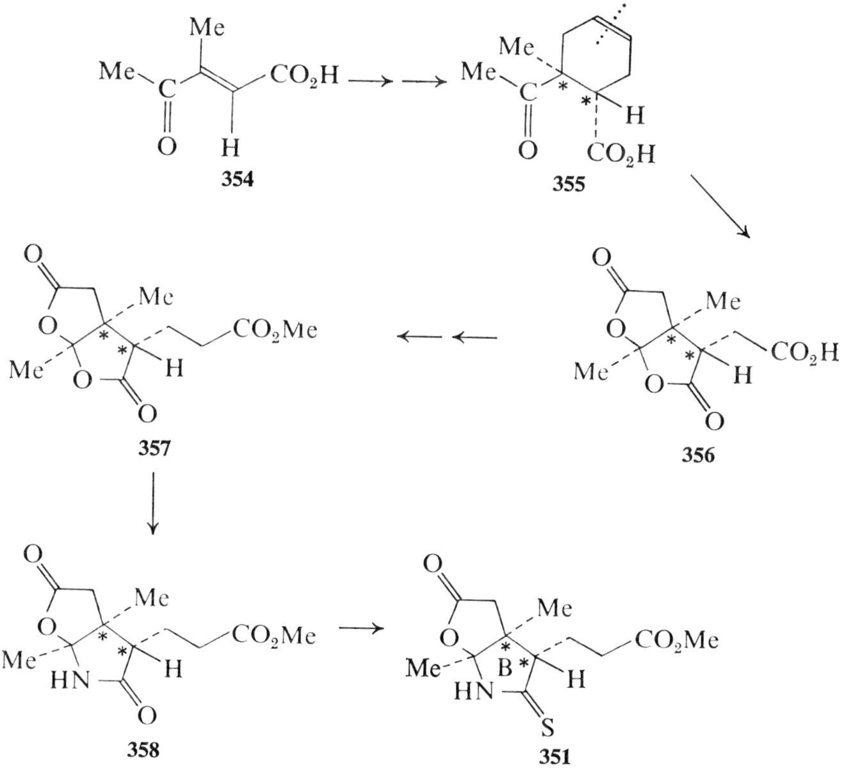

354 355 356 357 358 351

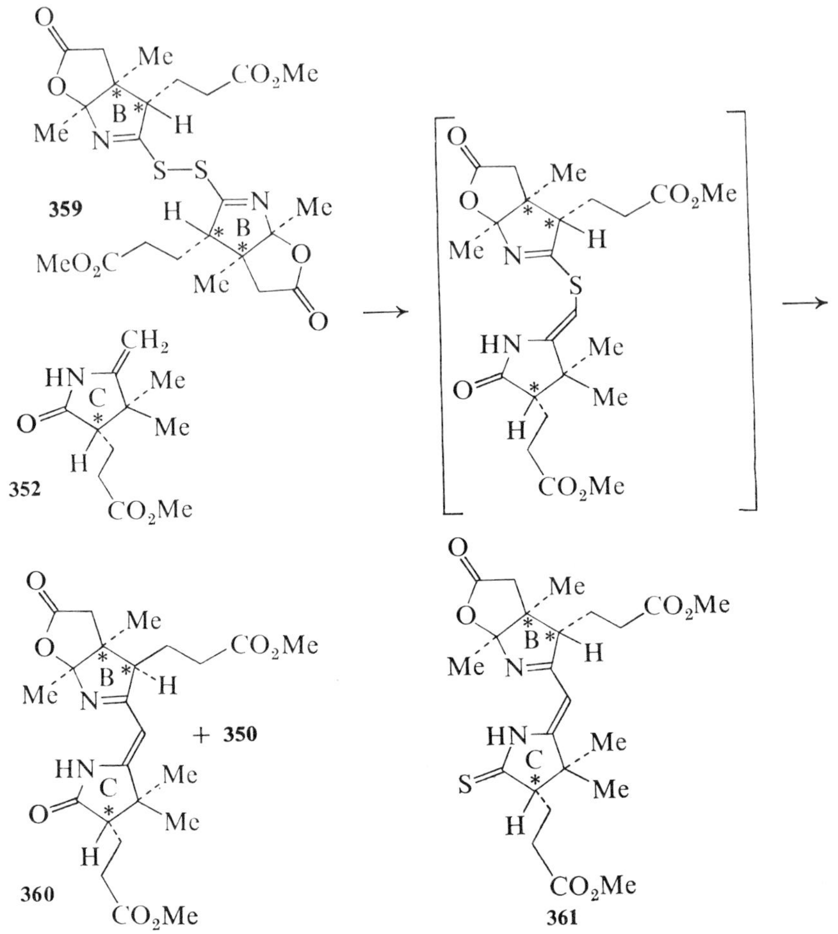

for coupling to ring *C* (**352**). Benzoyl peroxide oxidation of **351** in the presence of **352** gave initially the disulfide dimer **359**, which collapsed to a *B–C* sulfur bridged intermediate and was desulfurized directly by heating in triethyl phosphite to give **360** and its epimer **350** in 70% yield; the β-epimer **360** was readily obtained from this mixture in crystalline form. This material has the "wrong" stereochemistry in ring *B*, but this was not a disadvantage in light of the later discovery that several of the subsequent stages tended to epimerize this center, and furthermore it was known that the 8-position in vitamin B_{12} can be epimerized under suitable conditions.

Experience with the coupling of rings *B* and *C* had shown clearly that due to steric factors, it was unlikely that a straightforward iminoether condensation would be satisfactory for joining *B–C* to *D–A* and so the lactam in **350** was converted to the corresponding thiolactam derivative (**361**) as follows. Treatment with methyl mercury isopropoxide served to complex the nitrogen atoms so that reaction with Meerwein's salt effected *O*-alkylation specifically; subsequent decomposition of the intermediate with hydrogen sulfide gave the required activated *B–C* component (**361**).

The fact that the six contiguous chiral centers in the *A–D* component **349** exist in the most stable configurations, with large groups *trans* to each other, provided the ideal opportunity for "stereospecific synthesis by induction,"[207] given the ability to construct the required carbon skeleton.

The starting material, 6-methoxydimethylindole (**362**), though far removed from **349**, can be seen to contain certain elements of this component; for example, ring *A*, in a modified form, can still be discerned, with its two methyl groups, while the benzenoid ring also conceals the carbon skeleton of the ring *A* propionate side chain, with a carefully sited methoxyl substituent as a "handle" for its later manipulation by Birch reduction. The magnesium derivative of **362** gave the (±) indolenine when treated with propargyl bromide. Cyclization with mercuric oxide and boron trifluoride gave the key intermediate, the (±)-tricyclic ketone **364**, having the two methyl groups on

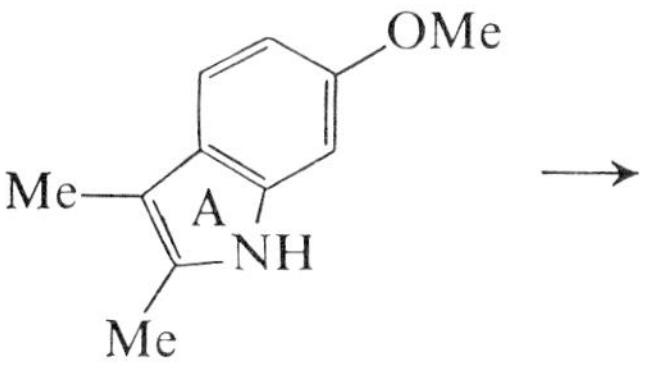

OMe
Me A NH
Me

362

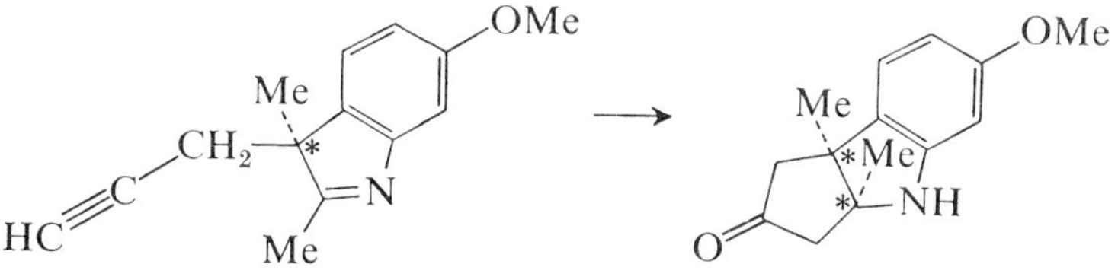

OMe
Me
CH_2 *
HC≡C
=N
Me

363

OMe
Me
*Me
* NH
O

364

OCONHMe
Me
NMe
N
Me H

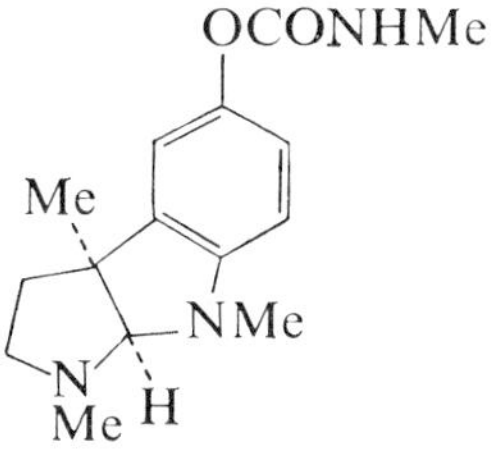

365

the same side of the molecule. The chirality of the second methyl group was dictated by that of the first, since the new five-membered ring must necessarily be *cis*-fused to the heterocyclic ring. A similar *cis* fusion must also occur between the two five-membered rings in the alkaloid physostigmine **365**, and its congeners, and this has been discussed in detail elsewhere.[208]

At this stage the tricyclic ketone was efficiently resolved through the agency of (+)-phenylethylisocyanate, and the absolute stereochemistries of the enantiomers defined by degradative comparison with derivatives of (+)-camphor. Experimentally, both of the enantiomers were used, the compounds of the "unnatural" series being utilized ingeniously as models with which to discover transformations to be applied to the "natural" substances* once they were established and optimized.

A suitable species (**366**) from which ring *D* might be fashioned was next assembled, in an optically pure form by absolute asymmetric synthesis from (−)-camphor as outlined in Scheme 15. (The corresponding enantiomer of **366**

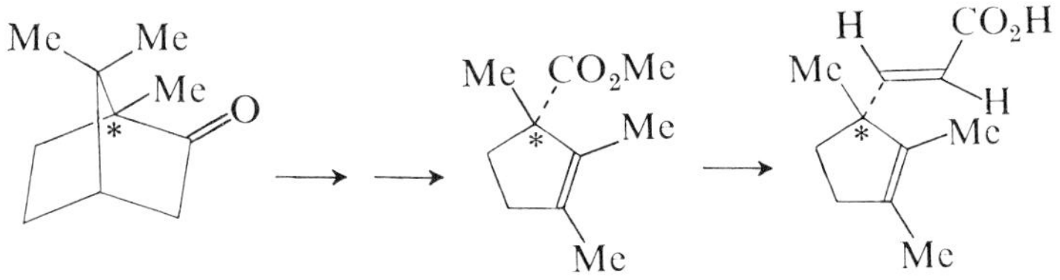

(−)-Camphor → → → **366**

Scheme 15

was likewise synthesized from (+)-camphor, for incorporation into the model "unnatural" series.)

The acid chloride from **366**, when treated with the ring-*A* precursor **364**, gave smoothly the required amide **367**, which was induced to cyclize with potassium *t*-butoxide in *t*-butanol to the pentacyclic lactam **368**. The two new chiral centers created in this cyclization were formed in the required sense by asymmetric induction and the large groups are all in a *trans* disposition to each other. This complex molecule **368** contains five of the required six contiguous asymmetric carbon atoms, and about 200 grams of this material and its enantiomer were available.

The aromatic ring, having served its purpose, was not merely to be discarded, for, as mentioned previously, Birch reduction of this benzenoid ring would eventually lead to the ring-*A* propionate side chain. Before this could be accomplished, however, it was necessary that certain labile functions within the pentacyclic lactam be protected. The ketonic carbonyl was easily masked

* All of the structures in this report are drawn in the "natural" sense to avoid complication.

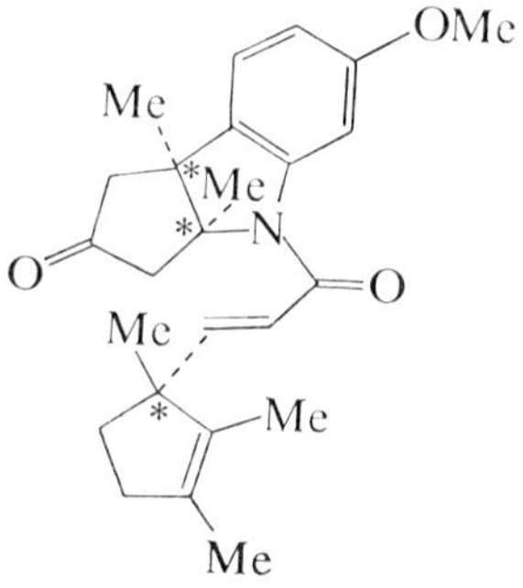

367 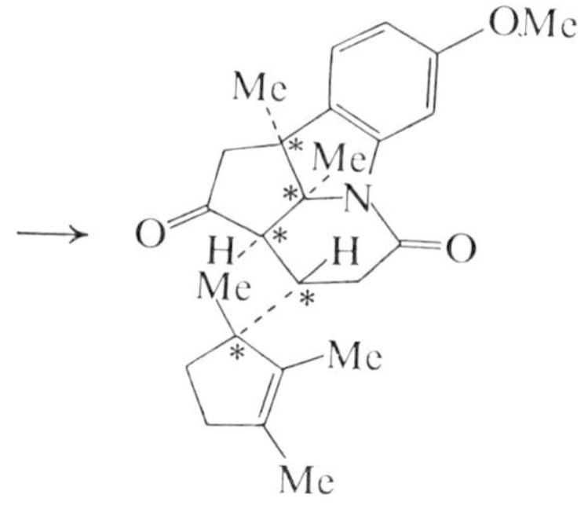 → 368

as its ketal **369** and the lactam group, which was liable to modification under the conditions of the reduction, also needed protection. This was carried out in a more circuitous, but subtly elegant, manner; thus Meerwein's salt gave the immonium salt **370**, which was converted with methoxide in methanol to the orthoamide derivative **371**. On heating, **371** furnished the methoxyenamine **372**, which withstood perfectly the rigors of Birch reduction with lithium and precisely determined amounts of liquid ammonia, *t*-butanol, and tetrahydrofuran. The resulting vinyl ether (**373**) was then readily hydrolyzed to the pentacyclenone **374**. Given a considerable amount of experimental skill this series of reactions could be carried out in remarkably high overall yields approaching 90% over *five* steps!

The newly created asymmetric center was, however, entirely in the opposite orientation to that required, although this was not realized until later. The oxygen atom of the spiroketal system of this highly concave molecule lies close to the carbon atom to which a proton must attach itself in the hydrolysis of the vinyl ether **373**. As a result of this steric hindrance to protonation the least thermodynamically stable epimer (**374**) is produced. Since this difficulty was not diagnosed at this point, the synthesis was continued initially without any attempt at rectification of the configuration in ring *A*.

Treatment of the pentacyclenone **374** with further acid liberated the ketone **375**, which was transformed via the dioxime to the monooxime **376**. Ozonolysis followed by treatment with periodic acid and esterification with diazomethane then gave **377**. This series of synthetic operations liberated the propionate side chain from the cyclohexenone ring as well as rupturing the cyclopentene ring to facilitate its conversion to a ring-*D* species. Quite remarkably, the oxime grouping survived the ozonolysis intact, no doubt due to its highly hindered steric situation; an otherwise unprotected oxime double bond would certainly have been cleaved by ozone. Pyrrolidine acetate-catalyzed cyclization of the diketone system gave **378** by activation of the least hindered carbonyl group through enamine formation. Mesylation

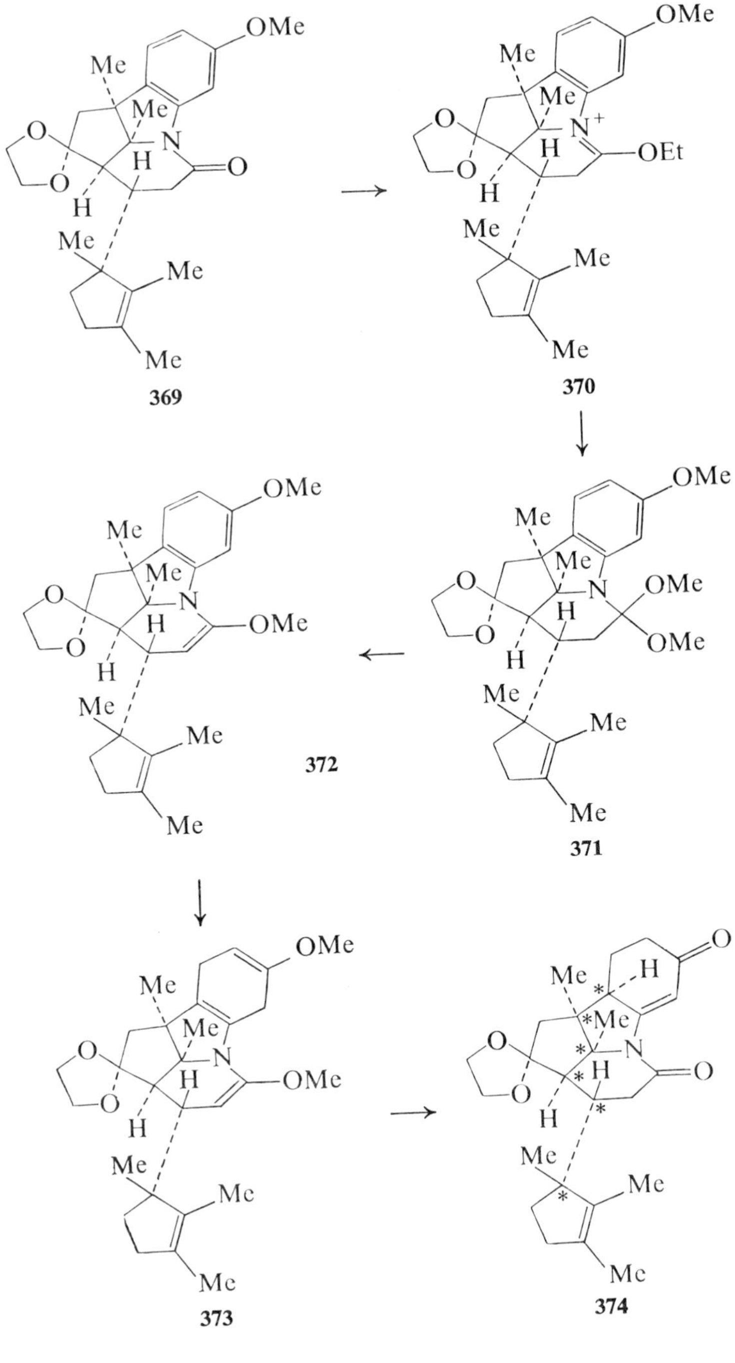

369

370

371

372

373

374

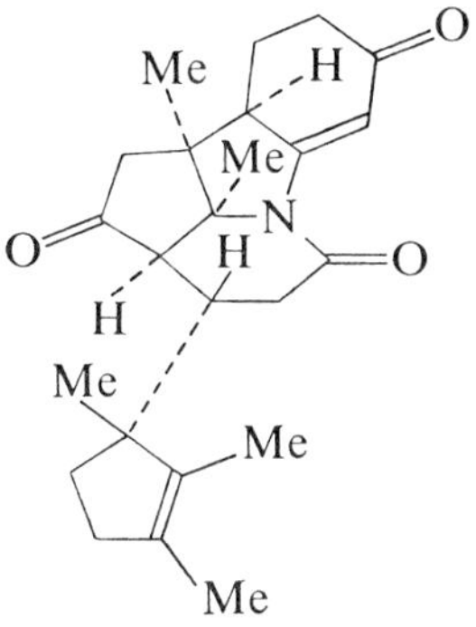

375

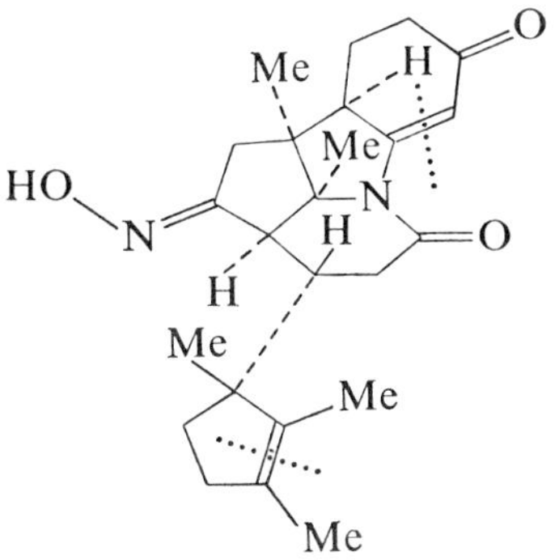

376

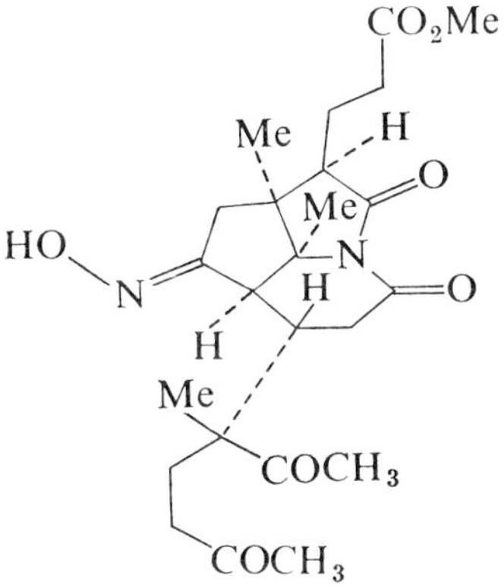

377

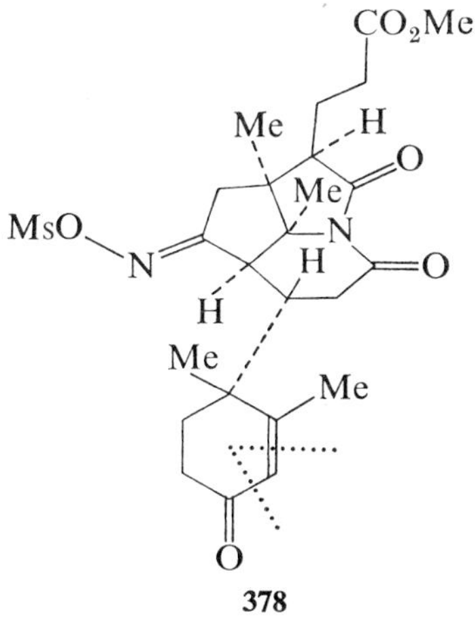

378

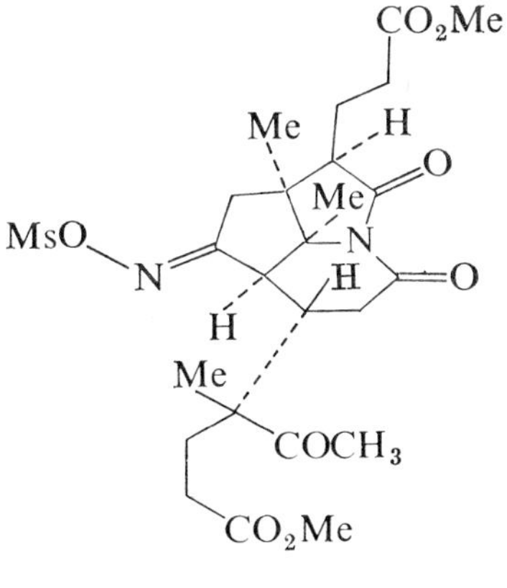

379

and ozonolysis followed by work-up with periodic acid and diazomethane gave **379**.

This molecule (**379**) now bore all of the essential features for conversion to a suitable *A–D* component, the *second* (ring-*D*) nitrogen atom of which was ingeniously masked by an oxime function. It was always a major feature of the synthetic strategy that this nitrogen atom should be introduced in this way and then moved into the correct position by means of a Beckmann rearrangement of the oxime. The ketone function of the cyclopentanone ring in the tricyclic ketone **364** was sited in the correct position to serve this purpose and the original ring can still be discerned in the highly complex intermediates at this late stage in the synthesis. Clearly, the strategy of this approach depended on the Beckmann reaction giving the required product, and the fact that failure at this late point could not be tolerated made this perhaps the most well-conceived transformation of the synthesis. Indeed, its experimental execution could not be faulted either, for in an extremely complex series of changes, resulting from treatment of the oxime mesylate **379** in methanol at 170° in the presence of a polystyrene sulfonic acid catalyst for 2 hours, the compound **380** was obtained. Not only had the required rearrangement taken place, but concomitantly with this the diacylamine system was cleaved, and the newly sited nitrogen atom attacked the carbonyl group of the acetyl function, which activated its α-methyl group toward formation of a new six-membered ring by condensation with its neighboring ester group. The product, given the very apt trivial name "α-corrnorsterone" (**380**), is on paper the immediate precursor of the required *A–D* species by hydrolysis of the lactam function (dots in **380** and rotation of ring *D* about 180°). However, it was found that a vast excess of base was required to effect hydrolysis of the amide function, and it was realized that the propionate side chain of ring *A* had been created in the wrong orientation. Hence when the lactam ring in α-corrnorsterone (**380**) was opened, the steric compression caused by the presence of two large groupings on the same face of ring *A* resulted in ring closure; less space was occupied by the groups concerned if the acetate side chain in this ring remained as a lactam grouping. In the Beckmann rearrangement a small amount of a compound isomeric with α-corrnorsterone was also produced, and the ease with which its lactam ring was opened with base showed that this material, β-corrnorsterone (**381**), must have had the required *trans* configuration in ring *A*, and the anomaly of the Birch reduction was confirmed. This observation, together with the relative rates of hydrolysis of the lactam groups, not only permitted an accurate prediction of the stereochemistries of ring *A* of the two corrnorsterones (which was subsequently verified by X-ray studies) but also gave a hint of a method by which α-corrnorsterone might be epimerized to its isomer. Clearly, if the lactam ring of α-corrnorsterone could be opened and equilibrated, then due to the

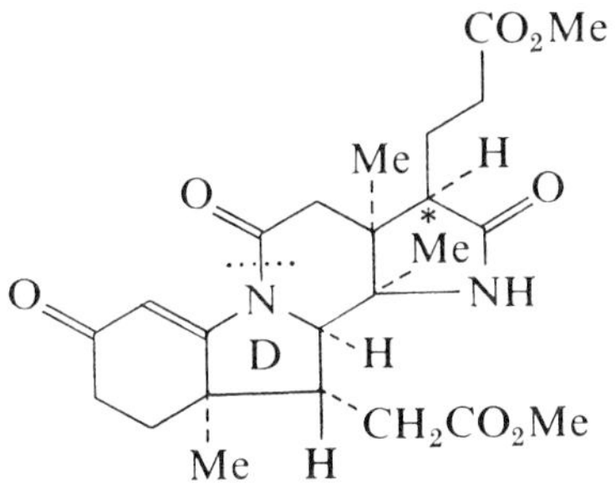

380

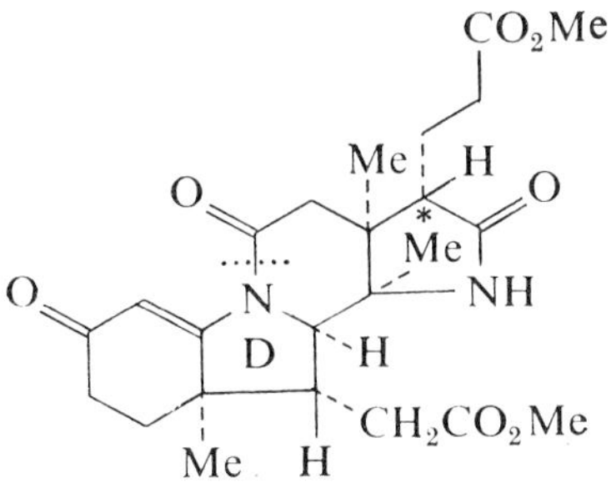

381

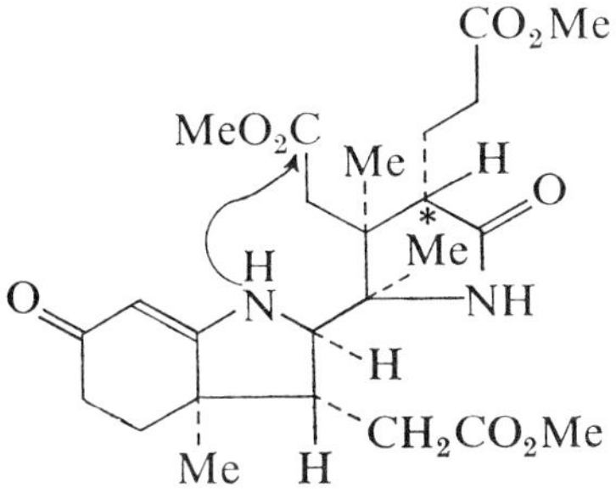

382

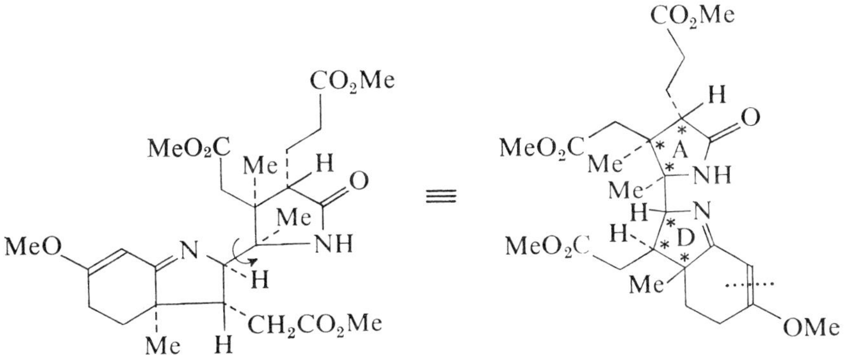

383

steric compression experienced by "opened" α-corrnorsterone, the major component of the equilibrium mixture must be the natural β-epimer. Thus heating of α-corrnorsterone with a large excess of base followed by acidification and treatment with diazomethane gave β-corrnorsterone (**381**) in about 90% yield and this was easily separated from the residual α-epimer. The road therefore seemed open to the *A–D* intermediate **382**, with the future ring-*D* propionate differentiated from the other esters by its incorporation into a cyclohexanone ring. However, when β-corrnorsterone was opened with base and treated with diazomethane in hope of obtaining **382**, the amine function of the ring-*D* vinylogous amide system reacted with the ring-*A* acetate group to give back β-corrnorsterone. Experimentally it was found impossible to achieve any synthetically acceptable conditions that would keep the lactam ring open, except for treatment with methanolic hydrogen chloride which gave the corresponding ring-opened compound, "hesperimine" **383**, featuring a ring-*D* vinylogus iminoether group. It was, however, apparent that the nucleophilic character of the vinylic position in hesperimine **383** was considerably less than would be expected from that in the corresponding vinylogous amide (**382**) were it to be accessible.

All the attempts to couple hesperimine **383** with the Zurich *B–C* component (**361**) were without reward, and so the **383** was ozonized to the aldehyde **384**; the imine function was not affected, no doubt due to the protection afforded to it by the vast array of substituents in its proximity, the aldehyde was then

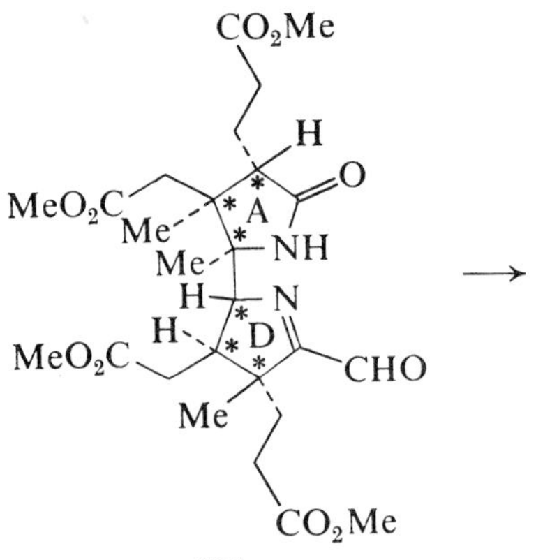

384 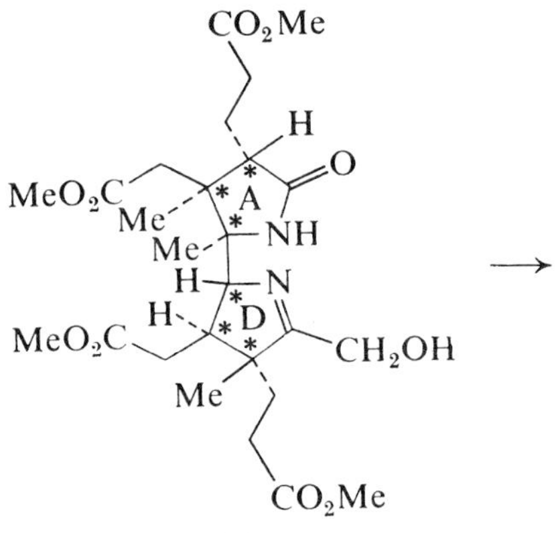 → **385** 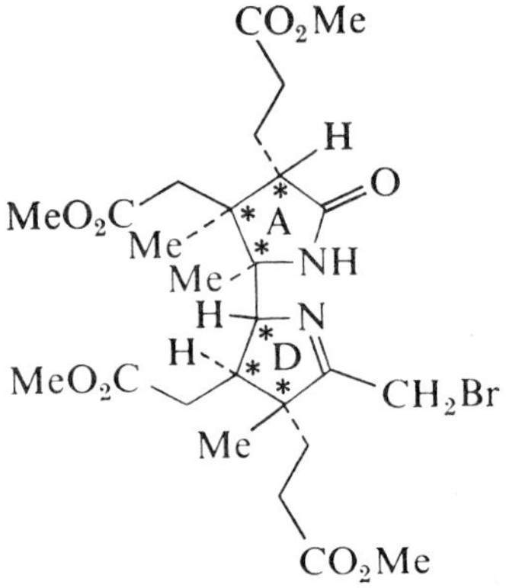 → **386**

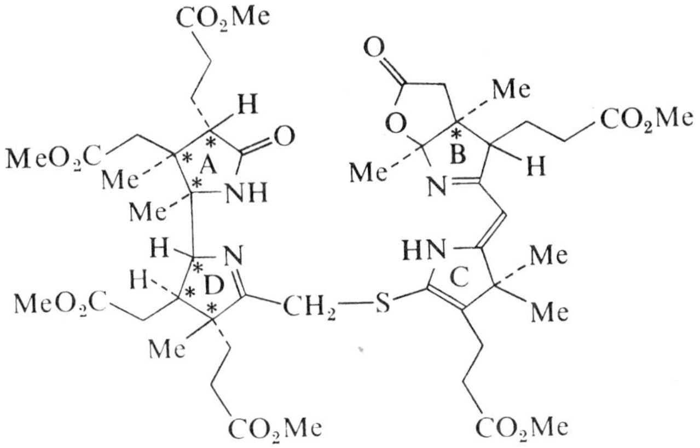

387

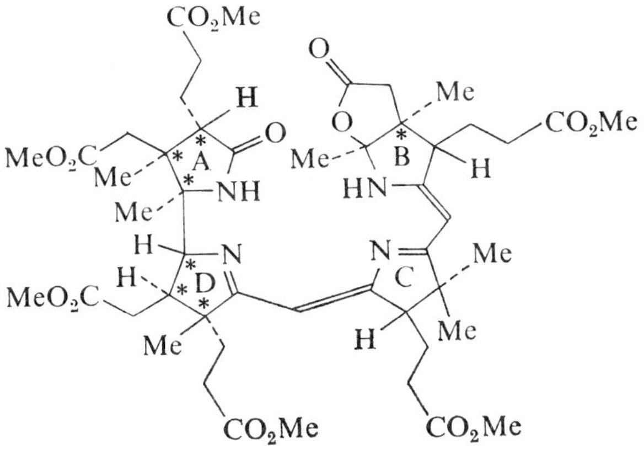

388

reduced with borohydride to the crystalline alcohol **385**. Mesylation with methanesulfonyl chloride in pyridine, followed by treatment with lithium bromide in dimethylformamide, gave the beautifully crystalline bromide **386**, the ultimate *A–D* intermediate.

Treatment with the anion of **361** gave the thioiminoether **387**, which was the most thermodynamically stable of no less than three such iminoethers. Acid-catalyzed sulfide contraction gave the *A–D–C–B* material, "corrigenolide" **388**, which unfortunately could not be crystallized, possibly as a result of epimerization of the ring-*B* and *C* propionate groups during the coupling process.

Phosphorus pentasulfide converted the ring-*A* lactam to the corresponding thiolactam, at the same time substituting a sulfur atom for the oxygen of the ring-*B* lactone, to give the dithio derivative **389**. Repetition of the series of changes established for the synthesis of metal-free corrins was shown by careful scrutiny of electronic absorption spectra to give a mixture of corrinoid compounds; in addition, their mass spectra exhibited the expected molecular ion. However, the relatively low yields experienced in this sequence stimulated experiments to find a more efficient access to the corrin macrocycle from **389** and these, both at Harvard and Zurich, have met with considerable success. Thus treatment of dithiocorrigenolide (**389**) with dimethylamine led to the labile substance **390**, which underwent chelation to **391** with zinc perchlorate. Oxidative coupling with iodine and potassium iodide gave, presumably, the intermediate **392**, which, on treatment with trifluoroacetic acid and triphenylphosphine in dimethylformamide, suffered sulfide contraction and loss of zinc ion; recomplexation with zinc furnished the well-characterized zinc chelate **393** in spectroscopic yields between 60 and 70%

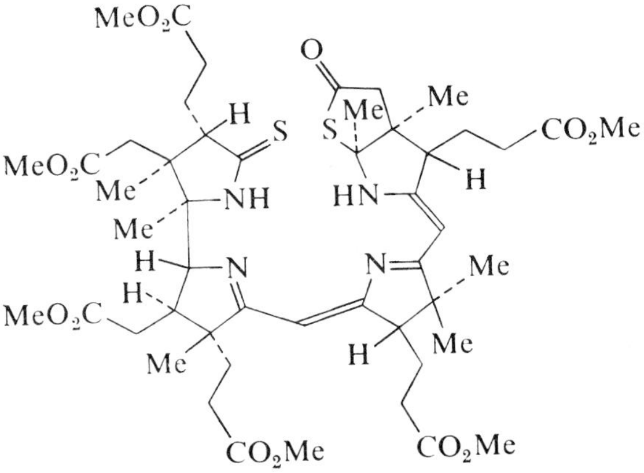

MeO$_2$C O Me H S S Me CO$_2$Me MeO$_2$C Me NH HN H Me H N N Me H MeO$_2$C Me H Me CO$_2$Me CO$_2$Me

389

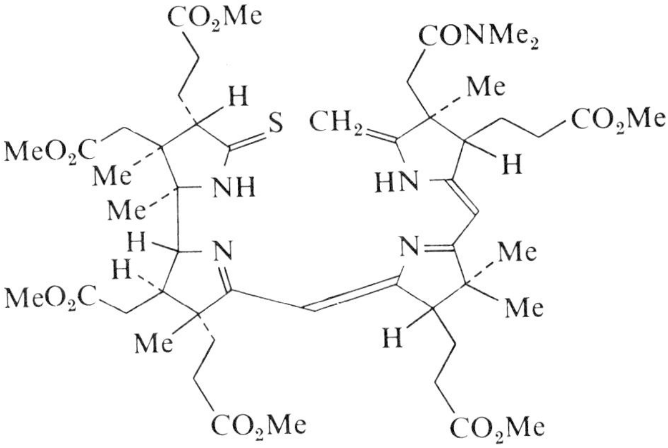

CO$_2$Me CONMe$_2$ H Me S CH$_2$ CO$_2$Me MeO$_2$C Me NH HN H Me H N N Me H MeO$_2$C Me H Me CO$_2$Me CO$_2$Me

390

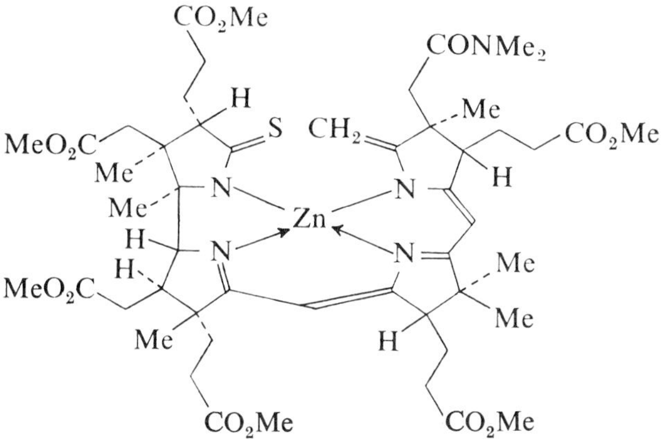

391

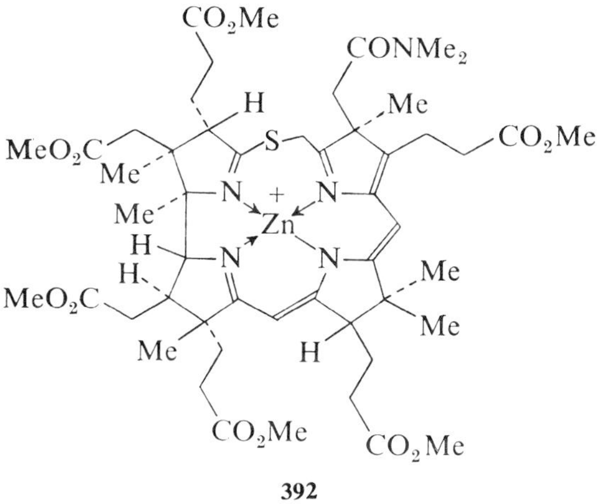

392

from **389**. The zinc in this material could be removed with acid and then replaced with cobalt through the agency of cobalt(II) chloride in tetrahydrofuran; cyanide ions and aerial oxidation then furnished the required chelate (**394**), which could also be prepared in similar yield by the more direct method of base-catalyzed thioiminoether cyclization of the appropriate precorrinoid dicyanocobalt(III) complex, although no details of this route are as yet available.

Meerwein's salt, followed by aqueous potassium bicarbonate, served to convert the dimethylamide **394** to the "beautifully crystalline"[202] complex **395** which had been shown to be identical with a sample of authentic material obtained[209] by removal of the 5- and 15-methyl groups from vitamin B_{12} by permanganate oxidation.

CO_2Me
$CONMe_2$
H
Me
CO_2Me
MeO_2C
Me
H
N
N
Me
Zn^+
H
N
N
Me
H
MeO_2C
Me
Me
H
CO_2Me
CO_2Me

393

CO_2Me
$CONMe_2$
H
Me
CO_2Me
MeO_2C
Me
CN
H
N
N
Me
Co
H
N
N
Me
H
CN
MeO_2C
Me
Me
H
CO_2Me
CO_2Me

394

CO_2Me
CO_2Me
H
Me
CO_2Me
MeO_2C
Me
N
CN
N
H
Me
Co
H
N
N
Me
H
CN
MeO_2C
Me
Me
H
CO_2Me
CO_2Me

395

The remaining problems to be overcome before the vitamin itself can be conquered are only three in number. First, the 5- and 15-methyl groups must be inserted in the macrocycle; considerable work on such reactions with simpler corrinoid compounds has already been reported,[210] though one might expect the unique steric restrictions of the vitamin itself to make this far from straightforward. Second, the propionate function at the 17-position must be differentiated from all of the other ester groupings. The original strategy had allowed for this, but ozonolysis of hesperimine **383** to the aldehyde **384** was an unexpected deviation, which, though an elegant solution to the immediate problems at the time, destroyed the potentially unique character of the substituent in question. Finally, the complicated problem of mixtures due to epimerization of the ring-*A*, *B*, and *C* propionate functions must be solved.

C. Cornforth's Approach

Although Cornforth has discussed his work many times in lectures,[211] which has clearly influenced parts of the Woodward-Eschenmoser synthesis, details have not hitherto appeared in print. His approach to the synthesis of vitamin B_{12} is based on the established, but previously little appreciated chemistry of the isoxazole ring system. The work was begun in the M.R.C. Laboratories in London in the late 1950s and continued in the Shell laboratories (Milstead) and at Warwick University.

Isoxazoles have long been known to undergo catalytic or chemical reduction to β-enamino-ketones, and ring opening by strong base affords β-keto nitriles (see Scheme 16). The ring system is also relatively stable to acid, but

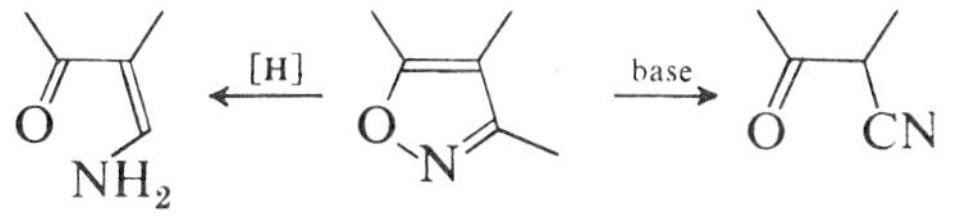

O NH$_2$ ← [H] — O N → base → O CN

Scheme 16

by virtue of its reductive ring opening, and because isoxazoles are commonly prepared from β-diketones, they may be regarded as a protected form of a β-diketone.

The basic strategy of the Cornforth approach involved attempts to construct a macrocyclic intermediate **396** containing three isoxazole units, which it was hoped would undergo reductive fission followed by recyclization to the corrin chromophore (**397**). The initial objectives, therefore, were to synthesize fragments of the molecule corresponding to the chiral centers of the molecule, in rings *A*, *B*, *C*, and *D*, each of which had the correct absolute stereochemistry. These were then to be joined by β-diketone type syntheses, and protected as isoxazoles during the initial construction of a macrocycle of type **396**. The first parts to be synthesized corresponded to rings *A* and *B* of the macrocycle; an additional advantage of the isoxazole approach lay in the fact that they could be synthesized from a common intermediate. The starting material, Hageman's ester (readily available from acetoacetic ester and formaldehyde) on treatment with cyanide in presence of magnesium acetate underwent 1,4-addition to give a mixture of the epimeric cyanoesters (**398**). The latter were separated as their semicarbazones and after regeneration with pyruvic acid, the *trans*-cyanoester, was methanolized to the *trans*-ketodiester (**399**). Formylation with methyl formate and base, or better with

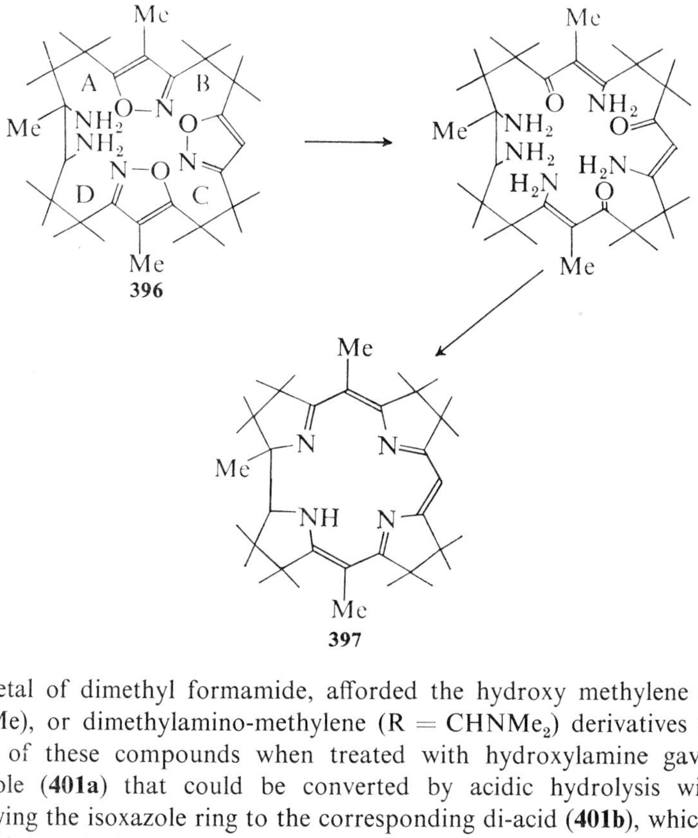

the acetal of dimethyl formamide, afforded the hydroxy methylene (R = CHOMe), or dimethylamino-methylene (R = $CHNMe_2$) derivatives (**400**). Either of these compounds when treated with hydroxylamine gave the isoxazole (**401a**) that could be converted by acidic hydrolysis without destroying the isoxazole ring to the corresponding di-acid (**401b**), which was the initial goal of this part of the synthesis. It was envisaged that alkaline fission at a later stage would result in cleavage of both rings as shown below to generate the acetic and propionic acid side chains of rings *A* and *B* (as in **403**); alternatively there was the option of opening the isoxazole ring only and converting the resulting keto-nitrile to the base stable, but acid labile, methoxyacrylonitrile (**404**). However, before continuing with the synthesis, it was essential to check the stereochemistry of the di-acid (**401b**), and after resolution as the quinine salt the (+)-enantiomer was correlated with a

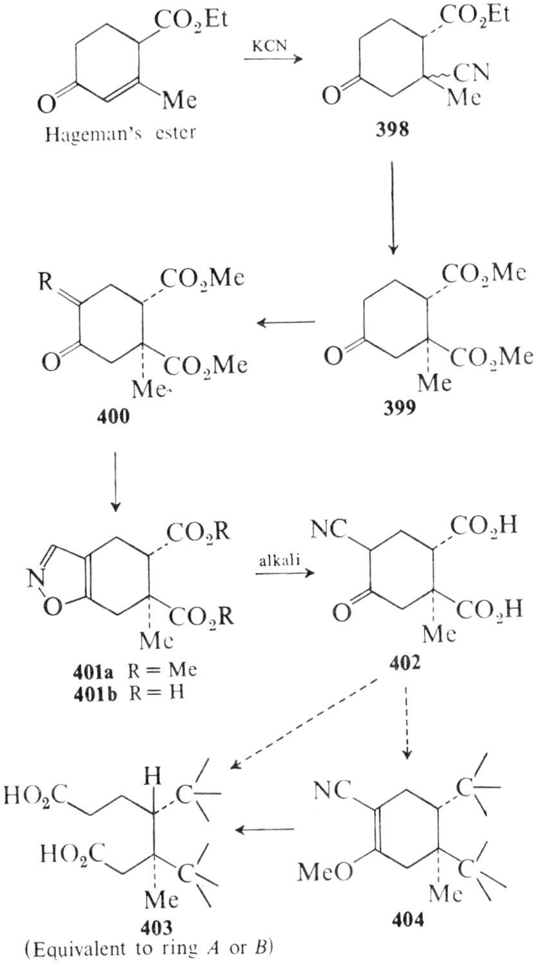

Hageman's ester

398

399

400

401a R = Me
401b R = H

402

403
(Equivalent to ring *A* or *B*)

404

steroid degradation product. Subsequently, the di-acid (**401b**) was converted into a mixture of the monoesters (**406a**) and (**406b**), and these were distinguished from each other by a synthesis in which radioactive cyanide was added to the initial Hageman ester; the methyl ester (**406b**) gave radioactive carbon dioxide on anodic decarboxylation, whereas the other ester (**406a**) did not. Treatment of the acid chloride derived from **406b** with cadmium dimethyl followed by hydrolysis then afforded the keto acid (**407**). It was later found that **407** could be derived directly and selectively from the intermediate anhydride (**405**) by treatment with cadmium dimethyl. Transformation of the keto acid (**407**) to the methylene amide (**408**) allowed addition of cyanide or nitromethylene groups as shown; the stereochemistry of the products **409** and **410** was assumed to be that shown, addition from the top face of **408** being governed by the angular methyl group. The relationship of these two products to their eventual role in the formation of ring *A* of the corrinoid macrocycle may be discerned from the schematic formulation **411** in which the isoxazole and cyclohexane rings have been assumed to be opened by alkaline hydrolysis.

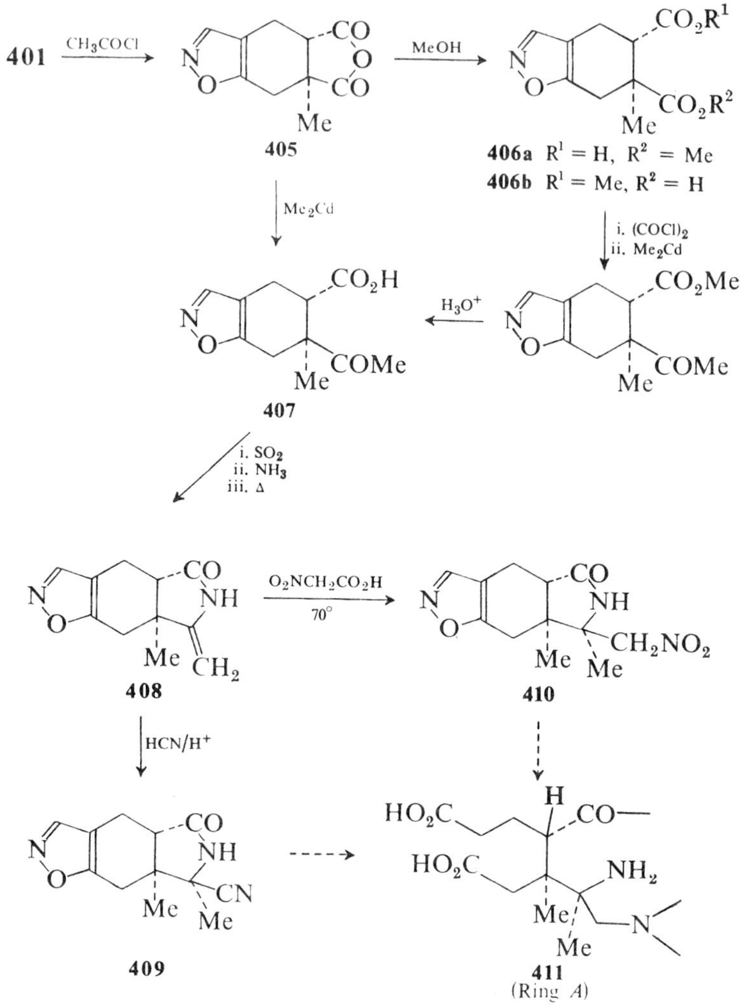

More recently, alternative syntheses of potential ring-*A* intermediates have been developed at Warwick. Base catalysed addition of nitroethane to Hageman's ester affords a mixture of the epimeric nitro compounds (**412**);

these were shown to differ only at the center bearing the nitro group by degradation of both compounds to the known ketodiester (**399**). The solid isomer **412** (35%) was brominated to give the bromonitroketone (**413**), and the corresponding ketal was also prepared. Further developments are awaited with interest, but it is envisaged that displacement of bromine from **413** with a suitable carbon nucleophile, followed by elaboration as shown schematically below could give rise to ring-*A* intermediates.

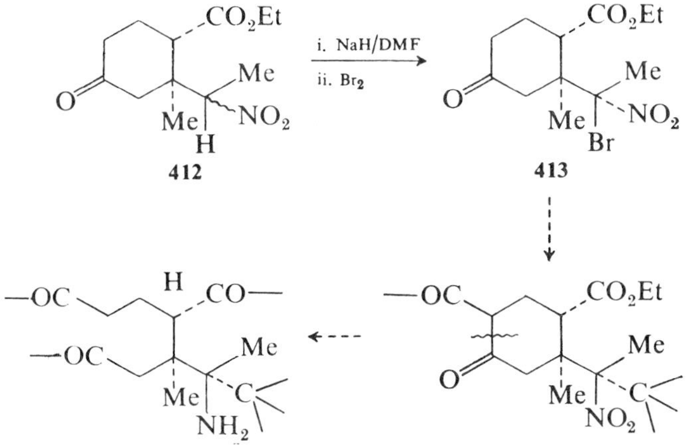

CO_2Et Me O Me NO_2 H **412** i. NaH/DMF ii. Br_2 CO_2Et Me O Me NO_2 Br **413**

—OC H CO— —OC Me Me C NH_2 Rings *A* and *B* —OC CO_2Et Me O Me C NO_2

Two approaches to the synthesis of ring *C* were studied in the original work between 1958 and 1961. The first route from a dibromoresorcinol dimethyl ether is illustrated below, a 5-carbon side-chain being introduced into the aromatic nucleus through the lithium aryl and chloroisobutyl methyl ketone. Ketalization of the resulting ketone (**414**) was effected with ethylene glycol in the presence of *p*-toluene sulphonic acid together with trimethyl orthoformate, a new procedure at the time, but one which has since become widely used. Lithium-ammonia reduction of the aromatic nucleus followed by oxidative ring opening with periodate then led to the desired ring-*C* intermediate (**415**).

The second approach to ring *C* started from camphor quinone (**353**) and is shown below. It was later adapted for use in the Woodward-Eschenmoser synthesis.

Work on the ring-*D* moiety was also begun in the late 1950s and also seems to have influenced parts of the Woodward-Eschenmoser synthesis. The α-methyl-α-carbethoxycyclopentanone (**416**) (available from adipic ester by base catalysed cyclization and alkylation) was reduced to the corresponding alcohol, dehydrated, and then hydrolysed to the acid (**417**). This was resolved with difficulty as the brucine and quinine salts, and the absolute configuration of the (+)-acid was shown by conversion to the ketone **418** and correlation

Me

$C(OH)CClMe_2$

Br Br Br

MeO OMe MeO OMe

i. PhLi

ii. $Me_2CClCOMe$

H_2O

$C(Me)_2C$ Me

$C(Me)_2COMe$

Br Br

MeO OMe MeO OMe

414

i. Li/NH_3

ii. 20% AcOH

Me Me

Me Me

Me HO_2C

Me COMe

O O OH CO_2H

$NaIO_4$

415

Me Me Me

Me MeCOO Me

O Me HN Me

* Me * H

Me

O H $CONH_2$ O

353

Me

CH_2 MeO Me O Me

HN Me HN Me O

C C HN * Me

O * Me O * Me H

H H

O

CO_2Me CO_2Me

352

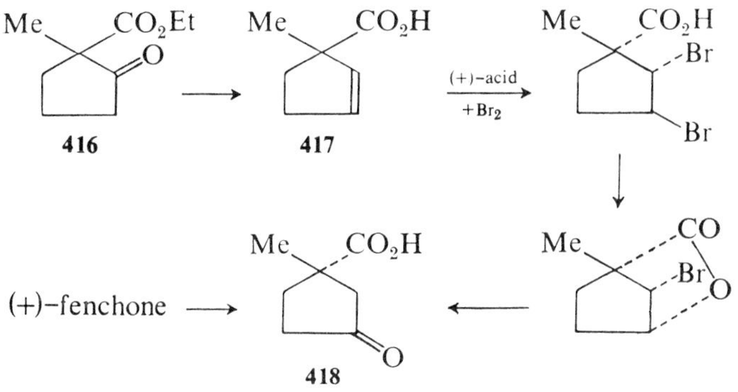

of this with (+)-fenchone. Subsequently, the corresponding homologous acid (**419**) was also prepared and resolved as the dehydroabietylamine. The additional methyl group in this compound (**419**) was eventually intended to be the C-15 bridge carbon atom of the corrin nucleus, and ring opening by cleavage of the double bond was expected to afford the ring-*D* propionic acid side-chain. The carboxylic group of this acid was destined to be C-18 of the

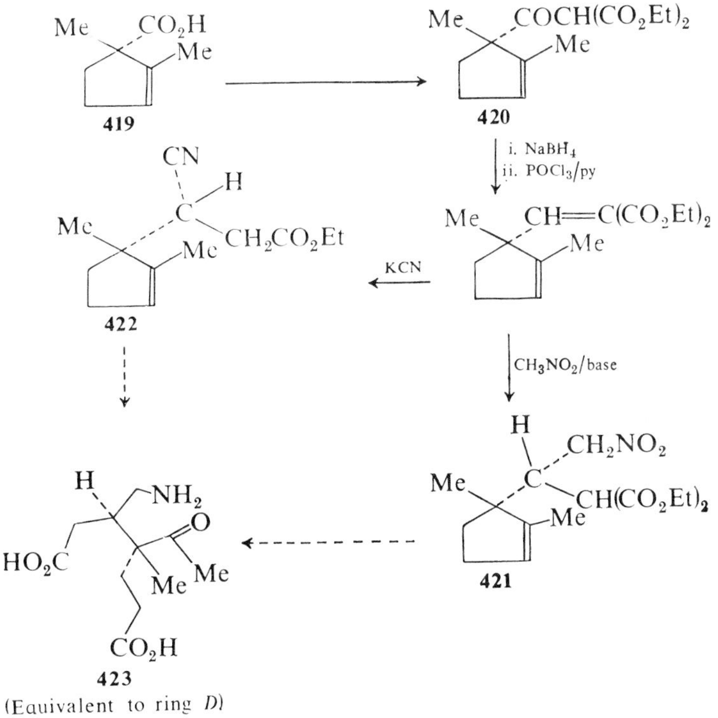

(Equivalent to ring *D*)

corrin nucleus, and various side-chain extension procedures were investigated; these included conversion to the corresponding aldehyde and alcohol, while the alkyl bromide was prepared from **414**. However, the most promising approach was chain extension by coupling the acid chloride with malonic ester to give the keto diester (**420**). Reduction and dehydration then gave the corresponding alkylidene malonate, which underwent addition of nitromethane in presence of base affording the nitromethyl derivative (**421**). Alternatively, treatment with potassium cyanide afforded the cyanoester (**422**) with concomitant decarboxylation; other potential intermediates were also synthesized from the alkylidene malonate. Possible routes from either of the two intermediates **421** or **422**, to a ring-*D* intermediate (**423**) can be discerned by assuming oxidative cleavage of the olefinic double bond as shown above.

Many attempts to construct intermediates corresponding to the *A–D* moiety were also investigated, and one of the more promising approaches is shown below (although this, like the others, failed to give the desired product). Attempts to couple the aldehyde corresponding to the acid **419** with the nitro

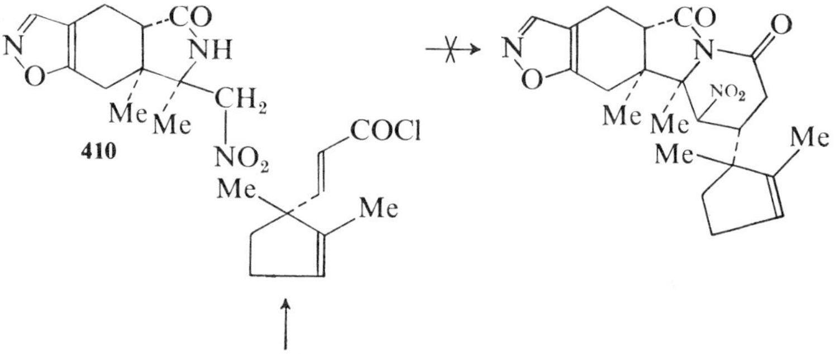

compound **410** also failed, and clearly, a considerable amount of further work is required in this area before the Cornforth approach can be considered a viable proposition. In addition, the precise manner in which the other intermediates are to be coupled also requires further exploration.

Very recently Stevens[212a] has described some experiments related to the synthesis of corrins, which also utilize isoxazoles; the preliminary reports[212a] include the preparation of model compounds ("semicorrins") analogous to the *BC* portion of the macrocycle involving not only isoxazoles as intermediates, but also γ-substituted butyrolactams.

The general potential of isoxazoles in organic synthesis has also been recognized by Stork[213] who has shown how they can be used in annelation reactions and in the synthesis of pyridine derivatives.

12. POSTSCRIPT

Since the original manuscript of this review was written, Stevens has reported his studies on the isoxazole approach to corrins[212b], Woodward[214] has described the completion of the joint Harvard-Zurich synthesis of cobyric acid (**348**), and Eschenmoser[215] has reported the adaption of his "new road" to a stereochemically controlled synthesis of the same acid. Although the yields in the penultimate stages of the last two investigations are as yet rather low, they constitute total syntheses of vitamin B_{12}, since the reconversion of cobyric acid to the vitamin has already been described.[206] At the time of writing the completion of the projects has only been described in lectures, and so only a very brief account can be given.

Eschenmoser's approach to the macrocycle involved the synthesis of an *A–D seco*-corrinoid following the route he had previously outlined;[202] rings *A*, *B*, and *C* were synthesized from the lactone (**424**), while the enantiomer could be converted into ring *D*; photochemical cyclization of the zinc or cadmium complex of the resulting *seco*-corrinoid (cf. **346** → **347**) followed by exchange of the metal for cobalt gave a mixture of stereoisomeric cobalt corrins related to **395**. Careful control of the reaction conditions ensured that the cyclization gave largely the "natural" series of isomers in which the ring-*D* acetic ester side chain was *trans* to the angular methyl group of ring *A*

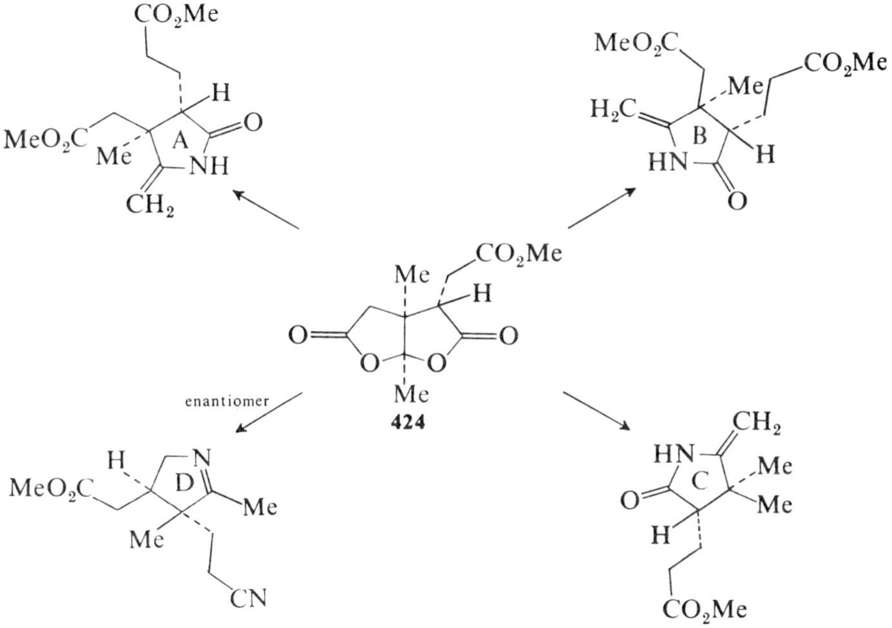

(cf. **348**). As in the joint Harvard-Zurich synthesis, a mixture of isomers was produced due to epimerization of the propionic ester side chains of rings *A*, *B*, and *C*. These were separated by high-pressure liquid chromatography, and the two heptamethylesters (**395**) prepared at Harvard and at Zurich were shown to be identical.

The remaining problems to be solved were the insertion of the two *meso*-methyl groups, differentiation of the 17-propionate function from the other esters, and conversion of the remaining esters to the corresponding amides. Alkylation of a derivative of the macrocycle with benzyloxymethyl chloride followed by brief treatment with thiophenol afforded the 5,15-di(phenylthiomethyl) derivative; alkylation at the 10-position in this derivative was sterically inhibited by the presence of an additional lactone ring (derived from the 7-acetic acid side chain of ring *B*) between the 7- and 8-positions of the macrocycle. Raney nickel desulphurization then gave a mixture of products from which the desired *meso*-dimethyl corrin could be separated by high-pressure liquid chromatography.

Differentiation of the 7-propionate function from the other propionate and acetate side-chains was achieved by use of a cyanoethyl group introduced before the corrin cyclization. Hydrolysis of the latter with concentrated sulphuric acid then afforded the corresponding amide, together with the 'neo'-compound due to partial epimerization of the propionate residue in ring *C*. After separation by high-pressure liquid chromatography, the appropriate amide was shown to be identical in all respects with naturally derived material. The yields at this stage are not yet satisfactory, but it is a tribute to the experimental skill of the investigators and the power of modern spectroscopic methods that fractions of a milligram of material were separated and crystallized and that the NMR spectrum with less than 100 μg was determined by the Fourier Transform method. Hydrolysis of the amide to the desired 17-propionic acid in presence of the other six ester functions required the development of new techniques, that is, careful nitrosation with N_2O_4 at Harvard and the use of a new reagent at Zurich. Finally, treatment of the hexamethylester mono-acid with ammonia in ethylene glycol gave cobyric acid (**395**).

The completion of this mammoth task, which has taken some 12 years, represents one of the major successes of modern organic synthesis. In some respects it may be compared with the ascent of Everest or the landing of man on the moon, for the actual achievement of the goal is of little direct value itself except as a symbol of man's striving for the ultimate in any field of endeavor. Perhaps more important are the new demands these syntheses have made on our chemical skills and ingenuity, to which we might have never otherwise aspired. The spin-off from this work has been immense and will have lasting effects on the future of organic chemistry, for example, the myriads of new synthetic methods that have been developed, and the recognition of the potential of earlier discoveries, for example, discoveries of iminoethers by Eschenmoser and of isoxazoles by Cornforth. However, the most profound results of the work arose from a redundant route, followed at Harvard, which led to a consideration of the effects of orbital symmetry on the course and stereochemistry of organic reactions.[216] These have been enshrined in the "Woodward-Hoffmann rules," and indeed without their aid Eschenmoser could not have contemplated his "new road" to corrins and ultimately to vitamin B_{12} itself.

ACKNOWLEDGEMENTS

We are very grateful to Dr. J. W. Cornforth for providing us with details of his work on the synthesis of vitamin B_{12} in advance of publication, and we also thank Dr. B. Golding and Professor V. M. Clark for letting us know of the associated work in progress at Warwick.

REFERENCES

1. H. Fischer and H. Orth, *Die Chemie des Pyrrols*, Vol. I. Akademische Verlag, Leipzig, 1934.
2. H. Fischer and H. Orth, *Die Chemie des Pyrrols*, Vol. II (i). Akademische Verlag, Leipzig, 1937.
3. H. Fischer and H. Stern, *Die Chemie des Pyrrols*, Vol. II (ii). Akademische Verlag, Leipzig, 1940.
4. T. S. Stevens, in E. H. Rodd, Ed., *Chemistry of Carbon Compounds*. Elsevier, Amsterdam, 1959, Vol. IVB, p. 1104.
5. R. L. N. Harris, A. W. Johnson, and I. T. Kay, *Q. Rev.*, **20**, 211 (1966).
6. H. H. Inhoffen, J. W. Buchler, and P. Jager, *Fortsch. Chem. Org. Naturst.*, **26**, 284 (1968).
7. G. S. Marks, *Heme and Chlorophyll*, Van Nostrand, London, 1969.
8. K. M. Smith, *Q. Rev.*, **25**, 31 (1971).
9. I.U.P.A.C. Rules for Nomenclature, *J. Am. Chem. Soc.*, **82**, 5582 (1960).
10. R. B. Woodward, *Ind. Chim. Belge*, **27**, 1293 (1962).
11. R. Lemberg and J. W. Legge, *Haematin Compounds and Bile Pigments*, Interscience, New York, 1949.
12. A. H. Corwin and E. C. Coolidge, *J. Am. Chem. Soc.*, **74**, 5196 (1952).
13. M. T. Cox, R. Fletcher, A. H. Jackson, G. W. Kenner, and K. M. Smith, *Chem. Comm.*, **1967**, 1141; A. H. Jackson, G. W. Kenner, and K. M. Smith, *J. Chem. Soc.* (*C*) **1971**, 502.
14. R. L. N. Harris, A. W. Johnson, and I. T. Kay, *J. Chem. Soc.* (*C*), **1966**, 22.
15. A. H. Jackson, G. W. Kenner, G. McGillivray, and G. S. Sach, *J. Am. Chem. Soc.*, **87**, 676 (1965).

16. A. H. Jackson, G. W. Kenner, and D. Warburton, *J. Chem. Soc.*, **1965,** 1328.
17. E. Baltazzi and L. I. Krimen, *Chem. Revs.*, **63,** 511 (1963).
18. P. Rothemund, *J. Am. Chem. Soc.*, **57,** 2010 (1935); P. Rothemund and A. R. Menotti, *J. Am. Chem. Soc.*, **63,** 267 (1941).
19. M. W. Roomi and S. F. MacDonald, *Can. J. Chem.*, **48,** 1689 (1970).
20. H. Fischer and E. Fink; *Z. physiol Chem.*, **280,** 123 (1944); **283,** 152 (1948).
21. G. G. Kleinspehn, *J. Am. Chem. Soc.*, **77,** 1546 (1955); G. G. Kleinspehn and A. H. Corwin, *J. Org. Chem.*, **25,** 1048 (1960).
22. E. Bullock, A. W. Johnson, E. Markham, and K. B. Shaw, *J. Chem. Soc.*, **1958,** 1430; A. W. Johnson and R. Price, *Org. Syn.*, **42,** 92 (1962).
23. W. G. Terry, A. H. Jackson, G. W. Kenner, and G. Kornis, *J. Chem. Soc.*, **1965,** 4389.
24. Y. K. Yurev, I. K. Korobitsyna and M. I. Kusnetsova, *Chem. Abs.*, **45,** 5680b (1951).
25. B. V. Gregorovitch, K. S. Y. Liang, D. M. Clugston, and S. F. MacDonald, *Can. J. Chem.*, **46,** 3291 (1968).
26. R. A. Chapman and S. F. MacDonald, see footnote in reference 25.
27. S. F. MacDonald and R. J. Stedman, *Can. J. Chem.*, **32,** 896 (1954).
28. S. F. MacDonald and K. H. Michl, *Can. J. Chem.*, **34,** 1768 (1956).
29. G. P. Arsenault, E. Bullock, and S. F. MacDonald, *J. Am. Chem. Soc.*, **82,** 4384 (1960).
30. E. J. Tarlton, S. F. MacDonald, and E. Baltazzi, *J. Am. Chem. Soc.*, **82,** 4389 (1960).
31. S. F. MacDonald, *J. Chem. Soc.*, **1952,** 4176.
32. R. G. Westall, *Nature*, **170,** 614 (1952).
33. T. E. Goodwin, Ed., *Porphyrins and Related Compounds*, Biochemical Society Symposium No. 28, Academic Press, London, 1968.
34. S. F. MacDonald and D. M. MacDonald, *Can. J. Chem.*, **33,** 573 (1955); A. H. Jackson, D. M. MacDonald, and S. F. MacDonald, *J. Am. Chem. Soc.*, **78,** 505 (1956); A. H. Jackson and S. F. MacDonald, *Can. J. Chem.*, **35,** 715 (1957); G. P. Arsenault and S. F. MacDonald, *Can. J. Chem.*, **39,** 2043 (1961).
35. H. Plieninger, P. Hess, and J. Ruppert, *Chem. Ber.*, **101,** 240 (1968).
36. B. Frydman, M. E. Despuy, and H. Rapoport, *J. Am. Chem. Soc.*, **87,** 3530 (1965); B. Frydman, S. Reil, M. E. Despuy, and H. Rapoport, *J. Am. Chem. Soc.*, **91,** 2338 (1969).
37. A. Albert, *Heterocyclic Chemistry*, University of London Press, 1959.
38. G. G. Kleinspehn and A. H. Corwin, *J. Am. Chem. Soc.*, **76,** 5641 (1954).
39. J. A. Ballantine, A. H. Jackson, G. W. Kenner, and G. McGillivray, *Tetrahedron*, *Suppl.* 7, 241 (1966).
40. A. F. Mironov, T. R. Ovsepyan, R. P. Evstigneeva, and N. A. Preobrazhenskii, *Zh. Obshch. Khim.*, **35,** 324 (1965).
41. H. Fischer, E. Baumann, and H. J. Reidl, *Ann.*, **475,** 237 (1929).
42. A. Hayes, G. W. Kenner, and N. R. Williams, *J. Chem. Soc.*, **1958,** 3779.
43. M. E. Flaugh and H. Rapoport, *J. Am. Chem. Soc.*, **90,** 6877 (1968).
44. J. M. Osgerby and S. F. MacDonald, *Can. J. Chem.*, **40,** 1585 (1962).
45. P. S. Clezy, A. J. Liepa, A. W. Nichol, and G. A. Smythe, *Aust. J. Chem.*, **23,** 589 (1970).

46. H. Fischer and A. Treibs, *Ann.*, **450,** 132 (1926).
47. W. Siedel and F. Winkler, *Ann.*, **554,** 162 (1943).
48. U. Eisner and R. P. Linstead, *J. Chem. Soc.*, **1955,** 3742.
49. U. Eisner, R. P. Linstead, E. A. Parkes, and E. Stephen, *J. Chem. Soc.*, **1956,** 1655.
50. U. Eisner, A. Lichtarowicz, and R. P. Linstead, *J. Chem. Soc.*, **1957,** 733.
51. A. Treibs and W. Ott, *Ann.*, **615,** 137 (1958).
52. A. W. Johnson, I. T. Kay, E. Markham, R. Price, and K. B. Shaw, *J. Chem. Soc.*, **1959,** 3416.
53. S. Krol, *J. Org. Chem.*, **24,** 2065 (1959).
54. E. Bullock, A. W. Johnson, E. Markham, and K. B. Shaw, *Nature*, **185,** 607 (1960).
55. I. T. Kay, *Proc. Nat. Acad. Sci.*, **48,** 901 (1962).
56. A. H. Jackson, P. Johnston, and G. W. Kenner, *J. Chem. Soc.*, **1964,** 2262.
57. D. Mauzerall, *J. Am. Chem. Soc.*, **82,** 2601 (1960).
58. E. Bullock, *Nature*, **205,** 70 (1965).
59. J. E. Falk, *Porphyrins and Metalloporphyrins*. Elsevier, Amsterdam, 1964.
60. S. Granick and D. Mauzerall, in D. M. Greenberg, Ed., *Metabolic Pathways*, Vol. II. Academic Press, New York, 1961, pp. 525–616.
61. L. Bogorad in M. B. Allen, Ed., *Comparative Biochemistry of Photoreactive Systems*, Academic Press, New York 1960, pp. 226–256; J. H. C. Smith, *idem.* pp. 257–277.
62. J. Lascelles, *Tetrapyrrole Biosynthesis and Its Regulation*. Benjamin, New York, 1964.
63. J. H. Mathewson and A. H. Corwin, *J. Am. Chem. Soc.*, **83,** 135 (1961).
64. J. Pluscec and L. Bogorad, *Biochem.*, **9,** 4736 (1970).
65. (a) H. Fischer and P. Halbig, *Ann.*, **448,** 193 (1926); (b) H. Fischer and G. Stangler, *Ann.*, **459,** 53 (1927).
66. P. A. Burbidge, M.Sc. Thesis, Liverpool, 1963.
67. A. Treibs and H. G. Kolm, *Ann.* **614,** 199 (1958).
68. J. S. Andrews, A. H. Corwin, and A. G. Sharp, *J. Am. Chem. Soc.*, **72,** 491 (1950).
69. J. Wass, Ph.D. Thesis, Liverpool, 1968.
70. R. P. Evstigneeva, V. N. Guryshev, A. F. Mironov, and G. Y. Volodarskaya, *Zh. Obshch. Khim.*, **39,** 2558 (1969).
71. G. M. Badger, R. L. N. Harris, and R. A. Jones, *Aust. J. Chem.*, **17,** 987 (1964).
72. R. B. Woodward, W. A. Ayer, J. M. Beaton, F. Bickelhaupt, R. Bonnett, P. Buchsacher, G. L. Closs, H. Dutler, J. Hannah, F. P. Hauck, S. Ito, A. Langemann, E. Le Goff, W. Leimgruber, W. Lwowski, J. Sauer, Z. Valenta, and H. Volz, *J. Am. Chem. Soc.*, **82,** 3800 (1960).
73. H. Fischer and J. Klarer, *Ann.*, **448,** 178 (1926).
74. H. Fischer, H. Friedrich, W. Lamatsch, and K. Morgenroth, *Ann.*, **466,** 147 (1928).
75. H. Fischer and A. Kirmann, *Compt. rend.*, **189,** 467 (1929).
76. H. Fischer, *Naturwiss.*, **17,** 611 (1929).
77. See reference 2, pp. 331 and 342.
78. H. Fischer and H. Helberger, *Ann.*, **480,** 235 (1930).
79. H. Fischer, H. Berg, and A. Schormuller, *Ann.*, **480,** 109 (1930).
80. F. Morsingh and S. F. MacDonald, *J. Am. Chem. Soc.*, **82,** 4377 (1960).

81. J. L. Archibald, D. M. Walker, K. B. Shaw, A. Markovac, and S. F. MacDonald, *Can. J. Chem.*, **44,** 345 (1966).

82. G. S. Marks, D. K. Dougall, E. Bullock, and S. F. MacDonald, *J. Am. Chem. Soc.*, **82,** 3183 (1960).

83. J. W. Purdie and A. S. Holt, *Can. J. Chem.*, **43,** 3347 (1965).

84. A. S. Holt, D. W. Hughes, H. J. Kende, and J. W. Purdie, *J. Am. Chem. Soc.*, **84,** 2835 (1962); A. S. Holt, J. W. Purdie, and J. W. F. Wasley, *Can. J. Chem.*, **44,** 88 (1966); A. S. Holt in T. W. Goodwin, Ed., *The Chemistry and Biochemistry of Plant Pigments*. Academic Books, London, 1965, pp. 3–28.

85. K. M. Smith, Ph.D. Thesis, Liverpool, 1967.

86. J. H. Mathewson, W. R. Richards, and H. Rapoport, *Biochem. Biophys. Res. Comm.*, **13,** 1 (1963); *J. Am. Chem. Soc.*, **85,** 364 (1963).

87. S. F. MacDonald, personal communication (cf. references 25 and 26).

88. P. S. Clezy and A. W. Nichol, *Aust. J. Chem.*, **18,** 1835 (1965).

89. P. S. Clezy, D. Looney, A. W. Nichol, and G. A. Smythe, *Aust. J. Chem.*, **19,** 1481 (1966).

90. R. Chong, P. S. Clezy, A. J. Liepa, and A. W. Nichol, *Aust. J. Chem.*, **22,** 229 (1969).

91. P. S. Clezy, A. J. Liepa, A. W. Nichol, and G. A. Smythe, *Aust. J. Chem.*, **23,** 589 (1970).

92. P. S. Clezy, A. J. Liepa, and G. A. Smythe, *Aust. J. Chem.*, **23,** 603 (1970).

93. P. S. Clezy and A. J. Liepa, *Aust. J. Chem.*, **23,** 2461 (1970).

94. P. S. Clezy and A. J. Liepa, *Aust. J. Chem.*, **23,** 2477 (1970).

95. R. Bonnett and M. J. Dimsdale, *Tetrahedron Lett.*, **1968,** 731; R. Bonnett, M. J. Dimsdale, and G. F. Stephenson, *J. Chem. Soc.* (*C*), **1969,** 564.

96. A. H. Jackson, G. W. Kenner, G. McGillivray, and K. M. Smith, *J. Chem. Soc.* (*C*), **1968,** 294.

97. H. Fischer and A. Schormüller, *Ann.*, **473,** 211 (1929).

98. D. Dolphin, A. W. Johnson, J. Leng, and P. Van den Broek, *J. Chem. Soc.* (*C*), **1966,** 880.

99. P. Bamfield, R. L. N. Harris, A. W. Johnson, I. T. Kay, and K. W. Shelton, *J. Chem. Soc.* (*C*), **1966,** 1436.

100. A. W. Johnson, *Chem. Brit.*, **3,** 253 (1967).

101. P. Bamfield, R. Grigg, R. W. Kenyon, and A. W. Johnson, *Chem. Comm.*, **1967,** 1029; *J. Chem. Soc.* (*C*), **1968,** 1259.

102. R. Grigg, A. W. Johnson, and M. Roche, *J. Chem. Soc.* (*C*), **1970,** 1928.

103. A. W. Johnson and I. T. Kay, *J. Chem. Soc.*, **1961,** 2418; R. Grigg, A. W. Johnson, R. Kenyon, V. B. Math, and K. Richardson, *J. Chem. Soc.* (*C*), **1969,** 176.

104. G. M. Badger, R. L. N. Harris, and R. A. Jones, *Aust. J. Chem.* **17,** 1013 (1964).

105. T. Lewis, Postdoctoral Fellowship Report, Liverpool, 1969.

106. A. H. Jackson, G. W. Kenner, and G. S. Sach, *J. Chem. Soc.* (*C*), **1967,** 2045.

107. A. H. Jackson, G. W. Kenner, and J. Wass, unpublished work.

108. R. P. Carr, P. J. Crook, A. H. Jackson, and G. W. Kenner, *Chem. Comm.*, **1967,** 1025; R. P. Carr, A. H. Jackson, G. W. Kenner, and G. S. Sach, *J. Chem. Soc.* (*C*), **1971,** 487.

109. A. H. Jackson, G. W. Kenner, and J. Wass, *Chem. Comm.*, **1967,** 1027.
110. P. J. Crook, Ph.D. Thesis, Liverpool, 1968. P. J. Crook, A. H. Jackson, and G. W. Kenner, *J. Chem. Soc.* (*C*), in press.
111. A. H. Jackson, G. W. Kenner, and K. M. Smith, *J. Chem. Soc.* (*C*), **1968,** 302.
112. (a) T. T. Howarth, Ph.D. Thesis, Liverpool, 1967; (b) M. T. Cox, Ph.D. Thesis, Liverpool, 1969.
113. M. T. Cox, T. T. Howarth, A. H. Jackson, and G. W. Kenner, *J. Am. Chem. Soc.*, **91,** 1232 (1969).
114. R. Fletcher, A. H. Jackson, and G. W. Kenner, unpublished work.
115. H. H. Inhoffen, H. Brockmann, K. M. Bliesner, *Ann.*, **730,** 173 (1969).
116. R. P. Carr, M. T. Cox, A. H. Jackson, and G. W. Kenner, unpublished work, cf. footnote 1 in reference 113.
117. G. Y. Kennedy, A. H. Jackson, G. W. Kenner, and C. J. Suckling, *FEBS Lett.*, **6,** 9 (1970); **7,** 205 (1970).
118. R. Lemberg, *Biochem J.*, **29,** 1322 (1935); R. Lemberg, B. Cortis-Jones, and M. Norrie, *Biochem. J.*, **32,** 171 (1938).
119. H. Fischer and H. Libowitzsky, *Z. Physiol. Chem.*, **255,** 209 (1938); **251,** 198 (1938); E. Stier, *Z. Physiol Chem.*, **272,** 239 (1942); **275,** 155 (1942).
120. Cf. R. Lemberg, *Rev. Pure Appl. Chem.* (*Aust.*), **6,** 1 (1956).
121. R. Bonnett, M. J. Dimsdale, and K. D. Sales, *Chem. Comm.*, **1970,** 962.
122. T. Kondo, D. C. Nicholson, A. H. Jackson, and G. W. Kenner, *Biochem. J.*, **121,** 601 (1971); cf. A. H. Jackson and G. W. Kenner in reference 33, p. 1.
123. M. T. Cox, A. H. Jackson, and G. W. Kenner, *J. Chem. Soc.* (*C*), **1971,** 1974.
124. M. E. Flaugh and H. Rapoport, *J. Am. Chem. Soc.*, **90,** 6877 (1968).
125. (a) A. Stoll and E. Wiedemann, *Fortschr. Chem. Forsch.*, **2,** 538 (1952); (b) L. P. Vernon and G. R. Seely, Eds., *The Chlorophylls*. Academic Press, New York, 1966.
126. H. H. Inhoffen, J. W. Buchler, and R. Thomas, *Tetrahedron Lett.*, **1969,** 1145.
127. H. Whitlock, R. Hanauer, M. Y. Oester, and B. K. Bower, *J. Am. Chem. Soc.*, **91,** 7485 (1969).
128. (a) H. Fischer and G. A. Von Kemnitz, *Z. Physiol. Chem.*, **96,** 309 (1915); (b) A. Treibs and E. Wiedemann, *Ann.*, **471,** 146 (1929); (c) reference 3, pp. 144–146; (d) H. Fischer and F. Balaz, *Ann.*, **553,** 166 (1942).
129. Cf. R. B. Woodward, *Angew. Chem.*, **72,** 651 (1960).
130. U. Eisner, *J. Chem. Soc.*, **1957,** 854.
131. U. Eisner and R. P. Linstead, *J. Chem. Soc.*, **1955,** 3749.
132. D. Mauzerall, *J. Am. Chem. Soc.*, **84,** 2437 (1962).
133. H. H. Inhoffen and P. Jäger, *Tetrahedron Lett.*, **1964,** 1317; **1965,** 3387.
134. H. H. Inhoffen, *Pure Appl. Chem.*, **17,** 443 (1968); see also reference 6.
135. D. Mauzerall, *J. Am. Chem. Soc.*, **82,** 2605 (1960).
136. M. Strell, A. Kalojanoff, and H. Koller, *Angew. Chem.*, **72,** 169 (1960); M. Strell and A. Kalojanoff. *Ann.*, **652,** 218 (1962).
137. A. Treibs and R. Schmidt, *Ann.*, **577,** 105 (1952); see also reference 128d.
138. M. Strell and A. Kalojanoff, *Ann.*, **577,** 97 (1952); *Angew. Chem.*, **66,** 445 (1954).
139. H. H. Inhoffen, *Angew. Chem. Int., Ed.*, **3,** 322 (1964).

140. H. Wolf, H. Brockmann, H. Biere, and H. H. Inhoffen, *Ann.*, **704,** 208 (1967); I. Fleming, *Nature*, **216,** 151 (1967); *J. Chem. Soc.* (*C*), **1968,** 2765; H. Brockmann, *Angew. Chem.*, **80,** 233 (1968).

141. H. Brockmann, *Angew. Chem.*, **80**, 234 (1968); H. Brockmann and I. Kleber, *Angew. Chem.*, **81,** 626 (1969).

142. J. J. Katz, G. D. Norman, W. A. Svec, and H. H. Strain, *J. Am. Chem. Soc.*, **90,** 6841 (1968); H. H. Strain and W. M. Manning, *J. Biol. Chem.*, **146,** 275 (1942).

143. R. C. Dougherty, H. H. Strain, W. A. Svec, R. A. Uphaus, and J. J. Katz, *J. Am. Chem. Soc.*, **88,** 5037 (1966); **92,** 2826 (1970).

144. W. Rüdiger, in reference 33, p. 121.

145. Z. Petryka, D. C. Nicholson, and C. H. Gray, *Nature*, **194,** 1047 (1962); P. O'Carra and E. Colleran, *FEBS Lett.*, **5,** 295 (1969); R. Bonnett and A. F. McDonagh, *Chem. Comm.* **1970,** 237.

146. P. O'Carra and E. Colleran, *Biochem. J.*, **115,** 13P (1969).

147. J. C. Kendrew, R. E. Dickerson, B. E. Strandberg, R. G. Hart, D. R. Davies, D. C. Phillips, and V. C. Shore, *Nature*, **185,** 422 (1960).

148. H. Fischer and A. Treibs, *Ann.*, **457,** 209 (1927); H. H. Inhoffen, J. H. Fuhrhop and F. v. d. Haar, *Ann.*, **700,** 92 (1966).

149. H. W. Siegelman, D. J. Chapman, and W. J. Cole, in reference 33, p. 107.

150. H. W. Siegelman, D. J. Chapman, and W. J. Cole, *J. Am. Chem. Soc.*, **89,** 3643 (1967); W. Rüdiger, P. O'Carra, and C. O'hEocha, *Nature*, **215,** 1477 (1967); H. L. Crespi, U. Smith, and J. J. Katz, *Biochem.*, **7,** 2232 (1968).

151. W. Siedel, *Z. Physiol. Chem.*, **245,** 257 (1937).

152. H. Fischer and H. Plieninger, *Z. Physiol. Chem.*, **274,** 231 (1942).

153. H. Fischer and H. Rose, *Z. Physiol. Chem.*, **82,** 391 (1912); *Berichte*, **45,** 1579 (1912); O. Piloty and S. J. Thannhauser, *Ann.*, **390,** 191 (1912).

154. H. Fischer and H. Rose, *Berichte*, **46,** 439 (1913); O. Piloty and S. J. Thannhauser, *Berichte*, **45,** 2393 (1912); O. Piloty, *Berichte*, **46,** 1000 (1913).

155. W. Siedel and H. Fischer, *Z. Physiol. Chem.*, **214,** 145 (1933).

156. H. Fischer and R. Hess, *Z. Physiol. Chem.*, **194,** 193 (1930).

157. W. Siedel, *Z. Physiol. Chem.*, **237,** 8 (1935).

158. H. Fischer and E. Adler, *Z. Physiol. Chem.*, **200,** 209 (1931).

159. H. Fischer and E. Adler, *Z. Physiol. Chem.*, **212,** 146 (1932).

160. Reference 2, p. 621 *et seq.*

161. W. Siedel, *Z. Physiol. Chem.*, **231,** 167 (1935).

162. H. Fischer and H. Höfelmann, *Z. Physiol. Chem.*, **251,** 187 (1937).

163. W. Siedel and E. Meier, *Z. Physiol. Chem.*, **242,** 101 (1936).

164. W. Siedel and H. Möller, *Z. Physiol. Chem.*, **264,** 64 (1940).

165. J. H. Atkinson, R. S. Atkinson, and A. W. Johnson, *J. Chem. Soc.*, **1964,** 5999.

166. D. R. Ridyard, M.Sc. Thesis, Liverpool, 1966.

167. H. Plieninger and M. Decker, *Ann.*, **598,** 198 (1956); H. Plieninger and J. Kurze, *Ann.*, **680,** 60 (1964).

168. H. Plieninger, Lecture to Société Vaudoise des Sciences Naturelles, Lausanne, October, 1963, abstracted in *Chimia*, **17,** 393 (1963); H. Plieninger and U. Lerch, *Ann.*, **698,** 196 (1966).

169. H. Plieninger and J. Ruppert, *Ann.*, **736,** 43 (1970).

170. C. H. Gray and D. C. Nicholson, *J. Chem. Soc.*, **1958,** 3085.

171. W. Siedel and E. Meier, *Z. Physiol. Chem.*, **242,** 121 (1936).

172. C. J. Watson, A. Moscowitz, D. A. Lightner, Z. J. Petryka, E. Davis, and M. Weiner, *Proc. Nat. Acad. Sci. U.S.*, **58,** 1957 (1967).

173. H. Plieninger and R. Steinsträsser, *Ann.*, **723,** 149 (1969).

174. A. H. Jackson, K. M. Smith, C. H. Gray, and D. C. Nicholson, *Nature*, **209,** 581 (1966); A. H. Jackson, G. W. Kenner, H. Budzikiewicz, C. Djerassi, and J. M. Wilson, *Tetrahedron*, **23,** 603 (1967).

175. H. Plieninger, K. Ehl, and A. Tapia, *Ann.*, **736,** 62 (1970).

176. G. Stork and R. Matthews, *Chem. Comm.*, **1970,** 445.

177. F. Wrede and A. Rothaas, *Z. Physiol. Chem.*, **215,** 67 (1933); **219,** 267 (1933); **222,** 203 (1933); **226,** 95 (1934).

178. H. Fischer and K. Gangl, *Z. Physiol. Chem.*, **267,** 201 (1941); A. Treibs and K. Hintermeier, *Ann.*, **605,** 35 (1957); A. J. Castro, A. H. Corwin, J. F. Deck, and P. E. Wei, *J. Org. Chem.*, **24,** 1437 (1959).

179. H. Rapoport and K. G. Holden, *J. Am. Chem. Soc.*, **84,** 635 (1962).

180. N. N. Gerber, *Appl. Microbiol.*, **18,** 1 (1969).

181. H. H. Wasserman, G. C. Rodger, and D. D. Keith, *Chem. Comm.*, **1966,** 825; K. Harashima, N. Tsuchida, and J. Nagatsu, *Agric. Biol. Chem.* (*Japan*), **30,** 309 (1966); K. Harashima, N. Tsuchida, T. Tanaka, and J. Nagatsu, *Agric. Biol. Chem.* (*Japan*), **31,** 481 (1967).

182. (a) H. H. Wasserman, G. C. Rodgers, and D. D. Keith, *J. Am. Chem. Soc.*, **91,** 1263 (1969); (b) H. H. Wasserman, D. D. Keith, and J. Nadelson, *J. Am. Chem. Soc.*, **91,** 1264 (1969).

183. N. N. Gerber, *Tetrahedron Lett.*, **1970,** 809.

184. U. V. Santer and H. J. Vogel, *Fed. Proc.*, **15,** 345 (1956); *Biochem. Biophys. Acta*, **19,** 578 (1956).

185. H. H. Wasserman, J. E. McKeon, L. Smith, and P. Forgione, *J. Am. Chem. Soc.*, **82,** 506 (1960).

186. H. Rapoport and C. D. Willson, *J. Am. Chem. Soc.*, **84,** 630 (1962).

187. D. W. Fuhlhage and C. A. Vanderwerf, *J. Am. Chem. Soc.*, **80,** 6249 (1958).

188. E. Mosettig, in *Organic Reactions*, Vol. VIII. Wiley, New York, 1954, p. 218; R. G. Jones and K. C. McLaughlin, *J. Am. Chem. Soc.*, **71,** 2444 (1949).

189. (a) H. Rapoport and N. Castagnoli, *J. Am. Chem. Soc.*, **84,** 2178 (1962); (b); J. Bordner and H. Rapoport, *J. Org. Chem.*, **30,** 3824 (1965).

190. R. Grigg, A. W. Johnson, and J. W. F. Wasley, *J. Chem. Soc.*, **1963,** 359; E. Bullock, R. Grigg, A. W. Johnson, and J. W. F. Wasley, *J. Chem. Soc.*, **1963,** 2326; R. Grigg and A. W. Johnson, *J. Chem. Soc.*, **1964,** 3315.

191. W. R. Hearn, M. K. Elson, R. H. Williams, and J. Medina-Castro, *J. Org. Chem.*, **35,** 142 (1970).

192. A. Ermili and A. J. Castro, *J. Heterocyc. Chem.*, **3,** 521 (1966); A. J. Castro, G. R. Gale, G. E. Means, and E. Tertzakian, *J. Med. Chem.*, **10,** 29 (1967).

193. R. Grigg, A. W. Johnson, and P. v. d. Broek, *Chem. Comm.*, **1967,** 502; I. D. Dicker, R. Grigg, A. W. Johnson, H. Pinnock, K. Richardson, and P. v. d. Broek, *J. Chem. Soc. (C)*, **1971,** 536.

194. E. Bertele, H. Boos, J. D. Dunitz, F. Elsinger, A. Eschenmoser, I. Felner, H. P. Gribi, H. Gschwend, E. F. Meyer, M. Pesaro, and R. Scheffold, *Angew. Chem.*, **76,** 393 (1964).

195. H. Meerwein, H. Hinz, P. Hoffmann, E. Kroning, and E. Pfeil, *J. Prakt. Chem.*, **147,** 17 (1937).

196. Cf. E. J. Corey, *Pure Appl. Chem.*, **14,** 19 (1967).

197. J. D. Dunitz and E. F. Meyer, *Proc. Roy. Soc. (A)*, **288,** 324 (1965).

198. A. Eschenmoser, R. Scheffold, E. Bertele, M. Pesaro, and H. Gschwend, *Proc. Roy. Soc. (A)*, **288,** 306 (1965).

199. I. Felner, A. Fischli, A. Wick, M. Pesaro, D. Bormann, E. L. Winnacker, and A. Eschenmoser, *Angew. Chem.*, **79,** 863 (1967).

200. J. I. Toohey, *Proc. Nat. Acad. Sci. U.S.*, **54,** 934 (1965); *Fed. Proc.*, **25,** 1628 (1966); K. Sato, S. Shimzu, and S. Fukui, *Biochem. Biophys. Res. Comm.*, **39,** 170 (1970).

201. A. Fischli and A. Eschenmoser, *Angew. Chem.*, **79,** 865 (1967).

202. A. Eschenmoser, *Q. Rev.*, **24,** 366 (1970).

203. Y. Yamada, D. Miljkovic, P. Wehrli, B. Golding, P. Loeliger, R. Keese, K. Müller, and A. Eschenmoser, *Angew. Chem.*, **81,** 301 (1969).

204. R. B. Woodward and R. Hoffmann, *Angew. Chem.*, **81,** 797 (1969).

205. R. Bonnett, *Chem. Revs.*, **63,** 573 (1963).

206. W. Friedrich, G. Gross, K. Bernhauer, and P. Zeller, *Helv. Chim. Acta*, **43,** 704 (1960).

207. R. B. Woodward, *Pure Appl. Chem.*, **17,** 519 (1968).

208. A. H. Jackson, Ph.D. Thesis, Cambridge, 1954; referred to by B. Robinson, *Chem. Ind.*, **1963,** 218; and by E. Coxworth in *The Alkaloids*, Vol. 8, R. H. F. Manske, Ed. Academic Press, New York. 1965, p. 27.

209. K. Bernhauer and F. Wagner, unpublished work.

210. D. Borman, A. Fischli, R. Keese, and A. Eschenmoser, *Angew. Chem.*, **79,** 867 (1967).

211. Cf. J. W. Cornforth (reported by P. B. D. de la Mare), *Nature*, **195,** 441 (1962).

212. (*a*) R. V. Stevens, L. E. DuPree, Jr., and M. P. Wentland, *Chem. Comm.*, **1970,** 821; R. V. Stevens and M. Kaplan, *idem*, **1970,** 822; (*b*) R. V. Stevens, C. G. Christensen, W. L. Edmonson, M. Kaplan, E. B. Reid and M. P. Wentland; *J. Amer. Chem. Soc.*, **93,** 6629 (1971); R. V. Stevens, L. E. DuPree Jr., W. L. Edmonson, L. L. Magid and M. P. Westland, *J. Amer. Chem. Soc.*, **93,** 6637 (1971).

213. G. Stork, S. Danishefsky, and M. Ohashi; *J. Am. Chem. Soc.* **89,** 5459 (1967); M. Ohashi, H. Kamachi, H. Kakisawa, and G. Stork, *idem*, **89,** 5460 (1967); G. Stork and J. E. McMurry, *idem*, **89,** 5463 and 5464 (1967).

214. R. B. Woodward, *XXIIIrd International Congress of Pure and Applied Chemistry*, Boston, July 1971, Vol. 2, Butterworths, London, 1972; *IUPAC Symposium on Natural Products*, Delhi, February 1972.

215. A. Eschenmoser, "W. Baker Lecture", Bristol, May 1972; cf. *XXIIIrd International Congress of Pure and Applied Chemistry*, Boston, July 1971, Vol. 2, Butterworths, London, 1972.

216. R. B. Woodward and R. Hoffmann, *The Conservation of Orbital Symmetry*, Verlag Chemie, Weinheim, 1970.

The Total Synthesis of Nucleic Acids*

SARAN A. NARANG

Biochemistry Laboratory, National Research Council of Canada, Ottawa, Canada

AND

ROBERT H. WIGHTMAN

Department of Chemistry, Carleton University, Ottawa, Ontario, Canada

1. INTRODUCTION

The hypothesis that the genes in all living organisms contain all the information required for the cell to reproduce is now more than 50 years old. In chemical terms the basic idea of this hypothesis has been that the deoxyribonucleic acid (DNA) in all living organisms directs the synthesis of all the protein required by that organism. This control is exerted through the intermediacy of ribonucleic acids (messenger or *m*RNA, transfer or *t*RNA), which transcribe and transmit the information originally present in DNA. Although most of the predictions of the "central dogma" have been verified using biological approaches, the most successful and direct confirmation has been obtained by the use of synthetic DNA molecules.[1] For many such investigations supplies of these molecules would be essential, but their pure form may not be derived from degradative procedures. Thus there have been attempts at a rational synthesis of such molecules not only because of the challenges of their chemical complexity but also for their usefulness in studying some cardinal biological processes.

Naturally occurring DNA consists of two long complementary polynucleotide chains twisted about each other in a regular helix. Each chain contains a large number of nucleosides joined through a $C_{3'} \rightarrow C_{5'}$ internucleotide phosphodiester bond. The two chains are held together by hydrogen bonds between pairs of bases in which guanine (purine) is always joined to cytosine (pyrimidine) and adenine (purine) always bonds to thymine (pyrimidine) or to uridine (pyrimidine) in RNA,* as illustrated in Fig. 1. Individual DNA molecules may be very large; in fact, the *E. coli* chromosome

* NRCC No. 11844

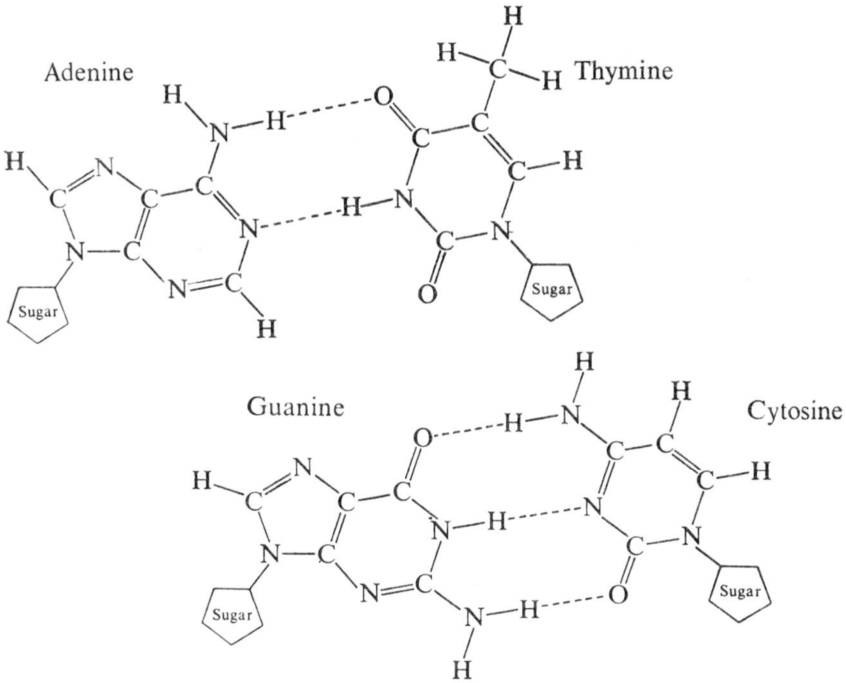

Figure 1. Base pairing in DNA due to hydrogen-bonding (note that in RNA thymine is replaced by uridine).

is probably a single DNA molecule with MW $\sim 2\text{–}4 \times 10^9$. Most DNA molecules correspond not to single genes but to a collection of genes. The average molecular weight is about 10^6 (i.e., 1500 nucleotide pairs).

Much preliminary work on the individual nucleosides (heterocyclic base plus sugar), nucleotides (base, sugar and phosphate), and small oligonucleotides (short chains) has shown that it is not feasible to attempt a purely

---

* The basic system of abbreviations used for polynucleotides and their protected derivatives is as used in *J. Biol. Chem.* Thus the single letters A, T, C, G, and U represent the nucleosides of respectively adenine, thymine, cytosine, guanine, and uridine. The letter *p* to the left of the nucleoside initial indicates a 5′-phosphomonoester group and the same letter to the right indicates a 3′-phosphomonoester group. Hence, in going from the left to the right the polynucleotide chain is specified in the $C_{3'} \rightarrow C_{5'}$ direction. The protecting groups on the purine or pyrimidine rings are designated by two-letter abbreviations added as superscripts after the nucleoside initial: thus A^{Bz} for *N*-benzoyldeoxyadenosine, C^{An} for *N*-anisoyldeoxycytidine, G^{Ac} for *N*-acetyldeoxyguanosine. The acetyl group at the 3′-hydroxyl group of a nucleoside is shown by —OAc added after the nucleoside initial: thus *d-p*T*p*G^{Ac}-OAc for the dinucleotide, 5′-*O*-phosphorylthymidylyl-(3′ → 5′)-*N*, 3′-*O*-diacetyldeoxyguanosine.

chemical synthesis of such a large molecule. The first concerted efforts to synthesize some of the simpler compounds were due to Todd and his co-workers.[2] Although these synthetic studies were primarily intended as final structure proofs, they did succeed in emphasizing some of the critical problems. Nucleotides and their polymers are sensitive molecules and because of their polar nature they are not amenable to the usual manipulations of organic chemistry. Also, methods for introduction of a phosphoric acid residue and specific formation of the internucleotide bond were relatively crude.

The modern phase of nucleic acid chemistry and synthesis rests on the development of appropriate condensing reagents, the application of ion-exchange chromatography for separation and purification of products, and the use of various enzymes as specific reagents for the characterization and synthesis of polynucleotides. The 1960s have witnessed tremendous strides in solving the problems associated with the synthesis of small chains of deoxyribo- (and ribo-) oligonucleotides. As a result much valuable information has been obtained concerning the genetic code, especially through the efforts of Khorana and his co-workers.[3] However, there seems little doubt that new concepts and approaches have to be explored to achieve more efficient and easier syntheses of the longer chains.

A few comprehensive articles have already been written concerning chemistry and synthesis in the nucleic acid field. These include an excellent book[4] and a small monograph,[5] both somewhat out-of-date; a reference text[6] pertaining primarily to manipulations of individual nucleosides and nucleotides; and some shorter reviews.[7] In addition much recent work has been reported on synthetic modifications[8] of the naturally occurring molecules. Some of these exhibit promising physiological activity.[9] Further, the very interesting "rare" bases present in *t*RNA are beginning to receive attention.[10]

With this review, which deals only with the development of procedures for the synthesis of deoxyribo- (and ribo-) oligonucleotides during the period 1960–1970, we hope to achieve two objectives. First, we attempt to mention all pertinent investigations and emphasize those approaches that have been most successful in practice. Second, we hope to illustrate some of the areas where improvements would prove most beneficial.

2. THE GENERALIZED PROBLEM

To achieve any success in nucleic acid synthesis one should be familiar with the organic chemistry of carbohydrates, nitrogen heterocycles, and phosphates esters as well as the handling of an increasing number of pertinent

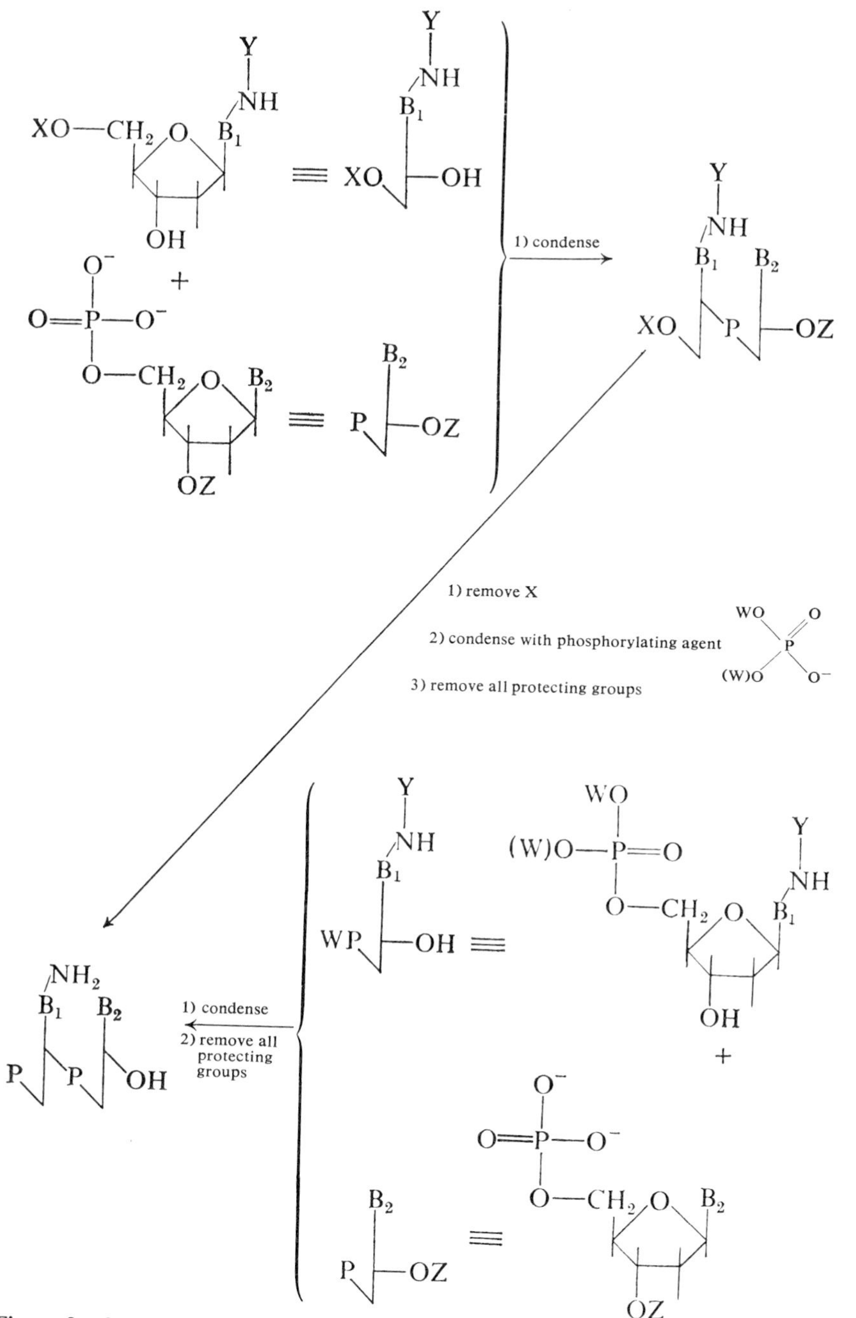

Figure 2. Schematic representation of possible approaches to a deoxydinucleotide diphosphate (B = heterocyclic base; W, X, Y, and Z = protecting groups for phosphate, primary hydroxyl, primary amino, and secondary hydroxyl respectively).

enzymes. The fundamental objective of the problem is efficient formation of the internucleotidic phosphodiester bond specifically between the $C_{3'}$ and $C_{5'}$ positions of two adjacent nucleosides. This is the natural bond occurring in DNA and RNA and is usually formed by reaction between a free phosphate group and a free hydroxyl group situated in the correct positions of the appropriate nucleosides or nucleotides. In order to perform this union most selectively it is often of paramount importance to protect any other functional groups, which may include (*a*) primary amino groups on the heterocyclic ring, (*b*) 5′-hydroxyl (primary), (*c*) 3′-hydroxyl (secondary) and/or 2′-hydroxyl in the *ribo* series, (*d*) phosphate groups at the 3′- or 5′-positions. Figure 2 illustrates two generalized approaches.

Thus a major portion of the synthetic work has been devoted to the following:

1. The improvement of methods for activating the phosphate ester group in a nucleotide so as to effect phosphorylation of the hydroxyl on another nucleotide or nucleoside.
2. The design of suitable protecting groups for the various functionalities; they must be quantitatively introduced and readily but selectively removed.
3. The development of· effective techniques for the purification and characterization of the resulting polynucleotides, since the condensation reactions currently in use are neither quantitative nor free of contaminating side products.
4. The elaboration of methods for the polymerization of mono- or block units.
5. The utilization of certain enzymes as specific reagents in the "multiplication-amplification" of polynucleotide templates or the linking of chains for the synthesis of longer oligonucletoides containing *hetero* sequences.

3. PHOSPHORYLATING AND CONDENSING REAGENTS

Although most simple nucleotides are now readily available from commercial sources it is occasionally necessary to introduce phosphate (sometimes containing P^{32}) into the molecule. Two reviews are available on phosphorylating agents[11] and several examples follow in later sections of this essay. Most commonly, two of the phosphate oxygens are blocked and this reagent is then activated, either directly (e.g., phosphorochloridates, **1** in Fig. 3) or indirectly via the condensing agent, for attack by an appropriate hydroxyl group from the nucleoside. Subsequent removal of the protecting group generates the free nucleotide which may then undergo a further reaction. Some of the more recent phosphorylating reagents include pyrophosphoryl chloride,[12a] *O*-phenylene phosphorochloridate,[12b] and trimetaphosphate.[12c]

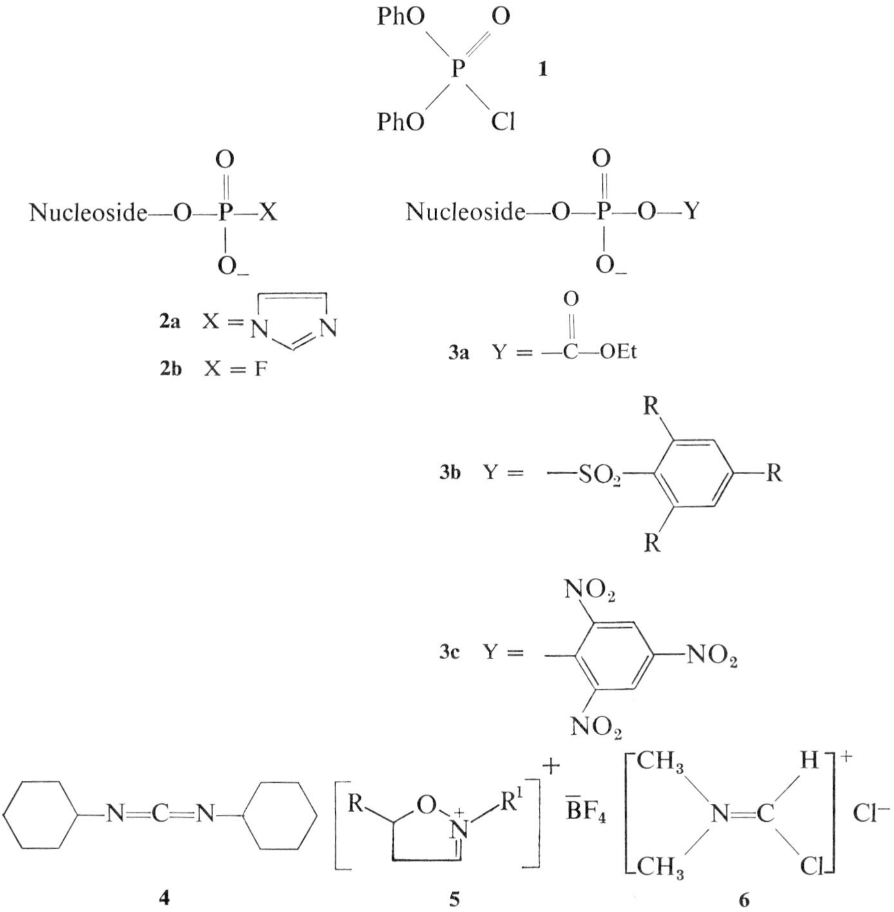

Figure 3. Condensing agents investigated for nucleic acid synthesis.

Condensation, or formation of the internucleotidic bond, is usually accomplished by reaction of a free hydroxyl function of one nucleoside or nucleotide with the reactive phosphate of another nucleotide. Various approaches have been reported for making the phosphate amenable to nucleophilic attack by hydroxyl. Some of these are illustrated in Fig. 3 and include:

1. Formation of derivatives such as imidazolide (**2a**)[13b] and phosphorofluoridate (**2b**).[14]

2. Formation of mixed anhydrides with acid chlorides, for example, ethyl chloroformate (**3a**),[15] mesitylene (or triisopropylbenzene) sulfonyl chloride (**3b**),[16] picryl chloride (**3c**).[17]

3. Reaction with dehydrating agents such as dicyclohexylcarbodiimide (**4**).[18]

4. Reaction with miscellaneous activating agents, for example, isoxazolium salts (**5**),[19] dimethylformamide chloride (**6**),[20] trichloroacetonitrile.[21]

A comparative study of some of these condensing reagents was reported in 1964[22] and since that time dicyclohexylcarbodiimide (DCC) and mesitylenesulfonyl chloride (MS) or triisopropylbenzenesulfonyl chloride (TPS) have been used almost exclusively. Since the hydroxyl group should be the poorest nucleophile available, all other functional groups are usually blocked and an excess of the phosphate-containing moiety is used when feasible. Under the usual experimental conditions, both reagents produce undesirable side products that often require extensive chromatographic separations while the DCC reaction is slow (2–3 days). Additionally, unless a large excess of the phosphate-containing component is used, there is a dramatic decrease in yield as the nucleotide chain grows larger than 8–10 units. Clearly, this is an area of nucleic acid chemistry that still requires considerable effort and ingenuity. For example, a recent report of new reagents for peptide synthesis[23] has promised to investigate their applicability for nucleic acids.

4. PROTECTING GROUPS

A. Primary Amino Function

Several of the heterocyclic bases (A, C, G) contain reactive primary amino functions which apparently must be blocked, preferably for an entire synthesis.* *N*-Acylation (acetyl, benzoyl, anisoyl) is most frequently employed as a method of protection. The general approach has been to fully acylate the mononucleotide or nucleoside and then selectively liberate the hydroxyl groups by taking advantage of the facile base hydrolysis of esters compared to amides. Commonly encountered conditions involve a brief treatment with 1*N* aqueous sodium hydroxide at room temperature or below. Additionally, aromatic amides are known to be more stable than aliphatic amides at high pH, apparently due to the ionization depicted in (**7**) (Fig. 4). This observation has led to a marked preference for the benzoyl or anisoyl groups, and excellent yields of the protected products, for example *N*-benzoyl deoxyadenosine (**8**), have been obtained[24] by the sequence outlined in Fig. 4.

* See, for instance, reference 5 (p. 105), reference 15 or reference 12 in P. T. Gilham and H. G. Khorana, *J. Am. Chem. Soc.*, **81,** 4647 (1959). However, for reactions without amino-protection, see A. M. Michelson, *J. Chem. Soc.*, **1959,** 1371 and reference 30.

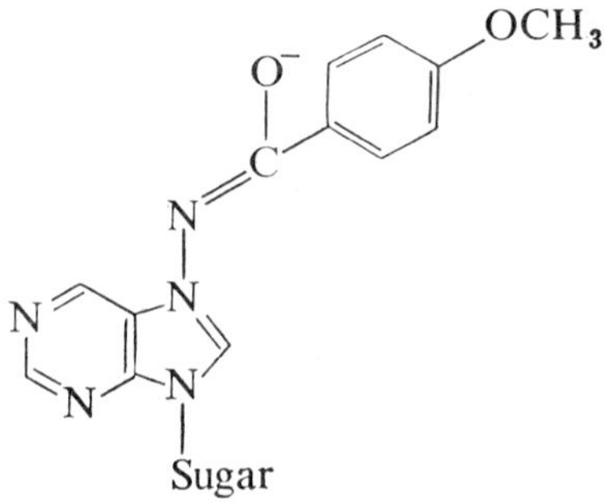

Figure 4. Protection of the amino function.

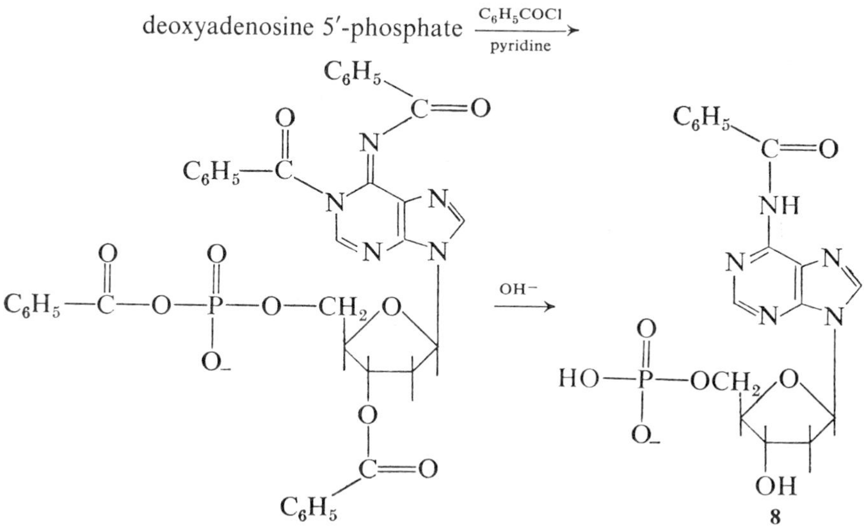

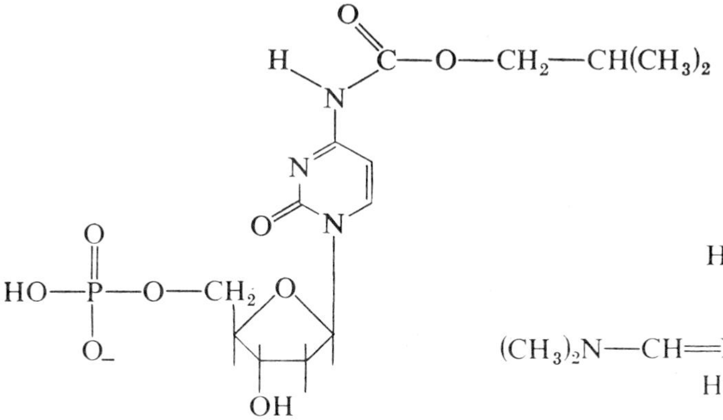

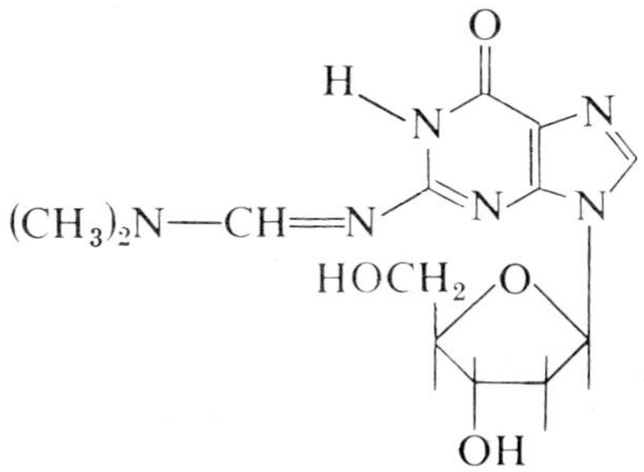

As well as acetyl, another nonaromatic protecting group recently introduced,[25] involved treatment with isobutyloxy chloroformate. After the sequence just outlined, good yields of the protected products, for example, *N*-isobutyloxycarbonyl deoxycytidine 5′-phosphate (**9**) (abbreviated pC^{BOC}), could be obtained. All of these groups can be removed readily by treatment with concentrated ammonium hydroxide for 2 hours at 50°C or 2 days at room temperature.

A selective protection of the amino group can also be accomplished by treatment with dimethylformamide dimethylacetal or the diethylacetal. Some complications involving transacetalization of the *cis*-2′,3′-diol system in ribonucleosides can be overcome by aqueous hydrolysis. Thus, *N*-dimethylaminomethylene derivatives of all the required nucleosides, such as deoxyguanosine (**10**), have been prepared and used in oligonucleotide syntheses.[26] In contrast to the acyl derivatives, this protecting group can be removed easily in either weakly alkaline or weakly acidic media.

B. For 5′-Hydroxyl Function (Primary)

Triphenylmethyl or trityl (**11a**), the standard group for the selective protection of a primary hydroxyl function, was found to be suitable for any synthesis involving only pyrimidines.[27] However, the conditions required for its removal (80% aqueous acetic acid at 100°C) caused extensive decomposition in the purine nucleotides due to the acid lability of the glycosyl bond. Therefore, more sensitive groups such as the mono-**11b** or dimethoxy-**11c** derivatives were developed.[28] These can be easily removed by treatment with 80% acetic acid at room temperature.

A bulky, base-labile group was reported earlier for thymidine[29] and has recently been reintroduced for the ribonucleotides.[30] Treatment of the nucleosides with pivaloyl chloride in pyridine causes selective protection of the 5′-hydroxyl (see **12**). This group can be readily removed by treatment with methanolic tetraethylammonium hydroxide for 2–3 hours at room temperature. Some alkoxy-substituted acetyl derivatives have also been tested on nucleosides in an attempt to achieve selectivity *at* the 5′-hydroxyl.[31]

C. For 3′-Hydroxyl Function (Secondary)

By far the most common method of protecting this hydroxyl function is simple acylation; acetylation (see **13a**) is most frequently encountered, followed by benzoylation.[32] The acetyl group which is readily introduced using acetic anhydride or acetyl chloride can be removed under very mild conditions (2*N* aqueous sodium hydroxide at 0°C for 5–10 minutes) and thus allows a high degree of selectivity.[33]

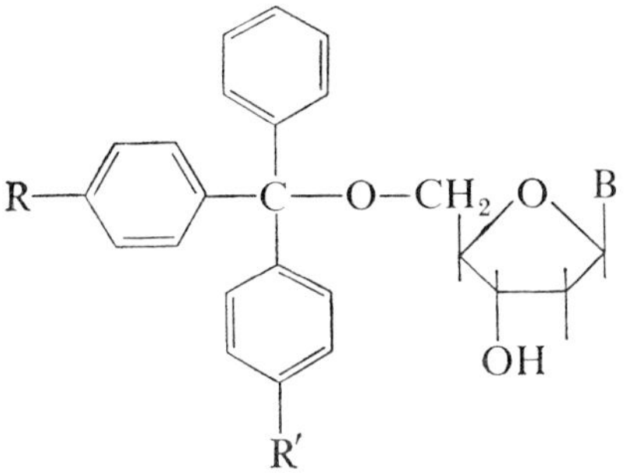

11a R = R′ = H
11b R = H; R′ = $-OCH_3$
11c R = R′ = $-OCH_3$

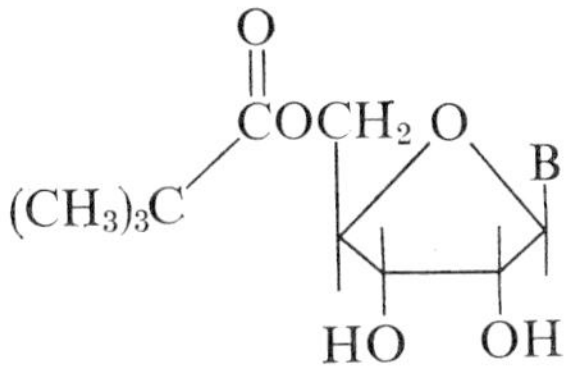

12

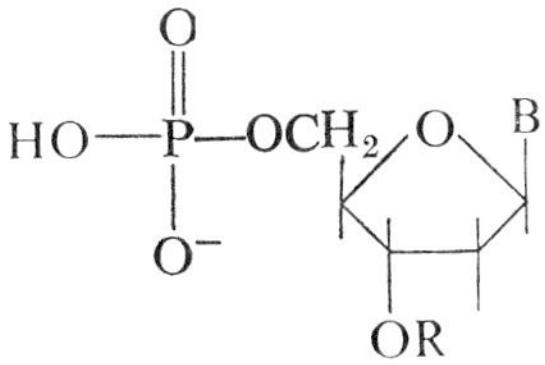

13a R = CH_3CO-
13b R = $O{=}C{-}CH_2CH_2{-}C(=O)Ph$
13c R = S–(2,4-dinitrophenyl) (NO_2, NO_2)

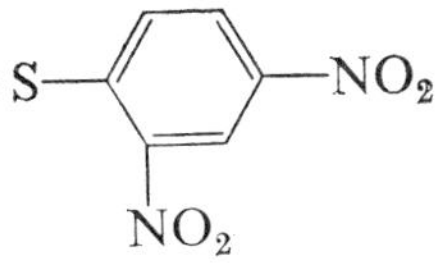

Figure 5. Protection of 5′-(primary) and 3′-(secondary) hydroxyl.

Recently, the β-benzoylpropionyl group (**13b**) has been introduced as a protecting group for this position.[34] It can be cleaved quantitatively within a few hours at room temperature by using dilute solutions of hydrazine hydrate in pyridine/acetic acid. However, the *N*-benzoyl groups of deoxycytidine and deoxyadenosine are also labile under these conditions. A dinitrophenylsulfenyl group (**13b**) susceptible to cleavage by thiophenol or Raney nickel was mentioned.[35]

In an attempt to achieve more selective removal, an increasing variety of other reagents suitable for protection of this functionality have been reported. Generally, they have been tested only on nucleosides and not thus far, employed in any extended synthesis. As a result, derivatives from reaction with β,β,β-tribromoethyl chloroformate have been cleaved using a zinc-copper couple,[36] while dihydrothiophene adducts are labile to silver ion.[37] Reaction with 2-chloroethyl orthoformate produces products which can be deblocked with 80% acetic acid at room temperature,[38] whereas chloroacetic anhydride[39] and *p*-nitro-chloroformate[40] provide derivatives which are most readily cleaved by organic amines at room temperature. Formyl[41] and benzoyl formyl[25b] nucleosides have also been mentioned as base-sensitive compounds. Esters of dihydrocinnamic acid can be specifically removed by the enzyme chymotrypsin.[42]

D. In the Ribonucleotides

An additional problem in ribooligonucleotide synthesis hinges on the protection of the 2′-hydroxyl function, preferably until completion of the synthesis. In addition, it must be released under conditions mild enough to

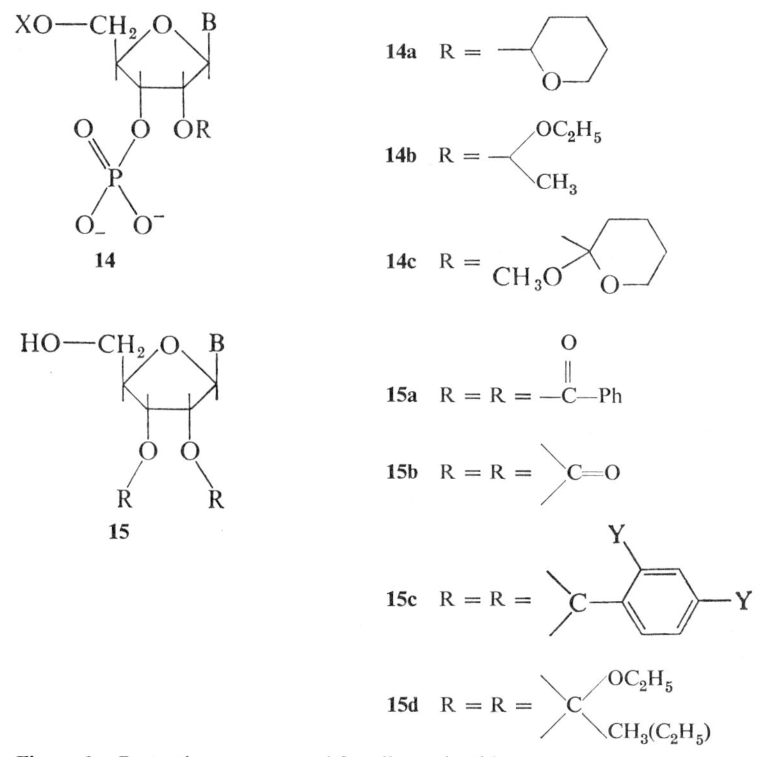

Figure 6. Protecting groups used for ribonucleotides.

ensure no isomerization of the $3' \rightarrow 5'$-phosphodiester linkage to the unnatural $2' \rightarrow 5'$ position. This means that any blocking group at this position must be stable to other manipulations required to extend the chain and hence its choice will determine the nature of all other protecting groups to be used in the synthesis. Three types of protecting group might be used: base-labile, acid-labile, or those labile to a specific conditions such as hydrogenolyzable benzyl ethers.

Acetates or benzoates[43] have been frequently employed as base-labile protecting groups in these positions, while the tetrahydropyranyl group (**14a**) is the most frequently encountered of the acid-labile blocking groups.[44] It has recently been shown that this group can be removed by 0.01 *N* aqueous hydrochloric acid in 3–4 hours at room temperature.[30] These conditions cause a negligible amount of phosphoryl migration. Various modifications of this latter group have also been reported, namely the 1-ethoxyethyl (**14b**)[45] and the 4-methoxytetrahydropyran-4-yl (**14c**)[46] derivatives.

Frequently, by blocking the *cis*-diol system in the 2′,3′-positions of the ribose molecule, the chain can be enlarged from the 5′- end. In addition to acetyl- and benzoyl-groups (**15a**),[47] the base sensitive cyclic carbonates (**15b**)[48] have been cited as potential blocking agents. Modifications to acid-sensitive groups include substituted benzylidenes (**15c**)[49] or the alkoxy methylene derivatives (**15d**) which can be converted into 2′- and 3′-*O*-acyl derivatives by treatment with acetic acid.[50]

E. For Phosphate

The blocking of a terminal phosphate from further reaction either intramolecularly to form cyclic phosphates or intermolecularly to form pyrophosphates is absolutely essential in nucleic acid synthesis. Formation of a β-cyanoethyl ester (**16a**) remains probably the most frequently employed method.[51] It can be quantitatively introduced (β-cyanoethanol and DCC for 2 days) and is removed by momentary treatment with mild alkali. This extreme lability is also a drawback since, frequently, reprotection of the phosphate must occur after every step.

The β-elimination mechanism for its removal is typical of a method frequently encountered for deblocking the phosphate group. Thus esters of 2-acetyl–2-methylethanol (**16b**) are found to be more sensitive than β-cyanoethyl derivatives,[52] whereas the 2-(α-pyridyl)-ethyl ester (**16c**) is most effectively removed by anhydrous sodium methoxide at 0°C after 48 hours.[53] The β,β,β-trichloroethyl group (**16d**) has also been used in oligonucleotide synthesis and is removed reductively with Zn/Cu in acetic acid or dimethylformamide.[54] However, these conditions are apparently not suitable for

N-benzoylcytosine.[55] A group of substituted phenyl hydracylamides (**16e**–**16f**) has also been developed. These are appreciably more stable to mild alkali and consequently permit selective hydrolysis of the 3′-*O*-acetyl.[56]

A direct displacement can be achieved under acidic conditions by using benzhydryl (**17a**)[57] or *t*-butyl esters (**17b**)[58]; certain substituted phenols (**17c**) are susceptible to aqueous alkali treatment.[59]

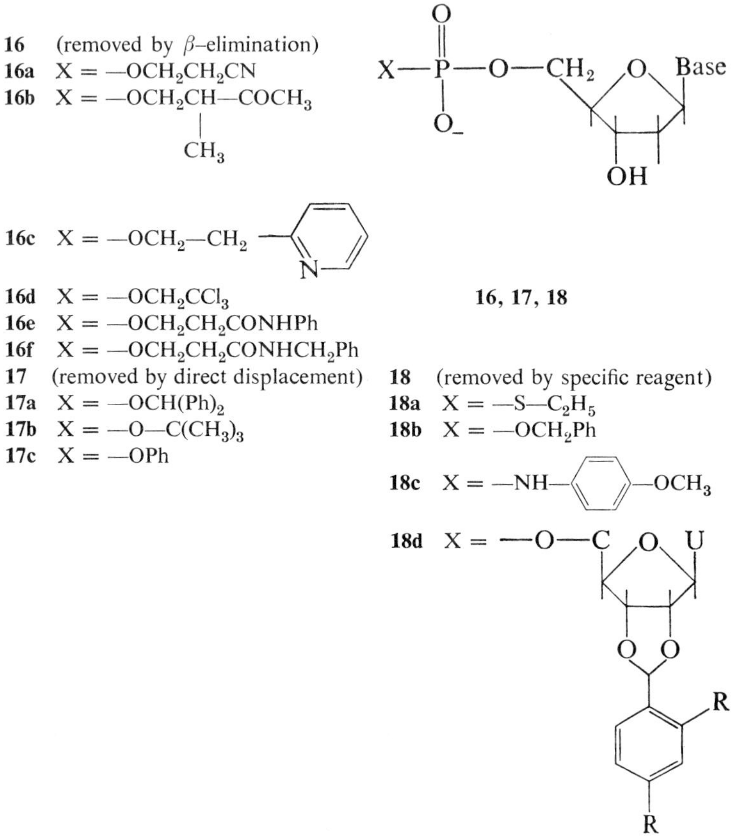

16 (removed by β-elimination)
16a X = $—OCH_2CH_2CN$
16b X = $—OCH_2CH(CH_3)—COCH_3$
16c X = $—OCH_2—CH_2$—(2-pyridyl)
16d X = $—OCH_2CCl_3$
16e X = $—OCH_2CH_2CONHPh$
16f X = $—OCH_2CH_2CONHCH_2Ph$
17 (removed by direct displacement)
17a X = $—OCH(Ph)_2$
17b X = $—O—C(CH_3)_3$
17c X = —OPh

X—P(=O)(O⁻)—O—CH₂ (nucleoside, Base, 3′-OH)
16, 17, 18

18 (removed by specific reagent)
18a X = $—S—C_2H_5$
18b X = $—OCH_2Ph$
18c X = —NH—C_6H_4—OCH_3
18d X = —O—C (uridine 2′,3′-O-benzylidene acetal, R-substituted phenyl)

Figure 7. Blocking groups for nucleotide phosphate.

A number of blocking groups are removed only under specific conditions. Ethylthioesters (**18a**) are selectively removed by oxidation with aqueous iodine;[55] this technique can also function as an activating procedure for oligonucleotide synthesis. Benzyl phosphates (**18b**) are split by hydrogenation,[60] but this is often accompanied by reduction of the pyrimidine ring.

However, benzyl phosphate triesters, formed by reaction of the internucleotidic phosphodiester bond, are debenzylated with sodium iodide in acetonitrile at 80°C.[61] Some recent investigations have shown that phosphoroanilidates (**18c**) can be specifically cleaved by isoamyl nitrite.[62] A 2′,3′-cyclic phosphate has also been employed as a method of protecting both the 3′-phosphate and the 2′-hydroxyl while adding to the chain from the 5′-hydroxyl end.[63] Perhaps the most curious report concerns the use of a substituted uridine (**18d**) as a phosphate-protecting group for deoxyoligonucleotides.[64] Deblocking is accomplished in a stepwise fashion in which the key step is oxidation of the *cis*-diol by periodate.

5. CHEMICAL SYNTHESES OF POLYNUCLEOTIDES

In the last decade, most attempts directed at the synthesis of oligonucleotides have involved the concepts and reagents outlined in the preceding section. Several variations of two fundamental approaches, polymerization and stepwise condensation, have been used and are outlined next.

A. Polymerization Method

Homopolymers

The availability of deoxyribopolynucleotides containing a single nucleotide unit aids in a variety of chemical, physicochemical, and enzymatic studies in the nucleic acid field. These compounds have been prepared by the polymerization of mononucleotides containing a free hydroxyl group. Initial studies, involving the reaction of thymidine 5′-phosphate with DCC,[65] produced a large number of polymeric products. As a result, the major problem was the development of satisfactory chromatographic methods for separation and characterization of the desired linear polynucleotides, which were isolated mainly by chromatography on columns of cellulose anion exchanger (ECTEOLA) and diethylaminoethyl (DEAE) cellulose. Usually two homologous series of polynucleotides are obtained. The first are the linear polynucleotides represented by the general structure (**19**), while the second series of compounds contain the cyclic oligonucleotides represented by (**20**); they arise by an intramolecular phosphodiester bond formation between the 5′-phosphomonoester group and the 3′-hydroxyl group. The formation of the cyclopolynucleotide compounds can be reduced by adding some 25% of 3′-*O*-acetylthymidine 5′-phosphate to the unprotected nucleotide at the start of polymerization.[66] The 3′-OAc nucleotides form the terminating unit of the greater portion of the resulting polynucleotide chains thus blocking the cyclization reaction and the acetyl group can subsequently be removed by mild alkali treatment. By using this technique and working in concentrated solution (1 M), the major products obtained were the linear

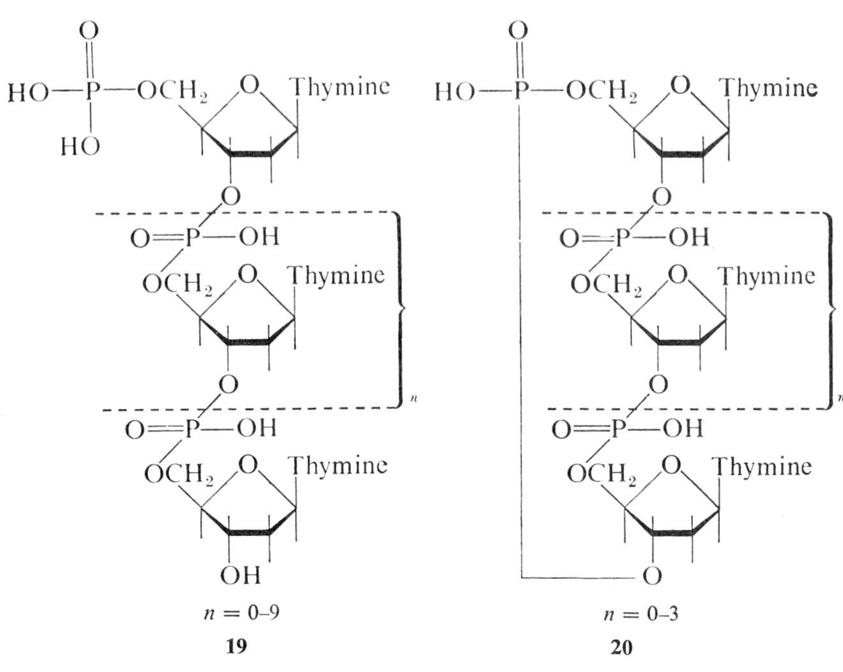

Figure 8. Major components (linear and cyclic oligonucleotides) of a polymerization reaction of *p*T. [Adapted from *J. Am. Chem. Soc.*, **83,** 675 (1961).]

polynucleotides, of which about 45% of the total polymeric mixture consisted of linear polynucleotides longer than tetranucleotides. Purification of components up to dodecanucleotides on a fairly large preparative scale was attained on a DEAE-Cellulose column (carbonate form) by using volatile triethylammonium bicarbonate as eluent (see Fig. 9).

The foregoing method for chemical polymerization by dicyclohexylcarbodiimide was also extended to N^6-benzoyldeoxyadenosine 5′-phosphate,[67a]

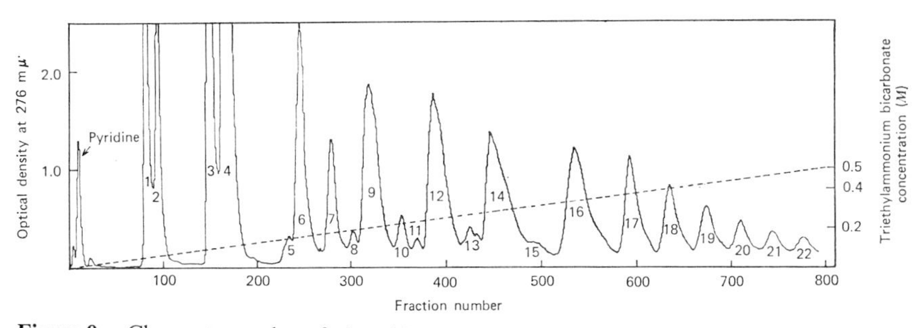

Figure 9. Chromatography of thymidine polynucleotides (total polymeric mixture) on DEAE-Cellulose (bicarbonate) column. [Adapted from *J. Am. Chem. Soc.*, **83,** 675 (1961).]

N^6-anisoyldeoxycytidine 5′-phosphate,[67b] and N^2-acetyldeoxyguanosine 5′-phosphate.[67c] Treatment of the reaction mixture with concentrated ammonia gave the expected homologous series of linear and cyclic polynucleotides.

Other commonly encountered side products from this type of polymerization were compounds containing mono- and oligonucleotides linked to each other by pyrophosphate bonds (see **21**). Their separation from the desired linear polynucleotides was extremely difficult; therefore before chromatography the entire polymerization reaction mixture was treated with excess acetic anhydride in pyridine to cleave the pyrophosphate bonds.[68]

Another series of minor contaminants was also present. The simplest member of the series contained a phosphomonoester group which could

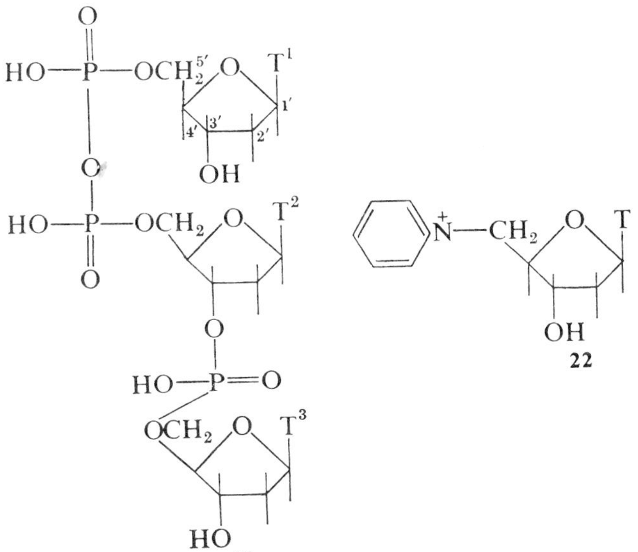

Figure 10. Typical contaminants (pyrophosphates and pyridinium nucleotides) of a polymerization reaction.

be dephosphorylated by incubation with bacterial alkaline phosphatase. The product thus obtained had an ultraviolet spectrum similar to that of an equimolar mixture of thymidine and *N*-methylpyridinium cation. From these results, it was tentatively concluded that the cation had the structure (**22**), while the phosphomonoester group was located at the 3′-position. Therefore it was assumed that a homologous series containing pyridinium-substituted terminal nucleotides were also present.

Various other reagents such as *p*-toluenesulfonyl chloride, 2,5-dimethylbenzenesulfonyl chloride, diphenylphosphorochloridate were also investigated but they did not compare favorably with DCC as a polymerization

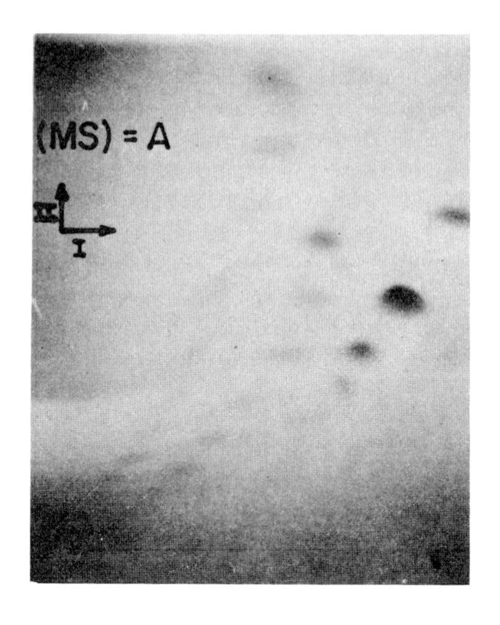

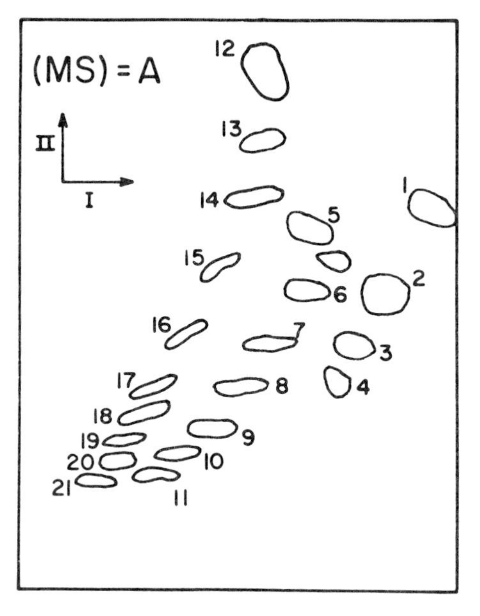

Figure 11. Two-dimensional TLC resolution of chemically polymerized *p*T on Avicel-Cellulose plate (20 × 20 cm, 0.1 mm thickness) (A) with Mesitylenesulfonyl chloride (MS) and (B) with dicyclohexylcarbodiimide (DCC). Solvent I, *n*-propanol: concentrated NH_3:H_2O (55:10:35) was used in the first dimension. Solvent II, Isobutyric acid: 1*M* NH_4OH:0.1*M* EDTA (100:60:1.6) was used in the second dimension. [Adapted from *Biochem. Biophys. Res. Comm.*, **41**, 1248 (1970).]

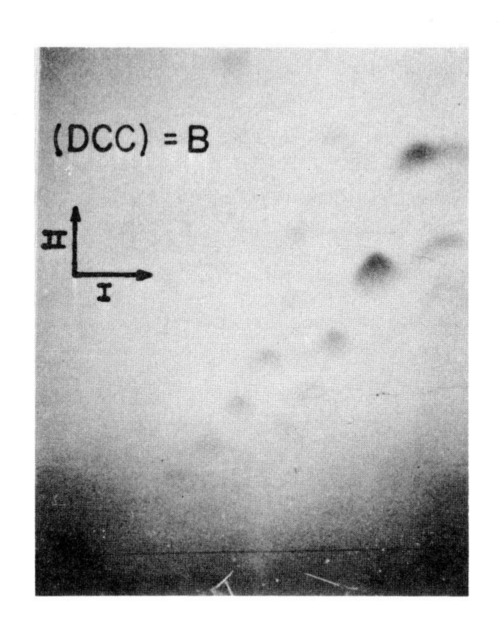
(DCC) = B
II
I

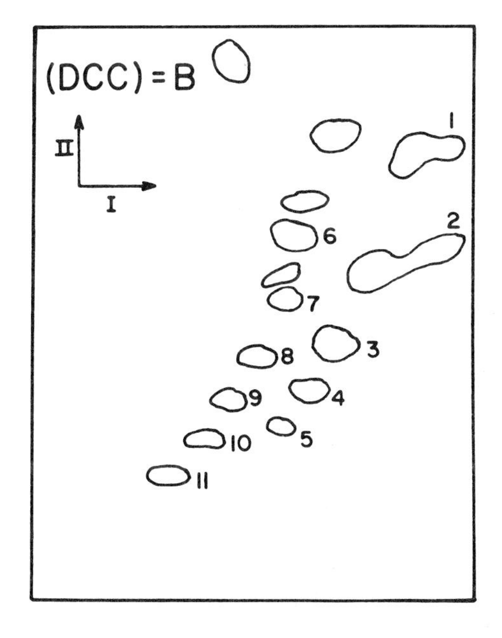
(DCC) = B
II
I
1
2
3
4
5
6
7
8
9
10
11

reagent.[68] It has been reported that a polythymidylic acid of 30 units was obtained by using picryl chloride[69] as the polymerizing reagent. More recently, β-imidazolyl-4(5) propanoic acid has been found to act as a specific catalyst in the polymerization of the unprotected 5′-deoxyribopolynucleotides.[70] Only the 3′ → 5′ internucleotide diester bond is apparently formed and the overall yield of di- and oligonucleotides is reported to be about 50%. The products were contaminated with 30–40% of the 5′-dephosphorylated oligonucleotides. Treatment of thymidine 5′-*S*-ethylphosphothioate with iodine in pyridine in the absence of any external nucleophile produced extensive self-condensation by attack of the 3′-hydroxyl group on the iodine-activated phosphorothioate.[55]

Recently a thin-layer chromatographic technique has been employed for the separation and characterization of various products of polymerization of thymidine 5′-phosphate.[71] This highly sensitive analytical technique clearly indicates not only the extensive polymerization occurring with MS and DCC treatment but also the various series of unwanted by-products (see Fig. 11).

Block Polymers

For the synthesis of polynucleotides containing repeating dinucleotide units, an approach has been used which is similar to that already described. It involved the polymerization of suitably protected dinucleotides such as (*d*-*p*T*p*C^{An}), (*d*-*p*T*p*G^{Ac}), (*d*-*p*C^{An}*p*A^{Bz}), and (*d*-A^{Bz}*p*G^{Ac}) with DCC in pyridine followed by the removal of protecting groups with alkali and separation of products by a combination of anion exchange and paper chromatography.[72] In addition to the desired compounds, two main types of side products were identified: those bearing a 3′-phosphomonoester group at one end of the chain and a 5′-phosphomonoester at the other end (**24**), and those bearing 5′-phosphomonoester groups at both ends with an unusual $C_{3'}$-$C_{3'}$ internucleotidic linkage within the chain (**26**). In all the contaminants, a common property which suggested the presence of two phosphomonoester groups per molecule was their striking increase in mobility on paper chromatography after treatment with phosphomonoesterase enzyme.

Formation of these side products could be explained by postulating various intermolecular and intramolecular triesters or pyrophosphates (see **23**, **25**, **27** in Fig. 12) as intermediates. These reactive triesters could then be broken down with pyridine from the reaction medium or hydroxide ion from the workup to give the observed products as well as the pyridinium-substituted nucleosides **(28)** observed earlier (see Fig. 10).

There is no doubt that polymerization methods available at present are inefficient and produce a plethora of contaminating side products which require extensive chromatography. The ready formation of these side

Figure 12. Contaminating side products from a polymerization of d-$p\mathrm{T}p\mathrm{C}^{\mathrm{An}}$, showing possible mechanisms for formation. [Adapted from *J. Am. Chem. Soc.*, **87**, 2956 (1965).]

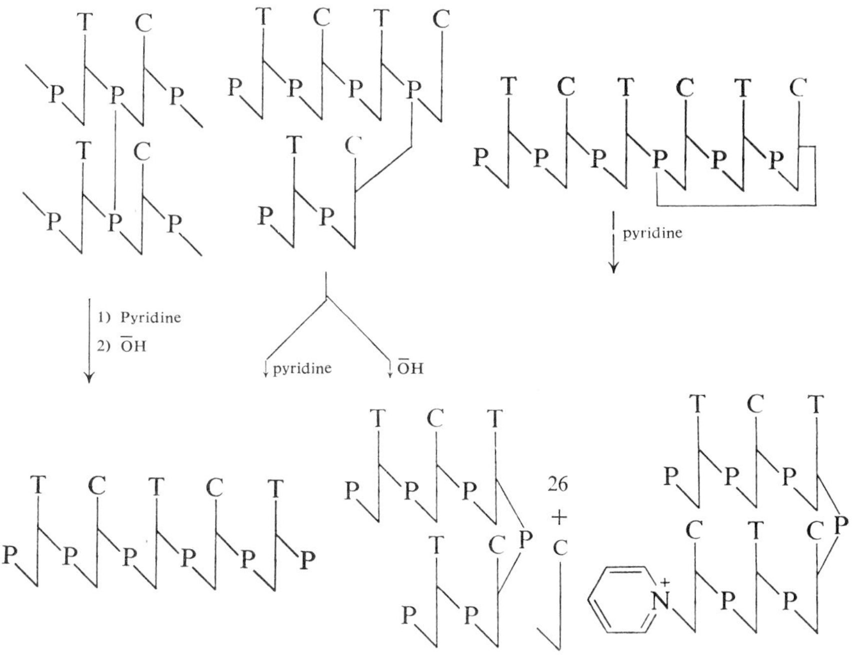

products illustrates very well the reactivity of the phosphate group and the multitude of sites available for reaction in these molecules. Nevertheless, this method does provide a rapid means of obtaining modest quantities of short-chain deoxyribopolynucleotides containing repeating sequences.

For the extensive polymerization of the trinucleotides (d-$p\mathrm{A}^{\mathrm{Bz}}p\mathrm{A}^{\mathrm{Bz}}p\mathrm{C}^{\mathrm{An}}$), ($d$-$p\mathrm{T}p\mathrm{T}p\mathrm{G}^{\mathrm{Ac}}$), ($d$-$p\mathrm{T}p\mathrm{A}^{\mathrm{Bz}}p\mathrm{G}^{\mathrm{Ac}}$), ($d$-$p\mathrm{A}^{\mathrm{Bz}}p\mathrm{T}p\mathrm{C}^{\mathrm{An}}$), ($d$-$p\mathrm{A}^{\mathrm{Bz}}p\mathrm{T}p\mathrm{C}^{\mathrm{An}}$), ($d$-$p\mathrm{A}^{\mathrm{Bz}}p\mathrm{T}p\mathrm{G}^{\mathrm{Ac}}$), ($d$-$p\mathrm{C}^{\mathrm{An}}p\mathrm{G}^{\mathrm{Ac}}p\mathrm{A}^{\mathrm{Bz}}$), ($d$-$p\mathrm{C}^{\mathrm{An}}p\mathrm{G}^{\mathrm{Ac}}p\mathrm{T}$), ($d$-$p\mathrm{C}^{\mathrm{An}}p\mathrm{C}^{\mathrm{An}}p\mathrm{T}$), ($d$-$p\mathrm{G}^{\mathrm{Ac}}p\mathrm{G}^{\mathrm{Ac}}p\mathrm{A}^{\mathrm{Bz}}$), ($d$-$p\mathrm{C}^{\mathrm{An}}p\mathrm{C}^{\mathrm{An}}p\mathrm{A}^{\mathrm{Bz}}$), ($d$-$p\mathrm{A}^{\mathrm{Bz}}p\mathrm{A}^{\mathrm{Bz}}p\mathrm{G}^{\mathrm{Ac}}$) and ($d$-$p\mathrm{T}p\mathrm{T}p\mathrm{C}^{\mathrm{An}}$), MS was found to be a more effective reagent.[73] Side products such as cyclic trinucleotides in minor amounts were always identified; this may be due to the use of excess condensing reagent. Also present were the usual contaminants containing phosphomonoester groups at both terminals and polynucleotides with a quaternary pyridine group at the 5′-carbon of the terminal nucleotides. These methods were subsequently extended for the polymerization of tetranucleotides such as (d-$p\mathrm{A}^{\mathrm{Bz}}p\mathrm{A}^{\mathrm{Bz}}p\mathrm{A}^{\mathrm{Bz}}p\mathrm{G}^{\mathrm{Ac}}$) and ($d$-$p\mathrm{A}^{\mathrm{Bz}}p\mathrm{T}p\mathrm{C}^{\mathrm{An}}p\mathrm{G}^{\mathrm{Ac}}$).[74]

B. Stepwise Condensation Methods for Polynucleotides without a Terminal Phosphate

1. The prevalent approach used in the stepwise synthesis of specific deoxyribopolynucleotides has involved the successive addition of mono-, di-, tri-, or tetranucleotide units to the 3′-hydroxyl end of a 5′-*O*-trityl protected mono- or oligonucleotide, as illustrated in Fig. 13. As the chain enlarges, an increasing excess of the incoming 3′-*O*-acetyl protected unit (**29**) or (**30**) must be used to maintain satisfactory yields with respect to the "growing" chain. The condensing agents in current use are MS, TPS, or DCC. In most of the synthetic work MS or TPS reagents offer the particular advantage that trialkylammonium salts can be used to aid solubilization of the nucleotidic components in the reaction medium.[75] Moreover, the presence of a trace of amine is a strong inhibitor of the DCC reaction. After each condensation step, the reaction mixture was submitted to a mild alkali treatment to remove the 3′-*O*-acetyl group and the desired product was obtained by anion-exchange chromatography on a DEAE-Cellulose column using volatile triethylammonium bicarbonate pH 7.5 buffer. By using this approach, the synthesis of a chain upto hexadecanucleotide $(TpTpApC)_4$ has been achieved.[76]

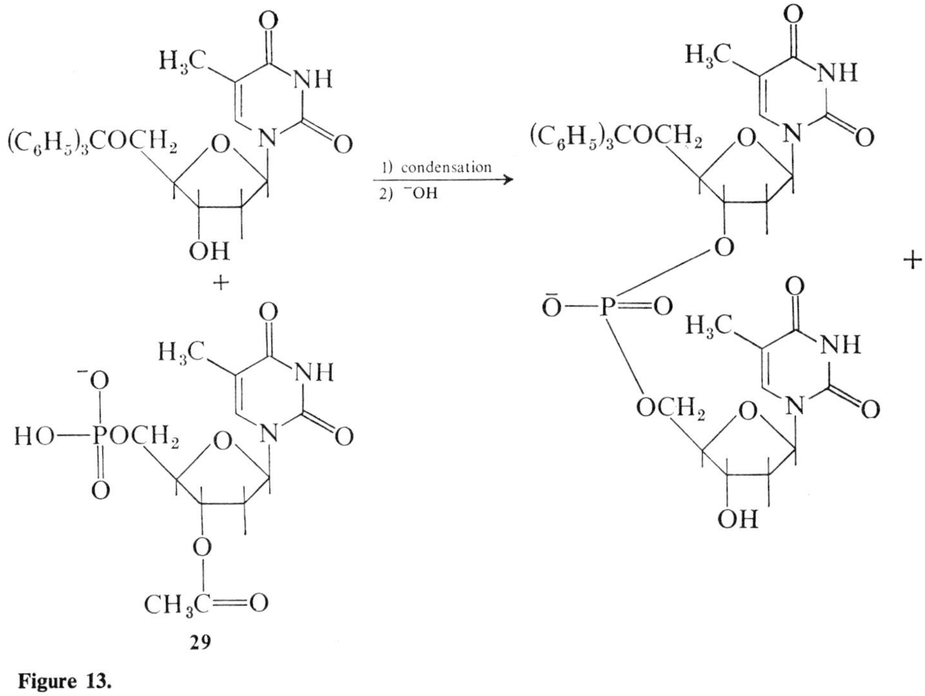

1) condensation
2) ^{-}OH

29

Figure 13.

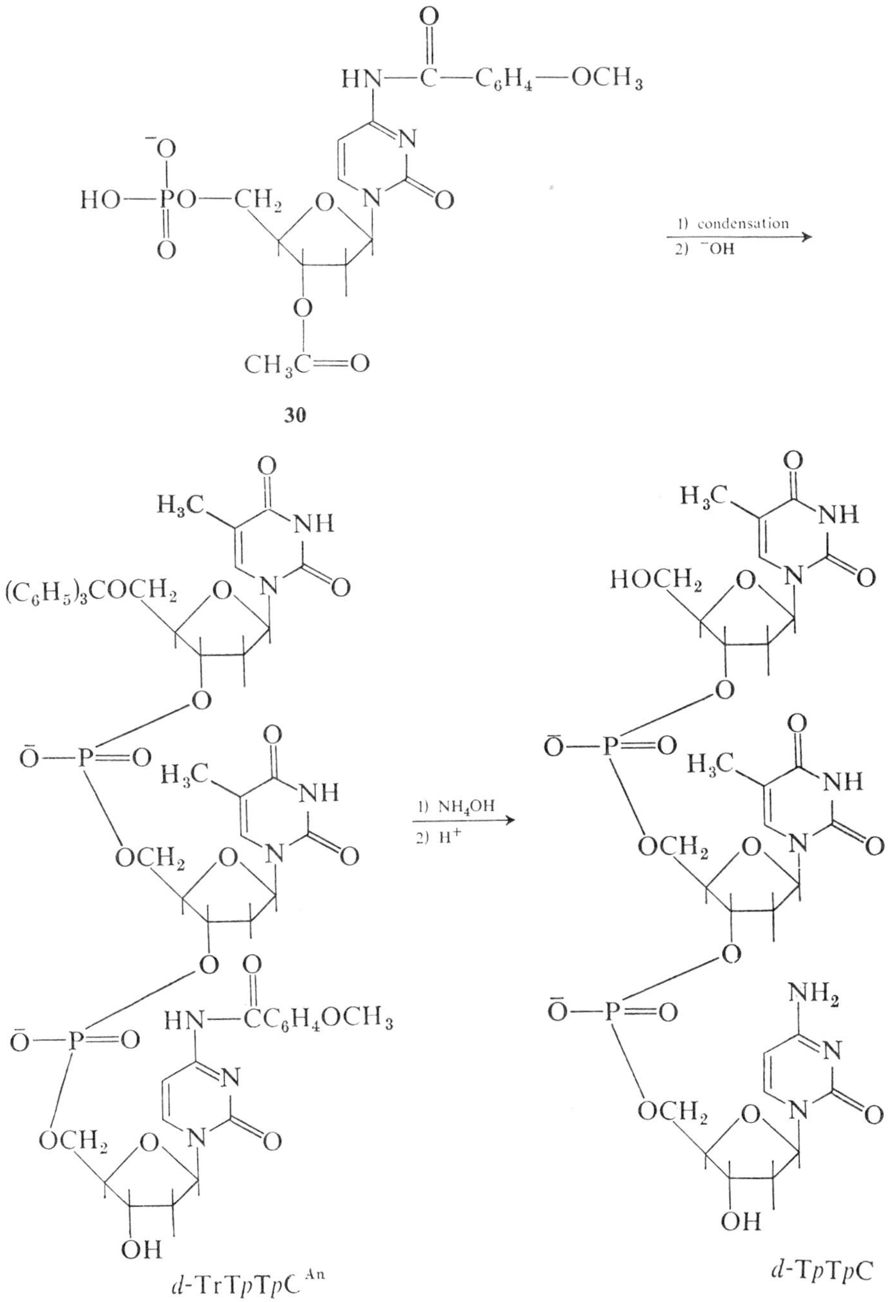

1) condensation
2) ^{-}OH

30

1) NH_4OH
2) H^+

d-TrT*p*T*p*C^{An}

d-T*p*T*p*C

Figure 13. A representative stepwise condensation to prepare an oligonucleotide without a terminal phosphomonoester group.

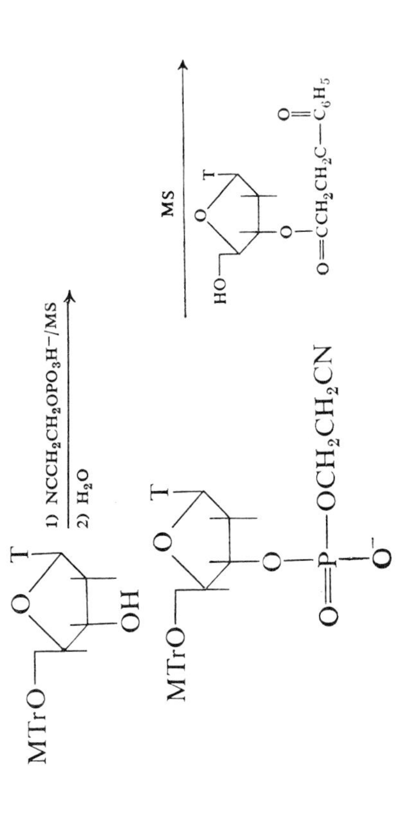
MTrO
T
O
OH
1) NCCH2CH2OPO3H−/MS
2) H2O
O=P—OCH2CH2CN
O−
MS
HO
O=CCH2CH2C—C6H5
O

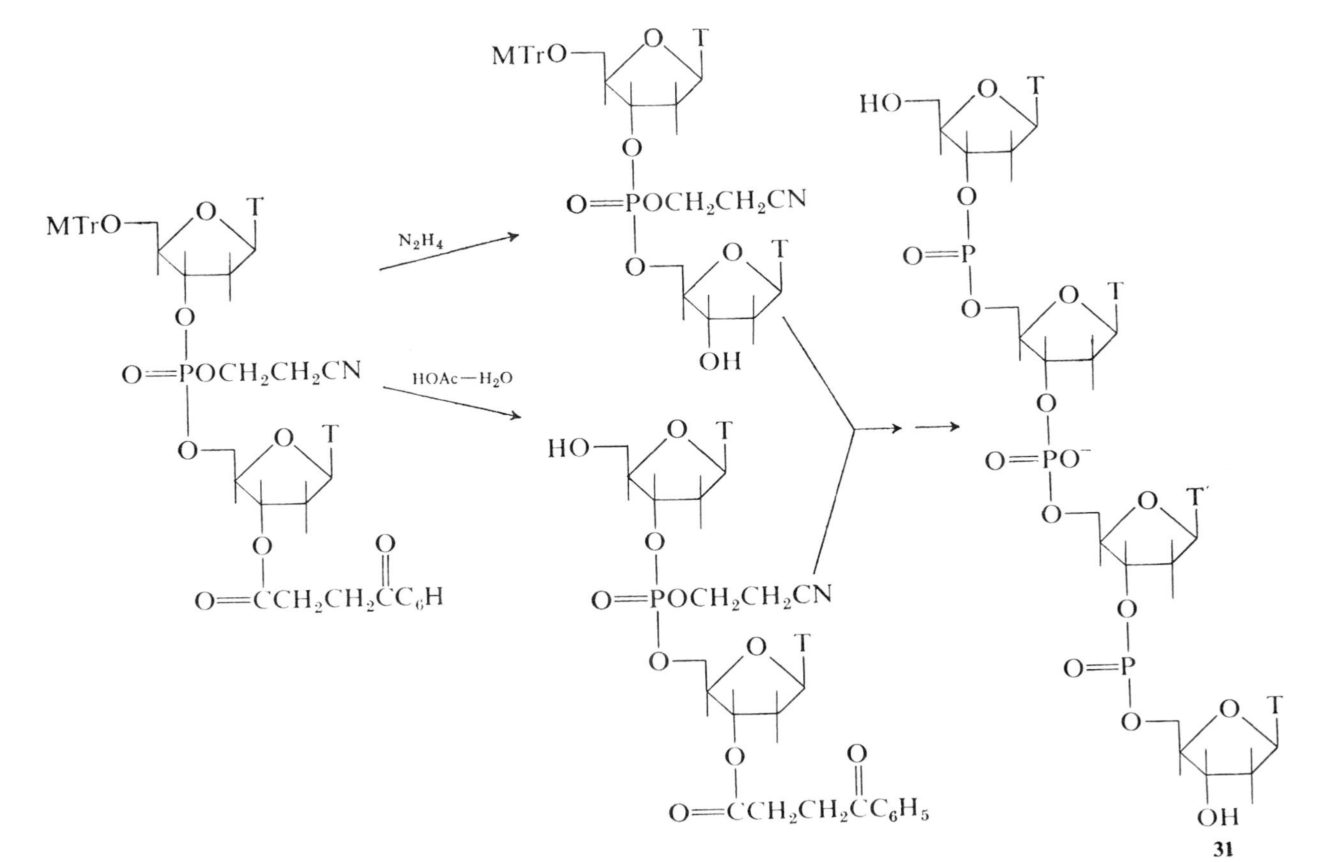

Figure 14. The triester approach to oligonucleotide synthesis. [Adapted from *J. Am. Chem. Soc.*, **91**, 3360 (1969).]

Side products were formed in small amounts in every condensation reaction. Two homologous series of compounds were positively identified.[77] The first group of compounds were Tr-T*p*, Tr-T*p*T*p*, and Tr-T*p*T*p*T*p* in the thymidine series; these correspond to products already discussed in the polymerization reactions, that is, a series in which the 3′-hydroxyl has been phosphorylated (see **24**), and another series containing pyridine substituted at the 5′-carbon of the sugar moiety (see **22** and **28**). In addition, stripping the anion-exchange column with 1 *M* salt at the end of gradient elution invariably yielded 5–20% of the total nucleotidic material. The formation of these uncharacterized "side products" with apparently a higher number of charges than those in the desired product is more serious; they may arise by phosphorylation of the heterocyclic ring by the activated nucleotides.[78]

This approach has two main disadvantages: (*a*) To maintain high yields in the condensation step, increasingly large excesses of the 5′-*O*-phosphomonoester component must be employed as the chain length of the oligonucleotide is increased. This requirement makes the approach very uneconomical from a practical point of view. (*b*) At each stage in the synthesis the products must be separated as salts by chromatography on DEAE-Cellulose columns with aqueous buffer solutions. Although efficient, such chromatography is time consuming (a typical separation requires several days) and yields the high molecular weight material—desired products—at the end of the elution pattern. However, this approach has been solely employed in the synthesis of all the deoxypolynucleotides required for building the gene corresponding to alanine *t*RNA.[79]

2. A synthetic approach involving β-cyanoethyl phosphotriesters for the synthesis of oligodeoxyribonucleotides has recently been reported.[80] As uncharged molecules, the phosphotriesters produced as intermediates would be expected to be soluble in organic solvents and amenable to the conventional techniques of separation and characterization of organic molecules. In addition, masking the active oxygen in the phosphodiester groups should prevent formation of the pyrophosphate bonds which are apparently responsible for chain fission and unwanted side products. The synthetic scheme, as outlined in Fig. 14, involves phosphorylation of the 3′-hydroxyl of a nucleoside or oligonucleotide derivative using pyridinium mono-(β-cyanoethyl) phosphate and MS or TPS. This phosphorylation is followed by condensation of the phosphodiester with an appropriate nucleoside, again employing an arenesulfonyl chloride as condensing agent. Workup of this mixture and chromatography on a silica-gel column with ethyl acetate and tetrahydrofuran afforded the desired product (e.g. **31**), after removal of protecting groups. By using this method the synthesis of a hexanucleotide, $(pT)_6$, has been achieved.[34] A trichloroethyl ester of phosphate has also been reported in a

phosphotriester approach;[81] also, benzyl[61] and phenyl[82] esters have been mentioned in triester-type syntheses.

3. An attempt has been made to prepare deoxyribooligonucleotides by activating the phosphate group to nucleophilic attack by a free hydroxyl which is ionized by strong base.[14] Only a modest yield of a tetranucleotide DMTr-T*p*T*p*T*p*T-MMTr was obtained, but it is interesting to note that no protecting groups were required for the amino functions and pyrophosphate-containing products were not detected. However, two major problems must be solved: appreciable cleavage of the glycosidic bond in the strong anhydrous base and the limited solubility of the oligonucleotide salts in dimethylformamide.

C. Stepwise Synthesis of Deoxyribopolynucleotides Bearing a 5′-Phosphomonoester End Group

1. For the synthesis of longer deoxypolynucleotides, it is necessary to have oligonucleotide units containing a terminal phosphomonoester end group, usually at the 5′-position. Chemically, such compounds have been synthesized by the stepwise condensation of a β-cyanoethyl protected mononucleotide with a 3′-*O*-acetyl mononucleoside 5′-phosphate.[83] The product was isolated by DEAE-Cellulose column chromatography, reprotected with a β-cyanoethyl group, and condensed with another appropriate mononucleotide, as outlined in Fig. 15. Insertion of phosphate at this position can also be accomplished enzymatically under certain conditions (see later section on enzymes).

2. By approximately doubling the oligonucleotide chain during each condensation step (see Fig. 16) the sequential method has been modified and extended for the synthesis of various dodecanucleotides.[84] In this approach, products and reactants differ substantially in molecular weight and therefore can be separated rapidly (products are usually off the column within 24 hours) and quantitatively by gel filtration through Sephadex gels with appropriate exclusion limits. An attractive feature of this separation technique is that the product peak emerges from the column before the starting material (see Fig. 17). However, the presence of the extremely labile β-cyanoethyl ester necessitated reblocking of the 5′-phosphate at each stage. In principle this difficulty could be overcome by using the recently developed β-β-β-trichloroethyl ester, *S*-ethyl phosphothioate, or anilidate-protecting groups which are more stable but can be cleaved selectively.

3. Recently a new approach, which utilizes substituted benzene derivatives (*a*) to block the phosphate group and (*b*) to increase dramatically the binding property of the nucleotidic fragment to benzoylated DEAE-Cellulose, has been developed for the synthesis of deoxyribooligonucleotides of defined sequence.[56,59] Hence after a condensation step the reaction mixture is passed through a column of benzoylated DEAE-Cellulose or benzoylated DEAE-Sephadex.[85] This technique offers the distinct advantage of effecting the complete removal of any products lacking aromatic residues. Thus, if non-aromatic protecting groups are used for other functionalities, a facile separation of unwanted contaminants such as pyrophosphates, cyclic phosphates, and the starting material containing the free phosphate is assured. The most useful blocking groups are phenyl-substituted hydracrylamides, which,

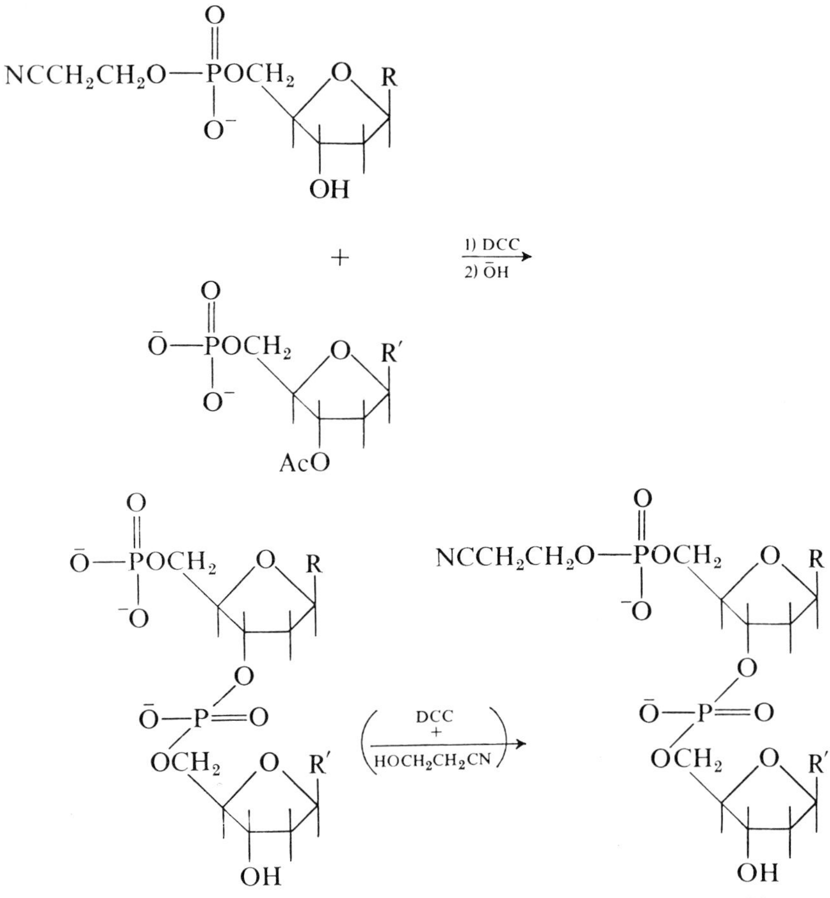

$NCCH_2CH_2O{-}P(=O)(O^-){-}OCH_2$ — sugar (O, R) — OH

\+

$\bar{O}{-}P(=O)(O^-){-}OCH_2$ — sugar (O, R′) — AcO

1) DCC
2) $\bar{O}H$

$\bar{O}{-}P(=O)(^-O){-}OCH_2$ — sugar (O, R) — O — $\bar{O}{-}P{=}O$ — OCH_2 — sugar (O, R′) — OH

DCC
\+
$HOCH_2CH_2CN$

$NCCH_2CH_2O{-}P(=O)(^-O){-}OCH_2$ — sugar (O, R) — O — $\bar{O}{-}P{=}O$ — OCH_2 — sugar (O, R′) — OH

31a

Figure 15. The stepwise approach to the synthesis of deoxyribooligonucleotide containing a 5′-phosphate. [Adapted from *J. Am. Chem. Soc.*, **89**, 2158 (1967).]

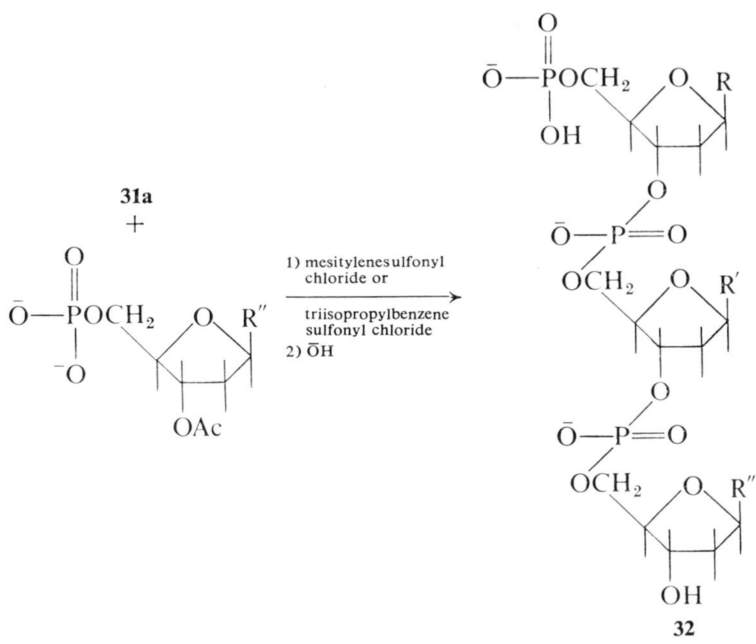

Figure 15. (*Continued*)

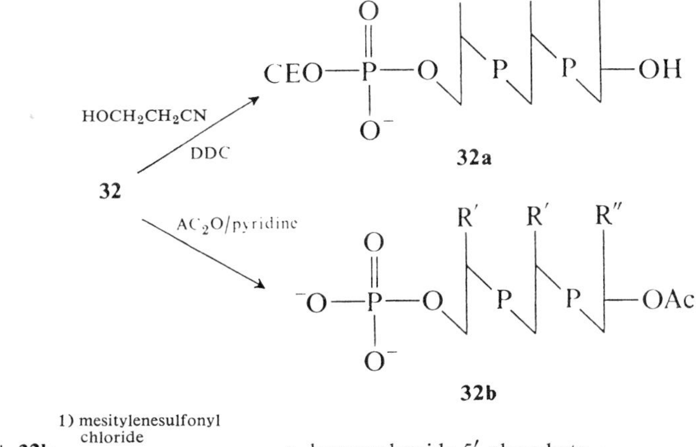

32a + **32b** $\xrightarrow[\text{2) OH}^-\text{ 3) gel filtration on Sephadex G-75 (Superfine)}]{\text{1) mesitylenesulfonyl chloride}}$ hexanucleoside 5′-phosphate

β-cyanoethyl ester of hexanucleotide + 3′-*O*-acetyl hexanucleoside 5′-phosphate $\xrightarrow[\text{2) OH}^-\text{ 3) gel filtration on Sephadex G-75 (Superfine)}]{\text{1) mesitylenesulfonyl chloride}}$ dodecanucleoside 5′-phosphate

Figure 16. Block synthesis of deoxyoligonucleotides possessing a 5′-phosphate group.

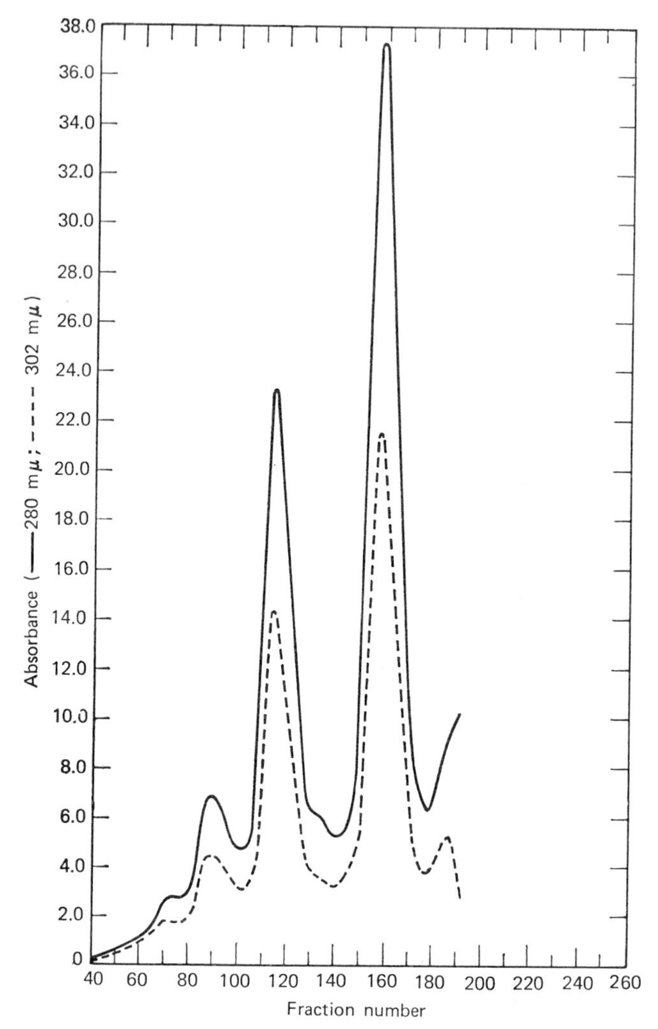

Figure 17. Chromatography of the reaction mixture (one-half portion) on a Sephadex G-75 (superfine), K25-100 column in the preparation of *d*-$pA^{Bz}pC^{An}pTpA^{Bz}pC^{An}pA^{Bz}$. Fractions of 2 ml were collected every 15 min. Fractions 108–128 contained the desired product. [Adapted from *Biochem.*, **8**, 3443 (1969).]

unlike the β-cyanoethyl group, are stable to conditions necessary for removing the 3′-*O*-acetyl group. Consequently, the phosphate does not have to be reprotected before each new condensation step, a substantial saving in time. Preparative scale TLC techniques on Avicel cellulose (1.0 mm thick) can also save much time in the isolation of synthetically useful amounts of pure oligonucleotides.[71]

D. Polymer-Support Synthesis

Some analogies can be drawn between the problems involved in nucleic acid synthesis and those of peptide synthesis. One of these is the necessity of rapid purification of product after each condensation step, a problem which has been elegantly circumvented in peptide work by Merrifield in the so-called solid-phase or polymer-support synthesis. Essentially, the method consists of attaching the growing chain to an insoluble and inert carrier by a covalent bond. After each condensation, the product is freed from incoming soluble component and reagents by simple filtration, thus eliminating any time-consuming chromatographic separations. During the last few years, several reports have appeared describing initial attempts at polymer-supported deoxyribooligonucleotide synthesis. Two of the special problems associated with these investigations are the polar nature of the nucleotides (necessitating polar solvents) and the extra functional groups that have to be protected.

Polystyrene supports having some reactive sites for attachment of the nucleotide (see below) have been used. Two general types have been reported: (*a*) *soluble* in the reaction medium, which is often anhydrous pyridine or dimethylformamide, and precipitated from water which is then used to wash out the reagents;[86] and (*b*) *insoluble* in the reaction medium.[87] These polymers are generally crosslinked with some divinyl benzene and, depending on the amount of crosslinking, they often swell, thus making elution of reagents more difficult.

Every functional group in the nucleotide has been used as a point of attachment to the polymer (see Fig. 18). Obviously the polymer will have some type of active site involving benzene rings and covalent linkages have been formed via:

1. A trityl (or derivative) ether bond to the 5′-hydroxyl of the nucleotide, **33**.[88]
2. A benzoic ester bond to the 5′- or 3′-hydroxyl, **34** and **36** respectively.[89]
3. A benzamide bond to the primary amine of the heterocyclic base, **37**.[90]
4. A phosphoramidate linkage to the 5′-phosphate, **35**.[91]

Both the diester and triester approach have been used to form the phosphate linkage between two nucleotides and all four different nucleosides have been employed in the investigations cited to date. With these methods deoxyribo-oligonucleotides of up to six units[89b] have been prepared but, depending on the conditions employed, various problems have been encountered.

Swelling of some supports often makes it difficult to remove all traces of starting material and reagents after each condensation step. The more insoluble polymers of course lead to heterogeneous reaction mixtures which can give rise to lower yields in the condensation step or alternately require longer than usual reaction times or more vigorous conditions. To remove the oligonucleotide chain from the support polymer frequently requires very vigorous conditions, and this difficulty increases with increasing chain lengths. A recent report describes the synthesis of hexa *d*T on an insoluble-type support in one week.[92] Many of the problems mentioned were apparently overcome; however, a fiftyfold excess of incoming mononucleotide was

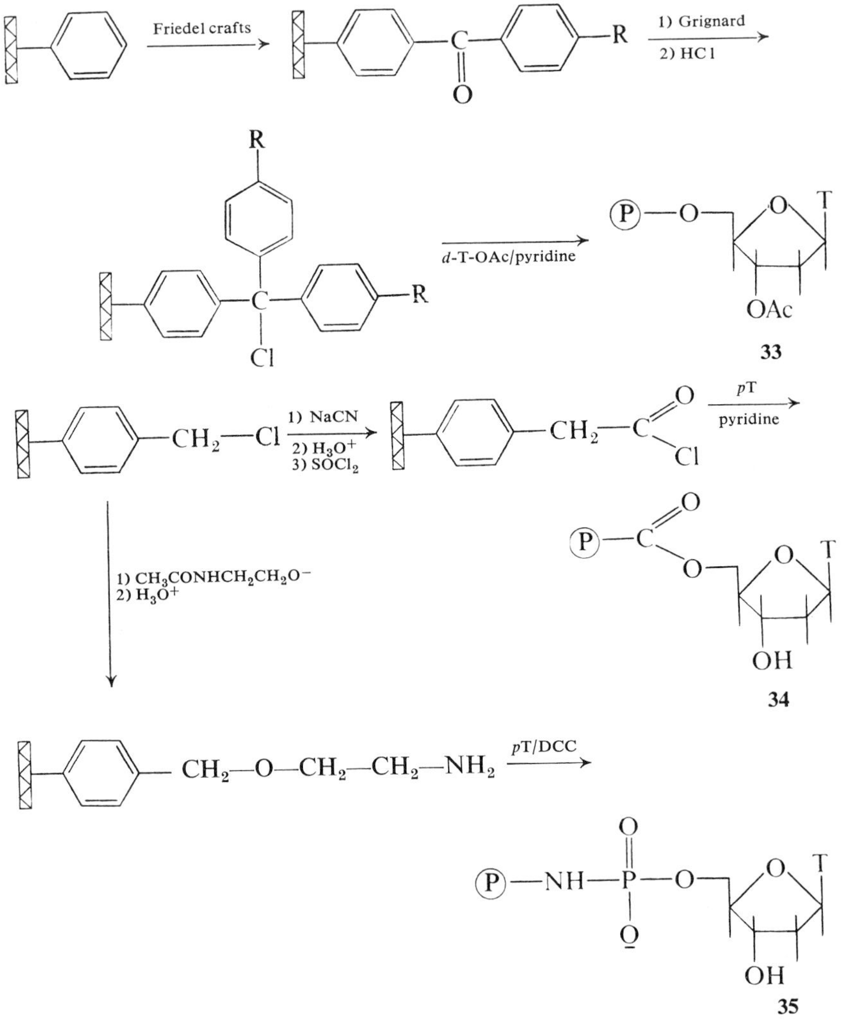

Figure 18. Preparation of various polymer supports and their attachment to the nucleotide.

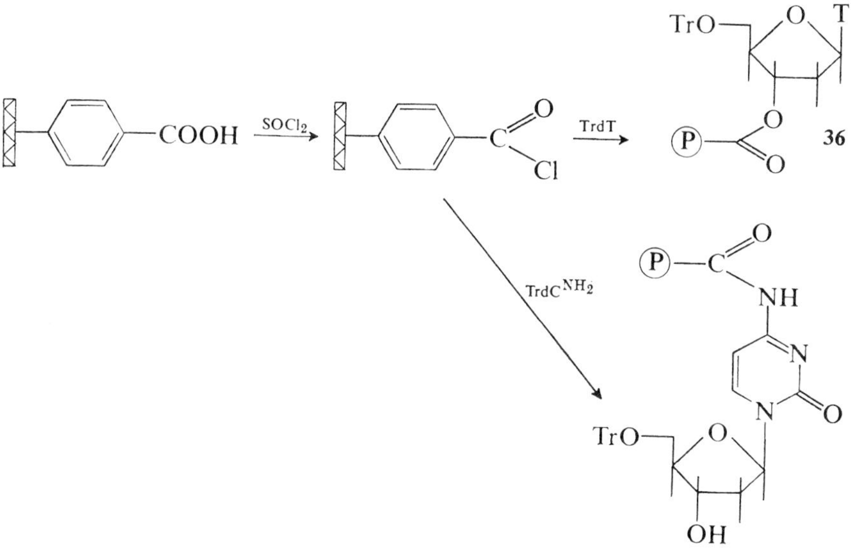

Figure 18. (*Continued*)

required at each condensation step and the experimental details were very sparse.

To be distinctly advantageous, the polymer support method requires a fast, almost quantitative condensation step and ready solubility of reagents. Both of these difficulties remain in nucleic acid syntheses.

E. Synthesis of Oligoribonucleotides

The synthesis of ribooligonucleotides is greatly complicated by the presence of the 2′-hydroxyl function on the ribose moiety. Since the natural phosphate bridge occurs between the 3′- and 5′-positions of two adjacent nucleosides, one must be able to differentiate between the two hydroxyl functions at the 2′- and 3′-positions, both of which are secondary. Additionally, the 2′-hydroxyl should have a protecting group that is not removed under conditions that deblock either the 3′- or 5′-positions prior to condensation; however, it must be readily deblocked at the conclusion of a synthesis and under conditions that will not isomerize the 3′ → 5′ linkage to 2′ → 5′. It has been shown that either acidic or basic conditions can accomplish this migration to the "unnatural" linkage. Although some small homopolymers have been synthesized from suitably protected monoribonucleotides[93] and some attempts have been made to achieve syntheses from cyclonucleoside precursors,[94] the main effort in the ribo-field has been directed toward a stepwise synthesis of chains containing defined sequences.[95] Several general approaches can be conceived and two of the most thoroughly investigated are illustrated in Fig. 19.

Initially, let us assume that any primary amino functions in the purine or pyrimidine rings can be satisfactorily protected and that there is no

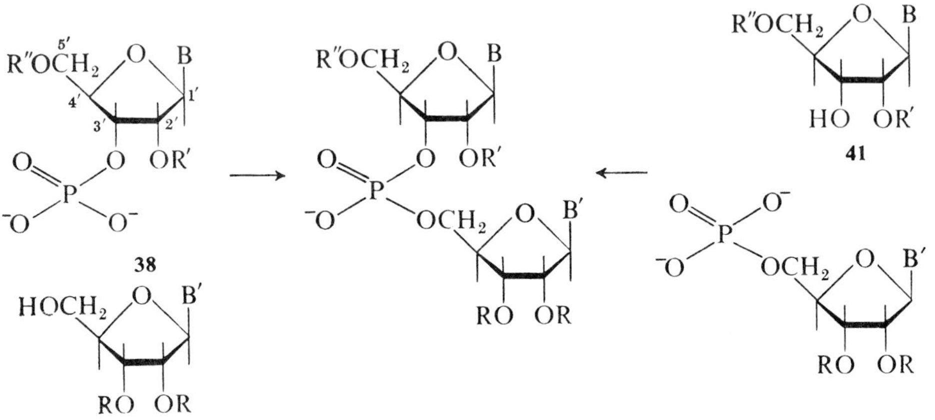

Figure 19. Two generalized approaches to the synthesis of oligoribonucleotides.

difference between effecting the internucleotide bond via a condensation of 3′-phosphate to 5′-hydroxyl or 3′-hydroxyl with 5′-phosphate. There are no known chemical methods for quantitatively distinguishing the 2′- and 3′-positions; however, with the aid of purification techniques compounds such as **38** and **41** can be obtained in pure form. Early efforts were focused on preparing **38** from selective, but not specific, basic hydrolysis of the 3′-5′-cyclic phosphate. Differentiation was achieved by using trityl at the 5′-position and tetrahydropyranyl for the 2′-hydroxyl.[28] This was later modified to the 5′-OTr and 2′-OAc.[43b] The use of pancreatic ribonuclease (RNase) to specifically and quantitatively cleave 2′,3′-cyclic phosphates to the 3′-phosphate, 2′-hydroxyl substitution pattern is exemplified by the preparation of 5′-OAc, 2′-OTHP nucleotides.[44a] Other vinyl ethers were introduced at the 2′-position to make deblocking easier.[45]

In order to be able to extend the chain, the 2′,3′-*cis* diol system of the entering nucleoside component, **39** must be protected with a group that is completely stable under conditions used for deblocking the 5′-hydroxyl. Thus acid-stable (acetyl, benzoyl) groups have been employed when trityl or derivatives are used to block the 5′-hydroxyl,[47] whereas base-stable alkylidene derivatives are employed when 5′-*O*-acetyl is present.[44]

The second approach, **41** plus **42**, requires differentiation of the 2′- and 3′-hydroxyls in the nucleoside component **41** and has been partially achieved by purely chemical methods.[30,46] Although isomerically pure samples could be isolated by crystallization techniques, the overall yield suffers from a somewhat lengthy sequence. However, several interesting points have been reported. The THP group protecting the 2′-hydroxyl can be removed under acidic conditions so mild that negligible isomerization to 2′-5′ occurs on deblocking. Furthermore, the bulky pivaloyl group affords essentially specific protection for the primary 5′-hydroxyl and is base-labile (acid-stable) in contrast to the commonly used trityl group.

Both of these procedures suffer from the same drawbacks. The final product **40** has no terminal phosphate, which would be necessary for chain elongation. If this discrepancy could be overcome, say by chemical or enzymatic phosphorylation, the chain could only be extended from the 5′- end since again the 2′- and 3′-hydroxyls of the 3′-terminus could not be specifically differentiated.

An attempt to overcome these difficulties was reported in which the first route, **38** + **39** → **40**, was used; however, the 2′- and 3′-hydroxyl groups of the nucleoside were "protected" by using the readily available 2′,3′-cyclic phosphate.[63] Condensation and workup produced a dinucleotide diphosphate (mixture of 2′- and 3′-phosphates at the 3′-terminus), which could be quantitatively cyclized with DCC and then specifically reopened to the 3′-phosphate using RNase. The chain was immune to enzymatic degradation since all the internal 2′-hydroxyl groups were protected.

Two major problems presented themselves. Although unwanted side products due to polymerization of the cyclic phosphate could be removed by chromatography, they presumably reduced the yield of the desired product. The enzymatic cleavage was practically useful only when no protecting groups were present on the heterocyclic base at the 3′-terminus. Therefore either that base must have no free amino group such as uridine, or it must be specifically deblocked—without affecting any other protecting groups—before the enzyme can function.

6. ENZYMES AS REAGENTS

The search for a more complete understanding, at the molecular level, of the mechanisms involved in the various reactions of DNA, for example, replication and transcription, has been greatly influenced by the various oligonucleotides fabricated by organic chemists. Paralleling these advances were the isolation and investigation of several enzymes. The interplay of these two factors has advanced the understanding of this vital biological process at an incredible rate during the last decade. In addition, the ready availability of several of these enzymes in pure form has allowed them to be used as routine—and highly specific—reagents for the synthesis and characterization of polynucleotides.

A. DNA-Polymerase

An enzyme has been isolated from *E. coli*[96] which catalyzes the synthesis of high-molecular-weight DNA in the presence of all four of the deoxynucleoside 5′-triphosphates, *d*-ATP, *d*-CTP, *d*-GTP, and TTP, plus a DNA template. No synthesis of DNA takes place in the absence of this template, which may consist of short-chain synthetic oligonucleotides or longer chains isolated from natural sources. For example, it has been demonstrated that short-chain synthetic deoxyribopolynucleotides containing alternating deoxyadenylate and thymidylate units serves as a template, in the presence of thymidine 5′-triphosphate and deoxyadenosine 5′-triphosphates, to induce the synthesis of a high-molecular-weight DNA-like polymer containing again deoxyadenylate and thymidylate units in alternating sequences (see Table 1, Reaction 1). Table 1 lists the types of reactions so far elicited from DNA-polymerase. These reactions are also very fast; polymers much larger than the original template can be produced in several hours.

Table 1. DNA-Polymerase Catalyzed Reactions

Reaction		Reference
1. $d\text{-(AT)}_3 + d\text{-ATP} + \text{TTP}$	→ Poly *d*-AT	97
2. $\text{T}_{11} + d\text{-A}_7 + d\text{-ATP} + \text{TTP}$	→ Poly *d*-A:T	98
3. $\text{T}_{11} + d\text{-A}_7 + d\text{-ATP}$	→ Poly *d*-A	98
4. $d\text{-(TG)}_6 + d\text{-(AC)}_6 + \begin{Bmatrix}\text{TTP}\\ d\text{ATP}\\ d\text{CTP}\\ d\text{GTP}\end{Bmatrix}$	→ Poly *d*-TG:CA[a]	98
5. $d\text{-(TTC)}_4 + d\text{-(AAG)}_3 + \begin{Bmatrix}\text{TTP}\\ d\text{ATP}\\ d\text{CTP}\\ d\text{GTP}\end{Bmatrix}$	→ Poly *d*-TTC:GAA[a]	99
6. $d\text{-(TATC)}_3 + d\text{-(TAGA)}_2 + \begin{Bmatrix}\text{TTP}\\ d\text{ATP}\\ d\text{CTP}\\ d\text{GTP}\end{Bmatrix}$	→ Poly *d*-TATC:GATA[a]	100

[a] All of the DNA-like polymers are written so that the colon separates the two complementary strands. The complementary sequences in the individual strands are written so that antiparallel base-pairing is evident.

Table 2. Nearest Neighbor Frequency Analysis of Poly *d*-TTC:GAA Templates: $d(TTC)_4 + d(AAG)_2$

	Radioactivity in Deoxynucleoside 3′-Phosphates							
	*d*A*p*		*d*G*p*		*d*C*p*		*d*T*p*	
α-P³²-Labeled Triphosphate	Count/Min	Percentage	Count/Min	Percentage	Count/Min	Percentage	Count/Min	Percentage
*d*ATP	12,836	50	12,851	50	0	0	0	0
*d*GTP	13,684	100	0	0	0	0	0	0
*d*CTP	0	0	0	0	0	0	9,623	100
*d*TTP	0	0	0	0	12,860	50.6	12,565	49.4

The most direct method for showing the chemical structure of the polymer is by the so-called nearest neighbor analysis technique,[101] in which one α-P^{32} labeled deoxyribonucleoside triphosphate is incorporated into the chain at a time. Degradation of the synthetic polymer to deoxyribonucleoside 3′-phosphates using enzymatic methods gives results which show complete accord with theoretical expectations from the sequence of the original template. Typically, the results from such a polymerization using templates with complementary repeating trinucleotide sequences is shown in Table 2.[99] Within experimental error, excellent agreement is observed concerning the incorporation of the various 5′-triphosphate monomers.

An additional property of these synthetic DNA-like polymers is their ability to reseed the synthesis by the DNA polymerase of more of the same product.[99] The importance of this finding can hardly be overstressed. In essence, it means that once the specific sequences have been put together by well-defined and unambiguous chemical synthesis, DNA-polymerase will ensure their permanent availability. Thus the well-known, dramatic feature of DNA-structure, its ability to guide its own replication, can be exploited at the molecular level. The total number of DNA-like polymers prepared so far is listed in Table 3.

Table 3. New DNA-like Polymers with Repeating Sequences

Repeating Dinucleotide Sequences	Repeating Trinucleotide Sequences	Repeating Tetranucleotide Sequences
	Poly *d*-TTC:GAA	
Poly *d*-TC:AG	Poly *d*-TTG:CAA	Poly *d*-TTAC:GTAA
Poly *d*-TG:AC	Poly *d*-TAC:GTA	Poly *d*-TATC:GATA
	Poly *d*-ATC:GAT	

The complete characteristics of the DNA-polymerase-catalyzed reactions can be summarized as follows: (*a*) chemically synthesized segments corresponding to both strands are required for reaction to proceed: (*b*) minimal size of the two complementary segments used as primers varies between 8 and 12 nucleotide units; (*c*) synthesis is extensive; (*d*) products are high molecular weight and are double stranded with sharp melting transitions; (*e*) nearest-neighbor analysis invariably indicates that the individual strands contain the appropriate repeating sequences; (*f*) high-molecular-weight products can be reutilized as primers for more synthesis.

An electron micrograph of poly *d*-TG:AC showed the average size to be 0.5 μ. This is indicative of a molecular weight in the range of one million.[98b]

B. Nucleotidyl Transferase Enzyme

A second DNA-polymerizing enzyme, the "end-addition enzyme," has been detected in a purified extract of calf thymus. Chromatography on a hydroxylapatite column permits its separation from DNA-dependent DNA-polymerase.[102] This particular enzyme extends oligodeoxyribonucleotides via the repeated addition of deoxyribonucleotide 5′-phosphate units to long homologous chains:

$$p\mathrm{T}p\mathrm{T}p\mathrm{T} + nd\text{-ATP} \xrightarrow{\text{end-addition enzyme}} p\mathrm{T}p\mathrm{T}p\mathrm{T}p\mathrm{A}p\mathrm{A}p\mathrm{A}p\mathrm{A} \cdots$$

The rate of this polymerization decreases in the sequence *d*-ATP > *d*-ITP > *d*-CTP ≫ *d*-TTP and *d*-GTP, while *d*-CTP is incorporated only in the presence of Co^{2+} ions.[103] The primer oligonucleotides must have a minimum length of three nucleotides at the same concentration of enzyme, while the length of the newly formed chain depends on the substrate/primer ratio. Thus DNA-like polymers having specific sequences at one end can be obtained for biological testing.

C. RNA-Polymerase (DNA-Dependent)

A DNA-dependent RNA-polymerase is present in bacteria and other tissues.[104] It is similar in action to DNA-polymerase and requires a double-stranded segment of DNA as a primer plus all four ribonucleoside 5′-triphosphates to be effective. The polyribonucleotides thus synthesized are single-stranded and of high molecular weight and their composition is determined by the composition of the DNA template. If synthetic polydeoxyribonucleotides with a known base sequence are used as templates, then a polyribonucleotide having a complementary defined base sequence is obtained. Typical results obtained with various synthetic polydeoxyribonucleotide primers for DNA-dependent RNA-polymerase are given in Table 4.

Such results emphasize the fact that only by providing in the reaction mixture a set of ribonucleotide 5′-phosphates which are appropriate for copying the template can the required single-stranded polyribonucleotide be obtained. In every case the single-stranded ribopolynucleotides thus prepared have been shown to contain the expected repeating sequences by the technique of nearest-neighbor analysis. A template of only 9 nucleotides (3 triplets) has

Table 4. DNA-Dependent RNA-Polymerase Reactions

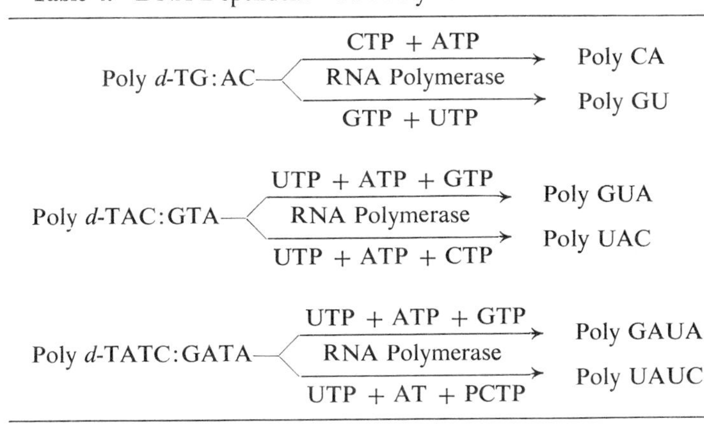

been shown to be sufficient for synthesizing a complementary polyribonucleotide with more than 150 nucleotide units.[105]

D. Polynucleotide Phosphorylase

Polynucleotide phosphorylase polymerizes nucleoside diphosphates to polynucleotides in which the bases are distributed at random:

$$n\text{XDP} \rightleftarrows (\text{XMP})_n + n\text{Pi}$$

This enzyme requires no template[106] and is specific for ribonucleoside 5′-diphosphate, even chemically modified 5′-diphosphates can be incorporated.[107] If several nucleoside 5′-diphosphates are present, they are incorporated into the polymer at random with the exception of GDP, which is incorporated preferentially.[108]

The nonspecificity of polynucleotide phosphorylase frequently can be circumvented by the action of other enzymes for the preparation of some definite characteristics. For example, random incorporation of a specific nucleotide into a polymer and subsequent cleavage at this nucleotide can give a polymer with a definite triplet at the 3′-end:[109]

UDP + GDP (30:1)

↓ polynucleotide phosphorylase

··· *p*U*p*U*p*U*p*G*p*U*p*U ··· *p*U*p*U*p*G*p*U ···

↓ T_1 ribonuclease

··· *p*U*p*U*p*U*p*G*p* + *p*U*p*U ··· *p*U*p*U*p*G*p* + U ···

↓ phosphatase

··· *p*U*p*U*p*U*p*G

Moreover, this enzyme can be used for the synthesis of short chains. With the primer-dependent enzyme from *M. lysodeicticus*, a dinucleotide phosphate is extended by only a few units if the reaction takes place in 0.4 *M* NaCl.[110] If the reaction is allowed to proceed until the few long chains formed initially have been converted by phosphorolysis into a large number of short chains, then the following sequence can be obtained:

A*p*U + ADP (1:4)

↓ polynucleotide phosphorylase (2 hr)

A*p*U*p*A*p*A ··· A (high molecular weight) + A*p*U (unused primer)

↓ polynucleotide phosphorylase (24 hr)

A*p*U (small amount) + A*p*U*p*A (*main product*) + A*p*U*p*A*p*A
+ A*p*U*p*A*p*A*p*A*p*A + a small amount of polynucleotide of high molecular weight

The trinucleoside diphosphates X*p*Y*p*Z can be separated chromatographically and thus is a purely enzymatic method suitable in principle for the synthesis of all 64 ribonucleotide triplets, which have been synthesized chemically.[47]

E. Ribonuclease Enzymes

Pancreatic ribonuclease has been mentioned as a reagent for opening 2′,3′-cyclic phosphates quantitatively to 3′-phosphate, 2′-hydroxyl in pyrimidine-containing ribonucleotides, U or C. This reaction is only one step in a sequence of reactions defining the ribonuclease's real function as an endonuclease, that is, to split any phosphodiester bond specifically to yield a pyrimidine 3′-phosphate. Thus, the enzyme has also received extensive use as a degradative tool in the sequencing of *t*RNA;[111] for example,

$$\mathrm{U}p\mathrm{G}p\mathrm{A}p\mathrm{C}p\mathrm{U}p\mathrm{U}p\mathrm{A}p\mathrm{C}p \cdots \xrightarrow{\text{pancreatic RNase}} \mathrm{U}p + \mathrm{G}p\mathrm{A}p\mathrm{C}p + \mathrm{U}p + \mathrm{U}p + \mathrm{A}p\mathrm{C}p \cdots$$

Additionally, T_1 ribonuclease[112] is specific for guanosine and several other RNase have been isolated with other characteristics. Also, these enzymes appear to require a free 2′-hydroxyl as well as a free amino function on the heterocyclic ring before they can be effective. Not all of them are commercially available and hence their use is limited.

F. Polynucleotide Kinase

Although introduction of phosphate can be accomplished chemically by using many types of phosphorylating agents, the recently purified[113] polynucleotide kinase can accomplish introduction specifically at the 5′-hydroxyl using only ATP. However, this enzyme requires the presence of phosphate in the 3′-position also, that is, a 3′-nucleotide or oligonucleotide chain, and is not commercially available. Consequently it has been employed only in a very limited sense often for incorporation of P^{32} for a crucial labeling experiment.

G. Alkaline Phosphatase (*E. Coli*)

This enzyme is used extensively for the nonspecific removal of any phosphomonoester group at alkaline *p*H.[114]

H. Phosphodiesterases (Spleen and Venom)

A phosphodiesterase is an enzyme which hydrolyzes the phosphodiester bonds linking the nucleotides in RNA or DNA. These two types of enzyme are commercially available and used routinely to characterize deoxy (or ribo) oligonucleotides. Venom diesterase hydrolyzes RNA to 5′-mononucleotides[115] and is also active in hydrolyzing the oligonucleotides produced by the action of deoxyribonucleases(I) on DNA to eventually yield deoxyribonucleoside 5′-phosphates. Spleen diesterase hydrolyzes RNA to 3′-mononucleotides and also acts on the mixture of oligonucleotides produced from DNA by spleen deoxyribonuclease(II) to yield deoxyribonucleoside 3′-phosphates.[116]

I. Polynucleotide Ligase

Recently, enzymes which catalyze the covalent joining of breaks in a single strand of bihelical DNA have been identified and purified from *E. coli*, phage-T_4 and phage-T_7 infected *E. coli*.[117] Given a DNA substrate containing single-stranded breaks, as shown below, the ligase enzyme accomplishes the repair by the esterification of an internally located 3′-hydroxyl group with the adjacent 5′-phosphomonoester. In the reaction ATP is cleaved to AMP and PPi.

```
3′ · · · ———————————————— · · · 5′
      |  |        |  |
      A  G        T  C          +ATP
      T  C        A  G         ———→
5′ · · · P P —OH P P P · · · 3′

                3′ · · · ———————————— · · · 5′
                          |  |  |  |
                          A  G  T  C        + AMP + PPi
                          T  C  A  G
                5′ · · ·  P  P  P  P  P · · · 3′
```

Studies on *E. coli* and T_4-induced ligases have revealed that they have many properties in common. Both enzymes are specific for the same DNA substrates and produce the same DNA end product. Furthermore, the reactions catalyzed by both enzymes are mediated by an enzyme-adenylate (enzyme AMP) complex. However, a distinguishing feature of the two enzymes is that the adenylate moiety for each of the two complexes is derived from different cofactors. The cofactor for *E. coli* enzyme is DPN, whereas that for the T_4-induced enzyme has been shown to be ATP.

Much has been deduced concerning the overall mechanism of the reaction. In both cases, the first step in the overall reaction consists of the transfer of an adenylate group from the cofactor to the enzyme to form a covalently linked enzyme-AMP intermediate. Once enzyme-AMP is formed by a reaction of the enzyme with DPN, there is a further transfer of the AMP to the 5′-phosphoryl terminus of a DNA chain to generate a new pyrophosphate bond linking the AMP and DNA. In the final step, the DNA phosphate of the pyrophosphate is presumably attacked by the 3′-hydroxyl group of the neighboring DNA chain, displacing the activating AMP and forming the phosphodiester bond.[118] However, specific characteristics made these enzymes of singular importance to the problem of the total synthesis of biologically specific high-molecular-weight DNA. Initial studies showed that a short deoxyribopolynucleotide carrying a 5′-phosphate group at one end and a 3′-hydroxyl group at the opposite end could be joined end-to-end in the presence of long complementary deoxyribopolynucleotides.[119] These experiments also showed that chains as short as hexa- or heptanucleotides could be joined. Further, if both the complementary polynucleotides are short, such as the hexadecanucleotide *d*$(TTAC)_4$, the octanucleotide *d*-$P^{32}(TAAG)_2$ can be added.

7. GENE SYNTHESIS

The fruition of many of the observations outlined in previous sections of this review were realized by the dramatic announcement, in 1970, of the total synthesis of the gene for an alanine transfer RNA from yeast[120] (see also reference 79). In defining the DNA sequence complementary to the *t*RNA sequence, the principle assumption was made that all the minor bases were produced from the four parent bases by modifications which occurred after transcription of the DNA gene with the four standard bases. The outstanding concept in the synthesis was to use polynucleotide ligase for joining relatively short, chemically synthesized, polynucleotide chains while they were held together in properly aligned bihelical complexes.

The total plan for the synthesis of this gene is outlined in Fig. 20. The gene was divided into three parts, shown as **A**, **B**, and **C** (or **C′**), and each part was to consist of several chemically synthesized segments. The chemically synthesized segments were phosphorylated using T_4-polynucleotide kinase, after heating to overcome any structural inhibitions of the phosphorylation reaction. The chains were elongated by stepwise addition of the appropriate segments using polynucleotide ligase. These segments were anchored in place by base-pairing with the overlapping end of the growing gene. The product from each addition could be separated from the starting materials on Agarose or Sephadex columns. The extent of the ligase-catalyzed joining reactions ranged from 40 to 70%. An analysis of the desired product hinged on the following steps: (*a*) resistance to phosphatase; (*b*) degradation to 3′-nucleotides using micrococcal and spleen phosphodiesterase; (*c*) hydrolysis to 5′-nucleotides using pancreatic deoxyribonuclease and venom phosphodiesterase. In a few cases repair with DNA polymerase was also used as further characterization.[121]

```
20 19 18 17 16 15 14 13 12 11 10  9  8  7  6  5  4  3  2  1

 G  A  U  U  C  C  G  G  A  C  U  C  G  U  C  C  A  C  C  A     (3′) ribo
┌──────────(4)──────────┐ ┌──────────────────(1)──────────────┐
—C—T—A—A—G—G—C—C  T—G—A—G—C—A—G—G—T—G—G—T                       (5′) deoxy   A
         |  |  |  |  |  |  |  |  |  |  |  |  |  |  |  |
         C—C—G—G—A—C—T—C—G—T  C—C—A—C—C—A                       (3′) deoxy
         └─────────(3)─────────┘ └──────(2)──────┘


50 49 48 47 46 45 44 43 42 41 40 39 38 37 36 35 34 33 32 31 30 29 28 27 26 25 24 23 22 21 20 19 18 17

Me₂                                     Me                                        H₂
—G  C  U  C  C  C  U  U  I  G  C  I  ψ  G  G  G  A  G  A  G  U  C  U  C  C  G  G  T  ψ  C  G  A  U  U     (3′) ribo
         ┌──────(9)──────┐ ┌───────────────────────────(6)───────────────────────────┐
        —G—A—A—T—C  G—T—A—C—C—C—T—C—T—C—A—G—A—G—G—C—C—A—A—G—                      (5′) deoxy   B
         |  |  |  |  |  |  |  |  |  |  |  |  |  |  |  |  |  |
—G—C—T—C—C—C—T—T—A—G—C—A—T—G—G—G  A—G—A—G—T—C—T—C—C—G—G—T—T—C—G—A—T—T           (3′) deoxy
└───────────────(8)───────────────┘ └─────(7)─────┘ └───────────(5)───────────┘
```

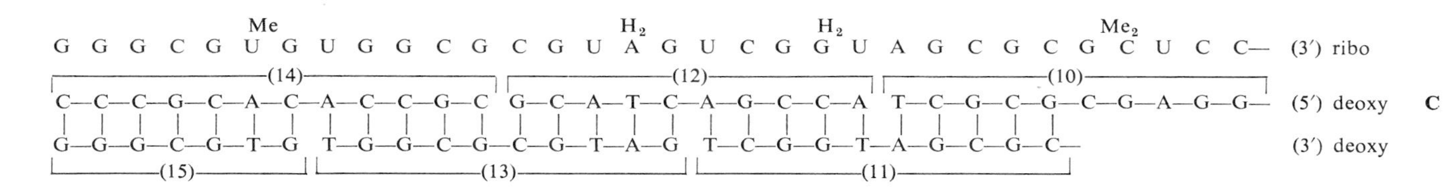

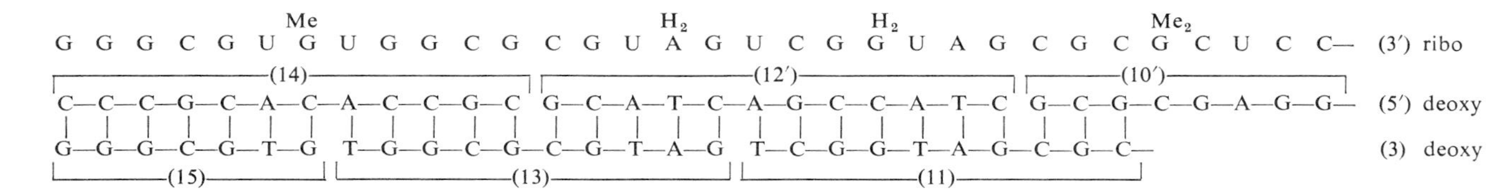

Figure 20. Total plan for the synthesis of a yeast alanine *t*RNA gene. The chemically synthesized segments are in brackets, the serial number of the segment being shown within the brackets. A total of 17 segments (including 10′ and 12′) varying in chain length from penta to icosanucleotides were synthesized. [Adapted from *Nature*, **227**, 27 (1970).]

With this successful application of the ligase enzyme in joining synthetic polynucleotides, it was considered essential to study the fidelity of the T_4-infected *E. coli* ligase enzyme. Recently it has been discovered[122] that this enzyme can also catalyze the covalent joining of interrupted deoxyribo-oligonucleotide strands with one mispaired base at the 3′-hydroxyl terminus of a bihelix ($P^{32}T_{11}C$ + Poly *d*A). Similar types of mismatched joining as well as end-to-end dimerizations have subsequently been reported to occur during the synthesis of alanine *t*RNA gene.[123]

A recent report showed that terminal crosslinking in DNA strands by an enzyme system consists of the T_4-ligase and another activity considered likely to be an exonuclease.[124]

8. CONCLUSION

Two distinct periods of nucleic acid chemistry have thus far emerged. In the pre-1960 period, some adventuresome organic chemists took up the challenges implicit in the structural complications of these macromolecules. The main impetus was provided by the Watson-Crick hypothesis of the early 1950s. In the period following 1960, as much of the basic chemistry was elaborated, more chemists and biologists began to contribute to this problem and increasingly complicated molecules were synthesized; this culminated in 1970, with the first announcement of the test-tube synthesis of a gene.

After this achievement, which undoubtedly constitutes a milestone in the development of new experimental techniques, it is probably safe to predict another period of prolific advancement in the 1970s. What areas need additional efforts or will experience growth? Some of the synthetic reagents require additional sophistication, two of the more important being those for effecting the formation of the internucleotidic bond and for differentiating the 2′-hydroxyl group in the ribonucleotides. The former is required to make the condensation faster and quantitative when using stoichiometric amounts of the two components to be condensed. There is always a need for faster separation techniques to relieve some of the tedious parts of the synthetic work in this field. But more important is the necessity for devising easier techniques for the sequence determination of the longer synthetic deoxypoly-nucleotides. The discovery of new enzymes and their ready availability in the pure form will undoubtedly assist nucleic acid chemists.

As techniques become better established, not only will other "genes" be synthesized but these will be incorporated into the natural DNA. In fact, at the rate this field is advancing such announcements may well be made before this review is published.

ACKNOWLEDGMENT

We are grateful to Mrs. Gloria Dumoulin for her patience in typing the manuscripts.

REFERENCES

1. H. G. Khorana, H. Büchi, H. Ghosh, N. Gupta, T. M. Jacob, H. Kössel, R. Morgan, S. A. Narang, E. Ohtsuka, and R. D. Wells, *Cold Spring Harbor Symp.*, **31**, 39 (1966).
2. A. M. Michelson and A. R. Todd, *J. Chem. Soc.*, **1955**, 2632.
3. H. G. Khorana, in D. Shugar, Ed., *Genetic Elements: Properties and Function*, Academic Press, New York, 1966, p. 209.
4. A. M. Michelson, *The Chemistry of Nucelosides and Nucleotides*, Academic Press, London, 1963.
5. H. G. Khorana, *Some Recent Developments in the Chemistry of Phosphate Esters of Biological Interest*, Wiley, New York, 1961.
6. W. W. Zorbach and R. S. Tipson, Eds., *Synthetic Procedures in Nucleic Acid Chemistry*, Interscience, New York, 1968.
7. (a) F. Cramer, *Pure Appl. Chem.*, **18,** 197 (1969); (b) H. G. Khorana, *Pure Appl. Chem.*, **17,** 349 (1968); (c) F. Cramer, *Angew. Chem. Int. Ed.*, **5,** 173 (1966).
8. For typical examples, see (a) R. L. Letsinger and W. S. Mungall, *J. Org. Chem.*, **35,** 3800 (1970); (*b*) G. H. Jones, H. P. Albrecht, N. P. Damodaran, and J. G. Moffatt, *J. Am. Chem. Soc.*, **92,** 5511 (1970); (c) D. A. Shuman, M. J. Robins, and R. K. Robins, *J. Am. Chem. Soc.*, **92,** 3434 (1970); (d) M. Ikehara, *Accts. Chem. Res.*, **2,** 47 (1969).
9. For typical examples, see (a) R. S. Goody, K. A. Watanabe, and J. J. Fox, *Tetrahedron Lett.*, **1970,** 293; (b) T. Y. Shen, *Angew. Chem. Int. Ed.*, **9,** 678 (1970); (c) D. Suhadolnik, *Nuceloside Antibiotics*, Wiley, New York, 1970.
10. For typical examples, see (a) R. H. Hall, M. J. Robins, L. Stasiuk, and R. Thedford, *J. Am. Chem. Soc.*, **88,** 2615 (1966); (b) H. G. Zachau, *Angew. Chem. Int. Ed.*, **8,** 711 (1969).
11. (a) D. Brown, in R. A. Raphael, E. C. Taylor, and H. Wynberg, Eds., *Advances in Organic Chemistry—Methods and Results*, Vol. 3, Interscience, New York, 1963, p. 75; (b) V. M. Clark, D. W. Hutchinson, A. J. Kirby, and S. G. Warren, *Angew. Chem. Int. Ed.*, **3,** 678 (1964).
12. (a) K. Imai, S. Fujii, K. Takanohashi, Y. Furukawa, T. Masuda, and M. Honjo, *J. Org. Chem.*, **34,** 1547 (1969); (b) T. A. Khawaja and C. B. Reese, *J. Am. Chem. Soc.*, **88,** 3446 (1966); (c) R. Saffhill, *J. Org. Chem.*, **35,** 2881 (1970).
13. F. Cramer and H. Neunhoeffer, *Chem. Ber.*, **95,** 1664 (1962).
14. R. von Tigerstrom and M. Smith, *Science*, **167,** 1266 (1970).

15. A. M. Michelson, *J. Chem. Soc.*, **1959,** 3655.
16. R. Lohrmann and H. G. Khorana, *J. Am. Chem. Soc.*, **88,** 829 (1966).
17. F. Cramer, R. Wittmann, K. Daneck, and G. Weimann, *Angew. Chem. Int. Ed.*, **2,** 43 (1963).
18. M. Smith, J. G. Moffatt, and H. G. Khorana, *J. Am. Chem. Soc.*, **80,** 6204 (1958).
19. F. Cramer, H. Neunhoeffer, K. H. Scheit, G. Schneider, and J. Tennigkeit, *Angew. Chem. Int. Ed.*, **1,** 331 (1962).
20. M. Ikehara and H. Uno, *Chem. Pharm. Bull. (Japan)*, **12,** 742 (1964).
21. (a) F. Cramer, W. Rittersdorf, and W. Böhm, *Justus Liebigs Ann. Chem.*, **654,** 180 (1962); (b) F. Cramer, H. J. Baldauf, and H. Küntzel, *Angew. Chem. Int. Ed.*, **1,** 54 (1962).
22. T. M. Jacob and H. G. Khorana, *J. Am. Chem. Soc.*, **86,** 1630 (1964).
23. Y. Wolman, S. Kivity, and M. Frankel, *Chem. Comm.*, **1967,** 629.
24. R. H. Ralph and H. G. Khorana, *J. Am. Chem. Soc.*, **83,** 2926 (1961).
25. (a) K. K. Ogilvie and R. L. Letsinger, *J. Org. Chem.*, **32,** 2365 (1967); (b) R. L. Letsinger and P. S. Miller, *J. Am. Chem. Soc.*, **91,** 3356 (1969).
26. (a) J. Zemlicka, S. Chládek, A. Holý, and J. Smrt, *Coll. Czech. Chem. Comm.*, **31,** 3198 (1966); (b) A. Hóly and J. Zemlicka, *Coll. Czech. Chem. Comm.*, **34,** 2449 (1969).
27. P. T. Gilham and H. G. Khorana, *J. Am. Chem. Soc.*, **80,** 6212 (1958).
28. M. Smith, D. H. Rammler, I. H. Goldberg, and H. G. Khorana, *J. Am. Chem. Soc.*, **84,** 430 (1962).
29. G. Weimann and H. G. Khorana, *J. Am. Chem. Soc.*, **84,** 4329 (1962).
30. B. E. Griffin, M. Jarman, and C. B. Reese, *Tetrahedron*, **24,** 639 (1968).
31. C. B. Reese and J. C. M. Stewart, *Tetrahedron Lett.*, **1968,** 4273.
32. R. K. Ralph and H. G. Khorana, *J. Am. Chem. Soc.*, **83,** 2926 (1961).
33. H. G. Khorana and J. P. Vizsolyi, *J. Am. Chem. Soc.*, **83,** 675 (1961).
34. R. L. Letsinger, K. K. Ogilvie, and P. S. Miller, *J. Am. Chem. Soc.*, **91,** 3360 (1969).
35. R. L. Letsinger and V. Mahaderan, *J. Am. Chem. Soc.*, **87,** 3526 (1965).
36. A. F. Cook, *J. Org. Chem.*, **33,** 3589 (1968).
37. L. A. Cohen and J. A. Steele, *J. Org. Chem.*, **31,** 2331 (1966).
38. T. Hata and J. Azizian, *Tetrahedron Lett.*, **1969,** 4443.
39. A. F. Cook and D. T. Maichuk, *J. Org. Chem.*, **35,** 1940 (1970).
40. R. L. Letsinger and K. K. Ogilvie, *J. Org. Chem.*, **32,** 296 (1967).
41. (a) F. Cramer, H. P. Bär, H. J. Rhaese, W. Saenger, K. H. Scheit, and G. Schneider, *Tetrahedron Lett.*, **1963,** 1039; (b) S. Chládek and J. Smrt, *Coll. Czech. Chem. Comm.* **29,** 214 (1964).
42. H. S. Sachdev and N. A. Starkovsky, *Tetrahedron Lett.*, **1969,** 733.
43. (a) Y. Lapidot and H. G. Khorana, *J. Am. Chem. Soc.*, **85,** 3852 (1963); (b) D. H. Rammler, Y. Lapidot, and H. G. Khorana, *J. Am. Chem. Soc.*, **85,** 1989 (1963).
44. (a) J. Smrt and F. Sörm, *Coll. Czech. Chem. Comm.*, **28,** 61 (1963); (b) D. H. Rammler and H. G. Khorana, *J. Am. Chem. Soc.*, **84,** 3112 (1962).
45. J. Smrt and S. Chládek, *Coll. Czech. Chem. Comm.*, **31,** 2978 (1966).
46. C. B. Reese, R. Saffhill, and J. E. Sulston, *Tetrahedron*, **26,** 1023, 1031 (1970).

47. R. Lohrmann, D. Söll, H. Hayatsu, E. Ohtsuka, and H. G. Khorana, *J. Am. Chem. Soc.*, **88,** 819 (1966); see also references 30 and 44b.

48. A. Hampton and A. W. Nichol, *Biochem.*, **5,** 2076 (1966); see also reference 40.

49. F. Cramer, W. Saenger, K. H. Scheit, and J. Tennigkeit, *Justus Liebigs Ann. Chem.*, **679,** 156 (1964); see also references 28, 44a.

50. (a) J. Zemlicka and S. Chládek, *Tetrahedron Lett.*, **1965,** 3057; (b) C. B. Reese and J. E. Sulston, *Proc. Chem. Soc.* (*London*), **1964,** 214; (c) F. Eckstein and F. Cramer, *Chem. Ber.*, **98,** 995 (1965); (d) S. Chládek and J. Zemlicka, *Coll. Czech. Chem. Comm.*, **32,** 1776 (1967); see also reference 31.

51. G. M. Tener, *J. Am. Chem. Soc.*, **83,** 159 (1961).

52. D. Söll and H. G. Khorana, *J. Am. Chem. Soc.*, **87,** 360 (1965).

53. W. Freist, R. Helbig, and F. Cramer, *Chem. Ber.*, **103,** 1032 (1970).

54. (a) F. Eckstein, *Chem. Ber.*, **100,** 2228 (1967); (b) A. Franke, F. Eckstein, K. H. Scheit, and F. Cramer, *Chem. Ber.*, **101,** 944 (1968).

55. A. F. Cook, M. J. Holman, and A. L. Nussbaum, *J. Am. Chem. Soc.*, **91,** 1522, 6479 (1969).

56. S. A. Narang, O. S. Bhanot, J. Goodchild, S. K. Dheer, and R. H. Wightman, *J. Chem. Soc.* (*D*), **1970,** 516.

57. F. Cramer and K. H. Scheit, *Justus Liebigs Ann. Chem.*, **679,** 150 (1964).

58. F. Cramer, H. P. Bär, H. J. Rhaese, W. Saenger, K. H. Scheit, and G. Schneider, *Tetrahedron Lett.*, **1963,** 1039.

59. S. A. Narang, O. S. Bhanot, J. Goodchild, and R. H. Wightman, *J. Chem. Soc.* (*D*), **1970,** 91; see also reference 82.

60. F. R. Atherton, H. T. Openshaw, and A. R. Todd, *J. Chem. Soc.*, **1945,** 382.

61. K. H. Scheit, *Tetrahedron Lett.*, **1967,** 3243.

62. (a) E. Ohtsuka, K. Murao, M. Ubasawa, and M. Ikehara, *J. Am. Chem. Soc.*, **91,** 1537 (1969); (b) E. Ohtsuka, M. Ubasawa, and M. Ikehara, *J. Am. Chem. Soc.*, **92,** 3441, 5507 (1970).

63. E. Ohtsuka, M. Ubasawa, and M. Ikehara, *J. Am. Chem. Soc.*, **92,** 3445 (1970).

64. F. Kathawala and F. Cramer, *Justus Liebigs Ann. Chem.*, **712,** 195 (1968).

65. (a) G. M. Tener, H. G. Khorana, R. Markham, and E. H. Pol, *J. Am. Chem. Soc.*, **80,** 6223 (1958); (b) A. F. Turner and H. G. Khorana, *J. Am. Chem. Soc.*, **81,** 4651 (1959).

66. H. G. Khorana and J. P. Vizsolyi, *J. Am. Chem. Soc.*, **83,** 675 (1961).

67. (a) R. K. Ralph and H. G. Khorana, *J. Am. Chem. Soc.*, **83,** 2926 (1961); (b) H. G. Khorana, A. F. Turner, and J. P. Vizsolyi, *J. Am. Chem. Soc.*, **83,** 686 (1961); (c) R. K. Ralph, W. J. Connors, H. Schaller, and H. G. Khorana, *J. Am. Chem. Soc.*, **85,** 1983 (1963).

68. H. G. Khorana, J. P. Vizsolyi, and R. K. Ralph, *J. Am. Chem. Soc.*, **84,** 414 (1962).

69. F. N. Hayes and E. Hansbury, *J. Am. Chem. Soc.*, **86,** 4172 (1964).

70. O. Pongs and P. O. P. Ts'o, *Biochem. Biophys. Res. Comm.*, **36,** 475 (1969).

71. S. A. Narang, O. S. Bhanot, S. K. Dheer, J. Goodchild, and J. J. Michniewicz, *Biochem. Biophys. Res. Comm.*, **41,** 1248 (1970).

72. E. Ohtsuka, M. W. Moon, and H. G. Khorana, *J. Am. Chem. Soc.*, **87,** 2956 (1965).

73. S. A. Narang, T. M. Jacob, and H. G. Khorana, *J. Am. Chem. Soc.*, **89,** 2167 (1967).
74. J. M. Jacob, S. A. Narang, and H. G. Khorana, *J. Am. Chem. Soc.*, **89,** 2177 (1967).
75. S. A. Narang and H. G. Khorana, *J. Am. Chem. Soc.*, **87,** 2981 (1965).
76. For a leading reference, see E. Ohtsuka and H. G. Khorana, *J. Am. Chem. Soc.*, **89,** 2195 (1967).
77. T. M. Jacob and H. G. Khorana, *J. Am. Chem. Soc.*, **87,** 368 (1965).
78. T. M. Jacob and H. G. Khorana, *J. Am. Chem. Soc.*, **87,** 2971 (1965).
79. K. L. Agarwal, H. Büchi, M. H. Caruthers, N. Gupta, H. G. Khorana, K. Kleppe, A. Kumar, E. Ohtsuka, U. L. Rajbhandary, J. H. Van de Sande, V. Sgaramella, H. Weber, and T. Yamada, *Nature*, **227,** 27 (1970).
80. R. L. Letsinger and K. K. Ogilvie, *J. Am. Chem. Soc.*, **91,** 3350 (1969).
81. T. Neilson, *J. Chem. Soc. (D)*, **1968,** 1139; see also reference 54.
82. C. B. Reese and R. Saffhill, *Chem. Comm.*, **1968,** 767.
83. S. A. Narang, T. M. Jacob, and H. G. Khorana, *J. Am. Chem. Soc.*, **89,** 2158 (1967).
84. S. A. Narang and S. K. Dheer, *Biochem.*, **8,** 3443 (1969).
85. J. J. Michniewicz, O. S. Bhanot, S. K. Dheer, J. Goodchild, R. Wightman, and S. A. Narang, *Biochim. Biophys. Acta*, **224,** 626 (1970).
86. (a) H. Hayatsu and H. G. Khorana, *J. Am. Chem. Soc.*, **89,** 3880 (1967); (b) F. Cramer, R. Helbig, H. Hettler, K. H. Scheit, and H. Seliger, *Angew. Chem. Int. Ed.*, **5,** 601 (1966).
87. (a) R. L. Letsinger and V. Mahadevan, *J. Am. Chem. Soc.*, **87,** 3256 (1965); (b) L. R. Melby and D. R. Strobach, *J. Am. Chem. Soc.*, **89,** 450 (1967).
88. L. R. Melby and D. R. Strobach, *J. Org. Chem.*, **34,** 421, 427 (1969); see also reference 86.
89. (a) R. L. Letsinger, M. H. Caruthers, and D. M. Jerina, *Biochem.*, **6,** 1379 (1967); (b) T. Kusama and H. Hayatsu, *Chem. Pharm. Bull. (Japan)*, **18,** 319 (1960); (c) T. S. Shmidzu and R. L. Letsinger, *J. Org. Chem.*, **33,** 708 (1968).
90. R. L. Letsinger and V. Mahadevan, *J. Am. Chem. Soc.*, **88,** 5319 (1966).
91. G. M. Blackburn, M. J. Brown, and M. R. Harris, *J. Chem. Soc. (C)*, **1967,** 2438.
92. F. Cramer and H. Köster, *Angew. Chem. Int. Ed.*, **7,** 473 (1968).
93. C. Coutsogeorgopoulus and H. G. Khorana, *J. Am. Chem. Soc.*, **86,** 2926 (1964).
94. For typical examples, see (a) K. K. Ogilvie and D. Iwacha, *Can. J. Chem.*, **48,** 862 (1970); (b) J. Nagyvary, *Biochem.*, **5,** 1316 (1966); (c) Y. Mizumo and T. Sasaki, *Tetrahedron Lett.*, **1965,** 4579; (d) P. C. Srivastava, K. L. Nagpal, and M. M. Dhar, *Experentia*, **25,** 356 (1969).
95. For typical examples, see (a) H. J. Rhaese, W. Siehr, and F. Cramer, *Justus Liebigs Ann. Chem.*, **703,** 215 (1967); (b) B. E. Griffin and C. B. Reese, *Tetrahedron*, **24,** 2537 (1968); (c) A. Holý and J. Smrt, *Coll. Czech. Chem. Comm.*, **31,** 3800 (1966); see also references 30, 43–50, and 63.
96. (a) A. Kornberg, I. R. Lehman, M. J. Bessman, and E. S. Simms, *Biochim. Biophys. Acta*, **21,** 197 (1956); (b) I. R. Lehman, M. J. Bessman, E. S. Simms, and A. Kornberg, *J. Biol. Chem.*, **233,** 163 (1958); (c) C. C. Richardson, C. L. Schildkraut, H. V. Aposhian, and A. Kornberg, *J. Biol. Chem.*, **239,** 222 (1964).
97. A. Kornberg, L. L. Bertsch, J. F. Jackson, and H. G. Khorana, *Proc. Nat. Acad. Sci. U.S.*, **51,** 315 (1964).

98. (a) C. Byrd, E. Ohtsuka, M. W. Moon, and H. G. Khorana, *Proc. Nat. Acad. Sci. U.S.*, **53,** 79 (1965); (b) R. D. Wells, E. Ohtsuka, and H. G. Khorana, *J. Mol. Biol.*, **14,** 221 (1965).

99. R. D. Wells, T. M. Jacob, S. A. Narang, and H. G. Khorana, *J. Mol. Biol.*, **27,** 237 (1967).

100. R. D. Wells, H. Büchi, H. Kössel, E. Ohtsuka, and H. G. Khorana, *J. Mol. Biol.*, **27,** 265 (1967).

101. H. Schachmann, J. Adler, C. Radding, I. Lehman, and A. Kornberg, *J. Biol. Chem.*, **235,** 3242 (1960).

102. (a) F. J. Bollum, E. Groeninger, and M. Yoneda, *Proc. Nat. Acad. Sci. U.S.*, **51,** 853 (1964); (b) M. Yoneda and F. J. Bollum, *J. Biol. Chem.*, **240,** 3385 (1965).

103. F. Bollum, *Fed. Proc.*, **24** (No. 2, Part I), 1207 (1965).

104. (a) J. Hurwitz, J. August, J. Davidson, and W. E. Cohen, *Progress in Nucleic Acid Research*, Vol. I, Academic Press, New York, London, 1963, p. 59; (b) J. Hurwitz, *Methods in Enzymology*, Vol. VI, Academic Press, New York, London, 1963, p. 23; (c) M. Chamberlin and P. Berg, *Proc. Nat. Acad. Sci. U.S.*, **48,** 81 (1963); (d) E. Fuchs, W. Zillig, P. H. Hofschneider, and A. Preuss, *J. Mol. Biol.* **10,** 546 (1964); (e) M. Chamberlin and P. Berg, *J. Mol. Biol.*, **8,** 297 (1964).

105. S. Nishimura, T. M. Jacob, and H. G. Khorana, *Proc. Nat. Acad. Sci. U.S.*, **52,** 1492 (1964).

106. S. Ochoa, *Angew. Chem.*, **72,** 225 (1960).

107. S. P. Colowick and N. O. Kaplan, *Methods in Enzymology*, Vol. VI, Academic Press, London, New York, 1962, p. 27.

108. M. F. Singer, R. J. Hilmoe, and M. Greenberg, *J. Biol. Chem.*, **235,** 2705 (1960).

109. F. Cramer, H. Küntzel, and J. H. Matthaei, *Angew. Chem. Int. Ed.*, **2,** 589 (1964).

110. R. E. Thach and P. Doty, *Science*, **184,** 632 (1965).

111. R. W. Holley, J. Apgar, G. A. Everett, J. T. Madison, M. Marquisse, S. H. Merrill, J. R. Penswick, and A. Zamir, *Science*, **147,** 1462 (1965).

112. F. Egami and K. Sato-Asaon, *Biochem. Biophys. Acta*, **29,** 655 (1958).

113. (a) C. C. Richardson, *Proc Nat. Acad. Sci. U.S.*, **54,** 158 (1965); (b) A. Novogrodsky, M. Tal, A. Traub, and J. Hurwitz, *J. Biol. Chem.*, **241,** 2933 (1966).

114. A. Garen and C. Levinthal, *Biochim. Biophys. Acta*, **38,** 470 (1960).

115. W. E. Razzell and H. G. Khorana, *J. Biol. Chem.*, **234,** 2105 (1959).

116. W. E. Razzell and H. G. Khorana, *J. Biol. Chem.*, **236,** 1144 (1961).

117. (a) S. B. Zimmerman, J. W. Little, O. K. Oshinsky, and M. Gellert, *Proc. Nat. Acad. Sci. U.S.*, **57,** 1841 (1967); (b) B. Olivera and I. R. Lehman, *Proc. Nat. Acad. Sci. U.S.*, **57,** 1426 (1967); (c) M. L. Gefter, A. Becker, and J. Hurwitz, *Proc. Nat. Acad. Sci. U.S.*, **58,** 240 (1967); (d) B. Weiss and C. C. Richardson, *Proc. Nat. Acad. Sci. U.S.*, **57,** 1021 (1967); (e) N. R. Cozzarelli, N. E. Melechen, T. M. Jovin, and A. Kornberg, *Biochem. Biophys. Res. Commun.*, **28,** 578 (1967).

118. B. M. Olivera, Z. W. Hall, Y. Anraku, J. R. Chien, and I. R. Lehman, *Cold Spring Harbor Symp.*, **33,** 27 (1968).

119. N. K. Gupta, E. Ohtsuka, H. Weber, S. H. Chang, and H. G. Khorana, *Proc. Nat. Acad. Sci. U.S.*, **60,** 285 (1968).

120. *The New York Times*, CXIX, 1 (June 3, 1970).
121. N. K. Gupta and H. G. Khorana, *Proc. Nat. Acad. Sci. U.S.*, **61,** 215 (1968).
122. C. M. Tsiapalis and S. A. Narang, *Biochem. Biophys. Res. Commun.*, **39,** 631 (1970).
123. V. Sgaramella, J. H. van de Sande, and H. G. Khorana, *Proc. Nat. Acad. Sci. U.S.*, **67,** 1468 (1970).
124. B. Weiss, *Proc. Nat. Acad. Sci. U.S.*, **65,** 652 (1970).

The Total Synthesis of Antibiotics

FRANCIS JOHNSON

The Dow Chemical Company Eastern Research Laboratory
Wayland, Massachusetts

1. INTRODUCTION

Organic substances that can inhibit or destroy one form of life or another are extremely common in nature. Their sources are found in almost all but the highest life forms, ranging from the lowly bacterium, such as *Clostridium botulinum*, which produces one of the most deadly toxins known to man, through the plant and insect kingdoms to the lower forms of aquatic and terrestrial life. In the latter categories one can mention the puffer fish, the eggs and ovaries of which contain one of the most potent toxic substances known—tetrodotoxin (LD_{50} in mice: 10 μg/kg, given ip.)[1a]—and Russell's viper, which harbors one of the deadliest venoms yet found [LD_{99} in mice: 0.13 mg/kg given iv.].[1b]

In many instances the real biological reasons for the existence of such substances are not known. However, it is obvious that frequently toxic substances are used defensively for self-protection and the protection of territory against predators and aggressors. Conversely, there are many instances where predators use such materials aggressively to render their victim harmless. In fact they constitute nature's arsenal for waging chemical warfare and taken in perspective this appears to be one of the commonest methods employed among the various species in their fight for continued existence.

Antibiotics are specific chemical compounds derived from or produced by living organisms that can, in small amounts, selectively inhibit the life processes of other organisms. In general, however, the term "antibiotic" frequently refers only to substances that inhibit microorganisms which in particular are pathogenic to man, other higher animals, or plants. Such microorganisms include bacteria, mycobacteria fungi, amebae, and more rarely viruses. For an antibiotic to be useful therapeutically it should have a high order of selective toxicity, that is, it should have little or no toxicity to the disease-bearing host at those levels at which it inhibits or kills the disease-producing organisms. Predicated on this requirement an antibiotic should be capable of being absorbed by the oral route (except where topical application is superior) and should remain at adequate levels in the blood for a reasonable period of time. Axiomatically perhaps, it should achieve wide distribution throughout the various tissues of the body.

Faced with such limitations very few organic compounds qualify as chemotherapeutic agents. Those that do largely are of microbial origin, although a few come from higher forms of life, a few are now made synthetically, and a number are actually hybrids of biochemical and laboratory syntheses. The latter are essentially laboratory modifications of existing antibiotics and such modified compounds are becoming increasingly necessary as new strains of microorganisms resistant to the parent antibiotic make their appearance. For example, in the cases of the penicillin and tetracycline antibiotics, substantially superior derivatives have been prepared by synthetic modification of the basic molecules. Again in the case of the rifamycins minor changes in structure led to a derivative that was not only less toxic to the host but also achieved much higher blood levels. Even here, however, there appears to be a limit because there is some evidence, albeit tenuous, that synthetic modification is only worthwhile when the microorganism(s) producing the antibiotic, is itself capable of producing a range of structural variants. When the substance produced is unique in structure, synthetic modification has failed, essentially, to produce a more effective antibiotic (e.g., in the cases of griseofulvin, fusidic acid, and chloramphenicol).

In this essay we are concerned only with the total synthesis of those antibiotics that are therapeutically useful (or whose derivatives are useful) and that are of microbiological origin. Also included are a number of drugs that look promising but are still under investigation. Table 1 lists alphabetically (within group classifications) the more important of the useful antibiotics *of known structure*, their microbial orgin, and their uses and indicates the status of the synthetic art surrounding them. A limited number of partial syntheses have also been documented where the chemistry is significant.

Antibiotics (e.g., emetine, the vinca alkaloids, and camptothecin) derived from higher plant life are not discussed, nor are antibiotics of nonbiogenetic origin such as the sulfonamides, the nitrofurans, or nalidixic acid.

Before proceeding to a discussion of the total syntheses accomplished so far it must be stated that only in rare cases has a totally synthetic product succeeded commercially, in supplanting the biosynthetically produced material. Generally this only happens when the molecule under consideration is extremely simple [such as aquamycin ($NH_2COC{=}CCONH_2$) used against rice blast] and can be produced in a few highly efficient and inexpensive steps. By comparison with life processes, laboratory methods are both clumsy and wasteful and rarely can be justified by the economics involved. Why then do chemists attempt the synthesis of what are for the most part highly complex and often unstable molecules? Part of the answer lies primarily

Table 1. Therapeutically Useful Antibiotics of Known Structure

Name	Source	Used Against	Total Synthesis
Basic Sugars			
Gentamicin	*Micromonospora purpurea*	Gram-positive and negative bacteria	Yes
Kanamycin	*Streptomyces kanamyceticus*	Gram-positive and negative bacteria; mycobacteria; leptospira	Yes
Kasugamycin	*Streptomyces kasugaensis*	Pyricularia oryzae (rice blast pathogen)	Yes
Lincomycin	*Streptomyces lincolnensis*	Gram-positive bacteria; mycobacteria	Yes
Neomycin-B	*Streptomyces fradiae*	Gram-positive and negative bacteria; mycobacteria	No
Paromomycin	*Streptomyces rimosus f. paromomycicus*	Gram-positive and negative bacteria; mycobacteria; amebic infection and murine leprosy	No
Spectinomycin	*Streptomyces spectabilis*	Gram-positive and negative bacteria	No
Streptomycin	*Streptomyces griseus*	Gram-positive and negative bacteria; mycobacteria	No
Streptozotocin	*Streptomyces achromogenes*	Gram-positive and negative bacteria; tumor cells	Yes
Macrolides			
Amphotericin-B	*Streptomyces nodosus*	Yeasts and fungi	No
Carbomycin	*Streptomyces halstedii*	Gram-positive bacteria rickettsias; protozoa; penicillin-resistant staphylococci	No
Chalcomycin	*Streptomyces bikiniensis*	Gram-positive and negative bacteria; mycobacteria	No
Erythromycin	*Streptomyces erythreus*	Gram-positive bacteria; rickettsias; protozoa; penicillin-resistant staphyloccoci	No
Filipin	*Streptomyces filipinensis*	Yeasts and fungi	No
Nystatin	*Streptomyces albulus*	Yeasts and fungi (mainly topical use)	No

Table 1. (continued)

Name	Source	Used Against	Total Synthesis
Oleandomycin	*Streptomyces antibioticus*	Gram-positive bacteria; rickettsias; protozoa; penicillin-resistant staphylococci	No
Spiramycin	*Streptomyces ambofaciens*	Gram-positive bacteria; rickettsias	No
Tylosin	*Streptomyces fradiae*	Gram-positive bacteria including mycobacteria	No
Nucleosides			
Angustmycin-A	*Streptomyces hygroscopicus*	Primarily mycobacteria	Yes
Blasticidin-S	*Streptomyces griseochromogenes*	*Piricularia oryzal*	No
Nucleocidin	*Streptomyces calvus*	Gram-positive and negative bacteria; tripanosomes	No
Polyoxins	*Streptomyces cacaoi var. asoensis*	*Piricularia oryzal*	No
Psicofuranine	*Streptomyces hygroscopicus*	Tumor cells	Yes
Puromycin	*Streptomyces alboniger*	Gram-positive bacteria; tumor cells; protozoa	Yes
Racemomycins-A and O	*Streptomyces racemochromogenus*	Gram-positive and negative bacteria; mycobacteria	No
Sangivamycin	*Unidentified streptomyces*	Leukemia cells	Yes
Septicidin	*Streptomyces fimbriatus*	Tumor cells; fungi	No
Toyocamycin	*Streptomyces toyocaensis*	Tumor cells	Yes
Tubercidin	*Streptomyces tubercidicus*	Tumor cells; mycobacteria	Yes
Penicillin group			
Cephalo-sporin-C	*Cephalosporium* species	Penicillin-resistant organisms, especially staphylococci	Yes
Penicillin	*Penicillium notatum*	Gram-positive bacteria; spirochetes; actinomycetes	Yes
Polypeptides and depsipeptides			
Actinomycin-D	*Streptomyces antibioticus*	Tumors	Yes

Table 1. (continued)

Name	Source	Used Against	Total Synthesis
Bacitracin	*Bacillus subtilis*	Gram-positive bacteria; amebae (mainly used topically)	No
Gramicidin	*Bacillus brevis*	Gram-positive bacteria (mainly topical application)	Yes
Polymixin-B_1	*Bacillus polymixa*	Gram-negative bacilli except proteus	Yes
Staphylomycin	*Streptomyces fradiae*	Gram-positive and mycobacteria	No
Tyrocidine	*Bacillus brevis*	Gram-positive bacteria (mainly topical application)	Yes
Tetracycline group			
Chlorotetra-cycline	*Streptomyces aureofaciens*	Gram-positive and negative rickettsias; large viruses	No
Oxytetra-cycline	*Streptomyces rimosus*	Gram-positive and negative bacteria; rickettsias; large viruses	Yes
Miscellaneous			
Anthramycin	*Streptomyces refuineus*	Tumor cells	Yes
Chloram-phenicol	*Streptomyces venezuelae*	Gram-positive and negative bacteria; rickettsias; large viruses	Yes
Cycloserine	*Streptomyces orchidaceus*	Mycobacteria	Yes
Coumermycin	*Streptomyces rishiriensis*	Gram-positive bacteria including mycobacteria; gram-negative bacteria	No
Cycloheximide	*Streptomyces griseus*	Phytopathic fungi; tumor cells	Yes
Daunomycin	*Streptomyces peucetius*	Tumor cells	No
Fumagillin	*Aspergillus fumigatus*	Amebae	No
Fusidic acid	*Fusideum coccineum*	Gram-positive bacteria; mycobacteria	No
Griseofulvin	*Penicillium griseofulvin*	Mycotic infections of hair, nails and skin	Yes
Mitomycin-C	*Streptomyces caespitosus*	Tumor cells	No

Table 1. (Continued)

Name	Source	Used Against	Total Synthesis
Monensin	*Streptomyces cinnamonensis*	Antibacterial (under investigation)	No
Novobiocin	*Streptomyces niveus*	Gram-positive and negative bacteria; penicillin-resistant staphylococci	Yes
Rifamycin-B	*Streptomyces mediterranei*	Primarily mycobacteria; gram-positive bacteria	No
Streptonigrin	*Streptomyces flocculus*	Tumor cells	No
Strepto-vitacin-A	*Streptomyces griseus*	Fungi; protozoa; tumor cells	No

in the challenge itself and secondarily in the attendant satisfaction and status that comes with success—the same reasons that led Hillary to attempt the scaling of Everest. Total synthesis is one of the highest aims a chemist can aspire to. As Woodward[2] has said: "There is excitement, adventure and challenge and there can be great art in organic synthesis." Apart from the human side, very few syntheses of complex molecules are accomplished without the concomitant development of either new ideas and/or new synthetic methods. These reasons alone are sufficient justification. In a few instances total syntheses are undertaken to provide a structure proof for a compound, but now that spectroscopic methods are so highly developed authentic cases of this are rare.

In the discussion that follows the antibiotics are grouped where possible, as they are in Table 1, according to structural similarity. However, the first to be considered is the penicillin group not only because of its importance but because of the prime position penicillin itself occupies in the history of antibiotics.

2. THE PENICILLIN GROUP

The term "penicillin" usually refers to compounds that have the general structure represented by **1** in which R is any acyl group. In the naturally occurring substances this acyl function is phenylacetyl (benzyl penicillin) or 5-carboxy–5-aminovaleroyl (penicillin N). Variations in the acyl function can be achieved in a limited number of cases by the addition of appropriate alcohols or acids to the fermentation medium. However, the development of both synthetic and biochemical methods[3] of preparing 6-aminopenicillanic

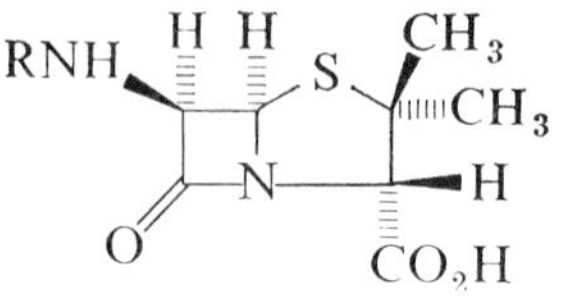

1

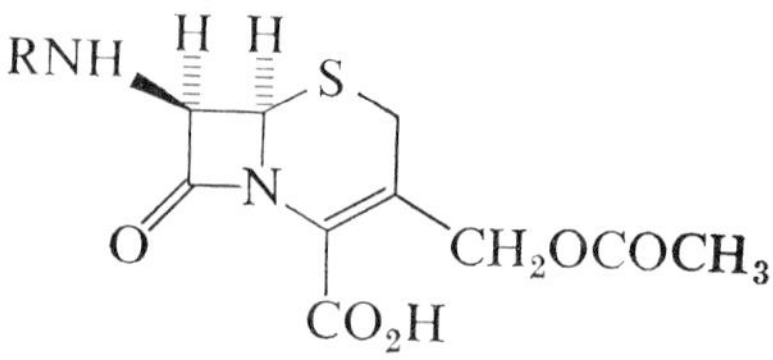

2

acid (**1**; R = H) has permitted access to a wide variety of penicillins with differing acyl groups.

From a therapeutic standpoint this has been of immense value since many of these new compounds are much move effective because of their resistance to cleavage by penicillinase. The latter is an enzyme, generated by some microorganisms, which inactivates the simpler penicillins by cleavage to **1** (R = H), which is itself only weakly antibiotic. The newer compounds[4] include methicillin (R = 2,6-dimethoxybenzoyl), ampicillin, (R = α-aminophenylacetyl), and nafcillin, (R = α-ethoxynaphthoyl), all of which are highly resistant to penicillinase. This is also true of the cephalosporins (**2**), the dominant naturally occurring form of which has R = D-5-amino–5-carboxyvaleroyl. The preferred[4] semisynthetic compound, cephaloridine, is the thienylacetyl derivative (2; R = [C_4H_3S]CH_2CO) of 7-aminocephalosporanic acid **2** (R = H).

A. The Penicillins

During World War II a prodigious amount of research[5] was devoted to solving the structure of penicillin and thereafter to attempting its synthesis. The major difficulty in the synthesis of penicillin lies in the fragility of the β-lactam. It has proved difficult to close such a ring and when closed, it is very easily cleaved again. In the cephalosporins the β-lactam is not quite so sensitive because the ring strain associated with it is not so great, by reason of its being attached to a six- rather than a five-membered ring.

The early wartime synthesis saw the preparation of benzyl penicillin (**12**) as its triethylammonium salt, in only minute yield.[6,7] The complete synthesis shown in Scheme 1 involves as the critical step the condensation of *D*-penicillamine **3** with 2-benzyl–4-methoxymethylene–5(4)oxazolone **4**.

The simplest and most practical synthesis that was devised[8] for the preparation of **3** involved the conversion of isobutyraldehyde to valine, by the Strecker reaction (60% yield) followed by chloroacetylation to give **5** in 86% yield. Treatment of **5** with acetic anhydride afforded the oxazolone **6** and the latter compound when treated with hydrogen sulfide in methanolic sodium methoxide yielded *N*-acetylpenicillamine (**7**). Acid hydrolysis then produced **3**, which was resolved as its *N*-formyl derivative by means of brucine.

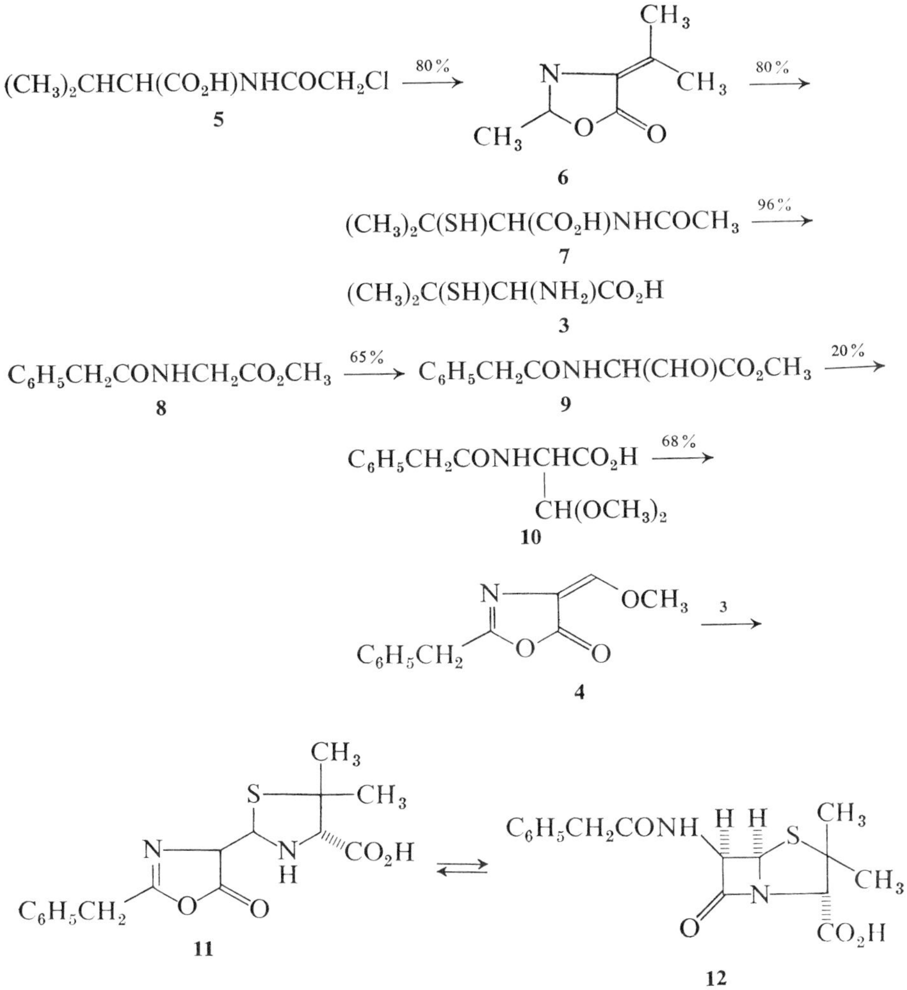

$$(CH_3)_2CHCH(CO_2H)NHCOCH_2Cl \xrightarrow{80\%} \mathbf{6} \xrightarrow{80\%}$$

5

6

$$(CH_3)_2C(SH)CH(CO_2H)NHCOCH_3 \xrightarrow{96\%}$$

7

$$(CH_3)_2C(SH)CH(NH_2)CO_2H$$

3

$$C_6H_5CH_2CONHCH_2CO_2CH_3 \xrightarrow{65\%} C_6H_5CH_2CONHCH(CHO)CO_2CH_3 \xrightarrow{20\%}$$

8 **9**

$$C_6H_5CH_2CONHCH(CH(OCH_3)_2)CO_2H \xrightarrow{68\%}$$

10

$$\mathbf{4} \xrightarrow{3}$$

4

$$\mathbf{11} \rightleftarrows \mathbf{12}$$

11 **12**

Scheme 1

The oxazolone **4** was prepared[9] by the acetic anhydride-induced cyclization of **10** itself obtained from **8** in three steps involving first formylation to give **9** followed then by ketalization and basic hydrolysis.

The condensation of **3** as its hydrochloride with **4** was carried out in pyridine in the presence of triethylamine at $\sim 0°$ and the reaction mixture was partitioned between chloroform and a phosphate buffer, after 10 minutes. The chloroform-soluble fraction was again subjected to treatment with a pyridine-triethylamine solution containing a small amount of pyridine hydrochloride and heated for 7 minutes at 130°. This was followed by a second phosphate

$NCH_2CO_2Bu^t$ $\xrightarrow{47\%}$ $NCHCO_2Bu^t$ (CHO) $\xrightarrow{3}$

13 14

N—CH—(CO—OBu^t) S, HN, CH_3, CH_3, CO_2H

15

$\left\{\begin{matrix} 24\%\ \alpha\text{-isomer} \\ 30\%\ \gamma\text{-isomer} \end{matrix}\right\}$ $\xrightarrow{82\%}$

NH_2CH (Bu^tO_2C) H S HN CH_3 CH_3 CO_2H $\xrightarrow{70\%}$

16

$C_6H_5OCH_2CONH$—CH (Bu^tO_2C) H S HN CH_3 CH_3 CO_2H $\xrightarrow{100\%}$

17

$C_6H_5OCH_2CONHCH$ (HO_2C) H S HN CH_3 CH_3 CO_2H $\xrightarrow{5\%}$

18

$C_6H_5OCH_2CONH$ H H S N O CH_3 CH_3 CO_2H

19

Scheme 2

buffer treatment after the pyridine-triethylamine mixture had been largely removed at low temperatures. Bioassay of the resulting material indicated a yield of ~0.07%. By extensive countercurrent partitioning and subsequent crystallization a pure sample (3.9 mg) of the triethylammonium salt of **12** was obtained. Its physical properties, including the infrared spectrum, were identical with those of the salt obtained from natural benzyl penicillin.

This synthetic approach was based on the thought that the base-catalyzed condensation would give an intermediate such as **11** which under the further influence of acid would cyclize to **12** to some degree. Other work had shown[10] that **12**, when treated with anhydrous HCl, gives the hydrochloride of **11**. Undoubtedly an equilibrium is involved that greatly favors **11**.

In 1957 a rational synthesis (depicted in Scheme 2) of penicillin V (**19**) was announced by Sheehan and Henery-Logan.[11] In their synthesis points worthy of note are (*a*) intermediates were used that would avoid azlactone formation and (*b*) again the construction of the β-lactam was delayed until the last step.

Formylation[12] of the phthalimide **13** afforded **14**, which, when treated with D-penicillamine in aqueous ethanol, yielded a mixture of two isomers (**15**), the more soluble of which (the α-isomer) corresponded to the natural product, in stereochemistry. When this isomer was split by hydrazine, **16** resulted. This was converted consecutively to **17** and then **18** by means of phenoxyacetyl chloride and anhydrous HCl, respectively. In the key step cyclization of **18** to penicillin V (**19**) was cleverly accomplished on the sodium salt in dilute aqueous dioxane by means of *N*,*N*-dicyclohexylcarbodiimide.

This synthesis was quickly followed by a general synthesis of penicillins[13] in which the key compound, 6-aminopenicillanic acid (**23**), was obtained as shown in Scheme 3. The amino group of **20** was first protected by means of a trityl function and the product, **21**, was cyclized to the β-lactam **22**, using *N*,*N*-diisopropylcarbodiimide. Saponification of **22** afforded the corresponding carboxylic acid (17% yield) and detritylation was accomplished in 32% yield by means of dilute hydrochloric acid. The product was identical with 6-aminopenicillanic acid (**23**), a compound that can be acylated to produce any desired penicillin as noted previously.

Recently an ingenious attempt[14] has been made to construct the β-lactam ring at an even earlier stage in the synthesis (Scheme 4). This involved the addition of azidoketene (from azidoacetyl chloride and triethylamine) to the thiazoline **24**. Unfortunately not only was the yield of product low but the compound proved to be the 6-epi-isomer **25**. Catalytic reduction of **25** gave **26** in moderate yield, which when acylated with phenoxyacetyl chloride afforded 6-epipenicillin V methyl ester (**26a**).

A number of other approaches to the penicillin nucleus have been reported[15,16] but as yet no new total synthesis has emerged.

H

$HClNH_2$—CH— S CH_3

| HN CH_3 ⟶

HO_2C CO_2CH_3

20

H

S CH_3

$(C_6H_5)_3CNH$—CH— CH_3 25% ⟶

HO_2C N CO_2CH_3

H

21

H H S CH_3 NH_2 H H S CH_3

$(C_6H_5)_3CNH$— CH_3 ⟶ CH_3

O N CO_2CH_3 O N CO_2H

22 23

Scheme 3

CH_3 N_3 H H S CH_3

N_3CH_2 S

| + CH_3 5–8% ⟶ CH_3 ⟶

COCl N CO_2CH_3 O N CO_2CH_3

24 25

H H

RHN S CH_3

CH_3

O N CO_2CH_3

26 R = H

26a R = $C_6H_5OCH_2CO$

Scheme 4

B. The Cephalosporins

The total synthesis of cephalosporin C **41** has been accomplished directly by Woodward and his associates[17] and the successful conversion of the penicillin nucleus to the cephalosporin nucleus has been reported by the Lilly group.[18–20]

In the total synthesis of cephalosporin the major difficulty that had to be overcome was the ease with which the cephalosporin analogs of penicillanic acid decarboxylate. The problem was avoided by using methods that avoided the generation of the free acid group. A critical intermediate was the β-lactam

33 and for its preparation (Scheme 5) the starting point chosen was the

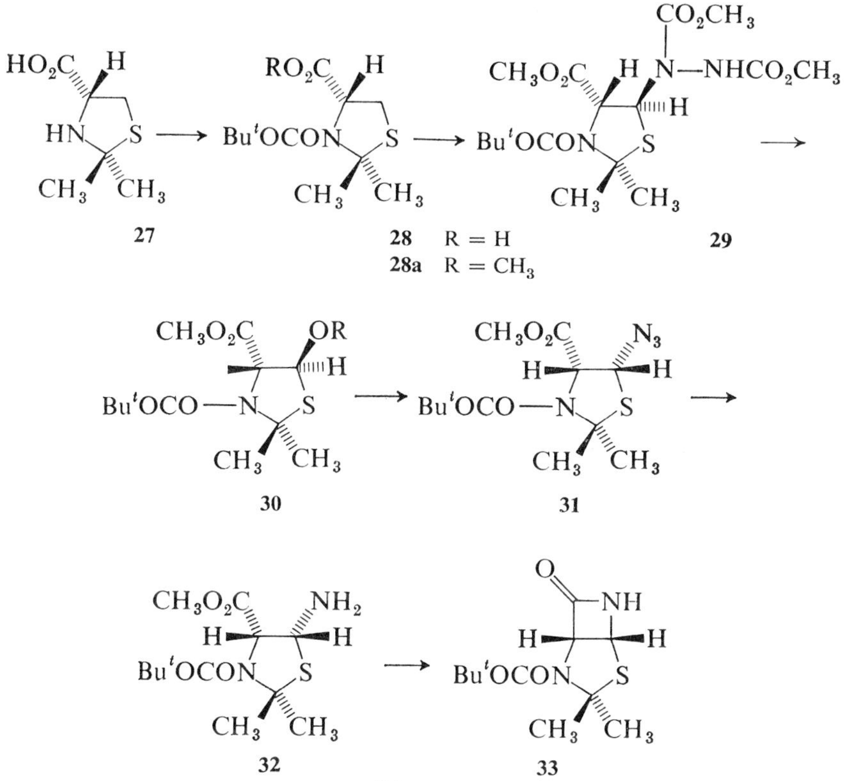

28 $R = H$
28a $R = CH_3$

Scheme 5

thiazolidine **27**, previously prepared[21] from L-(+)-cysteine. Treatment of **27** with *t*-butyloxycarbonyl chloride to protect the nitrogen atom afforded **28**, whose methyl ester **28a** underwent attack at the CH_2 group by dimethyl azodicarboxylate to give **29**. Oxidation of **29** by means of lead tetraacetate followed by treatment of the reaction mixture for 24 hours with sodium acetate under anhydrous conditions led to the *trans*-hydroxy ester **30** ($R = H$). The corresponding tosylate (**30**; $R = C_7H_7SO_2$) prepared *in situ* afforded the *cis*-azide **31** when treated with azide ion in water. Reduction of **31** at $-15°$ by aluminum amalgam gave the *cis*-amino ester **32**, which in a novel way was converted, by means of triisobutylaluminum in toluene, to the β-lactam **33**. The structures of both **32** and **33** were confirmed by X-ray crystallography at this stage.

In a parallel series of experiments (Scheme 6) di-β,β,β-trichloroethyl *d*-tartrate was oxidized by sodium metaperiodate to the glyoxylate hydrate

$CO_2CH_2CCl_3$
$CH(OH)_2$

34

$CO_2CH_2CCl_3$
CHOH
H
O H O

35

$CO_2CH_2CCl_3$
CH
C
OHC CHO

36

$CO_2CH_2CCl_3$
H
O N H O H
H H O
Bu^tOCON S
CH_3 CH_3

37

$CO_2CH_2CCl_3$
O CHO
H N
NH_2 S
H

38

$CO_2CH_2CCl_3$
O R
H N
$Cl_3CCH_2O_2CCH(CH_2)_3CONH$ S
H
$NHCO_2CH_2CCl_3$

39 R = CHO
39a R = CH_2OCOCH_3

$CO_2CH_2CCl_3$
O CH_2OCOCH_3
N
$Cl_3CCH_2O_2CCH(CH_3)_3CONH$ S
$NHCO_2CH_2CCl_3$

40

CO_2H
O CH_2OCOCH_3
NH_2 N
$HO_2CC(CH_2)_3CONH$ S

41

Scheme 6

34. The latter when condensed with sodium malondialdehyde afforded **35**, which when added to distilling octane underwent dehydration giving the highly reactive dialdehyde **36**.

Condensation of **36** with the β-lactam **33** in *n*-octane at 80° led to the adduct **37**, which in trifluoroacetic acid was smoothly converted in a masterful step, to the aminoaldehyde **38**. Acylation of **38** to give **39** was accomplished using *N*-β,β,β-trichloroethylethoxycarbonyl-D-(−)-α-aminoadipic acid in the presence of dicyclohexylcarbodiimide. Reduction in tetrahydrofuran by diborane followed by acetylation then led to the *iso*-ester **39a**, which was equilibrated with the normal ester **40** by means of pyridine at room temperature, ($K_{normal}/K_{iso} = \frac{1}{4}$). Reduction of **40** by zinc dust in acetic acid then removed all three trichloroethyl groups to give cephalosporin C (**41**), thus completing what must be regarded as a brilliant synthesis. In a separate series of experiments cephaloridine (cephalothin) (2; R = $[C_4H_3S]CH_2CO$) was also synthesized.

The partial synthesis of cephalosporin V (**49**; R = H) (Scheme 7) from penicillin V involves as the key step[18] the rearrangement of the penicillin sulfoxide methyl ester **42** in boiling acetic anhydride. This gives rise, undoubtedly via the intermediate **43**, to two products, the major one of which was assigned structure **45** and the minor one **44**. Treatment of **44** with mild base afforded the desacetoxycephalosporin **46** (R = CH_3), which also could be obtained more directly by heating **42** with a trace of acid in an inert solvent.

For the purposes of completing[19] the conversion, **46** (R = CH_3) was hydrolyzed and converted to the *p*-methoxybenzyl ester (**46**; R = CH_3OC_6-H_4CH_2). The latter was brominated by *N*-bromosuccinimide and the crude product treated with potassium acetate to give **47** (R = $CH_3OC_5H_4CH_2$), a mixture of the Δ^2- and Δ^3-isomers.

Oxidation of this mixture with *m*-chloroperbenzoic acid smoothly gave *only* the Δ^3-sulfoxide **48** (R = $CH_3OC_6H_4CH_2$), thus providing a method of converting all of the cephalosporin material present to the biologically active Δ^3-isomer. Reduction of **48** with sodium dithionite-acetyl chloride in DMF led to **49** (R = $CH_3OC_6H_4CH_2$—), which when cleaved by trifluoroacetic acid afforded cephalosporin V (**49**; R = H).

In a second approach Spry[20] prepared the sulfoxide **50** (R = *N*-phthalimidyl) by oxidation of **45** (R = *N*-pththalimidyl). Ring expansion of **50** by means of acetic anhydride and an acid catalyst gave a mixture of **51** (R = *N*-phthalimidyl) and **52** (R = *N*-phthalimidyl) in 30% yield. Conversion of **52** to **51** could be accomplished by acid-catalyzed dehydration. By using the *p*-nitrobenzyl ester in this sequence to protect the carboxyl group the conversion of 6-aminopenicillanic acid to 7-aminocephalosporanic acid was also achieved.

C_6H_5OCONH H H O S CH_3 CH_3 N CO_2CH_3 O

42

AcO^-

$C_6H_5OCH_2CONH$ H H $\overset{+}{S}$ CH_2 CH_3 N CO_2CH_3 O

43

$C_6H_5OCH_2CONH$ H H S $OCOCH_3$ + N CH_3 O CO_2CH_3

44

$C_6H_5OCH_2CONH$ H H S CH_2OAc CH_3 N CO_2CH_3 O

45

$C_6H_5OCH_2CONH$ H H S N CH_3 O CO_2R

46

$C_6H_5OCH_2CONH$ H H S N CH_2OAc O CO_2R

47

Scheme 7

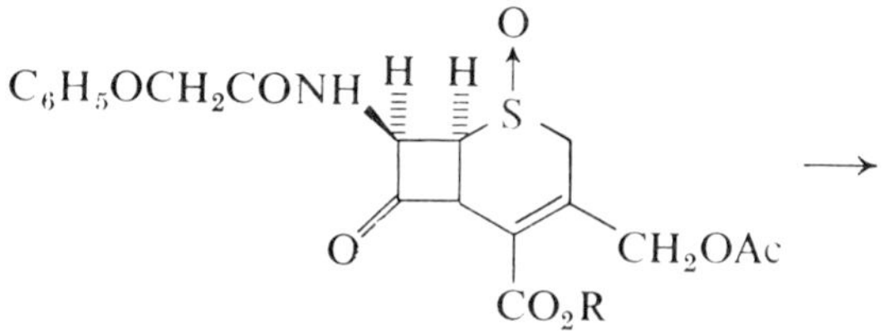

48

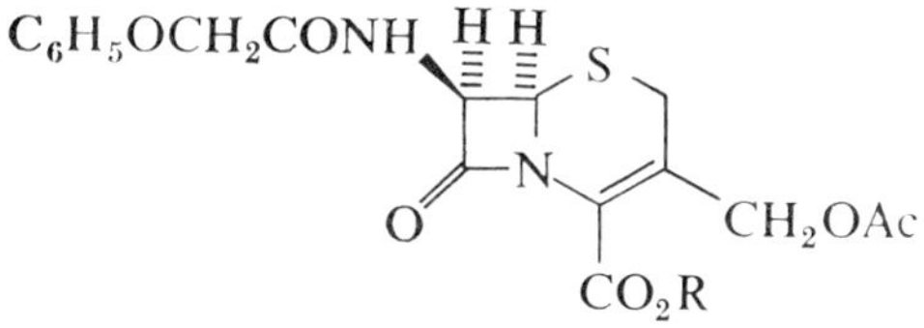

49

Scheme 7 (Continued)

An interesting approach to the cephalosporin nucleus has been published by French workers[22] and although it has not led to the synthesis of a naturally

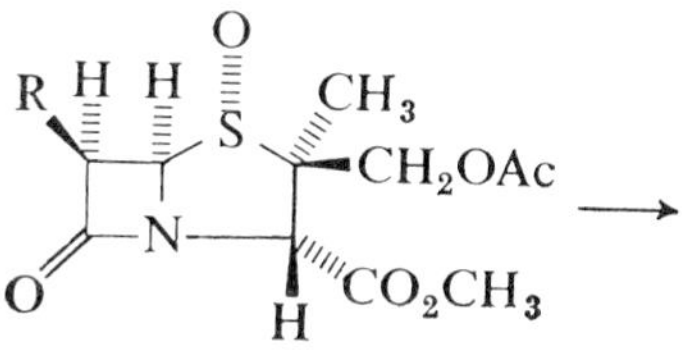

50

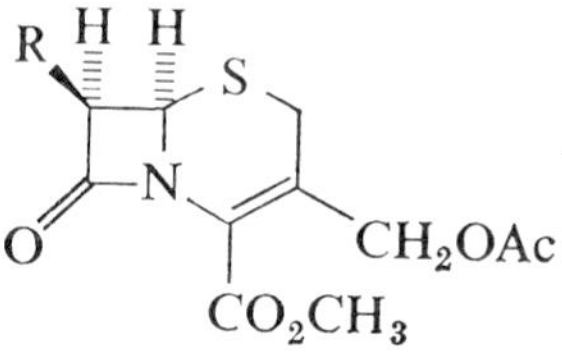

51

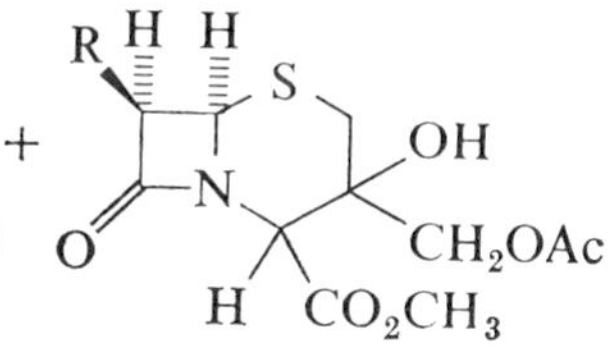

52

occurring compound, it is worthy of inclusion. It is predicated on the work of Sheehan and is outlined in Scheme 8.

The butenolide[23] **53** was converted to **54** by means of thiolacetic acid and **54** then was hydrolyzed under acid conditions to the unstable thiol **55**. Reaction of **55** with the phthalimido derivative **56** (itself prepared by the action of ammonium acetate on the corresponding formyl derivative) afforded the dihydrothiozine **57**. Removal of the phthaloyl group followed by tritylation of the amine led to **58**. The carboxyl group was then liberated

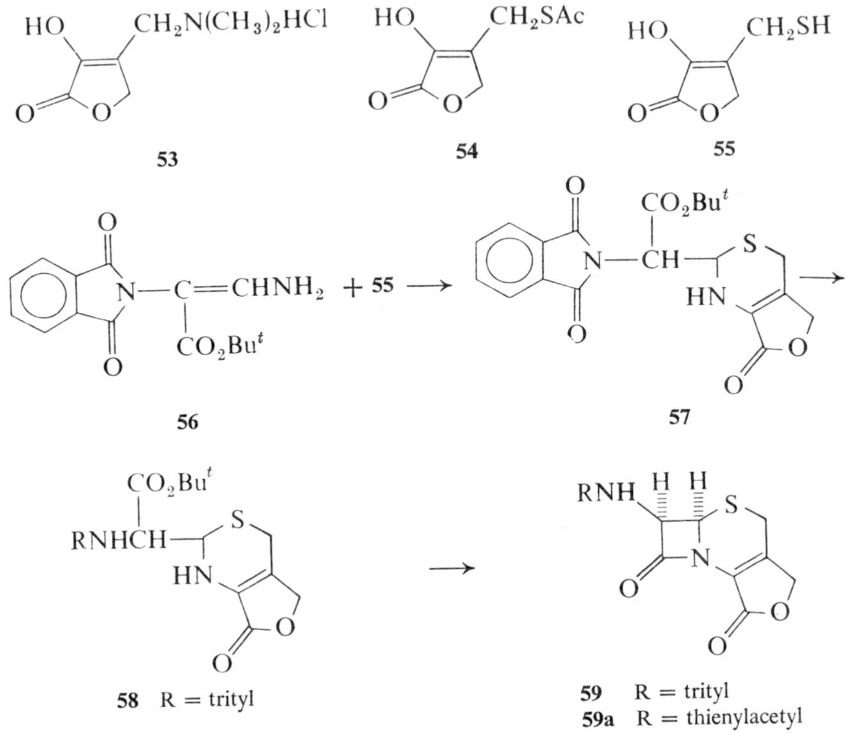

53

54

55

56

57

58 R = trityl

59 R = trityl
59a R = thienylacetyl

Scheme 8

by acid cleavage and the product treated with dicyclohexylcarbodiimide in nitromethane to give **59** in 70% yield. Standard procedures were then used to convert this product to **59a**, which proved to be identical with a sample prepared from cephalothine.[24]

3. THE TETRACYCLINES

The tetracyclines constitute an extremely useful and effective group of antibiotics. They are widely used in medical practice against a large variety of infective agents.

The basic hydronaphthacene structure is shown in **60** and investigations into structure-activity relationships have revealed that considerable variation in R_1, R_2, R_3, and R_4 can be made without much loss in its antibiotic activity. The naturally occurring derivatives of tetracycline (**60a**) are 7-chlorotetracycline (**60b**) and 5-hydroxytetracycline (**60c**). Two others are produced by a

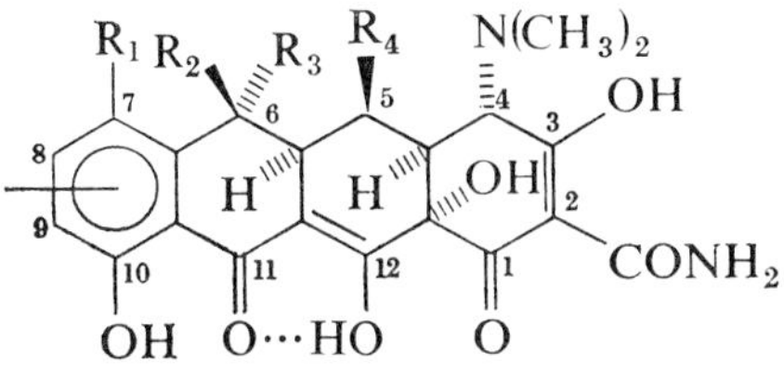

60a $R_1 = R_4 = H$; $R_2 = OH$; $R_3 = CH_3$
60b $R_1 = Cl$; $R_2 = OH$; $R_3 = CH_3$; $R_4 = H$
60c $R_1 = H$; $R_2 = R_4 = OH$; $R_3 = CH_3$
60d $R_1 = R_3 = R_4$; $R_2 = OH$
60e $R_1 = Cl$; $R_2 = OH$; $R_3 = R_4 = H$
60f $R_1 = R_2 = R_3 = R_4 = H$

mutant strain of the original *Streptomyces aureus*. These are 6-demethyltetracycline (**60d**) and 7-chloro–6-demethyltetracycline (**60e**) $R_1 = Cl$; $R_2 = OH$; $R_3 = R_4 = H$). The latter two compounds are very resistant to both acid and base degradation, thus making them valuable for oral use. Other semisynthetic derivatives are also in common use.

The most formidable of the synthetic problems posed by the tetracycline molecule have been discussed by Woodward.[26,27] They lie principally in ring *A*: every carbon atom of this ring bears at least one substituent and three of the potentially six asymmetric centers of the molecule fall in the consecutive chain C-4, C-4*a*, and C-12*a*. Added to these difficulties apart from the three remaining asymmetric centers is the fact that the most highly substituted derivatives having both methyl and hydroxyl groups at C-6 are very sensitive to acid and base. These obstacles have not discouraged synthetic efforts and to date three groups have reported success.

A. 6-Deoxy–6-Demethyltetracycline

The first success was reported by the Pfizer-Harvard group who in 1962 described[26,27] the total synthesis of *dl*-6-deoxy–6-demethyltetracycline (**60f**). Their synthesis was accomplished in two phases. The first phase constituted the development of a reasonably efficient route to the intermediate **61**, which

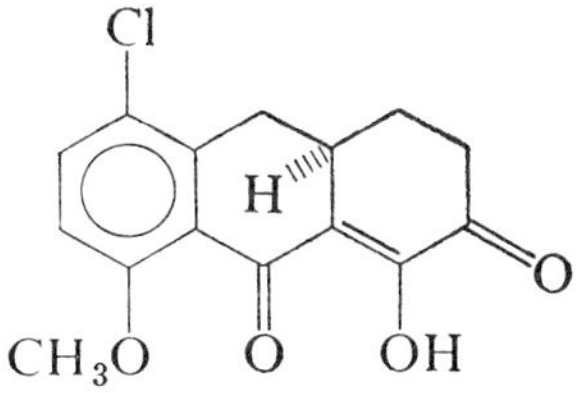

61

was prechosen on the basis of its expected chemical versatility for phase two, the elaboration of ring *A*.

The synthesis of **61**, shown in Scheme 9, began with the base-catalyzed condensation of methyl 3-methoxybenzoate with dimethyl succinate or with

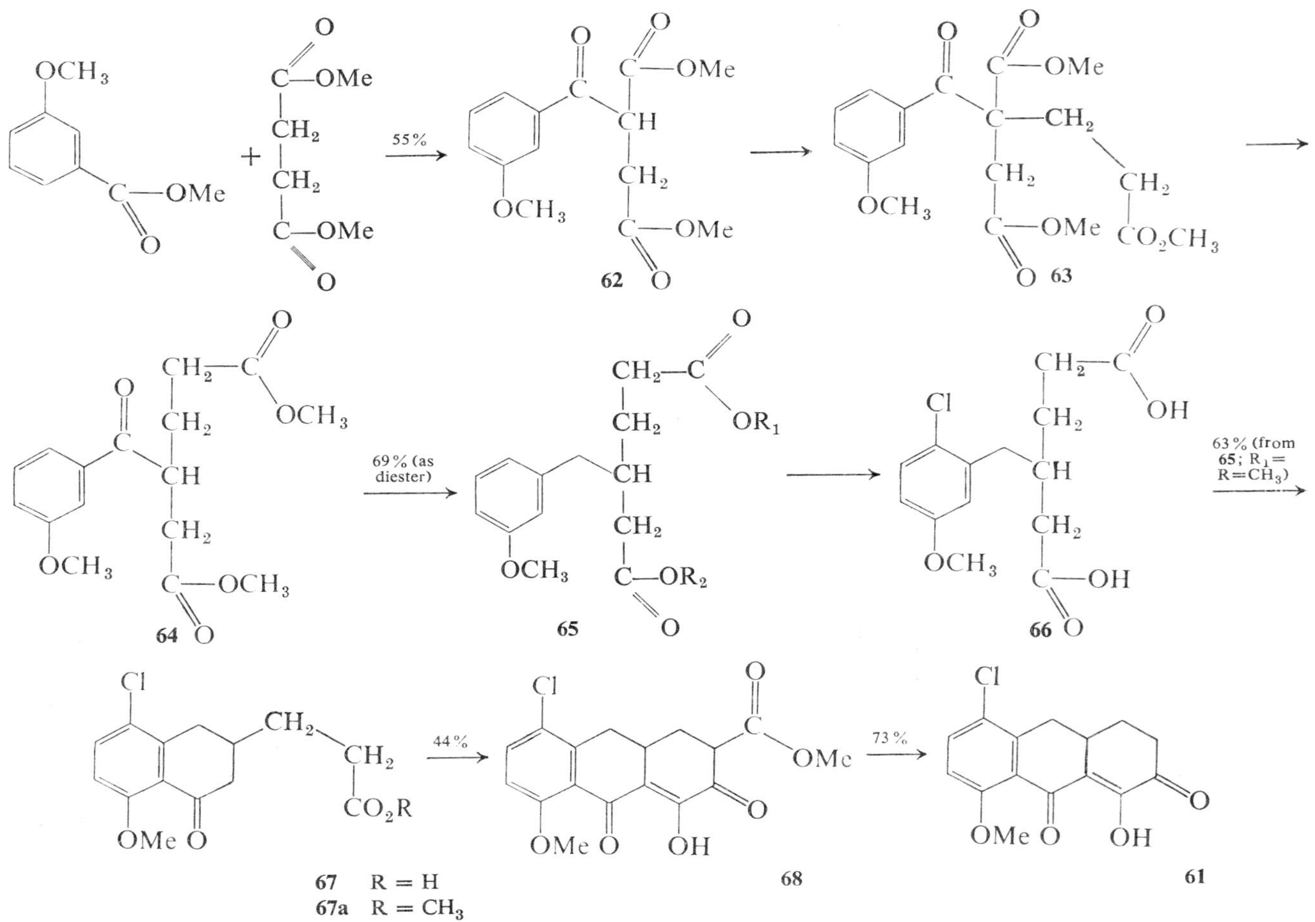

55%

62

63

69% (as diester)

64

65

66

63% (from **65**; $R_1 = R = CH_3$)

44%

73%

67 $R = H$

67a $R = CH_3$

68

61

Scheme 9

methyl acetate followed by alkylation of the intermediate β-ketoester with methyl bromoacetate. The product **62** was then condensed with methyl acrylate in a Michael reaction, using Triton-B as the catalyst, and afforded the keto-triester **63**. The latter material was boiled with aqueous sulfuric acid to extrude the tertiary carbomethoxy group then reesterified to give the β-aroyladipate **64**. Hydrogenolysis of the keto group of **64** using a palladium catalyst gave by way of an intermediate lactone the half acid-half ester **65** ($R_1 = H$; $R_2 = CH_3$) which was purified by esterification and distillation of the resulting diester (**65**; $R_1 = R_2 = CH_3$). Saponification then gave the pure diacid (**65**; $R_1 = R_2 = H$), which was chlorinated at the 6-position of the aromatic ring **66**, thus ensuring that in the upcoming cyclization step, mediated by hydrofluoric acid, ring formation would give exclusively **67**. The most crucial step in the sequence leading to **61** was the condensation of dimethyl oxalate with **67a** to give **68**. This was accomplished only after much experimentation and in view of the waywardness of the condensation must represent a kinetically controlled reaction. Success was achieved by using two equivalents of dimethyl oxalate almost four equivalents of sodium hydride and one equivalent of methanol in dimethyl formanide solution. Hot aqueous acid then smoothly transformed **68** into the much sought after triketone **61** and set the stage for the second phase (Scheme 10) of the synthesis.

Condensation of **61** with *n*-butyl glyoxylate in the presence of magnesium methoxide occurred smoothly to give **69**. Treatment of **69** at $-10°$ with dimethylamine afforded the Mannich base **70**, which easily lost dimethylamine and consequently was reduced immediately after its preparation at $-70°$ by sodium borohydride in dimethoxymethane. The product **71** was lactonized by means of a trace of acid in boiling toluene and the resulting compound **72** was reduced to **73** by means of zinc dust in formic acid (reaction time 1 minute) and then further to **74** by hydrogenation over a palladium catalyst.

The final steps in the formation of ring *A* were accomplished by condensing the mixed isopropyl carbonic anhydride **75** with a new derivative of malonic acid, ethyl *N*-*t*-butylmalonamate (**76**), in the presence of magnesium ethoxide in acetonitrile. The product **77** was difficult to purify and in the crude state was treated briefly with sodium hydride in dimethylformamide containing methanol. This afforded the tetracyclic derivative **78** as a crystalline material which when treated with hot aqueous hydrogen bromide afforded the demethylated primary amide **79**. The final step, the introduction of the 12a-hydroxyl group, was accomplished by means of carefully controlled oxygenation of **79** in the presence of cerous chloride in a buffered dimethylformamide-methanol solution. The product **60f** was isolated only after an extensive purification procedure, but its identity was established beyond doubt by comparisons with the behavior of an optically active specimen derived from natural sources.

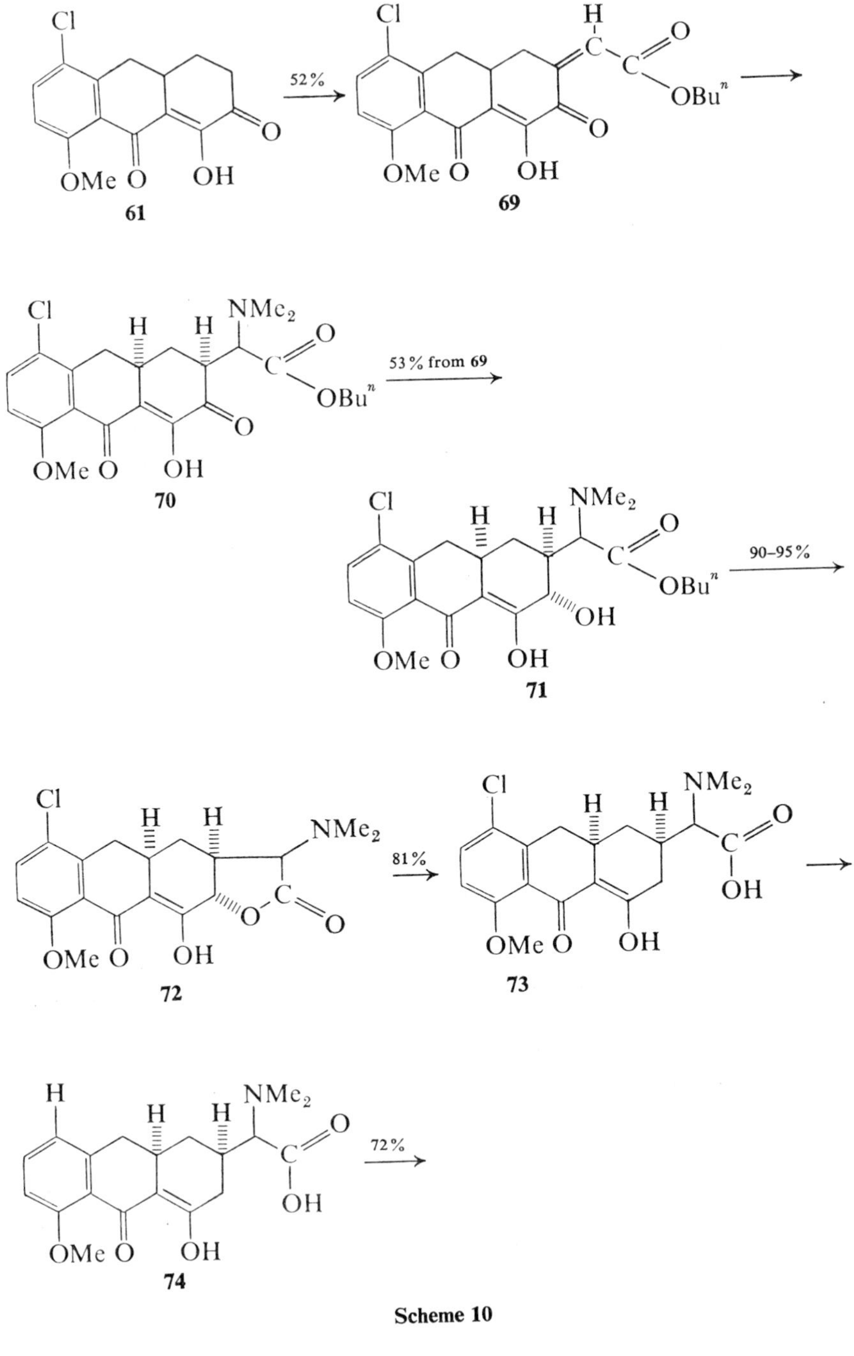

Cl, OMe, O, OH, O — **61**

52% →

Cl, H, C, C, O, OBu^n, OMe, O, OH, O — **69** →

Cl, H, H, NMe$_2$, O, C, OBu^n, O, OMe, O, OH — **70**

53% from **69** →

Cl, H, H, NMe$_2$, O, C, OBu^n, OH, OMe, O, OH — **71**

90–95% →

Cl, H, H, NMe$_2$, O, C, O, OMe, O, OH — **72**

81% →

Cl, H, H, NMe$_2$, C, O, OH, OMe, O, OH — **73** →

H, H, H, NMe$_2$, C, O, OH, OMe, O, OH — **74**

72% →

Scheme 10

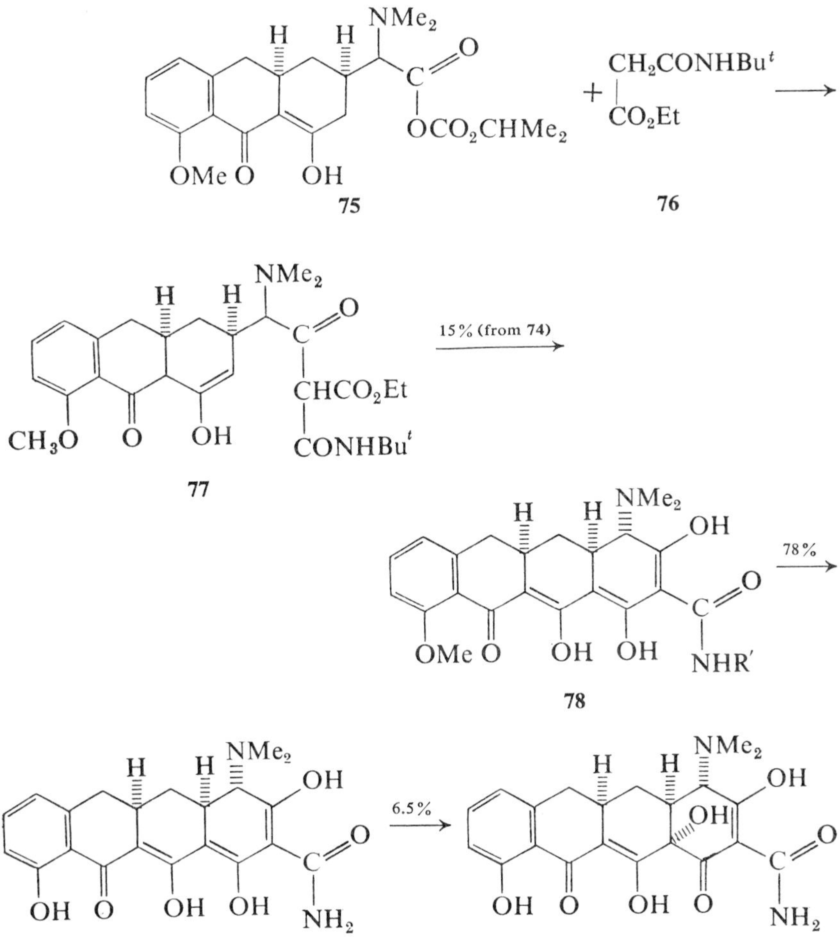

75 76

77

15% (from 74)

78

78%

79 60f

6.5%

Scheme 10 (Continued)

An alternate synthesis of *dl*-6-deoxy–6-demethyltetracycline has been accomplished by Muxfeldt and Rogalski.[28] They approached the problem by first constructing rings *C* and *D* and then simultaneously rings *A* and *B*. Their methods have the advantage of allowing the synthesis of tetracyclic compounds on a large scale, because of their relative simplicity.

The synthesis (Scheme 11) has as its starting point 1-chloro–2-bromo-methyl–4-methoxybenzene (**80**), which was used to alkylate the sodium salt of 1,1,2-*tris*(carbomethoxy)ethane. The resulting triester **81** was saponified to the corresponding acid which was decarboxylated at 160° to give the

Cl, Br, OCH_3

80

Cl, CO_2CH_3, C, CO_2CH_3, CH_2, CO_2CH_3, OCH_3

81

85% (from 80)

Cl, CO_2H, CO_2H, OCH_3

82

~100%

Cl, CO_2H, CH_3O, O

83

Cl, CHO, CH_3O, O, O

84

+ C=O, NH, CH_2CO_2H

Cl, N, ϕ, O, O, CH_3O, O, O

85

Cl, N, ϕ, O, O, CH_3O, O

86

+ H_3CO_2C, O, CONH-C(CH_3)$_3$

87

Cl, H, H, H, NHC, O, ϕ, OH, NHC(CH_3)$_3$, C, O, CH_3O, O, OH, O

88

Scheme 11

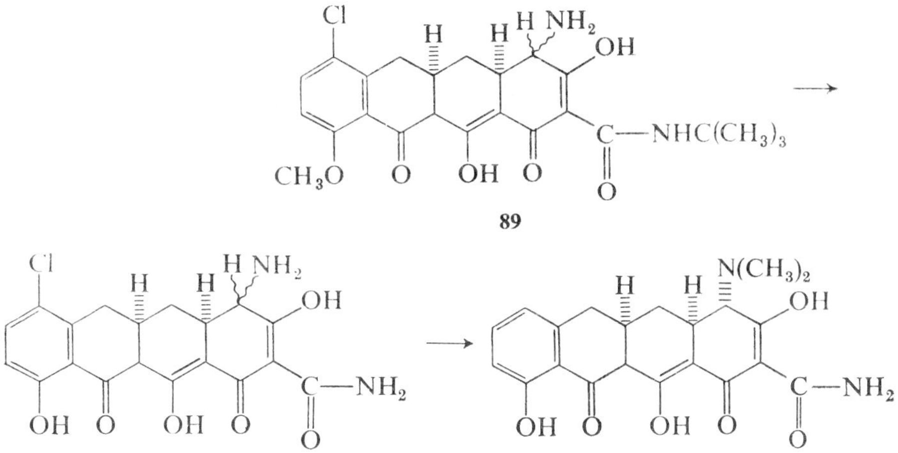

89

90 **79**

Scheme 11 (Continued)

succinic acid **82**. Polyphosphoric acid at 80° smoothly transformed **82** into **83**. This was then converted[29] to the aldehyde **84** in excellent yield by a standard series of reactions which included protection of the ketone as the dioxolane, conversion of the carbomethoxy group by a chain extension sequence to a CH_2CN group, then reduction of the latter to the aldehyde **84** with lithium aluminum hydride. Azlactone condensation of **84** with hippuric acid in the presence of lead acetate afforded **85**, which when carefully hydrolyzed with hydrochloric acid resulted in the formation of **86**, the key intermediate for the construction of rings *A* and *B*. This was accomplished in one step by condensing **86** with methyl *N*-*t*-butyl–3-oxoglutaramate (**87**) in the presence of two equivalents of sodium hydride in a mixture of ether and tetrahydrofuran. The product **88** proved to be a mixture of the two C-4 epimers, one of which could be isolated in a pure form. In order to remove the *N*-benzoyl group the mixture (**88**) was treated with triethyloxonium fluoborate and subsequently with aqueous acetic acid. This treatment yielded the pair of epimers **89**. Demethylation and debutylation were accomplished simultaneously by means of hot hydrogen bromide in acetic acid. The resulting epimeric mixture, **90**, was then methylated on the amine nitrogen by treatment with an excess of formaldehyde in methanol containing two equivalents of triethylamine and under the reducing conditions of a hydrogen atmosphere and a palladium catalyst. By this procedure, which also caused dehalogenation, **79** was obtained identical with the racemic material previously prepared by Conover et al.[26] Oxidation of **79** to 6-deoxy–6-demethyltetracycline (**60f**) was accomplished by means of oxygen and a freshly reduced platinum catalyst,[30] thus achieving the synthetic goal.

B. Tetracycline

Tetracycline (**60a**) formally has been synthesized by Russian workers.[31] In affect they synthesized racemic 12*a*-deoxy-5*a*,*b*-anhydrotetracycline **91**,

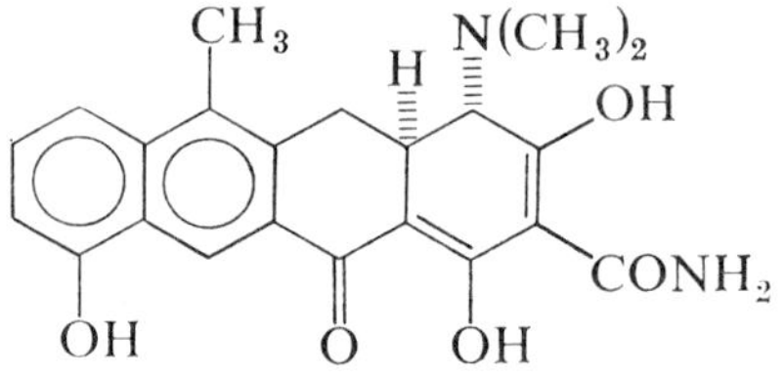

91

which in its levorotatory form had been converted previously, in two steps, to tetracycline itself. Their synthesis commenced with a tricyclic intermediate **92** containing basically rings *B*, *C*, and *D*, which originally had been synthesized by Inhoffen et al.[32] This was followed by modification of the functional groups of **92** to give the desired intermediate **93** on which was constructed the remaining elements of ring *A*.

The synthesis[32] of **92** is illustrated in Scheme 12. It was achieved in two steps from juglone **94**, first by a Diels-Alder reaction with 1-acetoxybutadiene, which gave dominantly isomer **95**, and then by reduction of the latter by one

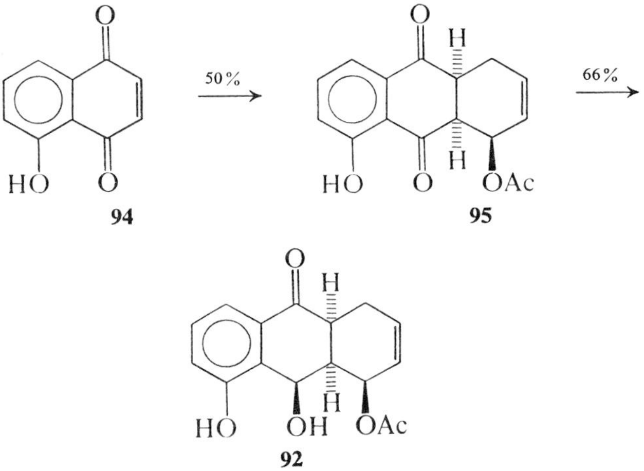

94 —50%→ **95** —66%→ **92**

Scheme 12

quarter equivalent of lithium aluminum hydride at $-60°$.

Modification[33] of **92** (Scheme 13) began by preparation of its benzyl ether **96**, which was then treated with excess methylmagnesium bromide

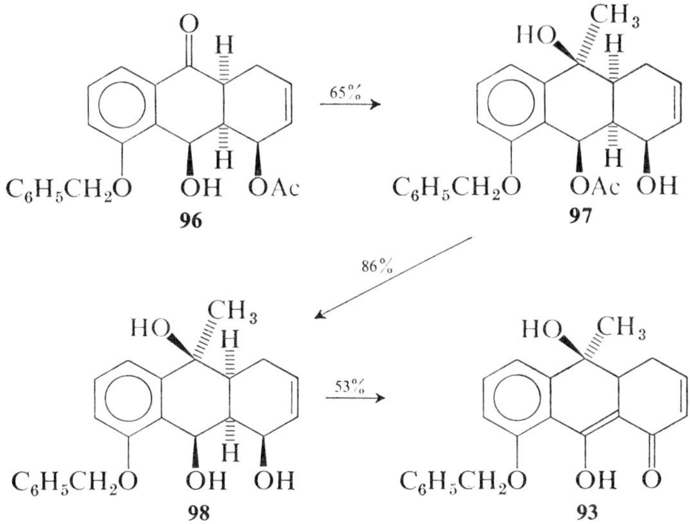

O H 65% HO CH$_3$ H

H H

$C_6H_5CH_2O$ OH OAc $C_6H_5CH_2O$ OAc OH

96 **97**

86%

HO CH$_3$ H HO CH$_3$

53%

H

$C_6H_5CH_2O$ OH OH $C_6H_5CH_2O$ OH O

98 **93**

Scheme 13

to give **97**, a base-catalyzed transfer of the acetate function also having taken place. Hydrolysis of **97** gave the triol **98**, which was easily converted to the desired dione **93**. Condensation of the diendiolone **93** (Scheme 14)

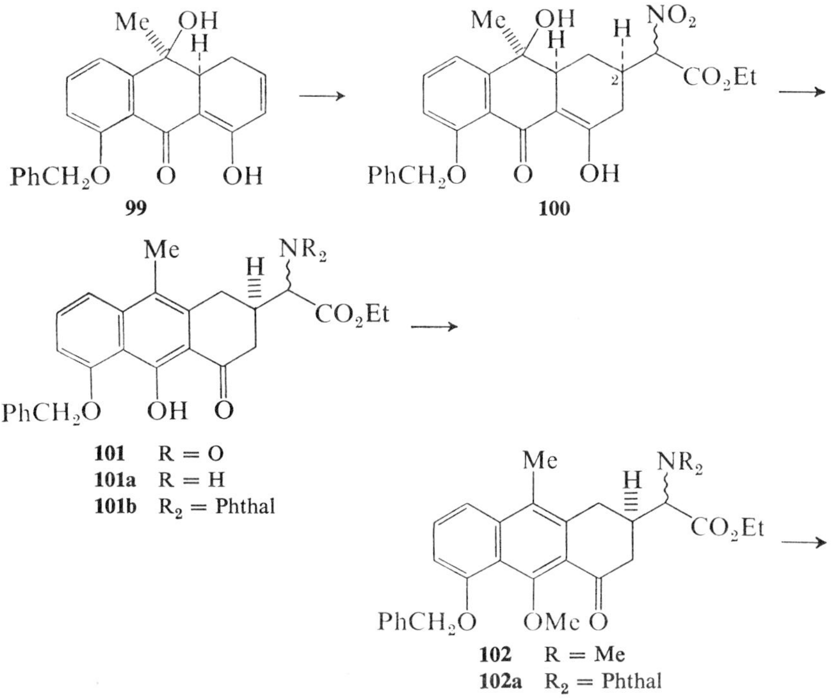

Me OH H Me OH H H NO_2

2 CO_2Et

$PhCH_2O$ O OH $PhCH_2O$ O OH

99 **100**

Me H NR_2

CO_2Et

$PhCH_2O$ OH O

101 R = O
101a R = H
101b R_2 = Phthal

Me H NR_2

CO_2Et

$PhCH_2O$ OMe O

102 R = Me
102a R_2 = Phthal

Scheme 14

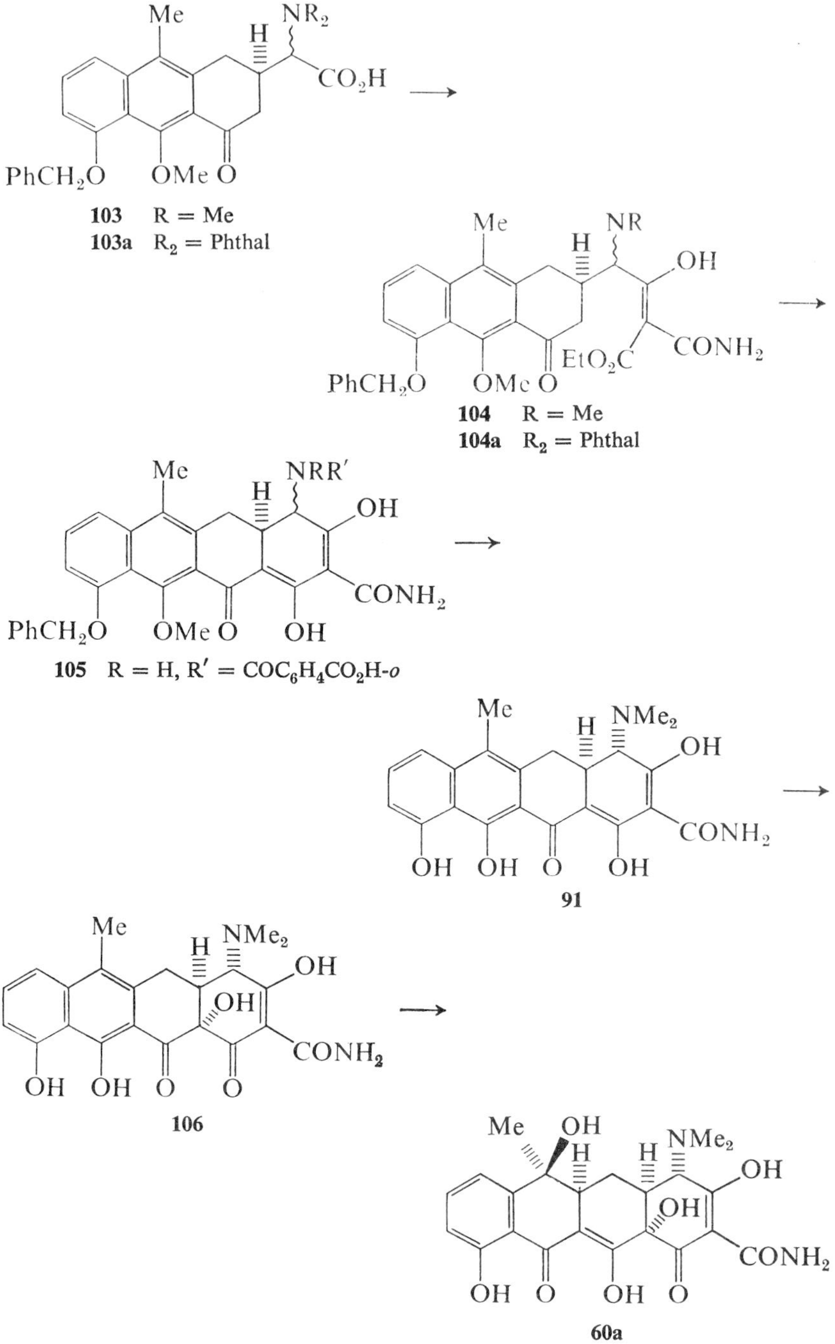

Me NR$_2$ H CO$_2$H

PhCH$_2$O OMe O

103 R = Me
103a R$_2$ = Phthal

Me NR H OH CONH$_2$ EtO$_2$C

PhCH$_2$O OMe O

104 R = Me
104a R$_2$ = Phthal

Me NRR′ H OH CONH$_2$

PhCH$_2$O OMe O OH

105 R = H, R′ = COC$_6$H$_4$CO$_2$H-*o*

Me H NMe$_2$ OH CONH$_2$

OH OH O OH

91

Me H NMe$_2$ OH OH CONH$_2$

OH OH O O

106

Me OH H H NMe$_2$ OH OH CONH$_2$

OH O OH O

60a

Scheme 14 (Continued)

with the triethylammonium salt of ethyl nitroacetate gave a mixture of two epimeric adducts **100**. These when treated with dilute alcoholic hydrogen chloride underwent dehydration and afforded a single (?) nitro compound **101**. Reduction of the latter by zinc in acetic acid then led to the amino ester **101a**.

At this stage the amino group was methylated with methyl iodide-silver oxide and the trimethylated product **102** hydrolyzed to the amino acid **103**. However, under no circumstances could this acid, as the chloride, as the isopropyl carbonate, or as the isobutyl carbonate, be condensed with the magnesium salt of ethyl malonamate to give **104**. In light of this the basic character of the amine function was for the time eliminated by the conversion of **101a** to the phthaloyl derivative **101b** by treatment with carboethoxy-phthalimide. Methylation as before then gave **102a**. When this compound was saponified by base, not only was the ester hydrolyzed, but the imide ring also was opened and had to be reclosed by heating in diglyme at 140°. The resulting acid **103a** was treated with phosphorus pentachloride in dimethylformamide to which was then added the ethoxymagnesium salt of ethyl malonamate. On this occasion the expected derivative **104a** was successfully obtained. Cyclization of this compound by means of dimsyl sodium in dimethyl sulfoxide then led to the tetracyclic derivative **105**, the imide ring again having opened. Hydrolysis by means of hot hydrogen bromide in acetic acid removed all of the protecting groups and when the resulting product was selectively methylated with methyl iodide in tetrahydrofuran the synthetic objective of the research 12*a*-deoxy-5*a*,6-anhydrotetracycline **91** was obtained.

Except for the question of resolution of **91**, this represents a formal synthesis of tetracycline because the conversion of **91** to **106** by oxygen and a platinum catalyst already had been reported[34] by the Russians themselves and the final step a photooxidation had been accomplished by Schach von Wittenau[35] using the procedure of Scott and Bedford.[36]

C. Oxytetracycline

The synthesis of racemic oxytetracycline or terramycin (**60c**), one of the most complicated and chemically sensitive of these antibiotics, was accomplished by Muxfeldt and his co-workers.[37] The synthesis is based on the general methods developed previously for this type of tetracyclic system and which had culminated[28] in the total synthesis of 6-deoxy-6-demethyltetracycline (**60**; $R_1 = R_2 = R_3 = R_4 = H$).

Oxytetracycline was assembled from three basic building blocks, the thiazolone **107**, methyl 3-oxoglutaramate (**108**), and the aldehyde **109**. The first two compounds are very easily prepared. The thiazolone **107** was

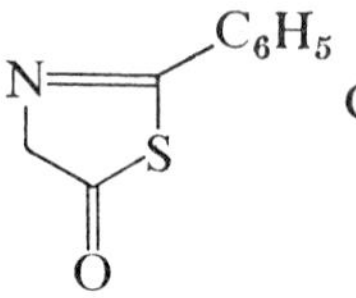

$CH_3O_2CCH_2COCH_2CONH_2$

107 108 109

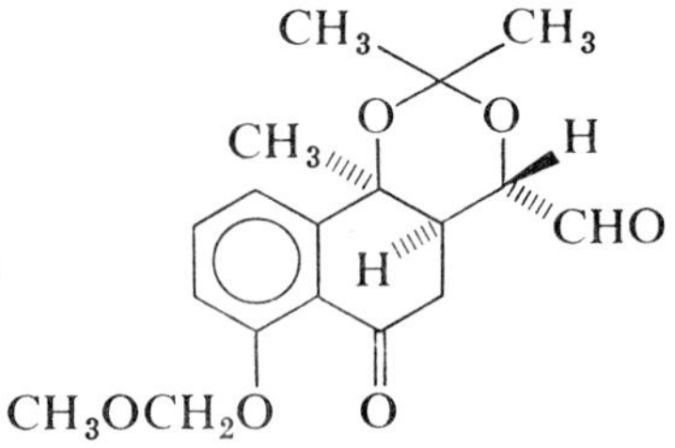

obtained as its hydrobromide by treatment of thiobenzoylglycine (**110**) with

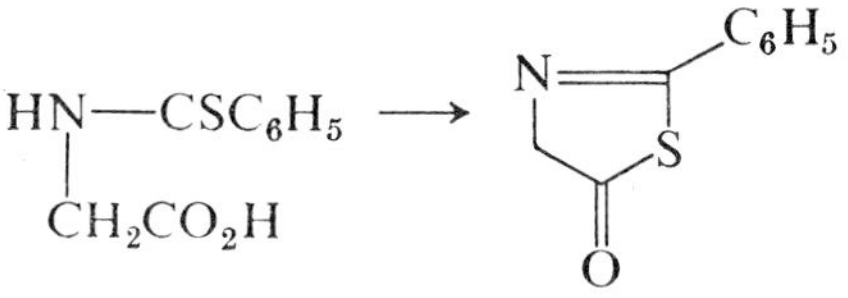

110 107

phosphorus tribromide. Neutralization with sodium acetate then produced the free base, which proved to be very unstable and was best used immediately.[38]

The keto-amide **108** was prepared in two steps from dimethyl 3-oxoglutarate first by careful treatment with ammonia, which led to methyl 3-aminoglutaconamate, followed by acid hydrolysis of the latter compound.

The synthesis (Scheme 15) of the aldehyde **109** had been achieved[39] in part at an earlier date from the product **110** of condensation of juglone acetate with 1-acetoxybutadiene. By contrast with juglone, which gives **95** (see above), juglone acetate gives as the major product of this Diels-Alder reaction the tetrahydroanthraquinone with the alcoholic oxygen at positions C-1 and C-5.

Treatment of **110** with methylmagnesium bromide results in selective attack at the C-9 carbonyl group to give product **111** with the desired stereochemistry. This selectivity is undoubtedly due to participation of the C-1 acetate function as an *ortho* ester magnesium salt. Basic hydrolysis of **111** also caused isomerization at C-4*a* and afforded the triol **112**, which was converted to the ketal **113** by means of acetone in the presence of anhydrous copper sulfate. The cyclohexene ring was next degraded by glycolation of the double bond by means of a potassium chlorate/osmium tetroxide combination, followed by lead tetraacetate cleavage of the product to give the dialdehyde **114**. Triethylammonium acetate then was used to catalyze the internal condensation of **114** to produce the unsaturated aldehyde **115**. Ozonolysis of **115** followed by treatment of the crystalline ozonide with

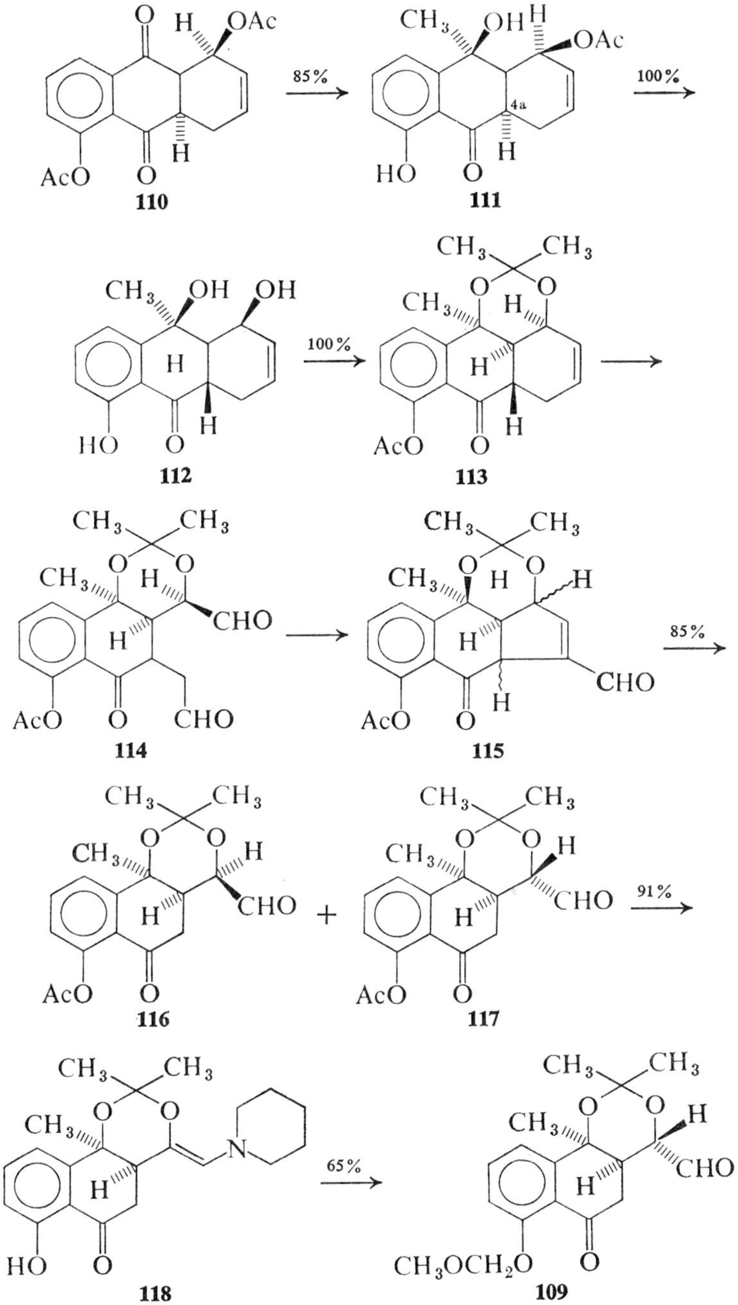

Scheme 15

aqueous sodium carbonate yielded a mixture of two aldehydes **116** and **117**, both of which could be isolated in a crystalline condition. That they are simply isomers about the carbon atom bearing the aldehyde group was shown by deuterium studies. In any event this mixture was treated with piperidine in boiling benzene and a single product, the enamine **118**, was isolated, hydrolysis of the phenolic acetate function having occurred concurrently. The sodium salt of **118** was selectively alkylated with chloromethyl ether on the phenolic oxygen atom to give an enamine, which when adsorbed on deactivated silica gel underwent selective hydrolysis of the amine function and afforded the desired intermediate **109** as an oil. This hydrolysis was not only selective but also stereospecific since **109** had an NMR spectrum consistent only with the structure depicted and in addition when treated with acetic acid regenerated **117** in almost quantitative yield.

The synthesis of oxytetracycline was now completed as shown in Scheme 16. The condensation of **109** with **107** occurred without any isomerization

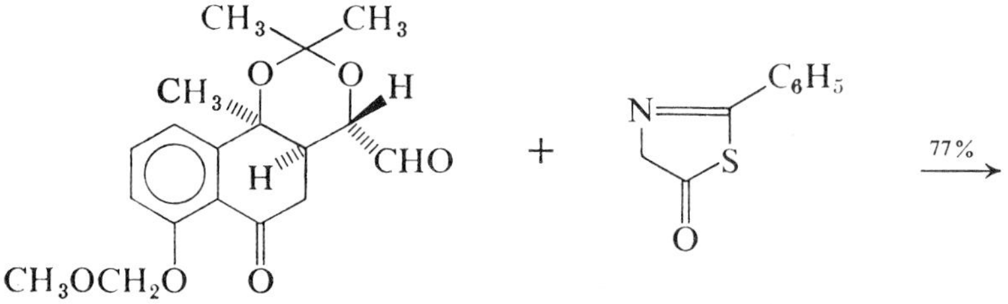

109 + **107** $\xrightarrow{77\%}$

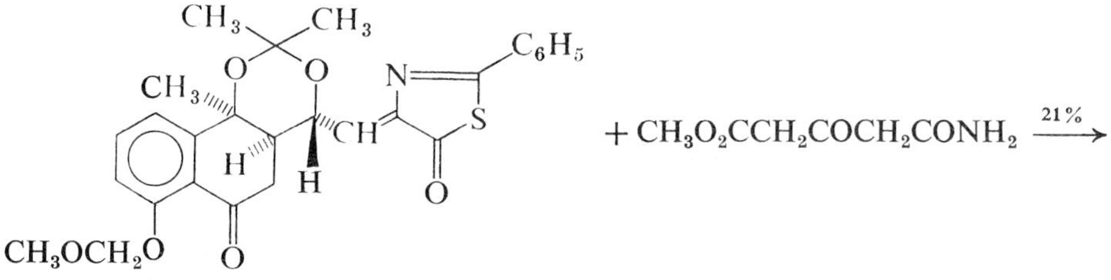

119 + $CH_3O_2CCH_2COCH_2CONH_2$ (**108**) $\xrightarrow{21\%}$

120 $\xrightarrow{32\%}$

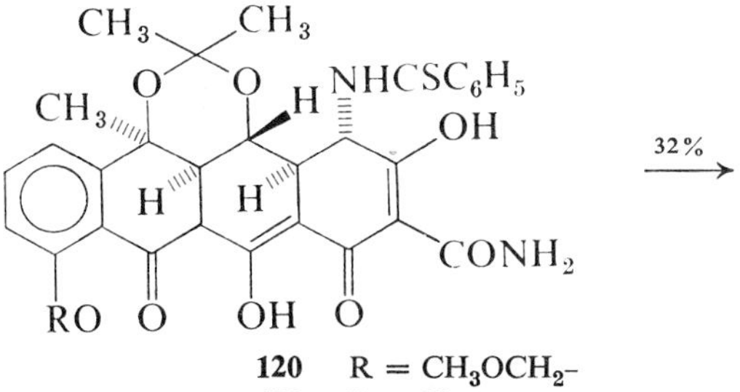

120 R = CH_3OCH_2–
120a R = H

Scheme 16

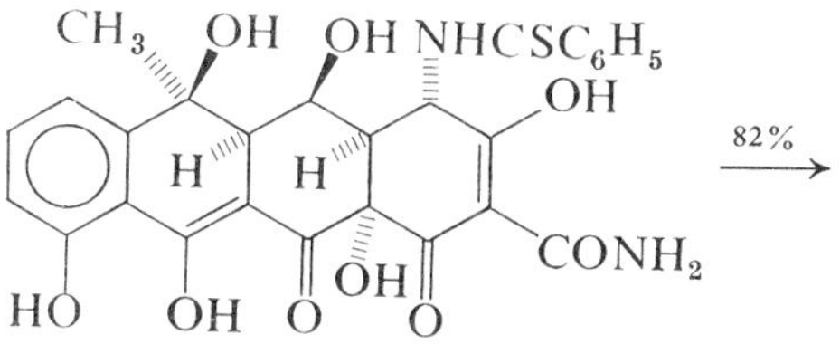

82%

121

33%

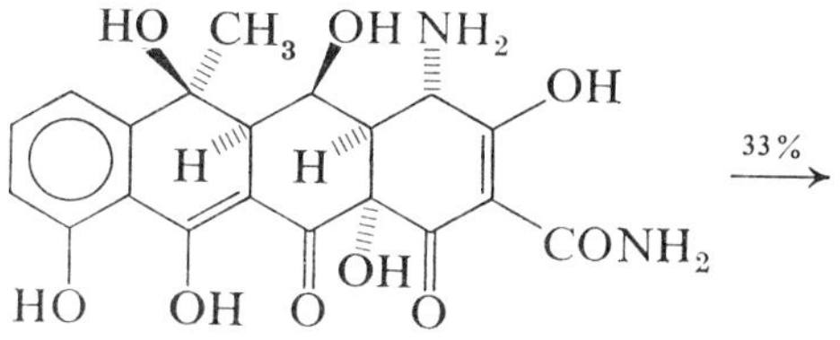

122

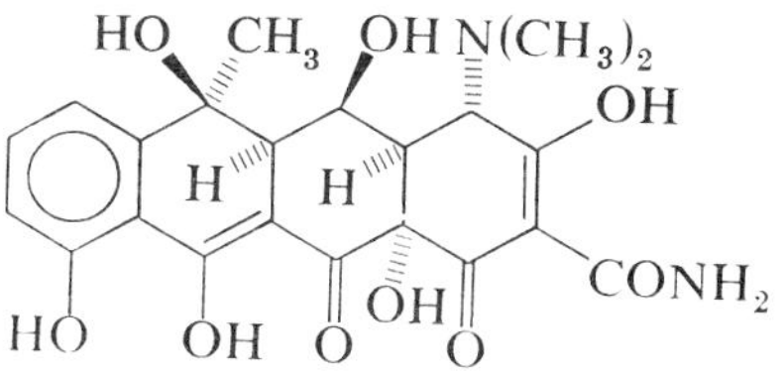

60c

Scheme 16 (Continued)

when basic lead acetate was used as the catalyst in tetrahydrofuran. In a masterful stroke the product **119** of this condensation was doubly condensed[40] with methyl 3-oxoglutaramate (**108**) using a combination of butyllithium and potassium *t*-butoxide as the catalyst to give the tetracyclic product **120**. Introduction of the 12a-hydroxyl group and the dimethylamino function were now carried out sequentially. The former was accomplished by oxygenation of **120a** (obtained from the parent methoxymethyl ether **120** by acetic acid catalyzed hydrolysis) in a basic medium, giving, after acid hydrolysis, the thioamide **121**. The latter compound was methylated on sulfur by treatment with methyl iodide at room temperature in tetrahydrofuran and the intermediate thioimino ether iodide hydrolyzed in acid without isolation to give *N*,*N*-bisdemethylterramycin (**122**) as the hydrochloride. The latter was immediately alkylated with a combination of methyl iodide and Hünig's base in tetrahydrofuran. Purification of the product by chromatography on a polyamide substrate then afforded *dl*-terramycin to mark the end of a brilliant and highly successful 10-year program directed to the synthesis of this compound.

Other groups have attempted total synthesis in this area. Most notable of these efforts are those due to Kende et al.,[41] which culminated in the synthesis of **123**, and the more recent work of Barton and his collaborators who have

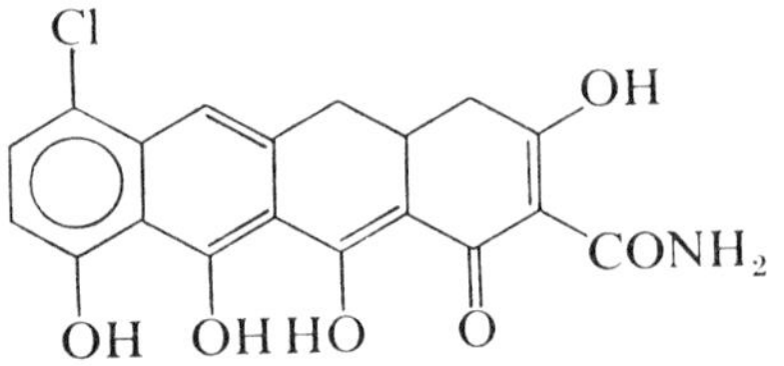

123

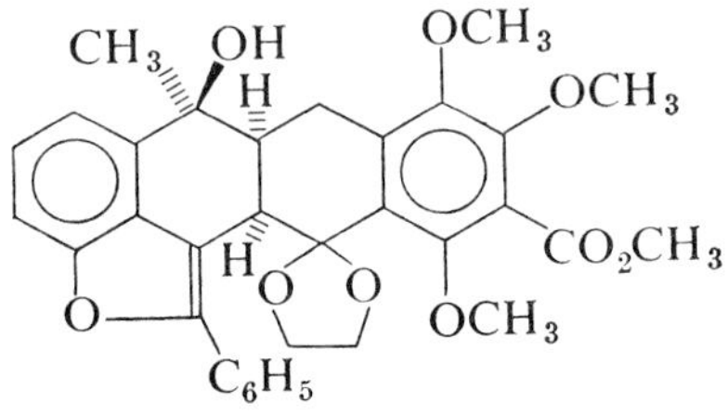

124

published[42] only very brief details of the synthesis of **124**. The latter differs from previous syntheses in that both rings *A* and *D* are aromatic throughout the elaboration of the tetracyclic nucleus. A reaction now needs to be found which will in effect introduce the 12*a*-hydroxyl group and eliminate the aromatic character of ring *A*.

4. THE BASIC SUGARS

Several basic sugars produced by differing species of *Streptomyces* have useful antibiotic properties. Some have similar activities while differing in toxicity, but generally they are most effective against gram-negative bacteria and tubercle bacilli in addition to gram-positive bacteria. Hepatotoxicity and ear damage are the chief drawbacks to their use.

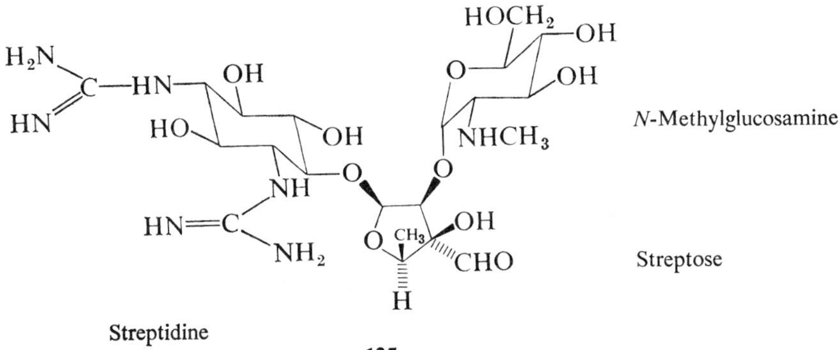

N-Methylglucosamine

Streptose

Streptidine

125

In structure there are some similarities but the one thing they all have in common is that they contain unusual amino sugars. The oldest of them in use is streptomycin-A (**125**) isolated[43] first in 1947. Despite this, the total synthesis of this molecule has not been achieved, although the syntheses of the individual components streptidine,[44] *N*-methylglucosamine,[45] and streptose[46] have been reported.

More closely related are gentamicin-A **126**, the kanamycins A (**127a**), B (**127b**), and C (**127c**), and the more complicated paromomycin II (**129a**) and neomycin B (**129b**). All of these antibiotics contain the cyclohexane derivative 2-deoxystreptamine. In addition the amino sugar glycosides of 2-deoxystreptamine, paromamine (**128a**) and neamine (**128b**), are contained,

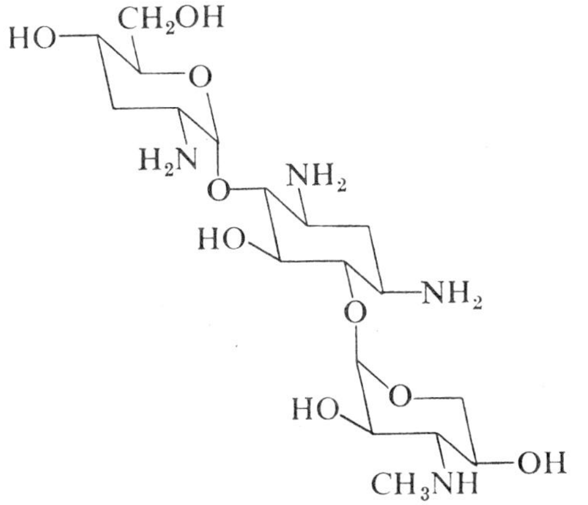

126

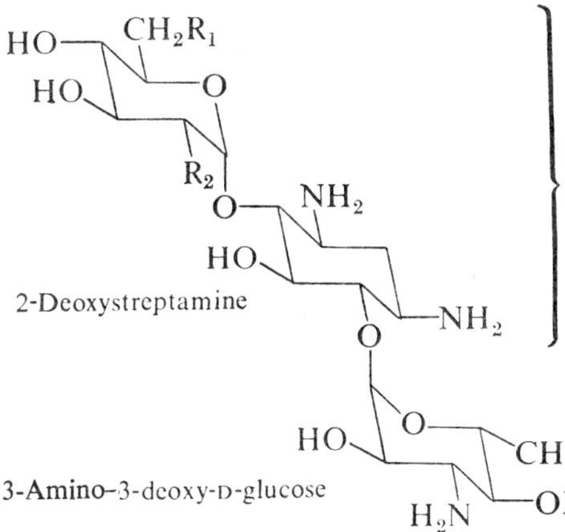

Paromamine (**128a**; $R_1 = OH$; $R_2 = NH_2$)
Neamine (**128b**; $R_1 = R_2 = NH_2$)

Kanamycin A (**127a**; $R_1 = NH_2$; $R_2 = OH$)
Kanamycin B (**127b**; $R_1 = R_2 = NH_2$)
Kanamycin C (**127c**; $R_1 = OH$; $R_2 = NH_2$)

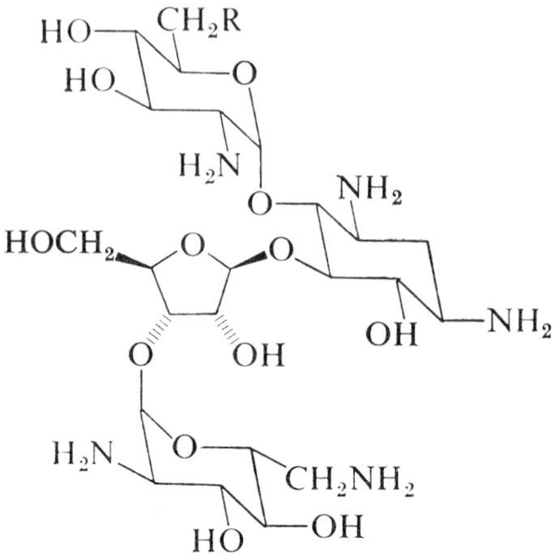

Paromomycin (**129a**; R = OH)
Neomycin (**129b**; R = NH_2)

respectively, in paromomycin II and kanamycin C, and neomycin B and kanamycin B. On the other hand, gentamicin A (**126**), which is the major component of a complex of at least 15 compounds, differs[47] only from kanamycin C in the 3-amino–3-deoxy-D-glucose ring. In this ring, in gentamicin A the 3-amino group is monomethylated and the hydroxymethyl group has been replaced by hydrogen. It is in fact a D-xylo-pyranose ring.

More distantly related to these compounds is spectinomycin[48] (**130**), which contains a streptamine ring, similar to that of streptomycin, but without the guanidino functions.

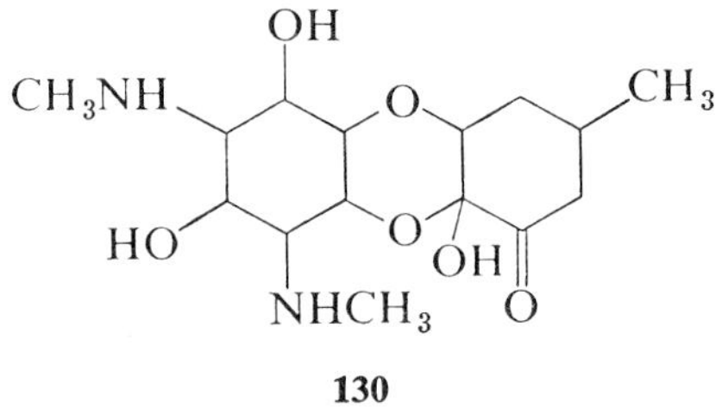

130

Still more remote in structure is kasugamycin (**131**), which sees use mainly against rice blast and has no useful therapeutic effect in mammalian species.

Completely unrelated to any of the foregoing compounds are lincomycin (**132a**) and streptozotocin (**133**). The former compound has potent antibacterial properties but its semisynthetic derivative (**132b**) is more useful

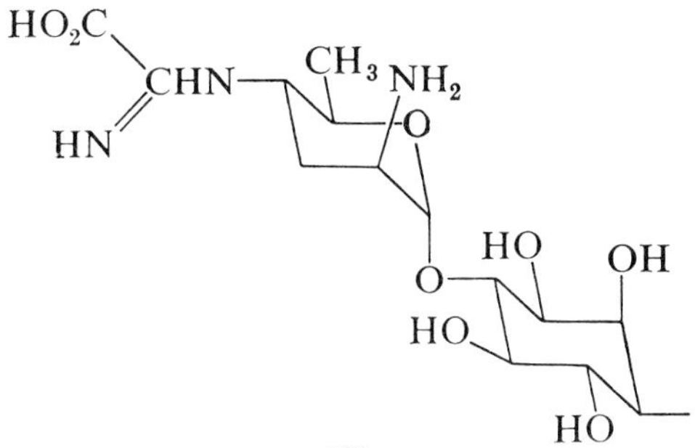

131

therapeutically, especially in the treatment of malaria. Streptozotocin, a glucosamine derivative, is used experimentally as an antibacterial and antitumor agent.

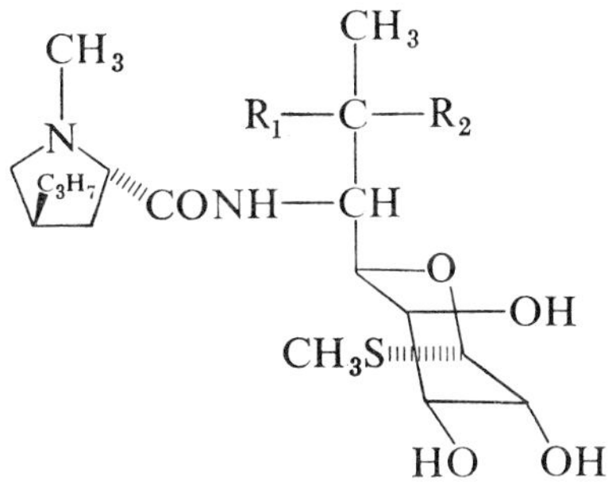

132a $R_1 = OH; R_2 = H$
132b $R_1 = H; R_2 = Cl$

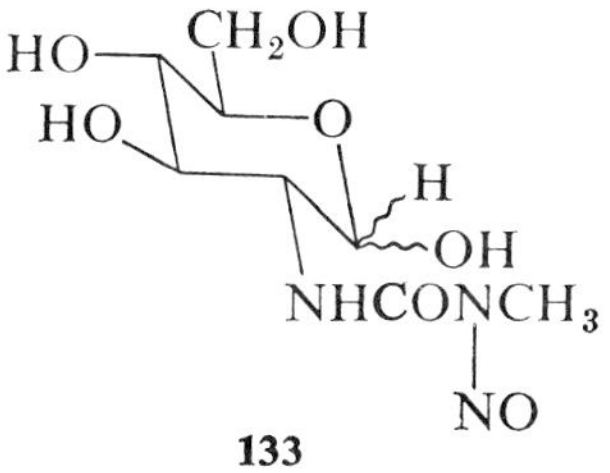

133

Of these antibiotics only the kanamycins, kasugamycin and streptozotocin have been synthesized. The extent to which protection of functionality must be taken in the synthesis of amino sugars is probably greater than that with any other group with the possible exception of the peptides. This represents both an experimental and psychological barrier to total synthesis in this area.

A. The Kanamycins

The key compound in the synthesis of the kanamycins is 2-deoxystreptamine (**134**). Its synthesis[49] was first reported in 1964 and is illustrated in Scheme 17. Hydrolysis[50] of the epoxy-diacetate **135** followed by acetylation afforded the tetraacetate **136**, which after subsequent treatment with bromine-water led to the bromohydrin **137** (via axial addition). Debromination of **137** with Raney

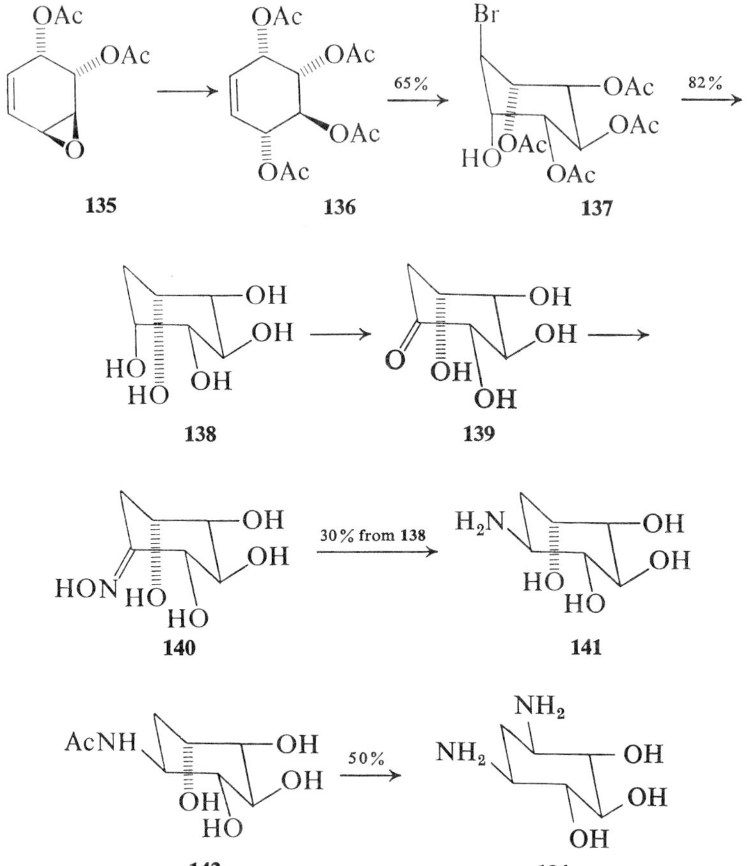

135 → 136 —65%→ 137 —82%→

138 → 139 →

140 —30% from 138→ 141

142 —50%→ 134

Scheme 17

nickel followed by ammonolysis to remove the acetyl groups gave 3-deoxy-epiinositol (**138**). Catalytic oxidation of **138** using oxygen and a platinum catalyst, specific for the oxidation of axial alcohols, yielded the monoketone **139** whose oxime **140** was reduced by means of sodium amalgam to **141** isolated as its pentaacetyl derivative. Ammonolysis of the latter afforded **142**, which was subjected to the same sequence of reactions used with **138**. This resulted in the pentaacetyl derivative of **134**. Racemic 2-deoxystreptamine (**134**) itself was liberated from this derivative by hydrolysis with 4*N* HCl and the subsequent use of an anion exchange column to obtain the free base.

Kanamycin A (**127a**) has been synthesized independently by Nakajima[51] and by Umezawa.[52] The former workers prepared a protected derivative of the 6-amino-6-deoxy-D-glucose moiety, coupled it with a masked derivative of

2-deoxystreptamine, and then forged the second glycosidic linkage by means of a protected form of the 3-amino–3-deoxy–D-glucose part of the molecule. Umezawa and his associates did just the reverse, but both groups needed roughly the same protected components.

The key intermediates chosen by Umezawa[53] for the first stage in the synthesis were 3-acetamido–2,4,6-tri-*o*-benzyl–3-deoxy–α-D-glucopyranosyl chloride (**143**) and the isopropylidene derivative of *N*,*N*′-bis(carbobenzyloxy)-2-deoxystreptamine (**144**).

The former compound was prepared (Scheme 18) by a method resting on some earlier work by Baer[54] in which methyl β-D-glucoside was oxidized to the dialdehyde **146** by means of periodate. Base-catalyzed condensation of **146** with nitromethane then produced the nitrosugar **147**, which when reduced catalytically gave the desired amino sugar **148**. Acetylation of **148** with acetic anhydride followed by benzylation with benzyl bromide in dimethylformamide in the presence of both barium oxide and hydroxide led to **149** in high overall yield. Hydrolysis of **149** with sulfuric acid followed by reacetylation afforded a mixture of anomers **150**, each of which gave the sought-after intermediate **143** in quantitative yield when treated with dry hydrogen chloride in dioxane containing acetyl chloride.

The second component, **144**, in its racemic form, was easily prepared by the action of 2,2-dimethoxypropane in the presence of an acid catalyst on *N*,*N*′-bis(carbobenzyloxy)-2-deoxystreptamine.

Condensation[55] of **143** with **144** in the presence of mercuric cyanide and Drierite in anhydrous dioxane/benzene gave a crude product in 80% yield which was hydrolyzed with 80% acetic acid to split the acetonide and hydrogenated over palladium to remove the benzyl groups. The residual material was dinitrophenylated and the product chromatographed to give two components, of which one was the desired diastereoisomer **151**. It was identified by comparison with a specimen prepared from a degradation product of kanamycin A. Hydrolysis of **151** with methanolic ammonia then afforded the free aminosugar **152**, signaling completion of the first phase of the synthesis.

In the second phase the amino groups of **152** were protected by carbobenzyloxylation and the derivative, **153**, was then converted to its bisacetonide. The latter when benzylated afforded **154**. Acetic acid cleavage of the ketal rings followed by treatment of the resulting tetrahydroxy derivative with 2,2-dimethoxypropane at 5° successfully yielded the monoacetonide **155**. Condensation[52] of **155** with 2,3,4-tri-*o*-benzyl–6-(*N*-benzylacetamido)–6-deoxy–α-D-glucopyranosyl chloride (**156**)[53,56] under Königs-Knorr conditions, as in the formation of the first glycosidic linkage gave a brown viscous mixture. This was treated with 80% acetic acid, then hydrogenated over palladium to remove *O*-benzyl groups, de-*N*-acetylated with barium

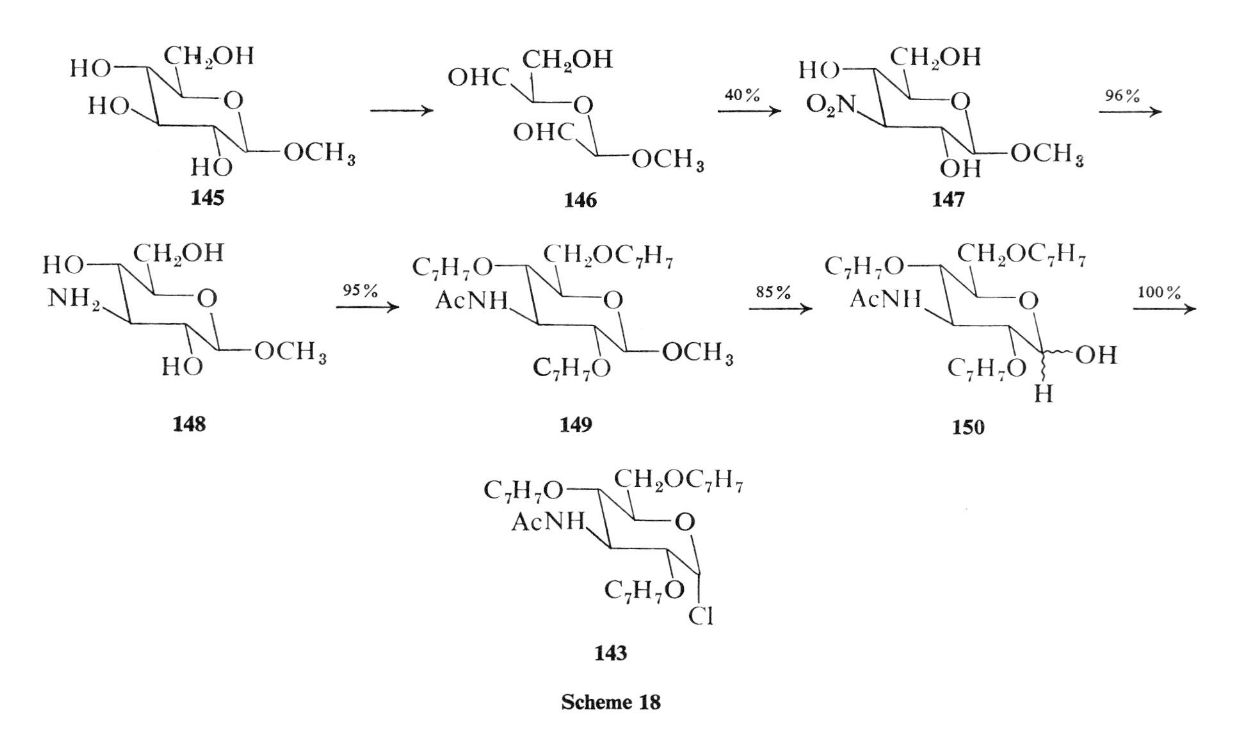

HO
CH2OH
HO
O
HO
OCH3
145
OHC
CH2OH
O
OHC
OCH3
146
40%
HO
CH2OH
O2N
O
OH
OCH3
147
96%
HO
CH2OH
NH2
O
HO
OCH3
148
95%
C7H7O
CH2OC7H7
AcNH
O
C7H7O
OCH3
149
85%
C7H7O
CH2OC7H7
AcNH
O
C7H7O
OH
H
150
100%
C7H7O
CH2OC7H7
AcNH
O
C7H7O
Cl
143

Scheme 18

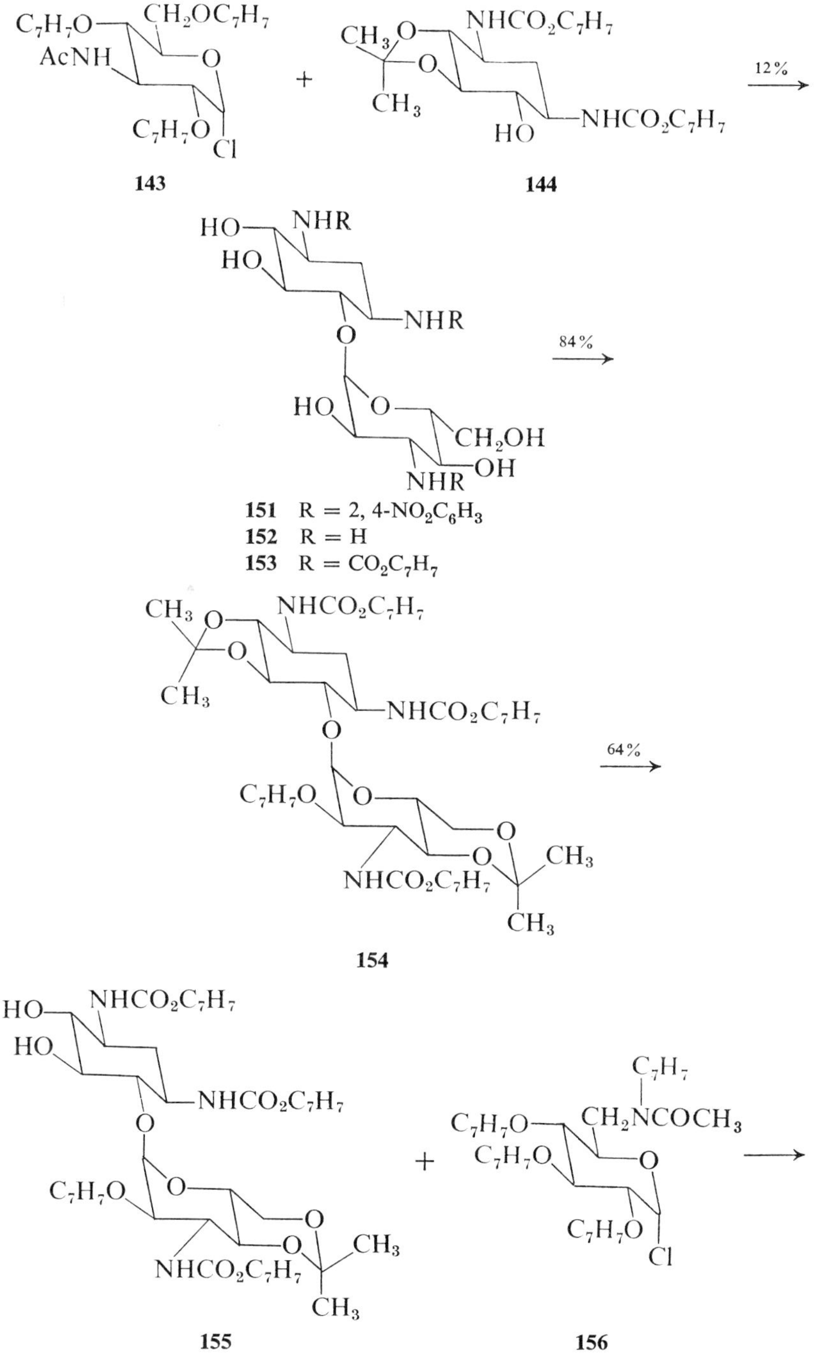

C_7H_7O $CH_2OC_7H_7$ AcNH O C_7H_7O Cl

143

+

CH_3 O $NHCO_2C_7H_7$ O CH_3 HO $NHCO_2C_7H_7$

144

12%
→

HO NHR HO NHR O HO O CH_2OH OH NHR

151 R = 2, 4-$NO_2C_6H_3$
152 R = H
153 R = $CO_2C_7H_7$

84%
→

CH_3 O $NHCO_2C_7H_7$ O CH_3 $NHCO_2C_7H_7$ O C_7H_7O O O O CH_3 $NHCO_2C_7H_7$ CH_3

154

64%
→

HO $NHCO_2C_7H_7$ HO $NHCO_2C_7H_7$ O C_7H_7O O O O CH_3 $NHCO_2C_7H_7$ CH_3

155

+

C_7H_7 C_7H_7O CH_2NCOCH_3 C_7H_7O O C_7H_7O Cl

156

→

Scheme 19

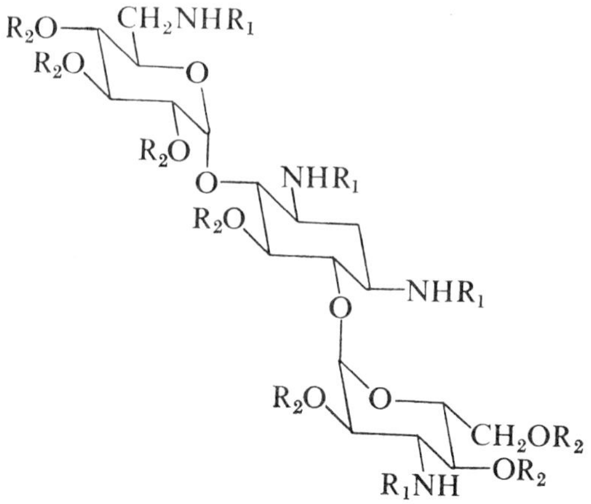

157 $R_1 = 2, 4\text{-}(NO_2)_2C_6H_4$; $R_2 = Ac$
127a $R_1 = R_2 = H$

Scheme 19 (Continued)

hydroxide solution, then hydrogenated again to remove the *N*-benzyl group. The product of these transformations was dinitrophenylated then *o*-acetylated The residual material, a six-component mixture, was chromatographed on a silica gel column and the main product **157** was isolated in 10.1% yield overall from **155**. When the latter material was treated with methanolic ammonia and the free base isolated by chromatography over an ion-exchange resin, crystalline kanamycin A was obtained, in all respects identical with the natural material.

In the alternative synthesis (Scheme 20) of **127a** due to Nakajima,[51] 6-acetamido–2,3,4-tri-*O*-benzyl–6-deoxy–α-D-glucosyl chloride (**158**) was condensed with racemic **144** in a modified Königs-Knorr reaction and after subsequent removal of the isopropylidene group afforded a mixture of the two diastereoismers **159** and **160** in 34 and 40% yield, respectively. A second condensation under essentially the same conditions this time between **159** and **143** produced the highly protected derivatives **161** and **162**. Removal of the benzyl and carbobenzyloxy groups from **161** was accomplished by sodium in liquid ammonia at −70°. This was followed by *N*-acetylation and led to tetra-*N*-acetylkanamycin A, completely identical with a specimen obtained from the natural product. Hydrolysis of the tetraacetyl derivative by barium hydroxide then afforded kanamycin A (**127a**) itself.

When **162** was subjected to the same treatment it also gave a tetraacetyl derivative isomeric with that derived from kanamycin A but whose NMR spectrum suggested that the new glycosidic linkage had the α-configuration,

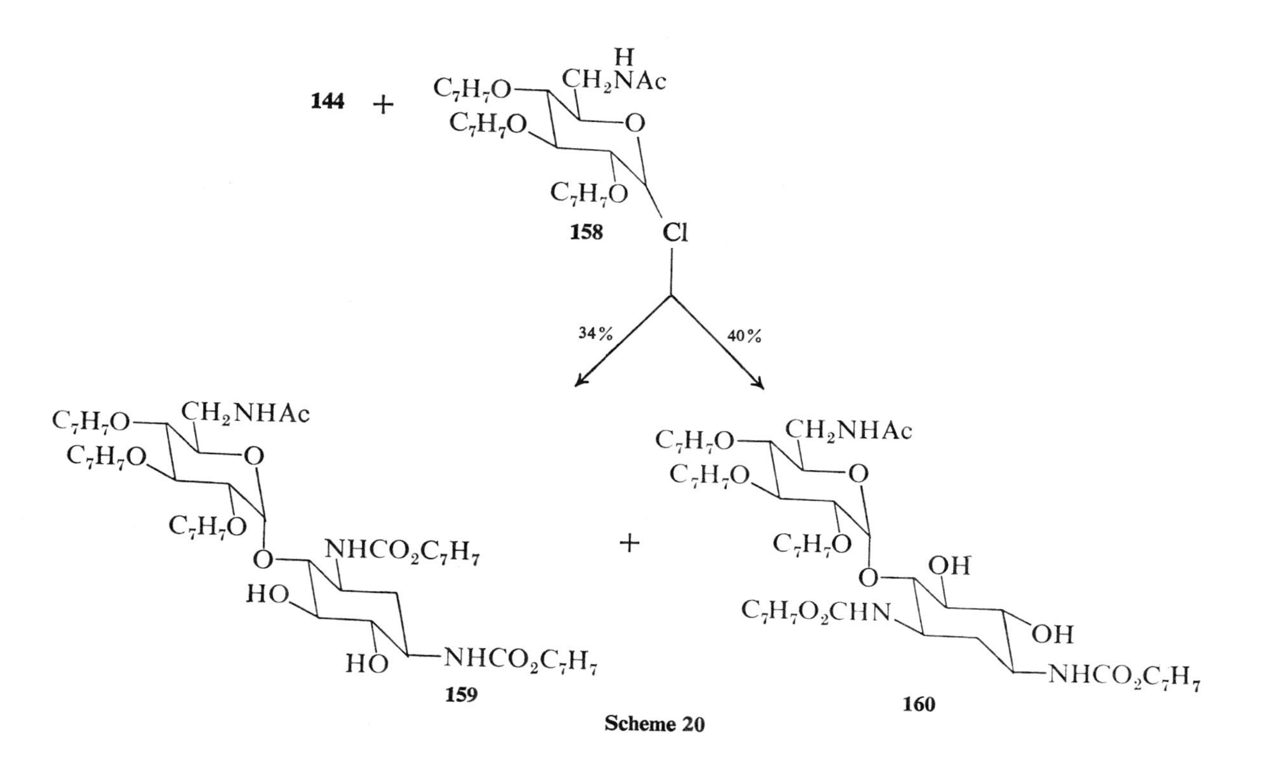
144
C_7H_7O
C_7H_7O
H
CH_2NAc
O
C_7H_7O
158
Cl
34%
40%
C_7H_7O
C_7H_7O
CH_2NHAc
O
C_7H_7O
O
$NHCO_2C_7H_7$
HO
HO
$NHCO_2C_7H_7$
159
+
C_7H_7O
C_7H_7O
CH_2NHAc
O
C_7H_7O
O
OH
$C_7H_7O_2CHN$
OH
$NHCO_2C_7H_7$
160
Scheme 20

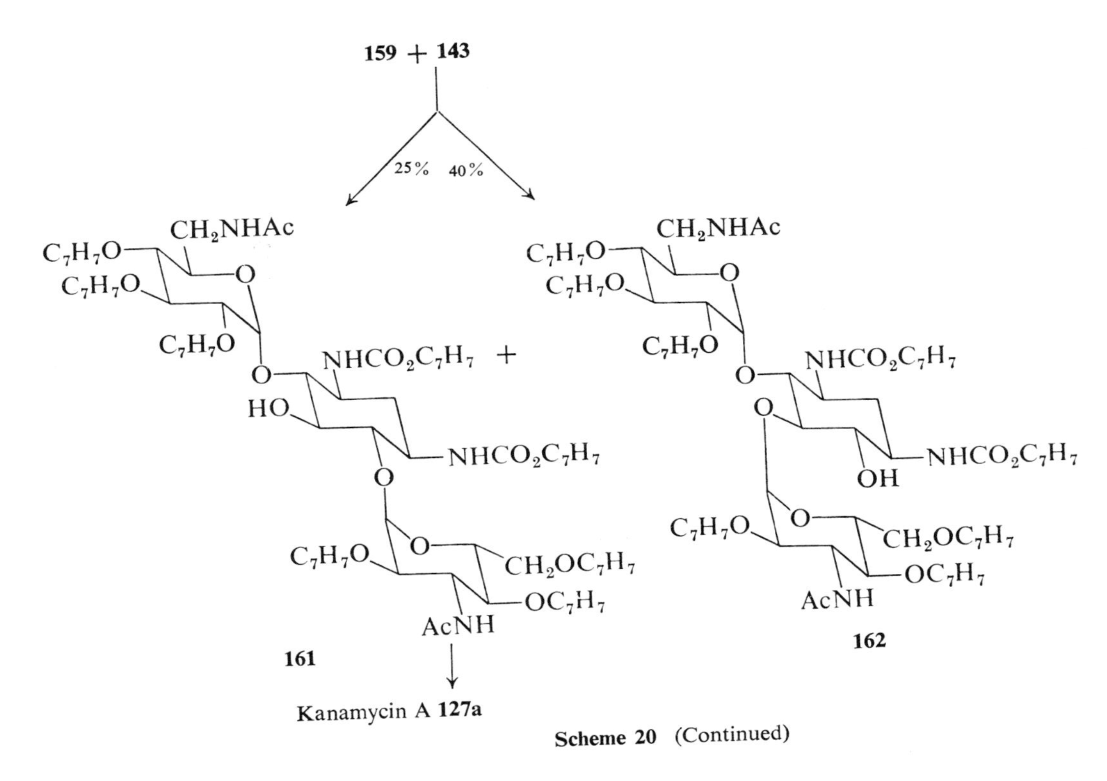

Scheme 20 (Continued)

and that it had formed at the C-5 oxygen of the 2-deoxystreptamine moiety.

The kanamycins B (**127b**) and C (**127c**) were both synthesized by Umezawa and his co-workers in a coupling sequence opposite to that used in their synthesis of kanamycin A. In addition both syntheses depend on the initial synthesis of paromamine (**128a**), which was used for the preparation of both kanamycin C (**127c**) and neamine (**128b**). The latter was then used to synthesize kanamycin B (**127b**).

The synthesis[57] (Scheme 21) of paromamine (**128a**) was carried out by

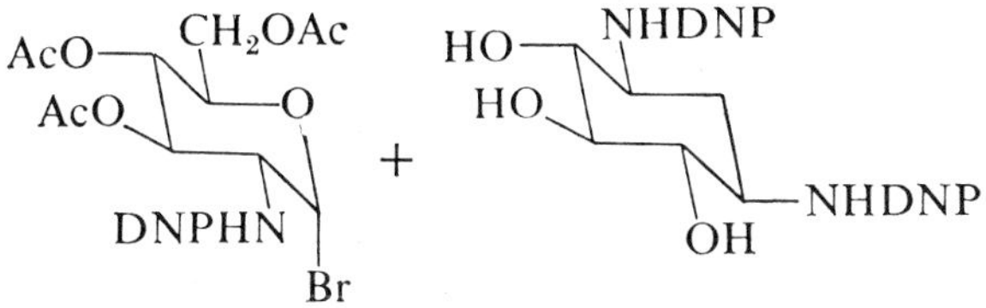

164 **163**

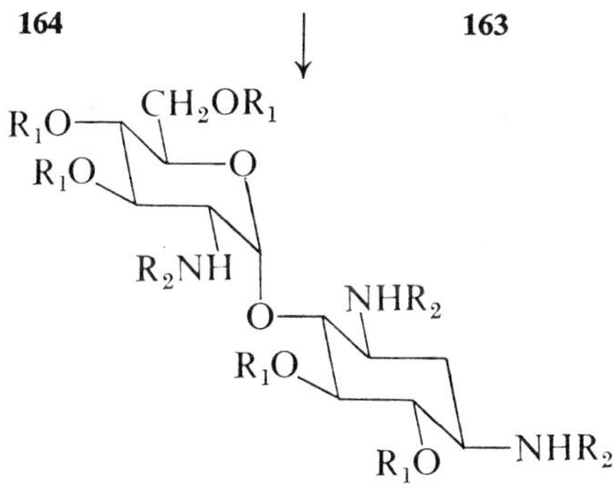

165 $R_1 = Ac; R_2 = 2,4\text{-}(NO_2)_2C_6H_3$
128a $R_1 = R_2 = H$

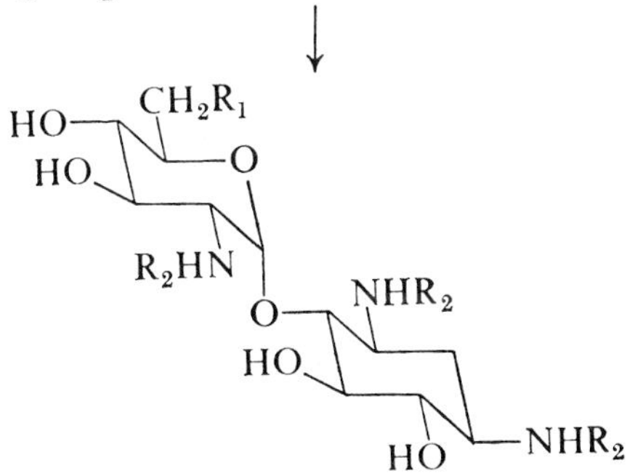

166a $R_1 = OH; R_2 = Ac$
166b $R_1 = OTs; R_2 = Ac$
166c $R_1 = NHAc; R_2 = Ac$
128b $R_1 = NH_2; R_2 = H$

Scheme 21

condensing bis-*N*,*N*′-(2,4-dinitrophenyl)–2-deoxystreptamine (**163**) with the protected α-glucopyranosyl bromide (**164**) derived[58] from glucosamine. The bromide was used here because of the comparative lack of reactivity of **163**, and the condensation was carried out in nitromethane at 95° for 15 hours in the presence of mercuric cyanide and bromide. This gave a mixture of glycosides which were acetylated and separated by chromatography. The correct diastereoisomer, **165**, was obtained in low yield. Hydrolysis of **163** with methanolic ammonia followed by treatment with an exchange resin gave crystalline paromamine (**128a**).

The conversion of paromamine (**128a**) to neamine (**128b**) proceeded via the tri-*N*-acetyl derivative **166a**, which was selectively monotosylated in 43% yield to give **166b**. The tosyloxy group of this compound was then successively converted to the azide, the amine, and the acetylamino compound **166c** by standard procedures. Deacetylation of **166c** by means of hydrazine then afforded, after the usual purification procedures, neamine (**128b**) in 72% yield.

The synthesis of kanamycin B (**127b**)[59] and C (**127c**)[60] were then completed by coupling, respectively, the protected derivatives **167** and **168** of neamine and paromamine with the α-D-glucopyranosyl chloride **143**. The syntheses parallel the methods used for the total synthesis of kanamycin A and will therefore not be discussed in further detail here.

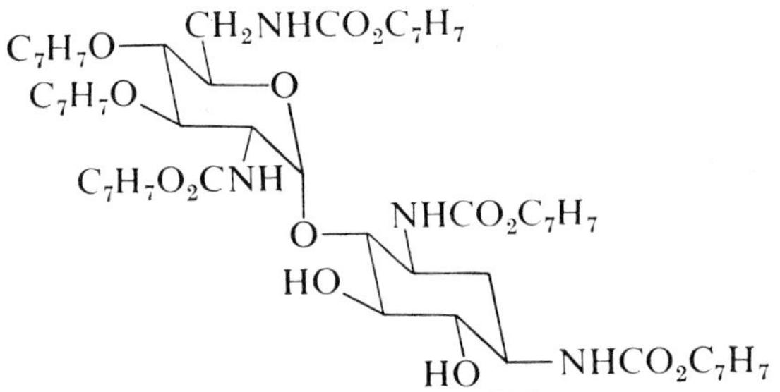

$CH_2NHCO_2C_7H_7$
C_7H_7O
C_7H_7O
O
$C_7H_7O_2CNH$
O
$NHCO_2C_7H_7$
HO
HO
$NHCO_2C_7H_7$

167

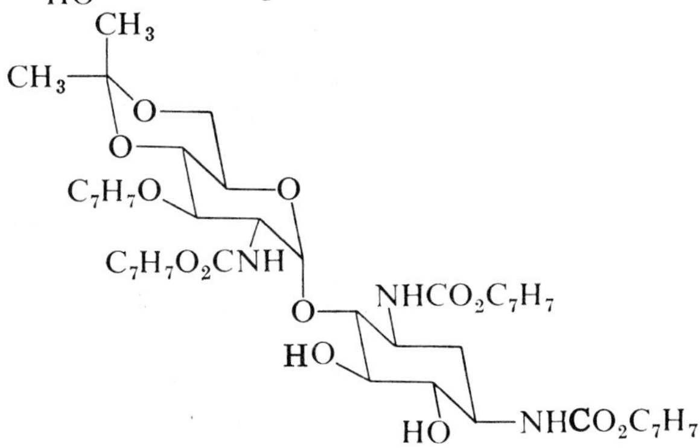

CH_3
CH_3
O
O
C_7H_7O
O
$C_7H_7O_2CNH$
O
$NHCO_2C_7H_7$
HO
HO
$NHCO_2C_7H_7$

168

B. Kasugamycin

The total synthesis of kasugamycin (**131**) has been accomplished by H. Umezawa[61] (not to be confused with S. Umezawa, whose group synthesized the kanamycins), who synthesized the aminosugar moiety beginning with a pyranone derivative. An alternate synthesis due to Nakajima reported[62] slightly earlier, started with a D-glucose derivative and afforded the naturally occurring optical isomer directly.

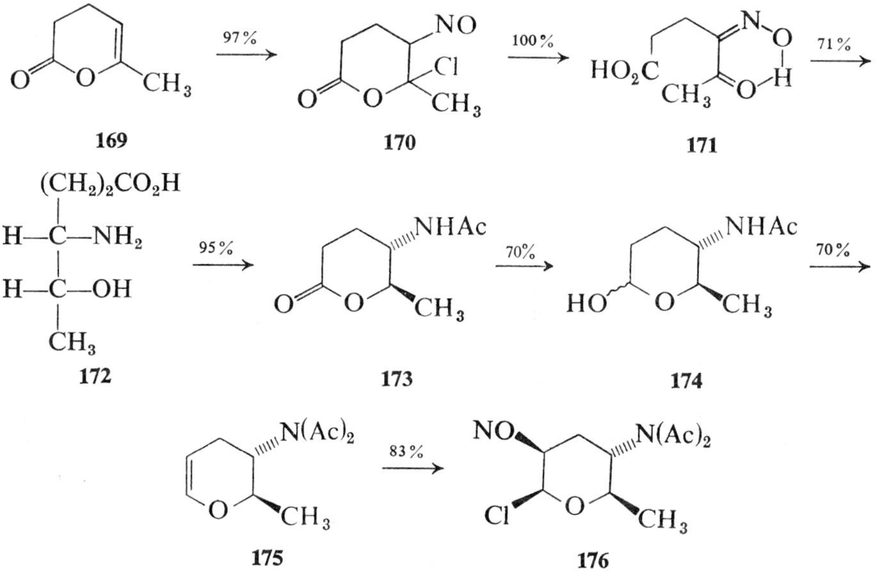

169 —97%→ **170** —100%→ **171** —71%→ **172** —95%→ **173** —70%→ **174** —70%→ **175** —83%→ **176**

Scheme 22

In the synthesis (Scheme 22) due to Umezawa and his group the starting dihydropyranone **169** was treated with nitrosyl chloride at −60° and afforded the dimer of **170**. Hydrolysis of **170** by water led easily to 4-oximino–5-oxohexanoic acid (**171**), which then was hydrogenated stereoselectively over a platinum catalyst to *dl*-erythro–4-amino–5-hydroxyhexanoic acid (**172**). Lactonization of **172** at room temperature with acetic anhydride afforded **173**, which was reduced to the hemiacetal **174** by lithium aluminum hydride. Refluxing with acetic anhydride converted **174** to **175**, and chloronitrosation of the latter compound under conditions similar to those used for **169**, afforded the dimer of **176**. Coupling of **176** with excess 1:2, 3:4-di-*o*-isopropylidene-(+)-inositol (**177**) was achieved in methylene chloride at 0° in the presence of silver carbonate and perchlorate and led to a mixture of diastereoisomers which were reduced over a platinum catalyst. The isopropylidene groups were removed by boiling 50% acetic acid and the

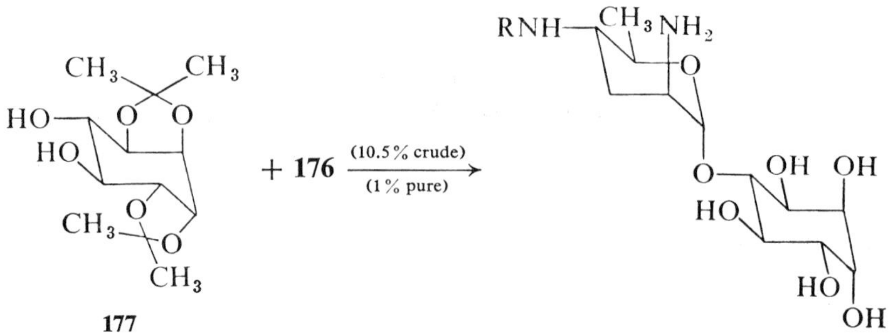

177 + **176** $\xrightarrow[\text{(1\% pure)}]{\text{(10.5\% crude)}}$

178a R = Ac
178b R = H
131 R = C(NH)CO_2H

Scheme 22 (continued)

reaction was product purified by chromatography over an ion-exchange resin and by crystallization to give **178a**. Hydrolysis of the acetylamino group by barium hydroxide then afforded optically active kasuganobiosamine (**178b**). The synthesis of **178a** is noteworthy not only from the point of view of the formation of the α-glycoside but also because of the resolution that took place during the isolation procedure.

The completion of the synthesis of **178b** in effect completed the synthesis of **131** because the final step, involving the reaction of **178b** with diethyl oxalimidate followed by acid hydrolysis under mild conditions, had been reported[63] previously.

Nakajima[61] in his synthesis (Scheme 23) of **178b** started with the 3-deoxy-glucopyranoside **179a** easily prepared[64] from D-glucose in five steps. Conversion of **179a** to **179b** was achieved in three steps involving a displacement with inversion of the tosyl group by azide ion followed by catalytic reduction over a platinum catalyst then acetylation. The benzylidene group of **179b** was removed by means of warm aqueous sulfuric acid and the resulting oil was monotosylated then acetylated to give **180a**. Replacement of the tosyl group by iodide and reduction with Raney nickel led to **180b**, which was deacetylated by sodium methoxide in methanol. The product **181** had its 4-hydroxyl group in the wrong configuration for further work and this was reversed by chromic acid oxidation to the 4-*oxo*-compound followed by platinum-catalyzed hydrogenation to give **182a**. The latter compound was converted by mesyl chloride to **182b**, which was then transformed into **183a** by the same procedures used for the conversion of **179a** to **179b**. Hydrolysis with 5*N* formic acid followed by acetylation led to **183b**, which was treated with a saturated solution of hydrogen chloride and acetic acid/chloroform to produce **184**. Coupling of **184** with **177** gave a crystalline glycoside **185**.

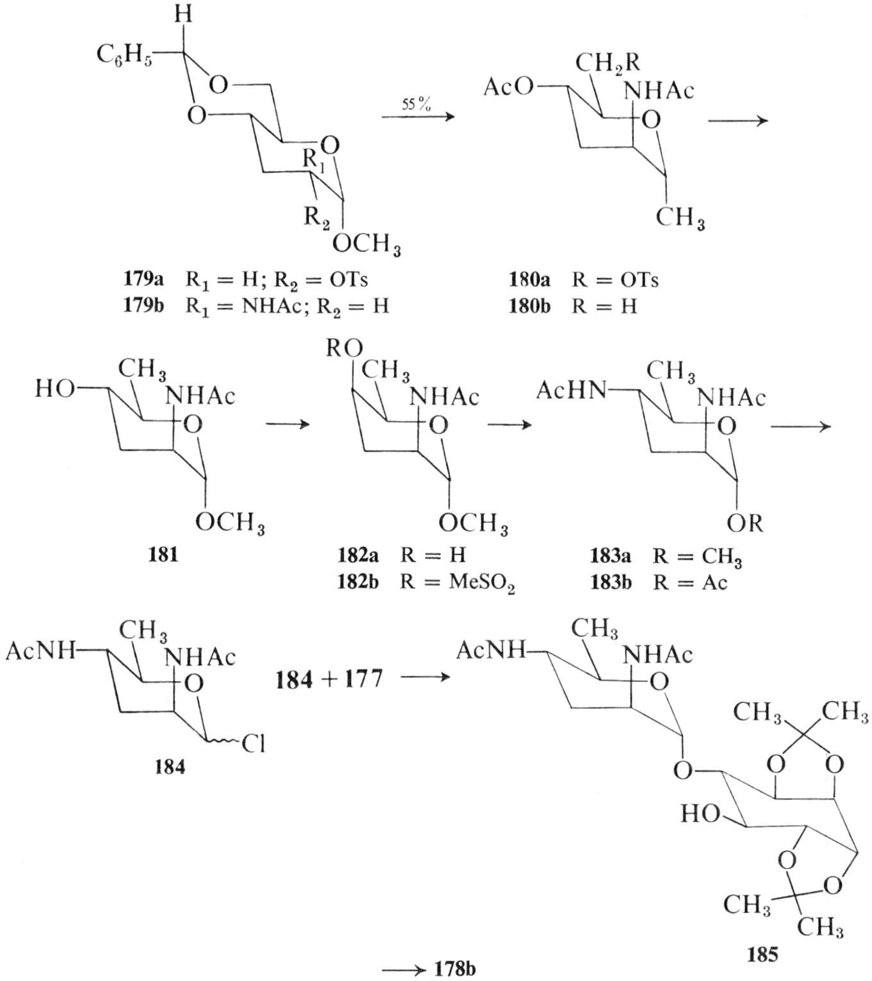

C_6H_5 O O R_1 O R_2 OCH_3 55% AcO CH_2R NHAc O CH_3

179a $R_1 = H$; $R_2 = OTs$
179b $R_1 = NHAc$; $R_2 = H$

180a $R = OTs$
180b $R = H$

HO CH_3 NHAc O OCH_3 **181**

RO CH_3 NHAc O OCH_3

182a $R = H$
182b $R = MeSO_2$

AcHN CH_3 NHAc O OR

183a $R = CH_3$
183b $R = Ac$

AcNH CH_3 NHAc O Cl **184**

184 + **177** ⟶

AcNH CH_3 NHAc O O HO CH_3 CH_3 O O O O CH_3 CH_3 **185**

⟶ **178b**

Scheme 23

Hydrolysis of **185** followed by acetylation provided a heptaacetyl compound, which when deacylated first by sodium methoxide to remove O-acetate groups then by boiling aqueous barium hydroxide to cleave the amide groups, led to kasuganobiosamine (**178b**), thus constituting, in view of previous work,[63] a total synthesis of **131**.

C. Streptozotocin

This antibiotic was first synthesized[65] at the time its structure was deduced. This was accomplished by treating tetra-*o*-acetylglucosamine hydrochloride

(**186a**) with methyl isocyanate to give the unsymmetrical urea **186b**. Nitros-

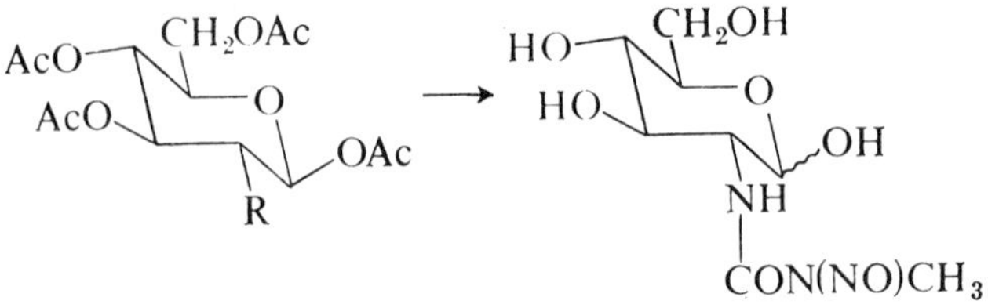

186a R = NH_2HCl
186b R = $NHCONHCH_3$
186c R = $NHCON(NO)CH_3$

133

ation of **186b** in pyridine produced **186c**, which was deacetylated by means of ammonia in methanol at −10°. This afforded streptozotocin (**133**) identical with an authentic sample.

A preparative method for the synthesis of **133** has also been published.[66] This involves a direct reaction of *D*-glucosamine with *N*-nitrosomethylcarbamylazide and gives streptozotocin in 31% yield. The nitrosoazide

$$CH_3N(NO)CON_3$$

was prepared in good overall yield from methylisocyanate by the following sequence:

$$CH_3NCO \rightarrow CH_3NHCOCl \rightarrow CH_3NHCOCl \rightarrow CH_3N(NO)CON_3$$

D. Lincomycin

This antibiotic was first synthesized[66a] in a partial manner from its cleavage products *trans*-1-methyl-4-*n*-propyl-*L*-proline (**187**) and the amino sugar methyl α-thiolincosaminide (**187a**). The latter had been isolated during previous degradation studies.[66b]

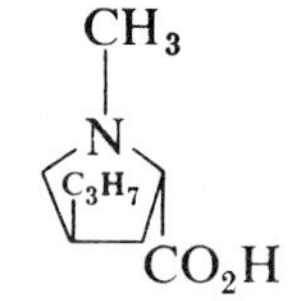

187 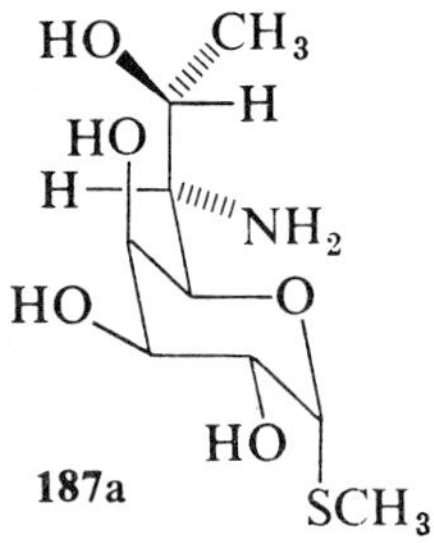 **187a**

These reactions are outlined below and utilize as the first step a variation of a procedure that Kenner et al.[66c] had used for the synthesis of 1-carbobenzoxy-4-methylene-*L*-proline. Hydrogenation of the intermediate propylidene derivative **187b** over a platinum catalyst on Dowex resin then gave a mixture of the desired *N*-blocked 4-*n*-propylprolines (**187c**) containing 25–35% of the desired *trans*-isomer. Catalytic reduction on other supports afforded little or none of this isomer. Coupling of **187c** with **187a** was accomplished using the mixed anhydride method (isobutyl chloroformate + triethylamine in dry acetonitrile) and led to **187d** in high yield. The blocking group was removed by hydrogenolysis, and the resulting product was reductively methylated using formaldehyde and a palladium-on-carbon catalyst in the presence of hydrogen. This gave a mixture of lincomycin **132a** and *cis*-lincomycin (**187e**) which were separated by careful chromatography.

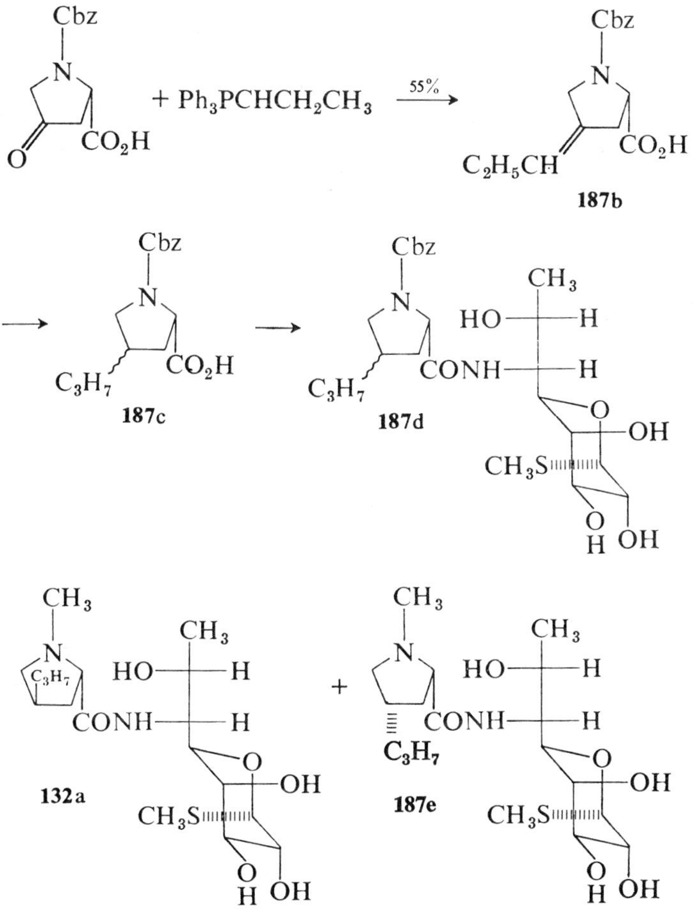

Cbz, N, O, CO_2H + $Ph_3PCHCH_2CH_3$ $\xrightarrow{55\%}$ Cbz, N, C_2H_5CH, CO_2H **187b**

→ Cbz, N, C_3H_7, CO_2H **187c** → Cbz, N, C_3H_7, CONH, CH_3, HO, H, H, O, OH, CH_3S, O H, OH **187d**

CH_3, N, C_3H_7, CONH, CH_3, HO, H, H, O, OH, CH_3S, O H, OH **132a** + CH_3, N, CONH, C_3H_7, CH_3, HO, H, H, O, OH, CH_3S, O H, OH **187e**

(Standard sugar nomenclature is used for the sugar side chain in these and subsequent diagrams of this section.)

Alternatively the mixture of these compounds could be prepared by deblocking **187c**, reductively methylating the nitrogen atom, followed by coupling with **187a** via the mixed anhydride method.

The total synthesis of **132a** then devolved simply on a synthesis of the amino sugar **187a**. This was accomplished both by Magerlein[66d] and by Szarek et al.,[66e] the latter workers devising two different syntheses of the molecule.

Contrary to the methods evolved by the Canadian group Magerlein introduced the S-methyl group at the outset. His synthesis of **187a** is outlined below. Treatment of D-galactose with methanethiol in hydrochloric acid afforded **187f**, which by a controlled tosylation reaction was converted to **187g** (R = H). Acetylation of the latter compound gave an oily triacetate **187g** (R = Ac), which underwent substitution with sodium iodide to give **187h**. Replacement of the 6-iodo by nitro was slow (sodium nitrite in dimethylformamide) but afforded about 20% of the desired compound **187i**. Repeated additions of acetaldehyde and sodium methoxide to a methanolic solution of **187i** then led to the nitro-alcohols **187j** in 50% yield. The latter

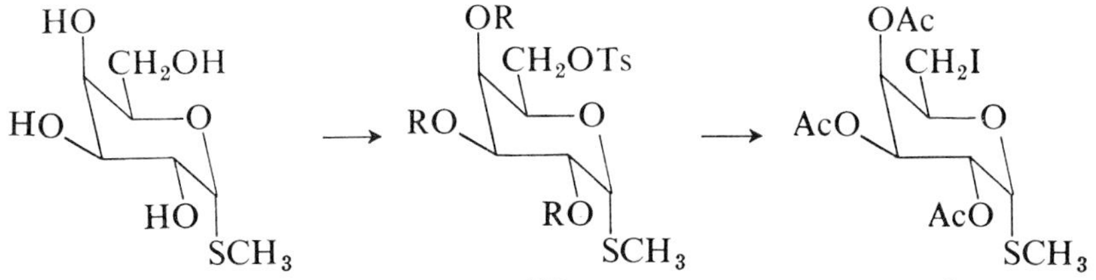

HO, CH_2OH, HO, O, HO, SCH_3 **187f** → OR, CH_2OTs, RO, O, RO, SCH_3 **187g** → OAc, CH_2I, AcO, O, AcO, SCH_3 **187h**

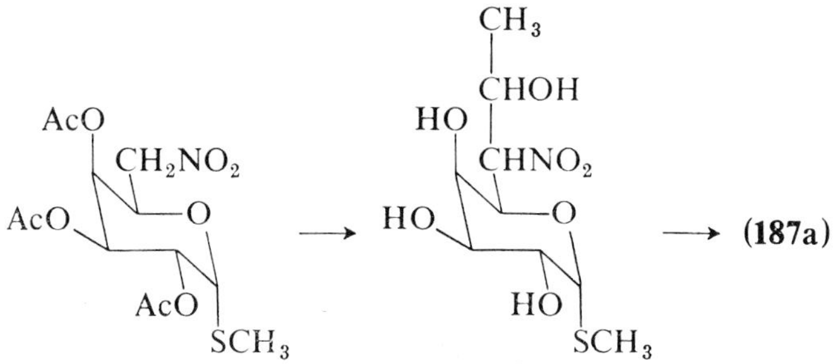

AcO, CH_2NO_2, AcO, O, AcO, SCH_3 **187i** → CH_3, CHOH, HO, $CHNO_2$, HO, O, HO, SCH_3 **187j** → (**187a**)

when reduced with lithium aluminum hydride in tetrahydrofuran afforded a mixture of amino sugars containing methyl α-thiolincosaminide. The desired isomer was separated as its pentaacetyl derivative by chromatography over silica gel.

In the alternative syntheses due to Szarek and his associates[66e] the thiomethyl group was introduced towards the end of the synthesis. The first method involved the sequence shown below, in which initially 1,2:3,4-di-*O*-isopropylidene-α-*D*-galactohexodialdopyranose-(1,5) was condensed[66f] with ethylidenetriphenylphosphorane to give predominantly the *cis*-isomer (**187k**) of the desired oct-6-enose. Treatment of the latter with aqueous

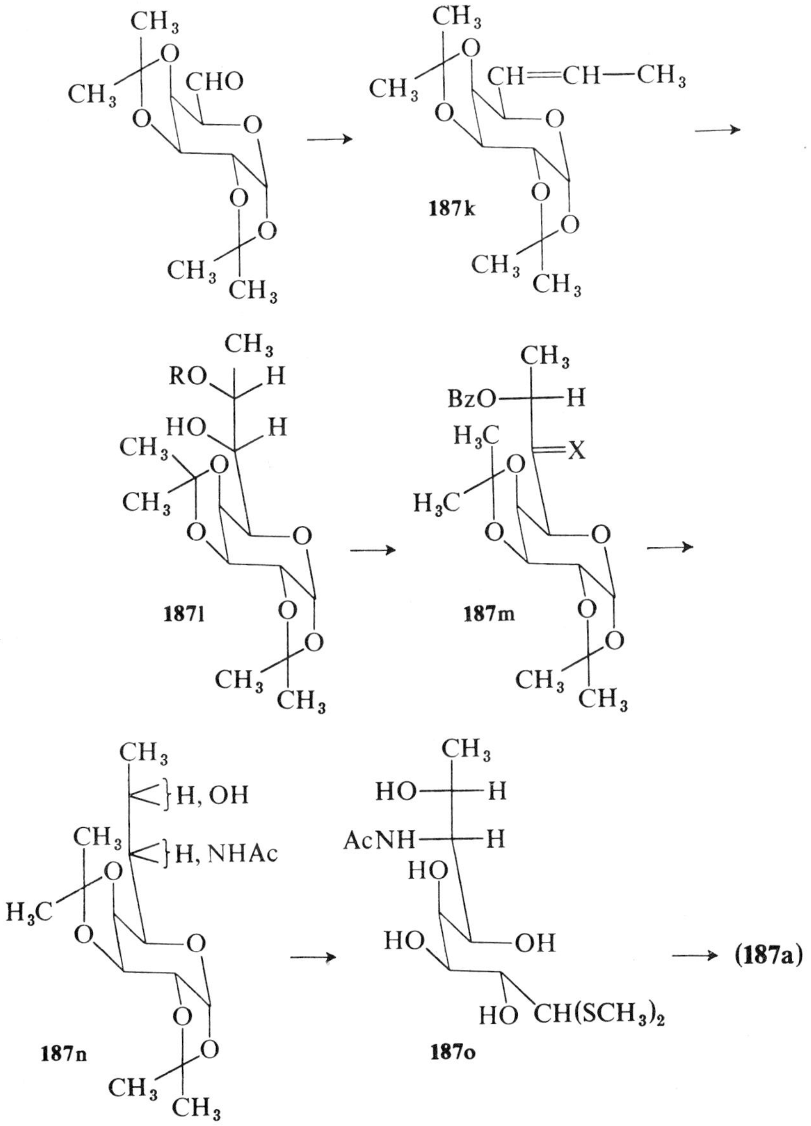

CH_3 O CH_3 CHO O O O O CH_3 CH_3 → CH_3 O CH_3 CH=CH—CH_3 O O **187k** O O CH_3 CH_3 →

CH_3 RO H HO H CH_3 O CH_3 O O **187l** O O CH_3 CH_3 → CH_3 BzO—H H_3C =X O H_3C O O **187m** O O CH_3 CH_3 →

CH_3 }H, OH CH_3 }H, NHAc O H_3C O O O O **187n** CH_3 CH_3 → CH_3 HO—H AcNH—H HO HO OH HO $CH(SCH_3)_2$ **187o** → (**187a**)

potassium permanganate gave the crystalline diol **187l** (R = H) from which the monobenzoate (**187l**; R = COPh) was easily prepared. Oxidation by the Pfitzner-Moffatt reagent then afforded the ketone **187m** (X = O), which was converted to a mixture of the oximes (**187m**; X = NOH) in which the *anti*-form predominated. Reduction of the latter isomer with lithium aluminum hydride led to a mixture of the *erythro* and *threo* forms of **187n**. Separation of the former in low yield was achieved with difficulty by means of chromatography. An alternate route to **187n** involved debenzoylation of the oxime **187m** (X = NOH) with a catalytic quantity of sodium methoxide in methanol followed by reduction of the oxime function by Raney nickel. Acetylation then afforded **187n**, but this constituted mainly the undesired *threo* isomer.

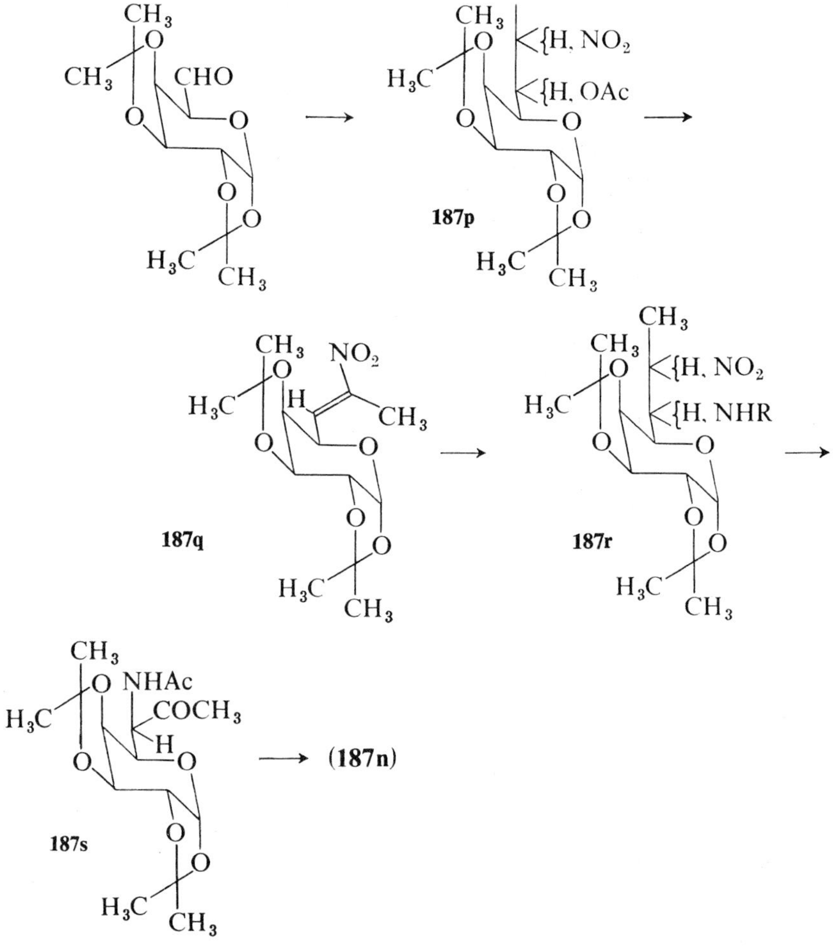

187p

187q

187r

187s → (**187n**)

Treatment of *erythro*-**187n** with methanethiol and concentrated hydrochloric acid afforded the dimethyl dithioacetal **187o** which treated with dilute acid yielded the *N*-acetyl derivative of thiolincosaminide. Deacetylation was achieved in high yield in refluxing 95% hydrazine hydrate and afforded **187a** identical with the material derived from lincomycin. The key intermediate in the above synthesis, namely the *erythro* form of **187n**, was also synthesized using the sequence described. The condensation of the starting aldehyde with nitroethane in the presence of sodium methoxide gave a mixture of β-nitro-alcohols, which were acetylated to give the corresponding β-nitro-acetates **187p** of which three were obtained pure by fractional crystallization. The predominant isomer when refluxed with triethylamine in benzene afforded the *cis*- and *trans*-isomers of **187q** in a ratio of 6:1. When the *cis*-isomer was treated with saturated methanolic ammonia, a mixture of the two stereoisomeric *vic*-nitroamines (**187r**; R = H) resulted. *N*-Acetylation of this mixture produced the corresponding acetyl derivatives (**187r**; R = Ac) in a 1:1 ratio. Oxidative denitration of this mixture with potassium permanganate led to two acetylamino-ketones the minor one of which (**187s**), when reduced with sodium borohydride gave two *vic*-acetamido alcohols. Fractional crystallization allowed their separation, and one proved to be identical with the previously prepared **187n**.

An alternate but somewhat lengthier procedure for the preparation of the intermediate *N*-acetyllincosamine, the hydrolysis product of **187n**, also has been published recently by Japanese workers.[66h] Again, the starting point in their synthesis was the 1,2,3,4-di-*O*-isopropylidene-α-D-*galacto*-hexodialdo-1,5-pyranose used by Sarek and his associates.[66e] The Japanese workers treated the latter with sodium cyanide in aqueous methanol and obtained a mixture of the *L*- and *D*-glycero cyanohydrins **187t**, which proved to be difficult to separate. Tosylation then gave the *O*-tosyl compounds, which were easily separated, and treatment[66i] of the *D*-glycero isomer **187u** with lithium aluminum hydride followed by acetylation afforded the aziridine **187v**. In contact with warm acetic acid **187v** afforded the 6-acetamido-7-*o*-acetate [**187w** (R — Ac)] exclusively. Deacetylation by means of sodium methoxide in methanol led to the 7-alcohol derivative **187w** (R = H). Oxidation of the latter by the Pfitzner-Moffatt procedure generated the aldehyde **187x** (R = Ts), which when treated with methyl magnesium bromide produced the undesired *threo* isomer **187y**. Conversion of this material to the desired isomer **187z** was achieved by oxidation of 7-hydroxyl group to the ketone by means of chromium trioxide in pyridine followed by sodium borohydride reduction. Hydrolysis of **187z** with aqueous acetic acid or the acid form of Amberlite IR-120 then afforded the desired *N*-acetyllincosamine.

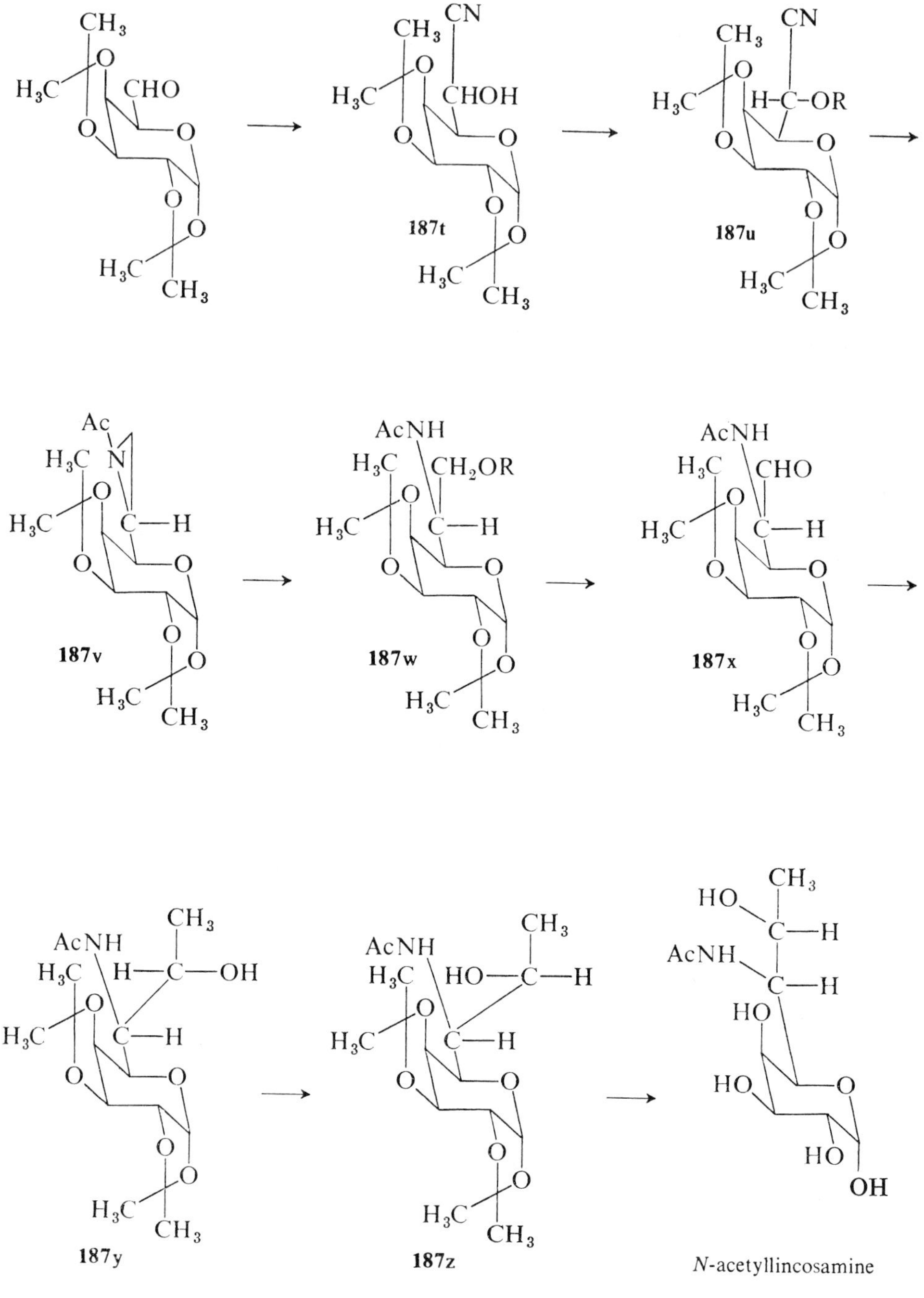

CH_3 H_3C O CHO O O O O H_3C CH_3 → CH_3 CN H_3C O CHOH O O O O H_3C CH_3 **187t** → CH_3 CN H_3C O H-C-OR O O O O H_3C CH_3 **187u** →

Ac N H_3C H_3C O C—H O O O O H_3C CH_3 **187v** → AcNH H_3C CH_2OR H_3C O C—H O O O O H_3C CH_3 **187w** → AcNH H_3C CHO H_3C O C—H O O O O H_3C CH_3 **187x** →

AcNH CH_3 H_3C H—C—OH H_3C O C—H O O O O H_3C CH_3 **187y** → AcNH CH_3 H_3C HO—C—H H_3C O C—H O O O O H_3C CH_3 **187z** → CH_3 HO C—H AcNH C—H HO HO O HO OH

N-acetyllincosamine

5. NUCLEOSIDE ANTIBIOTICS

About half of the therapeutically useful members of this group are closely related purine nucleosides. Angustmycin A (**189**) is the dehydration product of psicofuranine (**188**), both of which are derivatives of D-psicose

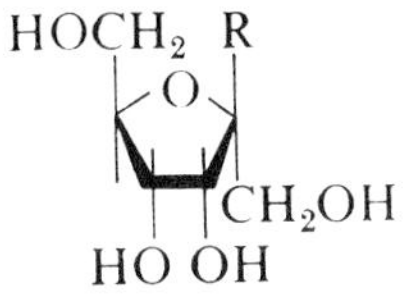

HOCH$_2$ R, O, CH$_2$OH, HO OH

188

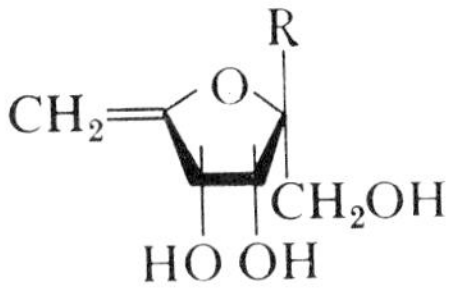

R, CH$_2$=, O, CH$_2$OH, HO OH

189

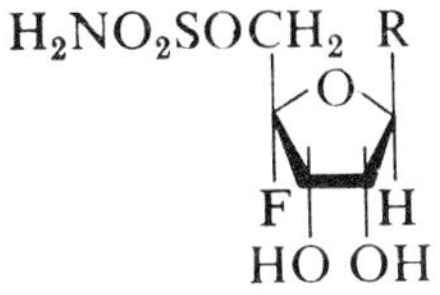

H$_2$NO$_2$SOCH$_2$ R, O, F, H, HO OH

190

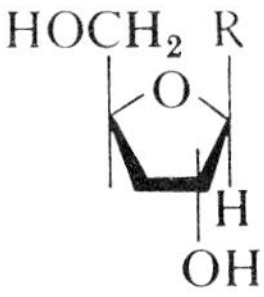

HOCH$_2$ R, O, H, OH

191

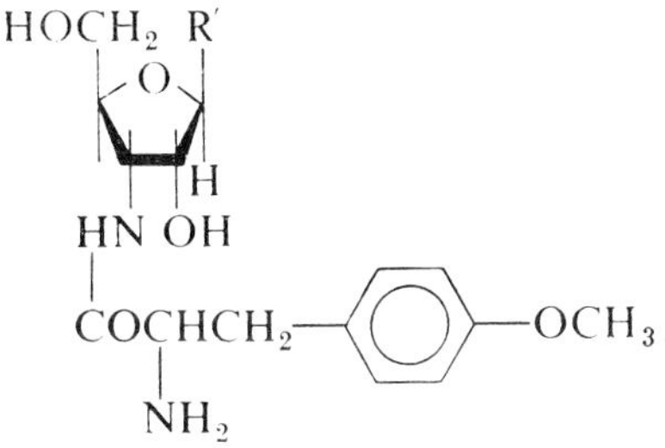

HOCH$_2$ R′, O, H, HN OH, COCHCH$_2$—C$_6$H$_4$—OCH$_3$, NH$_2$

192

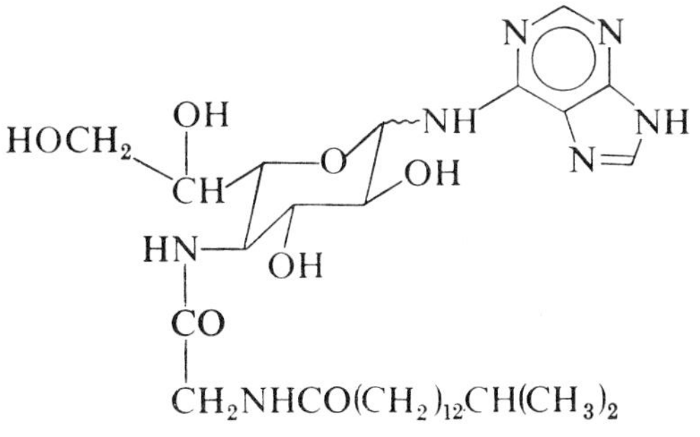

HOCH$_2$, OH, CH, O, NH, OH, HN, OH, CO, CH$_2$NHCO(CH$_2$)$_{12}$CH(CH$_3$)$_2$, N, N, NH, N

193

R = (adenin-9-yl: NH$_2$, N, N, N, N)

R′ = (6-dimethylaminopurin-9-yl: N(CH$_3$)$_2$, N, N, N, N)

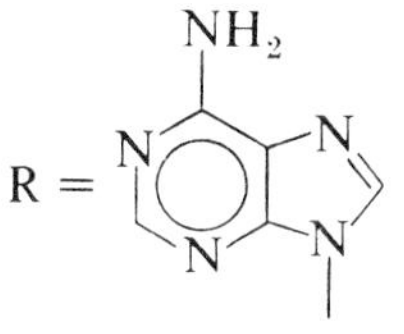

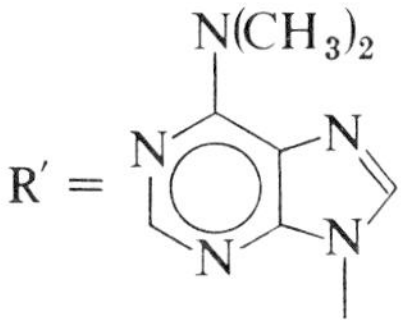

Nucleocidin[67] (**190**), on the other hand, is a ribofuranose derivative with the most unusual fluoro- and sulfamoyloxy groups on the sugar nucleus. It is related to cordycepin (**191**), an antibiotic of little medical value first isolated[68] from the culture fluids of *Cordyceps militaris* (Linn) link. In this group also is puromycin (**192**), a 3-amino–3-deoxyribofuranosyl purine. The most unusual member of the group is septacidin[69] (**193**), in which the 4-amino–4-deoxy-L-glycero-L-glucoheptose is attached at the 6-amino group of the adenine nucleus.

Three other compounds discussed in this section, tubercidin (**193a**), sangivamycin (**194**), and toyocamycin (**195**), are all closely related pyrrolopyrimidine nucleosides. These substances and the purine nucleosides described above all have been synthesized with the exception of nucleocidin and septacidin.

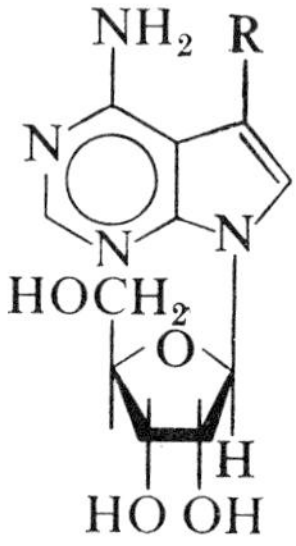

193a R = H
194 R = $CONH_2$
195 R = CN

The remaining compounds, excepting the racemomycins, are peptide nucleosides derived from pyrimidine. The polyoxins,[70] which comprise an antibiotic complex of 12 compounds, are nucleosides derived from uracil. Polyoxin A, for instance, has structure **196**. These compounds have great

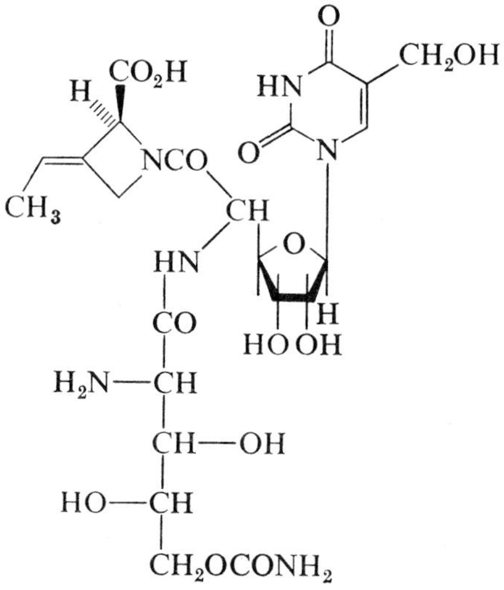

196

commercial importance in the Far East for the control of the sheath blight disease of the rice plant. None of the polyoxins has been synthesized but synthetic work on the sugar residue has been reported.[71] Blasticidin S (**197a**) is a cytosine nucleoside and is obviously related to gougerotin (**197b**), another nucleoside antibiotic,[71a] and to amicetin (**198**),[72] an antimicrobial

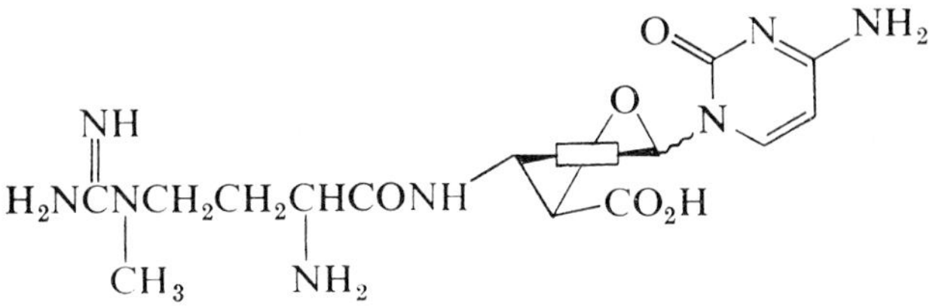

197a

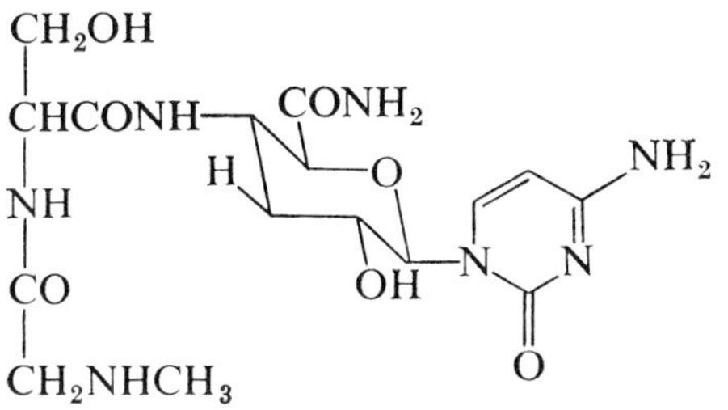

197b

compound having little or no therapeutic use at this time. Racemomycin A

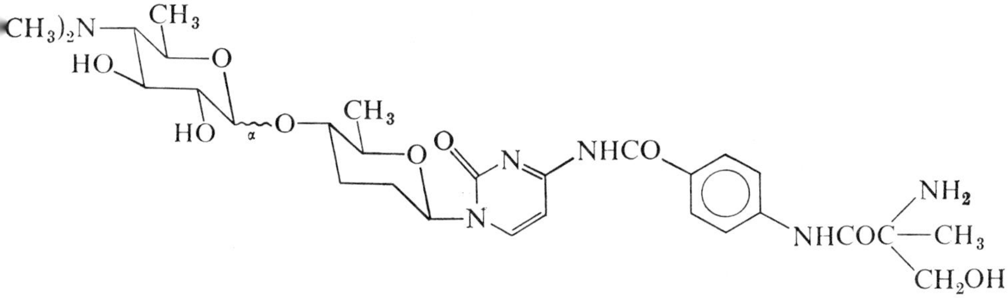

198

is extremely similar if not identical to streptolin[73] (**199**), perhaps the only question being the position of the carbamate group in the former. The structure of racemomycin O is again a variation of structure **199**, but the position of an additional $CH_3CH(OH)CH_2OCH_2CH_2CH<$ fragment has not been clarified.[74] No total synthesis work has been reported on these compounds.

In general the syntheses reported here have followed the well-established techniques for the synthesis of nucleosides wherein almost all of the work

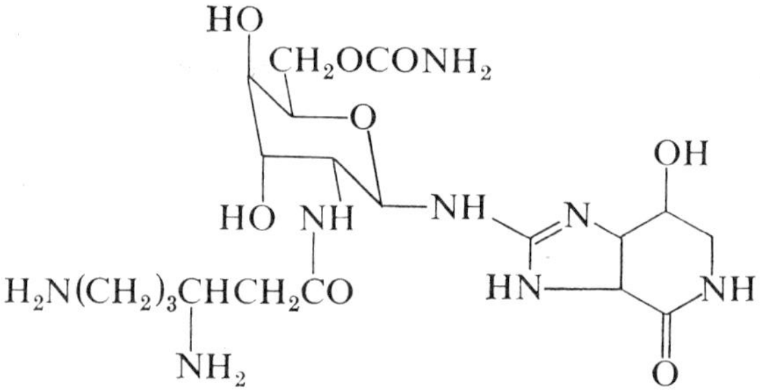

199

involves the modification of a readily available sugar (or derivative) via blocking-deblocking methods, to arrive at the required sugar moiety.

A. Purine Nucleosides

(1). *Puromycin*

Puromycin was the first of these antibiotics to be synthesized[75] and demanded that a suitably protected derivative (**214**) of 3-amino–3-deoxy–D-ribofuranose be prepared prior to the attempted formation of the nucleoside bond. This was accomplished[76] as shown in Schemes 24 and 25, starting with D-xylose (**200**), which was initially converted to a mixture of the methyl glycosides (**201**). Under very carefully controlled acid conditions (2×10^{-3} *N* sulfuric

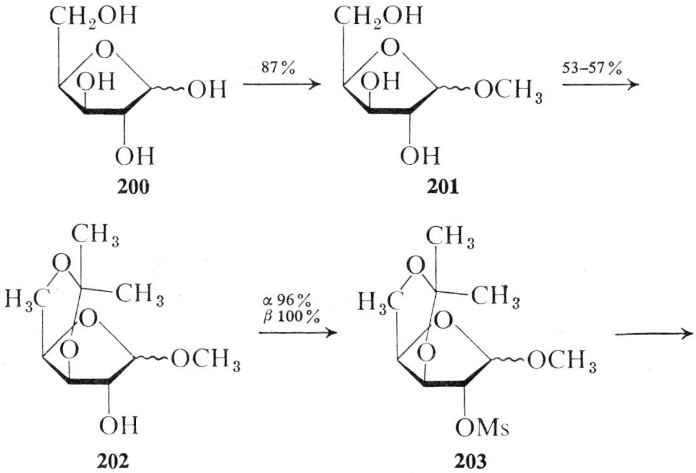

200 → (87%) 201 → (53–57%)

202 → (α 96%, β 100%) 203 →

Scheme 24

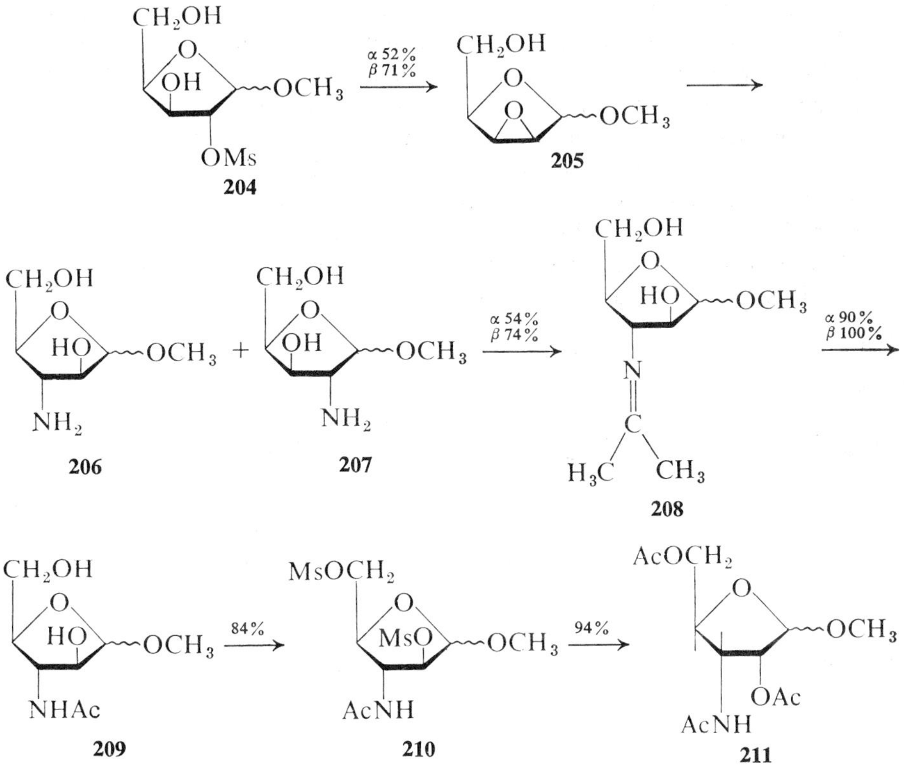

Scheme 24 (Continued)

acid) in an excess of acetone this mixture was transformed into the monoisopropylidene anomers **202**, which could be separated fairly easily (33 and 21% yield) because of a large difference in their boiling points. The remainder of the synthesis was carried out on each anomer individually, in an identical fashion.

Each series eventually afforded an anomer (**214**) that could be used in the final steps of the synthesis. Treatment of **202** with mesyl chloride afforded **203**, which was split to the diol **204** by means of warm 70% acetic acid. Sodium methoxide then was used to generate the anhydroglycoside **205**. The latter compound reacted with ammonia to give a mixture of the aminoglycosides **206** and **207**, but in the case of each anomeric series the addition of acetone caused the separation in crystalline form of the desired imine **208**. Aqueous acetic anhydride converted **208** to **209**, which had the 2-hydroxy group in the wrong orientation. This was reversed in two steps, first by conversion to the dimesylate **210** and then by treatment of **210** with sodium acetate in boiling 95% methyl cellosolve. The product of the reaction was in fact not **211** but the corresponding noncrystalline 2-hydroxy compound, which was converted to the highly crystalline **211** by means of acetic anhydride pyridine. The intermediate alcohol arises because of neighboring group participation of the acetylamino group which gives an oxazoline ring that subsequently is hydrolyzed by the water present in the reaction mixture.

In the next phase of the synthesis (Scheme 25) the triacetyl derivative was *O*-deacetylated by sodium methoxide in methanol to **212**, which under normal benzoylating conditions afforded the dibenzoate **213**. Either anomer of **213** when treated with hydrogen chloride in acetic acid under well-defined conditions led to the desired protected sugar **214**.

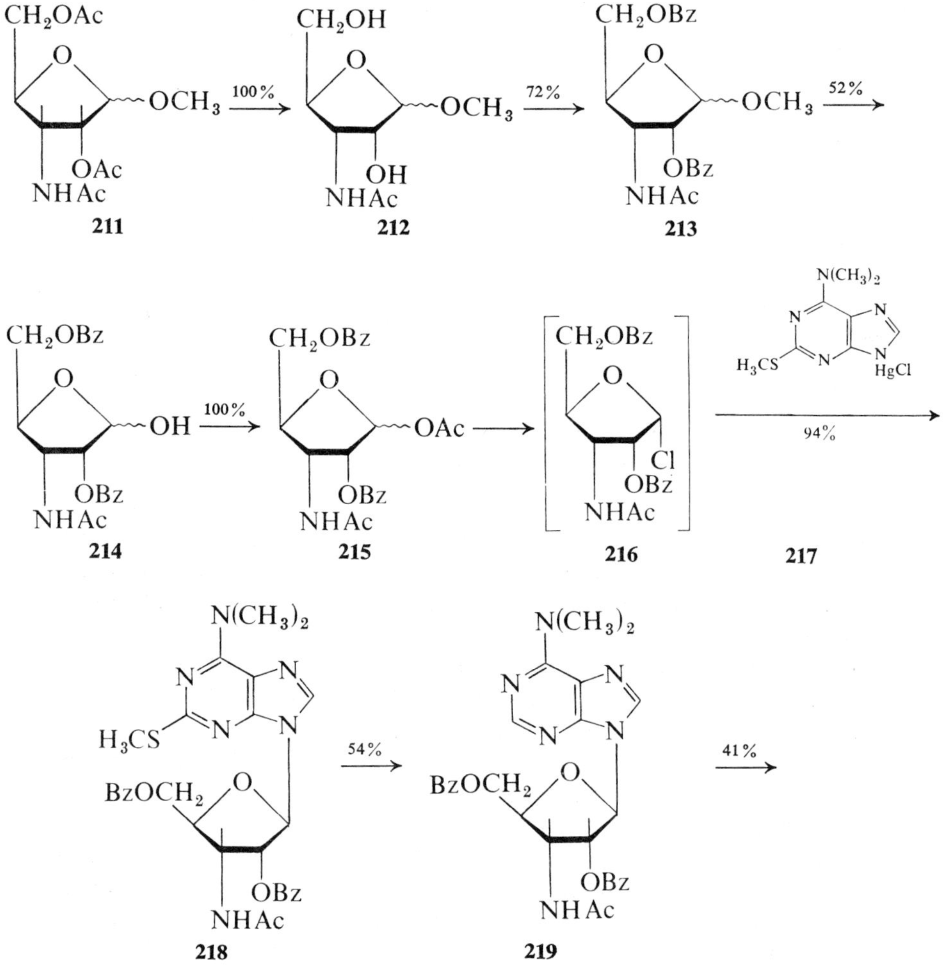

CH_2OAc O OCH_3 OAc NHAc **211** —100%→ CH_2OH O OCH_3 OH NHAc **212** —72%→ CH_2OBz O OCH_3 OBz NHAc **213** —52%→

CH_2OBz O OH OBz NHAc **214** —100%→ CH_2OBz O OAc OBz NHAc **215** → [CH_2OBz O Cl OBz NHAc] **216** —($N(CH_3)_2$ N N H_3CS N N HgCl **217**), 94%→

$N(CH_3)_2$ N N H_3CS N N $BzOCH_2$ O OBz NHAc **218** —54%→ $N(CH_3)_2$ N N N N $BzOCH_2$ O OBz NHAc **219** —41%→

Scheme 25

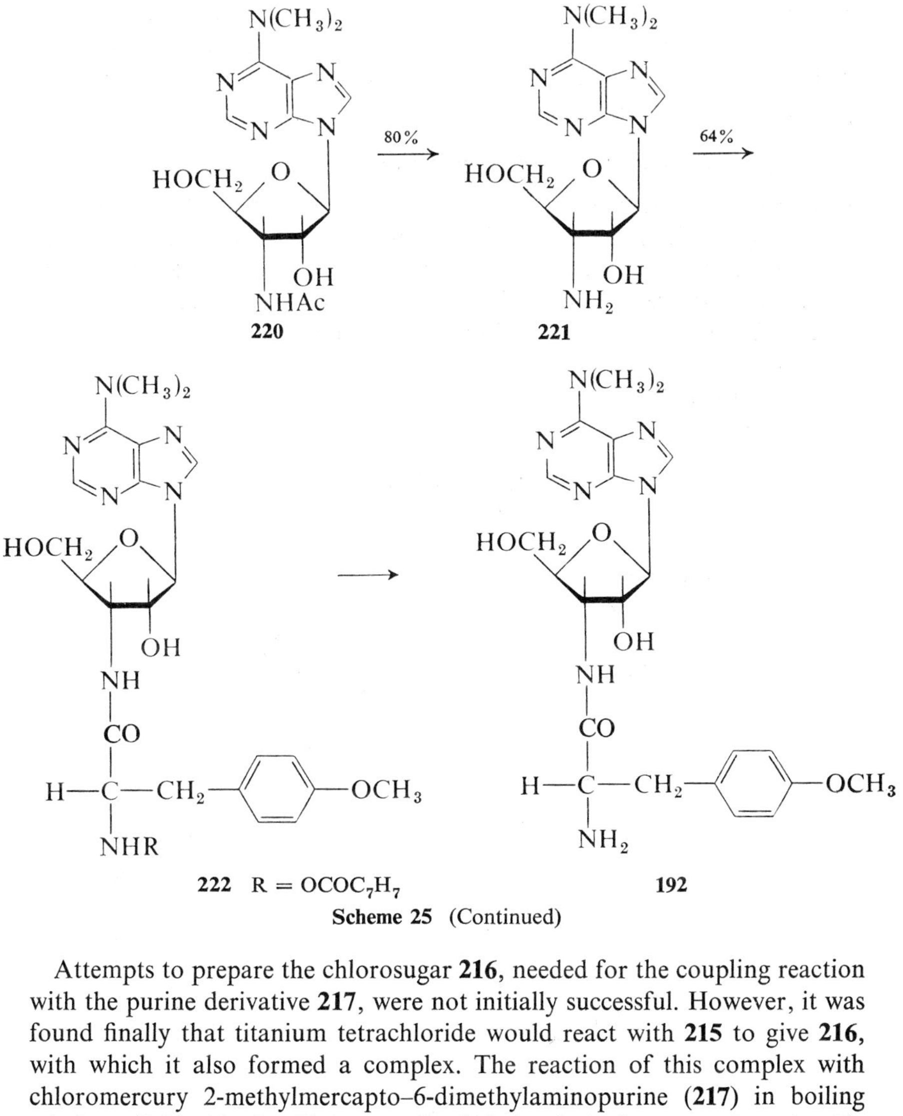

220 $\xrightarrow{80\%}$ 221 $\xrightarrow{64\%}$

222 R = $OCOC_7H_7$ ⟶ 192

Scheme 25 (Continued)

Attempts to prepare the chlorosugar **216**, needed for the coupling reaction with the purine derivative **217**, were not initially successful. However, it was found finally that titanium tetrachloride would react with **215** to give **216**, with which it also formed a complex. The reaction of this complex with chloromercury 2-methylmercapto–6-dimethylaminopurine (**217**) in boiling ethylene dichloride for 18 hours afforded the desired protected nucleoside **218** in high yield. Raney nickel desulfurization then eliminated the methythio group to give **219**. Removal of the sugar protective groups was achieved[77] by treatment first with sodium methoxide/methanol to give **220** then with aqueous barium hydroxide to effect *N*-deacetylation. The product was **221**, the necessary nucleosidic portion of puromycin. Introduction of the amino acid was then accomplished by the reaction of **221** with the carbethoxy mixed anhydride of *N*-carbobenzyloxy-*p*-methoxy-L-phenylalanine, which led to *N*-carbobenzyloxypuromycin (**222**). Hydrogenolysis of **222** with a palladium-on-charcoal catalyst in acetic acid then afforded puromycin (**192**), identical in all respects with the natural product.

(2). *Psicofuranine and Angustmycin A*

Psicofuranine **188** is a nucleoside comprised of adenine and *D*-psicose. It was first synthesized, essentially from these components, by the group of investigators that deduced its structure.[78] The D-psicosyl chloride tetraacetate **226** needed was prepared according to the procedure of Wolfrom et al.,[79] (Scheme 26) in which ribonyl chloride tetraacetate was treated with diazo-

$$\underset{\mathbf{223}}{H(CH_2OAc)_4COCl} \xrightarrow{81\%} \underset{\mathbf{224}}{H(CH_2OAc)_4COCHN_2} \xrightarrow{50\%}$$

$$\underset{\mathbf{225}}{H(CH_2OAc)_4COCH_2OAc} \xrightarrow{92\%}$$

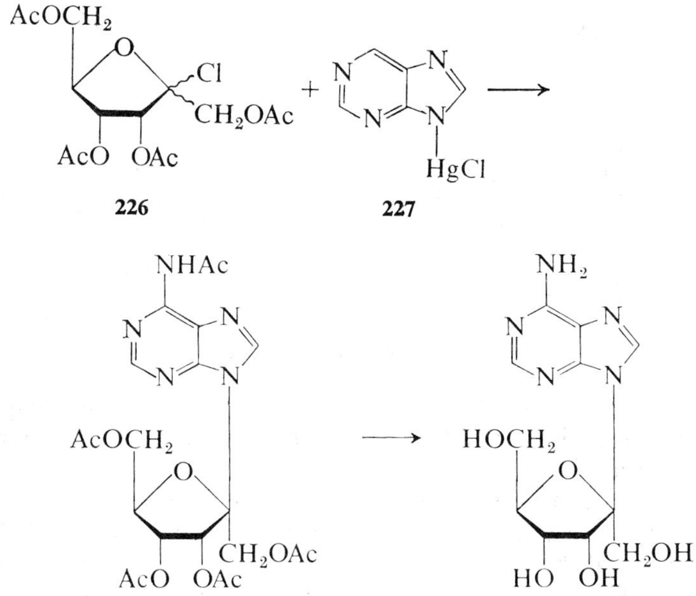

AcOCH₂ O Cl CH₂OAc AcO OAc + N N N N HgCl →

226 **227**

NHAc N N N N AcOCH₂ O CH₂OAc AcO OAc → NH₂ N N N N HOCH₂ O CH₂OH HO OH

228 **188**

Scheme 26

methane to give **224**. Reaction of **224** with acetic acid gave keto-D-psicose pentaacetate (**225**), which with hydrogen chloride in ether afforded the required chloride **226**. Condensation of **226** with chloromercuri-6-acetylaminopurine (**227**) provided **228** in unspecified yield. Deacetylation then afforded psicofuranine (**188**) identical with the natural product.

A second synthesis of **188** has also been recorded by Farkaš and Šorm.[80] In their case *D*-psicosyl bromide tetrabenzoate was condensed in dimethylacetamide with the chloromercury salt of 6-benzoylaminopurine and the product then debenzoylated by means of barium methylate solution. After chromatography of the crude material psicofuranine was obtained in 4.6% yield.

Angustmycin A (**189**) has been synthesized[81] partially from psicofuranine. In this transformation it was necessary to protect the 1,3- and 4-hydroxyl groups while a method was found for the dehydration of the 6-hydroxyl. This was accomplished as shown in Scheme 27.

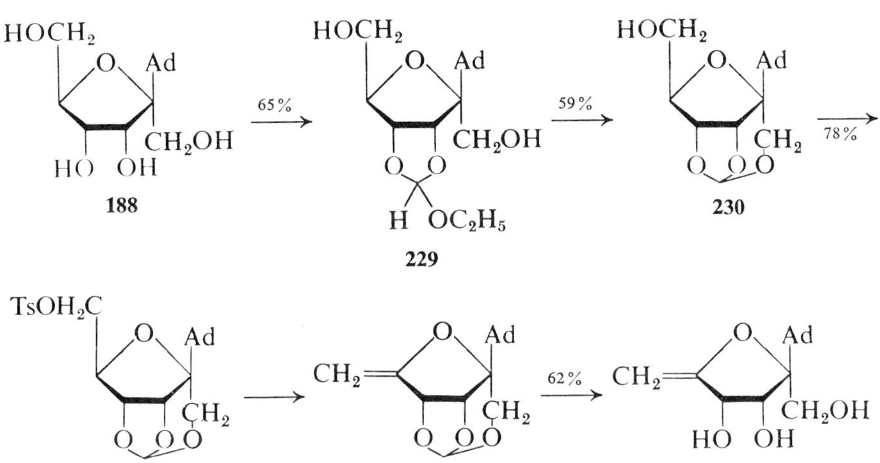

$HOCH_2$ O Ad CH_2OH HO OH **188** —65%→ $HOCH_2$ O Ad CH_2OH O O H OC_2H_5 **229** —59%→ $HOCH_2$ O Ad CH_2 O O O **230** —78%→

$TsOH_2C$ O Ad CH_2 O O O **231** —→ CH_2= O Ad CH_2 O O O **232** —62%→ CH_2= O Ad CH_2OH HO OH **189**

Ad = Adenine

Scheme 27

Treatment of psicofuranine (**188**) with triethyl orthoformate at room temperature afforded the ethoxymethylidine derivative **229**, which underwent a facile ring closure to the orthoester **230** when subjected to boron trifluoride etherate in dioxane. Reaction of the latter with tosyl chloride in pyridine gave the tosyl ester **231**, which when treated with potassium *t*-butoxide in *t*-butyl alcohol/pyridine led to the orthoformyl ester **232** of angustmycin A. Careful hydrolysis of **232** with warm aqueous acetic acid and subsequent treatment with alcoholic ammonia to remove the acetyl groups gave angustmycin A identical with an authentic specimen.

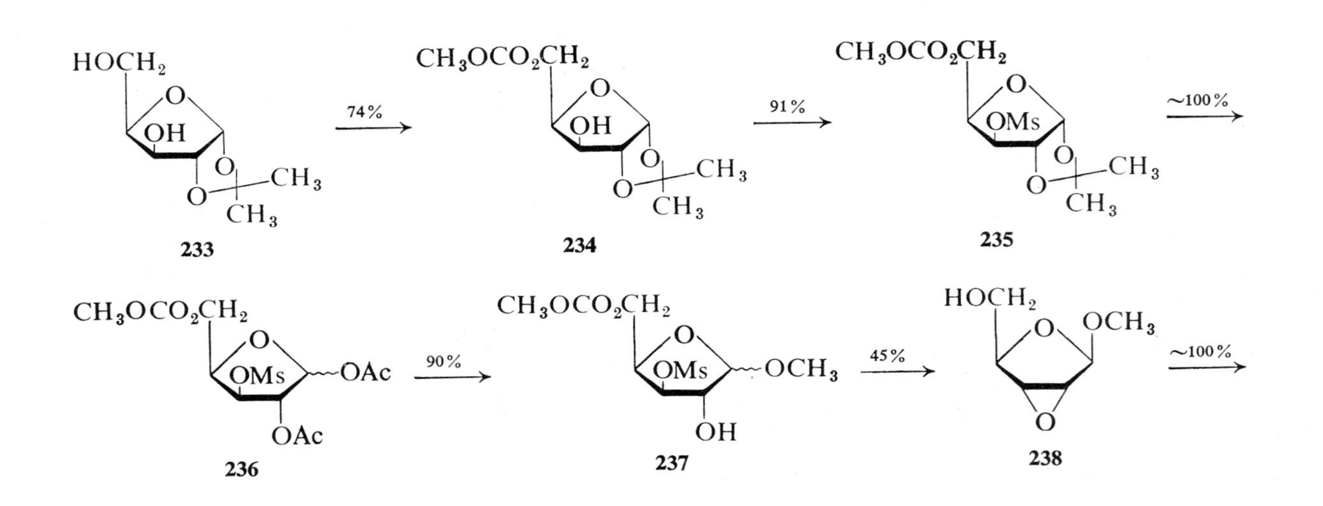
HOCH2
OH
O
CH3
233
74%
CH3OCO2CH2
234
91%
OMs
235
~100%
OAc
236
90%
OCH3
237
45%
238
~100%

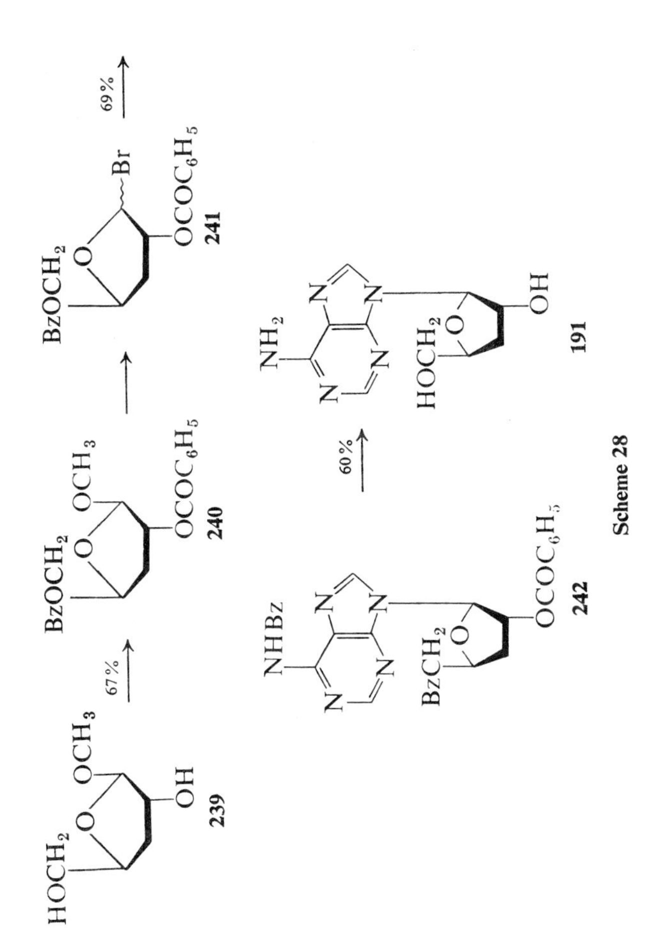

HOCH2
OCH3
OH
239
67%
BzOCH2
OCH3
OCOC6H5
240
BzOCH2
Br
OCOC6H5
241
69%
NHBz
BzCH2
OCOC6H5
242
60%
NH2
HOCH2
OH
191

Scheme 28

(3) *Cordycepin*

Cordycepin or 3′-deoxyadensosine was synthesized by Dalton and his associates.[82] The protected deoxyribofuranosyl bromide **241** needed for this synthesis was prepared in part according to the procedure of Baker[83] beginning with the readily available 1,2-*O*-isopropylidene-*D*-xylofuranose **233**. Treatment of **233** with excess methyl chloroformate in pyridine afforded almost exclusively the monoester **234**. Mesylation of **234** produced **235**, which when subjected to acetolysis in acetic acid/acetic anhydride containing 1% sulfuric acid generated a mixture of anomers (**236**). Methanolysis of **236** with 1% methanolic hydrogen chloride gave the corresponding α- and β-methyl glycosides (**237**), which under the influence of methanolic sodium methoxide yielded the anhydrosugars. These anomers were easily separated by distillation and the β-glycoside **238** was then hydrogenated over Raney nickel. This proceeded almost exclusively to methyl 3-deoxy-β-D-ribofuranoside (**239**), which was benzoylated to give **240**. Reaction of **240** with 30% hydrogen bromide in acetic acid for a few minutes then gave the crude bromide **241**, which was used directly in a condensation step with chloromercuri-6-benzoylaminopurine, which produced the protected nucleoside **242**. Finally debenzoylation was achieved by boiling methanolic sodium methoxide to give cordycepin (**191**).

B. Pyrrolopyrimidine Nucleosides (Tubercidin, Toyocamycin and Sangivamycin)

The only known naturally occurring pyrrolopyrimidine nucleosides have all been synthesized by Townsend and his group.[84] The synthetic scheme followed is shown in Scheme 29.

A previous attempt to synthesize toyocamycin (**195**) had failed[85] when no method could be found to desulfurize the penultimate compound **243**. The

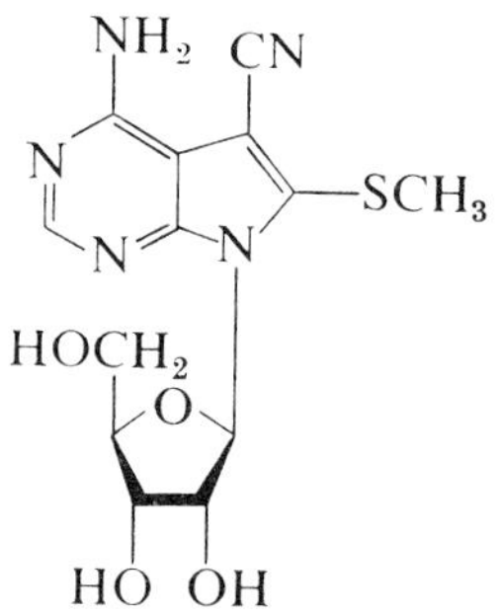

NH_2 CN

N

N N

SCH_3

$HOCH_2$

O

HO OH

243

NC CN H$_2$N N Br
244

51%

NH$_2$ CN N N N Br
245

94%

AcNH CN N N N Br
246

9%

AcNH CN N N N Br AcCH$_2$ O AcO OAc
247

92%

NH$_2$ CN N N N Br HOCH$_2$ O HO OH
248

NH$_2$ CN N N N HOCH$_2$ O HO OH
195

NH$_2$ CONH$_2$ N N N HOCH$_2$ O HO OH
194

NH$_2$ CO$_2$H N N N HOCH$_2$ O HO OH
249

13%

NH$_2$ N N N HOCH$_2$ O HO OH
193a

Scheme 29

new approach started with 2-amino–5-bromo–3,4-dicyanopyrrole **244**, which when refluxed with formamidine acetate in 2-ethoxyethanol afforded **245**. Acetylation of **245** led to **246** and the latter was then fused with 1,2,3,4-tetra-*O*-acetyl–*β*-D-ribofuranose in the presence of a catalytic amount of bis(*p*-nitrophenyl)phosphate[86] to give the tetraacetyl nucleoside **247**. Removal of the acetyl groups was accomplished by methanolic ammonia and yielded **248**, which was dehalogenated to toyocamycin, identical with the natural product, by catalytic reduction in the presence of a palladium catalyst. Since hydrolysis of **195** previously had been shown[87] to give the acid **249** (albeit in low yield), which underwent decarboxylation[84] when heated briefly at 238°, this constitutes a total synthesis of tubercidin (**193a**) also.

Sangivamycin (**194**) was in turn easily synthesized from toyocamycin (**195**) by oxidation with 30% hydrogen peroxide in concentrated ammonium hydroxide solution, thus completing total synthesis in this area. (Although several other subsidiary procedures also were reported[84] for the syntheses of these compounds, the methods just described appear to be the best, overall.)

C. Miscellaneous Nucleosides (Polyoxins, Blasticidin S)

Synthesis in the area of the polyoxins (e.g., **196**) has been limited so far to the sugar moiety of which a suitably protected derivative (**258**) has been prepared[71] (Scheme 30) starting with 1,2:5,6-di-*O*-isopropylidene–α-D-allofuranose (**250**). The latter material, easily synthesized[88] from D-glucose, was benzoylated and then treated with 1.5% sulfuric acid in aqueous ethanol to give **251** by selective hydrolysis of the 5,6-isopropylidene group. Mesylation of **251** followed by boiling the product in dimethyl formamide with sodium benzoate afforded the desired material **252** in which a reversal of configuration at the 5-position had been achieved. Debenzoylation of **252** by means of sodium methoxide in methanol, then treatment of the product with dimethoxypropane, yielded **253**. Benzoylation followed by selective hydrolysis led to **254**, which was in turn tritylated and mesylated to give **255**. The amino group desired at the 5-position was now achieved in two steps first by boiling **255** with sodium azide in dimethylformamide, which afforded **256** (configurational inversion at C-5), then catalytic reduction over a palladium catalyst to give the amine, which was benzoylated immediately then detritylated with *p*-toluenesulfonic acid in acetone. The product of these transformations, **257**, was next oxidized with potassium permanganate in acetone acetic acid solution and the resulting acid esterified with diazomethane to give **258**. This protected derivative of 5-amino–5-deoxy–α-D-allo-furanuronic acid, the sugar component of the polyoxins, is a suitable staging compound for further work toward the total synthesis of the polyoxins.

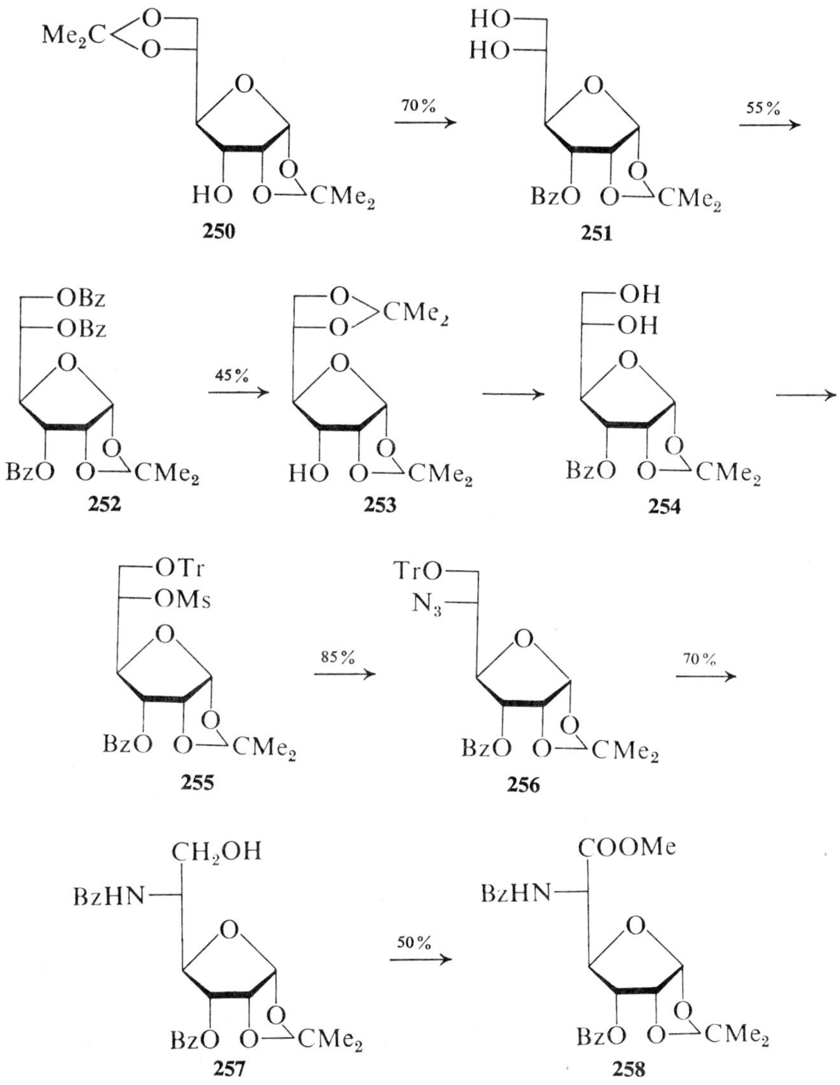

Scheme 30

The carbohydrate component **268** of blasticidin S (**197a**) also has been synthesized[89] recently, by the same group who synthesized both the nucleoside portion[90] ("C"-substance) of gougerotin (**197b**) and a nucleoside[90a] corresponding to that of blasticidin S in which there is a hydroxymethyl group in place of the carboxylic acid. The synthesis of **268** is reported here because of its potential for the ultimate synthesis of blasticidin S itself. Of the two methods described by Fox and his co-workers[89] for the synthesis of **268**, that shown in Scheme 31 appears to be the better.

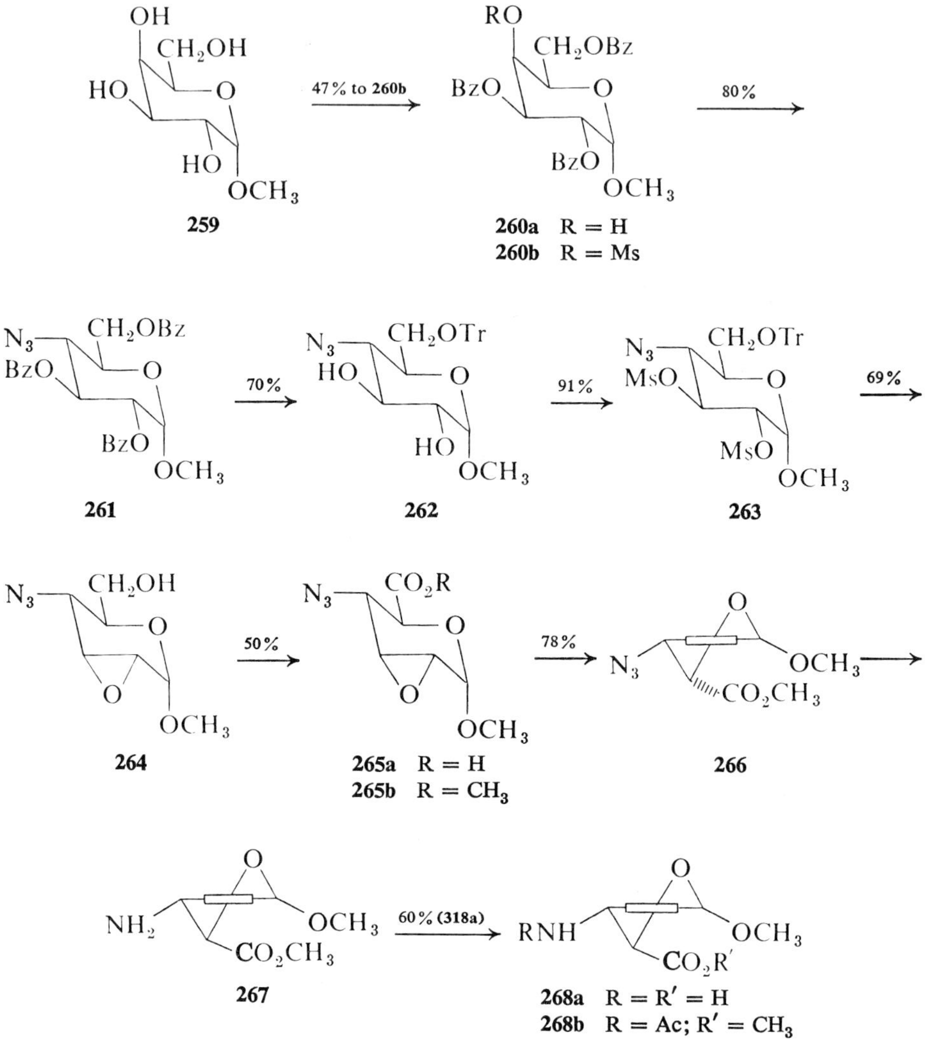

259 → (47% to **260b**) → **260a** R = H; **260b** R = Ms → (80%) → **261** → (70%) → **262** → (91%) → **263** → (69%) → **264** → (50%) → **265a** R = H; **265b** R = CH_3 → (78%) → **266** → **267** → (60% (318a)) → **268a** R = R′ = H; **268b** R = Ac; R′ = CH_3

Scheme 31

Methyl α-D-galactopyranoside (**259**) was benzoylated selectively at the equatorial hydroxyl groups to give **260a** according to the method of Reist et al.[90b] and then mesylated at the 4-position. The product **260b** was treated with sodium azide in hexamethylphosphoramide and gave **261**, which was debenzoylated with sodium methoxide in methanol then tritylated to give **262**. Mesylation of **262** afforded **263**, which when treated with sodium methoxide gave an epoxide. Detritylation of the latter led to **264** and oxidation of this material by potassium permanganate yielded the uronic acid **265a**. The methyl ester **265b** of this acid upon treatment with sodium iodide produced a mixture of iodohydrins that were mesylated in pyridine directly. Two products, which could be separated chromatographically, resulted from this reaction. One was the desired unsaturated azide **266** and the other an iodo mesylate that could be converted to **266** by a combination of tetramethylammonium chloride and zinc dust in pyridine. Reduction of **266** with sodium dithionite at *p*H 7 yielded the amino ester **267**, which without purification was saponified by methanolic sodium hydroxide to give crystalline methyl 4-amino-2,3,4-trideoxy-α-D-erythrohex-2-enpyranosiduronic acid (**268a**), the methyl glycoside of the carbohydrate moiety of blasticidin S. Hydrogenation of **266** was also carried out using a platinum catalyst. Acetylation of the crude product then afforded **268b** in almost quantitative yield.

Theoretically all that remains to complete the total synthesis of this antibiotic is to couple a suitable derivative such as the pyranosyl chloride from **268b** with cytosine (or a derivative) at the 1-position of the latter to give the cytosinine (**269a**) (after deblocking), since blasticidin S has already been reconstructed[91] (Scheme 32) from **269a** and blastidic acid (**270a**). In

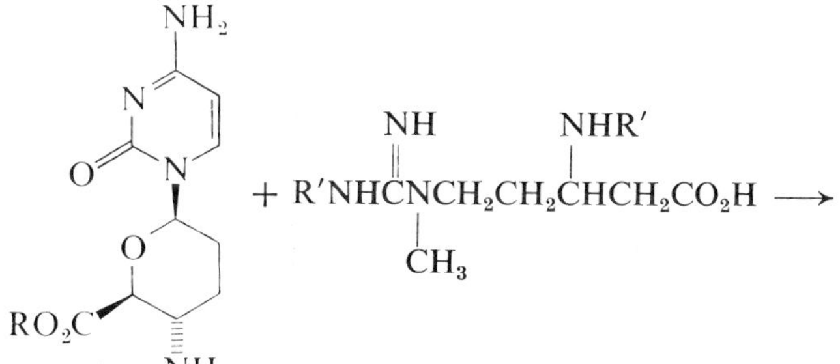

NH_2, N, O, N, O, RO_2C, NH_2

+ $R'NHC(=NH)N(CH_3)CH_2CH_2CH(NHR')CH_2CO_2H$ ⟶

269a R = H
269b R = CH_3

270a R = H
270b R = $C_7H_7OCO—$

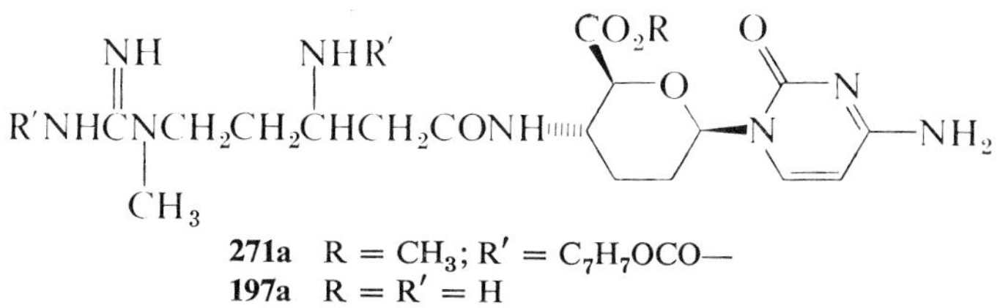

$R'NHC(=NH)N(CH_3)CH_2CH_2CH(NHR')CH_2CONH$, CO_2R, O, O, N, N, NH_2

271a R = CH_3; R′ = $C_7H_7OCO—$
197a R = R′ = H

Scheme 32

this partial synthesis the dihydrochloride of **269b** and *N,N′*-dicarbobenzyloxy-blastidic acid (**270b**) were coupled using *N,N′*-dicyclohexylcarbodiimide in methanol-acetonitrile containing triethylamine to give the protected derivative **271a**. It should be noted that the peptide bond formation occurs preferentially at the desired site because the latter is the most basic of those available. Removal of the carbobenzyloxy groups was then accomplished in dry methanolic hydrogen chloride. Saponification of the resulting methyl ester on a basic ion exchange resin afforded blasticidin S.

6. PEPTIDE AND DEPSIPEPTIDE ANTIBIOTICS

Many antibiotic substances are small polypeptides and those of known structure range from about 15 (gramicidin A) down to 2 (penicillin) in amino acid residues. Others of unknown structure obviously contain many more peptide linkages since molecular weights in this class range up to 14,000 (saramycetin[92]). Most, however, are less than 2000. Only a limited number of the 50 or more identified compounds are of known structure and only a few are of therapeutic value. Of these we may include (apart from the penicillins) the following: the bacitracins (topical use); actinomycin D (cancer therapy); capreomycin and viomycin (tuberculosis); the gramicidins and tyrocidins (topical use); the polymixins (urinary tract infections); the pristinamycins and staphylomycins (systemic infections).

Total syntheses have been achieved in the areas of the gramicidins, the tyrocidines, the polymixins, and the actinomycins (depsipeptides). The individual classes are discussed in this section. Much synthetic work has been carried out by Kaneko[93] in an attempt to synthesize bacitracin A (**272**), but so far no success has been recorded. This appears to be due in part to the

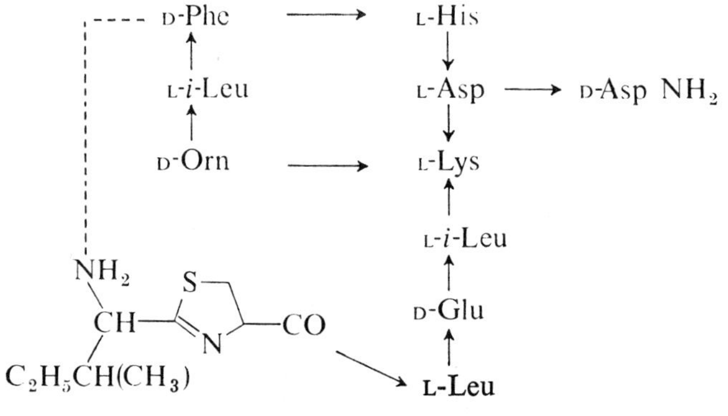

D-Phe → L-His → L-Asp → D-Asp NH_2

L-*i*-Leu → D-Phe

D-Orn → L-*i*-Leu; D-Orn → L-Lys

L-Asp → L-Lys

L-Leu → D-Glu → L-*i*-Leu → L-Lys

NH_2–CH($C_2H_5CH(CH_3)$)–thiazoline–CO → L-Leu

272 (broken line represents link in *cyclo* form)

few ambiguous points still unresolved about the structure[93,94] and in part to the racemization problems that are encountered when attempts are made to forge the peptide link between the thiazoline moiety and the L-leucine residue.[93]

Staphylomycin S (**273**) comprises only 5% of the crude staphylomycin complex. It is an unusual substance both in its amino acid content, which

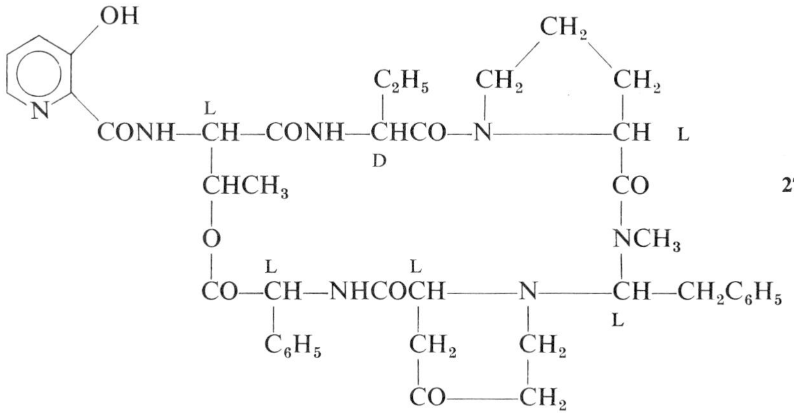

273

includes 4-piperidene–2-carboxylic acid, and in having one ester link in the macrocyclic ring. It is usually classified as a depsipeptide. Even more unusual in structure is its sister compound staphylomycin M[96] (**274**), which constitutes

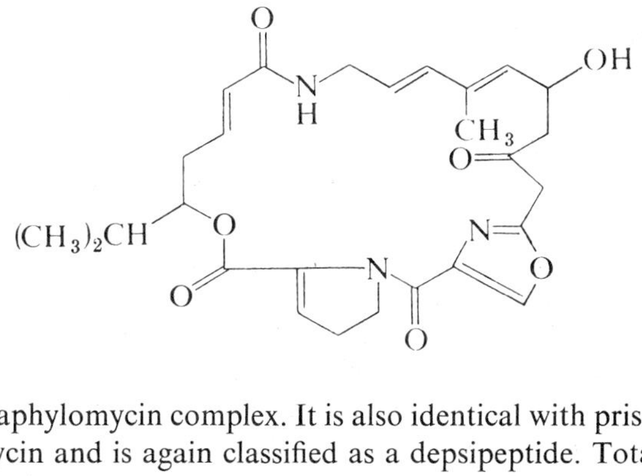

274

75% of the staphylomycin complex. It is also identical with pristinamycin IIA and ostreogrycin and is again classified as a depsipeptide. Total synthesis of either **273** or **274** has not been reported.

The structures of capreomycin and viomycin are not completely clear[97] and total synthesis is still out of the question.

Virtually none of the true macrocyclic depsipeptides is used therapeutically. Among the better known are the enniatins[98,99] A (**274a**), B (**274b**), and C

(**274c**), sporidesmolide-I,[100] serratamolide (**275**),[101] beauvericin (**276**),[102] and valinomycin (**277**).[103]

```
CH3        R    HC(CH3)2          R
   \       |    |                 |
    N—CHCO2CHCO—N————CHCO
    |  L     D       |       L     |
    |               CH3            |
    |                              |
    |          CH3                 |
    |  D     L  |    D             |
    COCHO2CCHNCOCH—————————O
       |      |      |
 (CH3)2CH     R      HC(CH3)2
```

274a $R = CH(CH_3)C_2H_5$
274b $R = CH(CH_3)_2$
274c $R = CH_2CH(CH_3)_2$
276 $R = CH_2C_6H_5$

```
                    CH2(CH2)5CH3
            CH2OH   |
            |       |
         HNCHCO2CHCH2CO              275
           |           |
          COCH2CHO2CCHNH
                |     |
     CH3(CH2)5CH2    CH2OH
```

```
[    CH3        CH(CH3)2   CH(CH3)2    CH(CH3)2  ]
     |          |          |           |         ]
[ —OCHCONHCH————CO2CHCONH—C—CO—               ]    277
     L          L          D           D         ]3
```

Valinomycin, which is the largest of these depsipeptides, contains a 36-membered ring[104] and is too toxic for use in mammalian species. It has in common with a number of these compounds the ability to complex with alkali metal cations and induce cation permeability of artificial and biological semipermeable membranes.

Most of the synthetic work in this area has been carried out by Shemyakin and his group at the Academy of Sciences in Moscow. A review[105] of much of the chemistry of the depsipeptides has appeared recently.

Syntheses of the depsipeptides, as with the macrocyclic and linear peptides discussed in this section, follow well-established patterns. In the area of the depsipeptides (apart from actinomycin S), the total syntheses of the antibiotic beauvaricin (**276**) and serratomolide (**275**) are discussed since they illustrate the two different approaches possible in this area.

A. Tyrothricin

The complex obtained by the extraction of acidified cultures of *Bacillus brevis* was termed "tyrothricin."[106] This material is easily resolved into two fractions, one of which is soluble in ether or acetone and is termed the "gramicidin" fraction constituting 20–40% of the initial mixture. The insoluble fraction is the "tyrocidine" fraction. The neutral gramicidin fraction[107] was fractionated[108] by countercurrent distribution and yielded a number of fractions, among which were three crystalline compounds, gramicidins A, B, and C. Although gramicidin A was thought to be cyclic originally, Witkop[109,110] has shown that it is in fact a mixture of two compounds Val[1]-gramicidin A (**278**) and Ileu[1]-gramicidin A (**279**). The former is a linear decapeptide formylated

OCH-L-Val-Gly-L-Ala-D-Leu-L-Ala-D-Val-L-Val-D-Val-
L-Try-D-Leu-L-Try-D-Leu-L-Try-D-Leu-L-Try-$NHCH_2CH_2OH$

278

at the NH_2-terminal residue and bearing ethanolamine at the CO_2H-terminal residue. In Ileu[1]-gramicidin A (**279**) the terminal valyl residue at the beginning of the chain is replaced by an isoleucine residue. Both have been synthesized by Witkop.[111] Gramicidin B again was shown[112] to be a mixture Val[1]-gramicidin B and Ileu-gramicidin B. These are very similar to the gramicidins A except that the L-Try[12] residue is replaced by an L-phenylalanine. In gramicidin C one tryptophane residue is replaced by tyrosine.

Gramicidin S (**279**) was isolated crystalline from *Bacillus brevis* by Gause and Brazhnikova.[113] The structure of gramicidin S was determined by Synge, Consden et al.[114] and its synthesis was accomplished by Schwyzer and Sieber.[115]

L-Val ⟶ L-Orn ⟶ L-Leu ⟶ D-Phe ⟶ L-Pro
↓ (L-Val) ... ↓ (L-Pro)
L-Pro ⟵ D-Phe ⟵ L-Leu ⟵ L-Orn ⟵ L-Val

279

The so-called gramicidins J_1 and J_2, which were originally thought to be cyclohepta- and cyclohexapolypeptides, recently have been shown to be identical[116] with gramicidin S.

The tyrocidine fraction of tyrothricin was separated by countercurrent distribution into three components, tyrocidines A, B, and C. Their structures were determined as **280**,[117] **281a**,[118] and **281b**,[119] respectively.

Syntheses of tyrocidines A,[120] B,[121] and C[122] have been accomplished by Izumiya and his group, who have also synthesized[123] a polypeptide antibiotic tyrocidine E, recently isolated by Kurahashi et al.,[124] from a culture of *Bacillus brevis* ATCC8185. This is the same as tyrocidine A except that the L-Tyr residue of the latter is replaced by L-Phe.

```
L-Val ⟶ L-Orn ⟶ L-Leu ⟶ D-Phe ⟶ L-Pro
  ↑                                 |
  |      NH2      NH2               |        280
  |       |        |                ↓
L-Tyr ⟵ L-Glu ⟵ L-Asp ⟵ D-Phe ⟵ L-Phe
```

```
L-Val ⟶ L-Orn ⟶ L-Leu ⟶ D-Phe ⟶ L-Pro
  ↑                                 |
  |      NH2      NH2               |
  |       |        |                ↓
L-Tyr ⟵ L-Glu ⟵ L-Asp ⟵  [X]  ⟵ L-Try
```

281a [X] = D-Phe

281b [X] = D-Try

In the area of the tyrothricin antibiotics the syntheses of Val[1]-gramicidin A, gramicidin S, and tyrocidine A are representative and are thus discussed individually. In these syntheses the following abbreviations are used: Z = carbobenzyloxy; Np = *p*-nitrophenyl; BOC = *t*-butyloxycarbonyl; Tos = *p*-methyltoluenesulfonyl; pmz = *p*-methoxy carbobenzyloxy and PHT = phthalyl.

Val[1]-gramicidin A (278)

The total synthesis of this molecule was carried out by Sarges and Witkop.[111] All of the amino acids in this molecule are neutral amino acids, so no particular difficulties were expected in their construction. The acid sensitivity of the tryptophane moiety, however, restricted the procedures that could be used. For this reason Sarges and Witkop planned to synthesize and join together the octapeptide Z-L-Val-Gly-L-Ala-D-Leu-L-Ala-D-Val-L-Val-D-Val-*R* **282** and the heptapeptide ethanolamide L-(H)-Try-D-Leu-L-Try-D-Leu-L-Try-D-Leu-L-Try-$NHCH_2CH_2OH$ **283**, either by the azide method or by coupling with dicyclohexylcarbodiimide at low temperature, methods that would largely avoid the bane of all such syntheses *viz.* racemization. The heptapeptide **283** was prepared by stepwise synthesis from the carboxyl end by the mixed anhydride method. This method (the general case is shown below) is fast and convenient and minimizes racemization. It was also used for the synthesis of the octapeptide **282** (R = OCH_3), but unfortunately the ester function could not be saponified to give the desired fragment **282**

$$\text{ZNHCH(R)CO}_2\text{H} + \text{ClCO}_2\text{Et} \xrightarrow{\text{Et}_3\text{N}} \text{ZNHCH(R)CO}_2\text{OCOEt}$$

$$\text{ZNHCH(R)CO}_2\text{OCOEt} + \text{NH}_2\text{CH(R')COR''} \longrightarrow$$

$$\text{ZNHCH(R)CONHCH(R')COR''} \xrightarrow{\text{cat.H}_2}$$

$$\text{NH}_2\text{CH(R)CONHCH(R')COR''} \longrightarrow \ldots$$

(R = alkoxyl, or amine, etc.)

(R = OH) and accordingly an alternate strategy was adopted. The ester function was converted to the hydrazide and coupled via the azide with the heptapeptide ethanolamide **283**. The product had the desired properties of the authentic *N*-carbobenzyloxydesformylgramicidin A but at the time simply because of lack of material, experimental conditions for the hydrogenolytic cleavage of the terminal carbobenzyloxy protecting group could not be found. Later it was found that the reaction could be accomplished in methanol over palladium black.

In the meantime a more convenient approach was explored which consisted of coupling the pentapeptide derivative Z-L-Val-Gly-L-Ala-D-Leu-L-Ala-OH (**284**) with the decapeptide ethanolamide D-(H) Val-L-Val-D-Val-L-Try-D-Leu-L-Try-D-Leu-L-Try-D-Leu-L-Try-$NHCH_2CH_2OH$ (**285**), which was prepared by the stepwise prolongation of the heptapeptide **283** (discussed earlier), again using the mixed anhydride method for the introduction of each amino acid residue. The pentapeptide **284** was also synthesized by this technique starting with methyl L-alaninate to give eventually the pentapeptide methyl ester which was saponified without difficulty. When **284** and **285** were coupled under the influence of dicyclohexylcarbodiimide the expected *N*-carbobenzyloxydesformylgramicidin A **286** was obtained.

286	Z-L-Val-Gly . . . D-Leu-L-Try-$NHCH_2CH_2OH$
287	L-(H)Val-Gly . . . D-Leu-L-Try-$NHCH_2CH_2OH$
288	OCH-L-Val-Gly . . . D-Leu-L-Try-$NHCH_2CH_2OCHO$
278	OHC-L-Val-Gly . . . D-Leu-L-Try-$NHCH_2CH_2OH$

Hydrogenation of **286** over palladium black in methanol afforded the desformyl polypeptide **287**, which when treated with formyl acetic anhydride led to *O*-formyl Val^1-gramicidin (**288**). Saponification of the latter with 1 *N* sodium hydroxide then gave the antibiotic itself identical in all biological and chemical respects with the naturally occurring material.

In a completely analogous fashion Z-L-Ileu-Gly-L-Ala-D-Leu-L-Ala-OH was coupled with the decapeptide **285** to give, after the three final steps, $Ileu^1$-gramicidin A.

Gramicidin S (279)

Although the linear sequence of gramicidin S had been synthesized in a number of laboratories, Schwyzer and Sieler[115] were the first to prepare the cyclic form. In fact this was the first synthesis of a cyclic peptide antibiotic and it succeeded only because of the suitable cyclization methods developed by these workers. Two syntheses were devised, both of which depended on the synthesis of the protected pentapeptide Z-L-Val-L-Orn (Tos)-L-Leu-D-Phe-L-Pro-OCH_3 (**289**). This was carried out by the method of activated esters (Scheme 33). The coupling of Z-L-Leu-ONp and H-D-Phe-OC_2H_5 was

L-Val	L-Orn	L-Leu	L-Phe	L-Pro	L-Val	L-Orn	L-Leu	D-Phe	L-Pro
		Z—ONp	H OC$_2$H$_5$						
		Z—	—OC$_2$H$_5$						
Z ONp	H—OCH$_3$ (Tos)	Z—	—OH						
Z—	—OCH$_3$ (Tos)	Z—	—OCH$_2$CN	H OCH$_3$					
Z—	—NHNH$_2$ (Tos)	Z—	—	—OCH$_3$					
Z—	—N$_3$ (Tos)	H—	—	—OCH$_3$					
Z—	—(Tos)	—	—	—OCH$_3$					

289

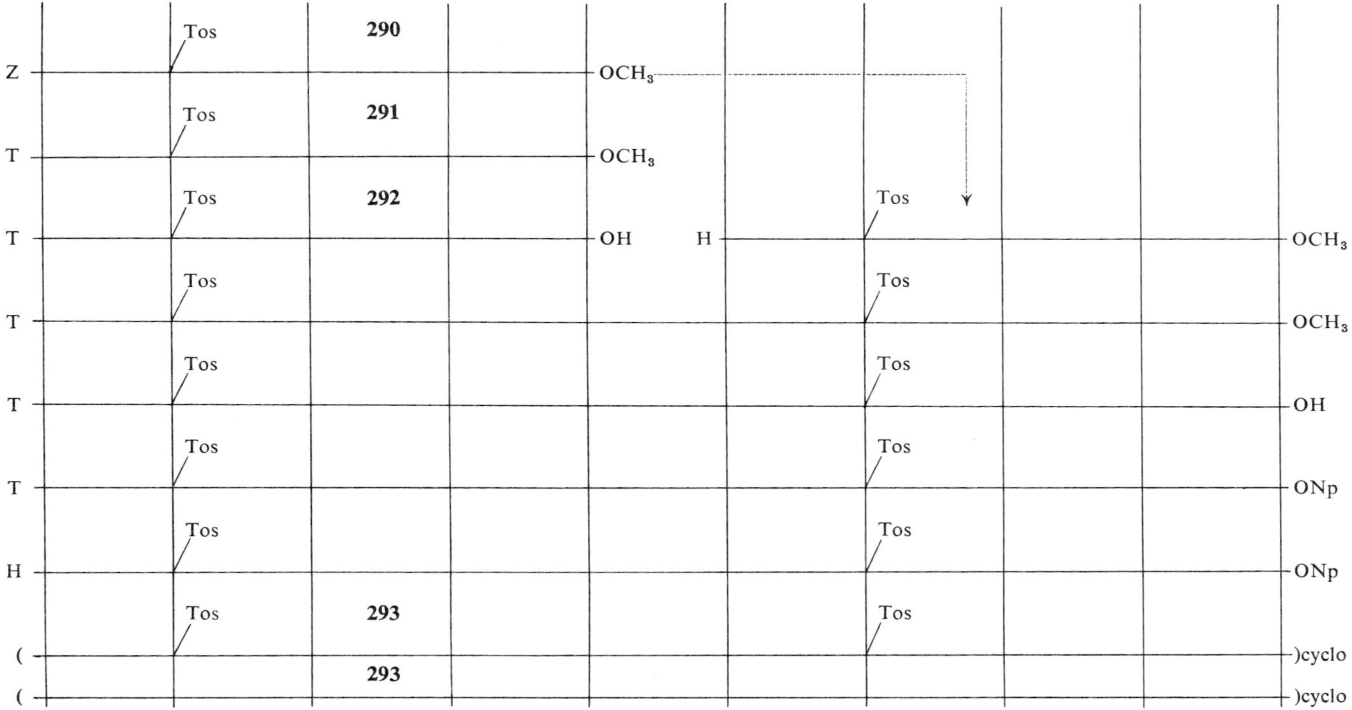

Z — Tos — 290 — OCH_3

T — Tos — 291 — OCH_3

T — Tos — 292 — OH   H — Tos — OCH_3

T — Tos — Tos — OCH_3

T — Tos — Tos — OH

T — Tos — Tos — ONp

H — Tos — Tos — ONp

(— Tos — 293 — Tos —)cyclo

(— 293 —)cyclo

Scheme 33

carried out 20° in tetrahydrofuran. The product was saponified to give Z-L-Leu-D-Phe-OH, which was converted to the corresponding cyanomethyl ester by means of chloroacetonitrile in the presence of triethylamine. Condensation of this active ester with L-proline methyl ester then afforded the tripeptide which was hydrogenolyzed to give H-L-Leu-D-Phe-L-Pro-OCH_3 suitable for further coupling. The dipeptide methyl ester *Z*-L-Val-L-Orn (Tos)-OCH_3 was prepared in the same way then converted to the hydrazide and subsequently to the azide Z-L-Val-L-Orn(Tos)-N_3, which when condensed with the tripeptide prepared above afforded the desired protected pentapeptide **289**. Catalylic reduction then removed the carbobenzyoxy group and the methyl ester **290** thus obtained was *N*-tritylated to give the *N*-trityl ester **291**. Saponification of **291** in a large excess of alkali then produced the *N*-protected peptidic acid **292**, from which two syntheses of gramicidin S were developed. In one,[115] **292** was coupled with the amino ester **290** by means of 1-cyclohexyl–3[2-morpholinyl-(4)-ethyl]carbodiimide. The resulting *N*-trityl decapeptide ester was saponified to the corresponding acid, which was then converted to the *p*-nitrophenyl ester by means of *p*-nitrophenyl sulfite in pyridine. Detritylation was accomplished by treatment with trifluoroacetic acid at −5° for 15 minutes and resulted in the trifluoroacetate of the decapeptide *p*-nitrophenyl ester. Cyclization was brought about by the dropwise addition of a dimethylformamide solution (containing a little acetic acid) of this compound to pyridine using a high dilution technique ($3 \times 10^{-3}M$ solution). Subsequent purification of the cyclic peptide afforded a 28% yield of crystalline material (**293**) identical with the bis-tosyl derivative of natural gramicidin S. The tosyl groups which had been used to protect the δ-amino groups of the ornithine residues were now removed by treatment of **293** with sodium in liquid ammonia. The product (70–90% yield) was identical in all respects with native gramicidin S (**279**).

In the second procedure[125] for the synthesis of **279** the *N*-tritylpentapeptide **292** was esterified by means of *p*-nitrophenyl sulfite then detritylated. Cyclization of the resulting amino ester then occurred with doubling to give **293** directly. This type of twinning was not unknown in other cases but linear pentapeptides do not always cyclize exclusively with doubling. However, in this case no cyclosemigramicidin could be detected, so the transition state for the formation of the cyclodecapeptide must be very favorable.

Tyrocidine A (280)

In the synthesis of tyrocidine A, Izumiya and his associates followed a pattern not dissimilar to that used by Schwyzer for gramicidin S. In general two pentapeptides were constructed and subsequently linked to give a decapeptide. Since Izumiya intended to cyclize the decapeptide by the use of the active *p*-nitrophenyl ester method of Schwyzer, he was guided in his choice of the two peptides by the limitations of the technique. This relates to the racemization that can occur at the amino acid moiety during ester formation, except when the amino acid residue is glycyl or prolyl. For this reason he chose to synthesize a decapeptide having at its terminus a proline group. The grid for the total synthesis of tyrocidine A is shown in Scheme 34. It needs little comment except to say that the isolated amino group of the ornithine moiety was protected up to the penultimate step as the carbobenzyloxy derivative as opposed to the tosyl derivative used by Schwyzer. This necessitated an alternate method of protecting the amino group end of the peptide chains during their synthesis. In the case of the pentapeptide L-Phe-D-Phe-L-Asp-L-Glu-L-Tyr-OH this was done using the *p*-methoxycarbobenzyloxy group, which can be removed by trifluoroacetic acid even in the presence of the sensitive *p*-nitrophenyl group. In the case of the pentapeptide L-Val-L-Orn-L-Leu-D-Phe-D-Pro-OH it was accomplished by means of a *t*-butyloxycarbonyl function which was eventually removed by hydrogen chloride in ethyl acetate. The couplings of Z-D-Phe-OH with H-L-Pro-OC_2H_5 and of Z-L-Leu-OH with H-D-Phe-L-Pro-OC_2H_5 and of pmZ-L-Phe-OH with H-D-Phe-OEt were all carried out by the mixed anhydride method. The dipeptide BOC-L-Val-L-Orn(Z)OC_2H_5 was made by dicyclohexylcarbodiimide coupling of BOC-L-Val-OH and H-L-Orn(Z)OEt using conventional procedures.

It is interesting to note that the first attempt to synthesize the decapeptide **296** was made by coupling the azide with the ester derived from **296** after removal of the BOC group. However, the Japanese workers were unable to hydrolyze the ester function in the coupled product—a result that parallels Witkop's experience[111] with the octapeptide **282** ($R = OCH_3$) discussed earlier. The synthesis scheme that they finally adopted produced pure crystalline tyrocidine A hydrochloride identical with the same salt of the natural material. The syntheses of the remaining members of this group were accomplished after the same fashion.

B. The Polymixins

The polymixins form a group of closely related cyclopolypeptides that are obtained from various strains of *Bacillus polymyxa*. They are potent antibiotics against gram-negative organisms but are somewhat limited by their toxicity. Structure determination in this area was for some time fraught with difficulty[126] and only after considerable synthetic and degradative work were the questions of peptide sequence and ring size solved.

The polymixins have the general structure **297** and are classified as A (**297a**), B (**297b**), B_2 (**297c**), C (**297d**), D_1 (**297e**), D_2 (**297f**), and M (**297g**). In polymixin the central Dab of the three Dab moieties in the ring has been

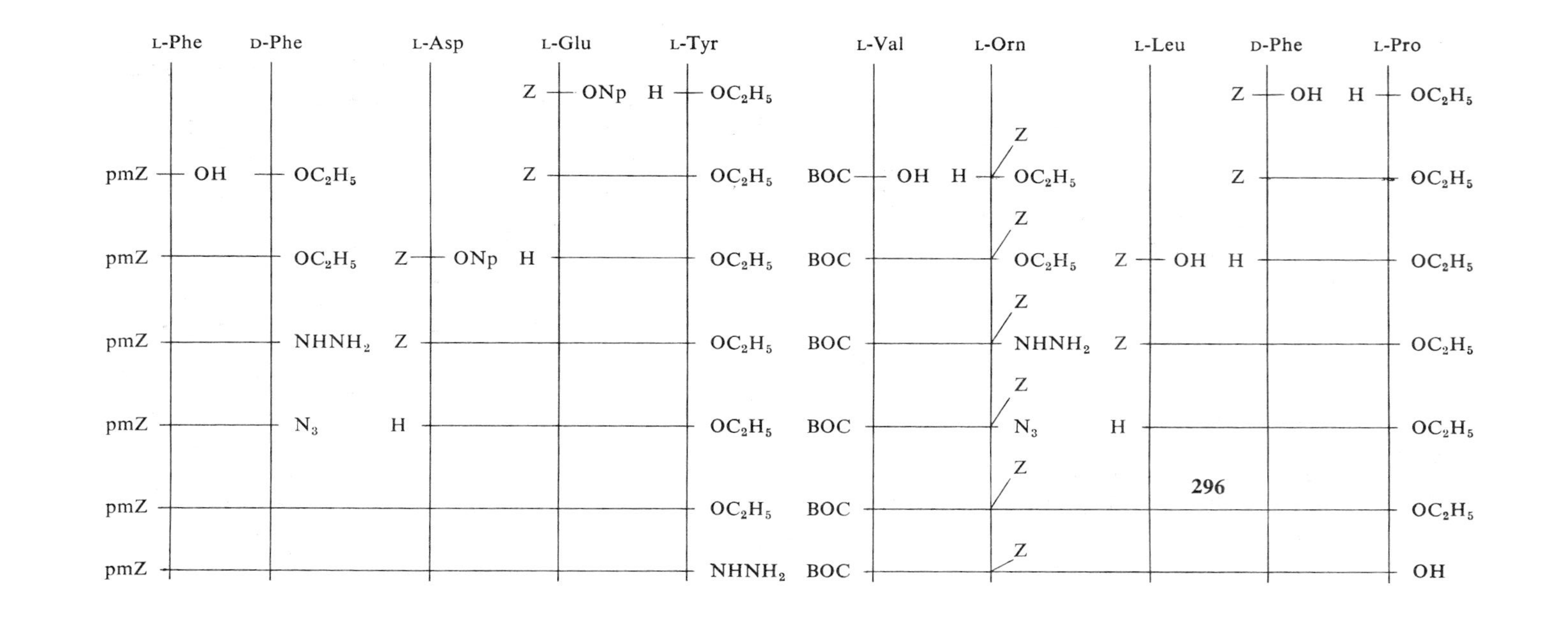

L-Phe
D-Phe
L-Asp
L-Glu
L-Tyr
L-Val
L-Orn
L-Leu
D-Phe
L-Pro
296

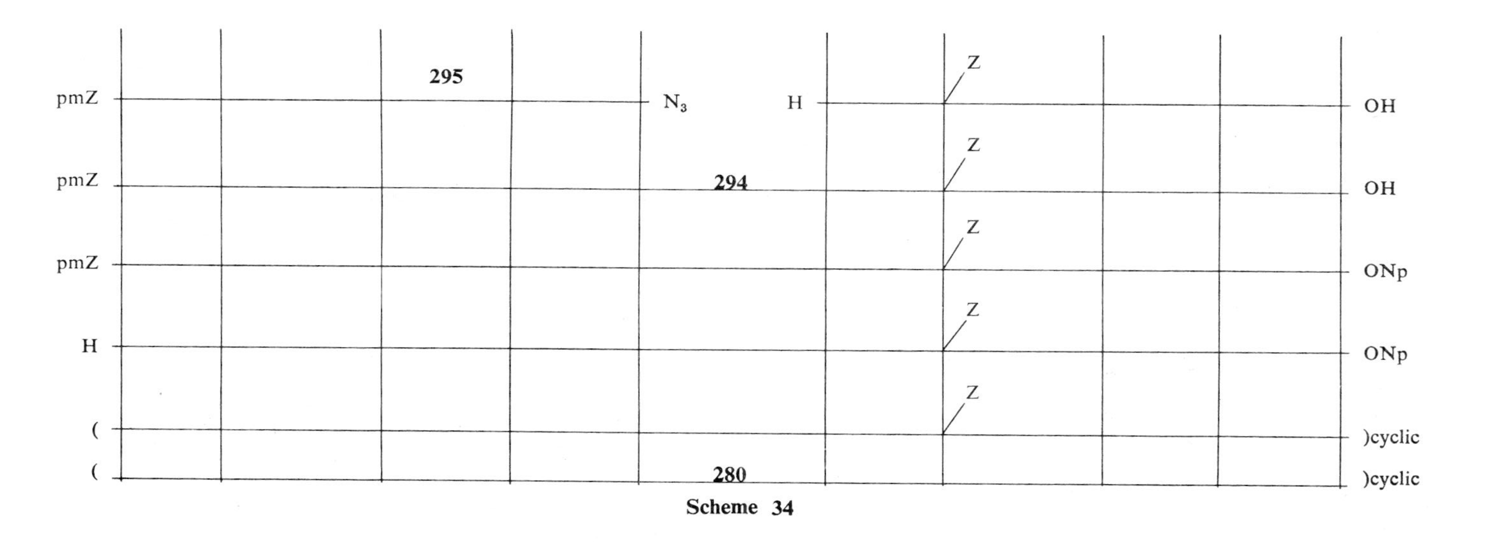

Scheme 34

exchanged for D-Leu. The circulins and the colistins also belong to this family. Polymixin E is identical with colistin A.

```
                                          L-Dab ──→ D-X ──→ L-Y
                                            ↑                 │
R-L-Dab ──→ L-Thr ──→ W ──→ L-Dab                 │
                                            ↑                 │
                                          L-Thr ←── L-Dab ←── L-Dab
```

297 Dab = α,γ-diaminobutyric acid

		R	W	X	Y
B_1	(**297b**)	(+) $CH_3CH_2CH(CH_3)(CH_2)_4CO$	L-Dab	D-Phe	L-Leu
B_2	(**297c**)	$CH_3CH(CH_3)(CH_2)_4CO$	L-Dab	D-Phe	L-Leu
C	(**297d**)	(+) $CH_3CH_2CH(CH_3)(CH_2)_4CO$	?	?	?
D_1	(**297e**)	(+) $CH_3CH_2CH(CH_3)(CH_2)_4CO$	D-Ser	D-Leu	L-Thr
D_2	(**297f**)	$CH_3CH(CH_3)(CH_2)_4CO$	D-Ser	D-Leu	L-Thr
M	(**297g**)	(+) $CH_3CH_2CH(CH_3)(CH_2)_4CO$	L-Dab	L-Thr	L-Dab

Only the structures just mentioned are known with certainty. One compound of this group has been synthesized,[127] polymixin B_1. A number of isomers of B_1 that do not occur naturally are also known. The synthesis of **297b** was accomplished by synthesizing the protected polypeptide moieties **298**, **299** and **300**, coupling them as discussed below, then cyclizing the heptapeptide chain after removal of the blocking groups.

The syntheses of these peptides are shown in Scheme 35. Standard methods were used to couple the amino acids in most of the steps and are obvious from the schemes. The introduction of the (+)-6-methyloctaroic acid (MOA) into the tetrapeptide **299** was carried out by causing the acid to react with H-L-Dab(NZ)-L-Thr-OCH_3 under the influence of carbonyldiimidazole. The necessary dipeptide **300** was obtained by condensing N^{α}-phthaloyl-N^{γ}-carbobenzyloxy-α,γ-diaminobutyric acid with D-phenylalanine *t*-butyl ester in the presence of dicyclohexylcarbodiimide. The phthaloyl group was then removed by means of hydrazine.

The coupling of the synthesized pepides was carried out as follows. The γ-BOC group of **299** was removed by means of trifluoroacetic acid and the resulting amine was condensed with the azide derived from **298**. The octapeptide that was produced was converted to the corresponding hydrazide and then to the azide which was coupled with the dipeptide **300** to give the

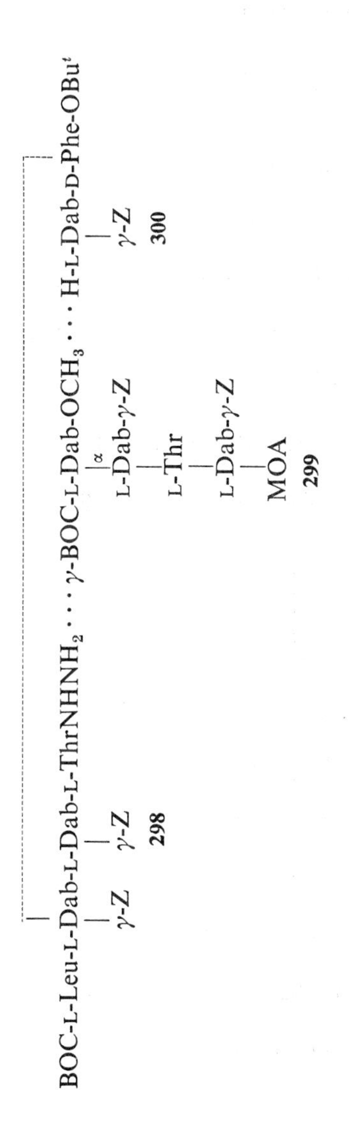
BOC-L-Leu-L-Dab-L-Dab-L-ThrNHNH2 · · · γ-BOC-L-Dab-OCH3 · · · H-L-Dab-D-Phe-OBu^t
γ-Z
γ-Z
298
α
L-Dab-γ-Z
L-Thr
L-Dab-γ-Z
MOA
299
γ-Z
300

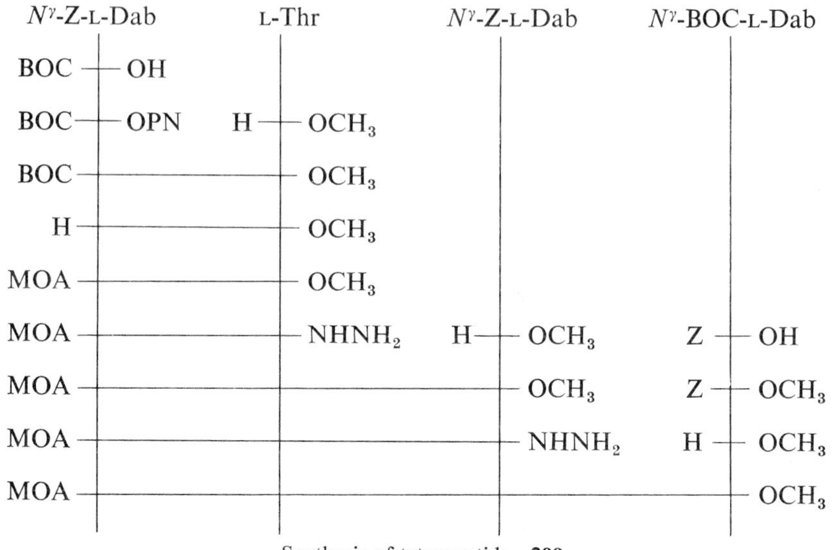

Synthesis of tetrapeptide **299**

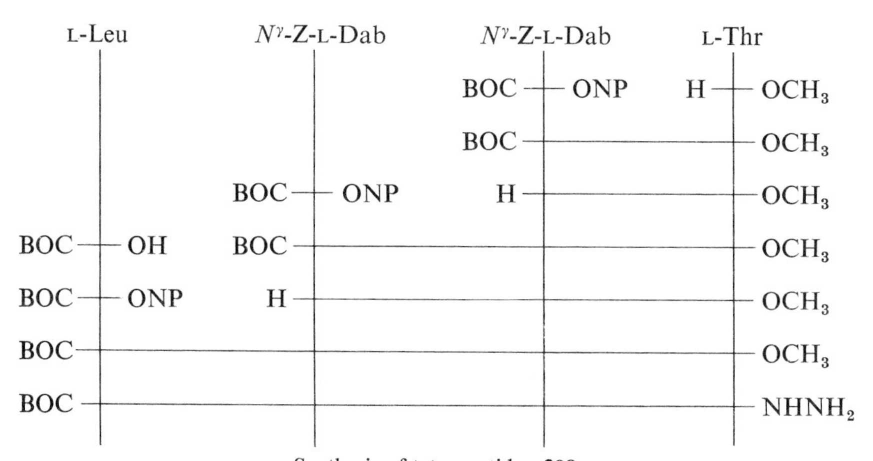

Synthesis of tetrapeptide **298**

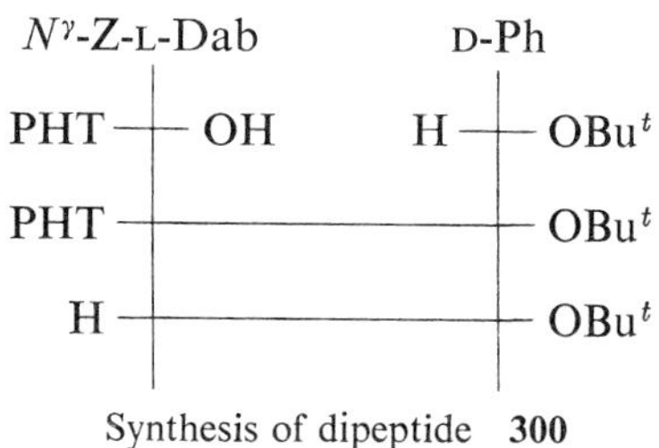

Synthesis of dipeptide **300**

Scheme 35

branched decapeptide **301**. The terminal *t*-butyloxycarbonyl and *t*-butyl

```
BOC-L-Leu-L-Dab-L-Dab-L-Thr-L-Dab-L-Dab-D-Phe-OBu^t
          |      |             |α     |
         γ-Z    γ-Z            |     γ-Z
                            L-Dab-γ-Z
                               |
                            L-Thr
                               |
                            L-Dab
                               |
                            MOA
```

301

groups were removed by trifluoroacetic acid at room temperature and the polypeptide amino acid that resulted was cyclized by means of dicyclohexylcarbodiimide in dimethylformamide-dioxane at high dilution. Finally the carbobenzyloxy groups were removed by cleavage with sodium in liquid ammonia. The product from this reaction after purification had all the physical and biological properties of polymixin B_1.

C. Cyclic Depsipeptides

Although actinomycin D contains only one ester linkage in the polypeptide half of the molecule and has also a large heterocyclic nucleus, it is still considered a depsipeptide and is included in this section along with the more authentic cases, serratamolide and beauvericin.

Actinomycin D

The actinomycins were first discovered by Waksman and Woodruff[127] who isolated actinomycin A from a species of *Streptomyces antibioticus*. The isolation of the various components of the complex, their structure elucidation, and their chemistry have been reviewed by Brockmann.[128] The major component of the complex is actinomycin D or C_1 (**302**). The first synthesis of the compound was reported by Brockmann[129] in 1964 and is shown in Scheme 36. The tetrapeptide **304** required for subsequent condensation was prepared by a series of carbodiimide condensations starting with *N*-formylvaline and the benzyl ester of L-proline. The resulting dipeptide was hydrogenated to cleave the benzyl group and the acid obtained used in a subsequent condensation. The formyl group of the tetrapeptide was removed by hydrolysis with a small amount of aqueous hydrochloric acid in benzyl alcohol, thus minimizing ester hydrolysis and allowing the formation of **304** (as its hydrochloride) in good yield. The other component **305** was

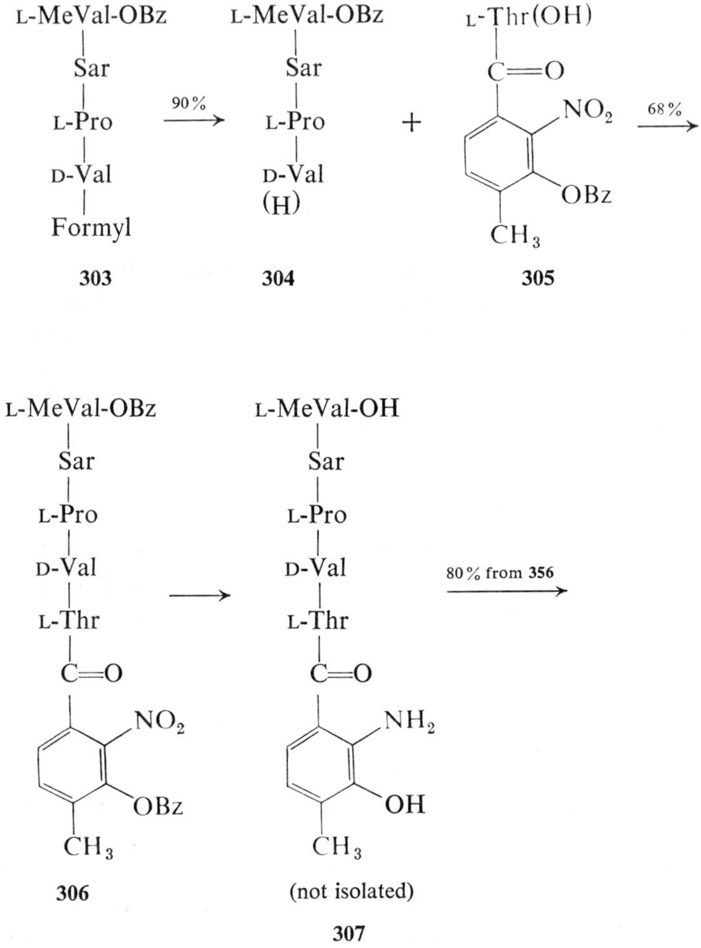

L-MeVal-OBz
|
Sar
|
L-Pro
|
D-Val
|
Formyl

303

$\xrightarrow{90\%}$

L-MeVal-OBz
|
Sar
|
L-Pro
|
D-Val
(H)

304

+

L-Thr(OH)
|
C=O
NO_2
OBz
CH_3

305

$\xrightarrow{68\%}$

L-MeVal-OBz
|
Sar
|
L-Pro
|
D-Val
|
L-Thr
|
C=O
NO_2
OBz
CH_3

306

$\longrightarrow$

L-MeVal-OH
|
Sar
|
L-Pro
|
D-Val
|
L-Thr
|
C=O
NH_2
OH
CH_3

(not isolated)

307

80% from **356** $\longrightarrow$

Scheme 36

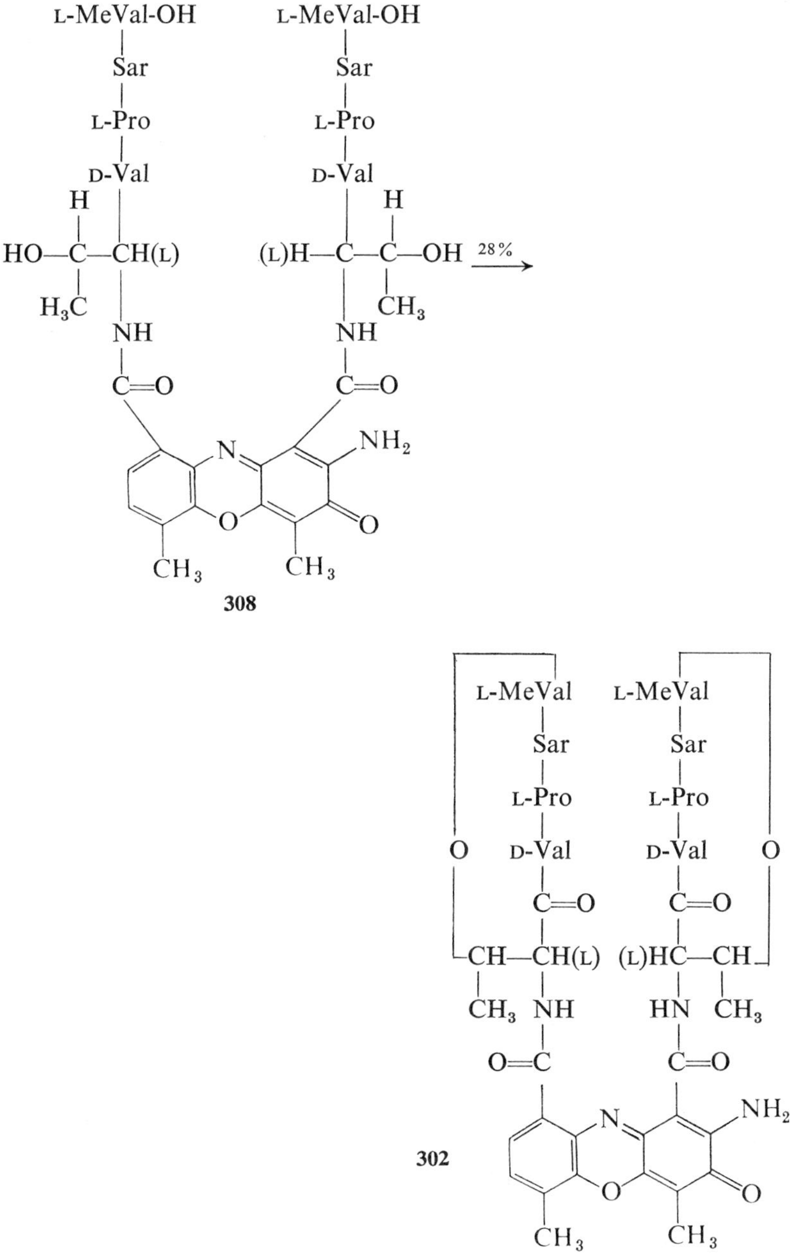

L-MeVal-OH L-MeVal-OH
Sar Sar
L-Pro L-Pro
D-Val D-Val
H H
HO—C—CH(L) (L)H—C—C—OH 28%
H_3C CH_3
NH NH
C=O C=O
N NH_2
O O
CH_3 CH_3

308

L-MeVal L-MeVal
Sar Sar
L-Pro L-Pro
O D-Val D-Val O
C=O C=O
CH—CH(L) (L)HC—CH
CH_3 NH HN CH_3
O=C C=O
N NH_2
O O
CH_3 CH_3

302

Scheme 36 (Continued)

prepared by allowing 2-nitro-3-benzyloxy-4-methylbenzoyl chloride to react with the sodium salt of threonine.[130] The condensation of **304** with **305** was accomplished by the use of *N*-ethyl-5-phenylisoxazolium-3′-sulfonate (Woodward's reagent) in nitromethane in the presence of triethylamine. The resulting compound (**306**) was submitted to catalytic hydrogenation to remove the two benzyl groups. Without characterization of the intermediate phenolic acid **307**, it was oxidized with potassium ferricyanide solution at *p*H 7.2 in an oxidative coupling reaction which was known in the case of simpler *o*-aminophenols. Actinomycin D (**302**) was now obtained by treating **308** with a mixture of acetyl chloride and *N*,*N*′-carbonyldiimidazole in tetrahydrofuran solution followed by chromatography of the crude product. The synthetic material was congruent with the natural product.

More recently a new synthesis of actinomycin D has appeared[131] which permits its preparation, and that of analogous substances, in substantial quantity. In this synthesis the difficult final lactone ring closure step of the Brockman synthesis is avoided and formation of a peptide linkage becomes the last reaction of the sequence. The key reaction, the cyclization to the pentapeptide lactone, was carried out by peptide bond formation between the prolyl and sarcosyl residues of **308a** using the *p*-nitrophenyl active ester

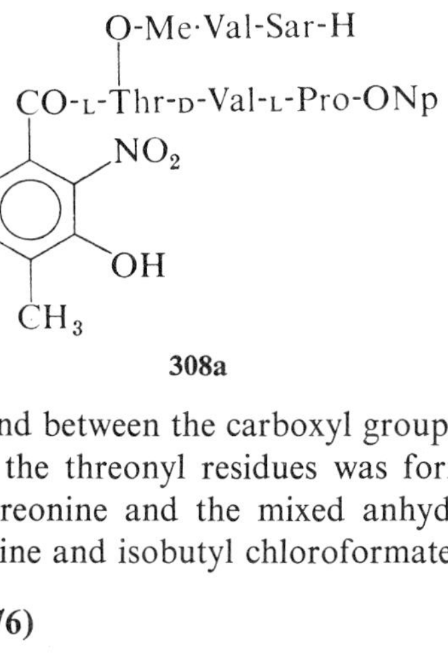

O-Me·Val-Sar-H

CO-L-Thr-D-Val-L-Pro-ONp

NO_2

OH

CH_3

308a

technique. The ester bond between the carboxyl group of *N*-methylvalyl and the hydroxyl group of the threonyl residues was formed by a reaction of *t*-butyloxycarbonyl-L-threonine and the mixed anhydride from benzyloxycarbonyl-L-*N*-methylvaline and isobutyl chloroformate.

Beauvericin (276)

The synthesis[132] of beauvericin is typical of this class of depsipeptide where peptide and ester functions alternate. The synthetic grid is shown in Scheme 37 and the methods are relatively simple because in this case only two residues are used to build the molecule, L-*N*-methylphenylalanine and D-2-hydroxyisovaleric acid. *N*-Carbobenzyloxy-*N*-methylphenylalanine was coupled with *t*-butyl D-α-hydroxyvalerate by means of carbonyldiimidazole. The product, **309** was in separate experiments (*a*) treated with trifluoroacetic

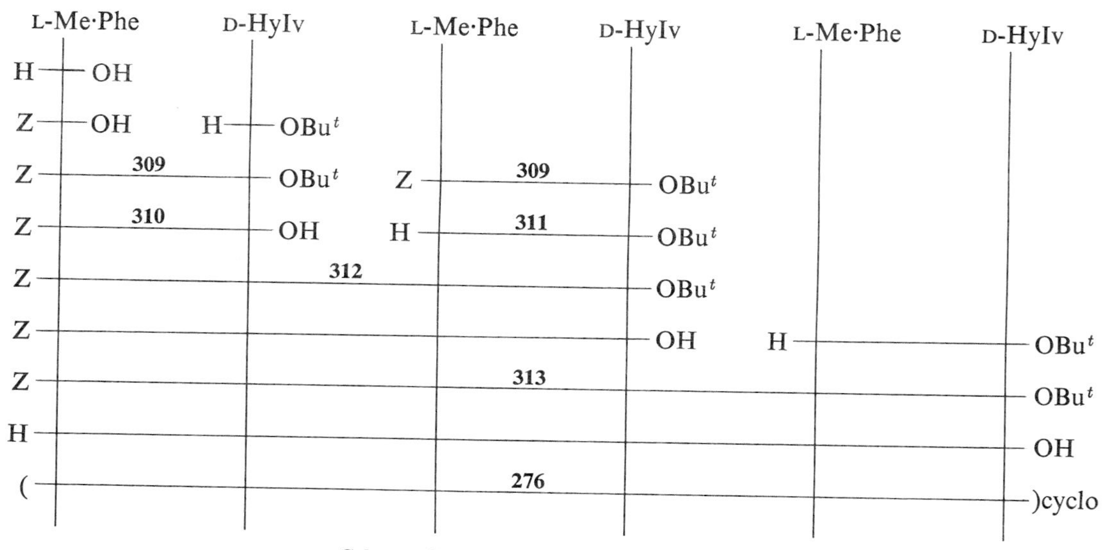

Scheme 37. The synthesis of beauvericin.

acid and (*b*) hydrogenated catalytically to give **310** and **311**, respectively, the two components of the next condensation. This condensation was accomplished by treating **310** with thionyl chloride in triethylamine and subsequently adding **311**. In a completely analogous fashion **312** was converted into **313**. The latter was treated with hydrogen bromide in acetic acid to remove the terminal protecting groups and the product was subjected to the action of thionyl chloride in triethylamine at high dilution. The synthetic material was essentially identical with the natural product. In each step of this synthesis the yield was better than 70% except the last which was 25%.

Serratamolide (275)

In his synthesis of the *O*,*O*′-diacetyl derivative of the antibiotic serratamolide[101] Shemyakin[132a] very cleverly made use of the known hydroxyacyl insertion reaction:

CH_2OBz | (PHT)NCHCO$_2$CH$_3$ **315**

CH_2OBz | L-(PHT)NCHCO$_2$H **314**

CH_2OBz | NH$_2$CHCO$_2$CH$_3$ **316**

CH_2OBz CH_2OBz | | (PHT)NCHCONHCHCO$_2$CH$_3$ **317**

318 R = Bz
319 R = Ac

For this synthesis D-β-benzyloxydecanoic acid and the diketopiperazine **319** were required. The latter was synthesized as follows: phthalylation of *O*-benzyl-DL-serine afforded *N*-phthalyl-*O*-benzyl-DL-serine, which was then resolved to give the L-isomer **314**. Methylation with diazomethane gave **315**, which was hydrazinolyzed in good yield to methyl *O*-benzyl-L-serine (**316**) isolated as the hydrochloride. Condensation of **314** with **316** in the presence of dicyclohexylcarbodiimide afforded the peptide **317**, which on treatment with hydrazine hydrate led to the dibenzyl ether **318**. Hydrogenolysis and acetylation then afforded the desired **319**. D-β-benzyloxydecanoic acid was prepared from the corresponding acid, silver oxide, and benzyl bromide. The acid chloride **320** when heated in boiling toluene with **319** led to **321**,

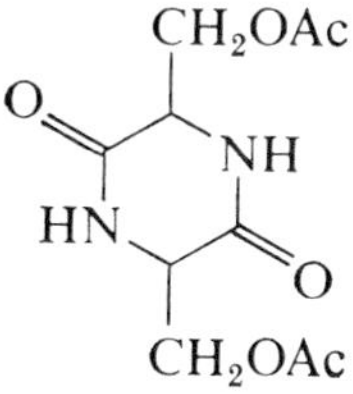
CH2OAc
O
NH
HN
O
CH2OAc

319

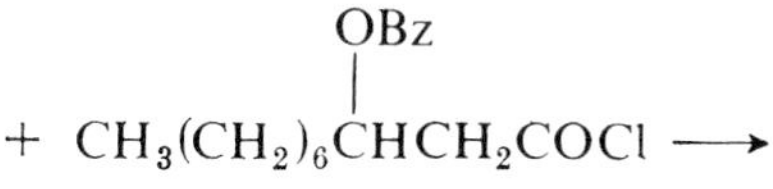
OBz
+ CH3(CH2)6CHCH2COCl

320

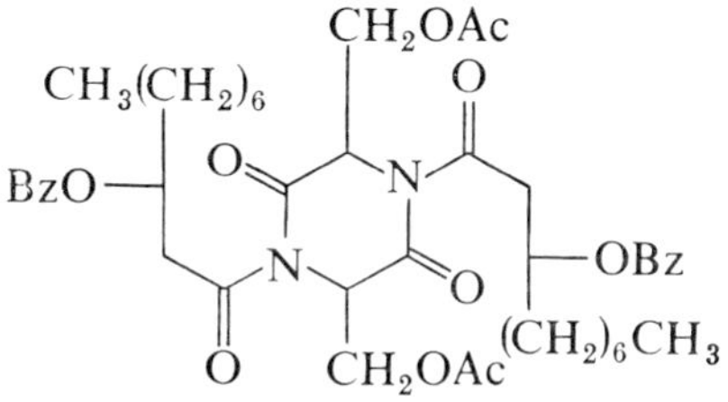
CH2OAc
O
CH3(CH2)6
O
N
BzO
N
OBz
O
O
CH2OAc
(CH2)6CH3

321

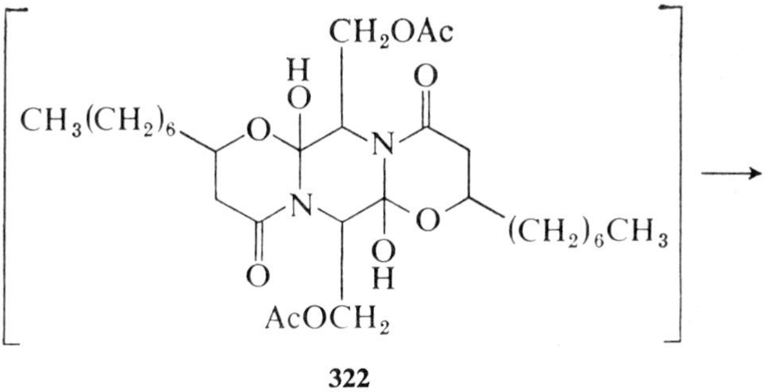
CH2OAc
H
O
O
CH3(CH2)6
O
N
N
O
(CH2)6CH3
O
O
H
AcOCH2

322

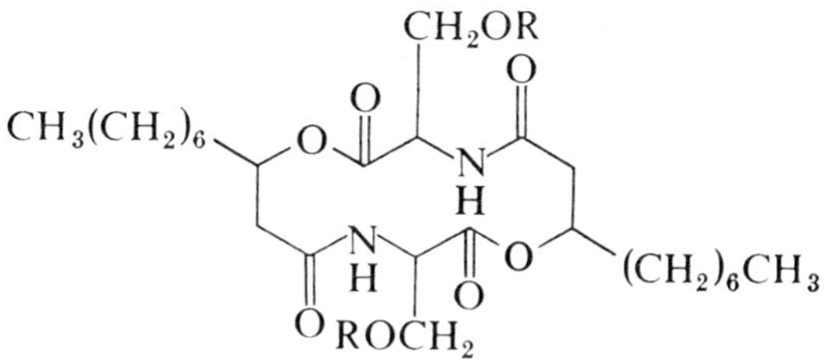
CH2OR
O
O
CH3(CH2)6
O
N
H
N
H
O
(CH2)6CH3
O
O
ROCH2

323 R = Ac
275 R = H

which was then hydrogenolyzed and afforded, via **322**, the *O,O'*-diacetyl-derivative **323** of serratamolide identical with that derived from the natural product.

In an attempt to synthesize serratamolide itself (**275**), the diketopiperazine **324** was prepared. However, hydrogenolysis of the four benzyl groups did

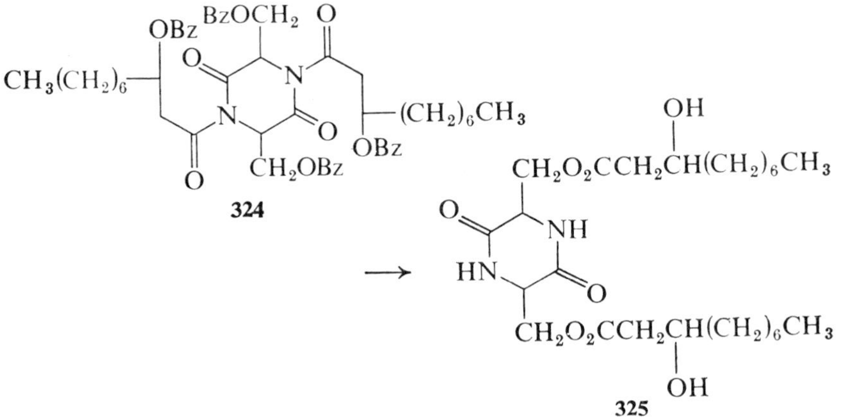

324

→

325

not give the desired material but only the product (**325**) of N → O acyl migration. For the hydroxyacyl insertion reaction to work in this system the hydroxyl groups of the diketopiperazine moiety must be protected.

7. MACROLIDE ANTIBIOTICS

The *complex* macrolide antibiotics can be subdivided into three classes based on both biological activity and structural features. The members of the first group, of which erythromycin[133] (**326**) and chalcomycin[134] (**327**) are typical

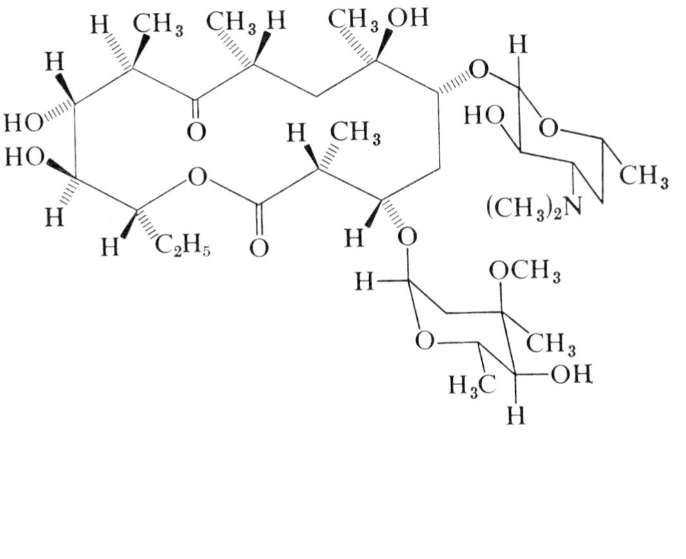

326

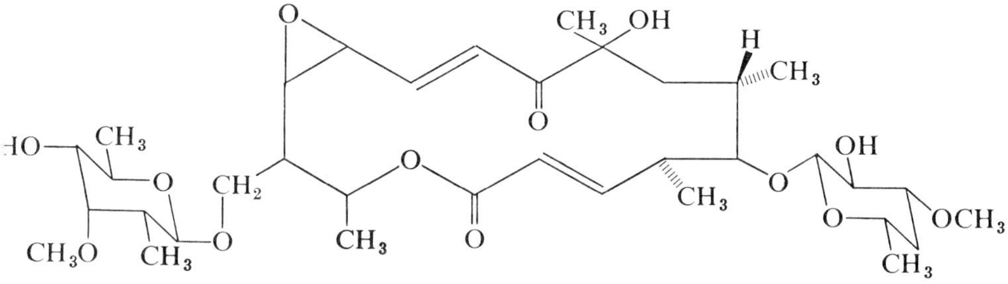

327

examples, have a biological activity similar to that of penicillin and have been used extensively in the past because of the high incidence of penicillin-resistant infections. Their importance has diminished recently with the availability of penicillins not affected by penicillinase.

The second group comprises such substances as filipin[135] (**328**), amphotericin B,[136] and nystatin.[137] This group is noted for its antifungal activity

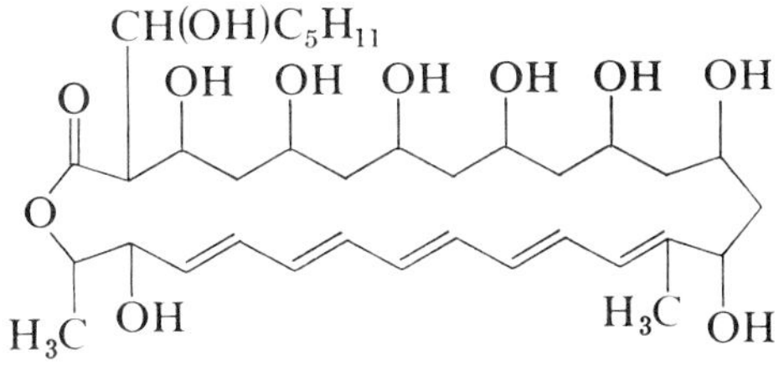

328

and, although mainly used topically, individual members such as amphotericin B are used intravenously for fungal infections of the blood. In comparison with the first group they have much larger lactone rings, which always contain extensive conjugated unsaturation. Sugars may be attached to the lactone ring. None of the members of either group have been synthesized. The presence in both groups of a large number of contiguous asymmetric centers in an aliphatic chain presents almost insuperable problems to the chemist bent on stereoselective synthesis. Certainly it seems that at least one probable way in which the synthesis of the lactone moiety, of say erythromycin, will be achieved will be by the synthesis of a perhydro suitably substituted tricyclic intermediate, followed by cleavage of the transannular bonds.

A third class of macrolide exists; by comparison with the two groups just mentioned the members are relatively simple in structure. Representatives of this class are curvularin (**329**), zearalenone (**330**), and radicicol[138] (**331**).

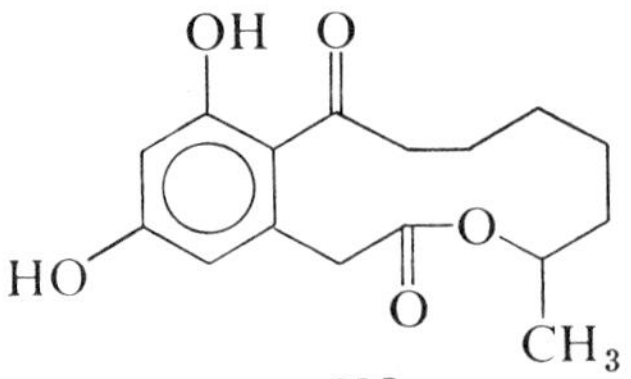

329 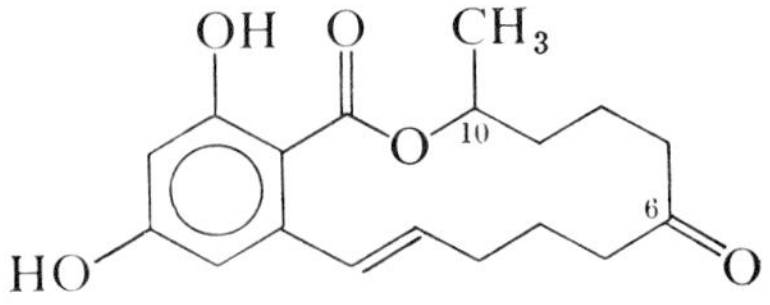 **330**

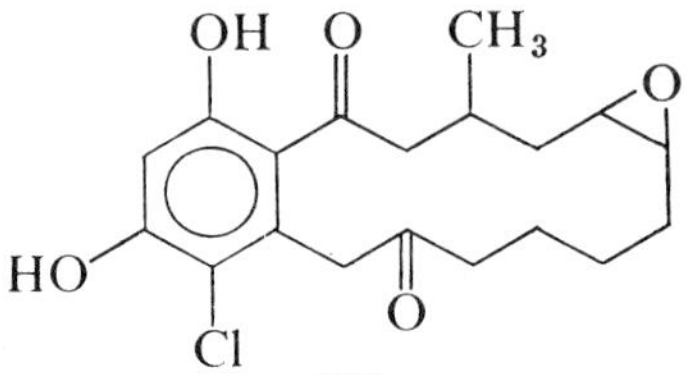

331

All of these compounds can be regarded as orsellinic acid derivatives generated without doubt from polyketide precursors. The syntheses of the di-*o*-methyl ether[139] of **329** and of **330** itself have been reported. These are described only because they represent rare examples of the synthesis of naturally occurring macrolides and not because of any outstanding biological activity.

A. Curvularin (329)

Initially, attempts were made to synthesize di-*o*-methylcurvularin (**335**) by the lactonization of **332**. However, all attempts to cyclize the compound with

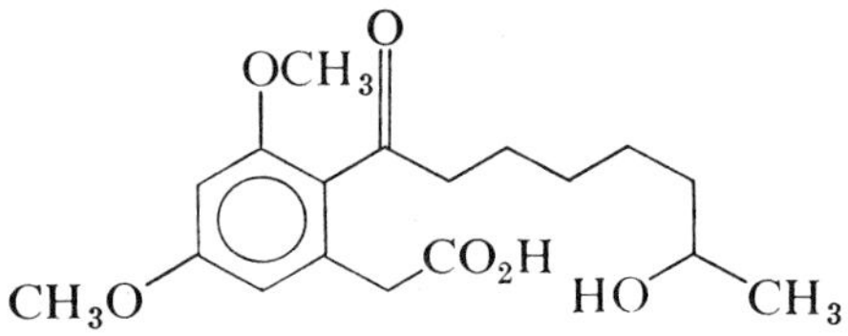

332

trifluoroacetic anhydride or dicyclohexylcarbodiimide failed. Lactones of this specific type are difficult to hydrolyze and the corresponding hydroxy acids are resistant to lactonization. This appears to be due to the steric hindrance offered by the methyl group and in part perhaps to the conformation of the macrocyclic ring. An alternate approach was therefore taken, using a Friedel-Crafts reaction. This type of reaction had been used successfully in the past for the synthesis of several medium-sized ring ketones.[140]

$$\text{3,5-}(CH_3O)_2C_6H_3CH_2COCl + CH_3CH(OH)(CH_2)_5CO_2Bz \xrightarrow{55\%}$$

333 R = Bz, **334** R = H $\xrightarrow{14\%}$ **335**

Benzyl 7-hydroxyoctanoate, prepared by benzylation and sodium borohydride reduction of 7-oxooctanoic acid, reacted smoothly at room temperature with 3,5-dimethoxyphenylacetyl chloride to yield the diester **333**. Hydrogenolysis afforded the acid **334**, which was cyclized to **335** by allowing it to stand in a dilute solution of trifluoroacetic anhydride and trifluoroacetic acid. Attempts to demethylate **335** to curvularin were not recorded.

B. Zearalenone (330)

The first synthesis of zearalenone was reported by Taub et al.[141] Zearalenone is produced by the fungus *Gibberella zeae* and has anabolic and uterotrophic activity.

As a prelude to the synthesis of **330** the recyclization of the *seco* acid (**336**) of di-*o*-methylzearalenone (**337**) was studied, since it was visualized that this reaction would be the penultimate stage in the synthesis of zearalenone. This was accomplished in 80% conversion yield, by 1 molar equivalent of trifluoroacetic anhydride in very dilute benzene solution at 6° for 18 hours. That ether cleavage of **337** to zearalenone by boron tribromide could be accomplished was also ascertained at this point. Having secured this information attention was turned to the total synthesis of the *seco* acid itself. It was envisaged that this could be accomplished by connecting the two components **338** and **339**. In **339** the potential functionalities of zearalenone at C-6 and C-10 are masked by internal ketal formation.

This compound was synthesized starting from 5-oxohexanoic acid by reduction with sodium borohydride followed by acid treatment which afforded **340**. The reaction of this tetrahydropyrone with 4-pentenyl magnesium bromide led to the keto acid **341**, which was converted thermally to **342**.

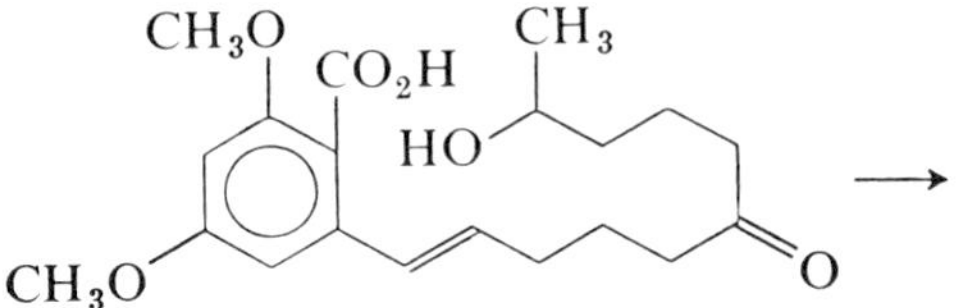

CH_3O CO_2H CH_3 HO CH_3O O

336

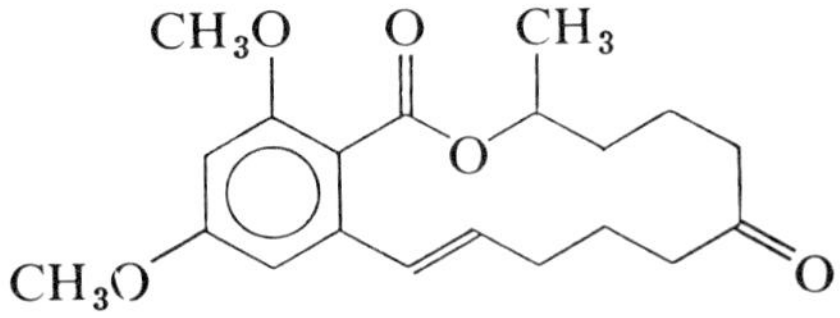

$\longrightarrow$

CH_3O O CH_3 O CH_3O O

337

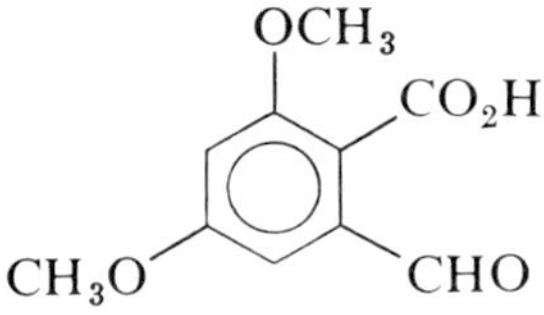

OCH_3 CO_2H CH_3O CHO

338

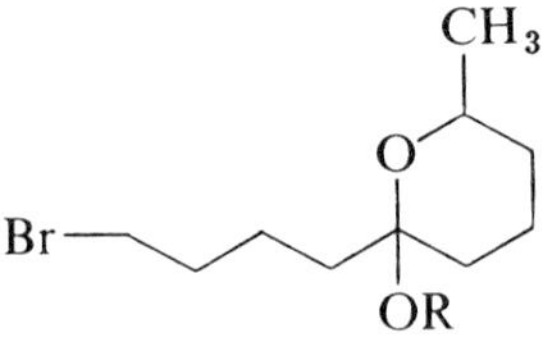

CH_3 O Br OR

339

$CH_3CO(CH_2)_3CO_2H$ $\longrightarrow$ CH_3 O O $\longrightarrow$

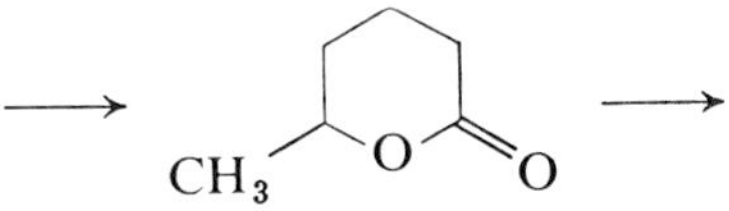

340

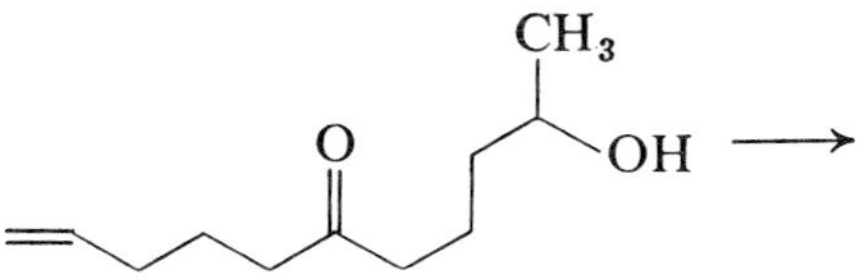

O CH_3 OH $\longrightarrow$

341

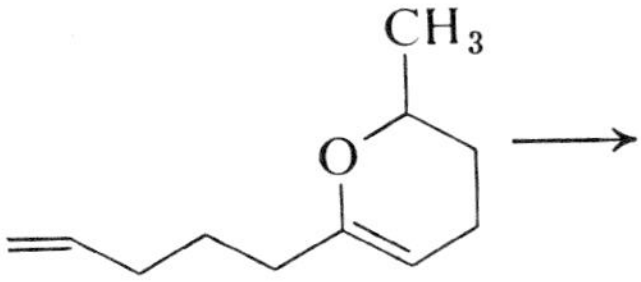

CH_3 O $\longrightarrow$

342

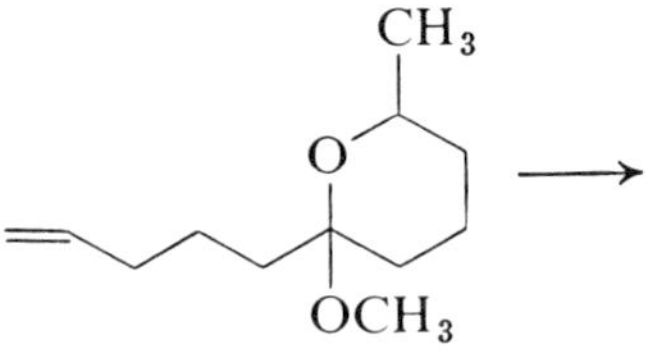

CH_3 O OCH_3 $\longrightarrow$

343

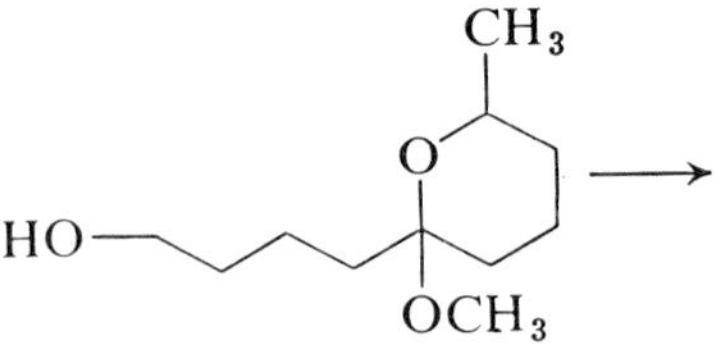

HO CH_3 O OCH_3 $\longrightarrow$

344

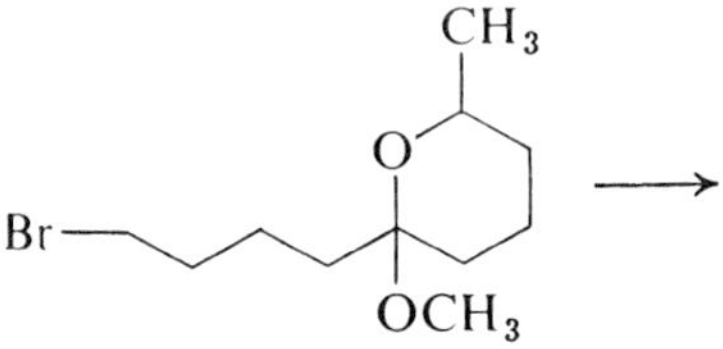

CH_3 O Br OCH_3 $\longrightarrow$

339

Br^- ϕ_3P^+ CH_3 O OCH_3

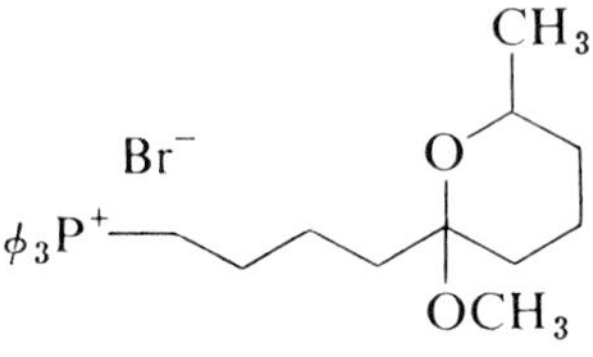

345

Methanolic hydrogen chloride then afforded **343**, which was ozonized at −60° in methanol and the ozonide reduced with sodium borohydride to the carbinol **344**. This carbinol was, however, extremely sensitive to acid, giving the spiran **346** with great ease. Under basic conditions then **344** was

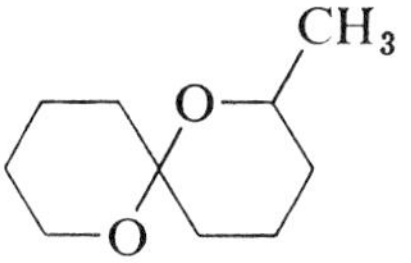

346

converted to the desired bromide via the tosylate. Triphenylphosphine in hot methanol then transformed **339** into the salt **345**. In both of the transformations from **344** to **345** the ketal group suffered some cleavage to the acyclic hydroxy ketone. This did not constitute a problem since the cyclic ketal was easily reformed in hot acidic methanol.

In preliminary experiments it was found that whereas the ylide

$$\phi_3P{=}CHCH_3$$

when condensed with the methyl ester of **338** gave complex results, condensation with the sodium salt of this acid in DMSO gave the desired product cleanly. When these conditions were used with the ylide from **345** condensation occurred smoothly to give a mixture of *cis* and *trans seco* acids **336** after work-up. Cyclization and demethylation under the previously mentioned conditions then afforded *dl*-zearalenone.

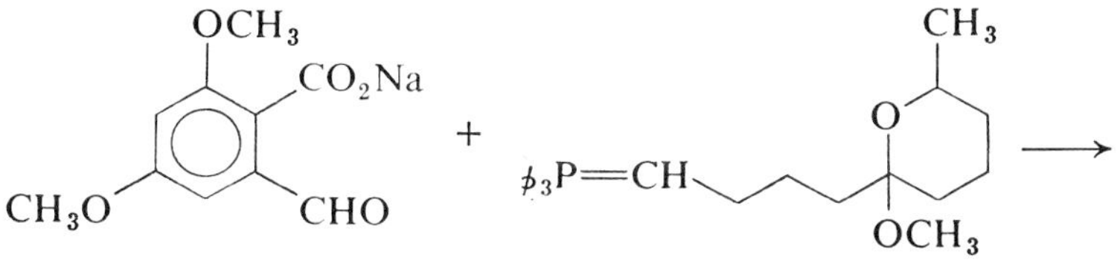

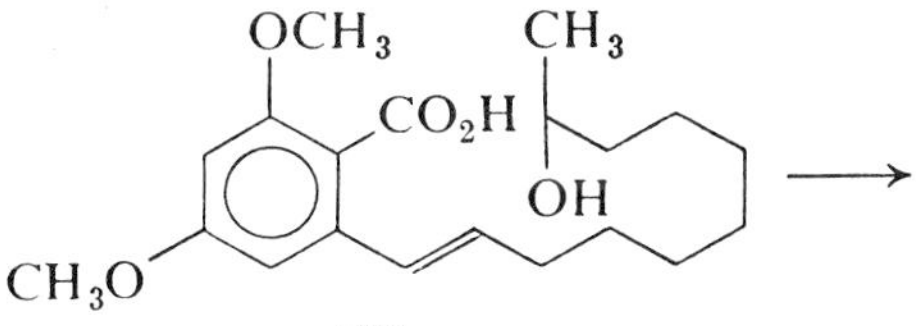

336

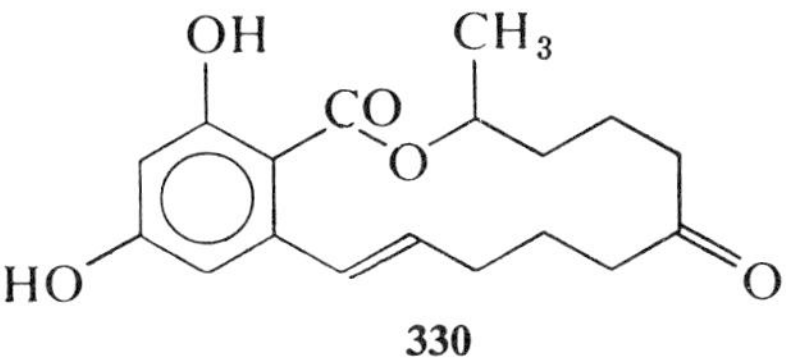

330

A second synthesis of this macrocycle has been reported[142] by a group at Syntex (Scheme 38). This again devolves on lactone formation in the penultimate step. The complete assembly of carbon atoms was smoothly elaborated by a Wittig reaction involving the ethyl ester of **338** with **347**.

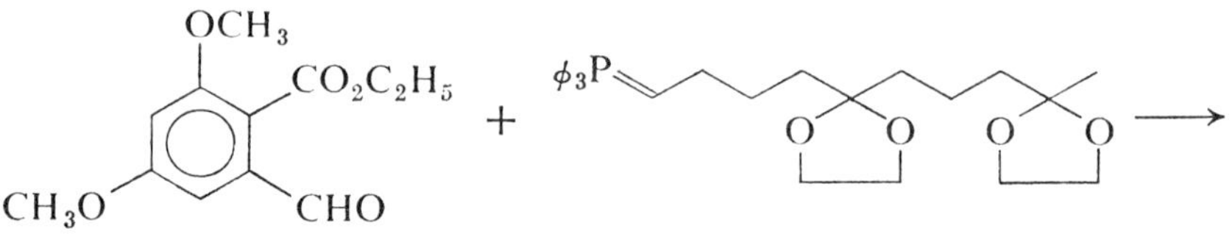

347

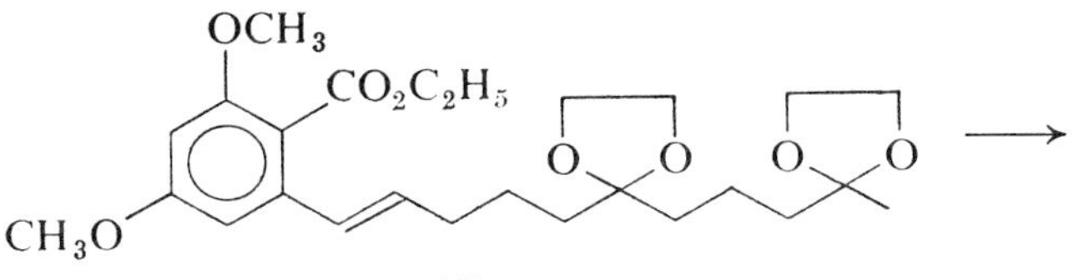

348

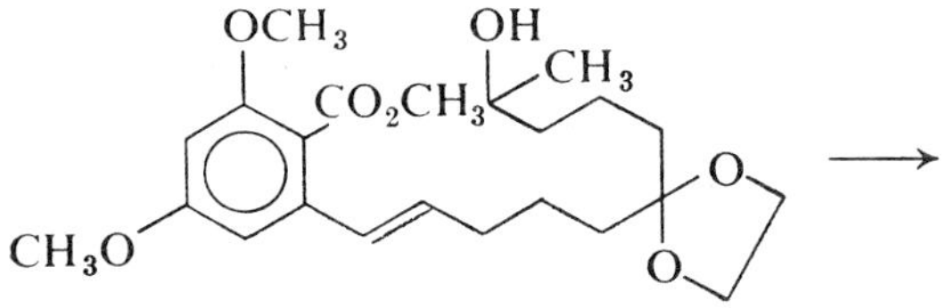

349

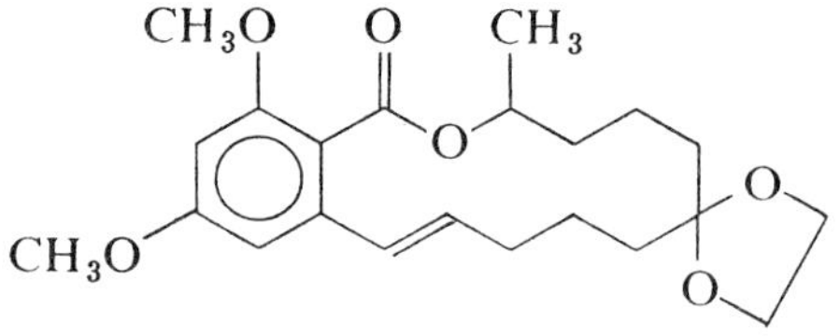

350

Scheme 38

This should be contrasted with the complex products obtained from the ethyl ester of **338** and the Wittig salt $\phi_3P^+C^-HCH_3$ mentioned previously. Selective cleavage of the terminal ketal group was possible with **348** and the ketone so obtained was reduced by borohydride to **349**. Base-catalyzed cyclization of **349** then led to **350**, which when hydrolyzed to remove the ketal group afforded *dl*-zearalenone.

Differentiation of the two ketone functions of **351**, the diketone derived from **348**, was also achieved in another way. Borohydride reduction of **351** led to **352**, which when treated with sodium hydride afforded the ether **353**.

Further treatment with the same reagent at elevated temperatures then gave the lactonic ether **354**, which under no circumstances could be modified to give zearalenone.

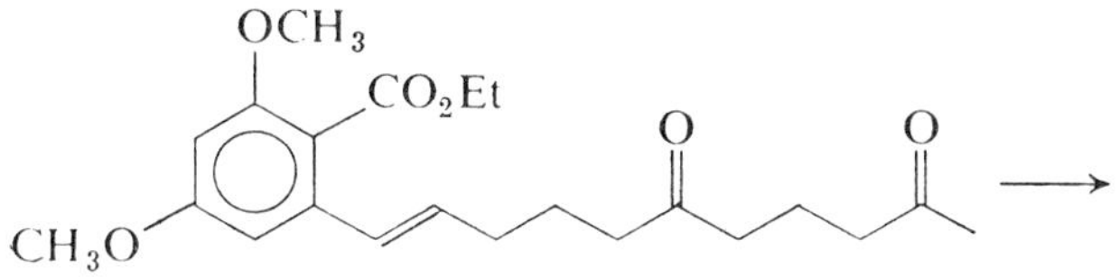

351

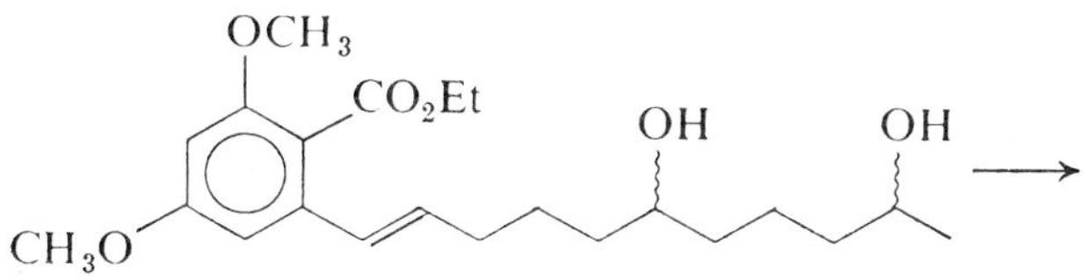

352

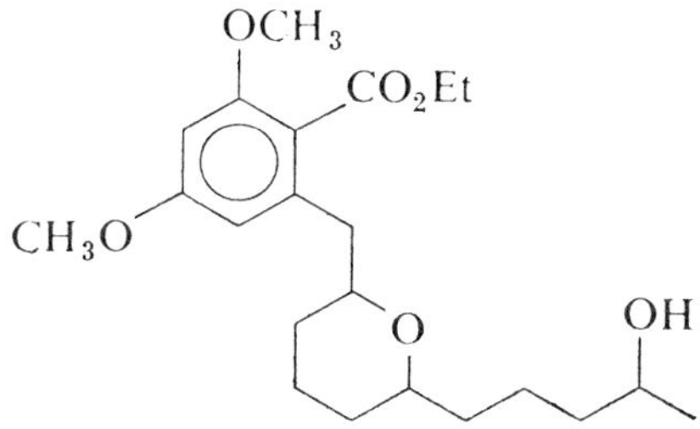

353

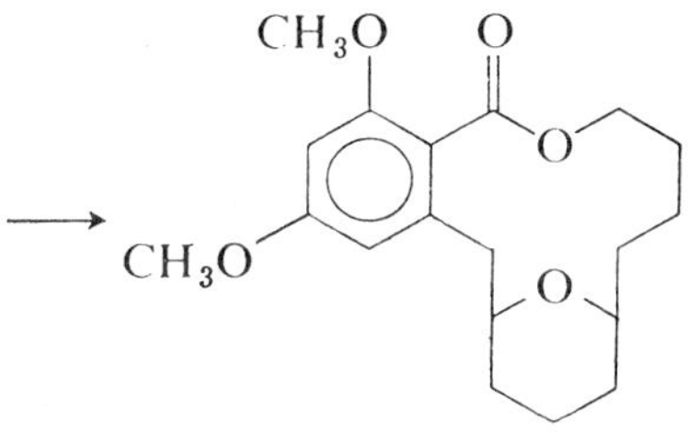

354

The foregoing methods for the synthesis of zearalenone use the third of the three conventional methods for the synthesis of macrolides. These methods are (*a*) Baeyer-Villiger oxidation of macrocyclic ketones,[143] (*b*) peracid oxidation of bicyclic enol ethers,[144] and (*c*) the direct cyclization of hydroxy acids and esters. Studies in connection with the synthesis of zearalenone and curvularin by the second of these methods have been reported by Immer and Bagli. In their investigations a series of compounds of the type **355** were synthesized and oxidized by means of *m*-chloroperbenzoic acid to **356**. No successful synthesis of zearalenone has yet emerged, although an

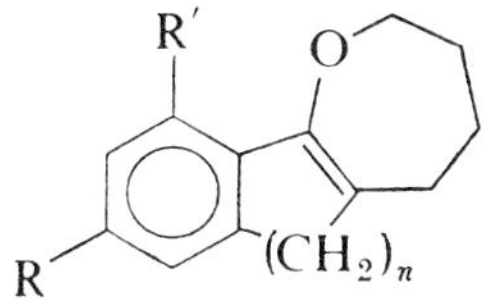

355

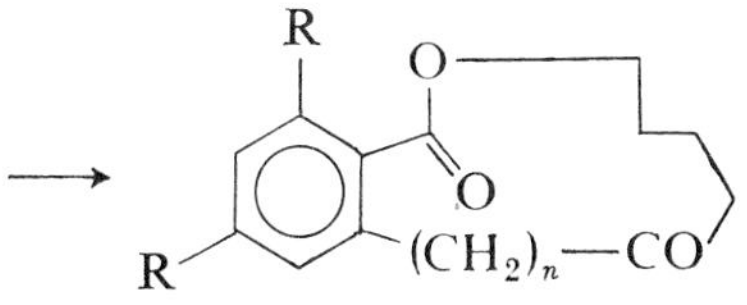

356

isomer of curvularin has been prepared.[145] Undoubtedly the presence of a double bond in the lactone ring of **355** will present complications to this method.

8. MISCELLANEOUS ANTIBIOTICS

There remains a group of antibiotics that in many instances are of very great therapeutic value but which do not fit into any one class and for convenience are simply assembled and discussed in this section. Of the 15 or so compounds that comprise this group only 6 have been synthesized, anthramycin (**357**), chloroamphenicol (**358**), cycloserine (**359**), cycloheximide (**360**), griseofulvin (**361**), and novobiocin (**362**). The synthesis of each of

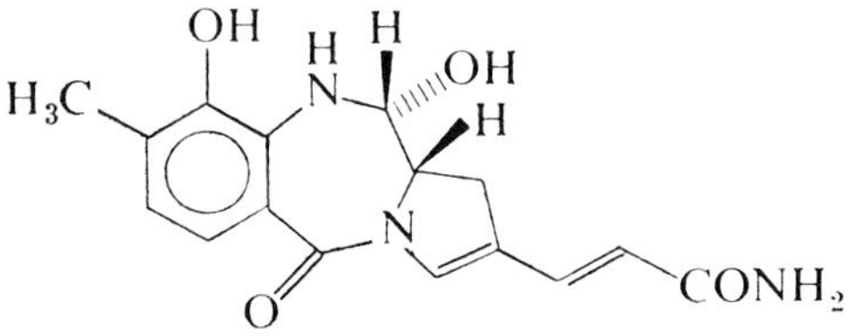

OH H H H_3C N OH H N O $CONH_2$

357

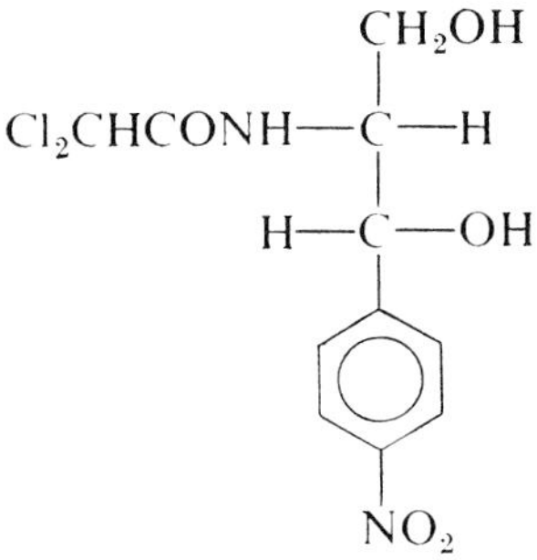

CH_2OH $Cl_2CHCONH—C—H$ $H—C—OH$ NO_2

358

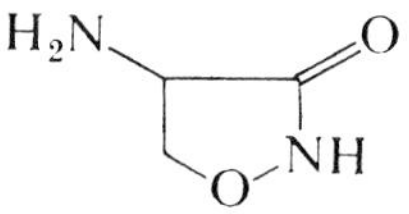

H_2N O NH O

359

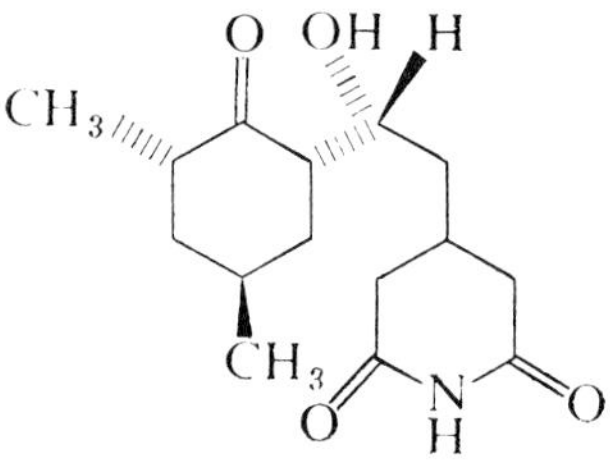

O OH H CH_3 CH_3 O N H O

360

OCH_3 CH_3O O OCH_3 Cl O O CH_3 H

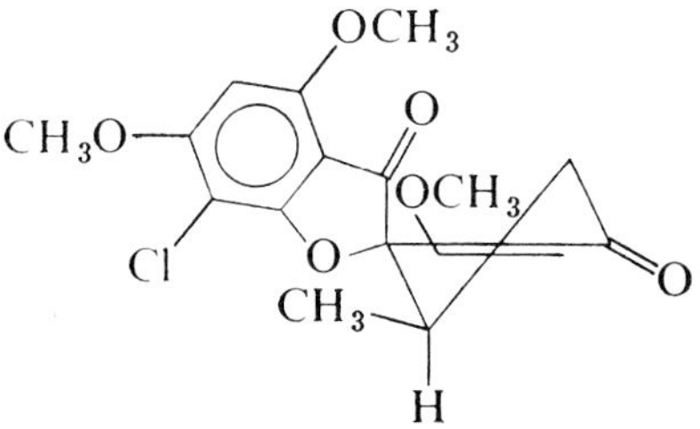

361

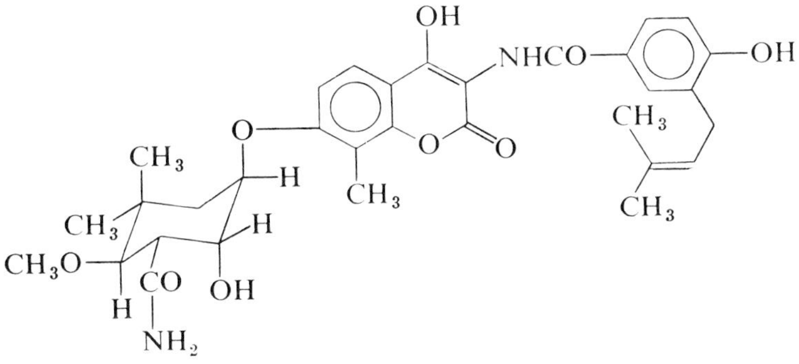

362

these molecules is discussed here with the exception of cycloserine whose preparation[146] presented no special problems.

Of the remaining compounds, coumermycin A,[147] (**363**) [which is closely related to novobiocin (**363**)], daunomycin[148] (**364**), fumagillin[149] (**365**), fusidic acid[150] (**366**), mitomycin C (**367**), monensin[151] (**368**), rifamycin B[152] (**369**), streptonigrin[153] (**370**), and streptovitacin A[154] (4*e*-hydroxycycloheximide), only fumagillin[155,156] and mitomycin C have seen any serious attempts being made toward their synthesis. The significant synthetic work dealing with mitomycin C is discussed briefly in this section.

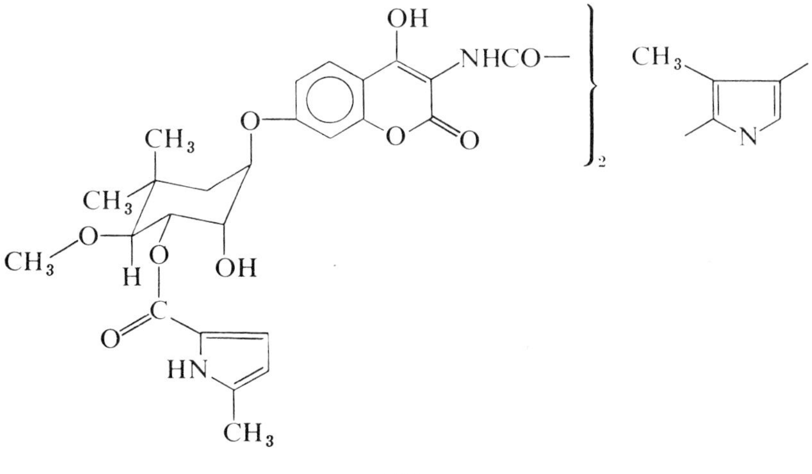

363

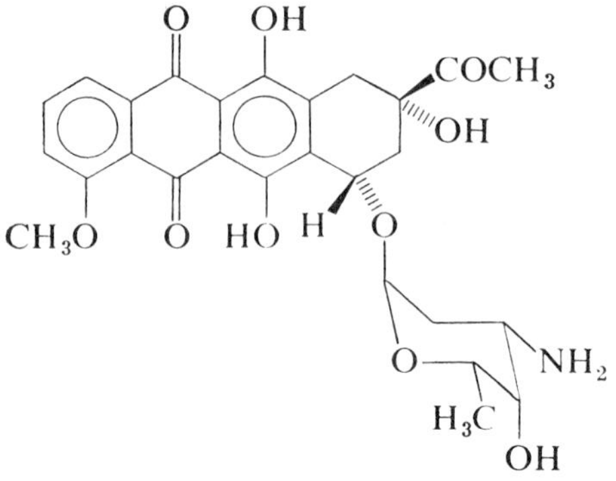

O OH
COCH3
OH
CH3O O HO H O
O NH2
H3C
OH

364

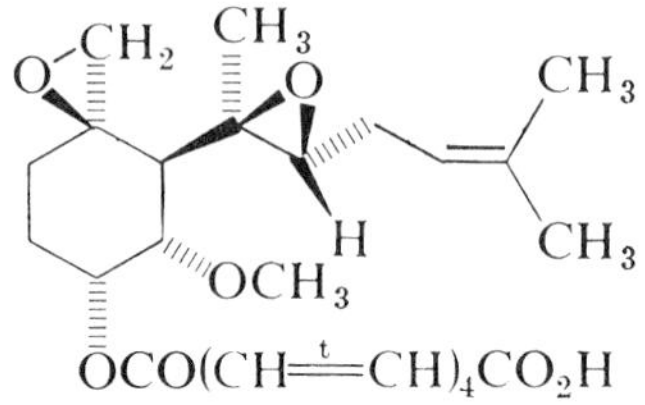

O CH2 CH3
O CH3
H
CH3
OCH3
OCO(CH=t=CH)4CO2H

365

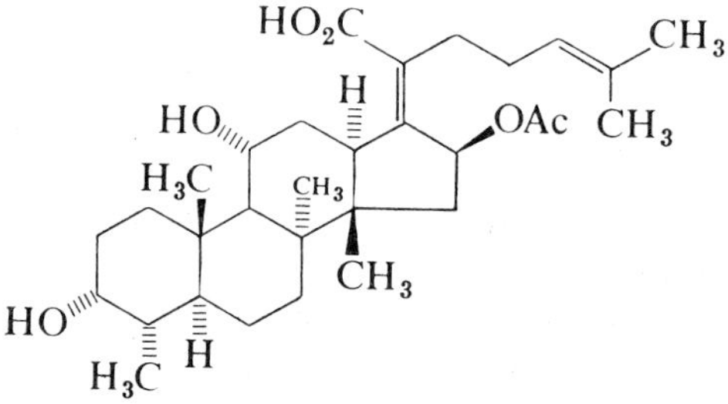

HO2C CH3
H
HO OAc CH3
H3C CH3
CH3
HO
H3C H

366

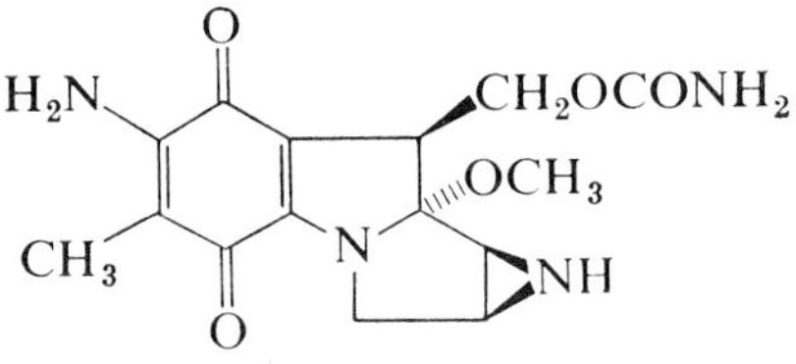

O
H2N CH2OCONH2
OCH3
CH3 N NH
O

367

CH3OCH3CH3
CH—CH—CH—CO2H
CH3 CH3 CH3
HOCH2 O
HO O O O CH3
C2H5 CH3
OH

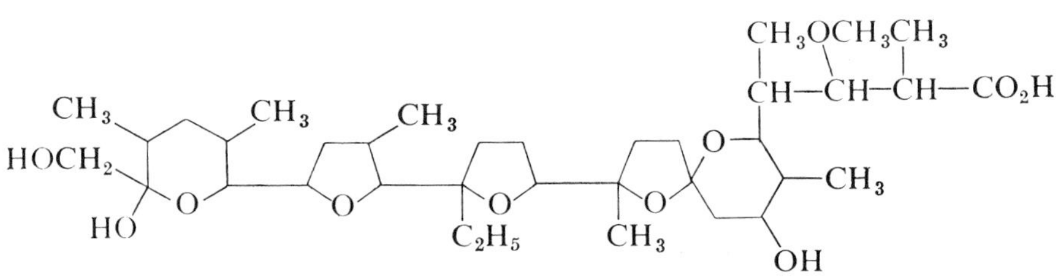

368

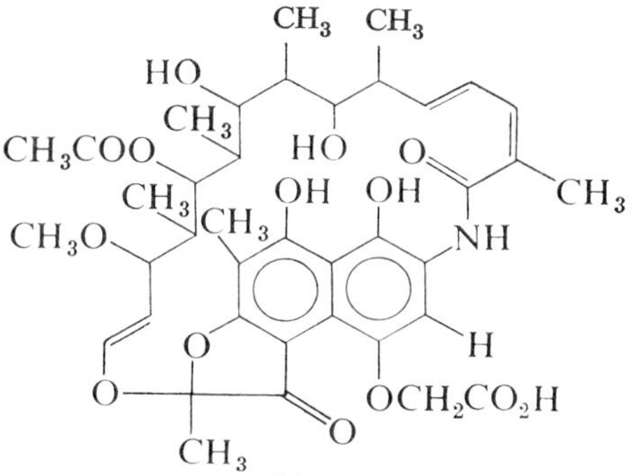

369

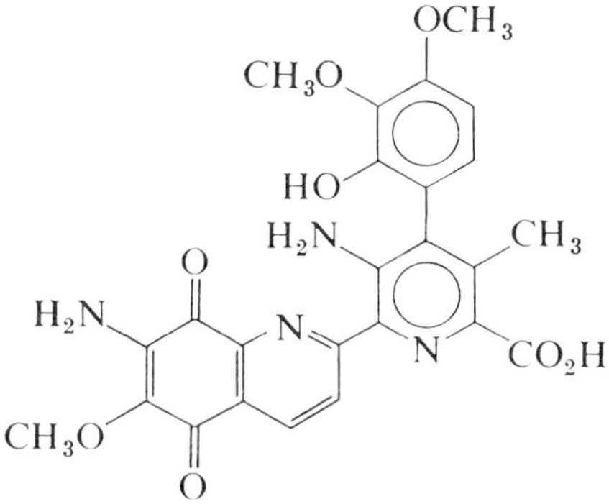

370

A. Anthramycin (357)

The total synthesis[157] of anthramycin, a potent antineoplastic agent, was carried out by the Hoffmann-LaRoche group who both isolated[158] the compound and proposed its structure.[159] During the structure work it had been shown that anthramycin methyl ether (**371**) could be converted, via anhydroanthramycin, to anthamycin itself. Since **371** was well characterized and comparatively stable, the problem was reduced itself to the synthesis of this compound. As a primary synthetic objective the corresponding cyclic amide (**372**) was chosen. A particularly attractive approach to this material

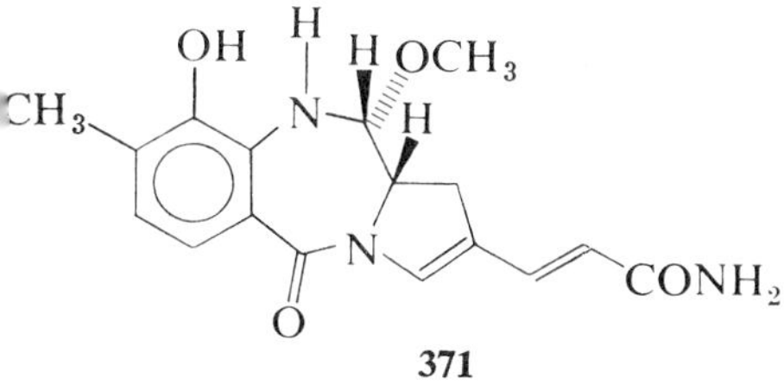

371

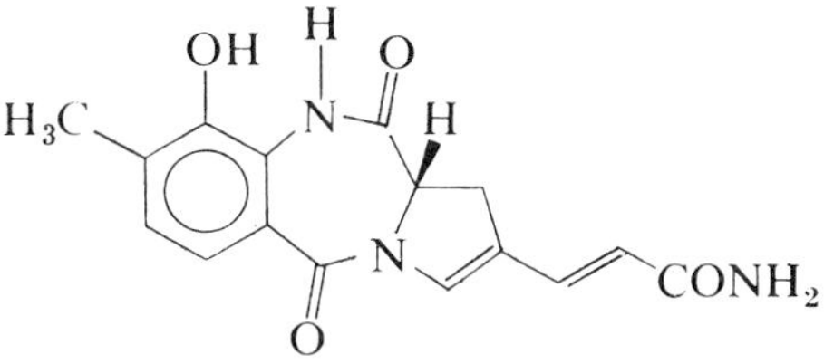

372

was to use a naturally occurring amino acid, which would result in the final product being optically active rather than racemic. The synthesis (Scheme 39) began therefore with the acylation of L-hydroxyproline methyl ester (**374**) with 3-benzyloxy–4-methyl–2-nitrobenzoyl chloride (**373**). The product **375** was reduced with sodium dithionite to the corresponding amine, which when heated with aqueous hydrochloric acid cyclized to the lactam **376**. Oxidation

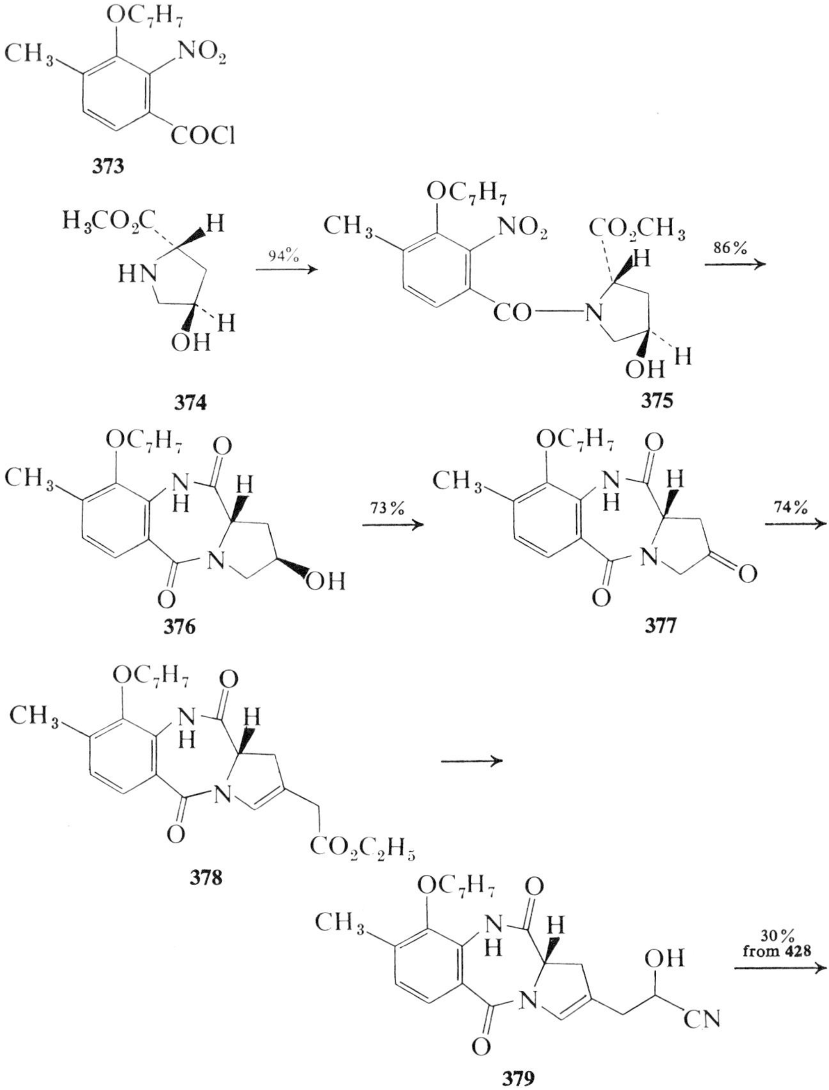

Scheme 39

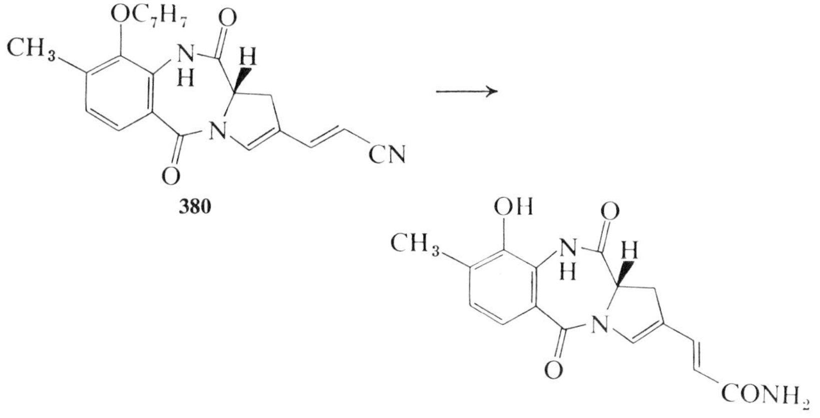

380

372

Scheme 39 (Continued)

of **376** with chromic acid led to **377**, which was condensed with the sodium salt of triethylphosphonoacetate at 0° to give as the major product the β,γ-unsaturated ester **378**. Reduction with diisobutylaluminum hydride at −60° gave a labile aldehyde which was immediately converted to the bisulfite adduct and then by treatment with potassium cyanide to a mixture of the epimeric cyanohydrins.[379] Treatment of this product with methanesulfonyl chloride in pyridine followed by boiling the mesylates in benzene with triethylamine afforded a *trans*:*cis*::4:1 mixture of the conjugated nitriles (**380**). These isomers could be separated by chromatography but were found to be easily interconvertible. For this reason the mixture was simply debenzylated by means of a combination of boron trifluoride etherate and trifluoroacetic acid at ambient temperatures. The resultant mixture of phenolic *cis*- and *trans*-nitriles was then heated in aqueous trifluoroacetic acid at reflux temperature and provided as the major product the desired intermediate **372**.

The remaining problem in the total synthesis consisted formally of specifically effecting reduction of the secondary amide group of **372**. Attempted reduction with an excess of lithium aluminum hydride left **372** unchanged, so it was decided that a derivative would have to be prepared to obtain the desired result. It was envisaged that **381** might be suitable because simple *N*-methylphenylamides are converted in good yield to aldehydes by lithium aluminum hydride. Unfortunately the desired intermediate could not be prepared from **372**. Nevertheless, the corresponding derivative **382** could be prepared from the *trans*-nitrile (itself obtained by debenzylation of **380**) by condensation with benzaldehyde dimethylacetal at 200°. This was converted to the amide **381** by hot polyphosphoric acid. Reduction of **381** with sodium

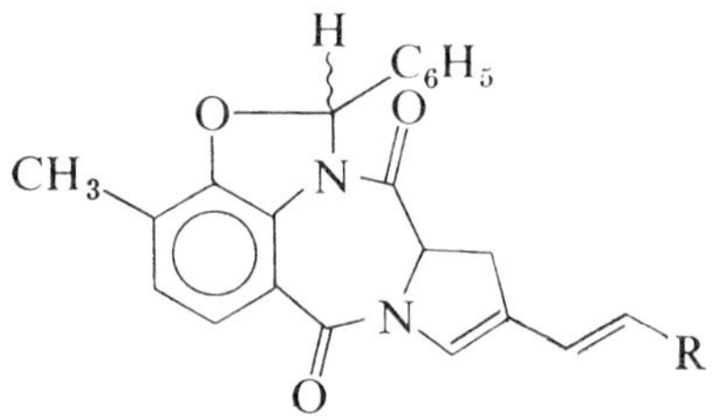

381 R = $CONH_2$
382 R = CN

borohydride in methanol at 5° followed by hydrolysis of the carbinolamine with very dilute acid in methanol then afforded anthramycin methyl ether (**381**) in 70% yield identical in every respect with an authentic sample.

A different approach to the anthramycin skeleton has recently been reported[160] but as yet a second synthesis of the antibiotic itself has not appeared.

B. Griseofulvin (361)

Griseofulvin was first discovered[161] in 1939, but its complete structure was not established[162] until 1952. Four different groups have completed a total synthesis of this molecule, three of them being announced initially in the same year (1960).

Scott and his associates[163] examined experimentally the suggestion by Barton and Cohen[164] that griseofulvin was derived biogenetically from the benzophenone **383** which underwent C-O oxidative coupling to give dehydrogriseofulvin **385** via the diradical **384**. Enzymatic reduction of **385** would then

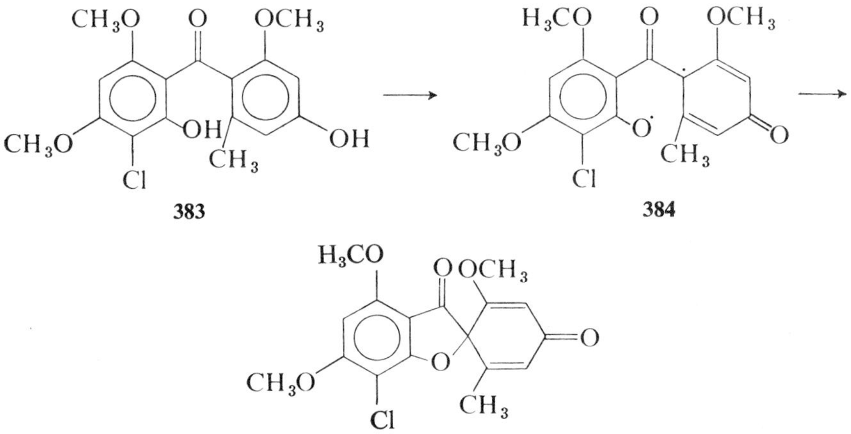

383

384

385

give **361**. The desired benzophenone was eventually synthesized, after many attempts, by the Friedel-Crafts reaction of the acid chloride **387** with the phenol **386**, followed by alkaline hydrolysis to remove the protecting methoxycarbonyl group. The overall yield was poor (~15%), but the benzophenone

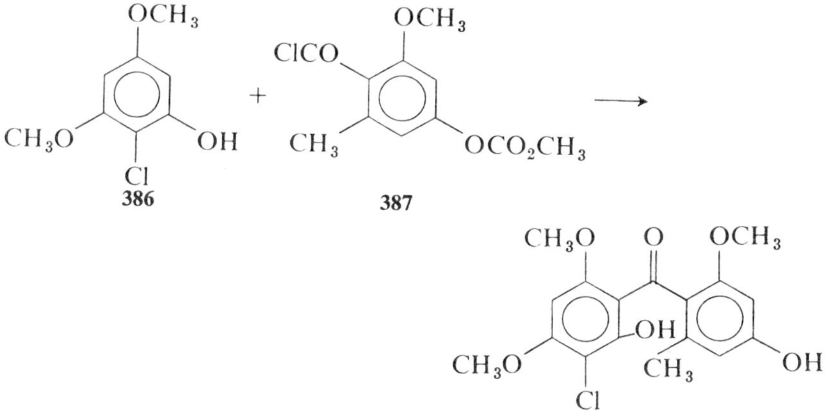

386 **387**

383

was available from griseofulvin itself by dehydrogenation with selenium dioxide to **385** followed by catalytic reduction over a platinum catalyst which simply caused hydrogenolysis of the C-O spiro bond. Internal coupling of **383** to regenerate dehydrogriseofulvin **385** proceeded smoothly as predicted (50–60% yield) in a mildly alkaline solution of potassium ferricyanide. Selective reduction of **385** to *dl*-griseofulvin was then accomplished by reduction over a rhodium-charcoal catalyst poisoned by the addition of 3% selenium dioxide. The yield was 30%, dropping to 2% in the absence of the added poison. Resolution of the racemic material was accomplished by acidic hydrolysis to *dl*-griseofulvic acid **388**, which was resolved as its quinine metho salt. The *d*-form of **388** had already been shown to regenerate

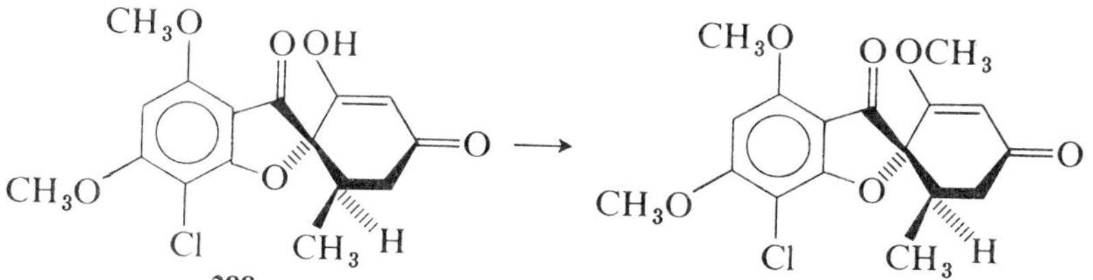

388 **361**

griseofulvin (**361**) when treated with diazomethane. Thus a formal synthesis of griseofulvin had been completed.

The synthetic pattern followed by the Merek group[165] for the synthesis of **361** paralleled that described by Scott and his associates[163] but was a much more thorough investigation. They found that the benzophenone **383** was best produced by allowing the free acid **389** and the phenol **386** to react in

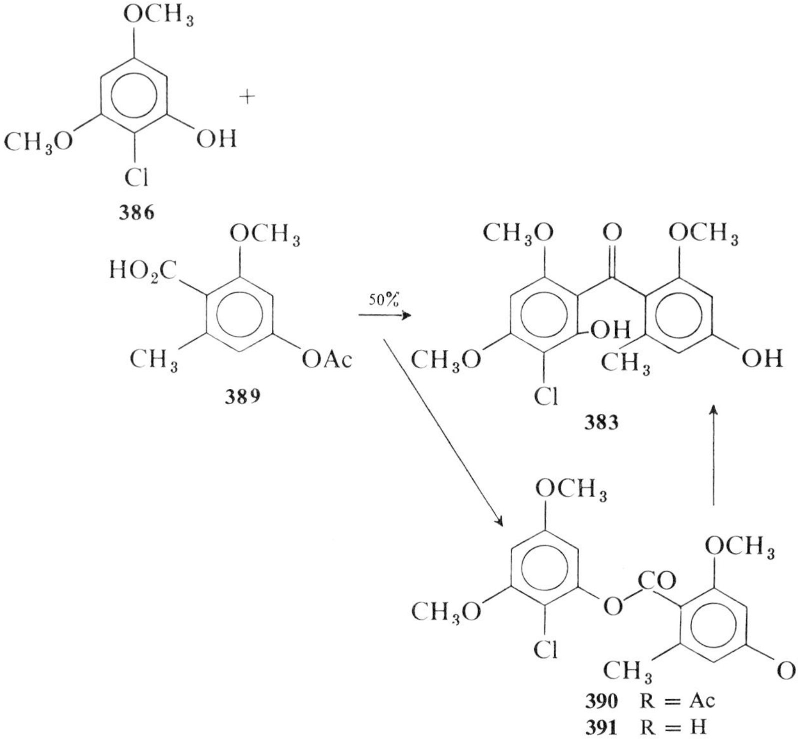

trifluoroacetic anhydride at 25°. A by-product (**390**), the ester formed from **386** and **389**, precipitated from the medium and mild alkaline hydrolysis of the soluble materials afforded **383**.

Alternative methods for the preparation of **383** consisted of either titanium tetrachloride (40–50% yield) or photoinduced (10–15% yield) Fries rearrangement of **391**. Improved methods for the oxidation of **383** to dehydrogriseofulvin (**385**) were also introduced. These involved carrying out the ferricyanide reaction with reverse addition or employing lead dioxide in acetone-ether solution. Both methods give 100% yields of product. Manganese dioxide in the same reaction gives 95–100% of **385**, whereas silver oxide affords only 5–10% of the dienone.

The reduction of **385** to griseofulvin itself was carried out over a large amount of a 10% palladium-on-charcoal catalyst but yields (~36%) were not much superior to those obtained by Scott.

A completely different approach to the synthesis of **361** was taken by Brossi and his group[166] Scheme (40). They *O*-alkylated the salicylic ester (**392**)

OCH_3
CO_2CH_3
CH_3O
OH
Cl
392

→

OCH_3
CO_2CH_3
CH_3O
$OCH_2CO_2CH_3$
Cl
393

→

OCH_3 O
CO_2CH_3
CH_3O
O
Cl
394

$\xrightarrow{57\%}$

CH_3O O
CO_2CH_3
CH_3O
O
CH_2COCH_3
CH_3
Cl
395

$\xrightarrow{(25\%)}$

CH_3O O O
O
CH_3O
O
Cl
CH_3 H
396

↓

OCH_3 CO_2R
CO_2R
CH_3O
O
Cl
CH_3
397 R = CH_3
398 R = H

CH_3O O OCH_3
O
CH_3O
O
CH_3
H
Cl
399

+

CH_3O O O
OCH_3
CH_3O
O
H
CH_3
400

Scheme 40

with methyl bromoacetate to give **393**, then subjected **393** to a Dieckmann condensation to obtain the coumaran-3-one **394**. Michael addition of **394** to *trans*-3-penten-2-one in methanol using Triton B as a catalyst afforded two compounds, m.p. 184° and 164°, in the ratio of ~5:1, respectively, both having structure **395**. Cyclization of the dominant isomer by means of sodium methoxide in methanol (sodium ethoxide in ethanol did not work) led to the spiro compound, *dl-epi*-griseofulvin (**396**), having the reverse configuration at the spiro-center with respect to griseofulvin itself. Minor amounts of the by-products **397** and **398** were also isolated from the mother liquors of this reaction. Methylation of **396** with diazomethane afforded a mixture of **399** and **400** from which **399** was separated by chromatography. Partial conversion of **399** to *dl*-griseofulvin was accomplished by using the base-catalyzed equilibration originally discovered by MacMillan[167] and for which he proposed as one possibility the mechanism of Scheme 41. The

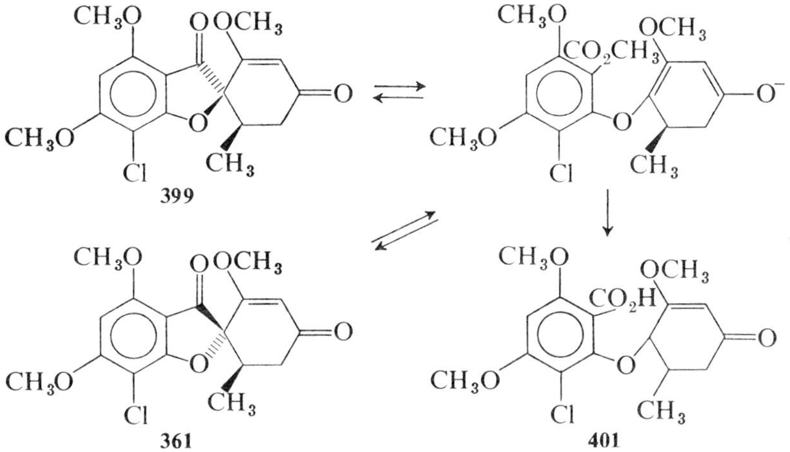

Scheme 41

isolation of small amounts of **401** suggests that this particular mechanism is correct. At equilibrium the system contains 60% of *dl*-epigriseofulvin (**399**) and 40% of *dl*-griseofulvin, which were separated by chromatography. *dl*-Griseofulvin was also synthesized by the base-catalyzed cyclization of the minor isomer of **395** followed by methylation of the product with diazomethane. However, the overall yield was extremely low. Resolution of *dl*-griseofulvin was carried out by using the brucine salt of griseofulvic acid followed by reconversion of the *d*-griseofulvic acid to **361** with diazomethane.

The griseofulvic acid **388** was generated by the methanolic acid-catalyzed isomerization of *dl*-griseofulvin to *dl*-isogriseofulvin (**402**), which was then hydrolyzed by mild aqueous base. This procedure avoids the problem of

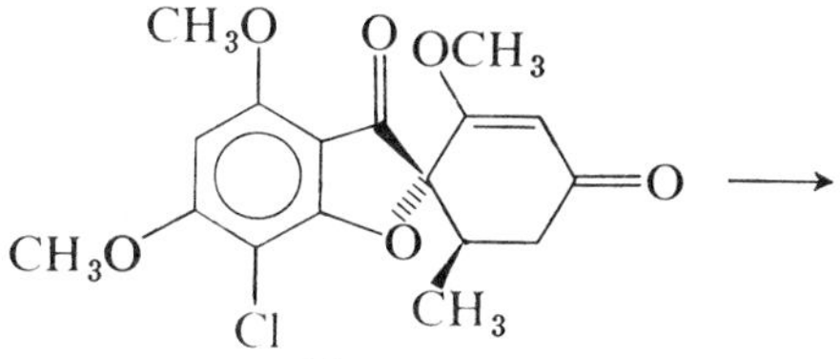

CH_3O O OCH_3 O → CH_3O O Cl CH_3

361

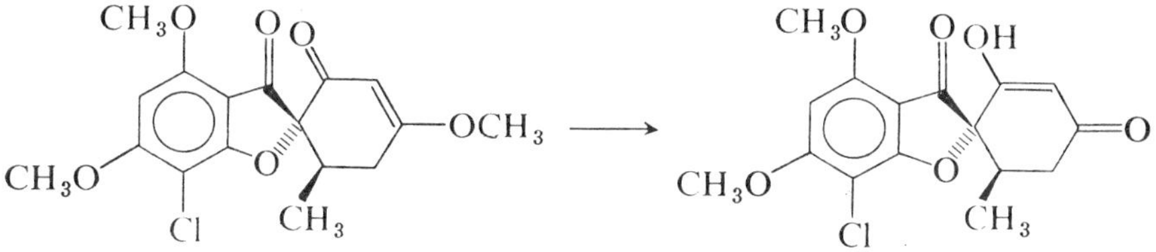

CH_3O O O OCH_3 → CH_3O O Cl CH_3

402

CH_3O O OH O CH_3O O Cl CH_3

388

generating some *epi*-griseofulvic acid which would almost certainly be produced, via the equilibration of **361** and **399**, if **361** were used directly in the base hydrolysis step.

Finally, a very elegant synthesis of *dl*-griseofulvin has been published by Stork and Tomasz.[168] They elected to synthesize the spiro-ring system of **361** by means of a double Michael addition of a cross-conjugated vinyl ethynyl ketone to an active methylene compound. Based on this concept they felt that the specific use of the coumaranone **403** and of methoxyethynyl propenyl ketone (**404**) would directly produce the griseofulvin structure as follows:

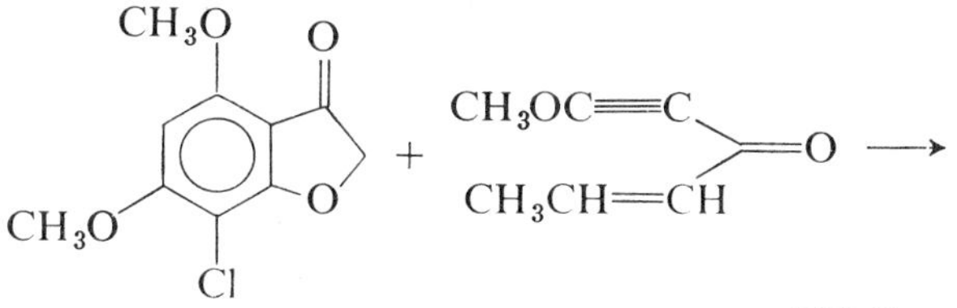

CH_3O O CH_3O O Cl + $CH_3OC{\equiv}C$ $CH_3CH{=}CH$ O →

403

404

CH_3O O OCH_3 O CH_3O O Cl CH_3

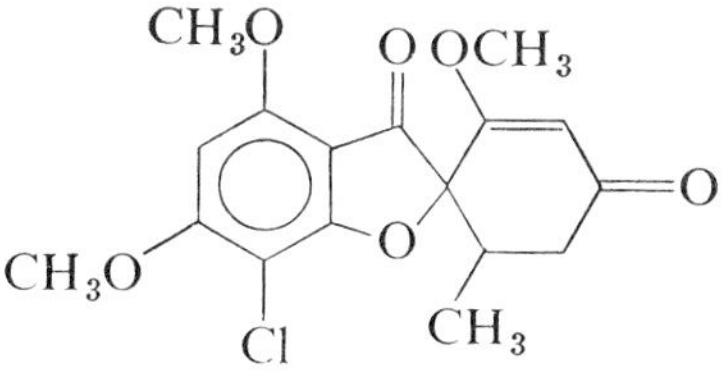

An attractive feature of this proposition is that should it work, it would lead to the proper enol ether without the ambiguities of the diazomethane method used in the previous syntheses. The real difficulty in this approach was that compounds of type **404** were previously unknown largely because of their great lability rather than to lack of a suitable preparative pathway. In fact the synthesis of **404** was solved in two steps. Condensation of the lithium salt of methoxyacetylene with crotonaldehyde at $-15°$ afforded the

$$Li{-}C{\equiv}C{-}OCH_3 + CH_3CH{=}CHCHO \xrightarrow{43\text{–}63\%} CH_3CH{=}CHCH(OH){-}C{\equiv}COCH_3 \;(\mathbf{405}) \xrightarrow{30\%} (CH_3OC{\equiv}C)(CH_3CH{=}CH)C{=}O \;(\mathbf{404})$$

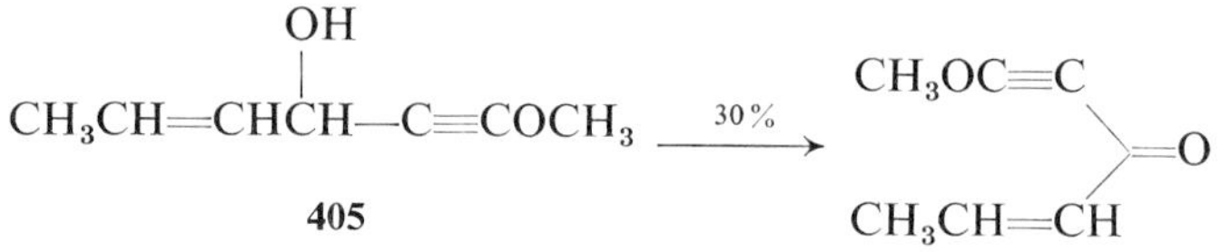

carbinol **405**, which was then oxidized with activated manganese dioxide in methylene chloride. Provided the manganese dioxide were washed to neutrality prior to use consistent results were obtained. Preliminary condensation experiments with the ethoxy analog of **404** and diethyl malonate in the presence of potassium *t*-butoxide (catalytic amount) led to a substituted cyclohexenone identified as **406**. Thus the initial concept was proven and

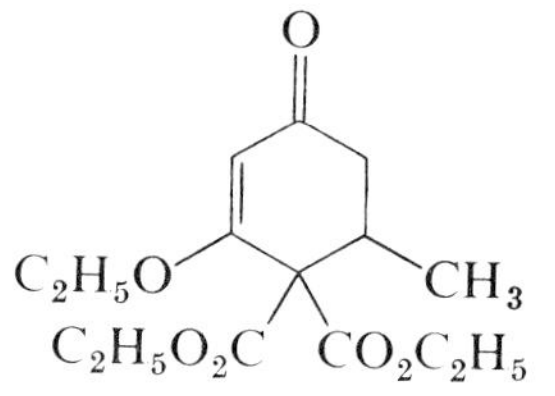

O, C_2H_5O, CH_3, $C_2H_5O_2C$, $CO_2C_2H_5$

406

the condensation of the coumaranone **403** with **404** was now attempted under essentially the same reaction conditions. Chromatography of the product then afforded *dl*-griseofulvin (**361**) in 7% yield. No *dl*-*epi*-griseofulvin (**399**) was detected in the reaction material, indicating that only the least stable isomer had been produced. The authors have suggested that the stereoselectivity observed here is the result of better overlap of the electron donor system in the transition state **407**, which leads to *dl*-griseofulvin, than in the transition state **408**, which leads to *dl*-*epi*-griseofulvin.

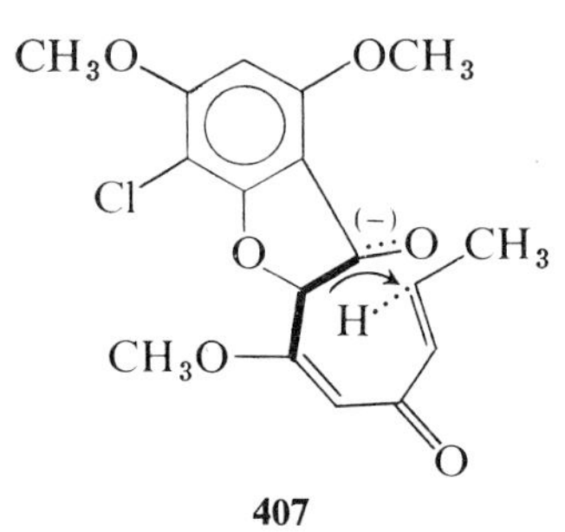

CH_3O, OCH_3, Cl, O, (−), O, CH_3, H, CH_3O, O

407

CH_3O, OCH_3, Cl, O, (−), O, CH_3, H, CH_3O, O

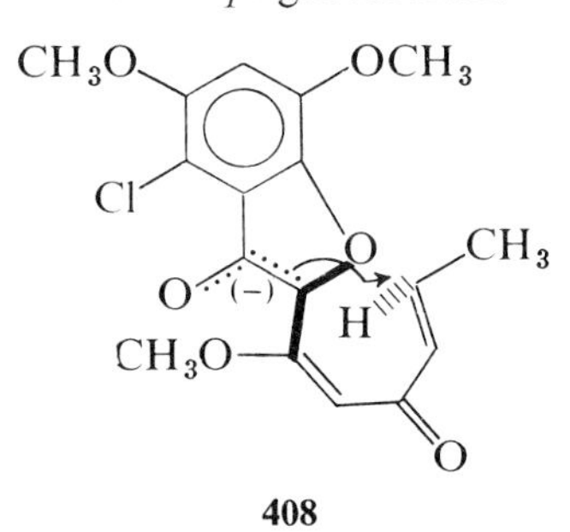

408

C. Novobiocin (362)

Novobiocin is an antibiotic used largely against gram-positive organisms. Its structure was first defined completely by Shunk et al.,[169] and Vaterlaus[170] reported its synthesis in 1964. Much of the difficulty in the synthesis of this molecule lay (*a*) in preparing a sugar moiety with the correct stereochemistry (Scheme 42) and (*b*) in determining proper conditions for the coupling of the protected sugar **419** to the coumarin nucleus.

The synthesis of **419** began with 3,5,6-tri-*O*-benzyl–2-*O*-methyl-D-glucofuranose (**409**), which was oxidized with *N*-bromosuccinimide in the presence of barium carbonate to the lactone **410**. Ring opening of **410** with methylamine afforded the amide **411**, whose mesylate (**412**) when refluxed in aqueous acetic acid afforded **413**, the 5-epimer of **410**. Treatment of **413** with methylmagnesium iodide gave the diol **414**. Benzoylation of the latter esterified the secondary hydroxyl only, and the product **415** on catalytic

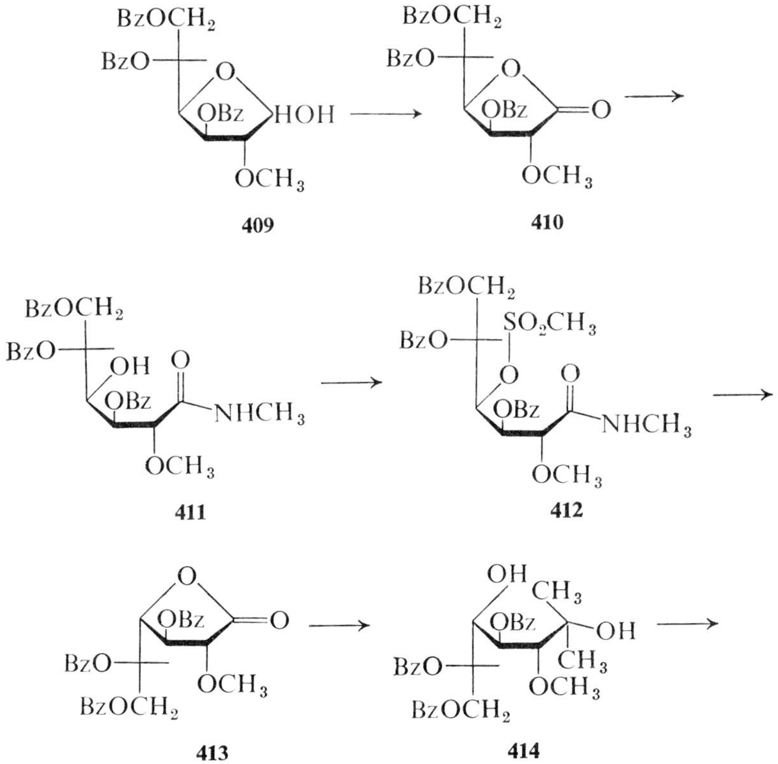

409 → **410** →

411 → **412** →

413 → **414** →

Scheme 42

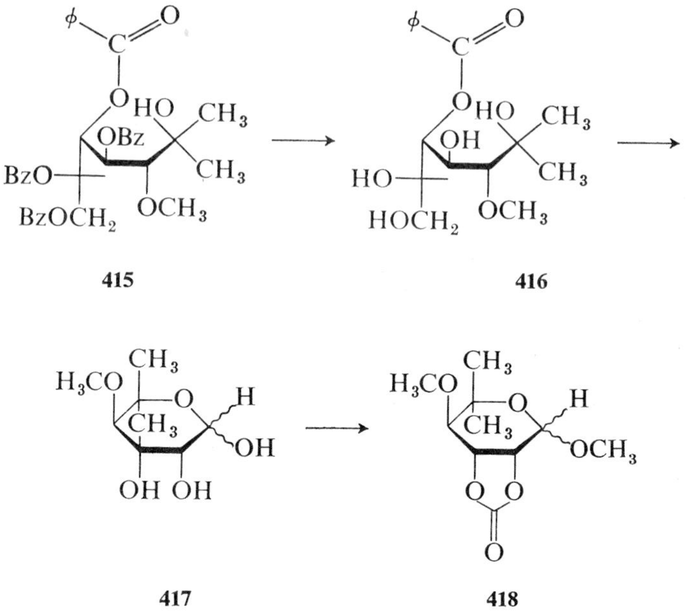

415

416

417

418

Scheme 42 (Continued)

reduction afforded the triol **416**. Lead tetraacetate cleavage of **416** followed by saponification of the benzoyl group produced noviose (**417**), the *des*-3-*O*-carbamoyl glycosidiccomponent of novobiocin. The methyl glycosides of **417** was treated by carbonyl chloride in pyridine to introduce the carbonate group. The mixture of glycosides (**418**) was then treated with acetyl chloride containing hydrogen chloride in nitromethane to give the key intermediate 2,3-*O*-carbonyl-noviosyl chloride **419**.

The required coumarin derivative **420** was synthesized by conventional methods (Scheme 43) and coupled successfully with **419** in quinoline in the presence of silver oxide and calcium sulfate to give the β-glycoside **421**. The latter was reduced catalytically to remove the protective benzyl group on the C-4 oxygen atom and coupled with phenyldiazonium chloride with a view to introducing the 3-amino function on the coumarin ring. The product **422** was again reduced catalytically over palladium and the desired goal was reached. Acylation of the amine **423** with the acid halide **424** led to the penultimate compound **425** of the complete synthetic sequence. The final step, ammonolysis of **425**, proceeded well, giving a mixture of novobiocin (**362**) and isonovobiocin (**426**) in which the former predominated. Fractional crystallization then gave pure novobiocin identical with the natural product.

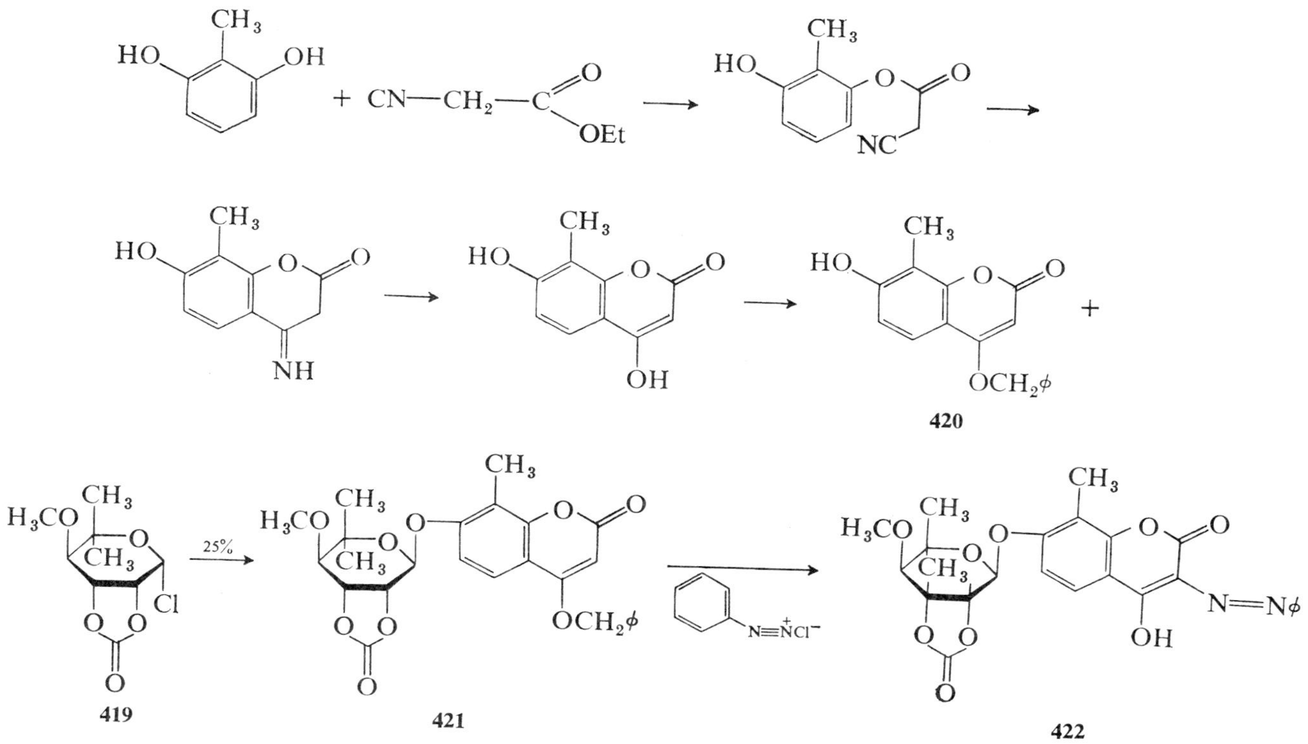

Scheme 43

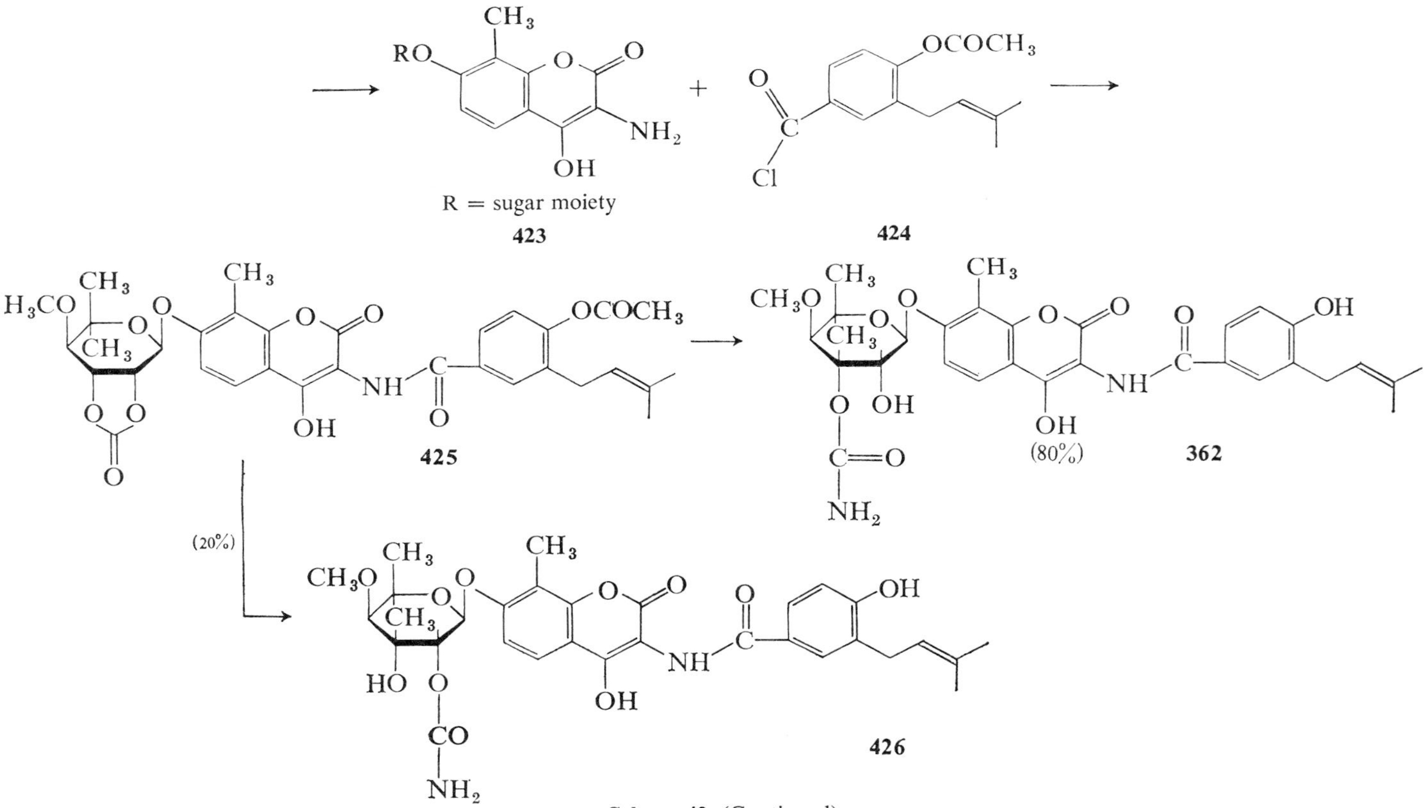

R = sugar moiety

423

424

425

(80%)

362

(20%)

426

Scheme 43 (Continued)

D. Cycloheximide (360)

Cycloheximide is a most interesting antibiotic from a biological point of view because it has such a broad spectrum of activity. In particular it has potent antifungal activity[171] and is sold commercially for the control of such plant diseases as "dollar spot", cherry leaf spot, and white pine blister rust. It is the most potent rodent repellent known, has antitumor activity, and is toxic to both protozoa and algae. Its activity springs from its ability to inhibit protein synthesis within the living cell. The structure of cycloheximide was determined by Kornfeld et al.[172] and the stereochemistry by Johnson and his associates.[173] The latter group also carried out the only total synthesis reported[174] so far for this molecule. In carrying out this synthesis stereoselectively, three major problems had to be solved: (*a*) the 4-methyl group of the cyclohexanone ring, which is remote from all other functionality, had to be built into the molecule in the unstable axial orientation; (*b*) the side-chain hydroxyl group had to be introduced in the (*R*)-configuration; and (*c*) the final steps of the synthesis had to be accomplished under fairly neutral conditions because of the sensitivity of cycloheximide to acid or base.

A method of establishing the methyl groups in at least a *trans* relationship was suggested by the prior work of Williamson.[175] He had proposed that enamines of 2-substituted cyclohexanones had the 2-substituent in the quasi-axial orientation and offered this as the explanation for the difficulty that is observed when alkylation of such enamines is attempted. Any reagent approaching the double bond of the enamine would necessarily be involved in a nonbonded interaction with the quasi-axial substituent at the 2-position.

The synthesis of cycloheximide required first a practical synthesis of glutarimide-3-acetic acid **429**, and an improved method (Scheme 44) for

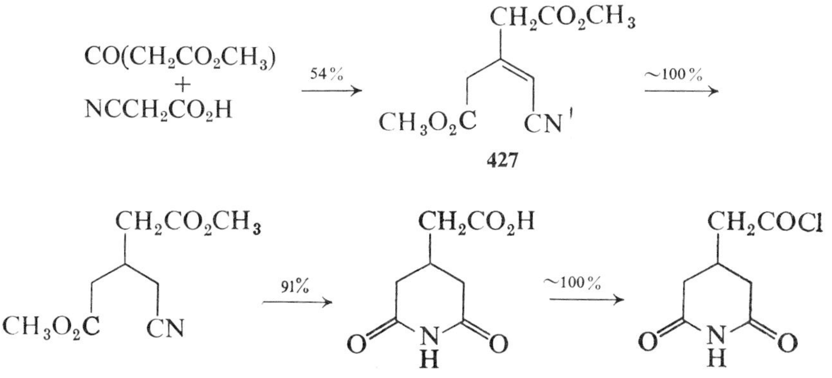

$CO(CH_2CO_2CH_3)$ + $NCCH_2CO_2H$ —54%→ **427** ($CH_2CO_2CH_3$, CH_3O_2C, CN) —~100%→ **428** ($CH_2CO_2CH_3$, CH_3O_2C, CN) —91%→ **429** (CH_2CO_2H, O, N, H, O) —~100%→ **430** (CH_2COCl, O, N, H, O)

Scheme 44

its preparation was reported by Johnson[176] in connection with a synthesis of actiphenol. Cope condensation of dimethyl acetonedicarboxylate with cyanoacetic acid led to the unsaturated nitrile, which when hydrogenated afforded **428**. When the **428** was boiled with moderately concentrated hydrochloric acid followed by raising the reaction temperature to 235° the highly crystalline acid **429** was produced. Conversion of this to the acid chloride occurred smoothly in the presence of thionyl chloride containing a trace of dimethylformamide.

In the critical step enamine formation with *cis*-2,4-dimethylcyclohexanone

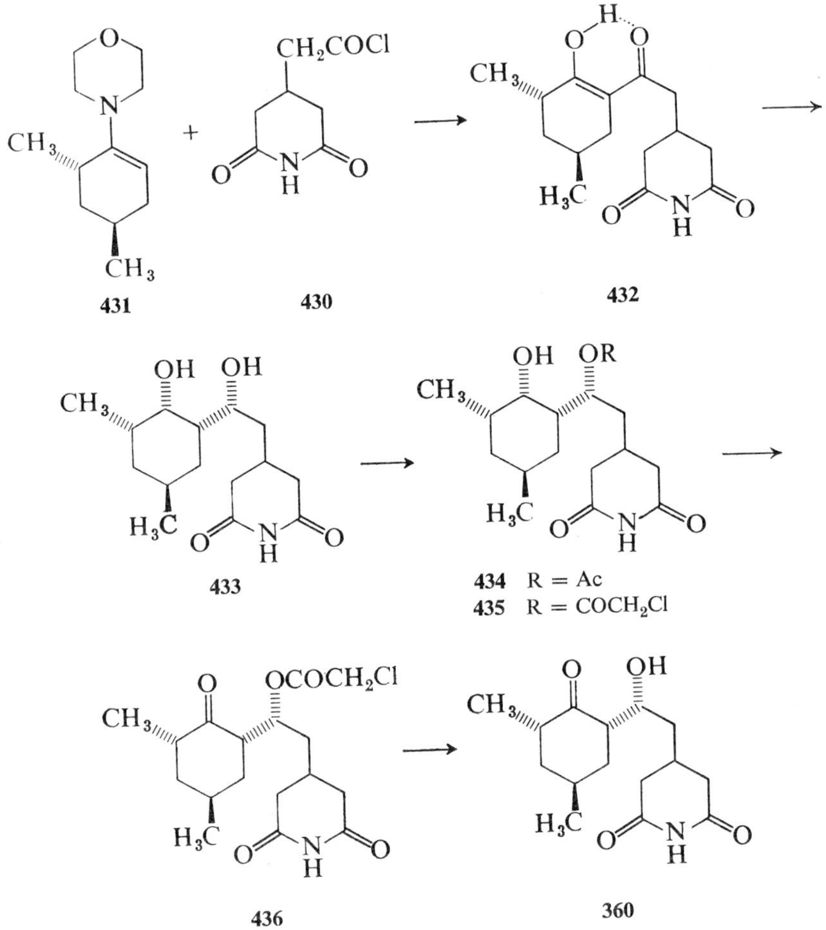

431 **430** **432**

433

434 R = Ac
435 R = $COCH_2Cl$

436 **360**

Scheme 45

led[177] to the enamine **431** of *trans*-2,4-dimethylcyclohexanone. Acylation of **431** with the acid chloride **430** (Scheme 45) gave, after hydrolysis with mild aqueous acid, dehydrocycloheximide (**432**). Reduction of **432** over a platinum catalyst very fortuitously occurred in such a way as to establish all five asymmetric centers of dihydrocycloheximide (**433**) stereoselectively (see below). An unexpected difficulty now arose. Although monoacetylation of **433** could be accomplished selectively at the side-chain hydroxyl group to give **434**, whose oxidation by chromium trioxide gave cycloheximide acetate, no method could be found to hydrolyze the acetate group without destroying the molecule. If monoacylation of **433** were carried out with trifluoroacetic anhydride, only the compound with a trifluoroacetyl group on the ring-hydroxyl could be isolated. Thus it appeared that if an unreactive monoester of **433** were prepared using a fairly weak acid, the ultimate product could not be hydrolyzed to give **360**. If a strong acid anhydride were used, acylation of the side-chain hydroxyl undoubtedly occurred first, but was accompanied by ester migration to the ring-hydroxyl. The problem was solved by rapidly acylating **433** with chloracetyl chloride at $\sim 0°$ in the presence of exactly one equivalent of pyridine. Under these conditions ester migration was minimized and **435** was produced in good yield. Oxidation of **435** with chromium trioxide gave cycloheximide chloroacetate **436**, which when hydrolyzed with potassium bicarbonate in aqueous methanol yielded cycloheximide (**360**). Both the *dl*- and *l*-forms of **360** were synthesized starting with the appropriate forms of *cis*-2,4-dimethylcyclohexanone.

The reduction of **432** to **433** in such a specific manner requires some explanation. The authors consider that of the two possible conformers **432a** and **432b** of this diketone, the topside surface of **432b** presented less steric hindrance to catalyst approach than any of the other three. Reduction from this direction would then lead to **437** having the desired ring stereochemistry (Scheme 46). On the other hand, reduction from the least hindered side of **432a** (i.e., from underneath) would generate **438**. Preferential reduction of **437** could be expected to take place from the topside of **437** for steric reasons and assuming that the conformation of the side chain is controlled by hydrogen bonding between the side-chain carbonyl groups and the ring hydroxyl, then **439** would be (and is in fact) the expected product of the reaction. By similar reasoning, **438** would lead to **439**.

Some evidence for a two-step reduction was obtained by halting the reaction after the absorption of one equivalent of hydrogen, when **437** was isolated.

Synthetic work in this area has also led to the synthesis of two isomers of cycloheximide, neocycloheximide[173] and α-*epi*-isocycloheximide,[178] but their description is not warranted in this essay.

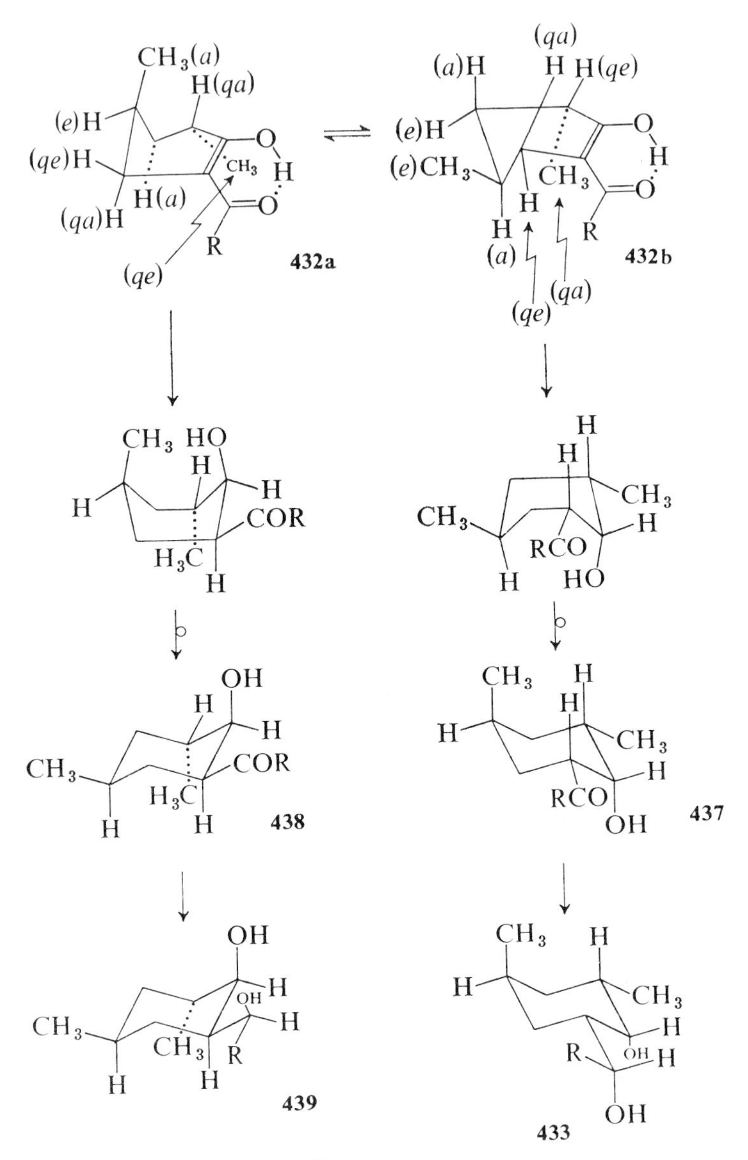
CH3(a)
H(qa)
(e)H
(qe)H
H(a)
(qa)H
CH3
H
O
R
(qe)
432a
(qa)
(a)H
H H(qe)
(e)H
(e)CH3
CH3
H
O
H
R
(a)
432b
(qa)
(qe)
CH3 HO
H
H
COR
H3C
H
CH3
RCO
HO
OH
CH3
COR
438
CH3 H
437
OH
OH
CH3
R
439
R
433

Scheme 46

E. Mitomycin C (367)

The mitomycins are a group of antibiotics that possess activity against both gram-positive and negative bacteria but are more noted for their antitumor properties. Only mitomycin C is used therapeutically. The structures of these compounds were deduced by Webb et al.[179] and are unusual not only in the pyrrolo(1,2-*a*)-indole ring that they contain but also because of the aziridine and aminobenzoquinone groups. Extensive synthetic work by a Lederle group[180] led to the elaboration of 7-methoxymitosene (**440**), but efforts to prepare the tetracyclic nucleus of **367** were not successful. The aziridine ring is very unstable.

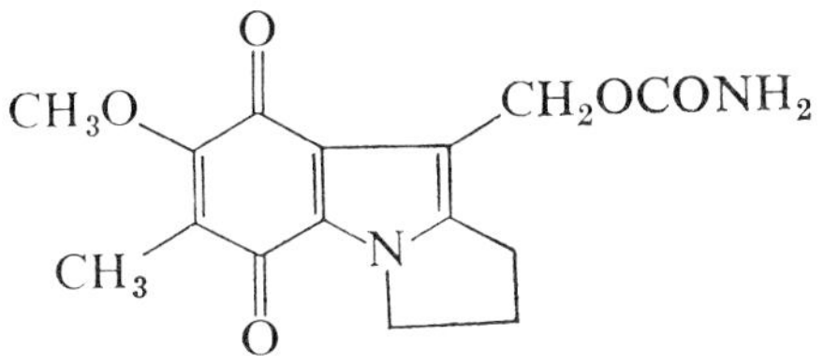

440

Recently a Japanese group has succeeded[181,182] in preparing a compound having the complete ring system of the mitomycins. Their synthesis is shown in Scheme 47. Conventional methods were used to convert the indole **441** to

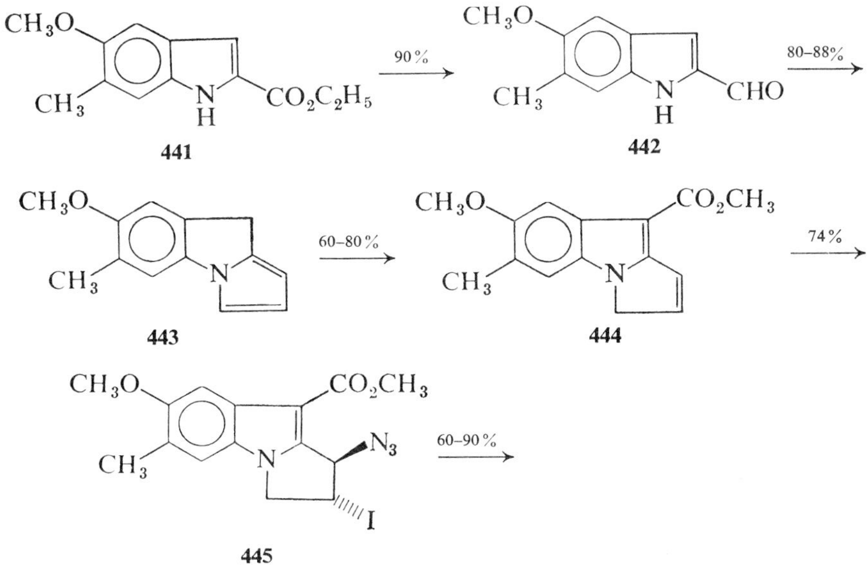

441 —90%→ **442** —80–88%→

443 —60–80%→ **444** —74%→

445 —60–90%→

Scheme 47

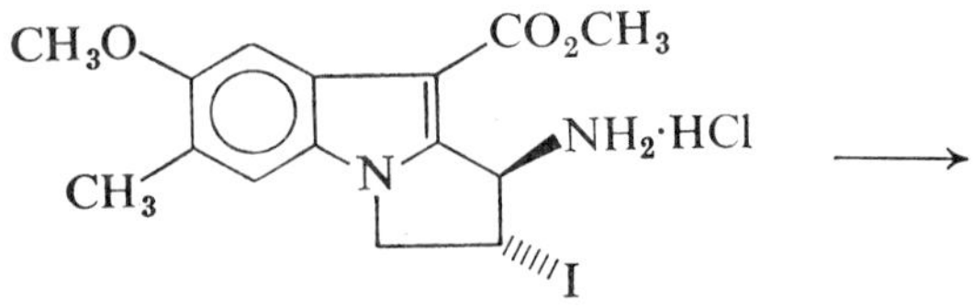

446

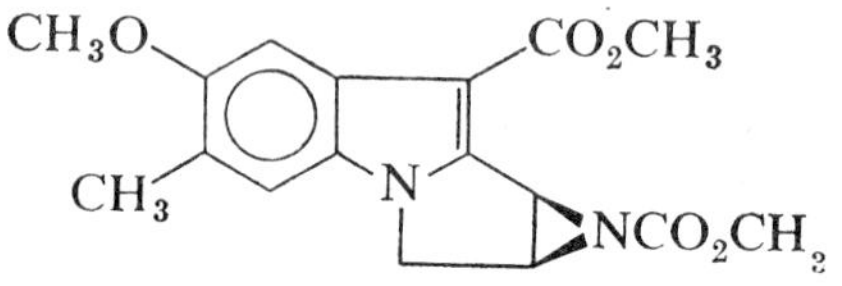

447

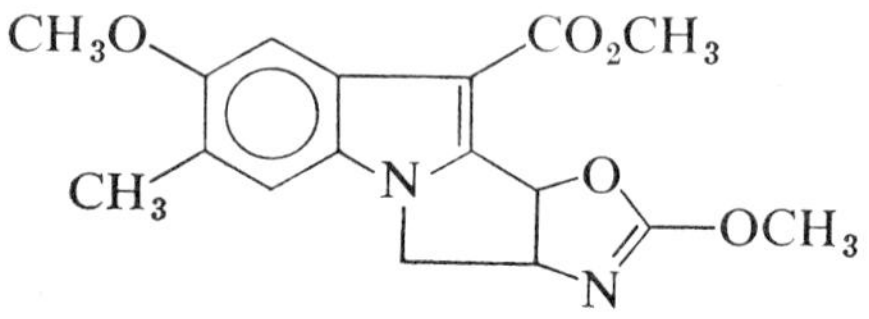

448

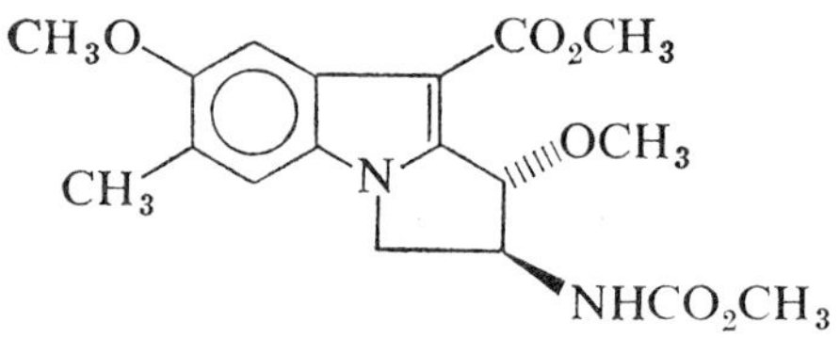

449

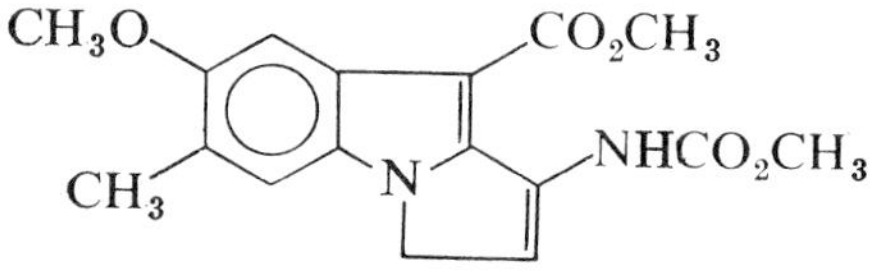

450

Scheme 47 (Continued)

the aldehyde **442**, which on treatment with vinyltriphenylphosphonium bromide and sodium hydride in tetrahydrofuran led to the pyrroloindole **443**. Carbomethoxylation with dimethyl carbonate and potassium *t*-butoxide afforded **444**, which was functionalized in the pyrroline ring with iodine azide preparatory to the introduction of the aziridine group. The iodo-azide **445** was reduced catalytically over palladium in methanol containing hydrogen chloride. The resulting amine hydrochloride **446** was cyclized by means of sodium methoxide in boiling methanol and yielded a crystalline mixture which was treated with methyl chlorocarbonate and triethylamine. From this reaction mixture the sought-after tetracyclic compound **447** was obtained by crystallization. There were also two by-products isolated to which structures **449** and **450** were assigned. The aziridino-pyrrolo(1,2-*a*) indole **447** proved to be thermally unstable and rearranged above 150° to the oxazoline **448**.

Further progress toward the synthesis of mitomycin C or related compounds has not been reported.

F. Chloramphenicol (358)

Although chloromycetin or chloramphenicol is the last antibiotic to be discussed, it certainly is not the least important. Structurally it is one of the simplest antibiotics in everyday use, yet it has a very broad spectrum of activity. A serious side effect, however, is that it causes blood dyscrasias. Its structure determination[183] and initial synthesis[184] were carried out by the same group at Parke Davis. Several other syntheses have since been developed.[185–189] The procedure discussed here, suitable for the large-scale production of **358**, was devised by Ehrhart and his associates.[187]

Condensation[190] of *p*-nitrobenzaldehyde with glycine methyl ester in methanolic hydrogen chloride gives exclusively the *threo* isomer of *p*-nitrophenylserine methyl ester (**451**) as its hydrochloride. Careful neutralization of this material with ammonia gave **451** itself, which was then acylated with dichloroacetyl chloride in aqueous sodium bicarbonate solution. The

$p\text{-}NO_2C_6H_4CHO + NH_2CH_2CO_2CH_3 \xrightarrow{50\%}$

$p\text{-}NO_2C_6H_4CH(OH)CH(NH_2)CO_2CH_3 \xrightarrow{81\%}$

451

$p\text{-}NO_2C_6H_4CH(OH)CH(NHCOCHCl_2)CO_2CH_3 \xrightarrow{83\%}$

452

$p\text{-}NO_2C_6H_4CH(OH)CH(NHCOCHCl_2)CONHNH_2 \xrightarrow{63\%}$

453

$p\text{-}NO_2C_6H_4CH(OH)CH(NHCOCHCl_2)CH_2OH$

358

amide ester **452** was converted to the hydrazide **453** by means of hydrazine hydrate in hot methanol. Treatment of **453** with nitrous acid led to the corresponding acyl azide, which without isolation was reduced with sodium borohydride in methanol to give *dl*-chloramphenicol **358**. The biologically active form of **358** (D-form) was obtained in essentially the same way by using the L(+)-threo isomer of **451**. Resolution of racemic **451** can be carried out using D(+)-tartaric acid.

Chloramphenicol is the only antibiotic of therapeutic value that is more economically produced by total synthesis than by a fermentation process. From the point of view then of the commercial use of sophisticated chemistry to effect practical total syntheses, the record in the field of antibiotics is somewhat sparse if not sad, and the situation seems unlikely to change in the immediate future!

REFERENCES

1. (a) C. Y. Kao, in F. E. Russell and P. R. Saunders, Eds., *Animal Toxins*. Pergamon Press, London, 1967, p. 109; (b) J. A. Vick, H. P. Ciuchta, and J. H. Manthei, p. 269.
2. R. B. Woodward, in A. Todd, Ed., *Perspectives in Organic Chemistry*. Interscience, New York, 1956, p. 156.
3. Biochemically 6-aminopenicillanic acid can be obtained in the absence of side chain precursors [F. R. Batchelor, F. P. Doyle, J. H. C. Nayler, and G. N. Rolinson, *Nature*, **183**, 257 (1959)]. It is also available from phenoxymethylpenicillin by treatment with kidney enzymes [U.S. Patent 3,070,511 (1962 to Lepetit)].
4. W. Cutting, *Handbook of Pharmacology*. Meredith Corporation, New York, 1969, pp. 26–37.
5. Summarized in *The Chemistry of Penicillin*, Princeton University Press, Princeton, N.J., 1949.
6. Reference 5, p. 1018.
7. V. du Vigneaud, F. H. Carpenter, R. W. Holley, A. H. Livermore, and J. R. Rachaele, *Science*, **104**, 431 (1946).
8. Reference 5, pp. 455–472.
9. Reference 5, pp. 806–807.
10. Reference 5, p. 909.
11. J. C. Sheehan and K. R. Henery-Logan, *J. Am. Chem. Soc.*, **79**, 1262 (1957); **81**, 3089 (1959).
12. J. C. Sheehan and D. A. Johnson, *J. Am. Chem. Soc.*, **76**, 158 (1954).
13. J. C. Sheehan and K. R. Henery-Logan, *J. Am. Chem. Soc.*, **81**, 5838 (1959).
14. A. J. Base, G. Spiegelman and M. S. Manhas, *J. Am. Chem. Soc.*, **90**, 4506 (1968).
15. I. Ugi, *Angew. Chem. Int. Ed. Engl.*, **1**, 8 (1962).
16. E. J. Corey and A. M. Felix, *J. Am. Chem. Soc.*, **87**, 2518 (1965).

17. R. B. Woodward, K. Heusler, J. Gosteli, P. Naegeli, W. Oppolzer, R. Ramage, S. Ranganathan, and H. Vorbrüggen, *J. Am. Chem. Soc.*, **88,** 852 (1966).
18. R. B. Morin, B. G. Jackson, R. A. Meuller, E. R. Lavingnino, W. B. Scanlon, and S. L. Andrews, *J. Am. Chem. Soc.*, **91,** 1401 (1969); **85,** 1896 (1963).
19. J. A. Webber, E. M. van Heyningen and R. T. Vasileff, *J. Am. Chem. Soc.*, **91,** 5675 (1969).
20. D. O. Spry, *J. Am. Chem. Soc.*, **92,** 5007 (1970).
21. G. E. Woodward and E. F. Schroeder, *J. Am. Chem. Soc.*, **59,** 1690 (1937).
22. R. Heymès, G. Amiard, and G. Nominé, *Compt. rend. C*, **263,** 170 (1966).
23. C. Mannich and M. Bauroth, *Ber.*, **57,** 1108 (1924).
24. R. R. Chauvette, E. H. Flynn, B. G. Jackson, E. R. Lavagnino, R. B. Morin, R. A. Mueller, R. P. Pioch, R. W. Roeske, C. W. Ryan, J. L. Spencer, and E. van Heyningen, *J. Am. Chem. Soc.*, **84,** 3401 (1962).
25. J. R. D. McCormick, E. R. Jensen, P. A. Miller, and A. P. Doerschak, *J. Am. Chem. Soc.*, **82,** 3381 (1960).
26. L. H. Conover, K. Butler, J. D. Johnston, J. J. Korst, and R. B. Woodward, *J. Am. Chem. Soc.*, **84,** 3222 (1962); **90,** 439 (1968).
27. R. B. Woodward, *Pure Appl. Chem.*, **6,** 561 (1963).
28. H. Muxfeldt and W. Rogalski, *J. Am. Chem. Soc.*, **87,** 933 (1965).
29. H. Muxfeldt, E. Jacobs, and K. Uhlig, *Chem. Ber.*, **95,** 2901 (1962).
30. H. Muxfeldt, G. Buhr, and R. Bangert, *Angew. Chem. Int. Ed. Engl.*, **1,** 157 (1962)
31. A. I. Gurevich, M. G. Karapetyan, M. N. Kolosov, V. G. Korobko, V. V. Onoprienko, S. A. Popravko, and M. M. Shemyakin, *Tetrahedron Lett.*, **1967,** 131.
32. H. H. Inhoffen, H. Muxfeldt, H. Schaefer, and H. Krämer, *Croat. Chem. Acta.*, **29,** 329 (1957).
33. M. N. Kolosov, S. A. Popravko, and M. M. Shemyakin, *Ann.*, **668,** 86 (1963).
34. A. I. Gurevich, M. G. Karapetyan, and M. N. Kolosov, *Khim. Prirodn. Sedin., Akad. Nauk. Uz. S.S.R.*, **2,** 141 (1966); *Chem. Abs.*, **65,** 13627 (1966).
35. M. Schach von Wittenau, *J. Org. Chem.*, **29,** 2746 (1964).
36. A. I. Scott and C. T. Bedford, *J. Am. Chem. Soc.*, **84,** 2271 (1962).
37. H. Muxfeldt, G. Hardtmann, F. Kathawala, E. Vedejs, and J. B. Mooberry, *J. Am. Chem. Soc.*, **90,** 6534 (1968).
38. H. Muxfeldt, J. Behling, G. Grethe, and W. Rogalski, *J. Am. Chem. Soc.*, **89,** 4991 (1967).
39. H. Muxfeldt, *Angew. Chem.*, **74,** 825 (1962).
40. Other examples of this type of condensation are given in reference 38.
41. A. S. Kende, T. L. Fields, J. H. Boothe, and S. Kushner, *J. Am. Chem. Soc.*, **83,** 439 (1961).
42. D. H. R. Barton, *Chem. Brit.*, **6,** 301 (1970). See also Barton and coworkers, *J. Chem. Soc. (C)* **1971,** 2164–2241.
43. A. Schatz, E. Bugie, and S. A. Waksman, *Proc. Soc. Exp. Biol. Med.*, **55,** 66 (1944).
44. M. L. Wolfrom, S. M. Olin, and W. J. Polglase, *J. Am. Chem. Soc.*, **72,** 1724 (1950).
45. F. A. Kuehl, Jr., E. H. Flynn, F. W. Holly, R. Mozingo, and K. Folkers, *J. Am. Chem. Soc.*, **69,** 3032 (1947).
46. J. R. Dyer, W. E. McGonigal, and K. C. Rice, *J. Am. Chem. Soc.*, **87,** 654 (1965).
47. H. Maehr and C. P. Schaffner, *J. Am. Chem. Soc.*, **92,** 1697 (1970).

48. H. Hoeksema, A. D. Argoudelis, and P. F. Wiley, *J. Am. Chem. Soc.*, **84**, 3212 (1962).

49. M. Nakajima, A. Hasegawa, and N. Kurihara, *Tetrahedron Lett.*, **1964**, 967.

50. M. Nakajima, I. Iomida, and S. Takei, *Chem. Ber.*, **92**, 163 (1959); M. Nakajima, A. Hasegawa and N. Kurihara, *Chem. Ber.*, **95**, 2708 (1962).

51. M. Nakajima, A. Hasegawa, N. Kurihara, H. Shibata, T. Ueno, and D. Nishimura, *Tetrahedron Lett.*, **1968**, 623.

52. S. Umezawa, K. Tatsuta, and S. Koto, *Bull. Chem. Soc. Japan*, **42**, 533 (1969).

53. S. Koto, T. Tsumura, Y. Koto, and S. Umezawa, *Bull. Chem. Soc. Japan*, **41**, 2765 (1968).

54. H. H. Baer, *Chem. Ber.*, **93**, 2865 (1960); *J. Am. Chem. Soc.*, **83**, 1882 (1961).

55. S. Koto, K. Tatsuta, E. Ketazawa, and S. Umezawa, *Bull. Chem. Soc. Japan*, **41**, 2769 (1968).

56. Compound **156** was prepared from 6-acetamido-6-deoxy-D-glucose [F. Cramer, O. Ollerbach and H. Springman, *Chem. Ber.*, **92**, 384 (1959)] in a manner completely analogous to that used for the conversion of **148** to **143**.

57. S. Umezawa and S. Koto, *Bull. Chem. Soc. Japan*, **39**, 2014 (1966).

58. P. F. Lloyd and M. Stacey, *Chem. Ind.*, **1955**, 917.

59. S. Umezawa, S. Koto, K. Tatsuta, H. Hineno, Y. Nishimura, and T. Tsumura, *Bull. Chem. Soc. Japan*, **42**, 537.

60. S. Umezawa, K. Tatsuta, and T. Tsumura, *Bull. Chem. Soc. Japan*, **42**, 529 (1969).

61. T. Suhara, F. Sasaki, K. Maeda, H. Umezawa, and M. Ohno, *J. Am. Chem. Soc.*, **90**, 6559 (1968).

62. M. Nakajima, H. Shibota, K. Kitahara, S. Takahashi, and A. Hasegawa, *Tetrahedron Lett.*, **1968**, 2271.

63. Y. Suhara, K. Maeda, H. Umezawa, and M. Ohno, *Tetrahedron Lett.*, **1966**, 1239.

64. E. Vis and P. Karrer, *Helv. Chim. Acta* **37**, 378 (1954).

65. R. R. Herr, H. K. Jahnke, and A. D. Argoudelis, *J. Am. Chem. Soc.*, **89**, 4808 (1967).

66. E. Hardegger, A. Meier, and A. Stoos, *Helv. Chim. Acta*, **52**, 2555 (1969).

66a. B. J. Magerlein, R. D. Birkenmeyer, R. R. Herr, and F. Kagan, *J. Am. Chem. Soc.*, **89**, 2459 (1967).

66b. H. Hoeksema, B. Bannister, R. D. Birkenmeyer, F. Kagan, B. J. Magerlein, F. A. MacKellar, W. Schroeder, G. Slomp, and R. R. Herr, *J. Am. Chem. Soc.*, **86**, 4223 (1964); R. R. Herr and G. Slomp, *J. Am. Chem. Soc.*, **89**, 2444 (1967).

66c. M. Bethell, G. W. Kenner, and R. C. Sheppard, *Nature*, **194**, 864 (1962).

66d. B. J. Magerlein, *Tetrahedron Lett.*, **1970** 33.

66e. G. B. Howarth, W. A. Szarek, and J. K. N. Jones, *J. Chem. Soc. (C)*, **1970**, 2218

66f. D. G. Lance, W. A. Szarek, J. K. N. Jones, and G. B. Howarth, *Can. J. Chem.*, **47**, 2871 (1969).

66g. G. B. Howarth, D. G. Lance, W. A. Szarek, and J. K. N. Jones, *Can. J. Chem.*, **47**, 75 (1969).

66h. H. Saeki and E. Ohki, *Chem. Pharm. Bull. (Tokyo)* **18**, 789 (1970).

67. G. O. Gorton, J. E. Lancaster, G. E. van Lear, W. Fulmor, and W. E. Meyer, *J. Am. Chem. Soc.*, **91**, 1535 (1969); D. A. Shuman, R. K. Robins, and M. J. Robins, *J. Am. Chem. Soc.*, **91**, 3391 (1969).

68. K. G. Cunninghan, S. A. Hutchison, W. Manson, and F. S. Spring, *J. Chem. Soc.*, **1951,** 2299.

69. J. D. Dutcher, M. H. von Saltza, and F. E. Pansy, *Antimic. Agts. Chemother.* **1963,** 83; H. Agahigian, G. D. Vickers, M. H. von Saltza, J. Reid, A. I. Cohen, and H. Gauthier, *J. Org. Chem.* **30,** 1085 (1965).

70. K. Isono, K. Asahi, and S. Suzuki, *J. Am. Chem. Soc.*, **91,** 7490 (1969).

71. T. Naka, T. Hashizume, and M. Nishimura, *Tetrahedron Lett.*, **1971,** 95.

71a. J. J. Fox, Y. Kuwada, T. Ueda, and E. B. Whipple, *Antimic. Agts. Chemother.*, **1964,** 518; J. J. Fox, Y. Kuwada, and K. A. Watanabe, *Tetrahedron Lett.*, **1968,** 6029; K. A. Watanabe, M. P. Kotick, and J. J. Fox, *Chem. Pharm. Bull.* (*Tokyo*), **17,** 416 (1969).

72. J. W. Hinman, E. L. Caron, and C. DeBoer, *J. Am. Chem. Soc.*, **75,** 5864 (1953); S. Hanessian, and T. H. Haskell, *Tetrahedron Lett.*, **1964,** 2451.

73. H. Taniyama and F. Muyoshi, *Chem. Pharm. Bull.* (*Tokyo*), **10,** 156 (1962); E. E. van Tamelen, J. R. Dyer, H. A. Whaley, H. E. Carter, and G. B. Whitfield, Jr., *J. Am. Chem. Soc.*, **83,** 4295 (1961); H. E. Carter, C. C. Sweeley, E. E. Daniels, J. E. McNary, C. P. Schnaffner, C. A. West, E. E. van Tamelen, J. R. Dyer, and H. H. Whaley, *J. Am. Chem. Soc.*, **83,** 4296 (1961).

74. S. Takemura, *Chem. Pharm. Bull.* (*Tokyo*), **8,** 574, 578 (1960).

75. B. R. Baker, R. E. Schaub, J. P. Joseph, and J. H. Williams, *J. Am. Chem. Soc.*, **77,** 12 (1955).

76. B. R. Baker, R. E. Schaub, and J. H. Williams, *J. Am. Chem. Soc.*, **77,** 7 (1955).

77. B. R. Baker, J. P. Joseph, and J. H. Williams, *J. Am. Chem. Soc.*, **77,** 1 (1955).

78. W. Schroeder and H. Hoeksema, *J. Am. Chem. Soc.*, **81,** 1767 (1959).

79. M. L. Wolfrom, A. Thompson, and E. F. Evans, *J. Am. Chem. Soc.*, **67,** 1793 (1945).

80. J. Farkaš and F. Šorm, *Tetrahedron Lett.*, **1962,** 813.

81. J. R. McCarthy, Jr., R. K. Robins, and M. J. Robins, *J. Am. Chem. Soc.*, **90,** 4993 (1968).

82. E. Walton, F. W. Holly, G. E. Boxer, R. F. Nutt, and S. R. Jenkins, *J. Med. Chem.*, **8,** 659 (1965).

83. C. D. Anderson, L. Goodman, and B. R. Baker, *J. Am. Chem. Soc.*, **80,** 5247 (1958).

84. R. L. Tolman, R. K. Robins, and L. B. Townsend, *J. Am. Chem. Soc.*, **91,** 2102 (1969).

85. R. L. Tolman, R. K. Robins, and L. B. Townsend, *J. Heterocyc. Chem.*, **4,** 230 (1967).

86. T. Hashizume and H. Iwamura, *Tetrahedron Lett.*, **1965,** 3095.

87. K. Ohkuma, *J. Antibiotics* (*Tokyo*), **13A,** 361 (1960).

88. O. Theander, *Acta Chem. Scand.*, **18,** 2209 (1964).

89. K. A. Watanabe, R. S. Goody, and J. J. Fox, *Tetrahedron*, **26,** 3883 (1970); M. P. Kotick, R. S. Klein, K. A. Watanabe, and J. J. Fox, *Carbohyd. Res.*, **11,** 369 (1969).

90. K. A. Watanabe, M. P. Kotick, and J. J. Fox, *J. Org. Chem.*, **35,** 231 (1970).

90a. K. A. Watanabe, I. Wempen, and J. J. Fox, *Chem. Pharm. Bull.* (*Tokyo*), **18,** 2368 (1970).

90b. E. J. Reist, R. R. Spencer, D. F. Calkins, B. R. Baker, and L. Goodman, *J. Org. Chem.*, **30,** 2312 (1965).

91. H. Yonehara and N. Otake, *Antimicrob. Agts. Chemother.*, **1965,** 855.
92. P. Baudet and E. Cherbuliez, *Helv. Chim. Acta*, **47,** 661 (1964).
93. Y. Hirotsu, T. Shiba, and T. Kaneko, *Bull. Chem. Soc. Japan*, **43,** 1870 (1970), and preceding papers.
94. D. L. Swallow and E. P. Abraham, *Biochem. J.*, **72,** 326 (1959); W. Stoffel and L. C. Craig, *J. Am. Chem. Soc.*, **83,** 145 (1961).
95. H. Vanderhaegl and G. Parmentier, *J. Am. Chem. Soc.*, **82,** 4414 (1960).
96. G. R. Delpierre, F. W. Eastwood, G. E. Gream, D. G. I. Kingston, P. S. Sarin, L. Todd, and D. H. Williams, *J. Chem. Soc. (C)*, **1966,** 1653.
97. B. W. Bycroft, D. Cameron, A. Hassanali-Walji, and A. W. Johnson, *Tetrahedron Lett.*, **1969,** 2539.
98. E. Schröder and K. Lübke, *Experientia*, **19,** 57 (1963).
99. P. Quitt, R. O. Studer, and K. Vogler, *Helv. Chim. Acta*, **47,** 166 (1964).
100. D. W. Russell, *J. Chem. Soc.*, **1962,** 753.
101. H. H. Wasserman, J. J. Keggi, and J. E. McKeon, *J. Am. Chem. Soc.*, **83,** 4107 (1961); **84,** 2978 (1962).
102. R. L. Hamill, C. E. Higgens, N. E. Boaz, and M. Gorman, *Tetrahedron Lett.*, **1969,** 4255.
103. H. Brockmann and H. Geeren, *Ann.*, **603,** 216 (1957).
104. M. M. Shemyakin, N. A. Aldanova, E. I. Vinogradova, and M. Yu Feigina, *Tetrahedron Lett.*, **1963,** 1921.
105. E. Schröder and K. Lübke, *The Peptides*, Vol. II. Academic Press, New York, 1966, pp. 396 ff.
106. R. D. Hotchkiss, *Adv. Enzymol.*, **4,** 153 (1944); R. L. M. Synge, *Q. Rev. (London)*, **3,** 245 (1949).
107. R. D. Hotchkiss and R. J. Dubos, *J. Biol. Chem.*, **132,** 79 (1940).
108. J. D. Gregory and L. C. Craig, *J. Biol. Chem.*, **172,** 839 (1948).
109. S. Ishii and B. Witkop, *J. Am. Chem. Soc.*, **85,** 1832 (1963); see also L. K. Ramachandran, *Biochem.*, **2,** 1138 (1963).
110. R. Sarges and B. Witkop, *J. Am. Chem. Soc.*, **87,** 2015 (1965).
111. R. Sarges and B. Witkop, *J. Am. Chem. Soc.*, **87,** 2020 (1965).
112. R. Sarges and B. Witkop, *J. Am. Chem. Soc.*, **87,** 2027 (1965).
113. G. F. Gause and M. G. Brazhnikova, *Lancet II*, **1944,** 715.
114. R. L. M. Synge, *Biochem. J.*, **39,** 363 (1945); R. Consden, A. H. Gordon, A. J. P. Martin, and R. L. M. Synge, *Biochem. J.*, **40,** xliii (1946); **41,** 596 (1947).
115. R. Schwyzer and P. Sieber, *Helv. Chim. Acta*, **40,** 624 (1957); *Angew. Chem.*, **68,** 518 (1958).
116. K. Kurahashi, *J. Biochem. (Tokyo)*, **56,** 101 (1964); S. Otani and Y. Saito, *J. Biochem. (Tokyo)*, **56,** 103 (1964).
117. A. R. Battersby and L. C. Craig, *J. Am. Chem. Soc.*, **74,** 4023 (1952); A. Paladini and L. C. Craig, *J. Am. Chem. Soc.*, **76,** 688 (1954).
118. T. P. King and L. C. Craig, *J. Am. Chem. Soc.*, **77,** 6627 (1955).
119. M. A. Ruttenberg, T. P. King, and L. C. Craig, *Biochem.*, **4,** 11 (1965).
120. M. Ohno, T. Koto, S. Makisumi, and N. Izumiya, *Bull. Chem. Soc. Japan*, **39,** 1738 (1966).

121. K. Kuromizu and N. Izumiya, *Bull. Chem. Soc. Japan*, **43,** 2199 (1970).

122. K. Kuromizu and N. Izumiya, *Bull. Chem. Soc. Japan*, **43,** 2944 (1970).

123. N. Mitsuyasu, S. Matsuura, M. Waki, M. Ohno, S. Makisumi, and N. Izumiya, *Bull. Chem. Soc. Japan*, **43,** 1829 (1970).

124. K. Fujikawa, T. Sakamoto, T. Suzuki, and K. Kurahashi, *Biochim. Biophys. Acta*, **169,** 520 (1968).

125. R. Schwyzer and P. Sieber, *Helv. Chim. Acta*, **41,** 2186 (1958).

126. For a summary of the structure work see reference 105, pp. 451–472. A review of the chemistry of these compounds has also appeared. K. Vogler and R. O. Studer, *Experientia*, **22,** 345 (1966); S. Wilkinson and L. A. Lowe, *Nature*, **188,** 311 (1966).

127. S. A. Waksman and H. B. Woodruff, *J. Bacteriol.*, **42,** 231 (1941).

128. H. Brockmann, *Angew. Chem.*, **66,** 1 (1954): **72,** 939 (1960); *Fortschr. Chem. Org. Naturst.*, **18,** 1 (1960); *Naturwiss.*, **50,** 689 (1963).

129. H. Brockmann and H. Lackner, *Naturwiss.*, **51,** 384 (1964); *Chem. Ber.*, **101,** 1312 (1968).

130. H. Brockmann, H. Lackner, R. Mecke, G. Troemel, and H. S. Petras, *Chem. Ber.*, **99,** 717 (1966).

131. J. Meienhofer, *J. Am. Chem. Soc.*, **92,** 3771 (1970).

132. Yu. A. Ovchinnikov, V. T. Ivanov, and I. I. Mikhaleva, *Tetrahedron Lett.*, **1971,** 159.

132a. M. M. Shemyakin, Yu A. Ovchinnikov, V. K. Antonov, A. A. Kiryushkin, V. T. Ivanov, V. I. Shchelokov, and A. M. Shkrob, *Tetrahedron Lett.*, **1964,** 47; A. A. Kiryushkin, V. I. Shchelokov, V. K. Antonov, Y. A. Ovchinnikov, and M. M. Shemyakin, *Khim. prirod.* Soedinenii, **3,** 267 (1967).

133. P. F. Wiley, K. Gerzon, E. H. Flynn, M. V. Sigal, Jr., O. Weaver, U. C. Quarck, R. R. Chauvette, and R. Monahan, *J. Am. Chem. Soc.*, **79,** 6062 (1957).

134. P. W. K. Woo, H. W. Dion, and Q. R. Bartz, *J. Am. Chem. Soc.*, **86,** 2726 (1964).

135. M. L. Dhar, V. Thaller, and M. C. Whiting, *J. Chem. Soc.*, **1964,** 842.

136. A. C. Cope, U. Axen, E. P. Burrows, and J. Weinlich, *J. Am. Chem. Soc.*, **88,** 4228 (1966).

137. A. J. Birch, C. W. Holzapfel, R. W. Rickards, C. Djerassi, M. Suzuki, J. Westley, J. D. Dutcher, and R. Thomas, *Tetrahedron Lett.*, **1964,** 1485.

138. R. N. Mirrington, E. Ritchie, C. W. Shoppee, W. C. Taylor, and S. Sternhell, *Tetrahedron Lett.*, **1964,** 365; F. McCapra, A. I. Scott, P. Delmotte, and J. Delmotte-Plaquée, *Tetrahedron Lett.*, **1964,** 869.

139. P. M. Baker, B. W. Bycroft, and J. C. Roberts, *J. Chem. Soc. (C)*, **1967,** 1913.

140. R. Huisgen and M. Reitz, *Tetrahedron*, **2,** 271 (1958).

141. D. Taub, N. N. Girotra, R. D. Hoffsommer, C. H. Kuo, H. L. Slates, S. Weber, and N. L. Wendler, *Tetrahedron*, **24,** 2443 (1968).

142. I. Vlattas, I. T. Harrison, L. Tokes, J. H. Fried, and A. D. Cross, *J. Org. Chem.*, **33,** 4176 (1968).

143. C. H. Hassell, *Org. Reactions*, **9** (1957).

144. I. J. Borowitz, G. Gonis, R. Kelsey, R. Rapp, and G. J. Williams, *J. Org. Chem.*, **31,** 3032 (1966).

145. J. F. Bagli and H. Immer, *Can. J. Chem.*, **46,** 3115 (1968).

146. C. H. Stammer, A. N. Wilson, F. W. Holly, and K. Folkers, *J. Am. Chem. Soc.*, **77**, 2346 (1955); Pl. A. Plattner, A. Boller, H. Frick, A. Fürst, B. Hegedüs, H. Kirchensteiner, St. Majnoni, R. Schläpfer, and H. Spiegelberg, *Helv. Chim. Acta*, **40**, 1531 (1957).

147. H. Kawaguchi, H. Tsukiura, M. Okanishi, T. Ohmori, K. Fujisawa, and H. Koshiyama, *J. Antibiotics*, **A18**, 1 (1965); H. Kawaguchi, T. Naito, and H. Tsukiura, *J. Antibiotics*, **A18**, 11 (1965).

148. F. Arcamone, G. Franceschi, P. Orezzi, G. Cassinelli, W. Barbieri, and R. Mondelli, *J. Am. Chem. Soc.*, **86**, 5334 (1964); F. Arcamone, G. Cassinelli, P. Orezzi, G. Franceschi, and R. Mondelli, *J. Am. Chem. Soc.*, **86**, 5335 (1964); F. Arcamone, G. Franceschi, P. Orezzi, and S. Penco, *Tetrahedron Lett.*, **1968**, 3349; F. Arcamone, G. Cassinelli, G. Franceschi, P. Orezzi, and R. Mondelli, *Tetrahedron Lett.*, **1968**, 3353.

149. D. S. Tarbell, R. M. Carman, D. D. Chapman, K. R. Huffman, and N. J. McCorkindale, *J. Am. Chem. Soc.*, **82**, 1005 (1960); N. J. McCorkindale and J. G. Sime, *Proc. Chem. Soc.*, **1961**, 331.

150. W. O. Godtfredtsen and S. Vangedal, *Tetrahedron*, **18**, 1029 (1962); D. Arigoni, W. von Daehne, W. O. Godtfredsen, A. Melera, and S. Vangedal, *Experientia*, **20**, 344 (1964); D. Arigoni, W. von Daehne, W. O. Godtfredsen, A. Marquet, and A. Melera, *Experientia*, **19**, 521 (1963).

151. A. Agtarap, J. W. Chamberlin, M. Pinkerton, and L. Steinrauf, *J. Am. Chem. Soc.*, **89**, 5737 (1967).

152. W. Oppolzer, V. Prelog and P. Sensi, *Experientia*, **20**, 336 (1964).

153. K. V. Rao, K. Biemann, and R. B. Woodward, *J. Am. Chem. Soc.*, **85**, 2532 (1963).

154. R. R. Herr, *J. Am. Chem. Soc.*, **81**, 2595 (1959); H. E. Hennis, L. G. Duquette, and F. Johnson, *J. Org. Chem.*, **33**, 904 (1968).

155. G. Büchi and J. E. Powell, Jr., *J. Am. Chem. Soc.*, **92**, 3126 (1970).

156. Unpublished work by the author.

157. W. Leimgruber, A. D. Batcho, and R. C. Czajkowski, *J. Am. Chem. Soc.*, **90**, 5641 (1968).

158. W. Leimgruber, V. Stefanović, F. Schenker, A. Karr, and J. Berger, *J. Am. Chem. Soc.*, **87**, 5791 (1965).

159. W. Leimgruber, A. D. Batcho, and F. Schenker, *J. Am. Chem. Soc.*, **87**, 5793 (1965).

160. M. Artico, G. De Martino, G. Filacchioni, and R. Giuliano, *Farm. Ed. Sci.*, **24**, 276 (1969).

161. E. Oxford, H. Raistrick, and P. Simonart, *Biochem. J.*, **33**, 240 (1939).

162. J. F. Grove, J. MacMillan, T. P. C. Mulholland, and M. A. T. Rogers, *J. Chem. Soc.*, **1952**, 3977.

163. A. C. Day, J. Nabney, and A. I. Scott, *J. Chem. Soc.*, **1961**, 4067.

164. D. H. R. Barton and J. Cohen, in *Festschrift A. Stoll*, Birkhauser, Basle, 1957, p. 117.

165. D. Taub, C. H. Kuo, H. L. Slates, and N. L. Wendler, *Tetrahedron*, **19**, 1 (1963).

166. A. Brossi, M. Baumann, M. Gerecke, and E. Kyburz, *Helv. Chim. Acta*, **43**, 2071 (1960).

167. J. MacMillan, *J. Chem. Soc.*, **1959**, 1823.

168. G. Stork and M. Tomasz, *J. Am. Chem. Soc.*, **86**, 471 (1964).

169. C. H. Shunk, C. H. Stammer, E. A. Kaczka, E. Walton, C. F. Spencer, A. N. Wilson, J. W. Richter, F. W. Holly, and K. Folkers, *J. Am. Chem. Soc.*, **78**, 1770 (1956).
170. B. P. Vaterlaus, J. Kiss, and H. Spiegelberg, *Helv. Chim. Acta*, **47**, 381 (1964); B. P. Vaterlaus, K. Doebel, J. Kiss, A. I. Rachlin, and H. Spiegelberg, *Helv. Chim. Acta*, **47**, 390 (1964); B. P. Vaterlaus and H. Spiegelberg, *Helv. Chim. Acta*, **47**, 508 (1964).
171. A. J. Whiffen, N. Bohonas, and R. L. Emerson, *J. Bact.*, **52**, 610 (1946).
172. E. C. Kornfeld, R. G. Jones, and T. V. Parke, *J. Am. Chem. Soc.*, **71**, 150 (1949).
173. F. Johnson, N. A. Starkovsky, and W. D. Gurowitz, *J. Am. Chem. Soc.*, **87**, 3492 (1965); F. Johnson, N. A. Starkovsky, and A. A. Carlson, **87**, 4612 (1965).
174. F. Johnson, N. A. Starkovsky, A. C. Paton, and A. A. Carlson, *J. Am. Chem. Soc.*, **88**, 149 (1966).
175. W. R. N. Williamson, *Tetrahedron*, **3**, 314 (1958).
176. F. Johnson, *J. Org. Chem.*, **27**, 3658 (1962).
177. F. Johnson and A. Whitehead, *Tetrahedron Lett.*, **1964**, 3825; H. J. Schaeffer and V. K. Jain, *J. Org. Chem.*, **29**, 2595 (1964).
178. F. Johnson, A. A. Carlson, and N. A. Starkovsky, *J. Org. Chem.*, **31**, 1327 (1966).
179. J. S. Webb, D. B. Cosulich, J. H. Mowat, J. B. Patrick, R. W. Broschard, W. E. Meyer, R. P. Williams, C. F. Wolf, W. Fulmor, C. Pidacks, and J. E. Lancaster, *J. Am. Chem. Soc.*, **84**, 3185, 3187 (1962).
180. G. R. Allen, Jr., J. F. Poletto, and M. J. Weiss, *J. Org. Chem.*, **30**, 2897 (1965); G. R. Allen, Jr., and M. J. Weiss, *J. Org. Chem.*, **30**, 2905 (1965); W. A. Remers, R. H. Roth and M. J. Weiss, *J. Org. Chem.*, **30**, 2910 (1965).
181. T. Hirata, Y. Yamada, and M. Matsui, *Tetrahedron Lett.*, **1969**, 19.
182. T. Hirata, Y. Yamada, and M. Matsui, *Tetrahedron Lett.*, **1969**, 4107.
183. M. C. Rebstock, H. M. Crooks, Jr., T. Controulis, and Q. R. Bartz, *J. Am. Chem. Soc.*, **71**, 2458 (1949).
184. J. Controulis, M. C. Rebstock, and H. M. Crookes, Jr., *J. Am. Chem. Soc.*, **71**, 2463 (1949).
185. L. M. Long and H. D. Troutman, *J. Am. Chem. Soc.*, **71**, 2469 (1949).
186. L. M. Long and H. D. Troutman, *J. Am. Chem. Soc.*, **71**, 2473 (1949).
187. G. Ehrhart, W. Siedel and H. Nahm, *Chem. Ber.*, **90**, 2088 (1957).
188. S. Umezawa and T. Suami, *Bull. Chem. Soc. Japan*, **27**, 477 (1954).
189. V. A. Mikhalev, M. I. Dorokhova, N. E. Smolina, A. M. Zhelokhovtseva, A. P. Skoldinov, A. I. Ivanov, A. P. Arendaruk, M. I. Galchenko, V. A. Skorodumov, and D. D. Smolin, *Antibiotiki*, **4**, 21 (1959).
190. E. D. Bergmann, H. Bendas, and W. Taub, *J. Chem. Soc.*, **1951**, 2673.

The Total Synthesis of Naturally Occurring Oxygen Ring Compounds

F. M. DEAN

University of Liverpool, England

1. INTRODUCTION

The types of compound discussed in this chapter form but a small proportion of those available, and the reader may wish to know the grounds on which the selections were made. Compounds or methods already multiply reviewed (e.g., flavones, oxidative cyclization) were rejected in favor of less well-documented areas of research provided these were currently lively. Then it was decided to discuss several syntheses within a few groups of compounds (rather than single syntheses from each of numerous groups) in order to maintain coherence and illustrate the subtler variations in problems and in methods of attack. Usually, the selected compounds contain more than one oxygen ring and all are derived from phenolic nuclei in order to bring out the fact that, not infrequently, problems in heterocyclic chemistry are quite as much problems in phenol chemistry, the heterocyclic part being the easier. And finally, of course, the choice was personal.

This chapter contains about 100 syntheses of compounds falling into 6 main classes: chromenes, dihydrofurobenzofurans, chromanochromanones, benzofurochromans, xanthones and pyronoquinones. It is proposed to rename this last group the amphipyrones.

2. CHROMENES

We are concerned here with compounds containing the nucleus of 2,2-dimethyl-2*H*-chromene **1**, of which a large number occur in the higher plants.[1,2] To produce this nucleus the biosynthetical pathways that, separately, produce terpenes or phenols must merge,[3] and one result is a very great range of complexity. In gambogic acid[4] **2**, for example, a phenolic kernel carries three C_5 residues and one C_{10} residue, each being modified in a different fashion thus presenting a considerable synthetical problem.

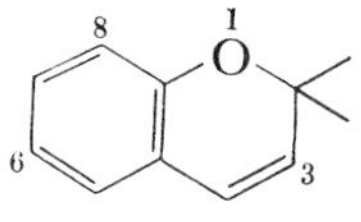

8 1 O 6 3

1

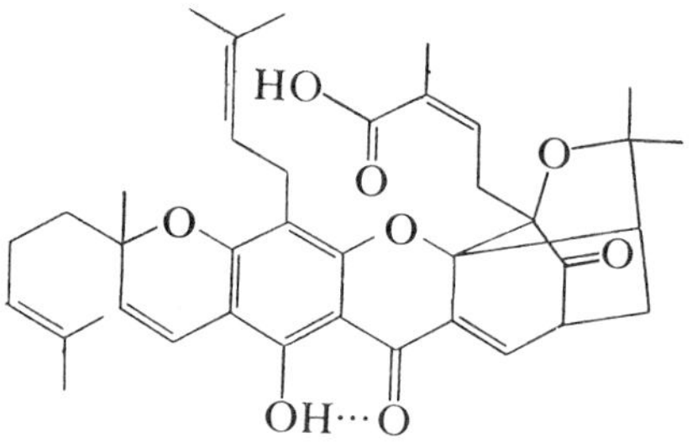

HO O O O O O OH···O

2 Gambogic acid

Good methods for synthesising chromenes are all recent discoveries. A synthesis of lapachenole[5] **3** shows the use of an older method,[6–9] still sometimes used, in which coumarins were treated with a Grignard reagent and the resulting alcohol cyclized:

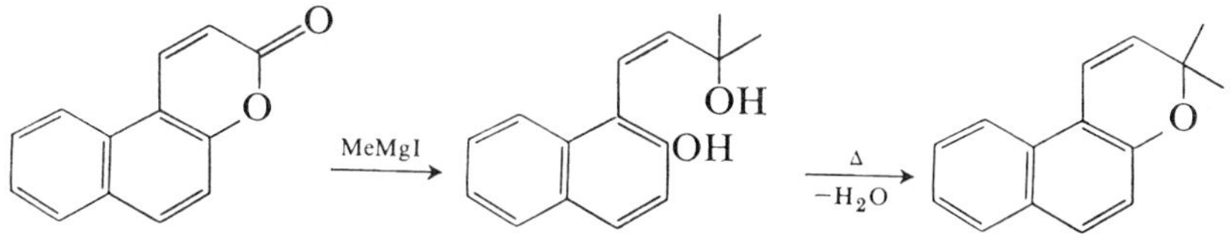

MeMgI

OH

OH

Δ

$-H_2O$

3 Lapachenole

The final cyclization has to be effected without the catalytic aid of acids. An ever-present difficulty in working with chromenes is their sensitivity to acids, especially protic acids, which, presumably, may lead to carbonium ions of types **4a** and **4b**, both of which are stabilized by interaction with the oxygen atom:

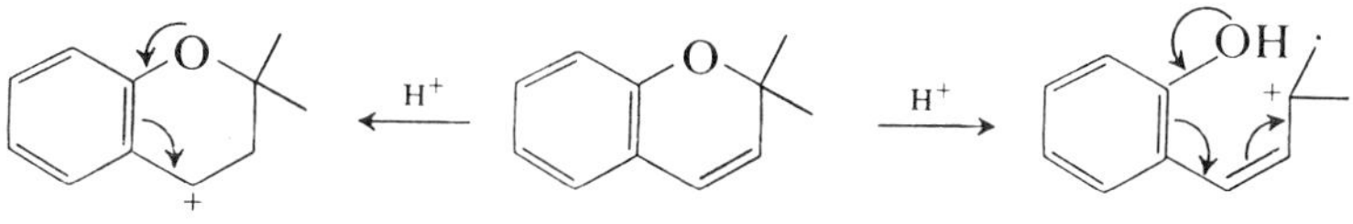

4a

4b

Ions of type **4a** must underlie the cyclization of gambogic acid **1** to gamboginic acid **5** and could explain the isolation from the essential oil of *Ageratum conyzoides* not only of ageratochromene **6** but also of a dimer[10] with structure **7**. On the other hand the complex polycyclic structures[5,8,9] obtained by treating lapachenole **3** and other chromenes with acids probably require the intervention of ions of type **4b**.

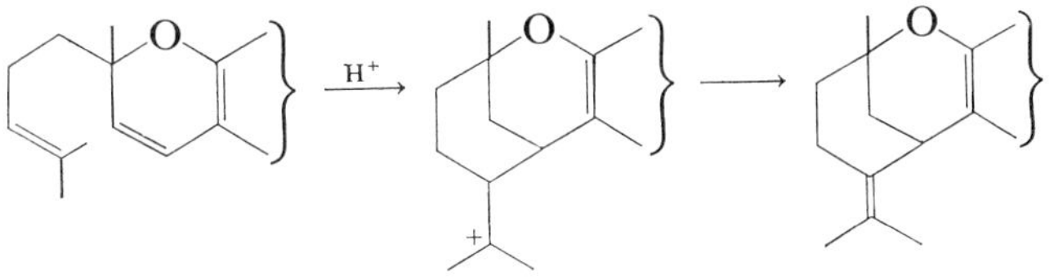

Gambogic acid

5 Gamboginic acid

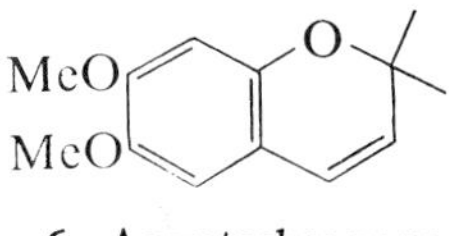

6 Ageratochromene

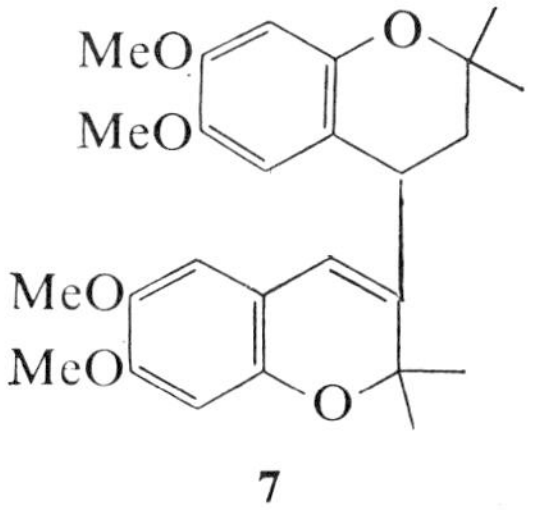

7

The most widely used chromene synthesis was that introduced by Späth[11] in which the requisite phenol was heated with 2-methylbut-3-yn-2-ol either alone or with some very mild Lewis acid catalyst such as zinc chloride. The method succeeded with seselin, xanthyletin, luvungetin, and some others, but it is enough to say that 2% was considered to be an acceptable yield. Efforts to use C_{10} acetylenes were even less rewarding,[12] and not all phenolic nuclei withstood the treatment. In a synthesis of pongachromene[13] **8** attempts to add the pyran ring to the flavone **9** failed entirely, whereas the more roundabout method gave a yield of ~0.1%:

9

8 Pongachromene

Allan-Robinson

C_5H_7OH, $ZnCl_2$

(rejected)

10 Lonchocarpine

11 Seselin

In a closely similar preparation of lonchocarpine **10**, Nickl[14] secured the catalytic effect of the cation in a medium of very low acidity by using zinc carbonate instead of zinc chloride, while other workers[15] have greatly improved the preparation of seselin **11** from umbelliferone (7-hydroxycoumarin) by omitting a catalyst altogether but using 1,2-dichlorobenzene as the solvent at 160°. But even so the yield was only 6%, and the technique is now wholly replaced by a modern version.

In this, the acetylenic alcohol is converted into 3-chloro–3-methylbutyne which is used to etherify the phenol in the presence of potassium carbonate with iodide catalyst, and the ether is heated in diethylaniline for some hours. Discovered by Iwai and Ide,[16] the method affords chromenes directly in yields of 80% or more. The illustration is for evodionol methyl ether **12** obtained thus by Hlubucek, Ritchie, and Taylor,[17] who prepared 7-methoxy–2,2-dimethylchromene, ageratochromene **6**, luvungetin, and seselin **11** in the same way.

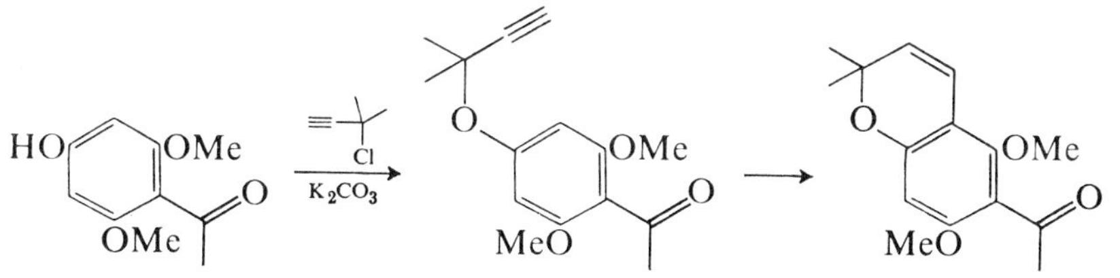

HO OMe O OMe — Cl, K_2CO_3 → O OMe O MeO → O OMe O MeO

12 Evodionol methyl ether

The rearrangement is very interesting since, if viewed as an allylic rearrangement, neither the parent acetylene nor the allenic transition state (or intermediate) appears to have a suitable geometry. Nevertheless, an allenic intermediate is indeed responsible and can be trapped internally by an electrocyclic reaction leading to a tricyclo-octane derivative[18] **13**:

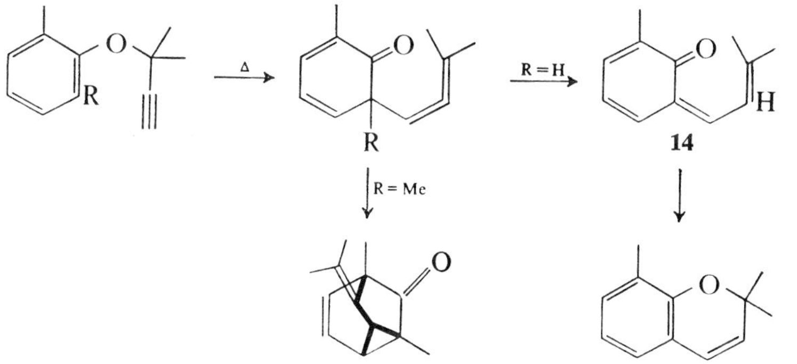

O R —Δ→ O R —R = H→ O H **14**

R = Me ↓ **13** (O) ; **14** ↓ (O)

13

The resemblance of **13** to the bridged part of gambogic acid **2** inspires speculations to be taken up in Section 7, but it is the electrocyclic ring closure of **14** that we have to notice particularly here, since it turns out to be a feature of nearly every chromene synthesis. A useful corollary is provided by the partial reduction of the acetylenic ethers to allylic ethers. These upon conventional Claisen rearrangement afford *o*-dimethylallylphenols* in excellent

* It is frequently convenient to refer to the 3-methylbut-2-enyl group as 3,3-dimethylallyl, or as dimethylallyl only, or even simply as prenyl.

overall yields. Osthenol **15**, demethylsuberosin, coumurrayin, and other compounds have been prepared thus,[19] and the method is generally much superior to most direct alkylation techniques:

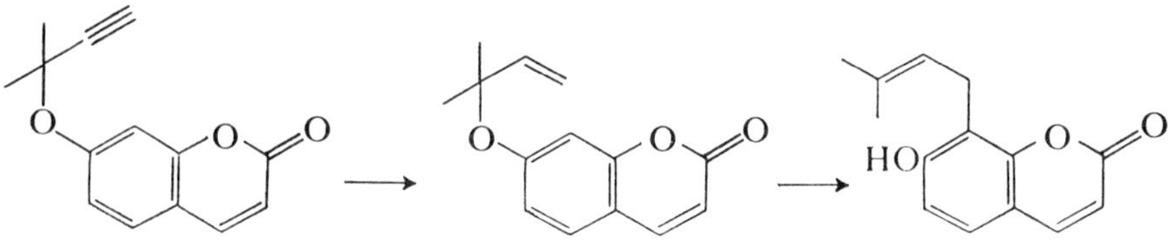

Osthenol **15**

Such *o*-3,3-dimethylallylphenols are very common, and it has long been recognized[1–3] that related dimethylallylphenols and 2,2-dimethylchromenes occur in the same plant or in closely allied species and believed that one structural type may be a biosynthetical precursor of the other. Some attention has been paid to the reduction of chromenes to allylphenols, a change that can be effected in simple cases by the Birch reduction,[20] or photochemically,[21] but much more has been devoted to the reverse process of oxidative cyclization, an idea first suggested by Ollis and Sutherland[1] and later modified by Turner[22] and others. Whatever the details, the central theme is the production of a quinone methide that cyclizes spontaneously:

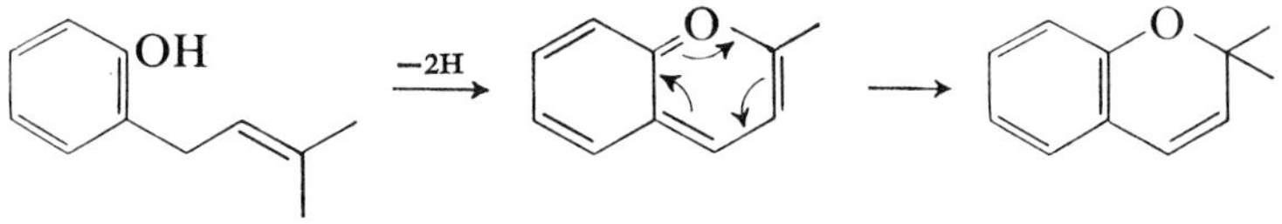

$\xrightarrow{-2H}$

Such a cyclization was first achieved *in vitro* during studies on mycophenolic acid,[23] the reagent being dichlorodicyanobenzoquinone (DDQ) in refluxing benzene. This system is known[24] to be able to operate by hydride abstraction from suitable hydrocarbon groupings, which suggests a sequence such as the following:

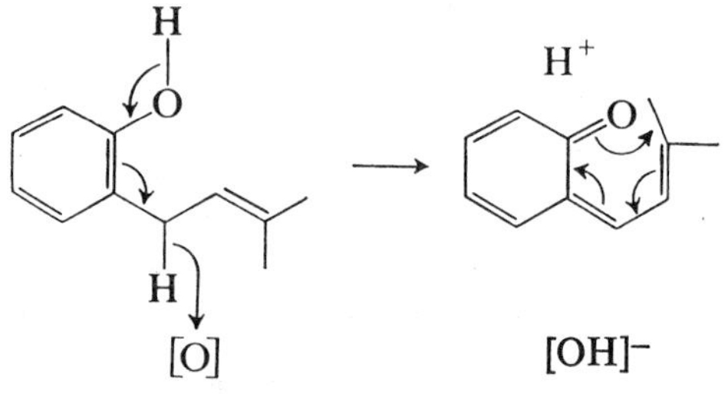

[O] $\qquad$ H^+ $\qquad$ $[OH]^-$

This question of mechanism is important, for it seems that cyclizations to chromenes would be on the same initial footing as other dehydrogenations which, consequently, might well interfere. Though little is known about the matter, the danger can be assessed from an example of dehydrogenation taken from the cannabinoid series:[25]

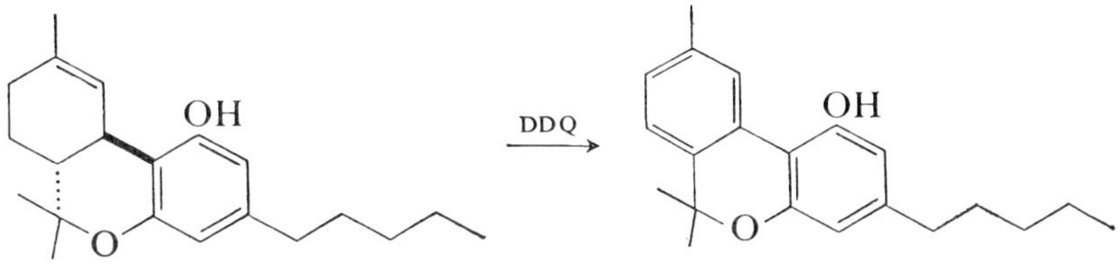

This reaction takes but 2 hours for 90% completion, whereas most chromene cyclizations require something between 1 and 16 hours.

Cyclizations by DDQ have been evaluated mainly by Italian workers[26] and usually give yields around 50%. Where the starting dimethylallylphenol is itself a readily available natural product such yields are not unsatisfactory, and the method is at its best in these circumstances. When the dimethylallylphenol has to be synthesized, the approach seems much less attractive, since the *C*-alkylation of phenols is often a confused affair not giving more than about 20% of the desired product. A common technique to introduce the dimethylallyl group is to employ the necessary phenol as its lithium salt, which is heated in benzene with 4-bromo–2-methylbut–2-ene. These conditions are designed to minimise *O*-alkylation, but there is a tendency for them to induce a concomitant cyclization to a chroman which, as explained later, may not be a viable intermediate. This criticism applies even more strongly to "biogenetically modeled" alkylations employing phosphates instead of bromides,[27] these often giving chromans in yields as high as 80%:

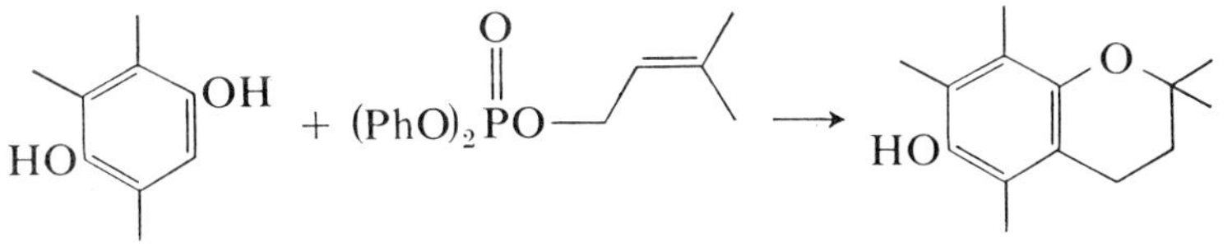

Examples[26] of the successful use of the route include the preparations of alloevodionol methyl ether **16** and of (±)-cannabichromene **17**, the latter exemplifying the insertion of a C_{10} side chain, and both exemplifying the directive effects of hydrogen bonded carbonyl groups:

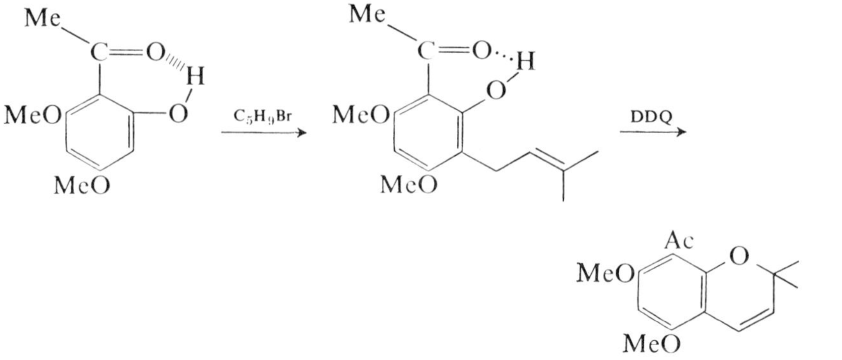

Alloevodionol methyl ether **16**

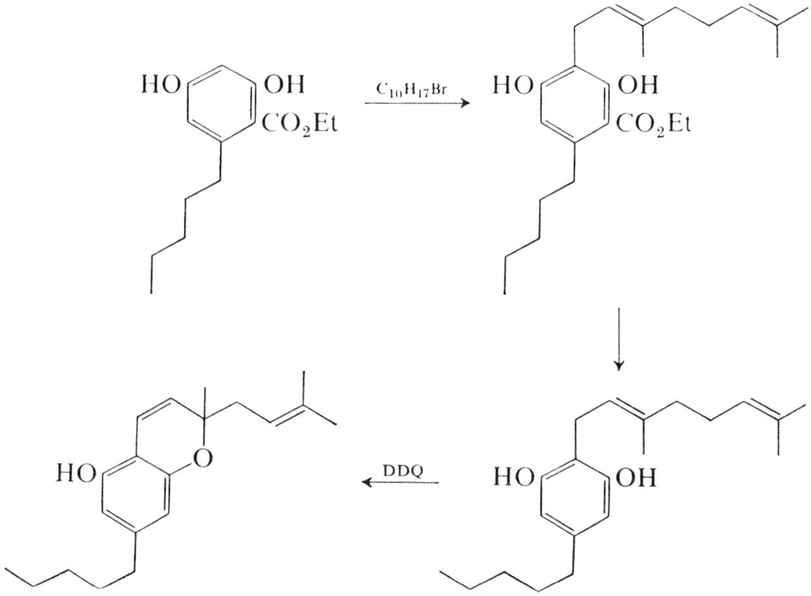

17 Cannabichromene

It must be noted, however, that ideas are changing rapidly in this area. It is now recognized that the cyclization of *o*-prenylphenols may be a relatively slow process, and that keeping the temperature as near to room temperature as possible and the reaction time as short as possible may be all that is necessary to obtain good yields of the desired alkenylphenols.[28] Clearly, the acidity of the medium is not as crucial as was thought originally, though the reaction is certainly acid catalyzed. Hence the success of the technique introduced by Bohlmann and Kleine[29] who report that treating phenols with 3-methylbut-2-enyl alcohol and boron trifluoride etherate in dioxan at about 25° gives good yields of their *C*-prenyl derivatives. The method is used increasingly,[30,31] as in a first synthesis[30] of evodionol (from *Evodia littoralis* and *Melicope simplex*):

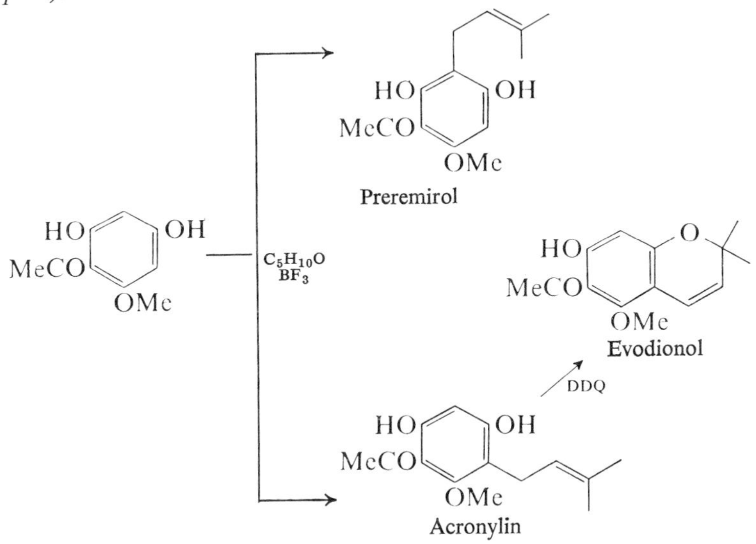

HO OH MeCO OMe — $C_5H_{10}O$ BF_3 → HO OH MeCO OMe Preremirol; → HO OH MeCO OMe Acronylin —DDQ→ HO O MeCO OMe Evodionol

If the dimethylallylation of the phenol leads inevitably to a chroman the DDQ reaction may still be applicable if there is a free hydroxyl group in a position allowing quinone-methide formation.[32] The point can be made with a synthesis of alloevodionol **18**. The unusual pyranochromene franklinone **19**, however, seems exceptional, for it is reported to be formed by DDQ oxidation of the corresponding tetrahydro-derivative with no free hydroxyl groups.[26] Previously, franklinone had evaded synthesis altogether. One last point of interest is that chelation will protect a phenolic hydroxy group to some extent and may therefore direct a DDQ cyclization as required for syntheses[26,33] of the trimethyl ether **20** of flemingin C:

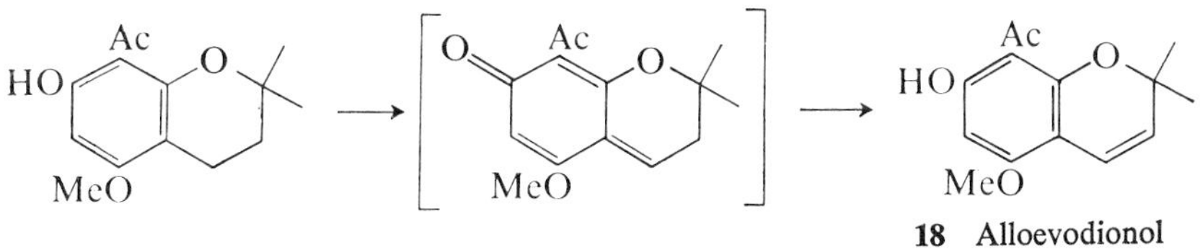

HO Ac O MeO → [O Ac O MeO] → HO Ac O MeO

18 Alloevodionol

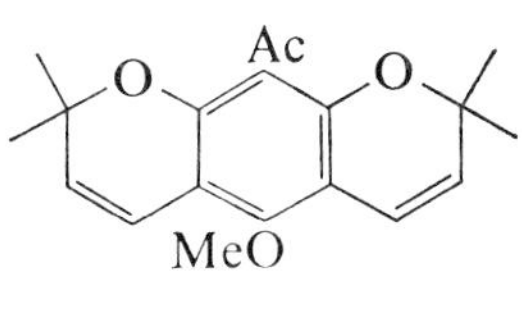

O Ac O MeO

19 Franklinone

MeO MeO MeO O O HO

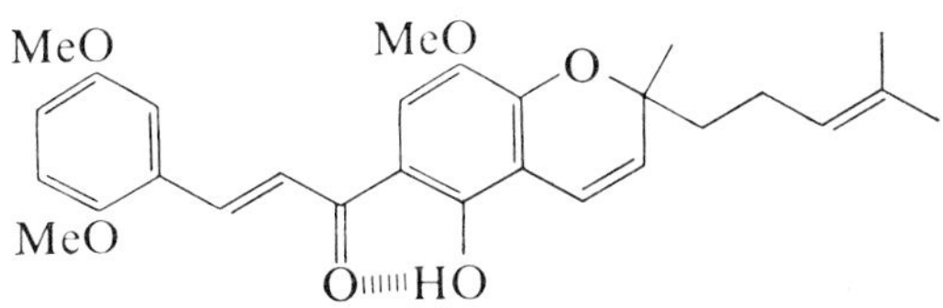

20 Flemingin C trimethyl ether

If the DDQ dehydrogenation fails to convert the chroman into a chromene it may still be possible to fall back on bromination-dehydrobromination techniques; one such instance is discussed near the end of Section 7.

Another form of oxidative cyclization is known in which a quinone nucleus can be regarded as an *internal* oxidant. It is catalyzed by bases and occurs with ease, and was discovered, largely accidentally, during a "purification" of ubiquinone on alumina columns that actually gave ubichromenol **21** instead:[34,35]

MeO MeO O O Me H 4 base MeO MeO O O Me $C_{20}H_{33}$

Ubiquinone

MeO MeO HO O Me H 4 MeO MeO O O Me $C_{20}H_{33}$

21 Ubichromenol

The mechanism outlined again emphasizes spontaneous cyclization in an *o*-quinone methide. Beside alumina, potassium hydroxide has been used as the catalyst and, much better, pyridine at its boiling point, this technique inducing a 92% conversion.[36] Sodium hydride in benzene has also been recommended.[37] Perhaps more significantly for plant life processes, the same change can be induced by irradiation with visible light,[38] a minor consequence of which is that TLC of such quinones must be conducted in the dark. If UV light is used, the quinone is mainly reduced to the quinol instead.

We turn now to a different type of chromene synthesis, one based on the fact that chromanones are rather easily accessible. Ageratochromene **22** provides our illustration of the approach,[39] (as shown on top of page 477). Meerwein-Ponndorf reduction may replace the hydride reduction. The last step, dehydration, is, however, critical, since the acid catalysts usually utilized prejudice the survival of the product.[40] One solution has already been noted. Another is to use anhydrous copper II sulfate,[40] but the problem has been better solved as in the case of acronycine **23** by the use of phosphorus oxychloride in refluxing pyridine.[32] In the case of xanthoxyletin **24** it was even possible to use sodium hydrogen sulfate provided the product was sublimed away from the reaction area.[41] This synthesis is illustrated on account of an interesting anomalous carbonyl reduction by borohydride, of clausenin

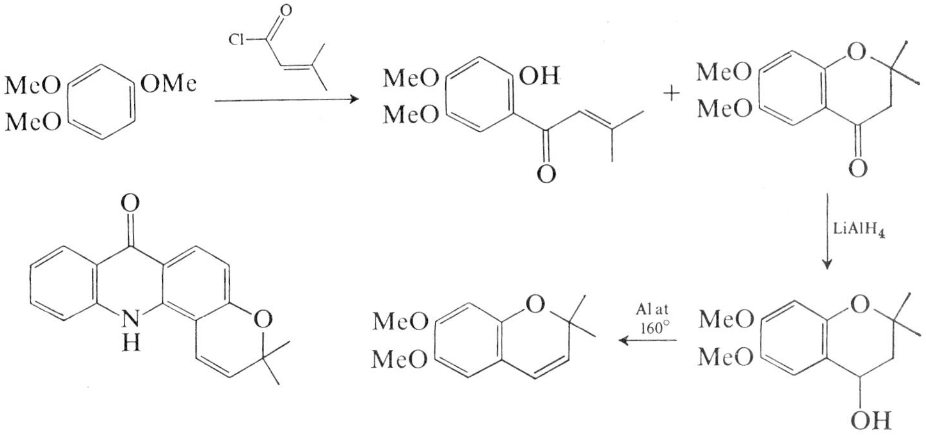

23 Acronycine

22 Ageratochromene

25 yielding the hydrocarbon as well as the expected alcohol. Presumably borohydride solutions are alkaline enough to open the lactone ring so that the way is open to quinone methide formation and reduction; upon work-up, the lactone ring would reclose spontaneously as is usual with coumarins. Of course, alcohol dehydration problems can be side-stepped altogether by pyrolyzing the acetates, as in a synthesis of lapachenole **3** via the related chromanone.[5]

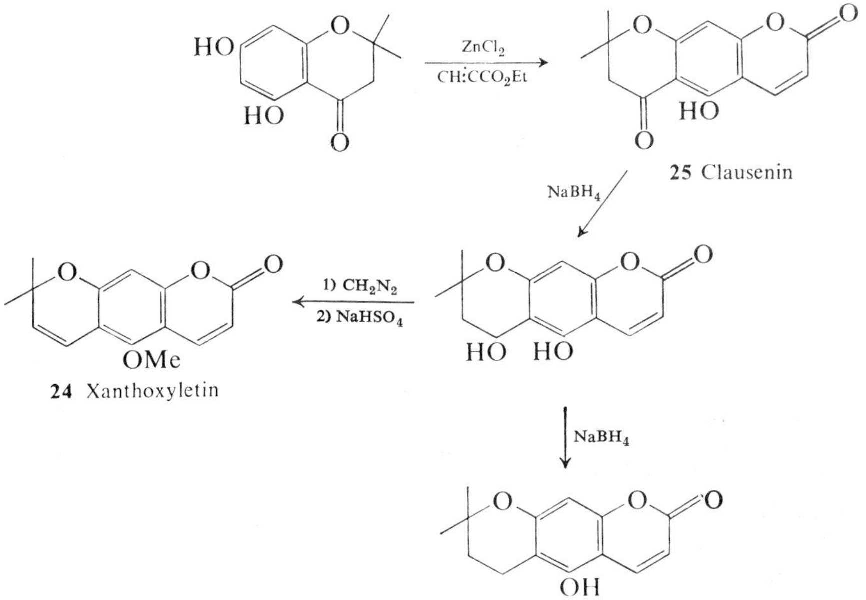

25 Clausenin

24 Xanthoxyletin

3-Hydroxychromans are available naturally, as reduction products of chroman-3-ones or, most recently, as thallic oxidation products of allyl aryl ethers,[41a] but acid-catalyzed dehydration is liable to induce skeletal rearrangements and prototropic shifts leading to isomers of the desired chromenes.[42] Examples occur in the chemistry of selinetin[43] (lomatiol) and jatmansinol.[44] The example shown concerns decursin[45] **26**, a coumarin constituent of an old Chinese medicinal herb; protic acids provoke a rearrangement leading to anhydronodakenetin **27** while the tosylate in collidine affords the chromene, xanthyletin **28** (as shown on opposite page).

26 Decursin

HO⁻ ; H+ ; 1) TsCl 2) collidine ; two prototropic shifts

27 Anhydronodakenetin

28 Xanthyletin

Unlike most chromens, siccanochromene A **30** is a fungal product. It comes from *Helminthosporium siccans*, a plant pathogen. An interesting synthesis of this compound by Nozoe and Hirai[46] contains steps of importance for the present discussion, and may be summarized as opposite.

The lithium derivative was used to attain an orientation not easily available by direct substitution reactions, but it is the spontaneous dehydration of intermediate **29** that invites attention since the contrast with the conditions for dehydrating 4-hydroxychromans is so marked. Again the formation and cyclization of a quinone methide allows a satisfactory rationalization for chromene ring formation.

OR
Li
+ RO
25°

OH
OR
RO

[R = (tetrahydropyran-2-yl)]

OH
O
O

aq. oxalic acid

O
OH
OR
RO

29

O
O
OH

(2 stereoisomers)

$CH_2 = PPh_3$

O
HO

30 Siccanochromene

If flindersine **31** be admitted to the honorary degree of chromene, then one synthesis[47] of it supplies another instructive variation upon the same theme:

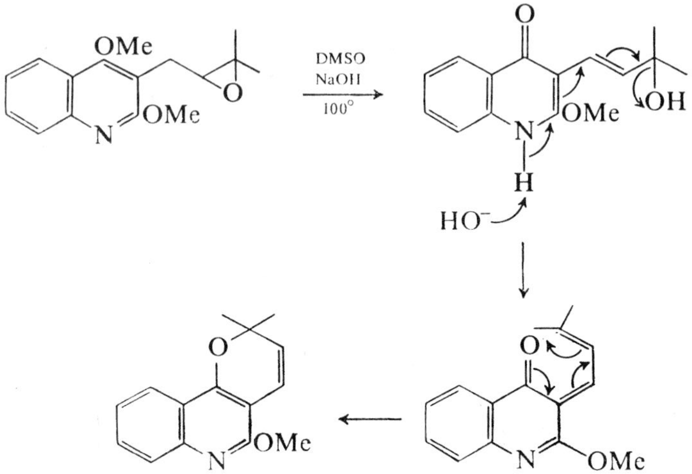

DMSO
NaOH
100°

HO⁻

31 Flindersine

The two preceding reactions serve to introduce one of the most powerful of all chromene syntheses. Essentially, the phenol (as phenoxide ion) reacts with an unsaturated aldehyde giving just that type of alcohol which, like **29**, spontaneously generates a quinone methide and therefore the chromene ring system.

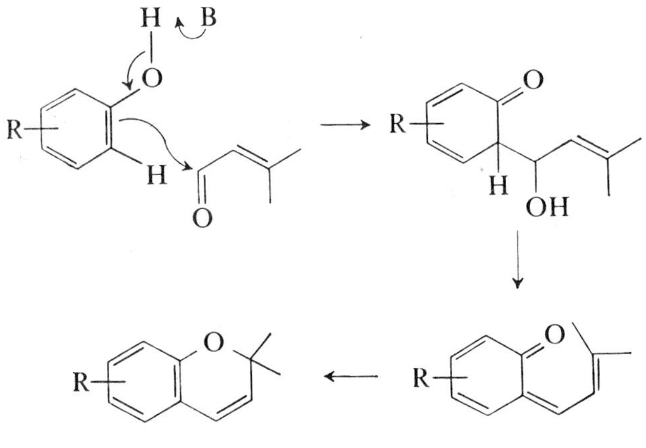

The reaction is induced by pyridine at 100–150° for several hours. It can be regarded as an aldol condensation occurring in the ketonic form of the phenol, and in fact it succeeds only with those phenols (resorcinol, phloroglucinol, and their derivatives) that are well known for this kind of tautomerism. Monohydric phenols and quinol are not amenable.

Announced by Crombie and Ponsford,[48a] one variation of the method was early used to synthesize certain constituents of hashish, the drug extracted from *Cannabis sativa*. Condensed with citral, olivetol yields several products,[48b] of which only the simplest, cannabichromene **32**, is produced in fair yield. This fact and the orientation can be attributed to the difficulty of inserting one large group *ortho* to another:

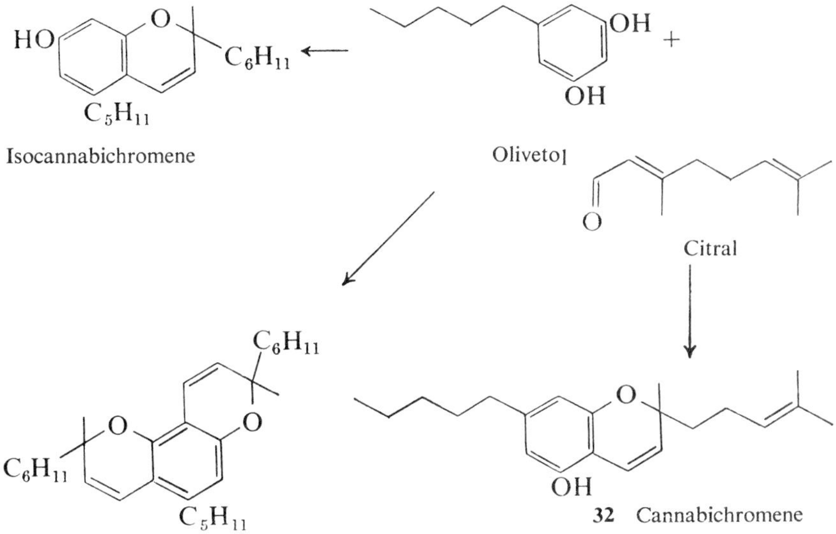

Isocannabichromene

Olivetol

Citral

32 Cannabichromene

Such reactions may go further, however, for some of these chromenes are sensitive to bases and the final results often depend upon the conditions used, especially the amount of pyridine. Although some details are not yet settled,[49,50] two cyclizations are formulated speculatively to emphasize yet again the central role of quinone methide chemistry as shown overleaf:

C_5H_{11} ^{-}O O H^+ → C_5H_{11} O O → C_5H_{11} O O

Citrylidene-cannabis

↑

C_5H_{11} HO O

Cannabichromene

↓

C_5H_{11} ^{-}O O → C_5H_{11} O O H $\xrightarrow{H^+}$ C_5H_{11} HO O

Cannabicyclol

A stepwise sequence is indicated for the cyclization to the cyclobutane derivative, cannabicyclol, by the steric factor which makes an electrocyclic reaction under Woodward-Hoffman rules very doubtful. A photochemical cyclization would be possible, of course, and is known.[51] The alkaloid mahanimbine **33** offers a nitrogen analog of the situation, and in fact this alkaloid can be synthesized and transformed similarly:[51,52]

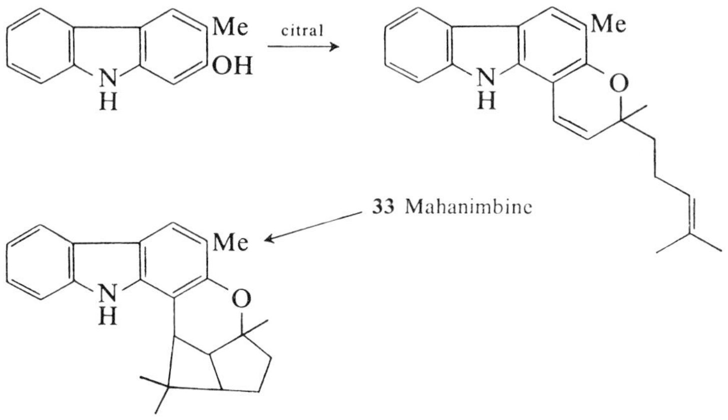

Me OH N H citral → Me O N H

33 Mahanimbine

Me O N H

Such further cyclizations and other misadventures can be rendered unimportant by masking the responsible hydroxy group by chelation. Thus the following condensation affords in 67% yield a chromene **34** suitable for elaboration into ethers of the flemingins A, B, and C:

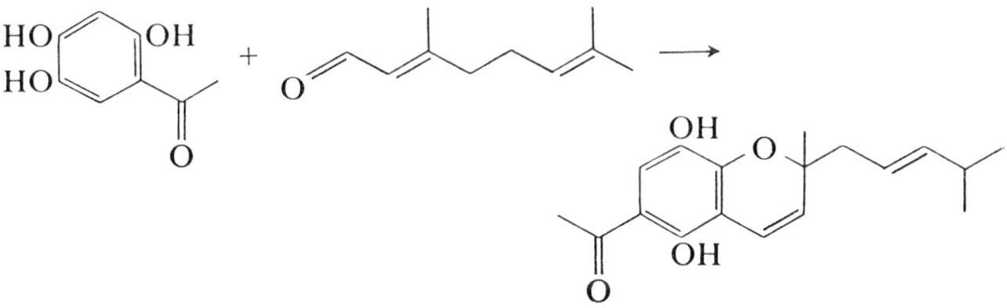

HO OH HO O + O →

OH O OH O

34

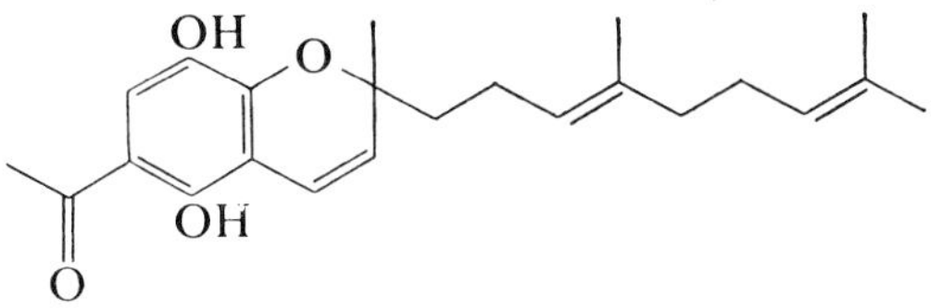

OH O OH O

35

Even the trienal, farnesal, undergoes a similar reaction[48a] giving 42% of the desired chromene **35**.

In a further refinement, unsaturation is avoided and the aldehyde replaced by an acetal function. Rather higher reaction temperatures seem necessary, but resins are much reduced and yields improved. Very satisfactory syntheses of evodionol methyl ether **12**, lonchocarpin **10**, and many other chromenes have been achieved thus.[53b] Illustrated are reactions leading to jacareubin **36** and to acronycine **37**, the latter, obtained from *Acronychia baueri*, being of particular interest for its antitumor activity. Further examples would include

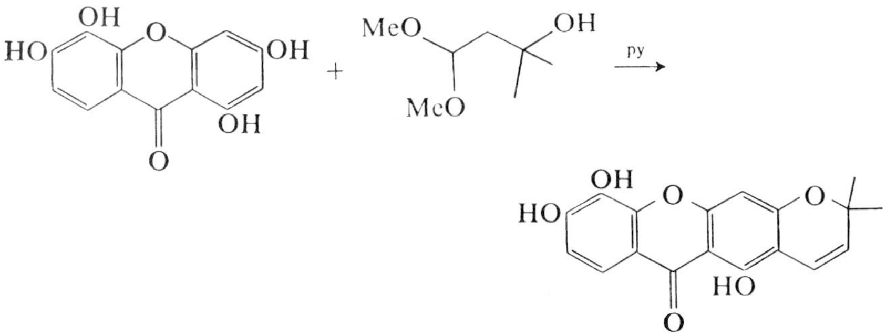

36 Jacareubin

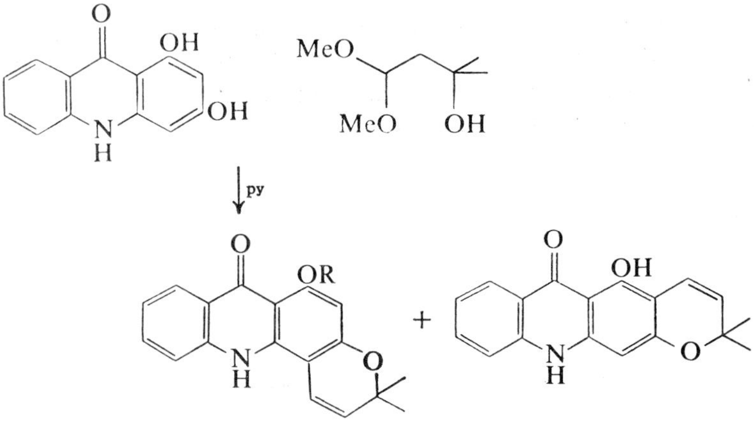

(R = H)

37 Acronycine (R = Me)

some of the pyranocoumarins[31] that are found in various species of *Mammea*. Another recent paper[54] describes additional cases of dual annelation as well as raising the old specter of dimerization in a new as well as the old forms—a claim that cyclization occurs at a chelated hydroxy group is perhaps not justified, however.

Of other methods of synthesizing chromenes, that of greatest interest centers on an ylide reaction. It has one advantage over the methods just oulined in that it begins with a salicylaldehyde derivative and therefore leaves no doubt as to the orientation of the product. The sodium salt of the derivative itself acts as the base initiating a Wittig reaction in dimethylformamide as solvent:

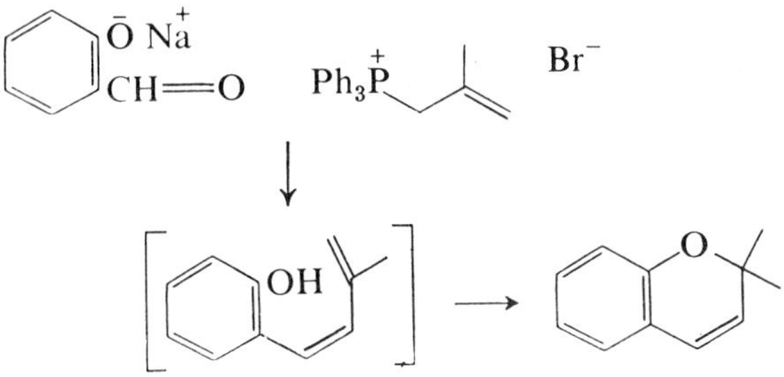

The primary product is believed to be the open-chain system shown, since hydrogenation at the early stage leads to the corresponding *o*-alkylphenol. The cyclization occurs at a later stage but exactly how is not clear. Yields of 40% are obtainable.[55a] The alkaloid girinimbine **38**, which occurs with mahanimbine in *Murraya koenigin*, has been synthesized in this fashion and its constitution defined thereby:[55b]

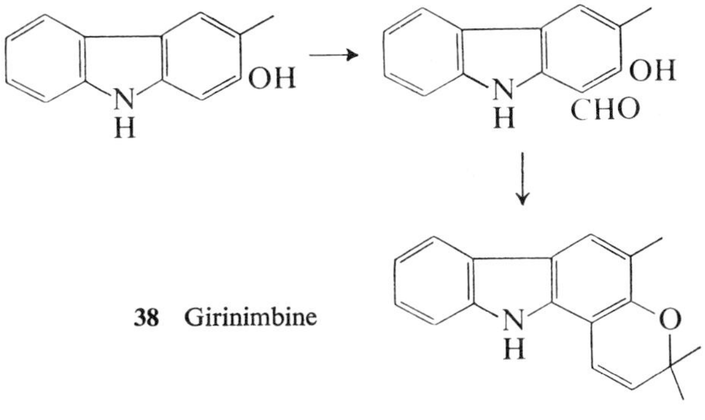

38 Girinimbine

3. STERIGMATOCYSTIN AND THE AFLATOXINS

A small but important group of fungal metabolites contain a furo[2,3-*b*]-benzofuran nucleus. In addition to this unusual structural feature, the high toxicity of all and especially the marked carcinogenic powers of some of these compounds has excited much interest in them.[56,57] Sterigmatocystin[58] (**39**), from some strains of *Aspergillus versicolor* (Vuill.) Tiraboschi, was the first member of the group to be recognized. The versicolorins are related anthraquinones,[59] and so is dothistromin,[60] a plant pathogen implicated in pine needle blight. The notorious carcinogens, the aflatoxins,[61] are produced by *Aspergillus flavus* and *A. parasiticus:* two have been synthesized, B_1 with structure **40** and B_2 with structure **41**. Another aflatoxin, aflatoxin M_1 **42**, comes from the milk of cows or the urine of sheep or the liver of rats that have ingested aflatoxin B_1 and has also been synthesized.[57] Some strains of *A. flavus* produce sterigmatocystin derivatives instead of, or as well as, aflatoxins, thus demonstrating a very close biosynthetical relationship between the two series. Other close relationships are believed to exist with the anthraquinone aversin, the methyl ether of which has very recently been synthesized as described at the end of this section.

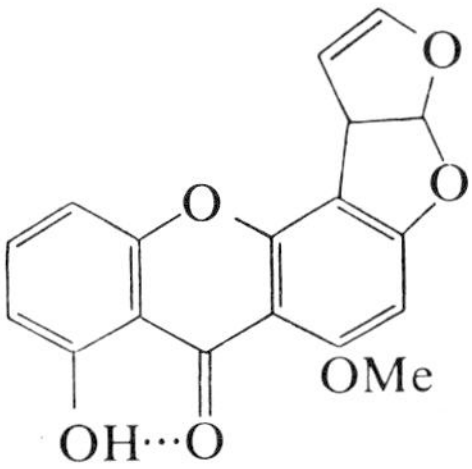

39 Sterigmatocystin

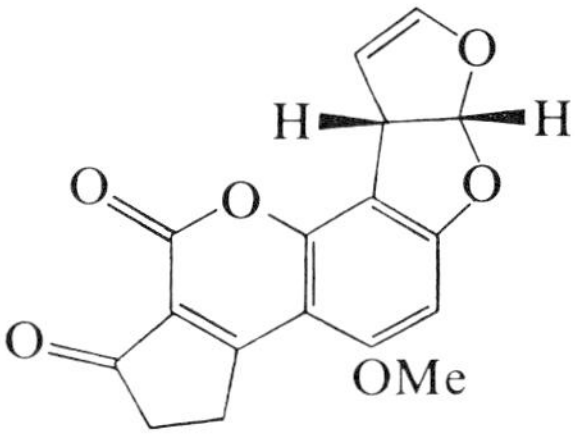

40 Aflatoxin B_1

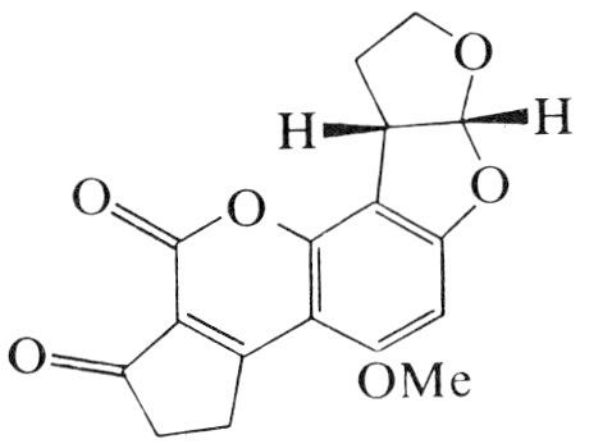

41 Aflatoxin B_2

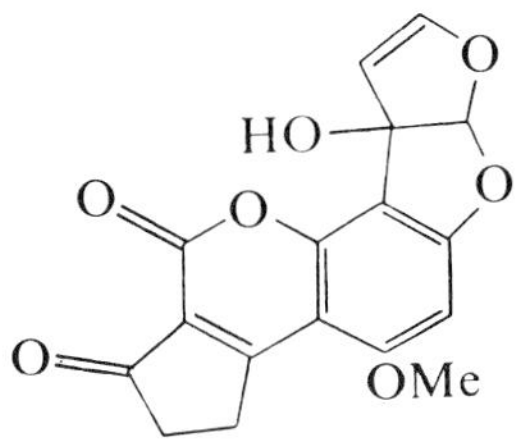

42 Aflatoxin M_1

Some direct methods of creating a furobenzofuran are known. Examples include (*a*) treatment of a suitable chloroethyl lactone with an amine:[62]

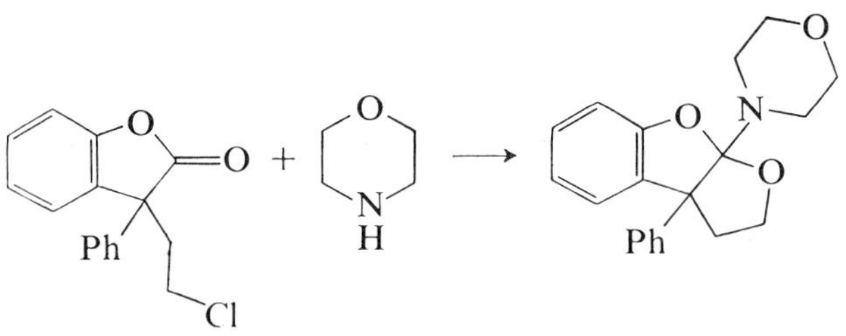

(*b*) condensation of glyoxal with a phenol:[63]

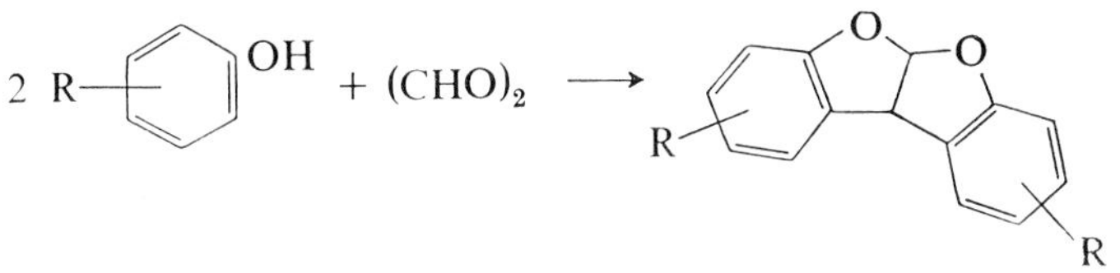

$$2\ R\text{—}C_6H_4\text{—}OH + (CHO)_2 \longrightarrow$$

and (*c*) condensation of dihydrofuran with acetylquinone:[64]

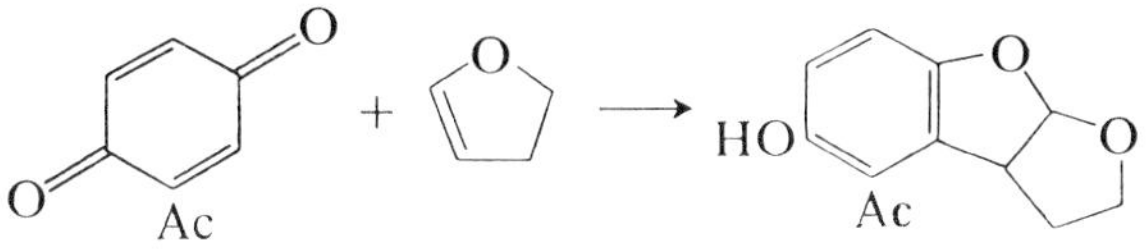

But since no direct method seemed sufficiently adaptable, relatively long sequences have been used in practice, often with 5,7-dihydroxy-4-methylcoumarin **43** as the starting point.*

Methylation[65] of **43** with dimethyl sulfate and base affords mainly the 7-methyl ether **44**. This selectivity depends on the fact that the ion **45** is *para* quinonoid and therefore more stable than the alternative, which is *ortho* quinonoid. Hence, provided that only *one* equivalent of base is used, the 7-oxide is chiefly present at the site of reaction. In agreement, the UV spectrum of the dihydroxycoumarin is nearly the same as that of the 5-methyl ether but different from that of the 7-methyl ether when all are examined in limited base.[66] By the same argument, on the other hand, the electrons of the 7-hydroxy group in the *unionized* dihydroxycoumarin are *less* readily available

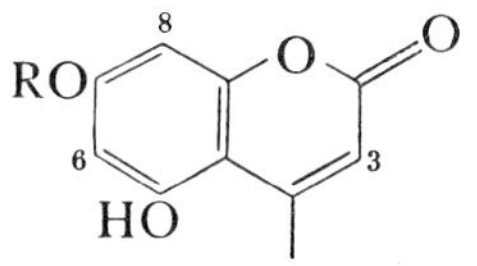

43 R = H
44 R = Me

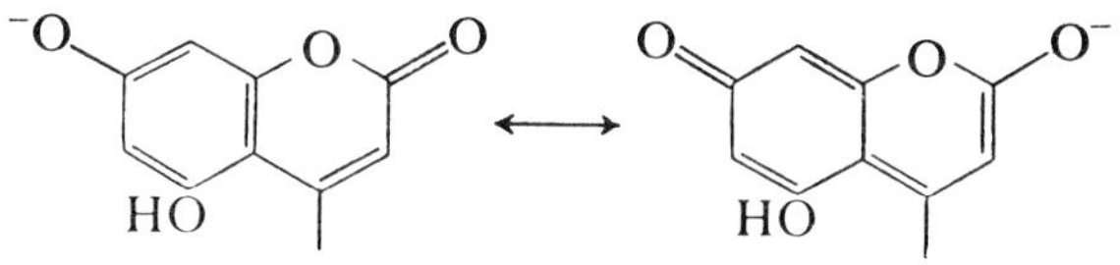

45

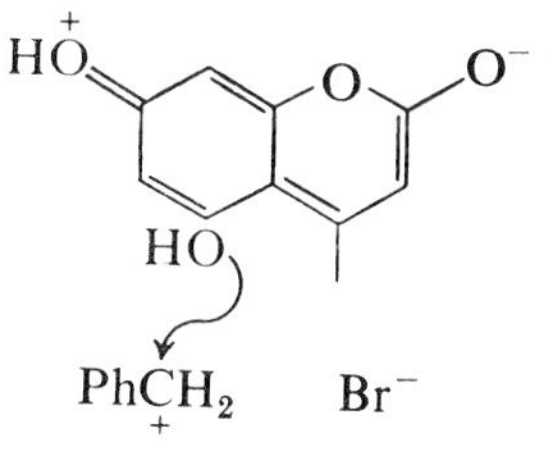

46

47

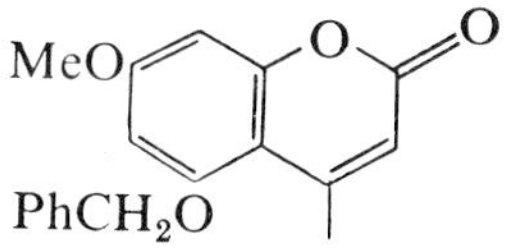

* The synthesis of the 3-(2-hydroxyphenyl)furan system and a cyclization by electrolytic oxidation have been achieved but details are not yet available.[64a]

than those of the 5-hydroxy group because of the *para*-quinonoid interaction indicated in **46**, so benzylation by sources of benzyl carbonium ion should occur elsewhere preferentially. This seems to explain the formation[66] of the 5-benzyl ether when the potassium carbonate-acetone technique is used.

Either way, the mixed ether **47** is accessible, and is oxidized[66,67] by selenium dioxide[68] to the yellow aldehyde **48** without damage to the benzyl methylene group. Acetal **49** is produced by the orthoformate method and then controlled hydrogenation saturates the olefinic bond taking the nucleus into the dihydrocoumarin series as **50**. At the same time the benzyl group is lost yet the acetal ether links survive, quite the reverse selectivity to the selenium dioxide reaction. Lithal reduction followed by acid hydrolysis now yields the tetrahydrofurobenzofuran derivative **51** as the racemate, but otherwise identical with (optically active) degradation product of sterigmatocystin.[67] The ring fusion is necessarily *cis* because of the considerable strain in the *trans* fused arrangement. Furthermore, Knight, Roberts, Roffey, and Sheppard[69] have condensed the furobenzofuran derivative **51** with ethyl cyclopentanone-2-carboxylate by Pechmann's method but with hydrogen chloride in ethanol instead of the usual sulfuric acid. (In phloroglucinol derivatives the electron-releasing effects of the three hydroxy groups reinforce one another so that the nucleus becomes very highly nucleophilic and only the mildest catalysts are needed to induce electrophilic substitutions, the more powerful ones merely inducing further, unwanted reactions.) The product **52** is the racemate corresponding to a reduction product of aflatoxin B_1 **40**.

The same workers[70] have also condensed the phenol **51** with ethyl 3-oxoadipate, thus obtaining a coumarinylpropionic ester and thence, by alkaline hydrolysis, the acid **53**. Alkaline hydrolysis leaves the acetal function untouched, of course, but hazards the lactonic system since this can open to a *cis*-cinnamic acid in which geometrical isomerization is known to occur with loss of viable material. In fact, losses were not serious, and the synthesis was continued by using oxalyl chloride to secure the acid chloride and aluminum chloride at $-5°$ to cyclize it to ($\pm$)-aflatoxin B_2 **54** in 33% yield. Such cyclizations are general.[61,70–72]

During the final stages great care was taken to ensure the survival of the acetal grouping under acidic attack. Acyclic acetals would probably not survive at all, the entropy factor being against it, but with cyclic ones such as these, ring opening does not, in itself, present a major difficulty since ring closure is always the preferred sequel.

More acute problems are posed by the synthesis of the unsaturated acetals, sterigmatocystin **39**, and aflatoxin B_1 **40**, in which acid-catalyzed additions to the vinylic ether group are a constant danger. This can be avoided by introducing the unsaturation at the last step, but even here conditions have

MeO O O

PhCH$_2$O CH=O

48

MeO O O

PhCH$_2$O CH(OEt)$_2$

49

MeO OH

OH

HO

EtO OEt

MeO O O

HO CH(OEt)$_2$

50

MeO OH

OH

HO

O

MeO OH

O

O

51

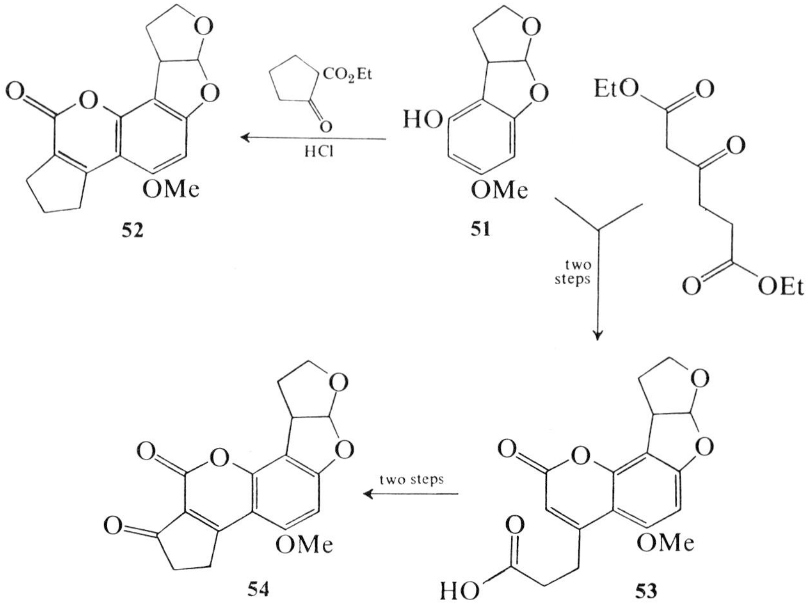

been found that make it possible to operate elsewhere in a molecule containing an exposed vinylic acetal group.

The synthesis of ($\pm$)-aflatoxin B_1 was announced by Büchi and his colleagues[66] in 1966. Their starting point was the aldehyde **48** alluded to earlier. Coumarins are rather easily reduced by zinc and acid, and in this one the double bond is activated further by the aldehyde carbonyl group so the reduction readily yielded the dihydrocoumarin aldehyde **55**. The next step was taken inadvertently, however, for dihydrocoumarins are rather easily hydrolyzed, here spontaneously exposing the latent aldehyde function thus allowing hemiacetal **56** formation and relactonization:

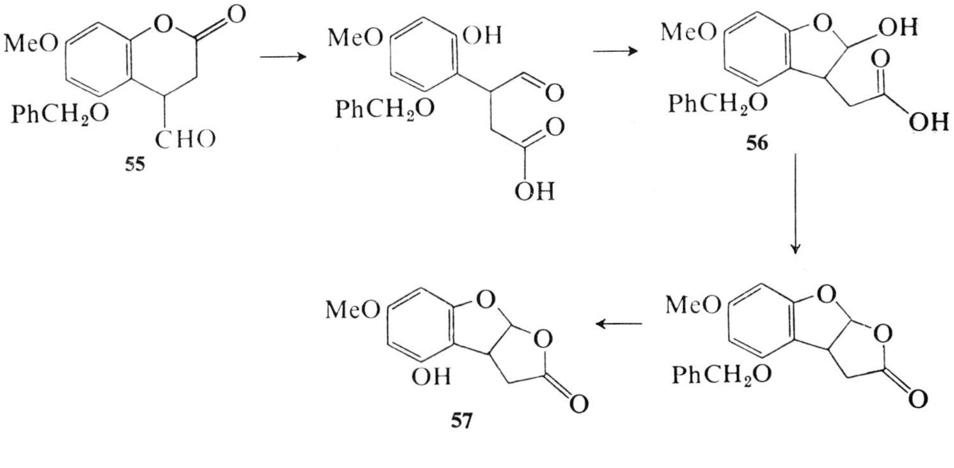

Such "β-acyl lactone" rearrangements are well known, having been extensively studied by Korte.[73]

Catalytic debenzylation produced the phenolic lactone **57**, and attempts were made to elaborate this into a coumarin using ethyl methyl 3-oxoadipate in a conventional Pechmann condensation under the influence of 86% sulfuric acid. This overpowerful reagent largely catalyzed ring-opening to a benzofuran **58**, thus exposing the system to a host of undesirable consequences, many of which took their toll. In a gentler procedure, the lactone **57** gave with methanolic hydrogen chloride the acetal-ester **59**, which was not isolated but allowed to react at once with ethyl methyl 3-oxoadipate thus smoothly producing a coumarin **60** in yields approaching 60% (see page 492). Next, hydrolysis gave the lactonic acid **61**, which, with oxalyl chloride and then aluminum chloride, afforded the cyclopentenocoumarin **62**. The crucial last stages were now at hand. Selective reduction of the γ-lactone carbonyl group giving the hemiacetal **63** and dehydration of this to give ($\pm$)-aflatoxin B_1 **40** were needed. Neither stage has proved to be particularly easy. Disiamylborane does attack the γ-lactonic function in preference to the others (the infrared band at 1790 cm^{-1} may perhaps indicate an electrophilicity higher than is usual in most lactones) and the hemiacetal **63** can be obtained in about 20% yield. Since dehydration cannot be accomplished directly, the acetate has to be pyrolyzed at 240° for 15 minutes, a process yielding 40% of ($\pm$)-aflatoxin B_1 **40**.

The racemate has not yet been resolved. Identities have been established by racemizing natural aflatoxin B_1 hydrate, which is easily obtained by treating the natural product with trifluoracetic acid and water. In base, the acetal-hemiacetal system opens liberating the aldehyde function; consequently, the adjacent optically active center* becomes labile and the compound racemises:

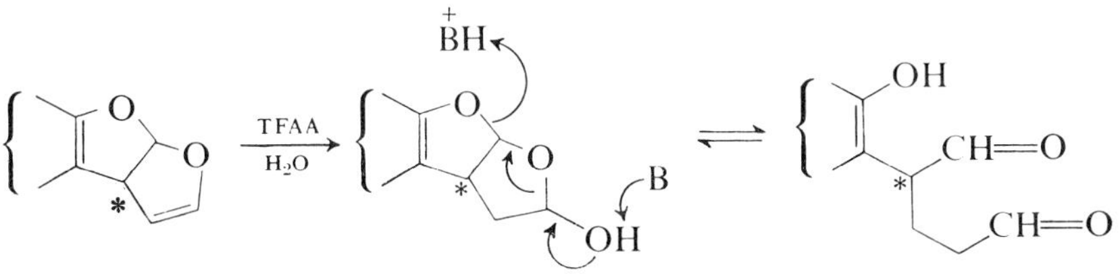

Subsequently, Büchi and Weinreb[74] evolved a most convincing synthesis of aflatoxin M_1 (**42**), provision for the angular hydroxy group being made at the very beginning by commencing with the coumaranone **64**. The methylene group was brominated using phenyltrimethylammonium perbromide in tetrahydrofuran, and the halogen replaced using benzyl alcohol in the presence of calcium carbonate (seemingly, this is an S_N1 reaction in which the

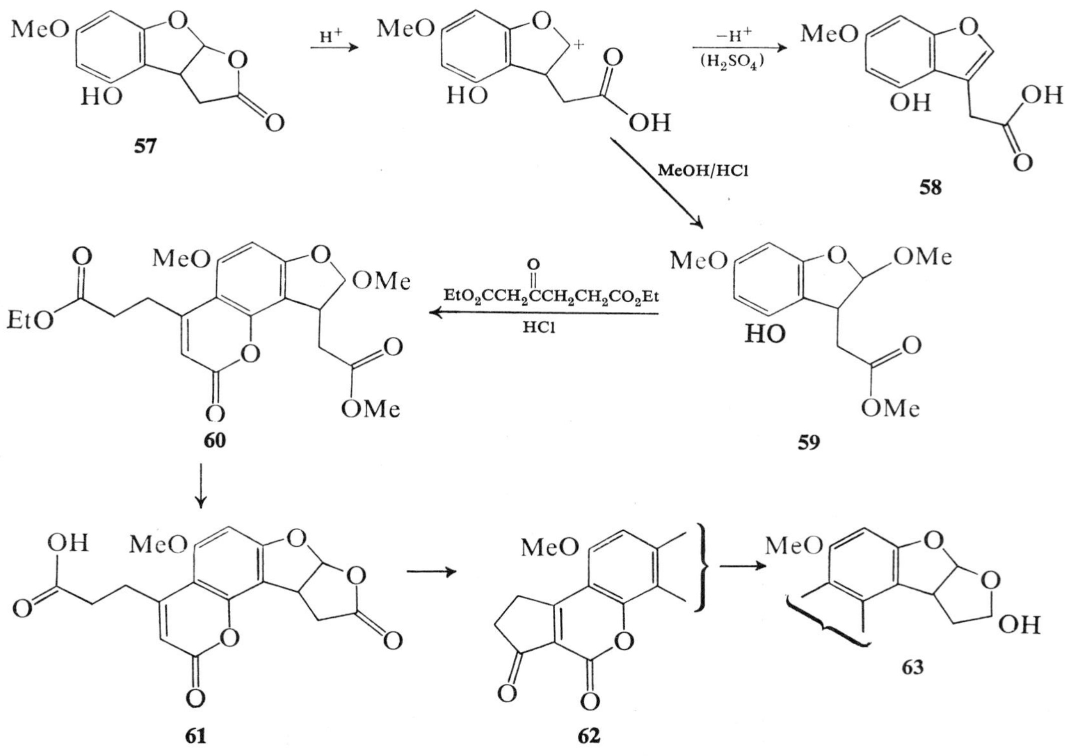
MeO
HO
57
H+
MeO
HO
OH
−H+
(H2SO4)
MeO
OH
OH
58
MeOH/HCl
MeO
OMe
HO
OMe
59
EtO2CCH2CCH2CH2CO2Et
HCl
EtO
MeO
OMe
OMe
60
OH
MeO
61
MeO
62
MeO
OH
63

carbonium center can be developed next to a carbonyl group, no doubt because of the considerable stabilizing effect of the ether oxygen atom). Treatment with allylmagnesium bromide and oxidation with the sodium periodate-osmium tetroxide reagent then supplied the aldehyde **65** as a mixture of stereoisomers:

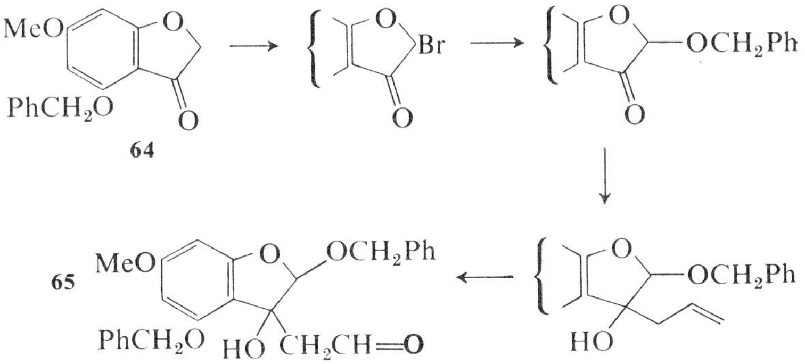

MeO, PhCH$_2$O, O, O, **64** → Br, O → OCH$_2$Ph, O ↓ OCH$_2$Ph, HO ← **65** MeO, O, OCH$_2$Ph, PhCH$_2$O, HO, CH$_2$CH=O

At this point the benzyl group had to be detached without disturbing either the aldehyde group or the angular hydroxy group, which is itself benzylic. This delicate manoeuvre was accomplished catalytically with the help of palladium-charcoal in ethyl acetate. Cyclization was spontaneous, as expected, and acetylation furnished the diacetate **66**, pyrolysis of which in a short-contact, continuous-flow system at 450° gave the vinylic ether **67** in 75% yield.

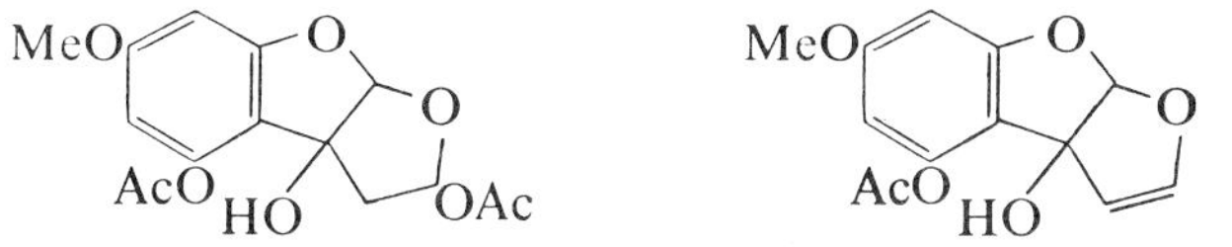

MeO, O, O, AcO, HO, OAc

66

MeO, O, O, AcO, HO

67

CO$_2$Et, O, R

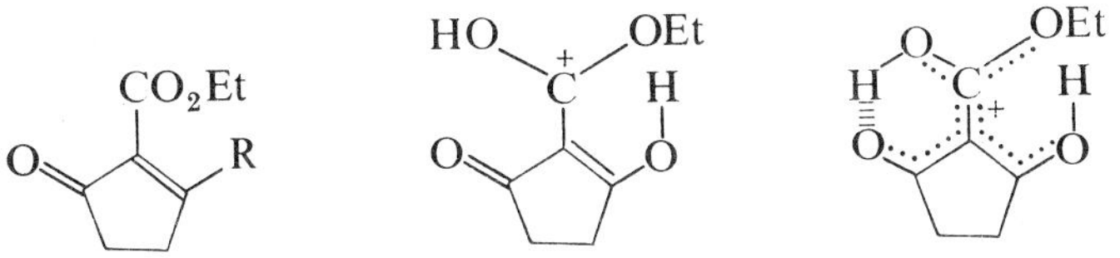

68a R = OH
68b R = Cl

HO, OEt, $\overset{+}{C}$, H, O, O

69

O, OEt, H, C, +, H, O, O

70

Phloroglucinol esters being susceptible to hydrolysis, the phenol was obtained from the acetate **67** and sodium hydrogen carbonate and attachment of the coumarin system investigated. At first sight all that is required is a Pechmann condensation with the cyclopentadienone ester **68a**, but this ester is well known to be inert in this respect.[75] We might speculate that protonation would yield ion **69** upon which hydrogen bonding and charge distribution as in **70** would confer considerable stability. However, the derived chloride **68b** readily condenses when zinc carbonate is used both as catalyst and acid scavenger and produces (±)-aflatoxin M_1 **42**, identical as nearly as may be with the natural product.

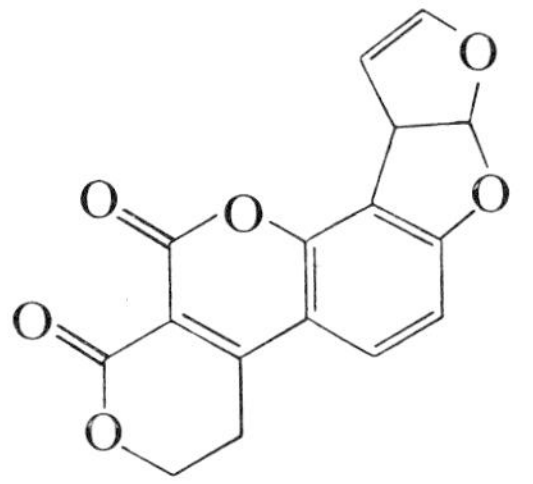

71
Aflatoxin G_2

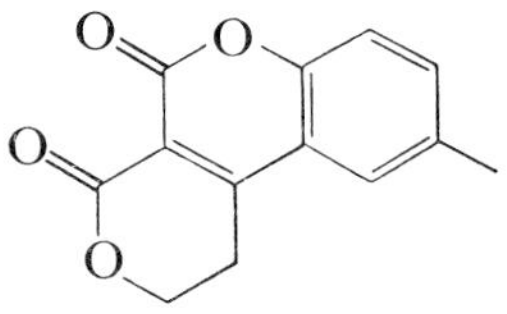

72

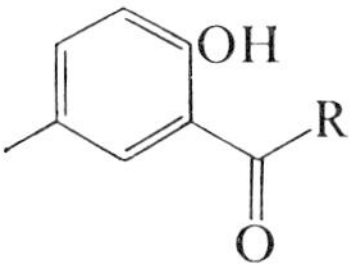

73a R = H
73b R = Me

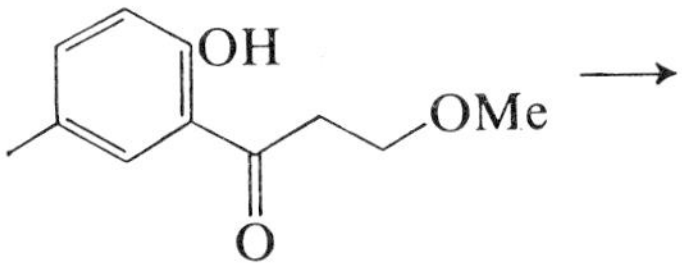

74

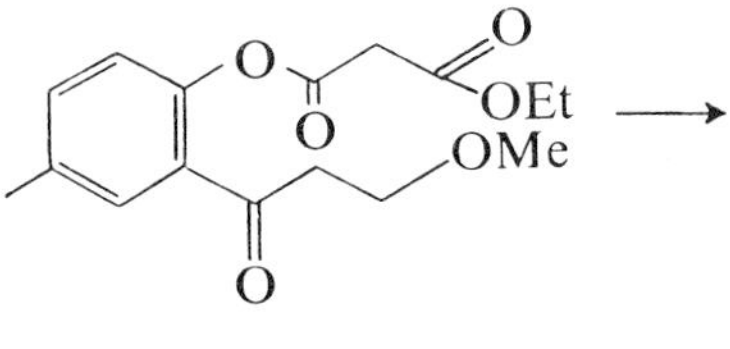

75

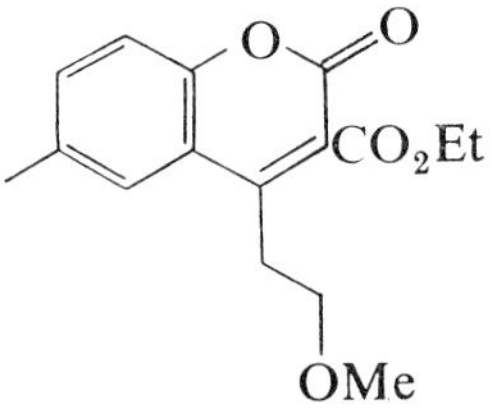

76

Another member of the aflatoxin series, aflatoxin G_2, has structure **71**. Bycroft, Hatton, and Roberts[76] have commenced explorations of routes to structures of this kind and succeeded in preparing lactone **72** as a model for the novel section of the G_2 molecule. Direct condensation between a salicylaldehyde derivative **73a** and, for example, ethyl acetoacetate, is a long-established method of preparing substituted coumarins, but is variable when applied to acetophenones such as **73b** and generally useless when applied to higher homologs. In the present case, the difficulty was cleverly overcome by arranging for entropy factors to counterbalance it. Phenol **74** was esterified by ethyl chlorocarbonylacetate and the product **75** readily cyclized to the coumarin **76** in the presence of piperidine. Cold concentrated sulfuric acid effected a second cyclization supplying the desired model **72**. Unfortunately, corresponding reactions were not observable upon transfer to the phloroglucinol or even to the resorcinol series.

In parallel studies, Rance and Roberts[77] have provided syntheses of (±)-*O*-methylsterigmatocystin **77** and of its dihydro derivative **78**, which was later found to be a natural product in its own right in an optically active form.[78] The only problems not already considered are those implicit in securing a xanthone nucleus. Phenol **51** condensed with methyl 2-bromo-6-methoxybenzoate giving the diphenyl ether **79** when the catalyst was cuprous chloride in refluxing pyridine. When the corresponding acid was treated with oxalyl chloride in dichloromethane it cyclized at once to the desired xanthone, (±)-dihydro-*O*-methylsterigmatocystin **78**, which is a tribute to the nucleophilicity of the phloroglucinol nucleus. For the synthesis of (±)-*O*-methylsterigmatocystin **77**, a xanthone ring was attached similarly to the acetal-ester **59**, the first stage giving **80**. Alkaline hydrolysis and acid treatment led to the lactone **81** converted by the oxalyl chloride method into the xanthone **82**. Then selective reduction by disiamylborane, acetylation, and pyrolysis served to complete the task.

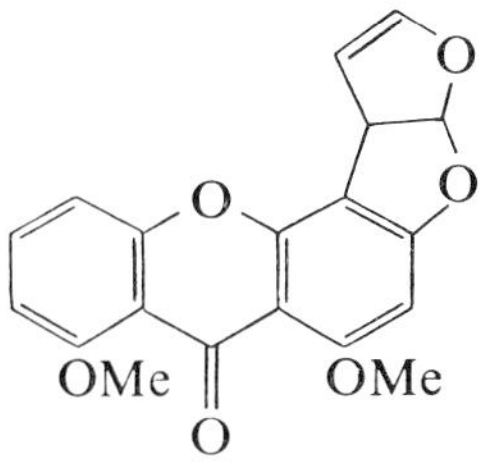

(±)-*O*-Methylsterigmatocystin
77

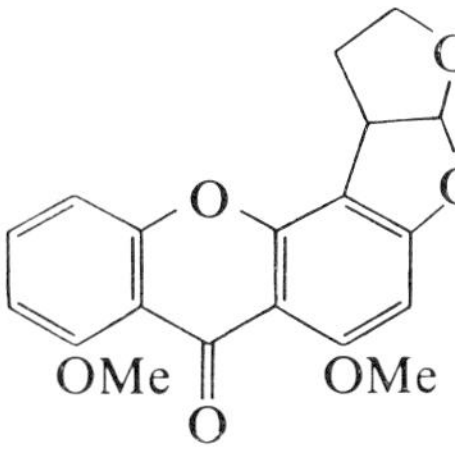

Dihydro-*O*-methylsterigmatocystin
78

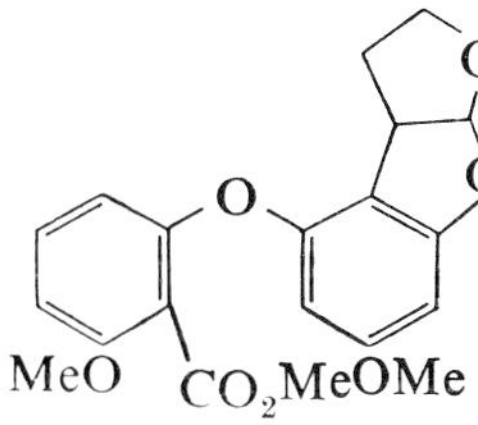

79

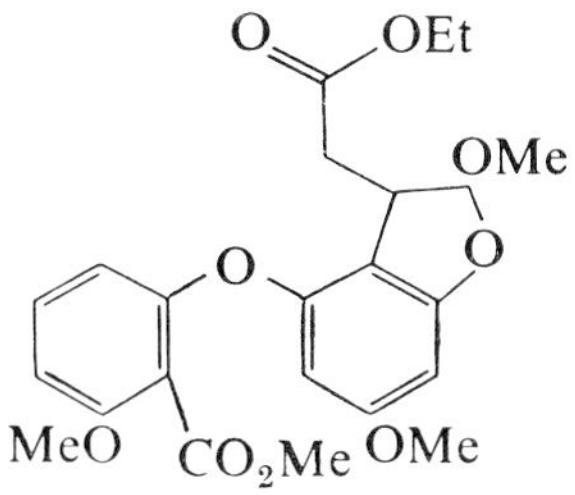

80

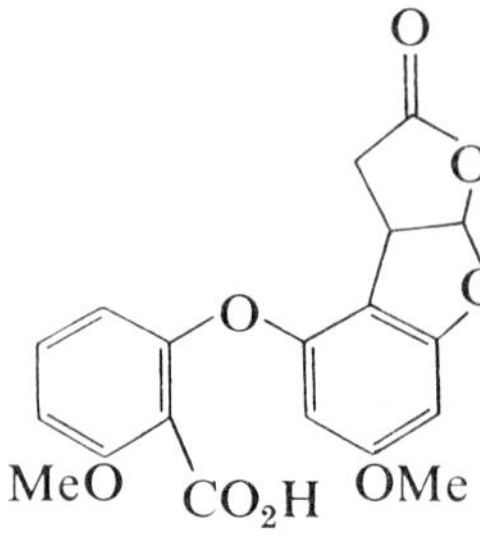

MeO CO_2H OMe

81

MeO OMe

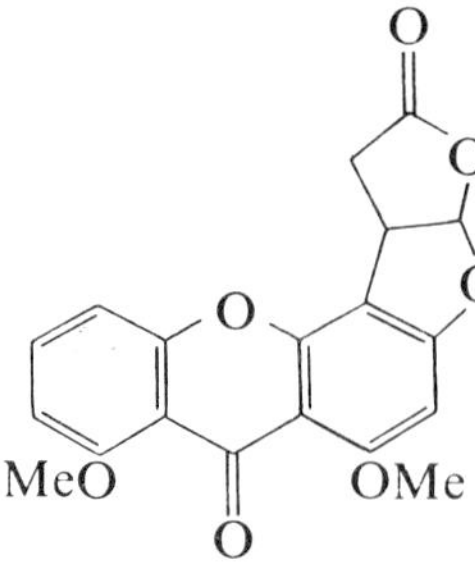

82

The anthraquinone, aversin **83a**, is of particular interest because anthraquinones of this type are likely to be the biogenetic precursors of the aflatoxins and sterigmatocystins.[79] It has been obtained[80] from the mold *Aspergillus versicolor* (Vuillemin) Tiraboschi, which also produces the versicolorins and divers related compounds, and its methyl ether **83b** has recently been synthesised[80] by a route that takes advantage of one of the newest techniques for building up hydroxyanthraquinone derivatives.[81]

Commonly, anthraquinones are prepared by Friedel-Crafts condensation of an aromatic substrate with a derivative of phthalic anhydride, the first stage being the formation of a benzoylbenzoic acid. This is often easily accomplished, but the next step, the cyclization forming the quinone ring, now demands electrophilic attack on an electron-deficient system and is often impossible without conditions of a stringency that precludes the method here. The new technique utilizes basic media,[81] and the present example starts at phenol **51** which, with oxalyl chloride in methylene chloride, furnishes the lactone **84**, treatment with methanol then giving the ester **85**, and methylation followed by hydrolysis the glyoxylic acid **86**. Such acids are not of great stability and in many of their reactions lose carbon monoxide. Thus thionyl chloride produces the acid chloride **87**, and a Friedel-Crafts acylation with 3,5-dimethoxybenzyl cyanide as substrate is now possible, the product being the benzophenone derivative **88**.

The cyclization to the cyanoanthracene is effected with methoxide ion in dimethylformamide or in dimethylsulfoxide. It is presumed that a carbanion is formed and extrudes a methoxy group by an addition-elimination mechanism made possible partly by the presence of the benzophenone carbonyl group and partly by the fact that the aromaticity of phloroglucinol derivatives renders them but little more stable than conjugated but nonaromatic analogs would be. Either methoxy group may be extruded, so the cyanoanthracene **89** is accompanied by the isomer with angular annelation. The isomers are readily distinguished by spectroscopic means, and the requisite

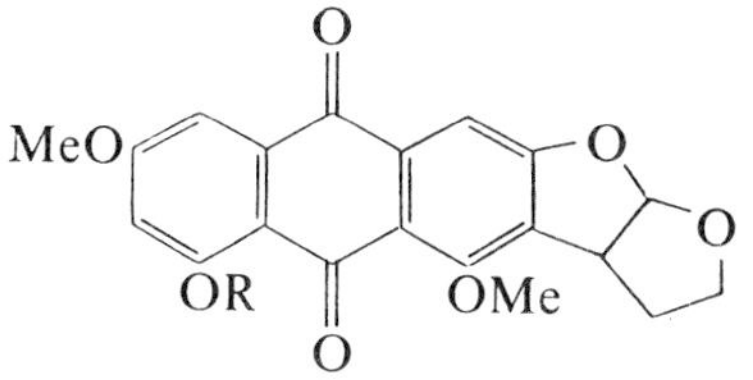

83a R = H; Aversin
83b R = Me

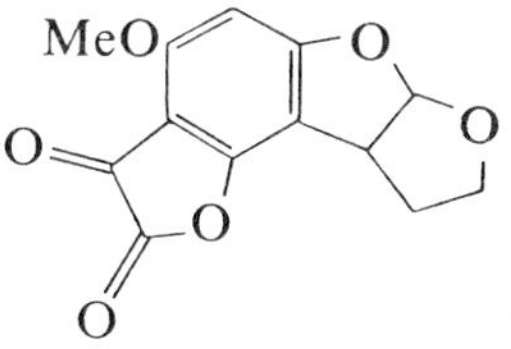

84

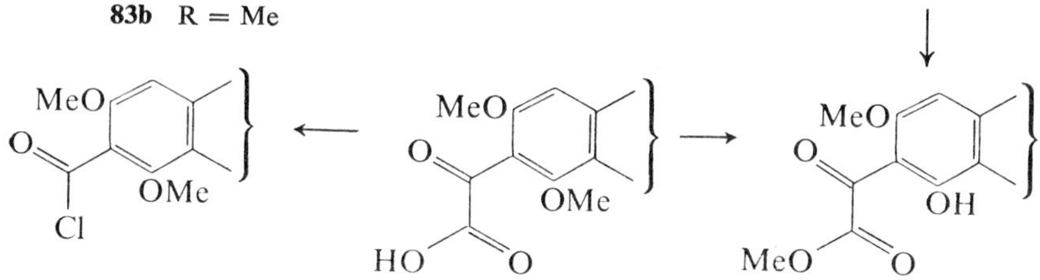

85

86

87

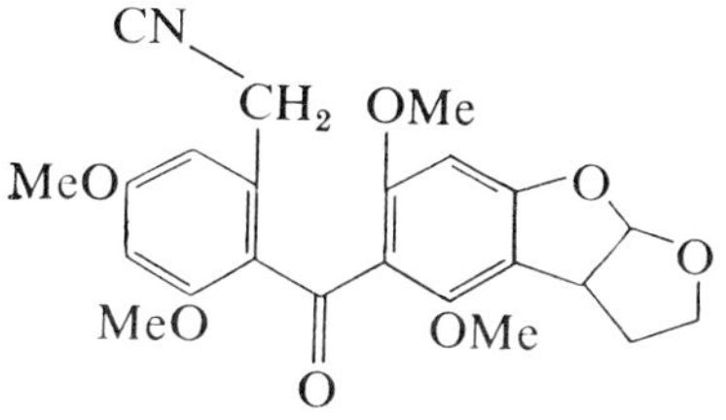

88

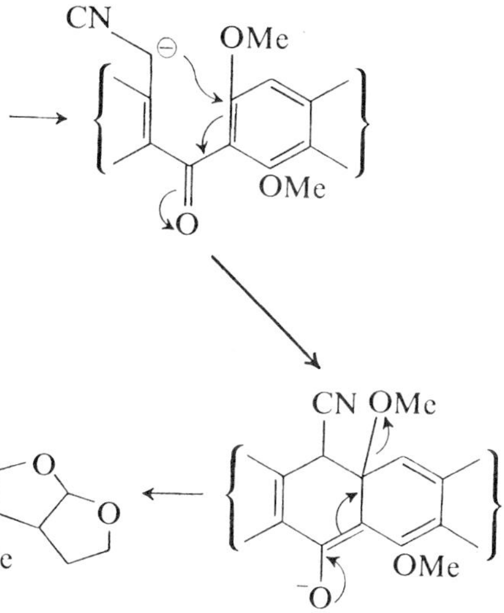

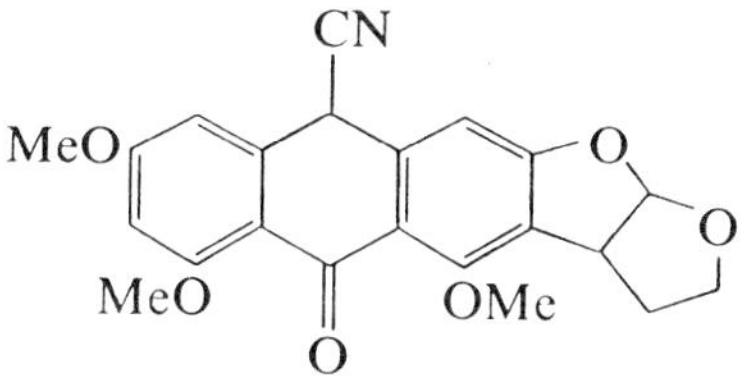

89

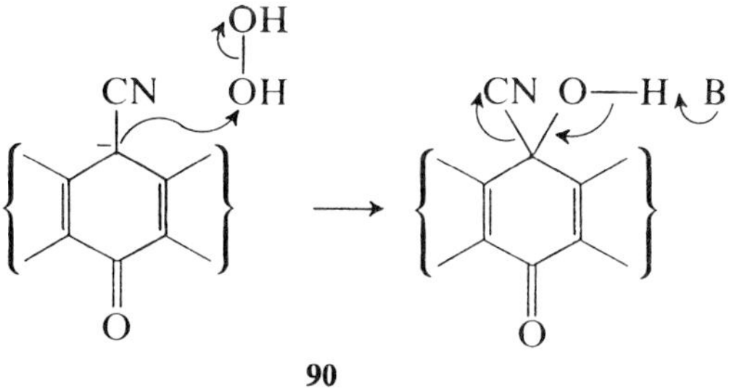

90

83a R = H; Aversin
83b R = Me

84

87 **86** **85**

88

89

90

one oxidized by alkaline hydrogen peroxide to aversin methyl ether **83b**. Probably the last reaction involves hydroxylation-elimination sequences as suggested in diagram **90**.

4. THE ROTENOIDS

With structure **91**, rotenone is at once the most important and nearly the most complex member of this group of plant products, munduserone **92** being the simplest, and all being characterized by what has usually been called a chromanochromanone nucleus.* The rotenoids are physiologically active, stunning fish and killing aphids and other plant pests, and are the active constituents of derris dust. Several synthetical routes are now available, although to date no fully satisfactory synthesis of rotenone itself has been described and no attempt has been made to synthesise amorphin **93**, the only glycoside yet discovered in this series.[82]

The foundations of rotenoid synthesis[84] were laid before 1940 by numerous contributions, often overlapping, from several distinguished groups of chemists; and we are forced to take much of this work for granted in order to concentrate on the more recent results. Throughout the earlier studies,

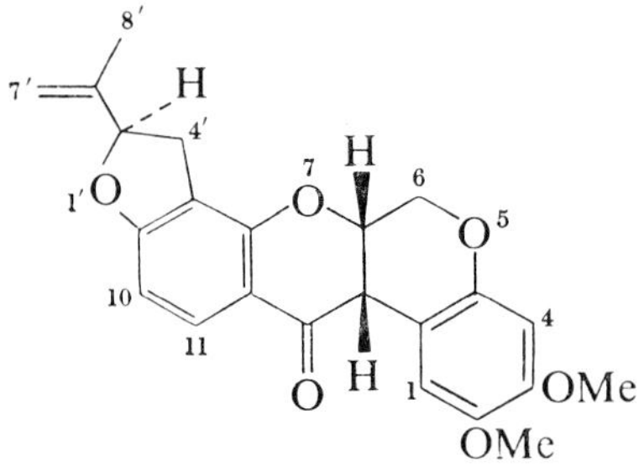

91
Rotenone

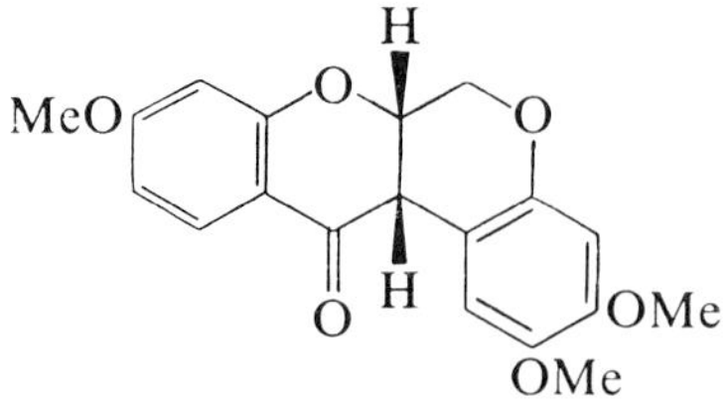

92
Munduserone

* For rotenone, *Chemical Abstracts* gives the name 1,2,12,12*a*-tetrahydro-2-isopropenyl-8,9-dimethoxy[1]benzopyrano[3,4-*b*]furo[2,3-*h*][1]benzopyran-6(6*aH*)-one. Another system of nomenclature is based upon the parent structure rotoxen:[83]

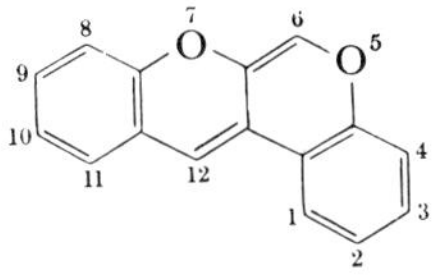

Rotoxen

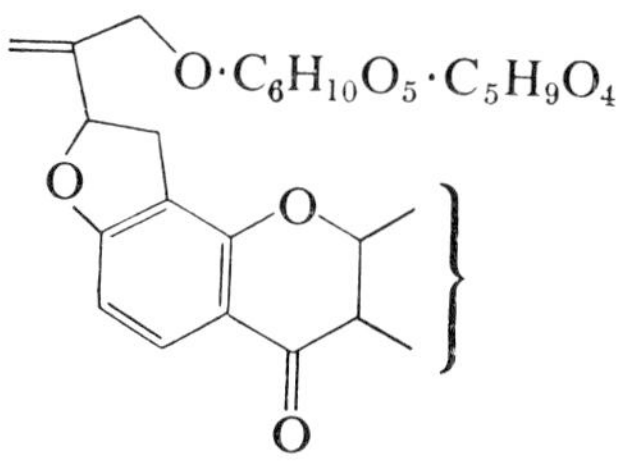

$O \cdot C_6H_{10}O_5 \cdot C_5H_9O_4$

93
Amorphin

ideas were dominated by the fact that gentle oxidation (iodine and sodium acetate is the favorite reagent, but manganese dioxide and ferricyanide are also used) of the chromanochromanone nucleus dehydrogenates it to the chromenochromone nucleus:

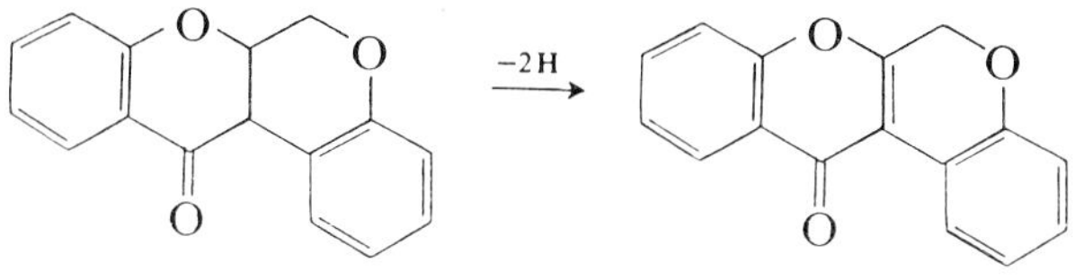

$\xrightarrow{-2H}$

The importance of this is that the new nucleus is readily dismantled by alkaline hydrolysis giving products that are relatively accessible synthetically:

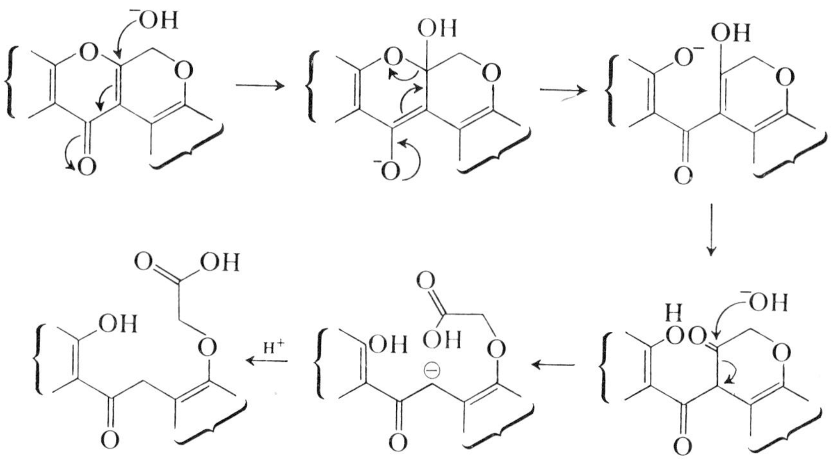

In reverse, such a sequence constitutes a general route to synthetic rotenoids. It has been usual to start with a suitably substituted salicylaldehyde and convert it via the azlactone synthesis into a benzyl cyanide:

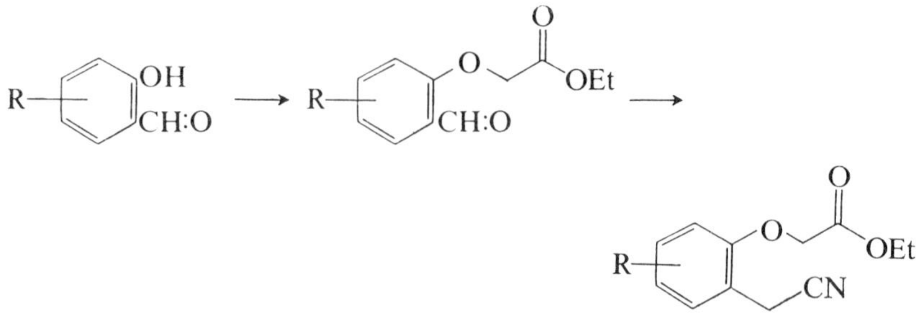

The next step, a particularly critical one, is the Hoesch condensation between the cyanide and a phenol which, since it has to be derived from resorcinol, phloroglucinol, or pyrogallol, is highly nucleophilic and reacts under the influence of hydrogen chloride with only mild catalysts such as zinc chloride or, sometimes, with no catalyst at all. The resulting imine is hydrolyzed by warm aqueous acid to the desired ketonic acid:

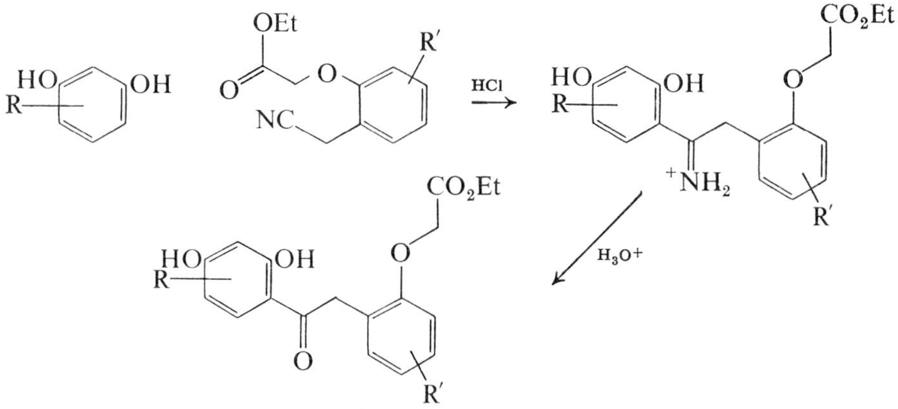

Although this approach to such ketonic acids is in some respects clumsy, it is fair to say that it is nearly always successful. An alternative, in which the phenylacetic acid itself is condensed with the phenol by means of polyphosphoric acid, is simpler and has been used as shown as part of a synthesis of munduserone,[85] though it is ambiguous and failed in another case:[86]

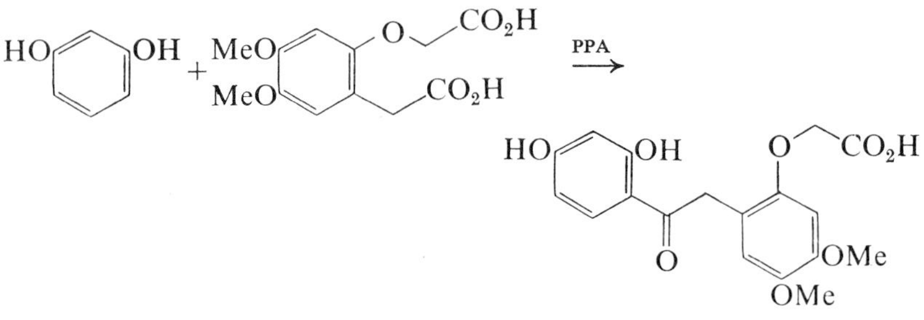

The dual cyclization necessary to convert the ketone into the chromenochromone is effected in one step by heating with acetic anhydride and sodium acetate: if the phenoxyacid group is esterified, the cyclization fails so it is likely that a mixed anhydride is an essential intermediate:[86,87]

Ac_2O, NaOAc

$-H_2O$

Dehydrodeguelin
94

95

In the absence of labile substituents the yields may be as high as 50%; and the method has sufficed for the synthesis[88] of racemic deguelin, for example, via dehydrodeguelin **94**. Occasionally the ester is cyclized instead of the acid, but the reagent is then sodium ethoxide and the reaction constitutes, in part, a Dieckmann condensation[85] giving the aroylchroman-3-one **95**, which cyclizes easily.

The problem of reducing the chromenochromone nucleus to the chromanochromanone nucleus had barred progress. Though hydrogenation can be successful[89,90] it is not easily controlled and, in any case, nearly all rotenoids contain other reducible double bonds that would be attacked. Matsui and Miyano[91] solved this problem with sodium borohydride, which reduces not only the carbonyl group but also the olefinic bond leaving a chromanol which is commonly oxidized to the required ketone by the Oppenauer method or, sometimes, by manganese dioxide used under conditions minimizing the further dehydrogenation leading back to the chromenochromone, (as shown on top of next page).

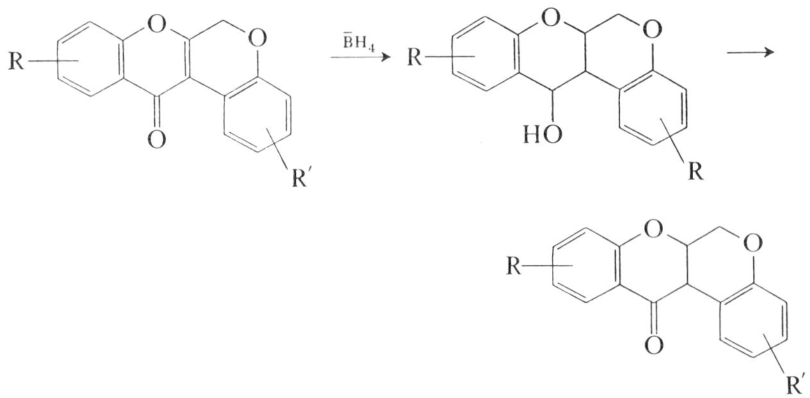

This reduction of an olefinic bond by borohydride ion may be understood in terms of a 1,4-addition which is greatly promoted because it leads to an enolic stilbene system with very extensive conjugation. Simple carbonyl reduction would result only in much decreased conjugation. After ketonization of the enol the phoenix carbonyl group can be reduced in a second step:

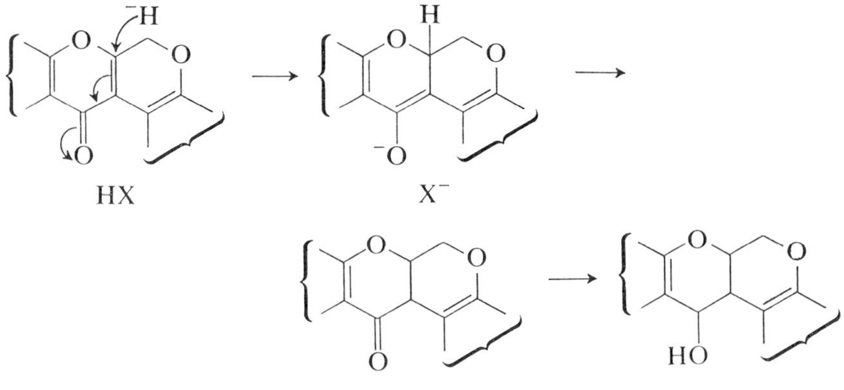

In theory, the ring fusion in the chromanochromanone nucleus can be *cis* or *trans*. However, the natural compounds are all *cis* fused and it is *cis* fusion only that is found in the synthetic materials. Models show[90,92] that in the *trans* arrangement there is an unavoidable collision between the carbonyl oxygen atom and the hydrogen atom at position 1 on the neighboring aromatic ring. In the more flexible *cis* arrangement there need be no such collision. Since there is probably little difference otherwise in energy between the *cis* and *trans* arrangements, this factor is enough to account for the somewhat unusual result.

To synthesize rotenone itself by this route, the half nitrile **96** of derric half ester has to be condensed with tubanol **97a** by means of hydrogen chloride, but attempts to do this directly fail because tubanol is sensitive to acids and can add hydrogen chloride, suffer scission of the allylic ether system, and isomerize into the more aromatic and more stable (but still sensitive) furan **97b**. The difficulty was minimized by the use, instead of tubanol, of the alcohol **98** prepared by the sequence shown from the benzofuran **99**. The Hoesch condensation then furnished a mixture of products from which the derrisic ester analog **100** was isolated in less than 8% yield and dehydrated with phosphorus tribromide in cold pyridine, this step giving the methyl ester **101** of (±)-derrisic acid in about 10% yield. The acid itself was needed for resolution, and the rather stringent conditions (potassium hydroxide in refluxing ethanol) reduced the yield yet further, only 10 mg of half-crystalline acid being obtained from 70 mg of crude ester. In these

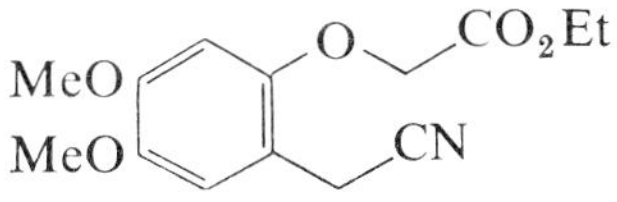

MeO MeO O CO_2Et CN OH OH

96 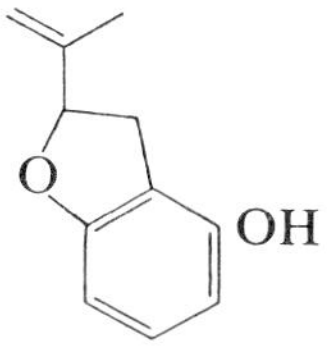 **97a** 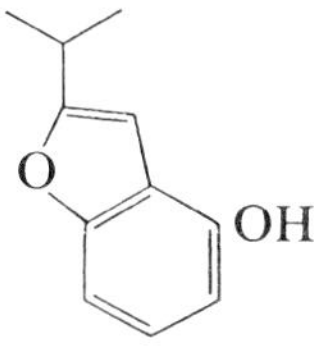 **97b**

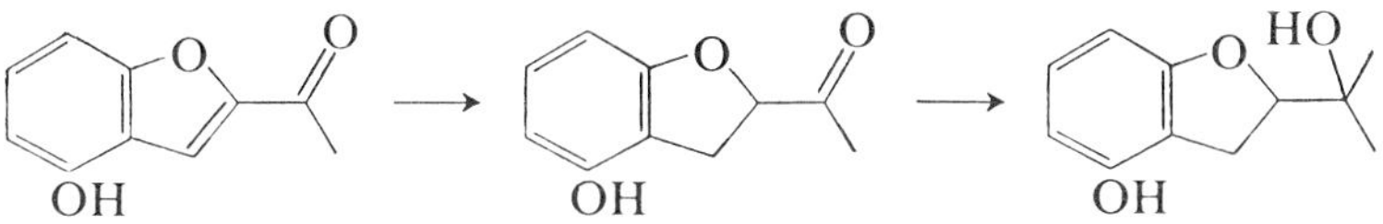

O O O O O HO OH OH OH

99 **98**

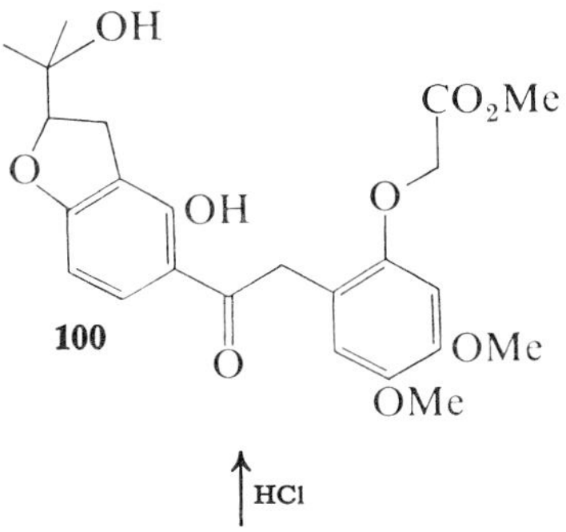

OH O OH O CO_2Me OMe OMe O **100**

↑ HCl

96 + 98

O OH O CO_2Me O OMe OMe

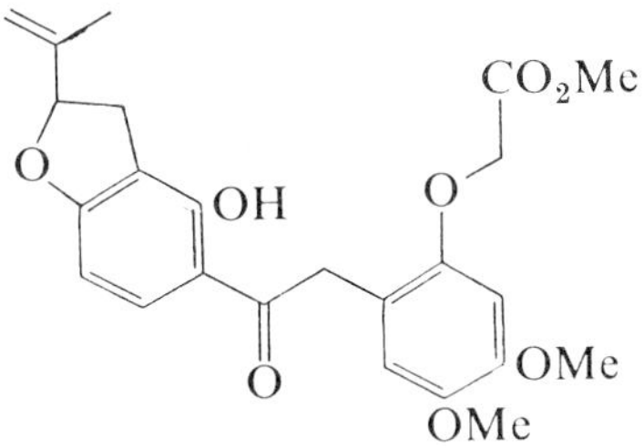

101

circumstances resolution could not be attempted, and the ingenious course was adopted of effecting an inverse resolution. Active derrisic acid, obtained by degradation of rotenone **91**, was observed to resolve (±)-1-phenylethylamine, one salt separating from ethanol. Consequently, the resolution of the synthetic derrisic acid by active phenylethylamine must also be possible, though it was not actually demonstrated. The final stages of rotenone synthesis had all been demonstrated earlier using materials from degradation; that is, cyclization of (optically active) derrisic acid to the chromenochromone (dehydrorotenone), borohydride reduction, and Oppenauer oxidation.

This synthesis of rotenone is hardly satisfying, and in a sense it is formal rather than actual. At the moment, however, it is the *only* synthesis, other routes not having been fully worked out. Moreover, the route has been greatly improved recently[93] by the use of dicyclohexylcarbodiimide (DCCD) to effect cyclization of derrisic acid **102** to dehydrorotenone **103**. No doubt this cyclization also involves a mixed anhydride[86] but it does not require acidic conditions so although a number of by-products complicate the issue, the propenyl grouping survives and the yields are good.

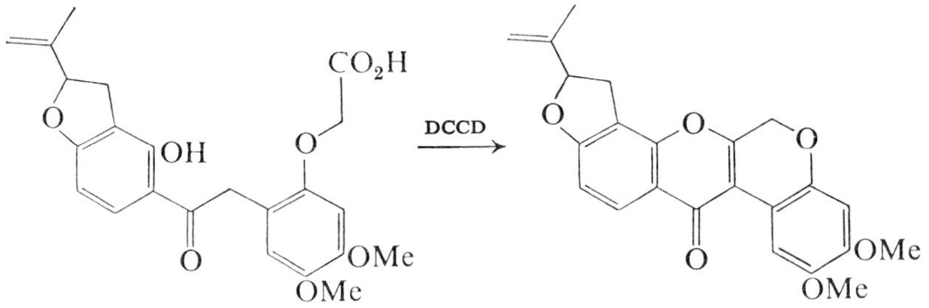

CO_2H OH O O OMe OMe DCCD O O O O OMe OMe

Derrisic acid

102

Dehydrorotenone

103

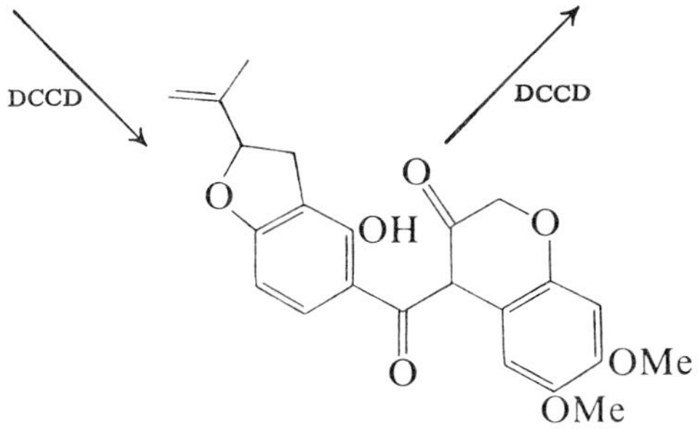

DCCD DCCD

O O OH O O OMe OMe

104

The postulated intermediate aroylchroman-3-one **104** above suggests a way of assembling the rotenoid skeleton so as to avoid entirely the Hoesch condensation and its attendant difficulties. Heated with sodium acetate and acetic anhydride, derric acid **105** readily cyclizes to the requisite chromanone **106**, which at once forms the enol acetate[94] **107**. Herbert, Ollis, and Russell[95] used piperidine to regenerate the chromanone from the acetate and simultaneously catalyze its condensation with 2-hydroxy–4-methoxybenzaldehyde. Catalytic hydrogenation completed a synthesis of munduseran **108**; unfortunately, no method of oxidizing this compound to the rotenoid

Derric acid
105

$-H_2O$
$-CO_2$

106

107

$C_5H_{10}NH$

108

Munduserone
109

munduserone **109** has yet been found. Of interest is the specificity in the direction of enolization and in the point of condensation. Conjugation with the benzene ring rather than the cyclic oxygen atom seems to control these reactions.

Here again DCCD has provided most striking improvements. In the form of its pyrrolidine enamine, dimethoxychroman-3-one **106** reacts with tubaic acid and DCCD giving dehydrorotenone directly;[96] and if the acid chloride **110** of tubaic acid is used with the enamine **111**, the yield of dehydrorotenone **112** runs as high as 15% instead of the disheartening 0.0023% for equivalent stages in the original method utilizing Hoesch condensations. Again an important feature is the avoidance of acids that would disrupt the propenyl grouping of rotenone.

O OH Cl C O $C_5H_{10}N$ O OMe OMe

Tubaic acid Chloride **110**

111

O O O O OMe OMe

Dehydrorotenone **112**

The thermal condensation techniques explored by Mentzer and his colleagues[97] provide another variation of this route that has not yet been evaluated fully. The essential points can be summarized schematically:

O CO_2Et CO_2Et $^-$OEt O CO_2Et O + O O CO_2Et

113 **114**

resorcinol at 160°; 25 hr

HO O O O + HO O O O

Phloroglucinol can replace resorcinol but, as always in this technique, it is best to etherify all hydroxy groups not actually needed for the reaction. A multiplicity of free hydroxy groups allows side reactions and tends to favor condensation modes leading to 2-pyrones (coumarins) at the expense of the required 4-pyrones. Another trouble is that the isomeric 3-oxo-esters **113** and **114** cannot be separated easily so a mixture is used and each component gives its own series of condensation products.

Originally suggested by taxonomic considerations and later established beyond doubt by experiments with tracers, the idea that the rotenoids are, biosynthetically, modified and elaborated isoflavones has stimulated a number of efforts to achieve a laboratory synthesis along similar lines. The chemistry of elliptone offers several examples. The first approach was to obtain a 2-bromomethylisoflavone containing a free hydroxy group at the 2′-position suitable for an internal etherification. Radical-catalyzed bromination of 2-methylisoflavones does yield the desired 2-bromomethyl compounds[98] but because in most cases the protection of sensitive substituents poses formidable problems we here illustrate a more general method reported by Mehta and Seshadri:[99]

HO, OH, OMe, O → ($EtOCH_2COCl$, py (2 steps)) → HO, O, CH_2OEt, OMe, O

↓ HBr

HO, O, CH_2Br, OH, O → (K_2CO_3, acetone) → HO, O, O

The product has been further elaborated by Kawase and Numata,[100] who used the Duff condensation to introduce a formyl substituent (the question of orientation is considered shortly) and the bromomalonate method to complete a furan ring as shown overleaf.

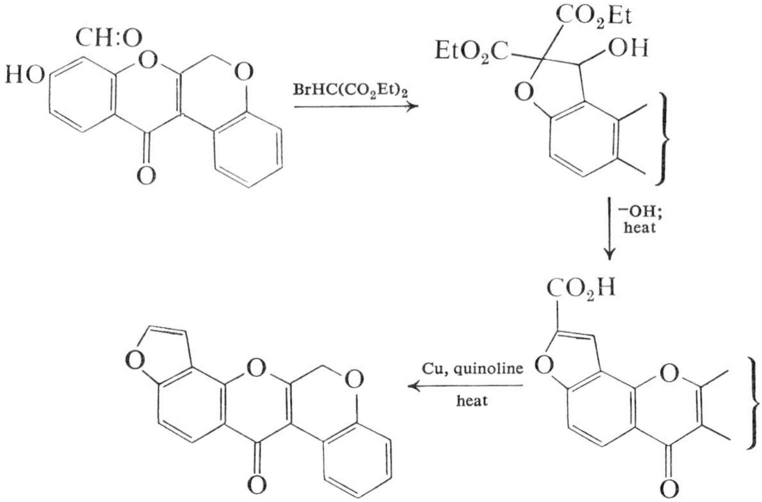

BrHC(CO_2Et)$_2$

−OH; heat

Cu, quinoline heat

This furan is a close relative of the rotenoid elliptone **115a**. The next synthesis from isoflavone precursors illustrates a superior method of providing for fused furan rings.[101] The Claisen migration in **116** is specific in direction and gives **117** because the transition state (or intermediate) represented by **118a** is less cross-conjugated and therefore energetically more favorable than the alternative **118b** needed for migration in the opposite direction. The same influences control similar migrations (and electrophilic

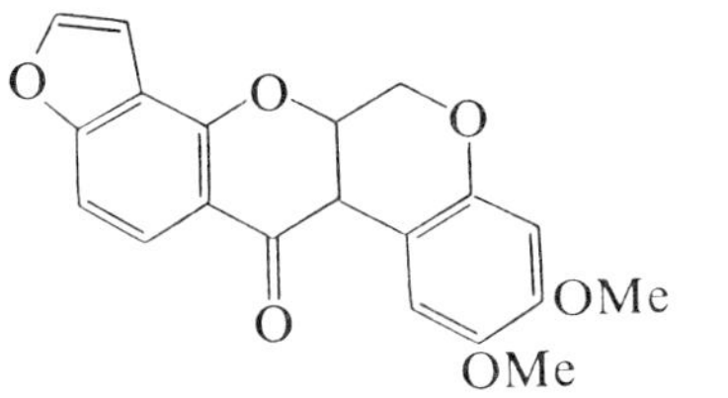

Elliptone **115a**

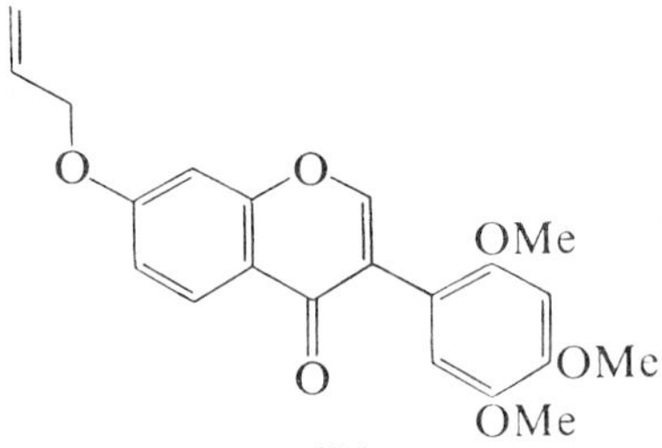

116

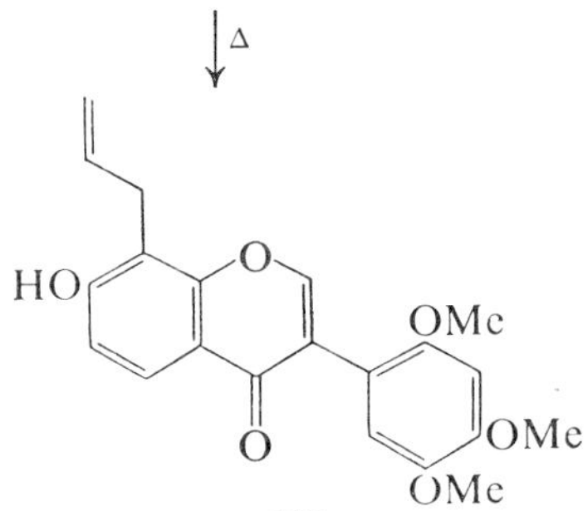

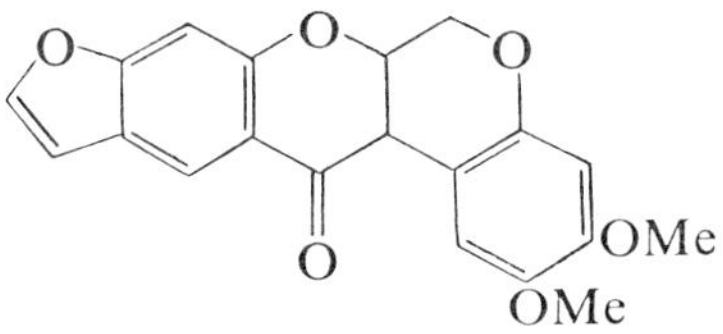

Δ

Isoelliptone **115b**

117

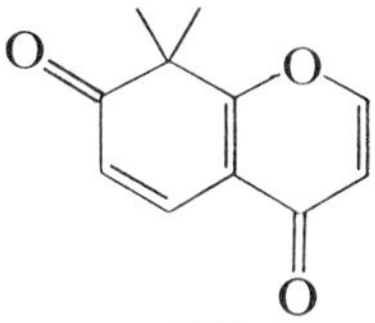

118a

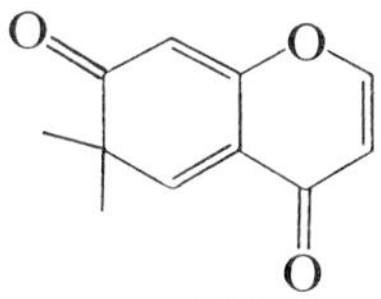

118b

substitutions) whether the second ring is 4-pyrone (as here), 2-pyrone, furan, or benzene, or even not a ring but certain substituents, and consequently it is usually difficult to construct with elegance compounds having annelation of the kind seen in isoelliptone **115b**. One solution is to work with non-aromatic systems and aromatize them at a late stage. Subsequent steps in elliptone synthesis are shown overleaf. An interesting feature is the selective demethylation of the 2′-methoxy group by aluminum chloride in refluxing methyl cyanide.[102] Other solvents are not satisfactory. Presumably the aluminum chloride forms a pyrylium salt with the pyrone grouping and the aluminum atom is then near enough to the 2′-methoxy group to coordinate with it and promote its demethylation, as indicated in diagram **119**. Recently, boron tribromide has been used for the same purpose.[103] The final steps (not shown) are the acetic anhydride-sodium acetate cyclization of elliptic acid **120** to the chromenochromone, dehydroelliptone, and the reduction of this to (±)-elliptone by the borohydride method. No resolution has been reported. Racemic forms of isoelliptone **115b** and of munduserone **92** have been obtained similarly and no doubt the method is of considerable applicability. Notwithstanding that, the method does not solve the problem convincingly since the isoflavone nucleus is used merely as a protective device for one hydroxyl group and is destroyed and has to be reconstituted. (In contrast, similar sequences form an excellent ingress to pterocarpans.) This deficiency has recently been made good by application of sulfur ylide chemistry.

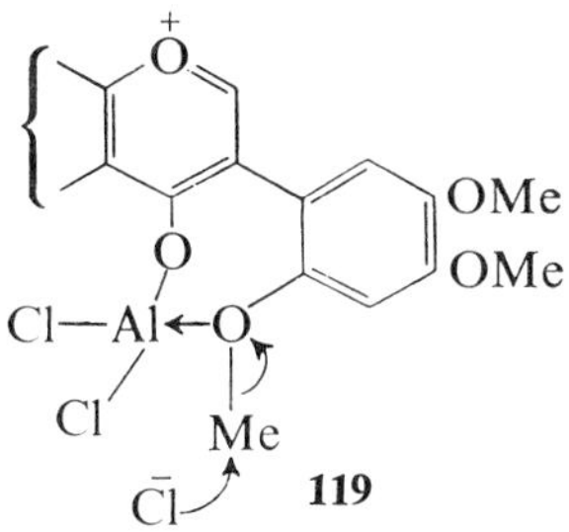

119

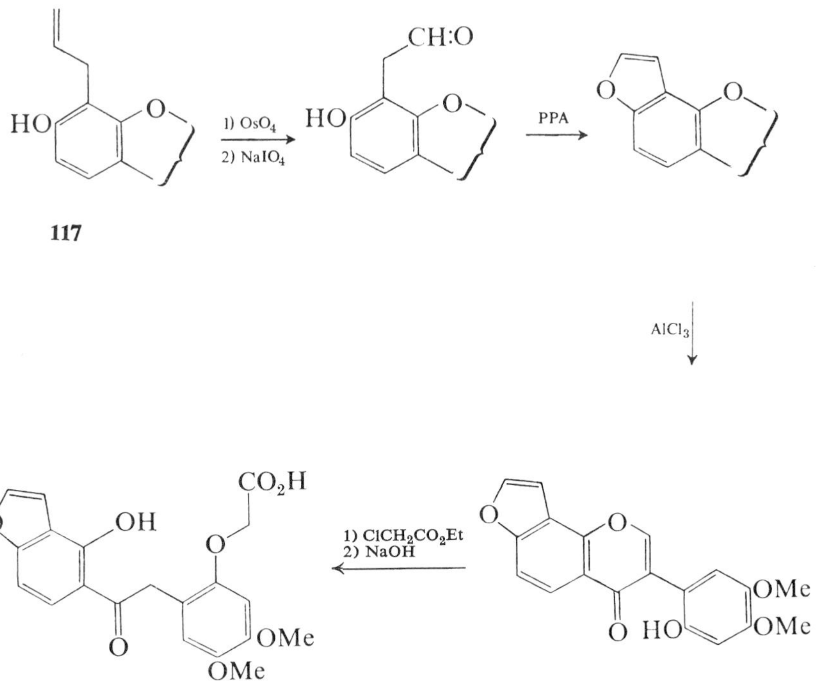

HO CH:O O 1) OsO_4 2) $NaIO_4$ HO O PPA O O

117

$AlCl_3$

CO_2H O OH O 1) $ClCH_2CO_2Et$ 2) NaOH O O OMe OMe O HO OMe O OMe

Elliptic acid

120

When dimethylsulfoxonium methylide adds to the 2,3-double bond of isoflavones it initiates a variety of reactions.[104,105] If an excess of reagent is avoided the result is simple, a cyclopropane derivative being formed:

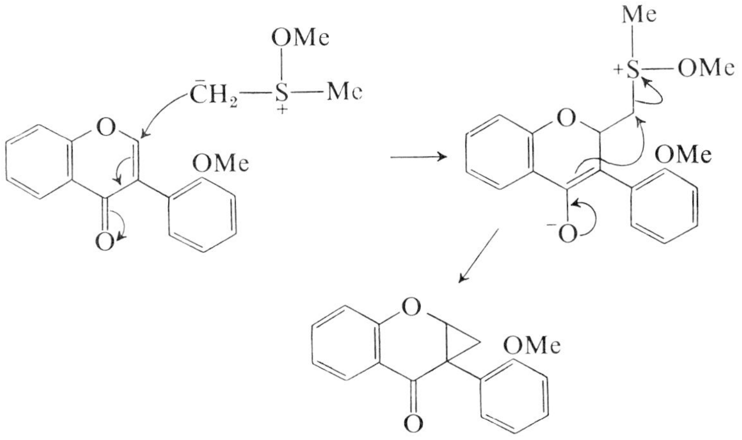

OMe $\bar{C}H_2$—$\overset{+}{S}$—Me O OMe O Me +S—OMe O OMe ^{-}O O OMe O

But if there is a free 2′-hydroxy group, a series of ring openings and closures leads to a vinylcoumaranone system, perhaps as follows, although this scheme allocates no particular role to the important 2′-hydroxy group itself:

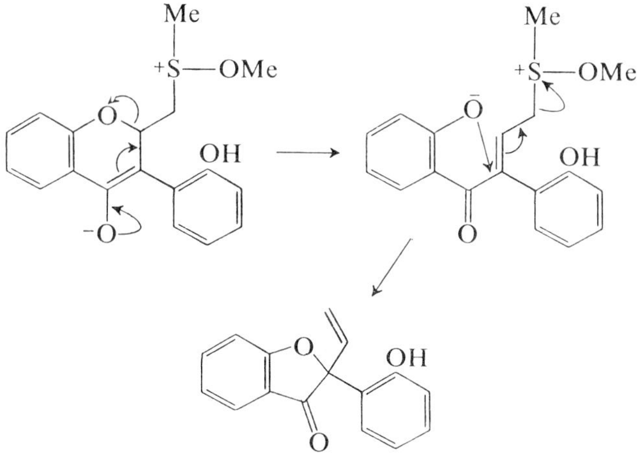

At first sight such vinylcoumaranones seem to be a long way from the desired goal, yet they are actually isomeric with the corresponding chromanochromanones and when heated with pyridine at 100° for 48 hours can be converted into them in high yields (~80%) as illustrated for isorotenone[106] **121**:

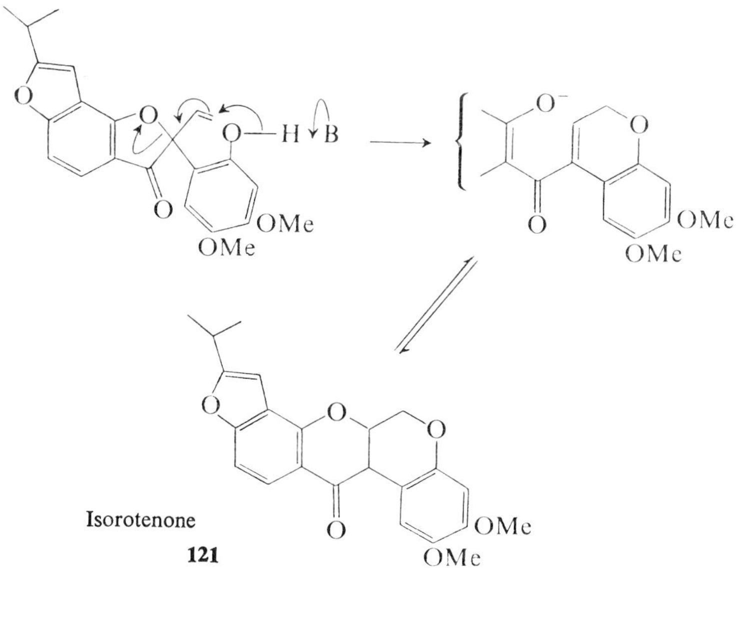

Isorotenone

121

There is still one problem to be solved. In the foregoing elegant method the sulfur ylide provides what eventually becomes the cyclic methylene group. In nature, however, the cyclic methylene group is perhaps provided by attaching the methyl group of the 2′-methoxy group to position 2 in the isoflavone nucleus. No comparable cyclization has been achieved in the rotenoid field, though something like it is known elsewhere.

Isoflavones are prerequisite for much of the work among rotenoids. They are generally obtained from phenol and phenylacetic acid derivatives by connecting these together in Friedel-Crafts style and effecting cyclization by Claisen condensations with ethyl oxalate or with ethyl orthoformate and piperidine[107] (the most popular method nowadays), though others are available. The thallium III oxidation of chalcones[108] is a different approach which was of additional interest when it was thought to simulate a biosynthetical route to isoflavones:

Ph O Ar O $\xrightarrow[\text{MeOH}]{\text{Tl}^{\text{III}}}$ Ph O $CH(OMe)_2$ Ar O $\xrightarrow{H_3O^+}$ O Ar O

An amusing isoflavone synthesis is provided by the following condensation of a chromanone with a quinone,[109] with its repeated prototropic shifts. The product is related to neotenone and the yam bean rotenoids.

$\bar{C}H_2SOMe$ H O O O MeO MeO O O → O O O MeO MeO O^- O

↓

B H O O MeO O $\overset{+}{H}B$ O $\xleftarrow{\text{Second } H^+ \text{ shift}}$ O O O MeO OH OH

Another way of reaching interesting isoflavones is an unusual variation of the Claisen allyl rearrangement applied in this example to an ether made from umtatin, a natural 2-hydroxychromone with the furanoid system found in rotenone[110]:

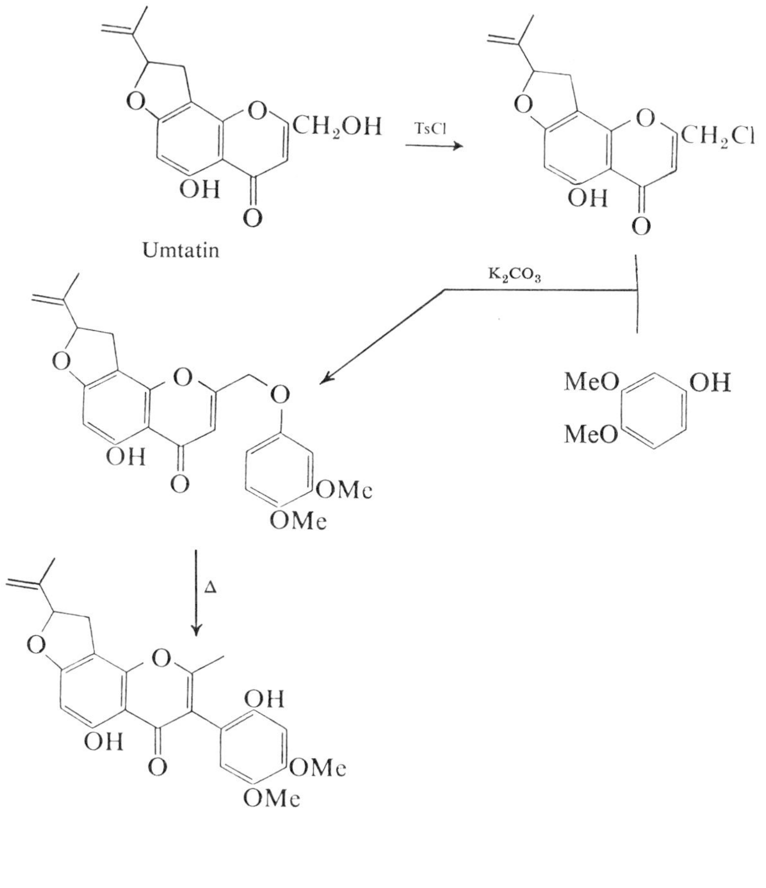

5. COUMESTANS AND PTEROCARPANS

Coumestan is the name of the fundamental ring system **122a** characteristic of a small, well-defined group of natural products related to the isoflavones. Pterocarpan describes the system **122b**, a reduced form of **122a**. The numbering of pterocarpan has been generally agreed for some time and is adhered to, but several systems have been used for coumestan and the one shown is perhaps the most common. Even the name "coumestan" itself is unsatisfactory, and it is surprising that so few authors prefer a more systematic name,* for example, benzofurano(3′,2′:3,4)coumarin.

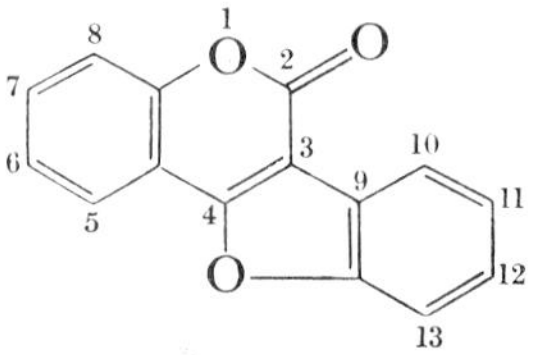

Coumestan

122a

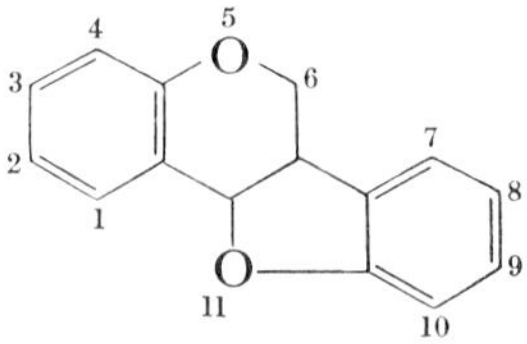

Pterocarpan

122b

It is convenient to begin with syntheses of coumestan derivatives, that of wedelolactone **123** illustrating the first success.[111,112] As with so many syntheses in the isoflavone field, this commences with a deoxybenzoin **124**, sodium powder and ethyl carbonate effecting a Claisen ester condensation and cyclization occurring spontaneously to give, after enolization, a 4-hydroxycoumarin **125** according to the technique introduced by Robertson.[113] The acidity of such coumarins is approximately that of acetic acid and ensures the production of the salt **126**, which is readily isolated as such. Various workers[113a] have described minor modifications of this method, while Deschamps-Vallet and Mentzer[113b] have shown how basic media can be dispensed with altogether if thermal condensation techniques are employed. The arrows show that the salt, notwithstanding its lactonic nature, is stable

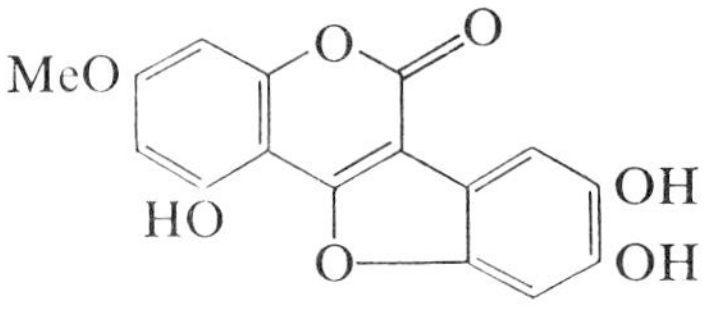

Wedelolactone

123

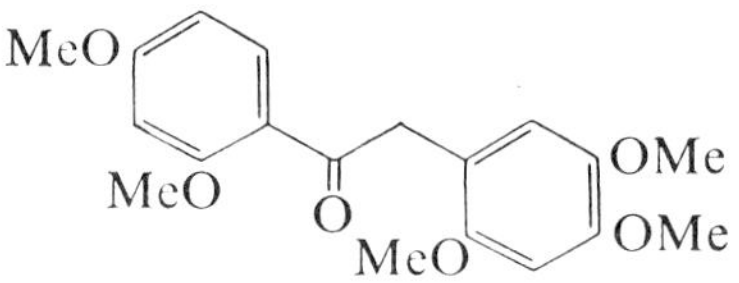

124

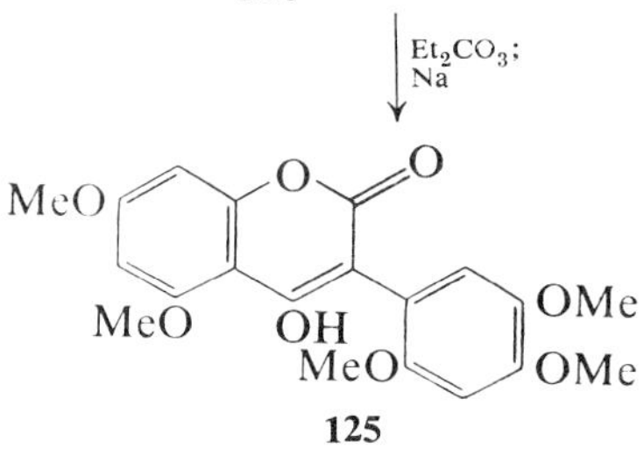

Et_2CO_3; Na

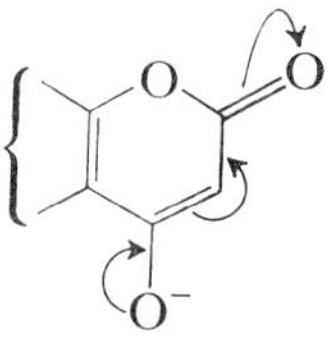

126

125

* *Chemical Abstracts* uses 6*H*-benzofuro[3,2-*c*][1]-benzopyran-6-one, with numbering as for pterocarpan above.

to nucleophilic attack and therefore to the large excess of base used for the reaction. Heated with hydrogen bromide, the 4-hydroxycoumarin **125** cyclizes to the tetramethoxycoumestan **127** directly, so a selective demethylation of one methoxyl group must occur. If a guess may be hazarded, the known basicity of 4-hydroxycoumarins would suggest formation of a pyrylium salt in which the arrangements for demethylation indicated in **128** seem particularly suitable. Careful treatment with hydrogen iodide composes the last step[112] in which **127** yields the monomethyl ether **123**. The interaction shown in **127** tends to protect the one methoxy group from protonation and hence from demethylation by iodide ion; other effects that probably promote demethylation at other positions are discussed on pp. 538, 540.

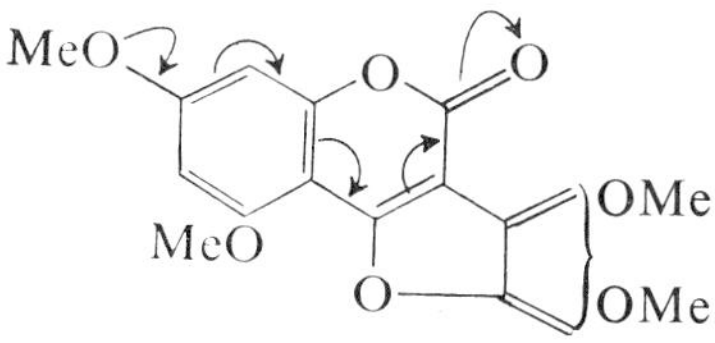

MeO MeO O O O OMe OMe

127

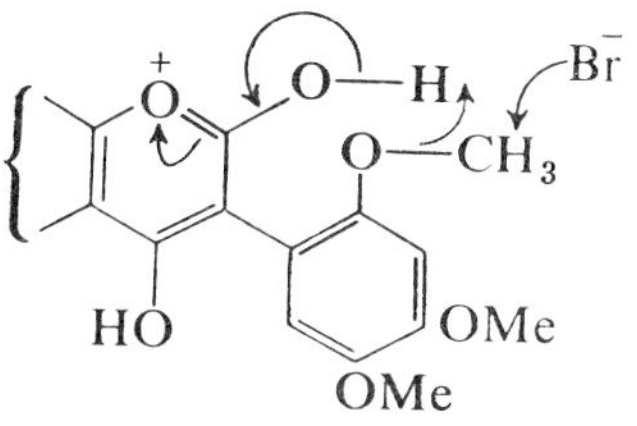

O O—H Br⁻ O—CH3 HO OMe OMe

128

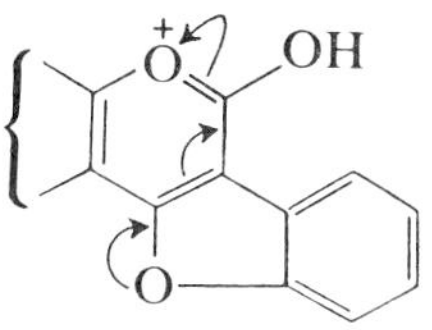

O OH O

129

Another point worth attention is that the wedelolactone kernel is indifferent to mineral acid despite the presence of a benzofuran nucleus which is usually easily protonated and may then undergo various reactions such as ring-opening with water giving a phenylacetaldehyde derivative rapidly resinified by acids. In wedelolactone, defense against such attacks is provided by the pyrone ring, which, as noted previously, is relatively basic. It can be assumed that the first proton is neutralized as the salt **129**; at this point the benzofuran oxygen atom is unable to participate in attaching any further proton except in extreme conditions.

Coumestrol **130** has been synthesized similarly, as have lucernol[114] **131** and derivatives of psoralidine[115] **132**, sativol[114] **133**, and others,[116] with an improvement in the use of the more reactive ethoxycarbonyl chloride instead of ethyl carbonate for preparing the 4-hydroxycoumarin derivatives, and of anilinium chloride for demethylation. Coumestrol is of interest because it has estrogenic activity (it contains a 4,4′-dihydroxystilbene nucleus). The more highly substituted coumestans appear to have little or none, but several have activity against plant pathogens and may be produced in response to fungal attack, and so on, and therefore can be classed as phytoalexins.

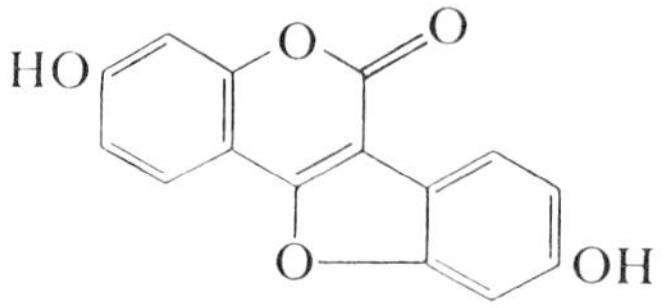

Coumestrol
130

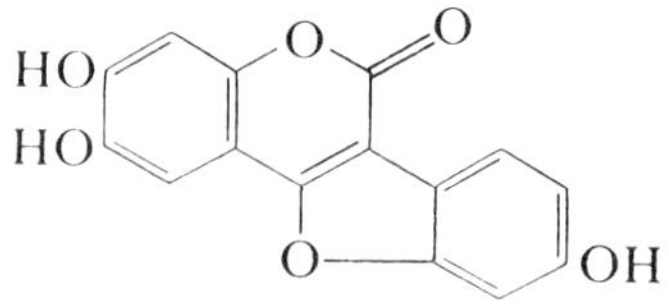

Lucernol
131

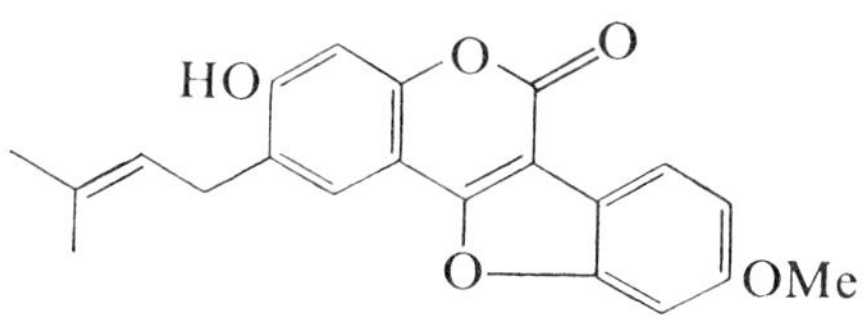

Psoralidine
132

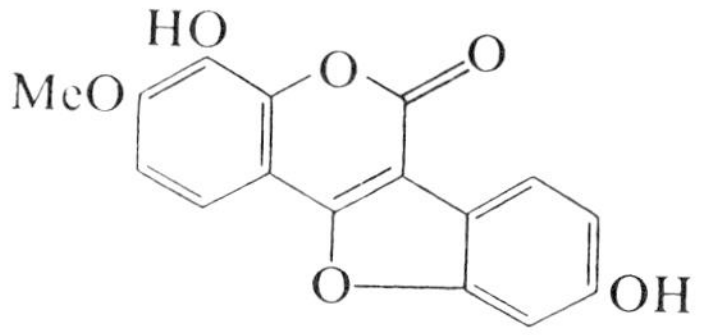

Sativol
133

These physiological activities have intensified interest in the coumestans and in their synthesis. Two elegant and powerful methods, one devised by Wanslick and one by Jurd, may be regarded as having replaced the method outlined.

Wanslick, Gritsky, and Heidepriem[117] demonstrated that oxidation of a mixture of 4,5-dihydroxy-7-methoxycoumarin and catechol with potassium iodate in acetate buffer led to wedelolactone in 85% yield:

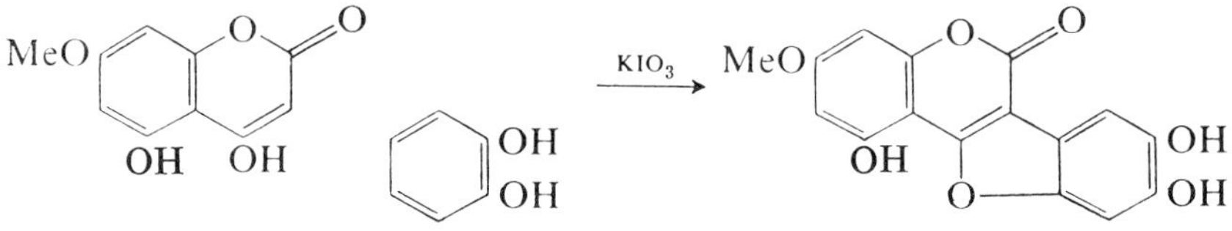

KIO_3

Wedelolactone

Enols other than 4-hydroxycoumarins can be used,[117] and examples have been described involving groups as sensitive as the amino group.[118] As to mechanism, one might exclude radical coupling on the grounds that high yields of unsymmetrical products could never be obtained thus. Since for one component a 1,2-dihydroxybenzene group is essential, 'nascent' 1,2-quinones may be intermediates[118a], the anion of the hydroxycoumarin adding to them; and with a simple variation on the same theme there is no difficulty in accounting for the remarkably ready cyclization that completes the sequence. The method is necessarily restricted to derivatives of 11,12-dihydroxycoumestan but has been used very successfully for numbers of these,[119–122] often with ferricyanide or hypoiodite as an alternative oxidant.

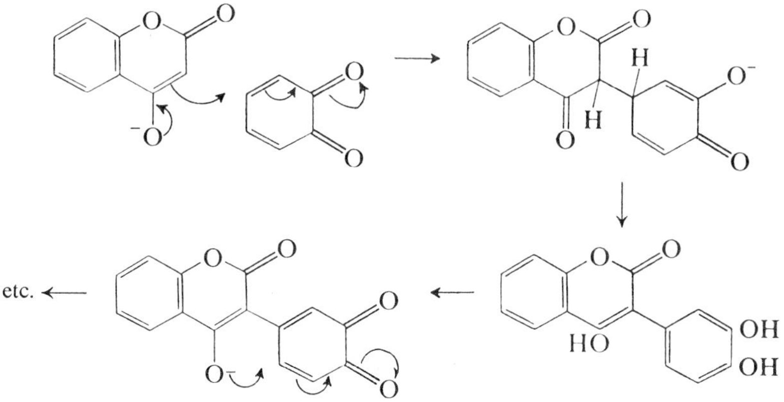

Originally studied in the hope of elucidating the structure of anthocyanins, the hydrogen peroxide oxidation of flavylium salts is a complex matter. The result for acetic acid solutions is merely a ring fission:[123]

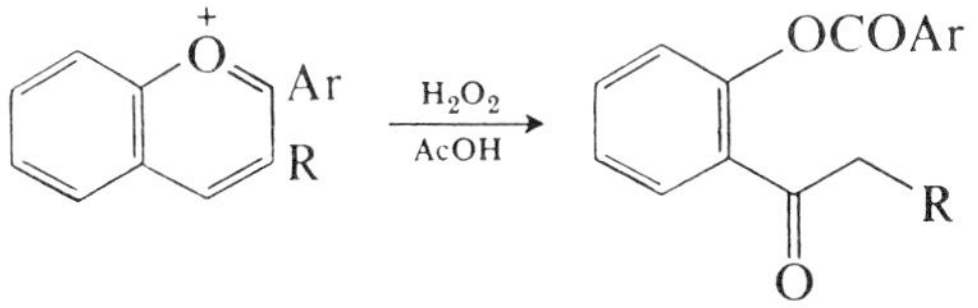

R = Ph, OMe, or $O{\cdot}C_6H_{11}O_5$

In aqueous methanol *p*H 5–7, however, the oxidation readily affords compounds derived from 2-phenylbenzofuran,[124] other studies conducted by Jurd[125] suggesting the following course of events:

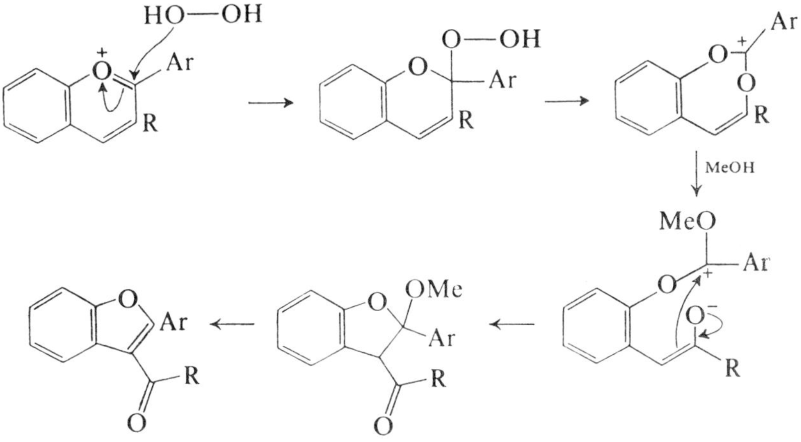

Applied to 3-methoxyflavylium salts containing a 2′-hydroxy group, the oxidation gives first an ester, but this usually rapidly cyclizes to a coumestan derivative:[124,126]

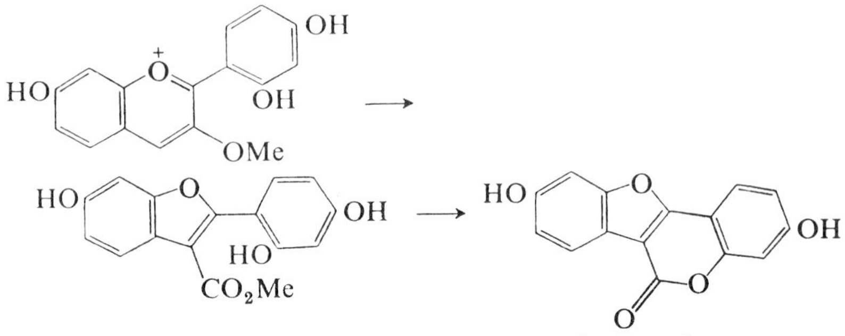

Coumestrol

Since flavylium salts are readily available in considerable variety, the method is a powerful and flexible one. The preceding synthesis supplies coumestrol in 50% yield.[124]

Syntheses of medicagol[127] are useful for comparing this technique with the Wanslick coupling reaction for obtaining certain substitution patterns. The coupling readily gives the trihydroxycoumestan **134**, but attempts at the selective methylenation (by CH_2I_2) required to give medicagol **135** afford, at best, only 5% of this compound.[122] For the flavylium salt route, the components **136** and **137** are needed; these condense very readily to form the salt **138**, which is debenzylated by hot aqueous acid and oxidized to form the coumestan. The oxidation affords a yield of about 40%, but of course the

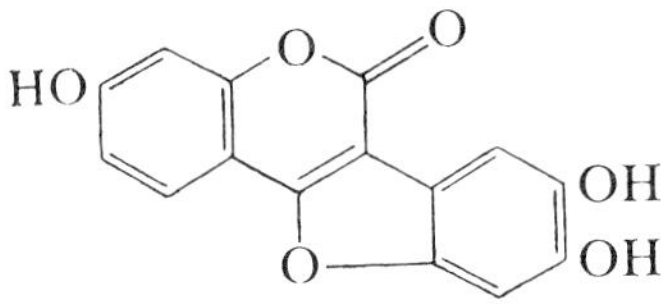

134

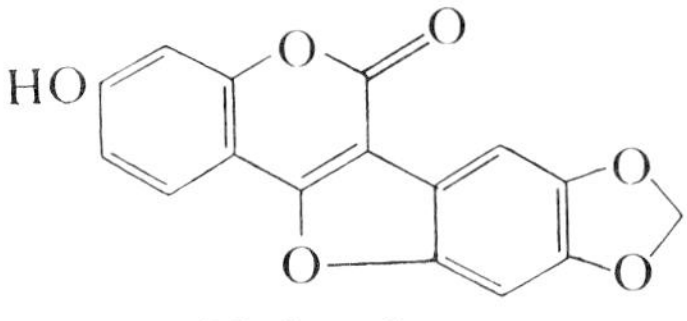

Medicagol
135

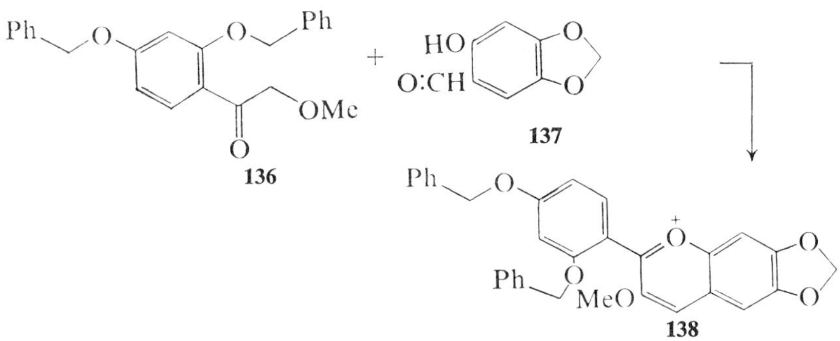

136

137

138

somewhat elaborate nature of the starting materials may be thought to detract from this advantage.

In general, Wanslick coupling offers the better route to derivatives of 11,12-dihydroxycoumestan (which are numerous) but is not adaptable to other hydroxylation patterns, an area where Jurd's flavylium salt oxidation excels.[128,129] Of course the scope of both methods is greatly increased by the proper use of protective devices as will be seen in the following examples, the first of which is a very much improved synthesis of medicagol **135** utilizing Wanslick oxidation:[130]

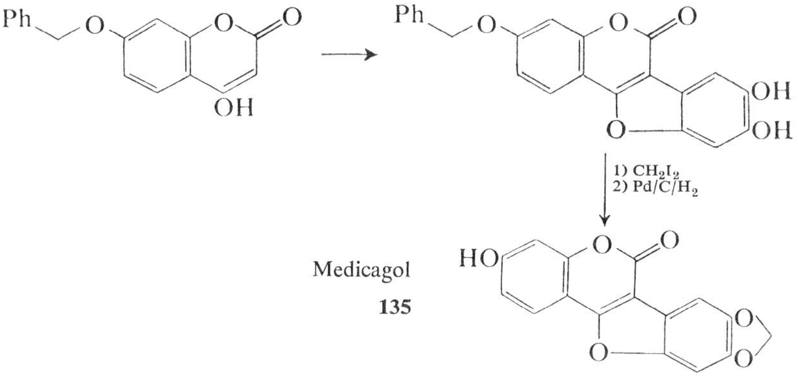

1) CH_2I_2
2) $Pd/C/H_2$

Medicagol
135

Trifoliol **139** presents a difficult problem elegantly solved by flavylium salt oxidation. The solution depends upon the ready availability of the benzoate **140**, which permits preparation of the flavylium salt **141** and thence the benzofuran **142**, which is methylated before the protective groups are removed.[131]

7-Hydroxy–11,12-dimethoxycoumestan **143**, which occurs in alfalfa, can be made[132] from the easily accessible 7,11,12-trihydroxycoumestan by means of a characteristic use of the relatively uncommon protective reagent, dichlorodiphenylmethane:[133]

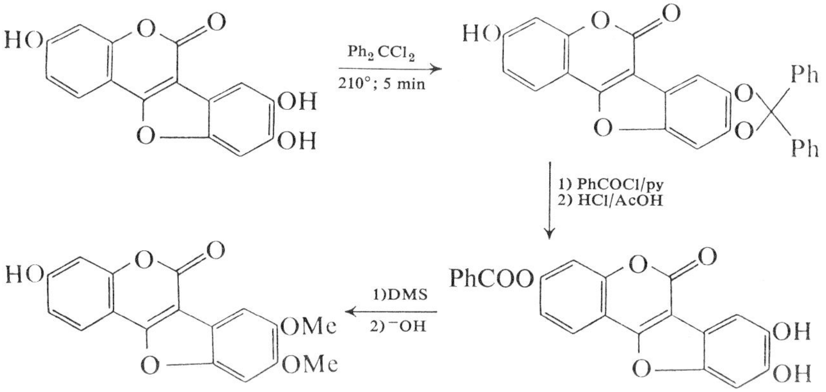

Ph_2CCl_2
210°; 5 min

1) PhCOCl/py
2) HCl/AcOH

1)DMS
2) ^{-}OH

143

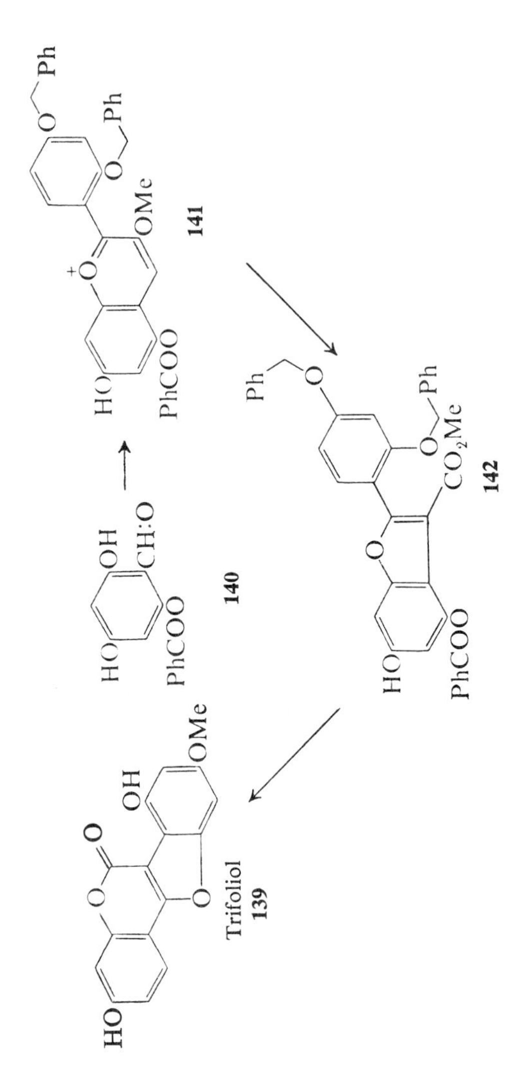
HO
OH
CH:O
PhCOO
140
HO
PhCOO
OMe
O
Ph
141
HO
PhCOO
O
Ph
CO_2Me
142
HO
O
OH
OMe
Trifoliol
139

Wanslick coupling is usually the method of choice if strongly acidic media have to be avoided, as in synthesizing erosnin[120] **144** with its benzofuran moiety:

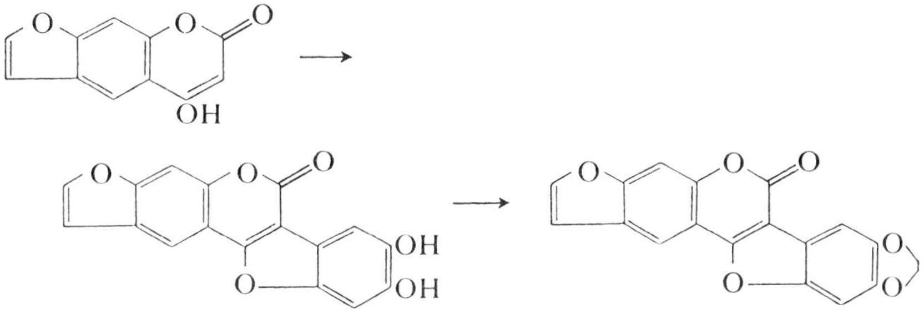

Erosnin

144

With the knowledge that coumestans are biogenetically isoflavanoids and not true coumarins, Grisebach and his colleagues[134] have investigated the formation of coumestrol **100** from 2′,4′,7-trimethoxyisoflavanone **145**. This material is heated at about 200° with pyridine hydrochloride in air, and a tiny amount (~0.2%) of coumestrol results. Almost certainly demethylation allows ring closure to a dehydropterocarpan derivative **146**, most of which would be destroyed because of uncontrollable oxidation of the phenolic rings, but some of which is oxidized at the methylene group to the coumarin. This is less sensitive to oxidation because of the electron-withdrawing effect of the carbonyl group, and a little survives. Radical oxidation (by air) of ring methylene groups in dibenzopyrans and related systems is known, having been investigated by Robertson and his colleagues,[135] and it may be significant that **146** has in fact been isolated during studies[136] on the heartwood constituents of *Swartzia madagascariensis*.

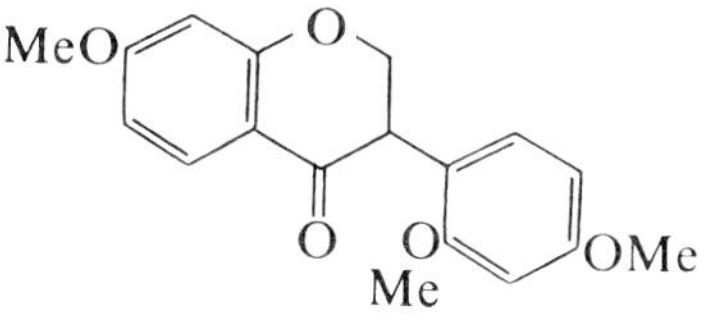

145

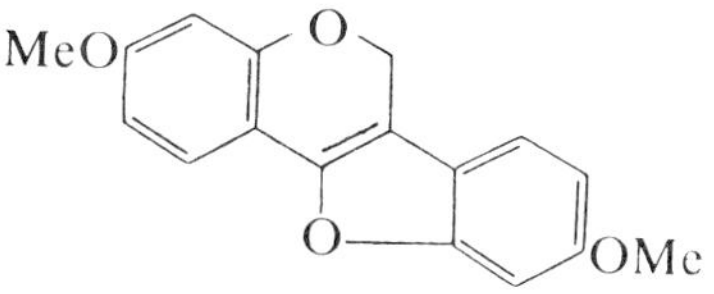

146

With these considerations we find ourselves discussing pterocarpan chemistry. There are but two independent routes to pterocarpans, and one of these leads through coumestan country. A case in point is the synthesis of maackiain, which begins with the benzyloxydihydroxycoumestan **147** intermediate in one synthesis of medicagol. From here, Fukui and his co-workers[130] effected methylenation and proceeded by lithium aluminum hydride reduction to the alcohol **148**, which they cyclized by dissolving it in boiling diethylene glycol. This is the standard method and gives very satisfactory yields of what are known as dehydropterocarpans. Usually, of course, such an etherification is effected using acid catalysts capable of producing carbonium ions from the allylic alcohol segment; here, however, acids have to be avoided on account of the sensitive benzofuran nucleus. Saturation of the olefinic bond in the dehydroptercarpan **149** is not easy but can be accomplished catalytically with palladium-on-charcoal at 70 atm and 70°. Rhodium catalysts are sometimes preferred. In the present example the benzyl group is lost simultaneously and a racemate corresponding to the natural product maackiain **150** is produced. It has not been resolved (none of the synthetic pterocarpans have been). The catalytic method ensures that the ring junction is *cis* as in all members of this series. Other ptercarpans obtained similarly

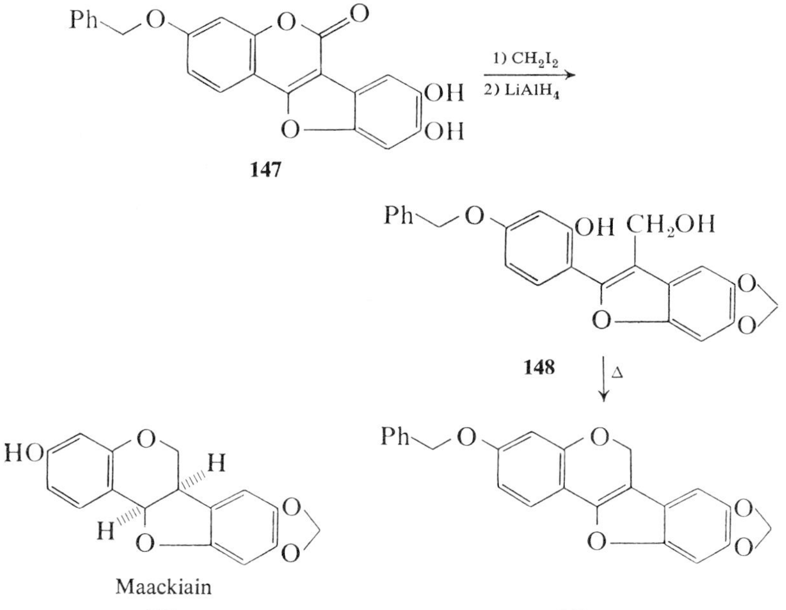

1) CH_2I_2
2) $LiAlH_4$

147

148

Δ

149

Maackiain

150

include 3-hydroxy–8,9-dimethoxypterocarpan,[137] 3,8,9-trimethoxypterocarpan,[138] (±)-pterocarpin[120] **151**, and (±)-4-methoxypterocarpin.[138] In a slightly modified version of this synthesis of 4-methoxypterocarpin **152**, diborane in tetrahydrofuran was used to reduce the coumarin carbonyl group to methylene without opening the lactone ring.[119] The yield is ~50%

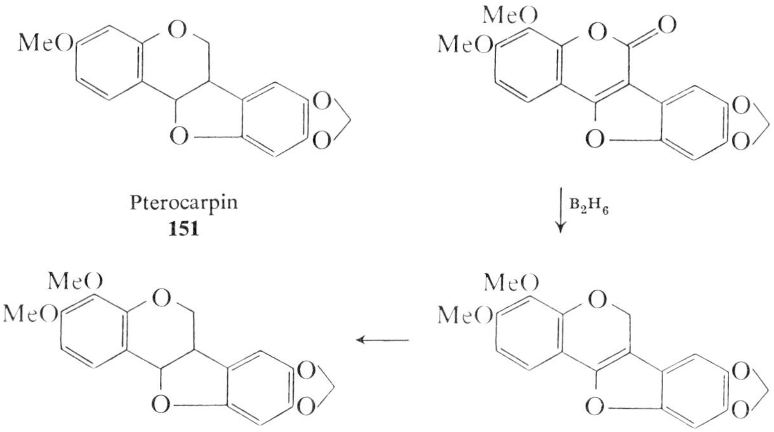

Pterocarpin
151

B_2H_6

152

and the idea worth imitation. (Another example occurs in Section 6.) Curiously, simpler coumarins do not behave in such a straightforward fashion. Coumarin itself gives 2-allylphenol.[139]

The other route to pterocarpan derivatives, which requires isoflavones as starting materials, was explored first by Suginome and Iwadare.[140] Another synthesis of maackiain[141] **153** is offered in illustration of the basis of the method:

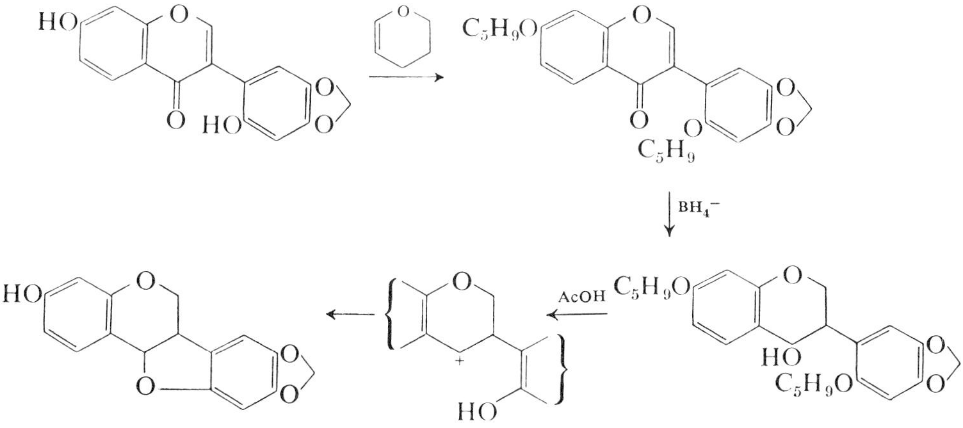

BH_4^-

AcOH

Maackiain

153

The reduction of isoflavones by borohydride has been discussed previously. The subsequent cyclization is conducted in hot 50% acetic acid so probably involves a benzylic carbonium ion as shown. That the requisite *cis* ring fusion results is merely a consequence of ring strain, which is somewhat greater in the *trans* fused arrangement. Racemic forms of pterocarpin,[142] inermin, 8-methoxyhomopterocarpin,[143] and other pterocarpans including analogs of neodulin[144] have been made by this method without significant variations. In all these examples a 2′-hydroxy group must be provided for in the isoflavone precursor, often by selective demethylation just as in some routes to the rotenoids as described earlier.

Protective techniques are also commonly used, and combined with a new method of obtaining isoflavones from enamine condensations they form a very flexible ingress to pterocarpan chemistry. An application to (±)-pterocarpin outlines the method:[145]

Ph
H
N
O
O
O
O
Ph
O
O
Cl
OMe
O
OAc
N
MeO
O
O
O
O
Ph
O
O
O
O
Ph
O
O
OMe
+N
OH
1) HCl
2) BH_4^-
3) AcOH
MeO
O
O
O
O
Pterocarpin

No attempt seems to have been made to synthesize trifolirhizin, the β-glucoside of 3-hydroxy-8,9-methylenedioxypterocarpan (inermin), although this would be of special interest in that it would probably also constitute a method of resolution in this series.

The last compound discussed here is pisatin **154**, curious in its hydroxy substituent, unique in the series, and in having rotation of the opposite sign to pterocarpin, from which it is formally derived. Pisatin is a phytoalexin of the common garden pea, whereas pterocarpin comes from a tropical heartwood—the plants concerned, however, both come from the same family Leguminosae, and even the same subfamily, the Papillionaceae. (±)-Pisatin has been synthesized, since it can be obtained from pterocarpin. The method requires a strictly controlled acid hydrolysis of the benzylic ether linkage in pterocarpin, 10*N* hydrochloric acid in ethanol being the catalyst. About 40% of the flavene **155** is obtainable, and with osmium tetroxide affords the black osmate **156**, which is decomposed by sodium carbonate in mannitol. The cyclic body is formed at once, notwithstanding the absence of the usual acid catalyst, and it is thought that a nucleophilic displacement occurs as the osmate collapses, as indicated.[146] Displacements at the benzylic position are usually very easily effected.

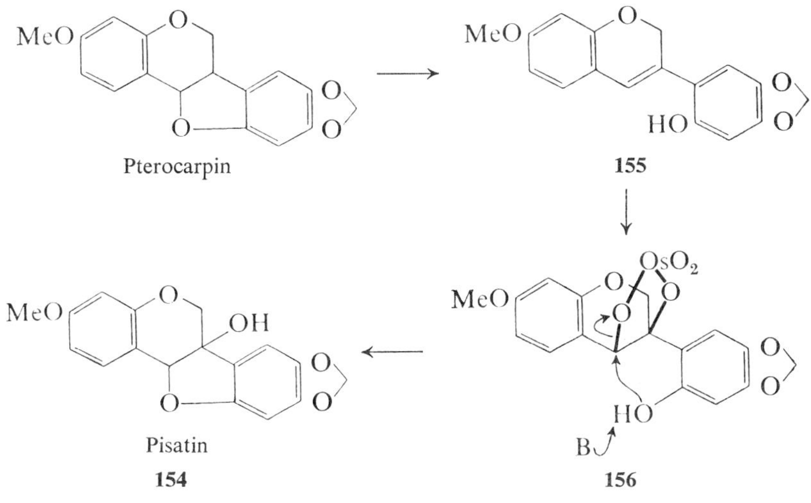

Pterocarpin

155

Pisatin

154

156

6. AMPHIPYRONES (PYRONOQUINONES)

The number of fungal metabolites containing or derived from nucleus **157** is steadily increasing.[147] This nucleus is often called the "pyronoquinone" nucleus, but the name is unfortunate in that it suggests a fusion of quinone and pyrone rings instead of the elision characteristic of **157**. The name "amphipyrone" is offered here as a substitute that indicates the correct number of oxygen atoms and their location at opposite sides of the ring system. At a time when none of these compounds had been allocated a structure the name "azaphilone" was coined for them because of their most striking reaction in which ammonia rapidly replaces an oxygen atom by the imino group. One of the first members to be extensively investigated was sclerotiorin[148] (**158**), a yellow pigment of *Penicillium sclerotiorum* van Beyma and other fungi, which with ammonia yields the red sclerotioramine **159**.

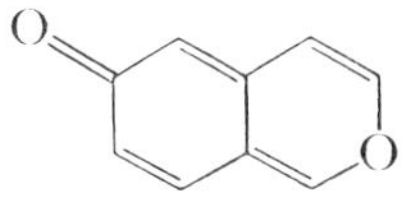

157

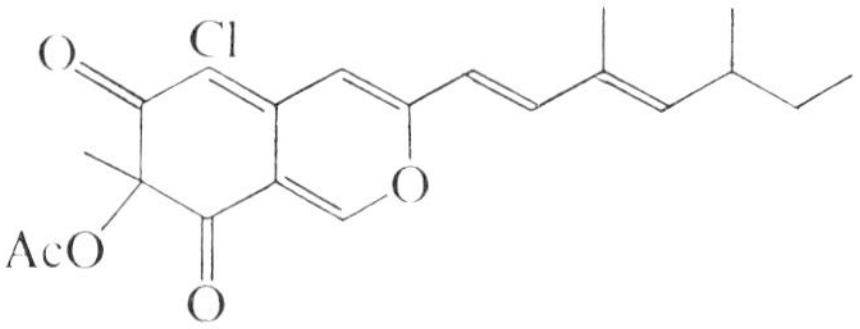

Sclerotiorin

158

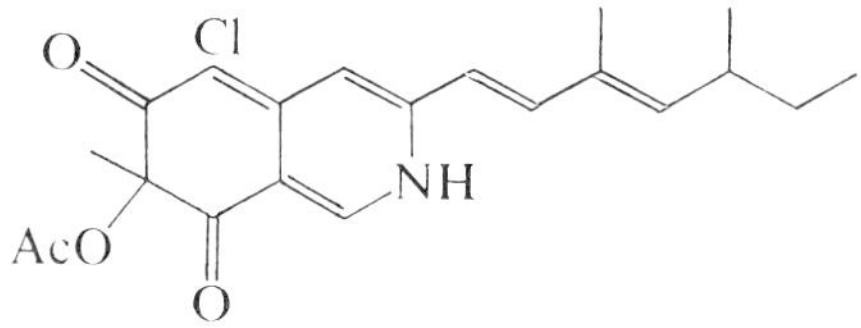

Sclerotioramine

159

Compounds with nucleus **157** are sensitive in other ways also, especially to nucleophiles, and their syntheses are correspondingly difficult. Only one reasonably simple synthesis of the system has been achieved, and although it has not yet been applied in the case of a natural product it is discussed here because of its intrinsic interest and especially because it does in fact constitute the extension of a simple pyrone.

Yamamura, Kato, and Hirata[149] heated 2,6-dimethyl-4-pyrone (**160**) with cyanoacetic acid in a large excess of acetic anhydride and obtained the extended pyrone **161**. There is limited precedent for such a condensation at the carbonyl group of a pyrone,[150] though these groups are usually virtually inert because of interactions indicated in **160**. The conditions suggest that (*a*) cyanacetic acid condenses with itself under the influence of acetic anhydride giving the dicyanoketone **162**, then (*b*) the enol of the dicyanoketone attacks position 4 in the pyrone whenever this is activated by protonation as in diagram **163**. Several other, rather similar schemes can be envisaged. However the reaction occurs, it gives the relatively good yield of 30%, and conversion into the diamide **164** is not difficult to accomplish with sulfuric acid. But what makes the method attractive is that polyphosphoric acid converts the extended pyrone **161** directly into the cyclized amphipyrone **165** in quantitative yield. Direct substitution into a pyrone ring is very rare, especially in acidic media where most pyrones would be expected to exist in a protonated form as pyrylium salts that are no more susceptible to electrophilic attack than are pyridinium salts in a like situation. Since the immediate substitution-cyclization product may well have structure **166**, it is reasonable to account for the special behavior in terms of the large increase in aromaticity, the nucleus of ion **166** being isoelectronic with naphthalene. This aromaticity is lost if a proton is removed and the amino group then becomes an enamine function, acid hydrolysis of which yields the enol **165**. With ammonia, **165** undergoes the characteristic replacement giving the isoquinoline derivative **167**.

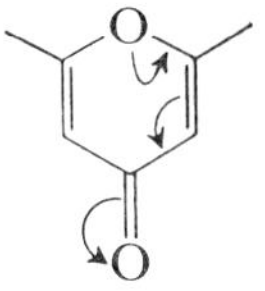

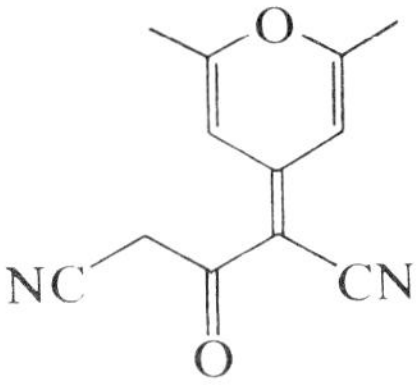

$CNCH_2COCH_2CN$
162

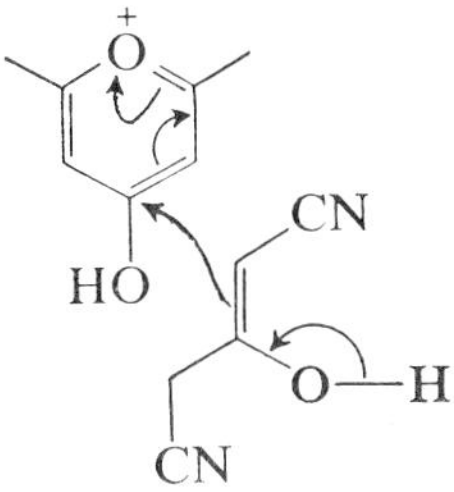

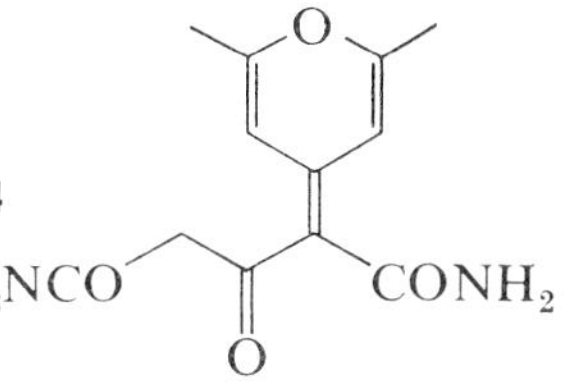

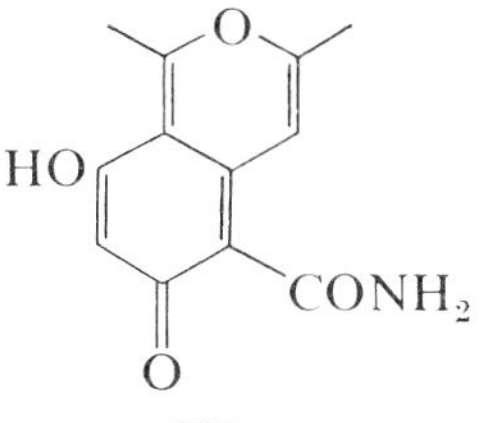

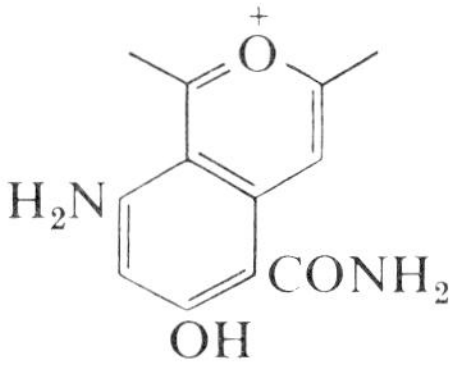

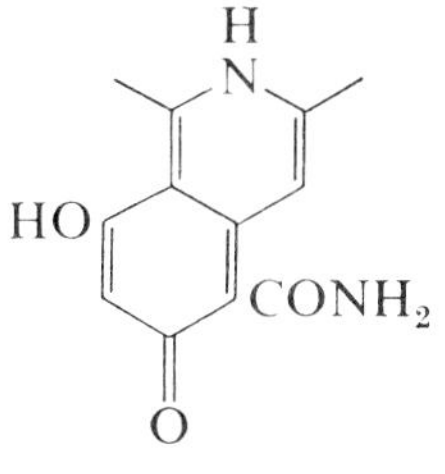

To date, however, all syntheses of the natural products have been planned along lines based on phenol chemistry and elaborated by Whalley and his colleagues who, taking up an earlier idea,[148] began[151] with a ranging shot fired at the relatively stable compound **168**, tetrahydrosclerotioramine, instead of the much more sensitive parent compound, sclerotiorin **158**. The phenolic ketone **169** was prepared by allowing the cadmium derivative of (+)-3,5(*S*)-dimethylheptyl bromide to act on 3,5-dimethoxy-4-methylphenylacetyl chloride and then demethylating the product. Usually the Gattermann aldehyde synthesis with hydrogen cyanide and hydrogen chloride converts resorcinols into aldimines, which are then hydrolyzed, but in this case the imine **170** inevitably cyclizes to the isoquinoline **171**. Chlorination with sulfuryl chloride in acetic acid gives **172**, and the last step is acetoxylation with lead(IV) acetate, a reaction extensively explored by von Wesseley.[152] Of course, this reagent is far from specific either positionally or stereochemically, and **168** is produced along with other compounds including the epimeric acetate, from which it seems not to have been separated. Nevertheless, the result establishes the method.

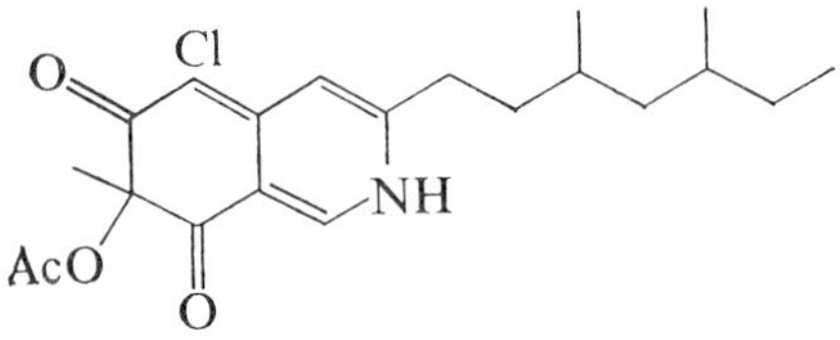

168

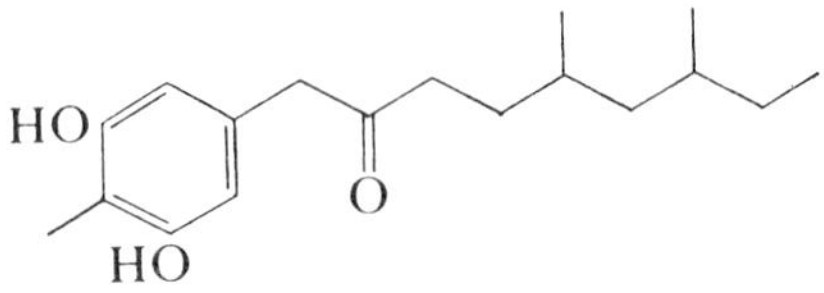

169

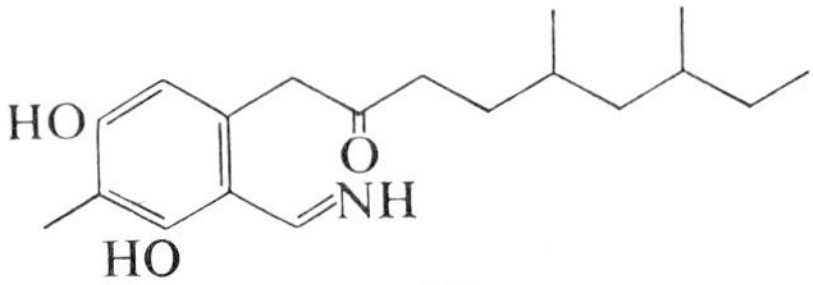

170

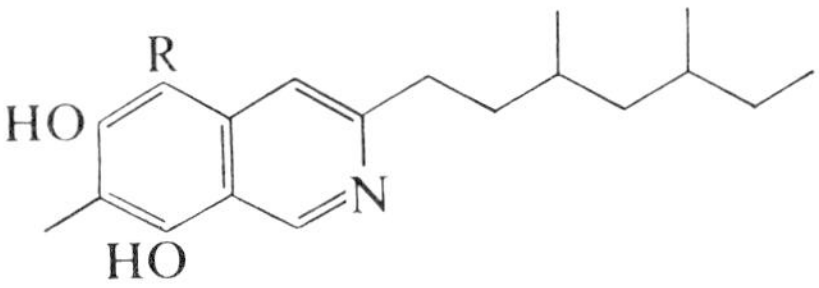

171 R = H
172 R = Cl

A different method of producing an amphipyrone appears in a synthesis, by Galbraith and Whalley,[153] of (±)-ascochitine **173**, a metabolite of *Ascochyta pisi* Lib. and *A. fabae* Speg. and one of the simplest of the natural amphipyrones. Methylation of 3,5-dimethoxybenzyl cyanide with sodium in liquid ammonia followed by iodomethane gave the starting material **174**, which was converted by standard methods into the phenolic ketone **175**.

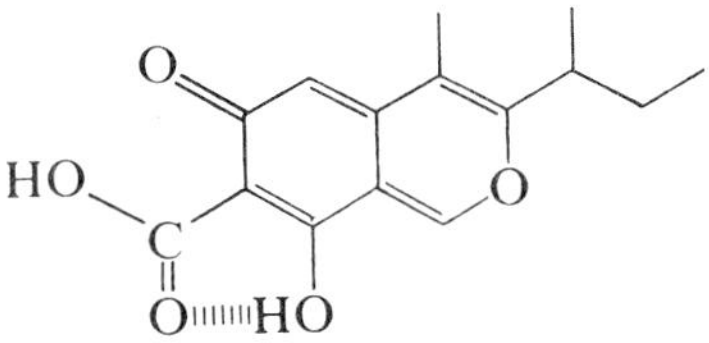

Ascochitine
173

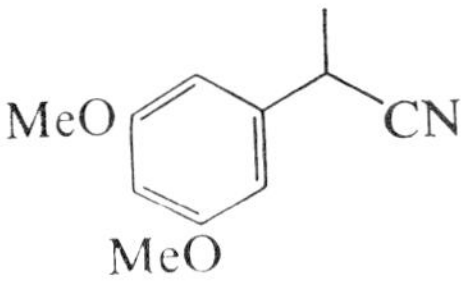

174

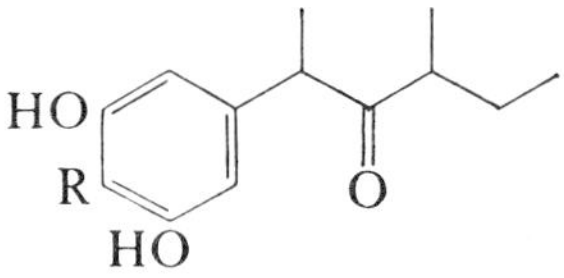

175 R = H
176 R = CO_2H

The carboxyl group was introduced between the hydroxy substituents by treatment at 180° with potassium hydrogen carbonate in glycerol, this orientation being usual where there is a 5-substituent bulky enough to shield positions 4 and 6. Heated with ethyl orthoformate and an acid catalyst (hydrogen chloride and toluene-*p*-sulfonic acid are both useful) the acid **176** gave ascochitine **173** in one step, though of course several stages must be involved—perhaps cyclization and other steps occur as indicated in diagrams **177** and **178**.

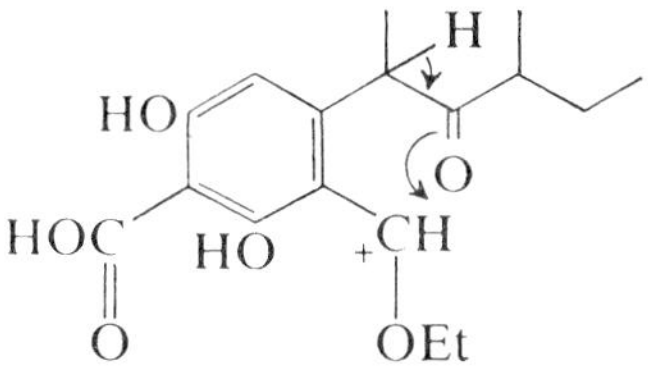

177

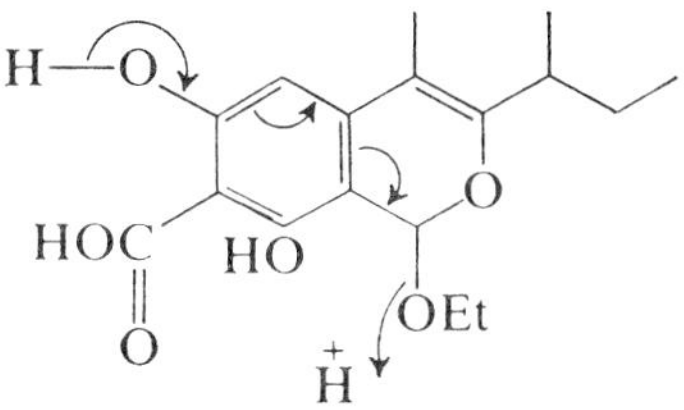

178

For other syntheses a return was made to the acetoxylation technique. Beginning again with phenolic ketone **169**, reaction with ethyl orthoformate affords the aldehyde **179**, and sulfuryl chloride (catalyzed by benzoyl peroxide) then gives the halogen derivative **180**. Here enolization, hemiacetal formation, and dehydration should give the amphipyrone **181**, but this seemed far too sensitive to be manipulated directly so it was used in the protonated form **182** obtained merely by treating the aldehyde **180** with hydrogen chloride in ether. As before, lead(IV) acetate supplied a mixture from which it was possible to isolate a mixture of tetrahydrosclerotiorin **183** and its epimer.

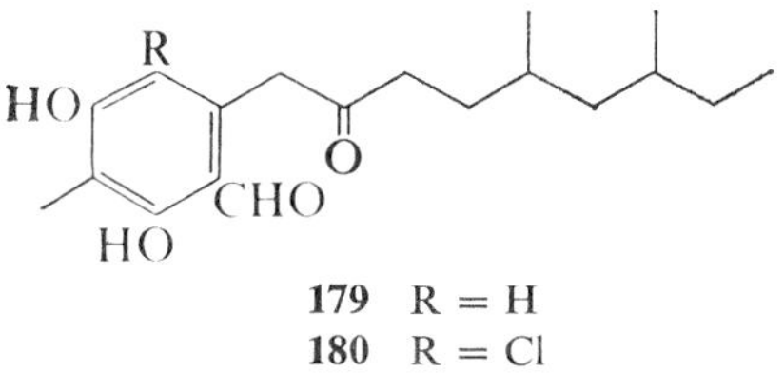

179 R = H
180 R = Cl

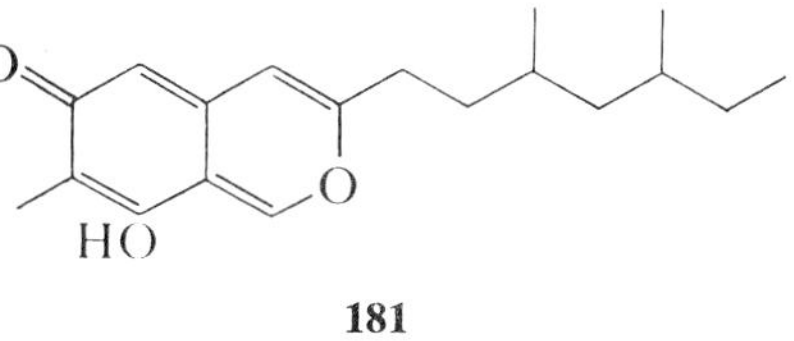

181

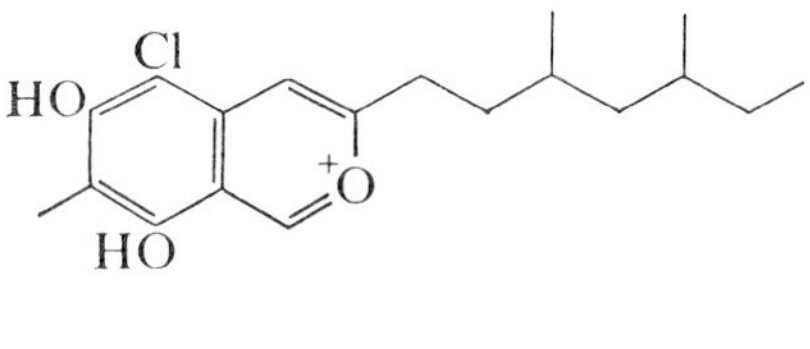

182

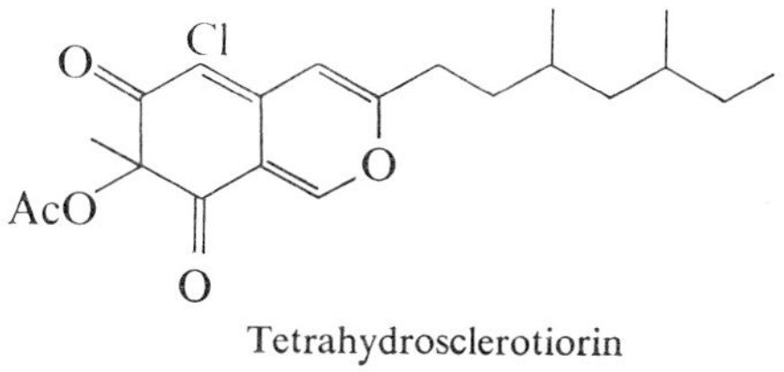

Tetrahydrosclerotiorin
183

These have not yet been separated, but the properties of the mixture leave no doubt as to the identity.[154]

A mixture of sclerotiorin epimers has been obtained similarly.[155] Standard methods suffice to convert acid **184** through the acid chloride and diazoketone into the bromomethyl ketone **185**, which is then debenzylated using boron tribromide and acetylated. The product **186** is transformed into a phosphorane under very mild conditions of minimal basicity by treatment with triphenylphosphine and methyloxirane at 25° for 16 hours. After removal of 2-bromopropan-1-ol the mixture is allowed to react with (±)-2,4-dimethylhex–*trans*-2-enal at 130° under nitrogen, and finally the acetate groups are removed by alkaline hydrolysis under nitrogen. This sequence would be expected to yield the phenolic ketone **187**, but the compound obtained is said to be a hydrate, which, rather surprisingly, is formulated as **187a**.

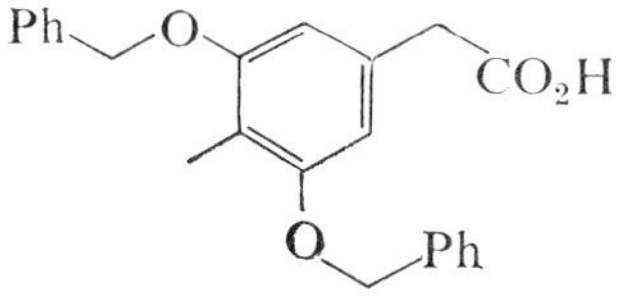

184

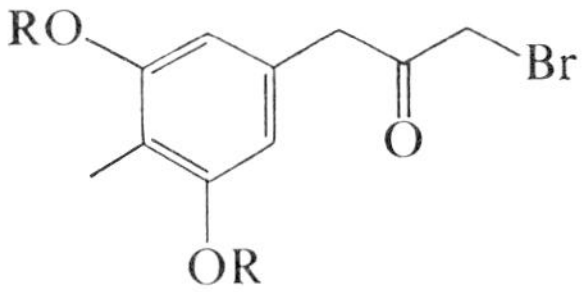

185 R = CH_2Ph
186 R = Ac

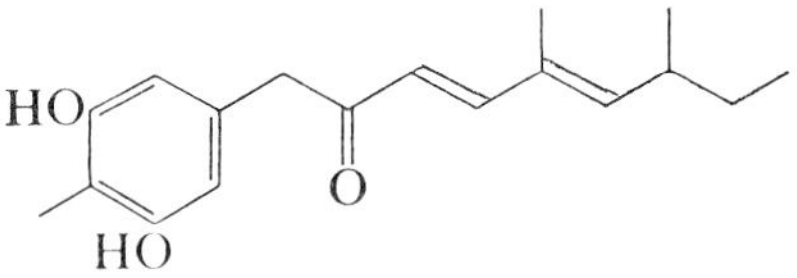

187

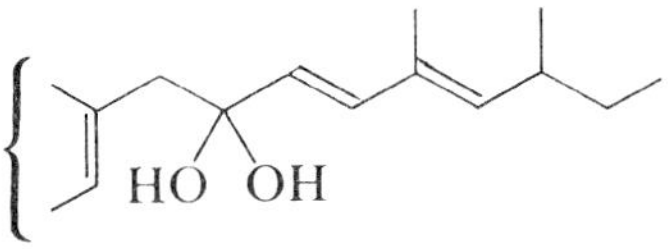

187a

The next steps are the now familiar ones where ethyl orthoformate introduces an aldehyde function and sulfuryl chloride a chlorine atom, the product being **188a**; a new aspect is the use of phosphorus pentoxide to induce formation of the amphiquinone **189** as a racemate corresponding to (+)-aposclerotiorin, a reduction product of the natural pigment. Natural (+)-aposclerotiorin forms a useful relay at this point, since even under nucleophilic attack as mild as that of acetate ion its heterocyclic ring opens to form the ketonic aldehyde **188a**, thus demonstrating the lability of the amphipyrone nucleus. Schematically, the fission takes the path shown in **188b** and **188c**. Acetoxylation of the natural compound **189** generates a mixture, and again a fraction can be isolated that contains both (+)-sclerotiorin **158** and an epimer too similar to be readily separated. The difficulty is always the same: there are two asymmetric centers only, but they are too far apart to have any influence on each other and too far for much hope of achieving asymmetric synthesis by induction.

188a

Aposclerotiorin
189

188b

188c

A further interesting elaboration is possible. Ethanolysis of (+)-sclerotiorin with sodium ethoxide at 0° gives the alcohol **190**, seemingly without collapse of the heterocyclic ring. It will be appreciated that now nucleophilic attack no longer leads to direct aromatisation as in **188b**. The alcohol reacts with diketene in hot pyridine-benzene in the presence of hydrogen chloride giving (presumably) first the acetoacetic ester **191** and then the cyclization product **192**. The yield is good, and the method[155] should offer a promising doorway to syntheses of the natural analogs[156] such as rotiorin, rubropunctatin[157] and monascorubrin[158] **193**.

190

191

192

Monascorubrin
193

Finally, a synthesis[159] of (±)-mitorubrin **194** illustrates some minor variations. The appropriate phenolic ketone with ethyl orthoformate and hydrogen chloride rapidly gives the pyrylium salt **195**, which is precipitated at once with ether. With ethanolic potassium acetate it immediately yields the ketonic aldehyde **196**. Since the amphipyrone is too unstable to withstand isolation, a solution of the aldehyde **196** in ethanol containing phosphorus pentoxide is used to obtain it in solution and it is then treated *in situ* with lead(IV) acetate to produce the more stable compound, **197**. The acetyl group is removed by sodium ethoxide and the alcohol esterified by 2,4-dibenzyloxy-6-methylbenzoic acid in the presence of trifluoroacetic anhydride. Then debenzylation with boron chloride at −70° gives (±)-mitorubrin **194**. The natural compound is the (−)-isomer and can be extracted from the phytotoxic fungus, *Penicillium rubrum*.[160]

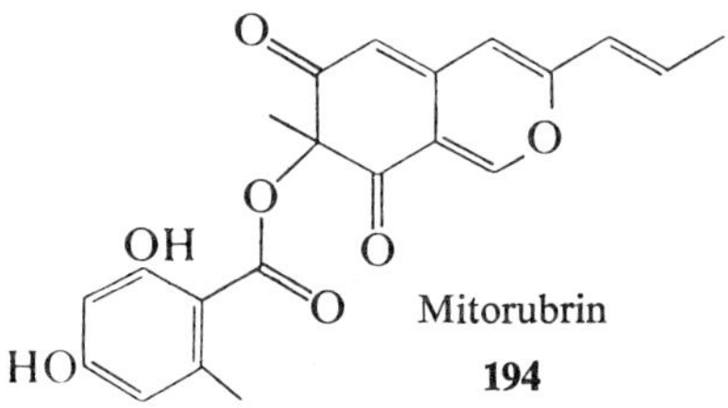

Mitorubrin

194

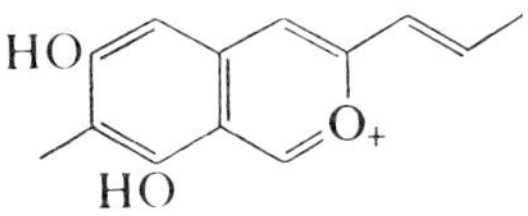

195

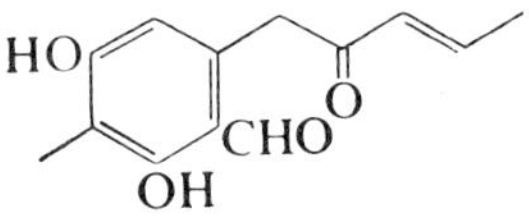

196

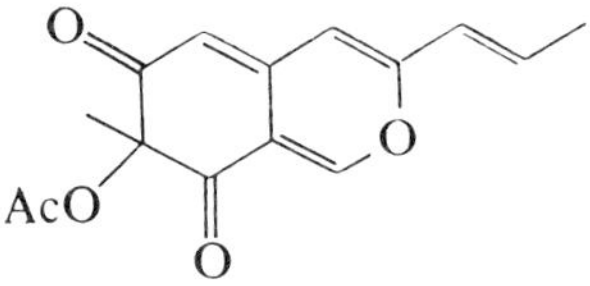

197

In one sense, amphipyrones are derived from isochromene and are consequently related to numerous compounds of the isocoumarin type. Most of these compounds contain true benzene rings and are therefore not of immediate interest here; but two others, citrinin and fuscin, both fungal metabolites, are nonaromatic and have been synthesized by methods indicative of the extent to which the quinonoid part of the system can be created by the oxidative techniques ignored so far.

The synthesis of citrinin **198** begins with the resorcinol derivative **199**, which, in Kolbé conditions, suffers carboxylation to the acid **200**, the formation of the isomer being prevented by the bulkily branched side chain. The Gattermann reaction[161] is often less sensitive than carboxylation to steric effects, so there is no difficulty in producing the aldehyde **201**. Finally, dehydration by sulfuric acid yields citrinin **198** as a racemate that can be resolved as the brucine salt.[162]

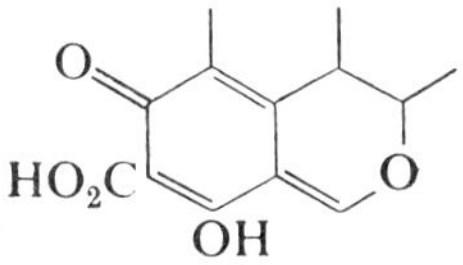

Citrinin

198

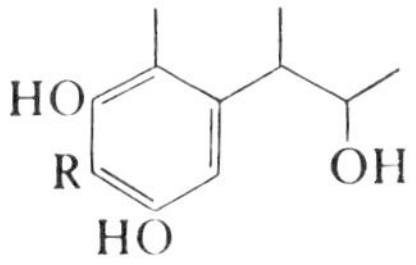

199 R = H
200 R = CO_2H

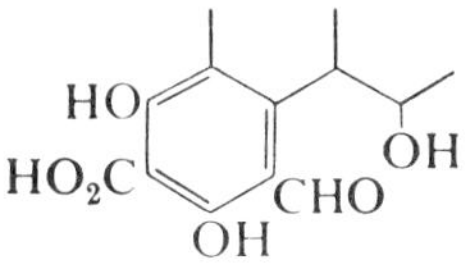

201

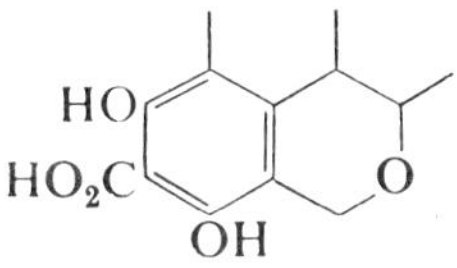

202

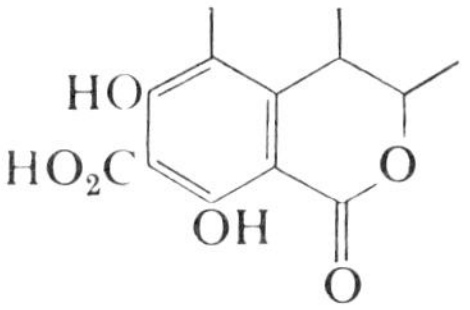

Dihydrocitrinone

203

The synthesis of dihydrocitrinin **202** is even easier, since formaldehyde condenses very readily with the methyl ester of acid **200** and the product cyclizes at once.[162] The gentlest of dehydrogenations (bromine or mercuric oxide) now gives the quinonoid citrinin.[163] The idea appears in various forms in isochromene chemistry. For example, Curtiss, Hassall, and Nazar[164] discovered that a fungal mutant no longer able to synthesize citrinin generated dihydrocitrinone **203** instead, and they used diborane to reduce this to dihydrocitrinin. In this work oxidation to citrinin was accomplished with manganese dioxide or simply with air. In strong contrast, the removal of further hydrogen to give a true amphipyrone seems very difficult to achieve without causing a deeper disruption—at all events, no such dehydrogenation has yet been described.

Fuscin **204** is another fungal metabolite. The description of amphipyrone can hardly be applied to this compound; however, in addition to its extended quinonoid system it does possess two modified pyran rings. The synthesis[165] starts with a standard chroman synthesis, methyl 3,4,5-trimethoxyphenylacetate and 3-methylbut-2-enoyl chloride giving the chromanone **205** and Clemmensen reduction the chroman **206**. Another acylation introduces an acetyl group, and borohydride reduction produces the corresponding alcohol which at once lactonizes, giving **207**. When demethylated, the product was simply left in air to suffer oxidation to the quinonoid fuscin **204**. In theory, the molecule could exist in any one of several tautomeric forms. That only **204** is observed is a consequence of the *para* quinonoid nature of this arrangement (the other possibilities are *ortho* quinonoid) and of the interactions indicated between certain parts of the molecule (such interactions are reduced or lost in tautomers).

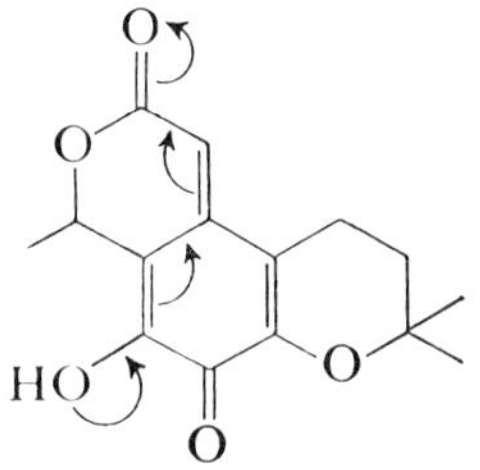

Fuscin

204

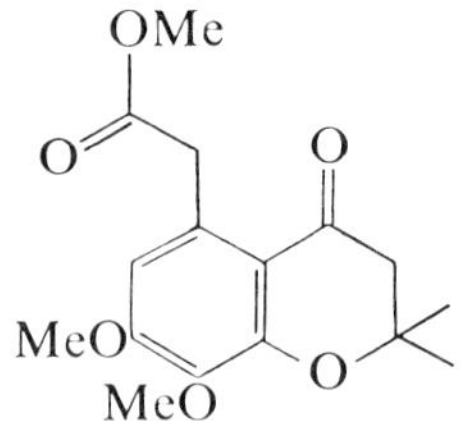

205

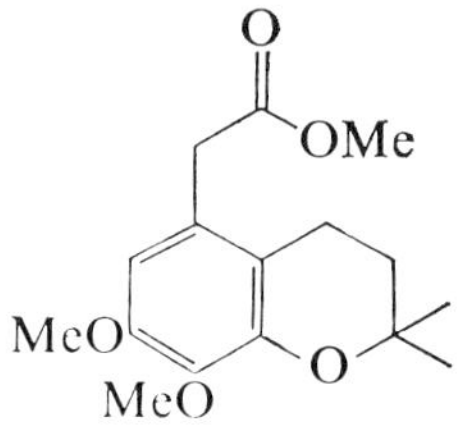

206

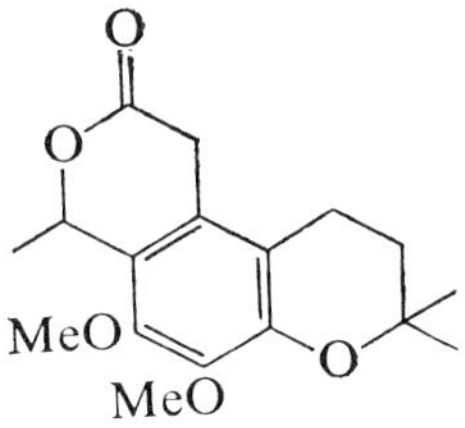

207

7. XANTHONES*

Xanthone occurs naturally as derivatives in a few families of higher plants, notably the Guttiferae and the Gentianaceae. Such compounds contain from one to five hydroxy and/or methoxy groups.[166,167] Sometimes C_5 isoprenoid side chains make an appearance. A few fungal xanthones are also known, and among these *C*-methyl groups and chlorine substituents are found.[166] The numbering of xanthone is now always as shown; letters are used in biosynthetical discussions,[167] A referring to the acetate and B to the shikimate-derived ring.

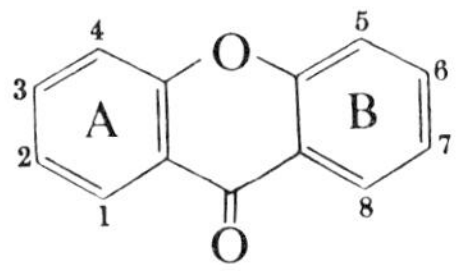

Xanthone

Although xanthone contains only four different locations for substituents, these being duplicated symmetrically, even with just five hydroxy and/or methoxy groups to dispose around the periphery one is faced with some hundreds of possibilities. About 70 of these are currently known as occurring

* The author is grateful to Dr. H. D. Locksley and Dr. F. Scheinmann for their help with the preparation of this section.

in plants, the great majority having been discovered since the last extensive reviews[168,169] were published around 1960. Clearly, selectivity and orientation control are the synthetical problems. It simplifies matters somewhat that while each benzene ring in xanthone interacts strongly with the central pyrone nucleus, there seems little transmission of effects across this nucleus from one benzene ring to the other. And it is a decided advantage that the xanthone nucleus is normally stable both to bases and to acids. The stronger acids merely protonate the carbonyl oxygen atom (reversibly) giving hydroxy-xanthylium salts, and ordinarily there is no danger of Wessely-Moser rearrangements.[169]

At present the xanthone synthesis introduced by Grover, Shah, and Shah[170] enjoys by far the greatest popularity.[171–180] It requires simple, usually accessible, materials: a salicylic acid derivative and a suitable phenol. It utilizes the simplest of techniques: the components are merely heated together with zinc chloride in phosphoryl chloride as solvent. In the following example the xanthone **208** formed by this method needs only selective methylation by diazomethane to yield lichexanthone (**209**), a metabolite of the lichen *Parmelia formosana:*

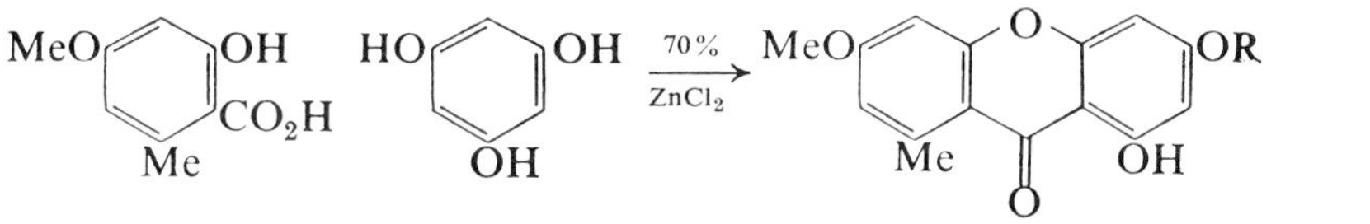

MeO, OH, CO_2H, Me + HO, OH, OH $\xrightarrow[ZnCl_2]{70\%}$ MeO, O, OR, Me, O, OH

208 R = H; Norlichexanthone
209 R = Me; Lichexanthone (CH_2N_2)

Reaction times vary from 0.5 to 24 hours. Sometimes the solvent is omitted and the temperature raised to 180° as in the Nencki reaction,[178] and sometimes polyphosphoric acid is preferred to the zinc chloride mix.[174,175]

Some acids (phloroglucinol carboxylic acid, resorcinol-2-carboxylic acid) are easily decarboxylated and require the mildest conditions for success, but gentisic acid (2,5-dihydroxybenzoic acid) provides great difficulty. It fails to react, which is a nuisance because quinol nuclei are common among xanthones and quinol itself is too resistant to electrophilic substitution to be used. Curiously, 2-hydroxy-5-methoxybenzoic acid is perfectly satisfactory in Grover-Shah-Shah (GSS) conditions,[170,176,178] and the next example is of its use in a preparation of 1,3-dihydroxy-7-methoxyxanthone[176] (**210**):

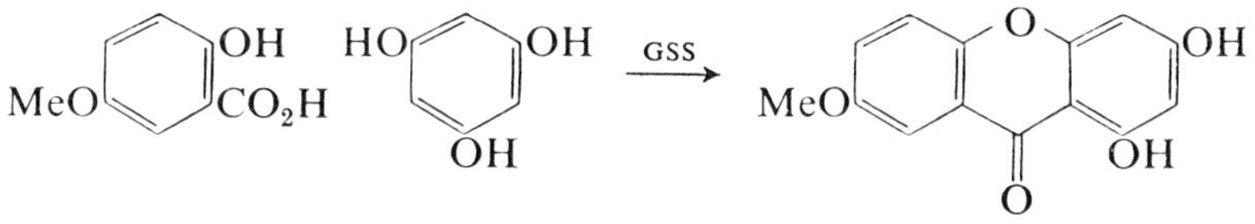

MeO, OH, CO_2H + HO, OH, OH $\xrightarrow{GSS}$ MeO, O, OH, O, OH

210

This striking effect of methylation has not been explained. It may be connected with the suppression of ionization in the acylium ion **211**, which is presumably the intermediate in what amount to Friedel-Crafts conditions. The first ionization would give a ketene **212** (there is analogy[181]), and a second would now greatly reduce the electrophilic character of the carbonyl group, as indicated:

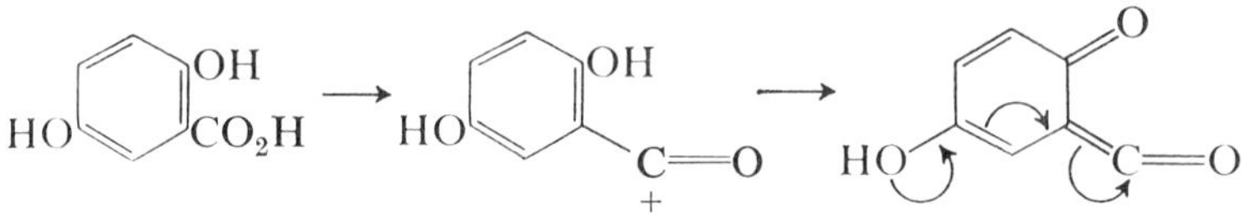

211 **212**

It follows, therefore, that free hydroxy groups at positions 4 and 6 in the salicylic acid would have no comparable effect and that a 3-hydroxy group would interfere but to a smaller extent; in all cases methyl ethers would give better results than the phenols do. No proper study of this kind has been undertaken, but qualitative comparisons appear to bear out these conclusions.

Strictly speaking, it is not xanthones that are usually formed in the general GSS reaction, but 2,2′-dihydroxybenzophenones that have to be cyclized in a separate step. Heating to about 200° in water in a sealed tube or an autoclave has been recommended, although lower temperatures suffice with acid catalysts:

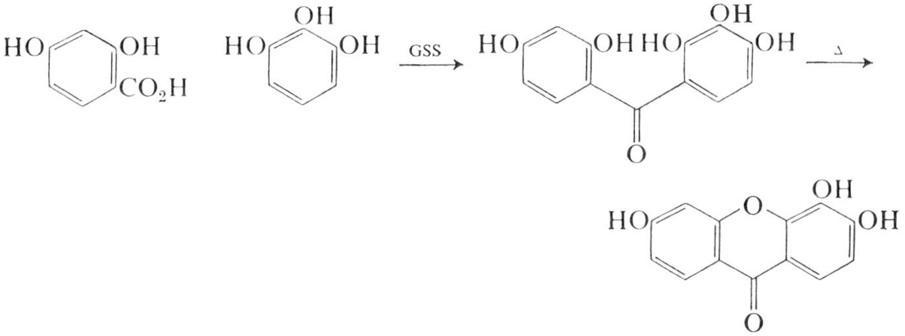

Xanthones are produced directly only if the benzophenone intermediate carries another hydroxy group at the 6 or the 6′-position, that is, if an alternative site for cyclization is available.[170] The following diagram **213** suggests that coordination with zinc ion might usually impede cyclization by forcing the molecule into an unfavorable conformation with the two hydroxy groups far apart; when the third hydroxy group is present cyclization is still possible—indeed, promoted, since the conformational restriction now insists that two hydroxy groups be mutually accessible as in diagram **214**.

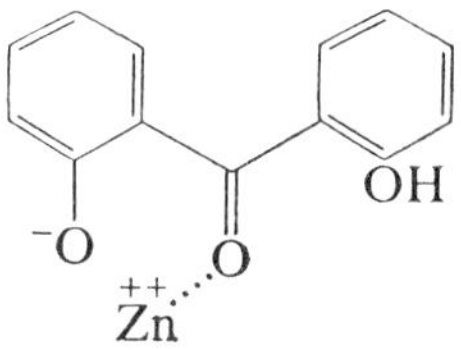

213

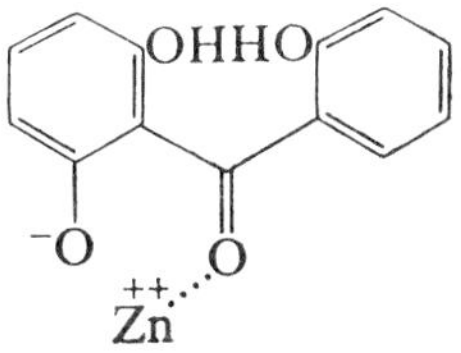

214

An important deviation from the general orientation rules is found in condensations with orcinol, the incoming carbonyl group becoming attached to the point between the hydroxy groups of this particular resorcinol derivative. Such behavior is common in the orcinol series and is to be attributed to the difficulty of pushing a bulky group (carbonyl coordinated to zinc ion) between one hydroxy group and the relatively large methyl group. Thus β-resorcylic acid and orcinol yield xanthone **215** directly, whereas orsellinic acid **216** and resorcinol yield a benzophenone and then the xanthone[170] **217**.

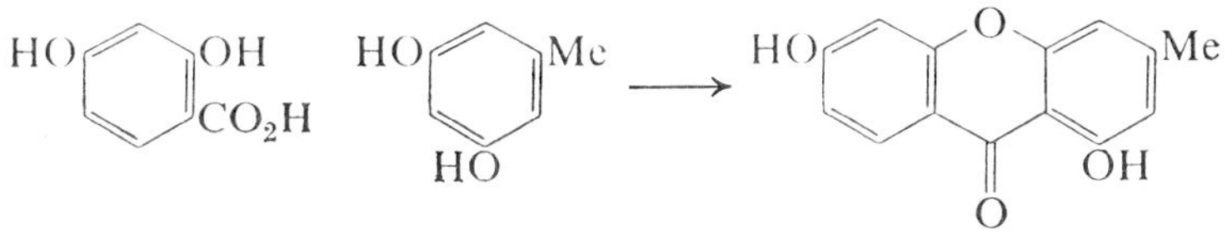

215

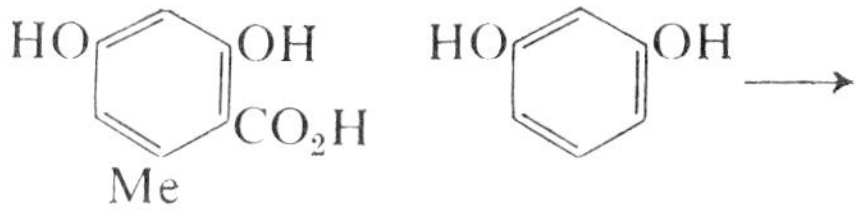

216

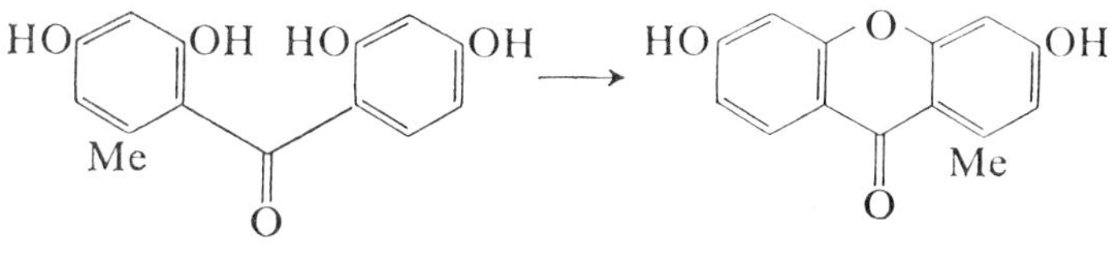

217

These idiosyncracies aside, the GSS reaction provides a wide variety of the simpler polyhydroxyxanthones and their ethers where these are not unduly sensitive to acids. The methoxy group (and methylenedioxy group[179]) is usually stable unless it arrives at position 1 (or 8) in the final xanthone, in which case it tends to be lost as it would be from any other *o*-alkoxycarbonyl compound. Mixtures may result, and demethylation can be completed, if desired, by aluminum chloride or boron chloride.[177,182] A good route[177] to xanthone **218** uses aluminum chloride in the GSS melt itself to open the way to xanthone formation and then to secure free hydroxy groups at position 1:

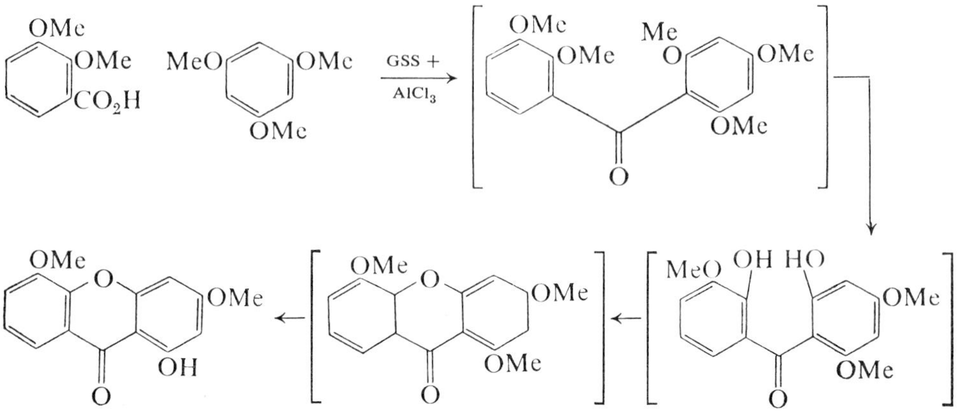

218

That at higher temperatures other methoxy groups may be affected is one of the defects of the older technique (Nencki reaction) in which the components were simply strongly heated with zinc chloride alone. With resorcinol derivatives differences in orientation result, too,[178,178a] presumably because at higher temperatures the products are determined thermodynamically rather than kinetically. Thus resorcinol and 2-hydroxy–5-methoxybenzoic acid under GSS conditions yield xanthone **219**, whereas under Nencki conditions the resorcinol is acylated at the 2-position so that, with concomitant demethylation, the product is the well known pigment euxanthone **220**.

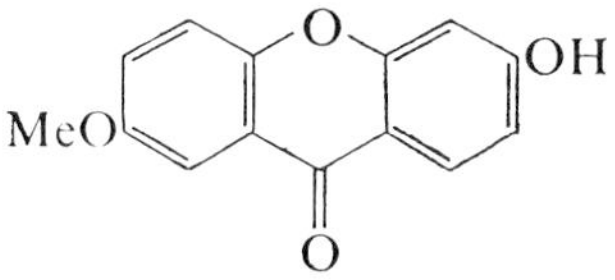

219

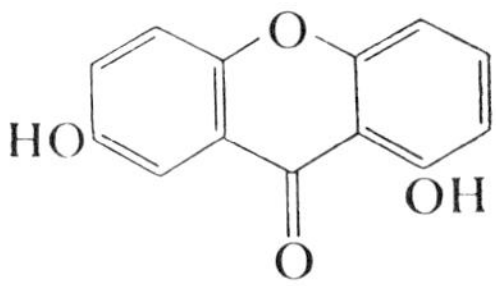

Euxanthone
220

An important demethylation occurs where pyrogallol nuclei are concerned.[180,183] Zinc chloride under GSS conditions converts pyrogallol trimethyl ether into 2,6-dimethoxyphenol in high yield. Sulfuric acid has the same effect, and the reaction often occurs in preference to the better known demethylation at position 1. As illustrations syntheses of xanthones **221** and **222** should suffice (as shown on top of facing page).

This selectivity is often a valuable feature for which buttress effects[184,185] have been held responsible,[180] since a group flanked on both sides by other (large) groups may be pushed out of the plane of the benzene ring and its behavior modified accordingly. Further examples will appear shortly.

MeO
MeO OH HO OH GSS → MeO HO O OH
CO_2H OH OH O
221

MeO
MeO O OMe H_2SO_4 50–55° → MeO HO O OMe
O OMe O OMe
222

However, this explanation may be accepted only cautiously at the moment, for there is a dearth of examples in which flanking groups other than methoxy groups have been studied. A possible alternative view assumes that coordination of metal ions or hydrogen bonding phenomena are responsible, the central methoxy group being always subject to a twofold attack, the outer ones to not more than a single attack. The idea is sketched in **223**, while a pleasing double application of the reaction has been described by Gottlieb and his colleagues,[183] who converted the ether **224** into the phenol **225**, which occurs in some species of *Kielmeyera*.

Me
O····Zn^{++}
OMe Cl^-
Zn^{++}
OMe
223

O OR
OR
OMe
O

224 R = Me
225 R = H
H_2SO_4

A different approach to xanthone synthesis is based on the cyclization of an *o*-phenoxybenzoic acid. The diphenyl ether link is forged by the Ullmann reaction as in a synthesis[186] of 2-methoxyxanthone **226**, used for identification of 2-hydroxyxanthone which is found in the seeds of *Mammea americana:*

Br KO
CO_2H OMe Cu →

O
CO_2H OMe → O OMe
O
226

The intramolecular acylation that concludes the synthesis is much more easily accomplished than its intermolecular counterpart and the weakest part of the sequence lies in the diphenyl ether formation. There can also be problems of orientation in the cyclization step; Goldberg and Wragg[187] have discussed some elementary situations in mechanistic terms. The *Kielmeyera* xanthones **227** and **228** cannot be prepared by the GSS method because of dealkylation but Dallacker and Damó offer the following solution[188] to the preparation of the former; the use of activated compounds to improve the Ullmann synthesis is noteworthy, as is the mildness of the cyclization conditions:

F, NO_2, CO_2Me + KO, OMe, O, O → DMF 80° → NO_2, O, OMe, CO_2Me, O, O

H_2/Pd; HNO_2 | Cu_2O

O, OMe, CO_2H, O, O ← KOH ← O, OMe, CO_2Me, O, O

AcCl, H_2SO_4 at 60°

O, OR, O, O, O

227 R = Me
228 R = H

The phenol **228** was made similarly, with the benzyloxy group instead of methoxy at the critical stages. However, the greatest successes of the route have been in the sterigmatocystin area (discussed earlier) where it was absolutely essential to employ none but the mildest reagents.

Recent innovations have again hinged upon benzophenone intermediates and their cyclization. Stout, Balkenhol, and their colleagues[189,190] combine methyl ethers (rather than phenols) with acid chlorides in true Friedel-Crafts reactions and cyclize the resulting benzophenone in *basic* conditions (tetramethylammonium hydroxide in pyridine). They have prepared a number of the highly methoxylated constituents of species of Gentianaceae in this manner; two examples will serve: (*a*) a synthesis of 1-hydroxy-2,3,4,7-tetramethoxyxanthone **229**, which is found in the root of *Frasera caroliniensis;* and (*b*) a synthesis of 1,2,3,7-tetramethoxyxanthone **230**:

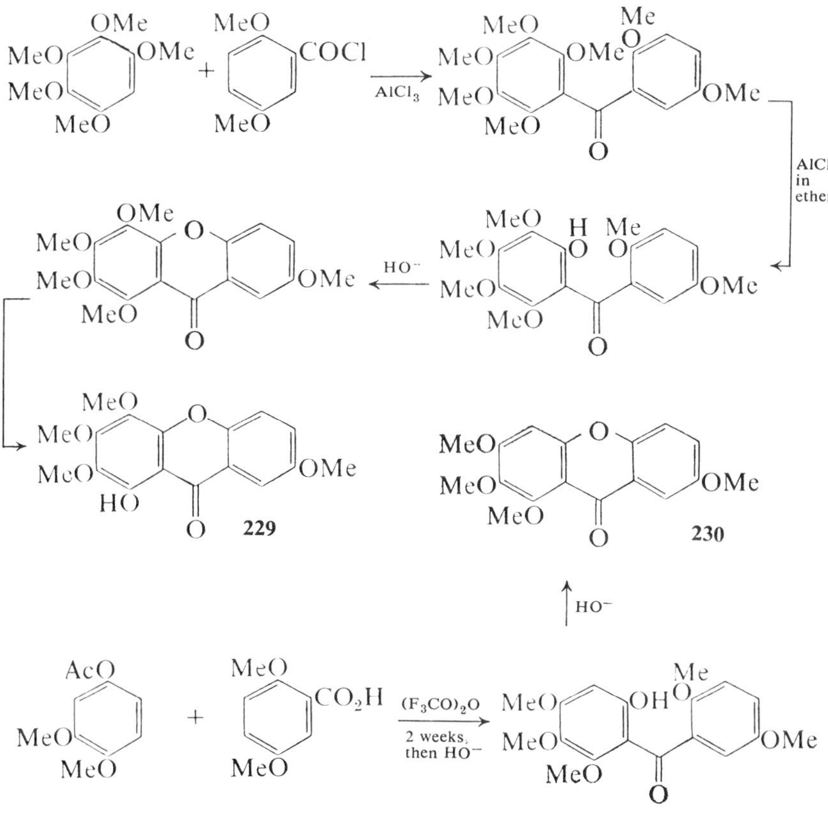

Both examples include selective demethylation by aluminum chloride of compounds with many methoxy groups at risk in so acid a medium. The gentlest possible technique is used (though boron chloride might serve the purpose better), and the only methoxy group affected is one subject to both coordination and to buttressing (or to a second incursion of the coordination phenomenon). We can now complete a series of flanking groups rendering methoxyl groups labile; in order of increasing efficiency this runs: one methoxy group, one carbonyl group, two methoxy groups, one methoxy with one carbonyl group, two carbonyl groups.

If desired, complete demethylation can be accomplished and the ring closed giving a polyhydroxyxanthone in one step. This is conveniently done using pyridinium chloride in a fashion prescribed by French authors.[191]

The most interesting feature is the base-catalyzed cyclization. The recognition of the possibility is due to Barton and Scott,[181] who pointed out that the benzophenone carbonyl group should promote nucleophilic addition-elimination reactions:

MeO O^- O → MeO O O^- → MeO^- O O

Considerable advances have been made following suggestions by Lewis[192] that many naturally occurring xanthones might arise by oxidative coupling in benzophenone precursors and his demonstration of the following reaction *in vitro*

HO OH OH O → HO O OH O

Several schools have elaborated upon the idea,[192–198] which receives impetus from tracer studies[194] showing that benzophenones can indeed act as xanthone precursors and from the (occasional) cooccurrence of both the xanthones that are theoretically possible when the cyclization can take alternative directions.[167] Chemical oxidation is usually reported as giving only the product of *para* coupling, but Atkinson and Lewis[195] have succeeded in isolating small amounts of the products of *ortho* coupling, as in the following instance in which two xanthones, **231** and **232**, found together in *Mammea americana* are formed simultaneously by ferricyanide oxidation:

a
b
pH 13
53%
10%
OH
O
231
232

Oxidative cyclization of 2,3′,6-trihydroxybenzophenone gave a similar pair of xanthones, permanganate proving the better oxidant in this case:

$KMnO_4$
32%
5%
OH
O

On the other hand, oxidation of the tetrahydroxybenzophenone **233** resulted only in the *para* coupled product, the plant xanthone gentisein **234**, a pigment of *Gentiana lutea*. In separate experiments it was demonstrated that the *ortho* coupled product **235** is so easily oxidized further that it would have been destroyed before detection.[195] This marks a rather serious limitation of the method, though it may perhaps be mitigated by the use of partly methylated benzophenones—interestingly, xanthone **235** has not been recognized as such in plants, although its ethers have been.

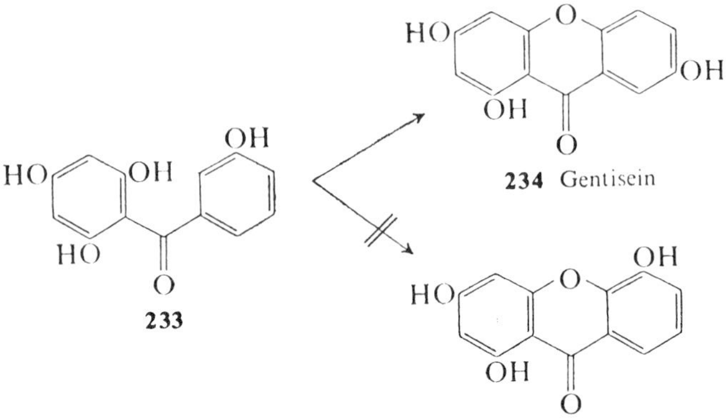

234 Gentisein

233

235

These oxidations are also known to be dependent not only on time of exposure to the reagent but also on the *p*H of the solution. Moreover, the enzymes peroxidase and laccase are capable of bringing about oxidative cyclization *in vitro*, *ortho*, and *para* coupling being observable.[195] Such cyclizations are generally held to have radical mechanisms demanding a phenolic hydroxyl group in the ring suffering substitution, a possible exception being DDQ oxidations which succeed with *para* methoxy groups suggesting electrophilic substitution and phenoxonium (phenoxylium) ion intermediates.[196]

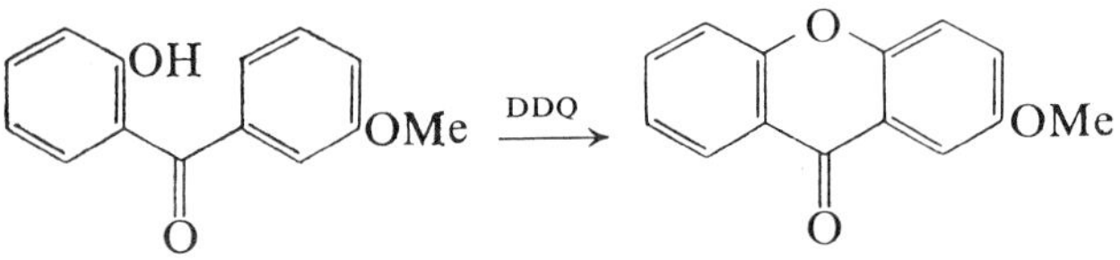

Somewhat similar studies by Ellis, Whalley, and Ball[197] disclose a different mechanism, oxidation to a quinone intervening. For example, ferricyanide oxidation converts **236** into the trihydroxyxanthone (**237**), presumably by a radical coupling mechanism, whereas some other oxidants supply the quinone **238**, which rapidly cyclizes in warm methanol to give the same product.

Another interesting variation has been provided by Jefferson and Scheinmann,[198] who discovered that the benzophenone maclurin **239** is best oxidised to the xanthone **240** (in 45% yield) by photochemical means. Although the product **241** of *ortho* coupling was not detected (it may have been destroyed concurrently), all three compounds do occur together in the plant *Symphonia globulifera*.[199] Similar photochemical cyclizations have given 1,6,7-trihydroxyxanthone, a constituent of *Garcinia eugenifolia* and *Mammea africana*. Unfortunately, the method has failed entirely in other cases even when ordinary chemical methods have been successful, and at present there seems to be no easy way of accounting for the differences.[200]

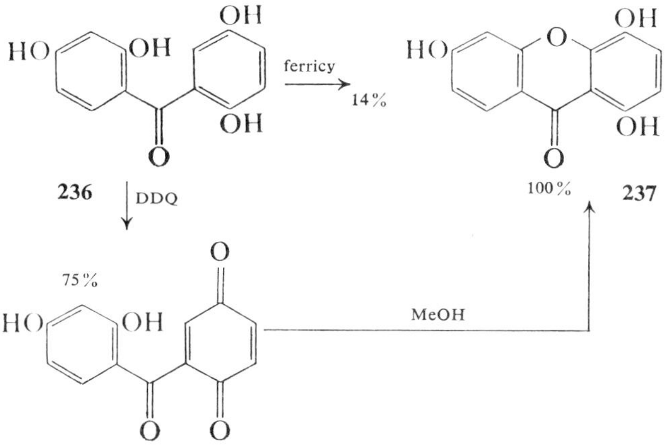

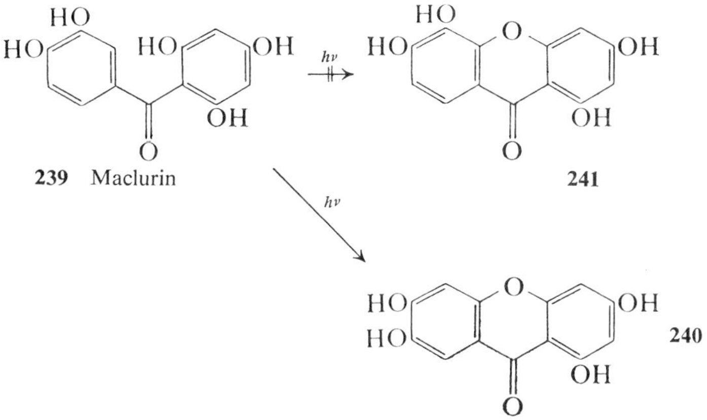

The benzophenones required for these studies can be made by standard techniques; that evolved by Usgaonkar and Jadhav[201] has proved particularly useful. For various reasons, however, some of which have already been touched on, derivatives of 2,6-dihydroxybenzophenone tend to be troublesome to prepare, yet, because they happen to be important, Locksley and Murray[200] have devised a general route to them (compare [46]). An example follows overleaf:

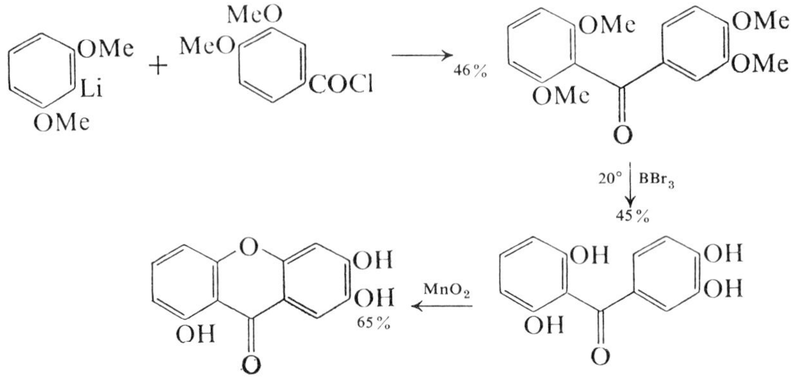

We can note three other methods briefly. That introduced by Tanase[202] utilizes xanthylium salt intermediates. Paquette's method[203] takas advantage of the synthetic powers of enamines, and Guyot and Mentzer's method[204] of the simplicity of noncatalyzed thermal condensations between suitably protected phenols and ethyl cyclohexanone-2-carboxylate. These last two methods require dehydrogenations for their completion.

We now leave the synthesis of the simpler xanthones to consider ways of building them up into more complex compounds. Scandinavian workers[205] find that chlorine in acetic acid converts norlichexanthone **208** into thiophanic acid **242**, a constituent of the lichen *Lecanora rupicola.* With a limited amount of chlorine the resorcinol ring is less attacked than the phloroglucinol ring (as would be predicted) so that some arthothelin **243** can be obtained. This is found in another lichen, *L. straminea*, and with diazomethane rapidly gives thuringion **244**. The selectivity of this reaction depends upon the very powerful acidifying effect that two *ortho* chlorine atoms have on a phenolic hydroxyl group. Finally, Raney nickel dehalogenation gave griseoxanthone C **245**, the sequence representing a most unusual way of effecting selective methylation in a xanthone derivative:

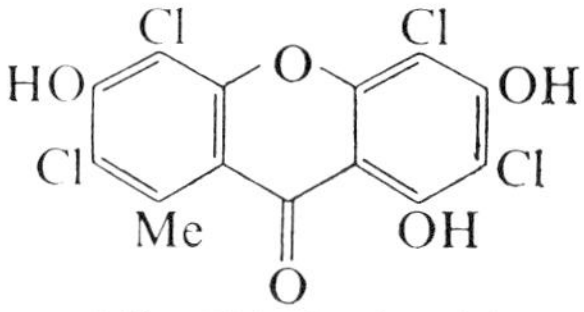

242 Thiophanic acid

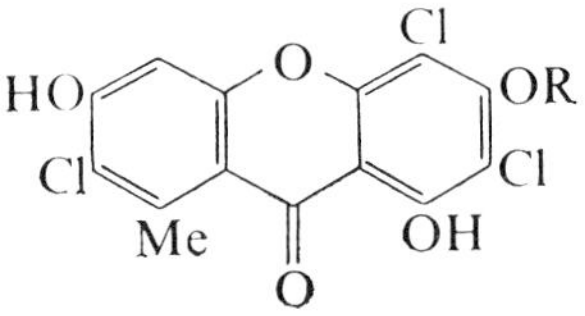

243 Arthothelin R = H
244 Thuringion R = Me

245 Griseoxanthone C

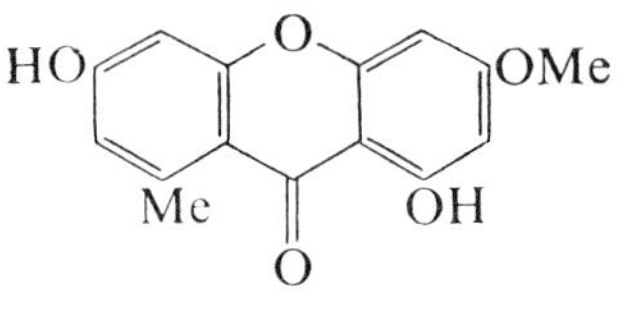

Belladifolin **246** occurs in *Gentiana belladifolia*. To synthesize it, Markham[206] neatly employed persulfate oxidation and selective methylation beginning with 1,3,8-trihydroxyxanthone. Neither reaction is truly selective, however, for oxidative attack on the phloroglucinol ring probably occurs and the methylation depends upon the enhanced acidity of hydroxyl *para* to a carbonyl group, an effect that spectroscopic studies[174,183] show to be much less clear cut in the xanthone than in other series.

$K_2S_2O_8$ DMS K_2CO_3

246 Belladifolin

The synthesis of polygalaxanthone-B **248** by Jain, Khanna, and Seshadri[207] began with xanthone **247** and illustrates a useful combination of Duff and Dakin reactions for providing hydroxyl groups often met with among other pyrones, especially the coumarins:[208]

247 $K_2S_2O_8$ DMS/K_2CO_3 Duff $(CH_2)_6N_4$ H_2O_2 DMS **248**

Formyl groups introduced by the Duff reaction or acetyl groups introduced by some form of Friedel-Crafts reaction have been used to build up furanoxanthones and other more complex combinations of rings despite the absence of any examples known to occur naturally.[209,210] We meet in the example one of the most important orientational specificities among hydroxyxanthones:

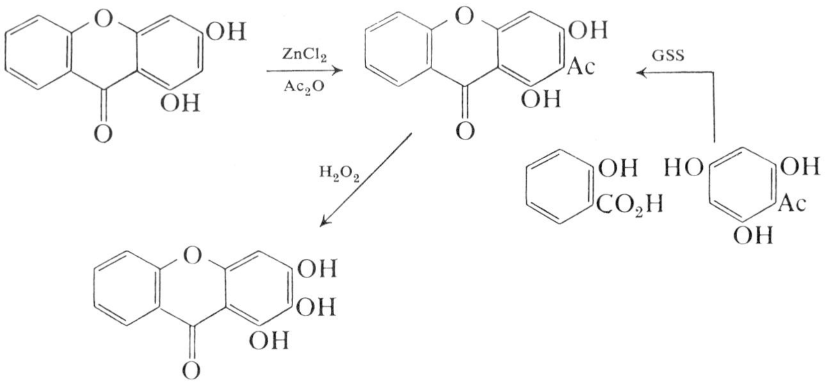

Specific 2-substitution characterizes most nuclear alkylation reactions of 1,3-dihydroxyxanthone derivatives. An important example is the synthesis of one of the more common xanthones, mangiferin **249**, with its unusual *C*-glucosyl substituent. This is introduced by treating 1,3,6,7-tetrahydroxyxanthone with 1-bromo-α-D-glucopyranosyl tetraacetate in methanolic sodium methoxide, with or without sodium iodide; not surprisingly, the yield is very poor.[211,212]

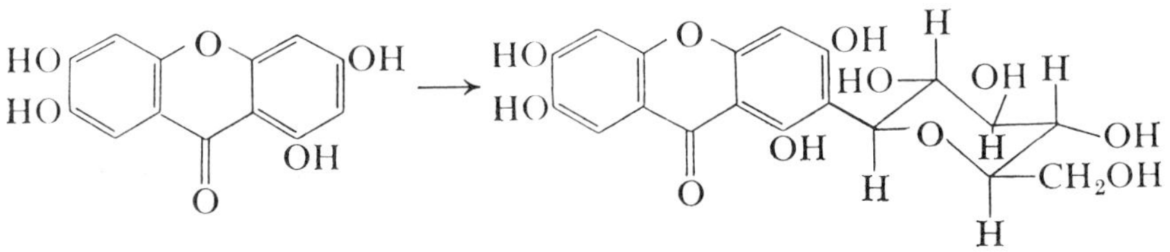

Mangiferin
249

Another example of this orientational effect occurs in routes to osajaxanthone **250**; as with the acetylxanthone mentioned previously, the orientation is observed whether the substituent is introduced into a preformed xanthone nucleus[213] or the xanthone ring is formed last.[172,179,214]

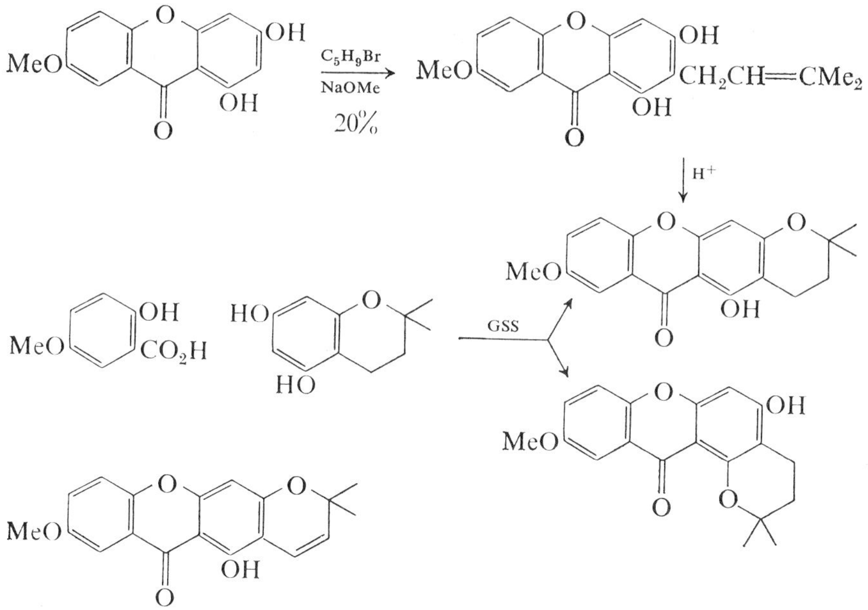

MeO OH OH O $\xrightarrow[\text{NaOMe}]{C_5H_9Br}$ 20% MeO OH $CH_2CH{=}CMe_2$ OH O

$\downarrow$ H^+

MeO OH O

MeO OH CO_2H HO HO GSS

MeO OH O

MeO OH O

Osajaxanthone

250

For the first fully definitive synthesis of a pyranoxanthone, jacareubin **255**, a pigment of the heartwoods of *Calophyllum brasiliense* Camb. and *C. sclerophyllum* Vesq., Jefferson and Scheinmann[215] began with **251**, having introduced the C_5 substituent by means of a combination of dimethylallyl bromide and silver oxide, which does not work well but is much better than the more usual bromide-sodium methoxide mixture (perhaps ion pairs are involved):

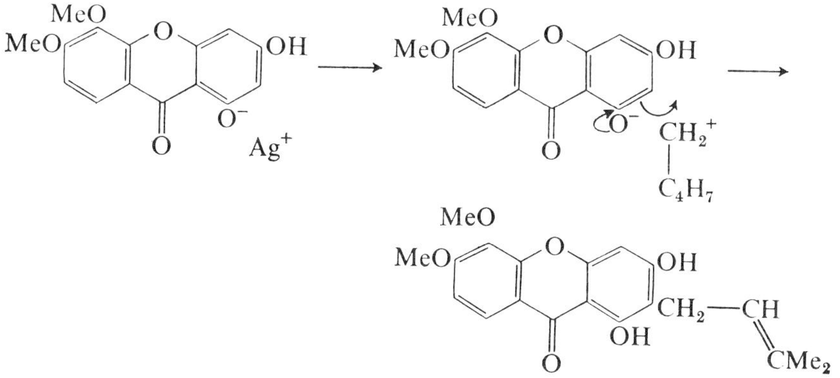

MeO MeO OH O O^- Ag^+ $\longrightarrow$ MeO MeO OH O O^- CH_2^+ C_4H_7 $\longrightarrow$

MeO MeO OH CH_2—CH CMe_2 OH O

251

Next, hydriodic acid deetherified the compound and cyclized it, both isomers (**252**, **253**) being obtained, although chelation of the 1-hydroxy group would be expected to oppose the formation of one. It may be that in the very strongly acidic solution the nucleus is protonated so that cyclization actually occurs in the salt **254** where chelation would be unimportant:

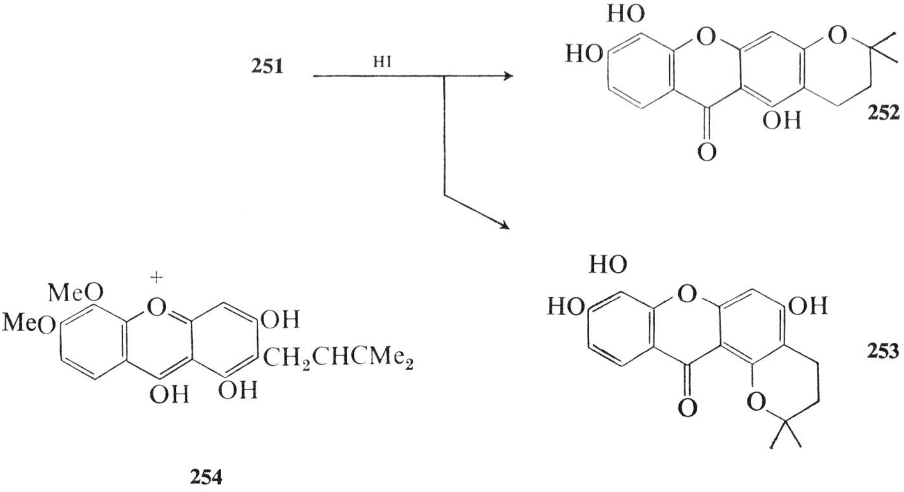

251 —HI→ 252 + 253

HO, HO, OH, O — **252**

HO, HO, OH, O — **253**

MeO, MeO, OH, CH_2CHCMe_2, OH, OH

254

In agreement, use of a weaker acid (formic acid) is known to favor the linear isomer in an equivalent case.[213] The final stages in the synthesis of jacareubin **255** can be summarized thus:

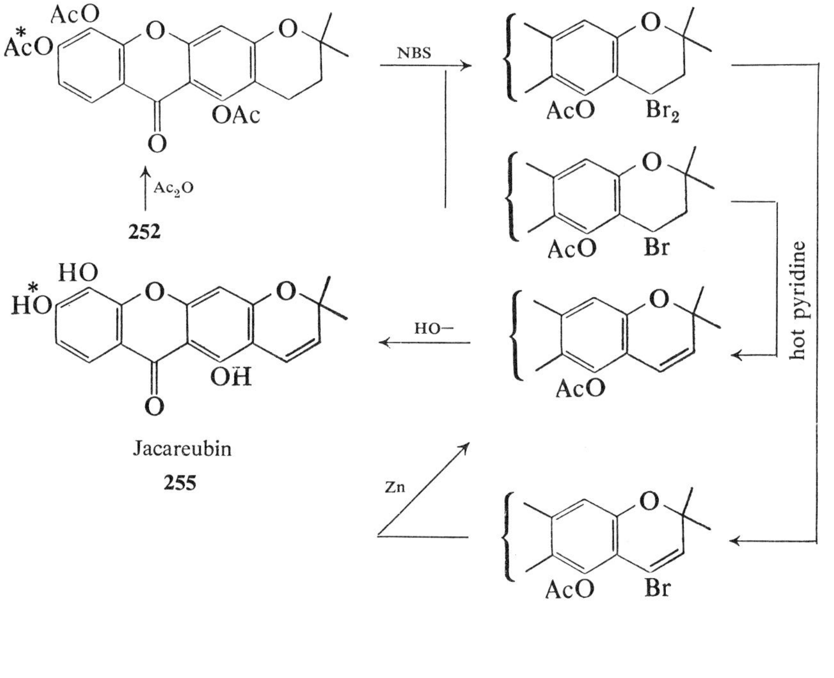

AcO, *AcO, OAc, O

↑ Ac_2O

252

NBS → AcO Br_2; AcO Br

hot pyridine → AcO; AcO Br

Zn

HO−

HO, *HO, OH, O

Jacareubin

255

The foregoing scheme includes bromination-dehydrobromination as a method of converting a chroman into a chromene, a practice going out of fashion now that gentler and more easily controlled syntheses are available; the dibromination sequence came to light during parallel work on deoxyjacareubin[177] (starred groups omitted). A direct synthesis of jacareubin **255** by Crombie's method was mentioned earlier but without reference to the orientational problem. This can be taken up now, routes to deoxyjacareubin offering the opportunity.

One approach to deoxyjacareubin **257** takes advantage of the Claisen rearrangement of the simple allyl ether **258**, which readily affords **259**, protection of the phenolic group and oxidative fission of the olefinic link following.[177] The resulting aldehyde **260** has many synthetic uses; sequences of this kind were originally introduced for the synthesis of furopyrones[216] and have recently been employed[174] in elaborating the side chain of scriblitifolic acid **266**. The present example[177] continues with the appropriate Wittig reaction needed to complete the desired dimethylallyl substituent, and after that boron chloride serves to remove one methyl group giving the phenol **262**.

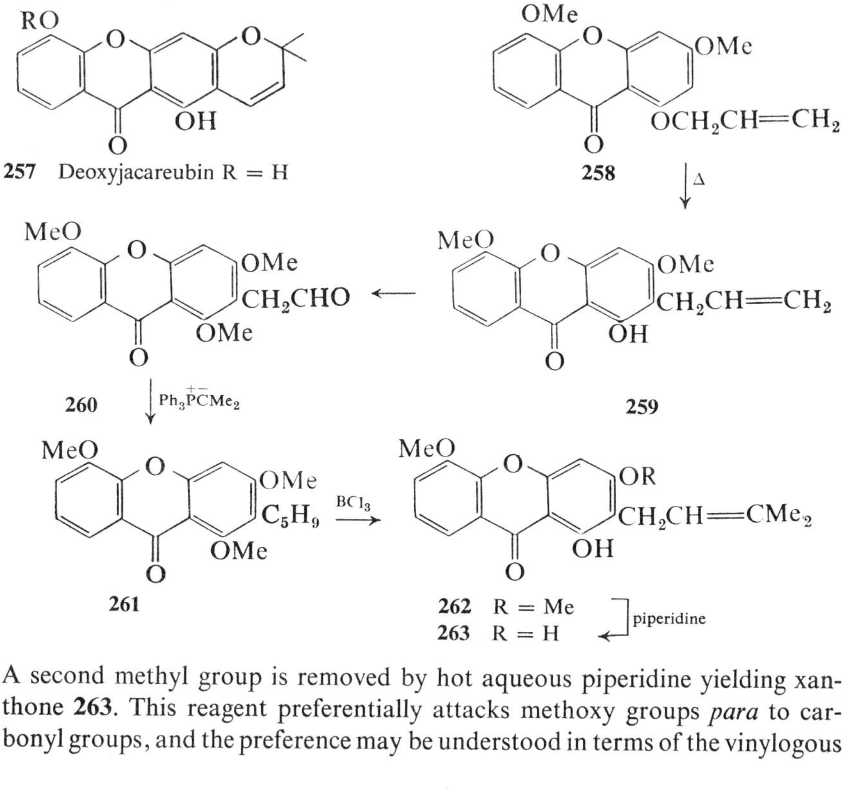

RO O O OH O

257 Deoxyjacareubin R = H

OMe O OMe O $OCH_2CH{=}CH_2$

258

Δ

MeO O OMe $CH_2CH{=}CH_2$ OH O

259

MeO O OMe CH_2CHO OMe O

260

$Ph_3\overset{+}{P}\overset{-}{C}Me_2$

MeO O OMe C_5H_9 OMe O

261

BCl_3

MeO O OR $CH_2CH{=}CMe_2$ OH O

262 R = Me
263 R = H

piperidine

A second methyl group is removed by hot aqueous piperidine yielding xanthone **263**. This reagent preferentially attacks methoxy groups *para* to carbonyl groups, and the preference may be understood in terms of the vinylogous ester system comprised by that arrangement. An *ortho* carbonyl group could serve similarly but, as is usually the case, *ortho* activation is less than *para* activation. The formation of buchanaxanthone[217] **265** from xanthone **264** is another example from the recent literature, but it must be noted that the degree of selectivity seems to vary considerably and unpredictably.

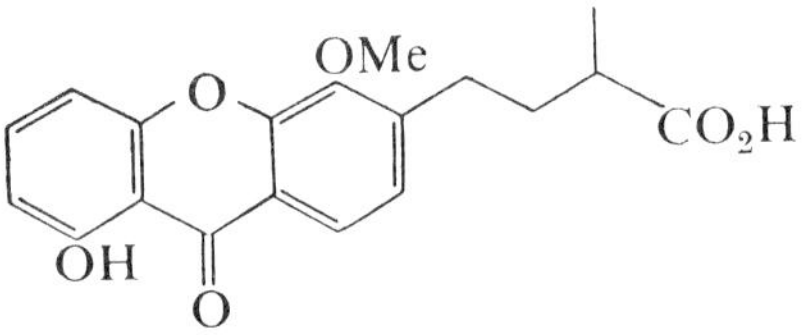

Scriblitifolic acid
266

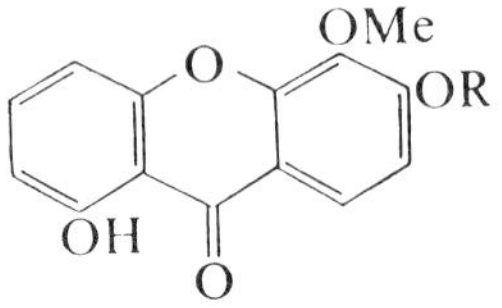

Buchanaxanthone **264** R = Me
265 R = H

The synthesis is completed with DDQ oxidation of **263** to give the desired linear product, deoxyjacareubin 5-methyl ether (**257**; R = Me), since the other hydroxyl group is protected by chelation. This result definitely orients the ring fusions in deoxyjacareubin, but this compound has not been prepared from the ether. Demethylation of the last methoxy group would not be easy and would endanger the pyran system. Deoxyjacareubin itself has been synthesized[177] by methods used for jacareubin.

Application of the Crombie chromene synthesis to the deoxyjacareubin problem discloses a curious anomaly. As Lewis and Reary have shown, the condensation of 1,3,5-trihydroxyxanthone with 3-methylcrotonaldehyde in boiling pyridine yields deoxyjacareubin **257** directly.[218] Other workers, however, find that 1,3-dihydroxy-5-methoxyxanthone **267** yields only the "wrong" isomer **268** in dilute solution, thus abrogating the orientation rule strictly obeyed in all the previous discussion. The secret lies in the concentration; for when this is high, the linear compound **269** does make an appearance.[177] One might guess that, since the boiling point of the concentrated

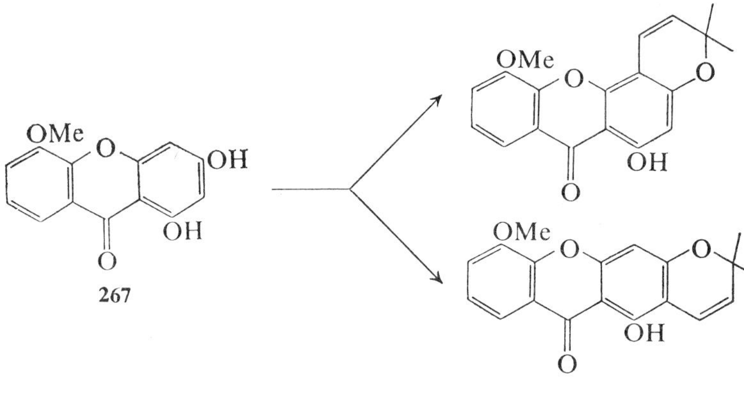

267

268

269

solution must be considerably higher than of the dilute solution, we may be witnessing here a changeover from kinetic to thermodynamic control. The appropriate studies have yet to be made.

Alvaxanthone **270** possesses a different type of prenyl group. An obvious synthetical plan would be based on a Claisen rearrangement in the ether **271**; this works, but not well.[180] The difficulty lies in fitting the bulky branched end of the C_5 group between the hydroxy and methoxy groups, and it has two consequences. The first is that the migrating group continues on its way, with another inversion, to the *para* position; and, since there is now no marked steric effect, **272** is the chief product. The second is that ring closure of the desired compound **273** is promoted, so the small yield is depleted further giving the furanoxanthone derivative **274**.

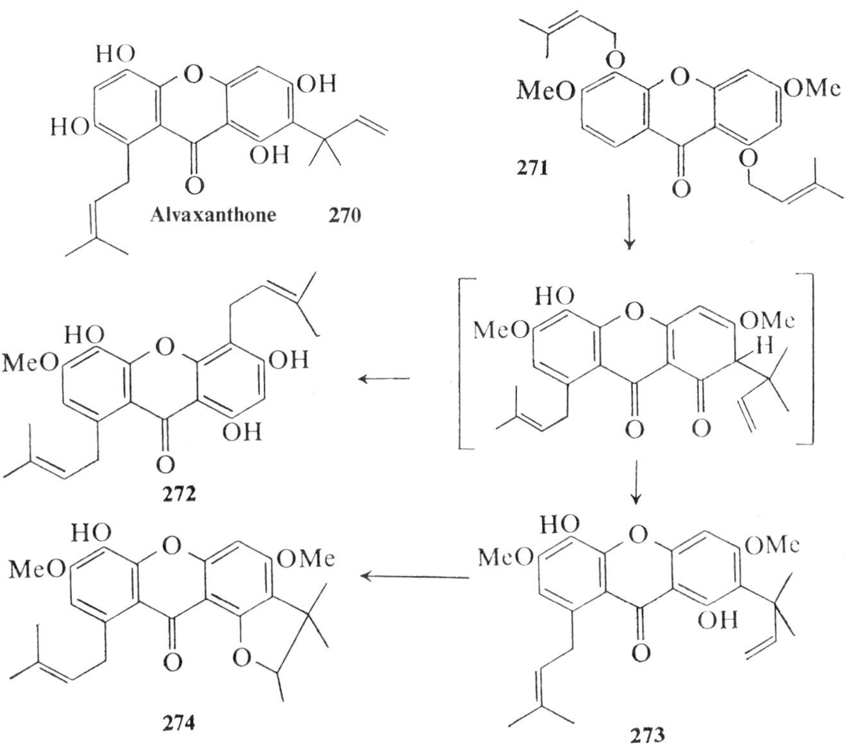

Alvaxanthone **270**

271

272

274

273

Another fact that has to be kept in mind is that in this series prenyl ethers may have an unexpected sensitivity to silica.[219] Attempts to purify **275** on silica columns lead to complex mixtures containing fair quantities of the 2-alkenylxanthone **276** not obtainable by Claisen rearrangement. It is believed that silica is acting as an acid* and inducing ion pair formation as in **277** (cf. p. 549). Whatever the explanation, the chromatographic analysis of natural materials has to be conducted with particular caution where allylic materials might be involved. Other workers have also noted the acidity of silica; xanthones sometimes show marked color changes on this column packing; and, because phenols are less basic than their ethers in the xanthone series, methylation induces a fall in R_f instead of the usual rise.[189]

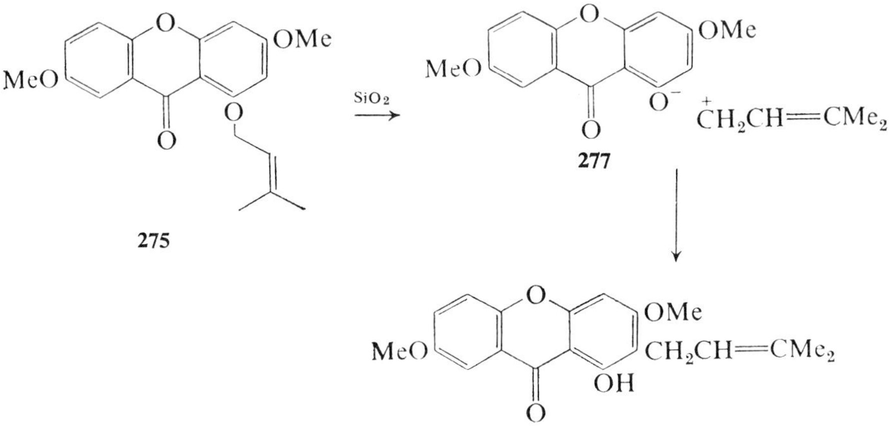

In conclusion, we consider the most novel recent advance in the chemistry of prenylxanthones. At the beginning of this chapter it was pointed out that some Claisen rearrangements might lead to bridged xanthone nuclei of the kind that characterize gambogic acid, the morellins, and bronianone. During the writing, the idea has been brought nearer actuality by Quillinan and Scheinmann,[220] who have demonstrated that the 5,6-diallyl ether **278** of jacareubin gives just such a bridged system **279** when heated in boiling decalin for 14 hours. (It should be noted that in 6-allyloxyxanthones migration is normally to position 5 as shown in the first stage.) The second stage is merely an internal Diels-Alder addition. Despite the severity of the reaction conditions in this instance, the reaction may well have biosynthetical implications.

A final note on spectroscopy. Xanthones have changed in the last few years from little known compounds to well studied ones, and considerable spectroscopical information is now available that helps enormously in deciding structures and particularly orientations in synthetical sequences. Quantities of

* In work not yet published[219] it was found that only prenyl ethers at position 1 are thus affected. This is entirely in accord with acid catalysis.

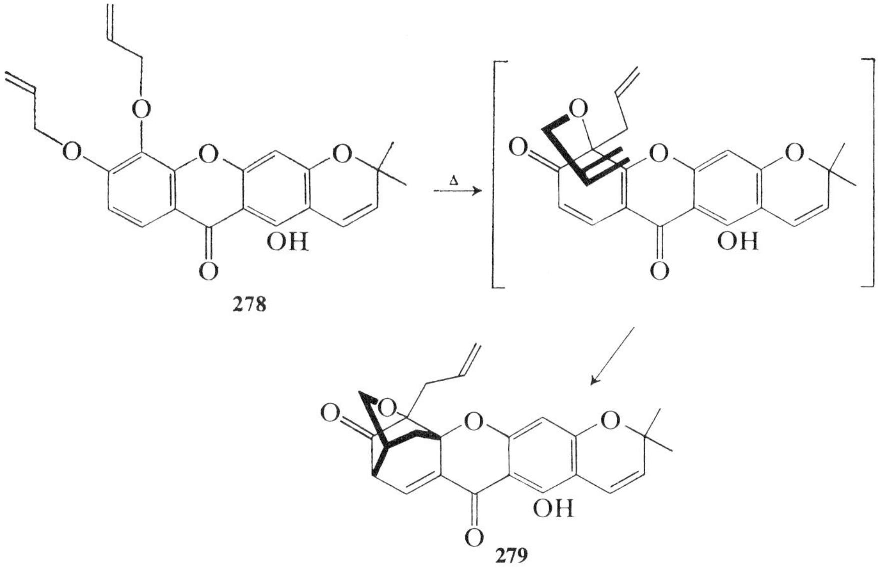

278

279

data are available for ultraviolet spectra[174,176,217,219,221,222] and for NMR spectra.[174,177,180,190,222,223]

REFERENCES

1. W. D. Ollis and I. O. Sutherland, in W. D. Ollis, Ed., *The Chemistry of Natural Phenolic Materials*, Pergamon, Oxford, 1961, Chapter 4.
2. F. M. Dean, *Naturally Occurring Oxygen Ring Compounds*, Butterworths, London, 1963, Chapter 7.
3. W. B. Whalley, in W. D. Ollis, Ed., *The Chemistry of Natural Phenolic Materials*, Pergamon, Oxford, 1961, Chapter 2.
4. W. D. Ollis, M. V. J. Ramsey, I. O. Sutherland, and S. Mongkolsuk, *Tetrahedron*, **21**, 1453 (1965).
5. R. Livingstone and R. B. Watson, *J. Chem. Soc.*, **1956**, 3701.
6. L. I. Smith and P. M. Ruoff, *J. Am. Chem. Soc.*, **62**, 145 (1940).
7. J. N. Chatterjea, *J. Indian Chem. Soc.*, **36**, 76 (1959).
8. J. Cottam, R. Livingstone, M. Walshaw, K. D. Bartle, and D. W. Jones, *J. Chem. Soc.*, **1965**, 5261.
9. C. S. Barnes, M. I. Strong, and J. L. Occolowitz, *Tetrahedron*, **19**, 839 (1963).
10. T. R. Kasturi and T. Manithomas, *Tetrahedron Lett.*, **1967**, 2573.
11. E. Späth and R. Hillel, *Ber. dtsch. chem. Ges.*, **72**, 963, 2093 (1939); E. Späth and H. Schmid, *Ber. dtsch. chem. Ges.*, **74**, 193 (1941).
12. G. Cardillo and L. Merlini, *Gazzetta*, **98**, 191 (1968).
13. S. K. Mukerjee, S. C. Sarkar, and T. R. Seshadri, *Tetrahedron*, **25**, 1063 (1969).

14. J. Nickl, *Chem. Ber.*, **91,** 1372 (1958).
15. H. D. Schroeder, W. Bencze, O. Halpern, and H. Schmid, *Chem. Ber.*, **92,** 2338 (1959).
16. I. Iwade and J. Ide, *Chem. Pharm. Bull. (Japan)*, **10,** 926 (1962); **11,** 1042 (1963).
17. J. Hlubucek, E. Ritchie, and W. C. Taylor, *Tetrahedron Lett.*, **1969,** 1369.
18. J. Zsindely and H. Schmid, *Helv. Chim. Acta*, **51,** 1510 (1968).
19. R. D. H. Murray, M. M. Ballantyne, and K. P. Mathai, *Tetrahedron Lett.*, **1970,** 243.
20. A. J. Birch and M. Maung, *Tetrahedron Lett.*, **1967,** 3275.
21. L. Edwards, J. Kolc, and R. S. Becker, *Photochem. Photobiol.* **13,** 423 (1971); J. Kolc and R. S. Becker, *J. Phys. Chem.*, **71,** 4045 (1967).
22. A. B. Turner, *Q. Rev.*, **18,** 347 (1964).
23. I. M. Campbell, C. H. Calzadilla, and N. J. McCorkindale, *Tetrahedron Lett.*, **1966,** 5107.
24. E. A. Brande, L. M. Jackman, R. P. Linstead, and G. Lowe, *J. Chem. Soc.*, **1960,** 3133, 3144.
25. R. Mechoulam, B. Yagnitinsky, and Y. Gaoni, *J. Am. Chem. Soc.*, **90,** 2418 (1968).
26. G. Cardillo, R. Cricchio, and L. Merlini, *Tetrahedron*, **24,** 4825 (1968).
27. J. A. Miller and H. C. S. Wood, *Chem. Comm.*, **39,** 40 (1965).
28. L. Jurd, K. Stevens, and G. Manners, *Tetrahedron Lett.*, **1971,** 2275.
29. F. Bohlmann and K. M. Kleine, *Chem. Ber.*, **99,** 885 (1966).
30. A. C. Jain and M. K. Zutshi, *Tetrahedron Lett.*, **1971,** 3179.
31. D. E. Games and N. J. Haskins, *Chem. Comm.*, **1971,** 1005.
32. J. R. Beck, R. Kwok, R. N. Booker, A. C. Brown, L. E. Patterson, P. Pranc, B. Rockey, and A. Pohland, *J. Am. Chem. Soc.*, **90,** 4706 (1968).
33. L. Merlini and R. Mondelli, *Tetrahedron*, **24,** 497 (1968).
34. J. Links, *Biochem. Biophys. Acta*, **38,** 193 (1960).
35. F. W. Hemming, R. A. Morton, and J. F. Pennock, *Biochem. J.*, **80,** 445 (1961).
36. D. McHale and J. Green, *Chem. Ind.*, **1962,** 1867; *J. Chem. Soc.*, **1965,** 5060.
37. B. O. Linn, C. H. Shunk, E. L. Wong, and K. Folkers, *J. Am. Chem. Soc.*, **85,** 239 (1963).
38. H. W. Moore and K. Folkers, *Ann.*, **684,** 212 (1965).
39. R. Huls, *Bull. Soc. Chim. Belg.*, **67,** 22 (1958).
40. F. Baranton, G. Fontaine, and P. Maitte, *Bull. Soc. Chim. France*, **1968,** 4203.
41. A. K. Ganguly, B. S. Joshi, V. N. Kamat, and A. H. Manmade, *Tetrahedron*, **23,** 4777 (1967).
41a. J. R. Collier and A. S. Porter, *Chem. Comm.*, **1972,** 618.
42. W. Bencze, J. Eisenbeiss, and H. Schmid, *Helv. Chim. Acta*, **39,** 923 (1956).
43. T. R. Seshadri and M. S. Sood, *Tetrahedron Lett.*, **1964,** 3367.
44. S. N. Shanbag, *Tetrahedron*, **20,** 2605 (1964).
45. K. Hata and K. Sano, *Tetrahedron Lett.*, **1964,** 3367.
46. S. Nozoe and K. Hirai, *Tetrahedron Lett.*, **1969,** 3017.
47. R. M. Bowman, M. F. Grundon, and K. J. James, *Chem. Comm.*, **1970,** 666.

48. L. Crombie and R. Ponsford, *Chem. Comm.*, (a) **1968**, 368; (b) **1968,** 894.

49. V. Kane and R. K. Razdan, *J. Am. Chem. Soc.*, **90,** 6551 (1968); *Tetrahedron Lett.*, **1969,** 591.

50. M. J. Begley, D. G. Clarke, L. Crombie, and D. A. Whiting, *Chem. Comm.*, **1970,** 1547.

51. L. Crombie, R. Ponsford, A. Shani, B. Yagnitinsky, and R. Mechoulam, *Tetrahedron Lett.*, **1968,** 5771.

52. S. P. Kureel, R. S. Kapil, and S. P. Popli, *Chem. Ind.*, **1970,** 958; *Chem. Comm.*, **1969,** 1120.

53. W. M. Bandaranyake, L. Crombie, and D. A. Whiting, *Chem. Comm.*, (a) **1969,** 58; (b) **1959,** 970.

54. W. J. G. Donnelly and P. V. R. Shannon, *Chem. Comm.*, **1971,** 76.

55a. E. E. Schweizer, J. Liehr, and D. J. Monaco, *J. Org. Chem.*, **33,** 2416 (1968); E. E. Schweizer, E. T. Schaffer, C. T. Hughes, and C. J. Berninger, *J. Org. Chem.*, **31,** 2907 (1966).

55b. N. S. Narazimhan, M. V. Paradkar, and A. M. Gokhale, *Tetrahedron Lett.*, **1970,** 1664.

56. K. Sargeant, R. B. A. Carnaghan, and R. Allcroft, *Chem. Ind.*, **1963,** 53; K. J. van der Merwe, L. Fourie, and de B. Scott, *Chem. Ind.*, 1963, 1660; F. Dickens and H. E. H. Jones, *Br. J. Cancer*, **19,** 392 (1965); G. N. Wogan, *Bacteriol Rev.*, **30,** 460 (1966); P. C. Spensley, *Endeavour*, **22,** 75 (1963); C. W. Hesseltine, O. L. Shotwell, J. J. Ellis, and R. D. Stubblefield, *Bacteriol. Rev.*, **30,** 795 (1966).

57. C. W. Holzapfel, P. S. Steyn, and I. F. H. Purchase, *Tetrahedron Lett.*, **1966,** 2799; I. F. H. Purchase, *Fed. Cosmet. Toxicol.*, **5,** 339 (1967).

58. E. Bullock, J. C. Roberts, and J. G. Underwood, *J. Chem. Soc.*, **1962,** 4179.

59. T. Hamasaki, M. Renbutsu, and Y. Hatsuda, *Agric. Biol. Chem.* (*Japan*), **31,** 11 (1967).

60. C. A. Bear, J. M. Waters, T. N. Waters, R. T. Gallagher, and R. Hodges, *Chem. Comm.*, **1970,** 1705.

61. T. Asao, G. Büchi, M. M. Abdel-Kader, S. B. Chang, E. L. Wick, and G. N. Wogan, *J. Am. Chem. Soc.*, **87,** 882 (1965).

62. H. E. Zaugg, F. E. Chadde, and R. J. Michaels, *J. Am. Chem. Soc.*, **84,** 4567 (1962).

63. E. C. M. Coxworth, *Can. J. Chem.*, **45,** 1777 (1967); B. S. Thyagarajan, K. K. Balasubramanian, and R. Bhima Rao, *Can. J. Chem.*, **44,** 633 (1966).

64. C. H. Eugster and P. Karrer, *Chimia*, **18,** 358 (1964).

64a. J. S. E. Holker, personal communication.

65. P. L. Sawhney and T. R. Seshadri, *Proc. Indian Acad. Sci.*, **37A,** 592 (1953).

66. G. Büchi, D. M. Foulkes, M. Kurono, G. F. Mitchell, and R. S. Schneider, *J. Am. Chem. Soc.*, **89,** 6745 (1967).

67. J. A. Knight, J. C. Roberts, and P. Roffey, *J. Chem. Soc.*, **1966,** 1308.

68. A. Schiavello and E. Cingolani, *Gazzetta*, **18,** 717 (1951).

69. J. A. Knight, J. C. Roberts, P. Roffey, and A. H. Sheppard, *Chem. Comm.*, **1966,** 706.

70. J. C. Roberts, A. H. Sheppard, J. A. Knight, and P. Roffey, *J. Chem. Soc. (C)*, **1968,** 22.

71. S. Brechbuhler, G. Büchi, and G. Milne, *J. Org. Chem.*, **32,** 2641 (1967).
72. J. G. Underwood and J. S. E. Holker, *Chem. Ind.*, **1964,** 1865.
73. C. L. Lange, H. Wanhoff, and F. Korte, *Chem. Ber.*, **100,** 2312 (1967).
74. G. Büchi and S. M. Weinreb, *J. Am. Chem. Soc.*, **69,** 5408 (1969).
75. S. Sethna and R. Phadke, *Org. Reactions*, **7,** 12 (1953).
76. B. W. Bycroft, J. R. Hutton, and J. C. Roberts, *J. Chem. Soc.* (*C*), **1970,** 281.
77. M. J. Rance and J. C. Roberts, *Tetrahedron Lett.*, **1969,** 277; **1970,** 2799.
78. R. J. Cole and J. W. Kirksey, *Tetrahedron Lett.*, **1970,** 3109
79. G. C. Elsworthy, J. S. E. Holker, J. M. McKeown, J. B. Robinson, and L. J. Mulheirn, *Chem. Comm.*, **1970,** 1069; M. Biollaz, G. Büchi, and G. Milne, *J. Am. Chem. Soc.*, **92,** 1035 (1970).
80. E. Bullock, D. Kirkaldy, J. C. Roberts, and J. G. Underwood, *J. Chem. Soc.*, **1963,** 829; G. M. Holmwood and J. C. Roberts, *Tetrahedron Lett.*, **1971,** 833.
81. J. S. Davies, V. H. Davies, and C. H. Hassall, *J. Chem. Soc.* (*C*), **1969,** 1873; C. H. Hassall and B. A. Morgan, *Chem. Comm.*, **1970,** 1345.
82. L. Crombie and R. Peace, *Chem. Comm.*, **1963,** 246.
83. G. Büchi, L. Crombie et al., *J. Chem. Soc.*, **1961,** 2861.
84. F. B. LaForge, H. L. Haller, and L. E. Smith, *Chem. Rev.*, **12,** 182 (1933); L. Feinstein and M. Jacobson, *Fortschr. Chem. Org. Naturst.*, **10,** 436 (1953); H. L. Haller, L. D. Goodhue, and H. A. Jones, *Chem. Rev.*, **30,** 33 (1942).
85. N. Nakatani and M. Matsui, *Agric. Biol. Chem.* (*Japan*), **32,** 769 (1968).
86. D. J. Ringshaw and H. J. Smith, *J. Chem. Soc.* (*C*), **1968,** 102.
87. F. B. LaForge and H. L. Haller, *J. Am. Chem. Soc.*, **54,** 810 (1932).
88. H. Fukami, M. Nakayama, and M. Nakajima, *Agric. Biol. Chem.*, **25,** 243 (1961).
89. O. Dann and G. Volz, *Justus. Liebigs Ann.*, **631,** 102, 111 (1960).
90. G. Büchi and L. Crombie, *J. Chem. Soc.*, **1961,** 2843.
91. M. Miyano and M. Matsui, *Chem. Ber.*, **92,** 1438, 2487 (1959); *Bull. Agric. Chem. Soc. Japan*, **24,** 540 (1960).
92. L. Crombie and J. W. Lown, *Proc. Chem. Soc.* (*London*), **1961,** 299.
93. M. Miyano, *J. Am. Chem. Soc.*, **87,** 3962 (1965).
94. A. Robertson and G. Rusby, *J. Chem. Soc.*, **1936,** 212.
95. J. R. Herbert, W. D. Ollis, and R. C. Russell, *Proc. Chem. Soc.*, **1960,** 177.
96. M. Miyano, *J. Am. Chem. Soc.*, **87,** 3958 (1965).
97. M. Baran-Marszak, J. Massicot, and C. Mentzer, *Compt. rend.*, **263C,** 173 (1966).
98. T. R. Seshadri and S. Varadarajan, *Proc. Indian Acad. Sci.*, **37A,** 784 (1953).
99. A. C. Mehta and T. R. Seshadri, *Proc. Indian Acad. Sci.*, **42A,** 192 (1955).
100. Y. Kawase and C. Numata, *Chem. Ind.*, **1961,** 1361.
101. V. Chandrashekar, M. Krishnamurti, and T. R. Seshadri, *Tetrahedron*, **23,** 2505 (1967).
102. K. Aghoramurthy, A. S. Kubla, and T. R. Seshadri, *J. Indian Chem. Soc.*, **38,** 914 (1961).
103. K. Fukui, M. Nakayama, and T. Harano, *Bull. Chem. Soc. Japan*, **42,** 1693 (1969).
104. G. A. Caplin, W. D. Ollis, and I. O. Sutherland, *J. Chem. Soc.* (*C*), **1968,** 2302.

105. L. Crombie, P. W. Freeman, and D. A. Whiting, *Chem. Comm.*, **1970,** 563.

106. L. Crombie, C. L. Green, and D. A. Whiting, *J. Chem. Soc. (C)*, **1968,** 3029.

107. V. R. Sathe and K. Venkataraman, *Current Sci.*, **18,** 373 (1949).

108. W. D. Ollis, K. L. Ormand, and I. O. Sutherland, *J. Chem. Soc. (C)*, **1970,** 119; W. D. Ollis, K. L. Ormand, B. T. Redman, R. J. Roberts, and I. O. Sutherland, *J. Chem. Soc. (C)*, **1970,** 125.

109. C. A. Weber-Schilling and H. W. Wanslick, *Tetrahedron Lett.*, **1964,** 2345.

110. E. K. Adesogan, F. M. Dean, and M. L. Robinson, unpublished observations.

111. W. J. Bowyer, A. Robertson, and W. B. Whalley, *J. Chem. Soc.*, **1957,** 542.

112. N. R. Krishnaswamy and T. R. Seshadri, *J. Sci. Ind. Res. India*, **16B,** 268 (1957).

113. A. H. Gilbert, A. McGookin, and A. Robertson, *J. Chem. Soc.*, **1957,** 3740.

113a. T. R. Govindachari, K. Nagarajan, and P. C. Parthasarathy, *J. Chem. Soc.*, **1957,** 548; O. H. Emerson and E. M. Bickoff, *J. Am. Chem. Soc.*, **80,** 4381 (1958); Y. Kawase, *Bull. Chem. Soc., Japan*, **32,** 690 (1959).

113b. C. Deschamps-Vallet and C. Mentzer, *Compt. Rendu.*, **251,** 736 (1960).

114. V. K. Kalra, A. S. Kubla, and T. R. Seshadri, *Tetrahedron Lett.*, **1967,** 2153.

115. D. Nasipuri and G. Pyne, *J. Sci. Indust. Res. India*, **21B,** 51 (1962).

116. K. Fukui and M. Nakayama, *J. Chem. Soc. Japan*, **85,** 790 (1964).

117. H.-W. Wanslick, R. Gritsky, and H. Heidepriem, *Chem. Ber.*, **96,** 305 (1963).

118. M. Darbarwar, V. Sundaramurthy, and N. V. Subba Rao, *Current Sci.*, **38,** 13 (1969).

118a. H. W. Wanslick, in W. Foerst, Ed., Newer Methods of Preparitive Organic Chemistry, Academic Press, New York, 1968, p. 139.

119. D. Bouwer, G. v. d. M. Brink, J. P. Engelbrecht, and G. J. H. Rall, *J. S. African Chem. Inst.*, **21,** 159 (1968).

120. K. Fukui and M. Nakayama, *Tetrahedron Lett.*, **16,** 1805 (1966); **1965,** 2559.

121. L. L. Simonova and A. A. Shamshurin, *Zh. Vsesoyuz. Khim. Obshch. im. D. I. Mendeleeva*, **11,** 252 (1966); *Chem. Abs.*, **65,** 10572 (1966); *Chem. Abs.*, **67,** 64380 (1967).

122. K. Fukui, M. Nakayama, and H. Sesita, *Bull. Chem. Soc., Japan*, **37,** 1887 (1964).

123. W. Dilthey and W. Hoschen, *J. Prakt. Chem.*, **246,** 42 (1933); P. Karrer, R. Widmer, A. Helfenstein, W. Hurliman, O. Nievergelt, and P. Monsarrat-Thomas, *Helv. Chim. Acta*, **10,** 729 (1927).

124. L. Jurd, *Tetrahedron Lett.*, **1963,** 1151; *J. Org. Chem.*, **29,** 2602 (1964).

125. L. Jurd, *Tetrahedron*, **1966,** 2913.

126. L. Jurd, *J. Org. Chem.*, **1964,** 3036.

127. A. L. Livingstone, S. C. Witt, R. F. Lundin, and E. M. Bickoff, *J. Org. Chem.*, **30,** 2353 (1965).

128. L. Jurd, *J. Pharm. Sci.*, **54,** 1221 (1965).

129. R. R. Spencer, B. E. Knuckles, and E. M. Bickoff, *J. Heterocycl. Chem.*, **3,** 450 (1966).

130. K. Fukui, M. Nakayama, H. Tsuge, and K. Tsuzuki, *Experientia*, **24,** 536 (1968).

131. A. L. Livingstone, E. M. Bickoff, R. E. Lundin, and L. Jurd, *Tetrahedron*, **20,** 1963 (1964).

132. R. R. Spencer, B. E. Knuckles, and E. M. Bickoff, *J. Org. Chem.*, **31,** 988 (1966).
133. L. Jurd, *J. Org. Chem.*, **27,** 872 (1962).
134. P. M. Dewick, W. Barz, and H. Grisebach, *Chem. Comm.*, **1969,** 466.
135. G. W. K. Cavill, F. M. Dean, J. F. E. Keenan, A. McGookin, A. Robertson, and G. B. Smith, *J. Chem. Soc.*, **1958,** 1544.
136. S. H. Harper, A. D. Kemp, and W. G. Underwood, *Chem. Comm.*, **1965,** 309.
137. K. Fukui, M. Nakayama, and T. Harano, *Bull. Chem. Soc. Japan*, **42,** 233 (1969).
138. K. Fukui, M. Nakayama, and K. Tsuzubi, *Experientia*, **25,** 122 (1969); *Bull. Chem. Soc. Japan*, **42,** 2395 (1969).
139. K. M. Biswas, L. E. Houghton, and A. H. Jackson, *Tetrahedron* Suppl., **22**(7), 261 (1966).
140. H. Suginome and T. Iwadare, *Bull. Chem. Soc., Japan*, **33,** 567 (1960).
141. M. Uchiyama and K. Ooba, *J. Agric. Chem. Soc. Japan*, **42,** 688 (1968).
142. K. Fukui and M. Nakayama, *Bull. Chem. Soc. Japan*, **42,** 1408 (1969).
143. V. K. Kalra, A. S. Kubla, and T. R. Seshadri, *Indian J. Chem.*, **4,** 201 (1966).
144. K. Fukui, M. Nakayama, and T. Harano, *Bull. Chem. Soc., Japan*, **42,** 1693 (1969).
145. M. Uchiyama and M. Matsui, *Agric. Biol. Chem.*, **31,** 1490 (1967).
146. C. W. L. Bevan, A. J. Birch, B. Moore, and S. K. Mukerjee, *J. Chem. Soc.*, **1964,** 5991.
147. W. B. Whalley, *Pure Appl. Chem.*, **7,** 565 (1963).
148. F. M. Dean, J. Staunton, and W. B. Whalley, *J. Chem. Soc.*, **1959,** 3004.
149. S. Yamamura, K. Kato, and Y. Hirata, *Tetrahedron Lett.*, **1967,** 1637.
150. L. L. Woods, *J. Am. Chem. Soc.*, **80,** 1440 (1958); M. Ohta and H. Kato, *Bull. Chem. Soc. Japan*, **32,** 707 (1959).
151. G. R. Birchall, M. N. Galbraith, and W. B. Whalley, *Chem. Comm.*, **1966,** 474.
152. F. Wessely, G. Lauterbach-Keil, and F. Sinwel, *Monatsh.*, **81,** 811 (1950); F. Wessely, E. Zbiral, and H. Sturm, *Chem. Ber.*, **93,** 2840 (1960).
153. M. N. Galbraith and W. B. Whalley, *Chem. Commun.*, **1966,** 620.
154. G. R. Birchall, R. W. Gray, R. R. King, and W. B. Whalley, *Chem. Comm.*, **1969,** 457.
155. R. Chong, R. R. King, and W. B. Whalley, *Chem. Comm.*, **1969,** 1512.
156. Reference 2, Chapter 15.
157. E. J. Haws, J. S. E. Holker, A. Kelly, A. D. G. Powell, and A. Robertson, *J. Chem. Soc.*, **1959,** 3598.
158. B. C. Fielding E. J. Haws, J. S. E. Holker, A. D. G. Powell, and A. Robertson, *Tetrahedron Lett.*, **1960,** 24.
159. R. Chong, R. W. Gray, R. R. King, and W. B. Whalley, *Chem. Comm.*, **1970,** 101.
160. G. Büchi, J. White, and G. N. Wogan, *J. Am. Chem. Soc.*, **87,** 3484 (1965).
161. W. E. Truce, *Org. Reactions*, **9,** 38.
162. D. H. Johnson, A. Robertson, and W. B. Whalley, *J. Chem. Soc.*, **1949,** 1563.
163. H. H. Warren, G. Dougherty, and S. E. Wallis, *J. Am. Chem. Soc.*, **79,** 3812 (1957).
164. R. F. Curtiss, C. H. Hassall, and M. Nazar, *J. Chem. Soc. (C)*, **1968,** 85.
165. D. H. R. Barton and J. B. Hendrickson, *J. Chem. Soc.*, **1956,** 1028.

166. O. R. Gottlieb, *Phytochem.*, **7**, 411 (1968).
167. I. Carpenter, H. D. Locksley, and F. Scheinmann, *Phytochem.*, **8**, 2013 (1969).
168. J. C. Roberts, *Chem. Rev.*, **61**, 591 (1961).
169. Reference 2, Chapter 9.
170. P. K. Grover, G. D. Shah, and R. D. Shah, *J. Chem. Soc.*, **1955**, 3982; *J. Sci. Ind. Res.*, **15B**, 629 (1956).
171. B. R. Samant and A. B. Kulkarni, *Indian J. Chem.*, **7**, 463 (1969).
172. P. R. Iyer and G. D. Shah, *Indian J. Chem.*, **8**, 691 (1970).
173. Y. S. Agasimundin and S. Rajagopal, *J. Org. Chem.*, **36**, 845 (1971).
174. B. Jackson, H. D. Locksley, and F. Scheinmann, *J. Chem. Soc. (C)*, **1967**, 785.
175. B. M. Desai, P. R. Desai, and R. D. Desai, *J. Indian Chem. Soc.*, **37**, 53 (1960).
176. M. L. Wolfrom, F. Komitsky, and J. H. Looker, *J. Org. Chem.*, **30**, 144 (1965).
177. H. D. Locksley, A. J. Quillinan, and F. Scheinmann, *Chem. Comm.*, **1969**, 1505; *J. Chem. Soc. (C)*, **1971**, in the press.
178. H. D. Locksley, I. Moore, and F. Scheinmann; *J. Chem. Soc. (C)*, **1966**, 430.
178a. J. S. H. Davies, F. Scheinmann, and H. Suschitzky, *J. Org. Chem.*, **23**, 307 (1958).
179. J. Moron, J. Polansky, and H. Pourrat, *Bull Soc. chim. France*, **1967**, 130.
180. E. D. Burling, A. Jefferson, and F. Scheinmann, *Tetrahedron*, **21**, 2653 (1965).
181. D. H. R. Barton and A. I. Scott, *J. Chem. Soc.*, **1958**, 1767.
182. F. M. Dean, J. Goodchild, L. E. Houghton, J. A. Martin, R. B. Morton, B. Parton, A. W. Price, and N. Somvichien, *Tetrahedron Lett.*, **1966**, 4153.
183. O. R. Gottlieb, M. Taveira Magalhães, M. Camey, A. A. Lins Mesquita, and D. de Barros Corrêa, *Tetrahedron*, **22**, 1777 (1966).
184. W. J. Horton and J. T. Spence, *J. Am. Chem. Soc.*, **80**, 2453 (1958).
185. G. H. Stout and V. F. Stout, *Tetrahedron*, **14**, 296 (1961).
186. R. A. Finnegan and P. L. Bachmann, *J. Pharm. Sci.*, **54**, 633 (1965).
187. A. A. Goldberg and A. H. Wragg, *J. Chem. Soc.*, **1958**, 4227.
188. F. Dallacker and Z. Damó, *Chem. Ber.*, **102**, 2414 (1969).
189. G. H. Stout and W. J. Balkenhol, *Tetrahedron*, **25**, 1947 (1969).
190. G. H. Stout, E. N. Christensen, W. J. Balkenhol, and K. L. Stevens, *Tetrahedron*, **25**, 1961 (1969).
191. R. Royer, J.-P. Lechartier, and P. Demerseman, *Bull. Soc. chim. France*, **1971**, 1707.
192. J. R. Lewis, *Proc. Chem. Soc.*, **1963**, 373.
193. J. R. Lewis and B. H. Warrington, *J. Chem. Soc.*, **1964**, 5074.
194. J. E. Atkinson, P. Gupta, and J. R. Lewis, *Chem. Comm.*, **1968**, 1386; *Tetrahedron*, **25**, 1507 (1969).
195. J. E. Atkinson and J. R. Lewis, *Chem. Comm.*, **1967**, 803; *J. Chem. Soc. (C)*, **1969**, 281.
196. J. W. A. Findlay, P. Gupta, and J. R. Lewis, *Chem. Comm.*, **1969**, 206.
197. R. C. Ellis, W. B. Whalley, and K. Ball, *Chem. Comm.*, **1967**, 803.
198. A. Jefferson and F. Scheinmann, *Nature*, **207**, 1193 (1965); *J. Chem. Soc. (C)*, **1966**, 175.
199. H. D. Locksley, I. Moore, and F. Scheinmann, *Tetrahedron*, **23**, 2229 (1967).

200. H. D. Locksley and I. G. Murray, *J. Chem. Soc.* (*C*), **1970,** 392.
201. U. R. Usgaonkar and G. V. Jadhav, *J. Indian Chem. Soc.*, **40,** 27 (1963).
202. Y. Tanase, *J. Pharm. Soc. Japan*, **61,** 341 (1941).
203. L. A. Paquette, *Tetrahedron Lett.*, **1965,** 1291, 3103.
204. M. Guyot and C. Mentzer, *Bull. Soc. chim. France*, **1965,** 2558.
205. J. Santesson and G. Sundholm, *Arkiv. Kemi*, **30,** 427 (1969); J. Santesson and C. A. Wachtmeister, *Arkiv. Kemi*, **30,** 449 (1969).
206. K. R. Markham, *Tetrahedron*, **21,** 1449 (1965).
207. A. C. Jain, V. K. Khanna, and T. R. Seshadri, *Current Sci.*, **37,** 493 (1968).
208. Reference 2, Chapters 6–12.
209. Y. S. Agasimundin and S. Rajagopal, *Chem. Ber.*, **100,** 383 (1967).
210. F. Lamb and H. Suschitsky, *Tetrahedron*, **5,** 1 (1959); J. S. H. Davies, *J. Chem. Soc.*, **1956,** 2140; **1958,** 1790.
211. P. E. Knott and J. C. Roberts, *Phytochem.*, **6,** 1597 (1967).
212. V. K. Bhatia and T. R. Seshadri, *Tetrahedron Lett.*, **1968,** 1741.
213. A. C. Jain, V. K. Khanna, and T. R. Seshadri, *Tetrahedron*, **25,** 2787 (1969).
214. M. L. Wolfrom, E. W. Koos, and H. B. Bhat, *J. Org. Chem.*, **32,** 1058 (1967).
215. A. Jefferson and F. Scheinmann, *J. Chem. Soc.* (*C*), **1966,** 175.
216. Reference 2, Chapters 6, 8, 9, 10, and 12.
217. B. Jackson, H. D. Locksley, I Moore, and F. Scheinmann, *J. Chem. Soc.* (*C*), **1968,** 2579.
218. J. R. Lewis and J. B. Reary, *J. Chem. Soc.* (*C*), **1970,** 1622.
219. H. D. Locksley, I. Moore, and F. Scheinmann, *J. Chem. Soc.* (*C*), **1966,** 2265.
220. A. J. Quillinan and F. Scheinmann, *Chem. Comm.*, **1971,** 966.
221. A. A. Lins Mesquita, D. de Barros Corrêa, O. R. Gottlieb, and M. Taveira Magalhães, *An. Acad. brasil. Clenc.*, **1968,** 40; *Anal. Chim. Acta*, **42,** 311 (1968).
222. H. D. Locksley, I. Moore, and F. Scheinmann, *J. Chem. Soc.* (*C*), **1966,** 2186.
223. D. Barraclough, H. D. Locksley, F. Scheinmann, M. Taveira Magalhães, and O. R. Gottlieb, *J. Chem. Soc.* (*B*), **1970,** 603.

Subject Index

Page numbers in bold face indicate the synthesis of this type of compound.

Reaction Index